Roy J. Glauber

Quantum Theory of Optical Coherence

1807–2007 Knowledge for Generations

Each generation has its unique needs and aspirations. When Charles Wiley first opened his small printing shop in lower Manhattan in 1807, it was a generation of boundless potential searching for an identity. And we were there, helping to define a new American literary tradition. Over half a century later, in the midst of the Second Industrial Revolution, it was a generation focused on building the future. Once again, we were there, supplying the critical scientific, technical, and engineering knowledge that helped frame the world. Throughout the 20th Century, and into the new millennium, nations began to reach out beyond their own borders and a new international community was born. Wiley was there, expanding its operations around the world to enable a global exchange of ideas, opinions, and know-how.

For 200 years, Wiley has been an integral part of each generation's journey, enabling the flow of information and understanding necessary to meet their needs and fulfill their aspirations. Today, bold new technologies are changing the way we live and learn. Wiley will be there, providing you the must-have knowledge you need to imagine new worlds, new possibilities, and new opportunities.

Generations come and go, but you can always count on Wiley to provide you the knowledge you need, when and where you need it!

William J. Pesce
President and Chief Executive Officer

Peter Booth Wiley
Chairman of the Board

Roy J. Glauber

Quantum Theory of Optical Coherence

Selected Papers and Lectures

WILEY-VCH Verlag GmbH & Co. KGaA

The Author

Roy J. Glauber
Lyman Laboratory
Harvard University
Cambridge, MA 02138
USA

Cover

Illustration by Michael Buser and
Kathleen F. Dodson-Schleich.

Photograph of the Author on Backcover:
© Stu Rosner, Cambridge, Mass.

All books published by Wiley-VCH are carefully produced. Nevertheless, authors, editors, and publisher do not warrant the information contained in these books, including this book, to be free of errors. Readers are advised to keep in mind that statements, data, illustrations, procedural details or other items may inadvertently be inaccurate.

Library of Congress Card No.:
applied for

British Library Cataloguing-in-Publication Data:
A catalogue record for this book is available from the British Library.

Bibliographic information published by the Deutsche Nationalbibliothek
The Deutsche Nationalbibliothek lists this publication in the Deutsche Nationalbibliografie; detailed bibliographic data are available in the Internet at <http://dnb.d-nb.de>.

© 2007 WILEY-VCH Verlag GmbH & Co. KGaA, Weinheim

Every effort has been made to trace copyright holders and secure permission prior to publication. If notified, the publisher will rectify any error or omission at the earliest opportunity.

All rights reserved (including those of translation into other languages). No part of this book may be reproduced in any form – by photoprinting, microfilm, or any other means – nor transmitted or translated into a machine language without written permission from the publishers. Registered names, trademarks, etc. used in this book, even when not specifically marked as such, are not to be considered unprotected by law.

Printed in the Federal Republic of Germany
Printed on acid-free paper

Typesetting Dr. Michael Bär, Wiesloch
Printing Strauss GmbH, Mörlenbach
Bookbinding Litges & Dopf Buchbinderei GmbH, Heppenheim

ISBN: 978-3-527-40687-6

Contents

1   The Quantum Theory of Optical Coherence   *1*
1.1   Introduction   *1*
1.2   Elements of Field Theory   *2*
1.3   Field Correlations   *7*
1.4   Coherence   *10*
1.5   Coherence and Polarization   *15*
   Appendix   *18*
   References   *20*

2   Optical Coherence and Photon Statistics   *23*
2.1   Introduction   *23*
2.1.1   Classical Theory   *27*
2.2   Interference Experiments   *30*
2.3   Introduction of Quantum Theory   *35*
2.4   The One-Atom Photon Detector   *38*
2.5   The n-Atom Photon Detector   *46*
2.6   Properties of the Correlation Functions   *51*
2.6.1   Space and Time Dependence of the Correlation Functions   *54*
2.7   Diffraction and Interference   *56*
2.7.1   Some General Remarks on Interference   *58*
2.7.2   First-Order Coherence   *59*
2.7.3   Fringe Contrast and Factorization   *64*
2.8   Interpretation of Intensity Interferometer Experiments   *66*
2.8.1   Higher Order Coherence and Photon Coincidences   *67*
2.8.2   Further Discussion of Higher Order Coherence   *70*
2.8.3   Treatment of Arbitrary Polarizations   *71*
2.9   Coherent and Incoherent States of the Radiation Field   *75*
2.9.1   Introduction   *75*
2.9.2   Field-Theoretical Background   *77*
2.9.3   Coherent States of a Single Mode   *80*

Quantum Theory of Optical Coherence. Selected Papers and Lectures. Roy J. Glauber.
Copyright © 2007 WILEY-VCH Verlag GmbH & Co. KGaA, Weinheim
ISBN: 978-3-527-40687-6

2.9.4	Expansion of Arbitrary States in Terms of Coherent States *86*
2.9.5	Expansion of Operators in Terms of Coherent State Vectors *89*
2.9.6	General Properties of the Density Operator *92*
2.9.7	The *P* Representation of the Density Operator *94*
2.9.8	The Gaussian Density Operator *100*
2.9.9	Density Operators for the Field *104*
2.9.10	Correlation and Coherence Properties of the Field *109*
2.10	Radiation by a Predetermined Charge–Current Distribution *117*
2.11	Phase-Space Distributions for the Field *121*
2.11.1	The *P* Representation and the Moment Problem *123*
2.11.2	A Positive-Definite "Phase Space Density" *124*
2.11.3	Wigner's "Phase Space Density" *127*
2.12	Correlation Functions and Quasiprobability Distributions *132*
2.12.1	First Order Correlation Functions for Stationary Fields *134*
2.12.2	Correlation Functions for Chaotic Fields *136*
2.12.3	Quasiprobability Distribution for the Field Amplitude *139*
2.12.4	Quasiprobability Distribution for the Field Amplitudes at Two Space-Time Points *145*
2.13	Elementary Models of Light Beams *148*
2.13.1	Model for Ideal Laser Fields *153*
2.13.2	Model of a Laser Field With Finite Bandwidth *156*
2.14	Interference of Independent Light Beams *164*
2.15	Photon Counting Experiments *170*
	References *181*

3	**Correlation Functions for Coherent Fields** *183*
3.1	Introduction *183*
3.2	Correlation Functions and Coherence Conditions *184*
3.3	Correlation Functions as Scalar Products *186*
3.4	Application to Higher Order Correlation Functions *189*
3.5	Fields With Positive-Definite *P* Functions *191*
	References *195*

4	**Density Operators for Coherent Fields** *197*
4.1	Introduction *197*
4.2	Evaluation of the Density Operator *199*
4.3	Fully Coherent Fields *205*
4.4	Unique Properties of the Annihilation Operator Eigenstates *209*
	References *216*

5	**Classical Behavior of Systems of Quantum Oscillators** *217*
	References *220*

6	**Quantum Theory of Parametric Amplification I** *221*
6.1	Introduction *221*
6.2	The Coherent States and the *P* Representation *223*
6.3	Model of the Parametric Amplifier *227*
6.4	Reduced Density Operator for the A Mode *233*
6.5	Initially Coherent State: *P* Representation for the A Mode *234*
6.6	Initially Coherent State; Moments, Matrix Elements, and Explicit Representation for $\rho_A(t)$ *238*
6.7	Solutions for an Initially Chaotic B Mode *241*
6.8	Solution for Initial *n*-Quantum State of A Mode; B Mode Chaotic *244*
6.9	General Discussion of Amplification With B Mode Initially Chaotic *249*
6.10	Discussion of *P* Representation: Characteristic Functions Initially Gaussian *252*
6.11	Some General Properties of $P(\alpha,t)$ *258*
	Appendix *260*
	References *261*

7	**Quantum Theory of Parametric Amplification II** *263*
7.1	Introduction *263*
7.2	The Two-Mode Characteristic Function *265*
7.3	The Wigner Function *267*
7.4	Decoupled Equations of Motion *271*
7.5	Characteristic Functions Expressed in Terms of Decoupled Variables *273*
7.6	*W* and *P* Expressed in Terms of Decoupled Variables *275*
7.7	Results for Chaotic Initial States *278*
7.8	Correlations of the Mode Amplitudes *283*
	References *286*

8	**Photon Statistics** *287*
8.1	Introduction *287*
8.2	Classical Theory *288*
8.3	Quantum Theory: Introduction *290*
8.4	Intensity and Coincidence Measurements *293*
8.5	First and Higher Order Coherence *297*
8.6	The Coherent States *300*
8.7	Expansions in Terms of the Coherent States *307*
8.8	Characteristic Functions and Quasiprobability Densities *313*
8.9	Some Examples *319*
8.10	Photon Counting Distributions *322*
	References *329*

9 Ordered Expansions in Boson Amplitude Operators *331*

- 9.1 Introduction *331*
- 9.2 Coherent States and Displacement Operators *333*
- 9.3 Completeness of Displacement Operators *337*
- 9.4 Ordered Power-Series Expansions *345*
- 9.5 s-Ordered Power-Series Expansions *353*
- 9.6 Integral Expansions for Operators *358*
- 9.7 Correspondences Between Operators and Functions *366*
- 9.8 Illustration of Operator–Function Correspondences *375*
- Appendix A *377*
- Appendix B *378*
- Appendix C *379*
- Appendix D *380*
- References *380*

10 Density Operators and Quasiprobability Distributions *383*

- 10.1 Introduction *383*
- 10.2 Ordered Operator Expansions *385*
- 10.3 The P Representation *389*
- 10.4 Wigner Distribution *393*
- 10.5 The Function $\langle \alpha | \rho | \alpha \rangle$ *399*
- 10.6 Ensemble Averages and s Ordering *402*
- 10.7 Examples of the General Quasiprobability Function $W(\alpha,s)$ *408*
- 10.8 Analogy with Heat Diffusion *416*
- 10.9 Time-Reversed Heat Diffusion and $W(\alpha,s)$ *418*
- 10.10 Properties Common to all Quasiprobability Distributions *420*
- References *423*

11 Coherence and Quantum Detection *425*

- 11.1 Introduction *425*
- 11.2 The Statistical Properties of the Electromagnetic Field *426*
- 11.3 The Ideal Photon Detector *428*
- 11.4 Correlation Functions and Coherence *429*
- 11.5 Other Correlation Functions *432*
- 11.6 The Coherent States *434*
- 11.7 Expansions in Terms of Coherent States *437*
- 11.8 A Few General Observations *439*
- 11.9 The Damped Harmonic Oscillator *440*
- 11.10 The Density Operator for the Damped Oscillator *444*
- 11.11 Irreversibility and Damping *447*
- 11.12 The Fokker–Planck and Bloch Equations *449*

11.13	Theory of Photodetection. The Photon Counter Viewed as a Harmonic Oscillator *453*
11.14	The Density Operator for the Photon Counter *459*
	References *462*

12 Quantum Theory of Coherence 463

12.1	Introduction *463*
12.2	Classical Theory *468*
12.3	Quantum Theory *471*
12.4	Intensity and Coincidence Measurements *474*
12.5	Coherence *487*
12.6	Coherent States *495*
12.7	The P Representation *499*
12.8	Chaotic States *514*
12.9	Wavepacket Structure of Chaotic Field *521*
	References *530*

13 The Initiation of Superfluorescence 531

13.1	Introduction *531*
13.2	Basic Equations for a Simple Model *532*
13.3	Onset of Superfluorescence *534*
	References *536*

14 Amplifiers, Attenuators and Schrödingers Cat 537

14.1	Introduction: Two Paradoxes *537*
14.2	A Quantum-Mechanical Attenuator: The Damped Oscillator *542*
14.3	A Quantum Mechanical Amplifier *548*
14.4	Specification of Photon Polarization States *558*
14.5	Measuring Photon Polarizations *561*
14.6	Use of the Compound Amplifier *563*
14.7	Superluminal Communication? *565*
14.8	Interference Experiments and Schrödinger's Cat *569*
	References *575*

15 The Quantum Mechanics of Trapped Wavepackets 577

15.1	Introduction *577*
15.2	Equations of Motion and Their Solutions *578*
15.3	The Wave Functions *581*
15.4	Periodic Fields and Trapping *584*
15.5	Interaction With the Radiation Field *587*
15.6	Sum Rules *590*
15.7	Radiative Equilibrium and Instability *592*
	References *594*

16 Density Operators for Fermions 595

- 16.1 Introduction *595*
- 16.2 Notation *597*
- 16.3 Coherent States for Fermions *597*
- 16.3.1 Displacement Operators *597*
- 16.3.2 Coherent States *599*
- 16.3.3 Intrinsic Descriptions of Fermionic States *600*
- 16.4 Grassmann Calculus *601*
- 16.4.1 Differentiation *601*
- 16.4.2 Even and Odd Functions *601*
- 16.4.3 Product Rule *602*
- 16.4.4 Integration *602*
- 16.4.5 Integration by Parts *603*
- 16.4.6 Completeness of the Coherent States *604*
- 16.4.7 Completeness of the Displacement Operators *604*
- 16.5 Operators *605*
- 16.5.1 The Identity Operator *605*
- 16.5.2 The Trace *606*
- 16.5.3 Physical States and Operators *606*
- 16.5.4 Physical Density Operators *607*
- 16.6 δ Functions and Fourier Transforms *608*
- 16.7 Operator Expansions *610*
- 16.8 Characteristic Functions *612*
- 16.8.1 The s-Ordered Characteristic Function *613*
- 16.9 s-Ordered Expansions for Operators *614*
- 16.10 Quasiprobability Distributions *616*
- 16.11 Mean Values of Operators *618*
- 16.12 P Representation *619*
- 16.13 Correlation Functions for Fermions *620*
- 16.14 Chaotic States of the Fermion Field *621*
- 16.15 Correlation Functions for Chaotic Field Excitations *624*
- 16.16 Fermion-Counting Experiments *626*
- 16.17 Some Elementary Examples *628*
- 16.17.1 The Vacuum State *628*
- 16.17.2 A Physical Two-Mode Density Operator *629*
- References *631*

Index *632*

Foreword

It was in the first days of January 1944 that a young man of nineteen years climbed out of the train at the lonely train station in Lamy, New Mexico, where the Santa Fe Railroad comes closest to the city of Santa Fe. The young man was not the only one getting out of the train; he was joined by a pudgy gentleman. However, it was only when both approached 109 East Palace Avenue in Santa Fe that they realized they were heading in the same direction. The older man registered with Dorothy McKibbin, who was running the place, using the name of Jonny Newman. The younger one wrote the name Roy Glauber into the book. The first was the famous mathematician, physicist and computer scientist John von Neumann, the second one had used his real name. Their common destination was Los Alamos.

Why, you may well ask, is a nineteen year old being recruited for the Manhattan project? Simple – Roy Glauber was a child prodigy. He constructed his first telescope as a pre-teenager, won the Westinghouse Prize and entered Harvard at the age of sixteen. There he studied physics under several notables and by the age of nineteen had learned most of what was known in physics at the time.

Nuclear Physics and First Experience

During his one and a half years in Los Alamos Roy worked in the group of Robert Serber on neutron diffusion. He was also close to Richard Feynman who was directing a small theory department. In this way Roy had first-hand experience of many of Feynman's pranks which are nowadays common knowledge due to the book *Surely you are joking Mr. Feynman*.

On the occasion of the Trinity test shot near Alamogordo, NM, there was only room for the senior staff (such as Fermi). So our resourceful hero Roy teamed up with a group to view the event on Sandia peak some 100 miles from the test site. The appointed hour came and they saw – nothing. After having waited a while longer, they decided that the bomb hadn't worked after all. So they started packing up for the trip back to Los Alamos. Just then the southern sky lit up with an eerie purple glow; you know the rest of the story.

Quantum Theory of Optical Coherence. Selected Papers and Lectures. Roy J. Glauber.
Copyright © 2007 WILEY-VCH Verlag GmbH & Co. KGaA, Weinheim
ISBN: 978-3-527-40687-6

After his time in Los Alamos, Roy started to work on his PhD thesis on meson theory under the guidance of Julian Schwinger (Nobel Prize in physics 1965) at Harvard. After graduating in the fall of 1949 he received a grant to go to the Institute of Advanced Studies in Princeton. A couple of months earlier he had realized, in a hotel in Berkeley, an interesting relation concerning the multiplication of two exponential operators which contain creation and annihilation operators. This identity impressed the director of the institute, Robert Oppenheimer, enormously. Also the book Messiah's on quantum mechanics refers to Glauber in this context. Later, this relation turned out to be very important in the context of coherent states.

During his first year in Princeton, Roy met Wolfgang Pauli who had come to visit the Institute. Pauli offered Roy a position for the summer of 1950. He accepted and traveled to Europe with the luxury liner *Île de France*. Even today Roy keeps talking about the impressive life that was taking place on this ship: champagne, cocktails and an unusual wealth of dishes at any time of the day or night and a plentitude of young women on their pilgrimage to Europe in the Holy Year of 1950. His first stop in the Old World was Paris, where a large European physics conference was held at the Institute Poincaré. On this occasion he had his first experience of Pauli's direct way of dealing with people. Roy had caught a cold during his first days in Paris and had lost his voice. When he saw Pauli in Paris he could barely say "I have lost my voice". Pauli with a huge smile answered "Yeh, you have lost your voice and you don't look so good either".

After the meeting, Roy took a few days off and enjoyed the life in Paris. Since he had never been to this town before, "The postcards came to life", as he used to call it later. For this reason he arrived in Zürich a few weeks after Pauli. His mother had been worried about her son since she hadn't heard from him since he had left America. For this reason she had written a letter to Prof. Pauli asking him about Roy. Pauli immediately forced Roy to write a letter to his mother. Later Pauli used every opportunity to tease Roy about his mother's concern. He often interrupted his lectures when he saw Roy entering the class room to ask "and how is your mother doing?".

Another example of the impish nature of Pauli showed up during an excursion of his department to Stansstad. In the late afternoon the group played soccer and some people were swimming in the lake. From time to time Pauli kicked the ball purposefully into the lake and the students and assistants had to swim out to get it back. Finally he took aim at Roy, who had carried around a large camera all afternoon and was now taking pictures of Pauli during the soccer game. There exists a remarkable photograph of Pauli as he kicks the ball towards the camera. Moments later Roy was knocked down to the floor because of the transfer of momentum.

A New Scattering Approximation

In the summer of 1951, Roy got his first permanent position. Caltech had just hired Feynman from Cornell. However, he wanted to spend his first year on a Sabbatical in

Brazil. It is well known that Feynman learned how to play the bongo drums during this visit. Roy substituted for Feynman and took over his class in quantum mechanics. On this occasion he also collaborated, in the spring of 1952, with the quantum chemistry group of Linus Pauling. Here he became interested in the scattering of electrons from molecules. The experiments of V. Schomaker showed some rather remarkable results which could not be explained by first-order perturbation theory, i.e. by the Born approximation. For this reason Roy developed a new approximation which is also valid at short wavelengths. This theory has found applications in many branches of physics and is known in scattering theory as "Glauber approximation".

In the meantime, Schwinger had negotiated with the administration at Harvard that an assistant should be attached to him. Roy was offered this position and returned to Harvard. After the first year he became Assistant Professor and stayed at Harvard until today. Since 1976 he is Mallinckrodt Professor. In the following years Roy dedicated much of his time to complete his scattering theory. Of central importance in this context is his discovery of diffractive dissociation.

The Birth of Quantum Optics

In the early sixties Roy was drawn into a completely different topic. He developed the quantum theory of optical coherence and photon detection, for which he was finally awarded the Nobel Prize in 2005. In this way he laid the foundations for modern quantum optics.

The quantum theory of radiation had already been developed in the early days of quantum mechanics by Max Born, Pascual Jordan, Werner Heisenberg, and Paul A. M. Dirac. Nevertheless, this theory could not provide a quantitative description of physical processes since it contained singularities. It was not until 1947 and the experiments of Willis E. Lamb and Polykarp Kusch on the shift of the energy levels of the hydrogen atom and the anomalous magnetic moment of the electron that it was generally accepted that light needs to be treated quantum-mechanically. These revolutionary experiments led to the development of quantum electrodynamics. However, conventional wisdom still believed that quantization of the radiation field had little relevance for optical processes.

In the mid-fifties, interferometric measurements of the size of distant stars accomplished by Robert Hanbury Brown and Richard Q. Twiss brought a dramatic turn of events. They used intensity correlations of photo currents from two spatially separated detectors and observed an enhancement when the difference of the optical wavelengths of the two signals disappeared. Our hero Roy interpreted this enhancement as a quantum effect. Thermal light comes in clumps and is therefore bunched. The probability to detect a photon immediately after another one has been found is higher than at a later moment. This photon bunching experiment with a thermal light source can be considered the birth of quantum optics.

Quantum Theory of Optical Coherence

After the discovery of the maser and the laser in the early sixties new ideas for quantum effects of the radiation field were in the air. However, there was no theory for their observation. It was only in 1963 that Roy had developed the quantum theory of optical coherence. Here the concept of a coherence state plays a central role. Coherent states had been proposed for the first time by Erwin Schrödinger in 1927 in order to show that a wavepacket needs not always be bound to spread. The coherent state became the crucial tool for Roy's theory of optical coherence. In particular, he could show that for coherent fields all correlation functions factorize. As a consequence, the intensity correlations are independent of the delay time. This effect is also observed using laser light.

However, besides photon bunching there is also the opposite effect, i.e. anti-bunching. For short delay times the intensity correlations vanish. The probability to detect two photons right after each other vanishes. Anti-bunched light appears for example in the resonance fluorescence of a single atom and was detected independently by the groups of Leonhard Mandel (Rochester) and Herbert Walther (Garching). In these cases, the light is in a non-classical state. The characterization of non-classical light fields relies on coherent states and the introduction of distributions in the quantum mechanical phase space. They provide a bridge between the classical and the quantum mechanical description of the radiation field.

The Glauber representation of the density operator in terms of coherent states was also important for understanding the laser. The development of the quantum theory of the laser pursued by the groups of Willis Lamb and Marlan O. Scully (Yale), Melvin Lax (Bell Labs), and Hermann Haken with Hannes Risken (Stuttgart) and even the theory of the atom laser benefits from the early work of Roy. Indeed, the quantum theory of coherence is not restricted to light but can be applied to any bosonic field and, in particular, to a Bose–Einstein-Condensate (BEC). Recently, intensity correlations of a BEC have been measured and are in full agreement with theory. Moreover, with the help of Grassmann algebras Roy could also extend his theory to fermionic fields which can be tested using fermionic atoms. First experimental results have been reported.

The Old Question: What is a Photon?

During the Les Houches summer school of 1964, Lamb, who had received the Nobel Prize in 1955 for discovering the shift of the energy levels in a hydrogen atom, lectured on his semi-classical theory of the laser. In this formalism light is treated by using classical electrodynamics and matter by using quantum mechanics. In Lamb's theory there is no need for the concept of the photon. Despite this fact, many scientists in Les Houches were using the word "photon" even when they referred to an effect whose explanation did not rely on the quantum theory of radiation. This misuse of the word "photon" annoyed Lamb and he introduced a licence which entitled its owner to use the word "photon". Scientists without licence were not allowed to even mention photons. Roy was one of the very few colleagues who received such a licence from Lamb.

The problem of the photon is best summarized by a 1951 quote from Albert Einstein: "In 50 years of pondering, I still have not come closer to an answer for the question what light quanta are. Today every Tom, Dick and Harry thinks he knows, but he is wrong". Roy's theory of a photodetector has, however, furnished considerable progress in this direction. He could connect a light quantum in the field with a click in the detector. The relevant theory had to take into account that the absorption of a photon in the detector changes the state of the radiation field. This condition leads to the normal ordering of creation and annihilation operators of the field. Roy once jokingly summarized his theory of photo detection by the sentence: "I don't know anything about photons, but I know one when I see one". This sentence is reminiscent of the American Supreme Court Justice Potter Steward who in 1964 was asked to define pornography and said: "I know it when I see it".

The Teacher and the Human Being

Roy can be proud to have an impressive list of PhD students and Postdocs who themselves have started famous schools in theoretical physics. His PhD student Dan Walls who unfortunately died very early had established a highly successful school of quantum optics in New Zealand and Australia. In Germany it was Fritz Haake in Essen and Maciej Lewenstein in Hannover who propagated the Glauber fame. Roy has had always strong ties to Germany. He was a Humboldt awardee at the Max Planck Institute for Quantum Optics and has spent many summers at the University of Ulm.

We are also proud of Roy's dedication to his children, Val and Jeff. As a single parent, Roy did all the hard work of running the home, being a soccer Dad and bringing up children. Doing all this while carrying his duties as Harvard Professor and father figure to us all, is indeed reason enough for a "Noble Prize".

Roy has also a great sense of humor. For this reason he is often asked to give the after-dinner speech at conference banquets. An excellent illustration is his after-dinner speech at the *Nato Summer School on Gravitation and Squeezed States* in 1981. This meeting took place in the city of Bad Windsheim close to Nuremberg. Roy had noticed that the town had been promoted to the status of a spa after World War II and now carried the name "Bad Windsheim". In his speech he remarked, "I have no idea when this city went bad, but we have had such a great time here that from now on we will call it Good Windsheim".

September 2006
Marlan Scully, College Station, TX
Wolfgang P. Schleich, Ulm

1
The Quantum Theory of Optical Coherence[1]

1.1
Introduction

Correlation, it has long been recognized, plays a fundamental role in the concept of optical coherence. Techniques for both the generation and detection of various types of correlations in optical fields have advanced rapidly in recent years. The development of the optical maser, in particular, has led to the generation of fields with a range of correlation unprecedented at optical frequencies. The use of techniques of coincidence detection of photons[1,2] has, in the same period, shown the existence of unanticipated correlations in the arrival times of light quanta. The new approaches to optics, which such developments will allow us to explore, suggest the need for a fundamental discussion of the meaning of coherence.

The present paper, which is the first of a series on fundamental problems of optics, is devoted largely to defining the concept of coherence. We do this by constructing a sequence of correlation functions for the field vectors, and by discussing the consequences of certain assumptions about their properties. The definition of coherence which we reach differs from earlier ones in several significant ways. The most important difference, perhaps, is that complete coherence, as we define it, requires that the field correlation functions satisfy an infinite succession of coherence conditions. We are led then to distinguish among various orders of incomplete coherence, according to the number of conditions satisfied. The fields traditionally described as coherent in optics are shown to have only first-order coherence. The fields, generated by the optical maser, on the other hand, may have a considerably higher order of coherence. A further difference between our approach and previous ones is that it is constructed to apply to fields of arbitrary time dependence, rather than just to those which are, on the average, stationary in time. We have also attempted to develop the discussion in a fully quantum theoretical way.

It would hardly seem that any justification is necessary for discussing the theory of light quanta in quantum theoretical terms. Yet, as we all know, the successes of classical theory in dealing with optical experiments have been so great that we feel no hesitation in introducing optics as a sophomore course. The quantum theory, in other

---

[1] Reprinted with permission from R. J. Glauber, *Phys. Rev.* **130**, 2529–2539 (1963). Copyright 2006 by the American Physical Society.

Quantum Theory of Optical Coherence. Selected Papers and Lectures. Roy J. Glauber.
Copyright © 2007 WILEY-VCH Verlag GmbH & Co. KGaA, Weinheim
ISBN: 978-3-527-40687-6

words, has had only a fraction of the influence upon optics that optics has historically had upon quantum theory. The explanation, no doubt, lies in the fact that optical experiments to date have paid very little attention to individual photons. To the extent that observations in optics have been confined to the measurement of ordinary light intensities, it is not surprising that classical theory has offered simple and essentially correct insights.

Experiments such as those on quantum correlations suggest, on the other hand, the growing importance of studies of photon statistics. Such studies lie largely outside the grasp of classical theory. To observe that the quantum theory is fundamentally necessary to the treatment of these problems is not to say that the semi-classical approach always yields incorrect results. On the contrary, correct answers to certain classes of problems of photon statistics[3] may be found through adaptations of classical methods. There are, however, distinct virtues to knowing where such methods succeed and where they do not. For that reason, as well as for its intrinsic interest, we shall formulate the theory in quantum theoretical terms from the outset. Quite a few of our arguments can easily be paraphrased in classical terms. Several seem to be new in the context of classical theory.

We shall try to construct this paper so that it can be followed with little more than a knowledge of elementary quantum mechanics. Since its subject matter is, in the deepest sense, quantum electrodynamics, we begin with a section which describes the few simple aspects of that subject which are referred to later.

1.2
Elements of Field Theory

The observable quantities of the electromagnetic field will be taken to be the electric and magnetic fields which are represented by a pair of Hermitian operators, $\boldsymbol{E}(\boldsymbol{r}t)$ and $\boldsymbol{B}(\boldsymbol{r}t)$. The state of the field will be described by means of a state vector, $|\ \rangle$, on which the fields operate from the left, or by means of its adjoint, $\langle\ |$, on which they operate from the right. Since we shall use the Heisenberg representation, the choice of a fixed state vector specifies the properties of the field at all times. The theory is constructed, by whatever formal means, so that in a vacuum the field operators $\boldsymbol{E}(\boldsymbol{r}t)$ and $\boldsymbol{B}(\boldsymbol{r}t)$ satisfy the Maxwell equations

$$\nabla \cdot \boldsymbol{E} = 0,$$

$$\nabla \times \boldsymbol{E} = -\frac{1}{c}\frac{\partial \boldsymbol{B}}{\partial t},$$

$$\nabla \times \boldsymbol{B} = \frac{1}{c}\frac{\partial \boldsymbol{E}}{\partial t},$$

$$\nabla \cdot \boldsymbol{B} = 0.$$

(1.1)

1.2 Elements of Field Theory

We omit the source terms in the equations since, for the present, we are more interested in the fields themselves than the explicit way in which they are generated or detected. It follows from the Maxwell equations that the electric field operator obeys the wave equation

$$\left(\nabla^2 - \frac{1}{c^2}\frac{\partial^2}{\partial t^2}\right)\boldsymbol{E}(\boldsymbol{r}t) = 0, \tag{1.2}$$

and the magnetic field operator does likewise.

One of the essential respects in which quantum field theory differs from classical theory is that two values of the field operators taken at different space-time points do not, in general, commute with one another. The components of the electric field, which is the only field we shall discuss at length, obey a commutation relation of the general form

$$[E_\mu(\boldsymbol{r}t), E_\nu(\boldsymbol{r}'t')] = D_{\mu\nu}(\boldsymbol{r}-\boldsymbol{r}', t-t'). \tag{1.3}$$

That the tensor function $D_{\mu\nu}$ has as arguments the coordinate differences $\boldsymbol{r}-\boldsymbol{r}'$ and $t-t'$ follows from the invariance of the theory under translations in space and time. We shall not need any further details of the function $D_{\mu\nu}$, but may mention that it vanishes when the four-vector $(\boldsymbol{r}-\boldsymbol{r}', t-t')$ lies outside the light cone, i.e., for $(\boldsymbol{r}-\boldsymbol{r}')^2 > c^2(t-t')^2$. The vanishing of the commutator, for points with spacelike separations, corresponds to the fact that measurements of the stated field components at such points can be carried out to arbitrary accuracy. Such accuracy is attainable since no disturbances can propagate through the field enough to reach one point from the other.

An important element of the discussion in this paper will be the separation of the electric field operator $\boldsymbol{E}(\boldsymbol{r}t)$ into its positive and negative frequency parts. The separation is most easily accomplished when the time dependence of the operator is represented by a Fourier integral. If, for example, the field operator has a representation

$$\boldsymbol{E}(\boldsymbol{r}t) = \int_{-\infty}^{\infty} \boldsymbol{e}(\omega, \boldsymbol{r})\, e^{-i\omega t}\, d\omega, \tag{1.4}$$

where the Hermitian property is secured by the relation $\boldsymbol{e}(-\omega, \boldsymbol{r}) = \boldsymbol{e}^\dagger(\omega, \boldsymbol{r})$, then we define the positive frequency part of $\boldsymbol{E}$ as

$$\boldsymbol{E}^{(+)}(\boldsymbol{r}t) = \int_0^{\infty} \boldsymbol{e}(\omega, \boldsymbol{r})\, e^{-i\omega t}\, d\omega, \tag{1.5}$$

and the negative frequency part as

$$\boldsymbol{E}^{(-)}(\boldsymbol{r}t) = \int_{-\infty}^0 \boldsymbol{e}(\omega, \boldsymbol{r})\, e^{-i\omega t}\, d\omega, \tag{1.6}$$

$$= \int_0^{\infty} \boldsymbol{e}^\dagger(\omega, \boldsymbol{r})\, e^{-i\omega t}\, d\omega. \tag{1.7}$$

It is evident from these definitions that the field is the sum of its positive and negative frequency parts,

$$E(rt) = E^{(+)}(rt) + E^{(-)}(rt) . \qquad (1.8)$$

The two parts, regarded separately, are not Hermitian operators; the fields they represent are intrinsically complex, and mutually adjoint,

$$E^{(-)}(rt) = E^{(+)\dagger}(rt) . \qquad (1.9)$$

In the absence of a Fourier integral representation of $E(rt)$, the positive and negative frequency parts of the field may be defined more formally as the limits of the integrals,

$$E^{(+)}(rt) = \lim_{\eta \to +0} \frac{1}{2\pi i} \int_{-\infty}^{\infty} \frac{E(r, t-\tau)}{\tau - i\eta} d\tau , \qquad (1.10)$$

$$E^{(-)}(rt) = -\lim_{\eta \to +0} \frac{1}{2\pi i} \int_{-\infty}^{\infty} \frac{E(r, t-\tau)}{\tau + i\eta} d\tau . \qquad (1.11)$$

It follows from the intrinsically different time dependences of $E^{(+)}(rt)$ and $E^{(-)}(rt)$ that they act to change the state of the field in altogether different ways, one associated with photon absorption, the other with photon emission. In particular, the positive frequency part, $E^{(+)}(rt)$, may be shown[4] to be a photon annihilation operator. Applied to an n-photon state it produces an $(n-1)$-photon state. Further applications of $E^{(+)}(rt)$ reduce the number of photons present still further, but the regression must end with the state in which the field is empty of all photons. It is part of the definition of this state, which we represent as $|vac\rangle$, that

$$E^{(+)}(rt)|vac\rangle = 0 . \qquad (1.12)$$

The adjoint relation is

$$\langle vac|E^{(-)}(rt) = 0 . \qquad (1.13)$$

Since the operator $E^{(+)}(rt)$ annihilates photons, its Hermitian adjoint, $E^{(-)}(rt)$, must create them; applied to an n-photon state it produces an $(n+1)$-photon state. In particular, the state

$$E^{(-)}(rt)|vac\rangle$$

is a one-photon state.

It has become customary, in discussions of classical theory, to regard the electric field $E(rt)$ as the quantity one measures experimentally, and to think of the complex fields $E^{(\pm)}(rt)$ as convenient, but fictitious, mathematical constructions. Such an attitude can only be held in the classical domain, where quantum phenomena play no essential role. The frequency ω of a classical field must be so low that the quantum energy $\hbar\omega$ is negligible. In such a case, we can not tell whether a classical test charge emits or absorbs quanta. In measuring a classical field strength, $\boldsymbol{E}(\boldsymbol{r}t)$, we implicitly sum the effects of photon absorption and emission which are described individually by the fields $\boldsymbol{E}^{(+)}(\boldsymbol{r}t)$ and $\boldsymbol{E}^{(-)}(\boldsymbol{r}t)$.

Where quantum phenomena are important the situation is usually quite different. Experiments which detect photons ordinarily do so by absorbing them in one or another way. The use of any absorption process, such as photoionization, means in effect that the field we are measuring is the one associated with photon annihilation, the complex field $\boldsymbol{E}^{(+)}(\boldsymbol{r}t)$. We need not discuss the details of the photoabsorption process to find the appropriate matrix element of the field operator. If the field makes a transition from the initial state $|i\rangle$ to a final state $|f\rangle$ in which one photon, polarized in the μ direction, has been absorbed, the matrix element takes the form

$$\langle f | E_\mu^{(+)}(\boldsymbol{r}t) | i \rangle . \tag{1.14}$$

We shall define an ideal photon detector as a system of negligible size (e.g., of atomic or subatomic dimensions) which has a frequency-independent photoabsorption probability. The advantage of imagining such a detector, as we shall show more explicitly in a later paper, is that the rate at which it records photons is proportional to the sum over all final states $|f\rangle$ of the squared absolute values of the matrix elements (1.14). In other words, the probability per unit time that a photon be absorbed by an ideal detector at point $\boldsymbol{r}$ at time t is proportional to

$$\sum_f \left| \langle f | E_\mu^{(+)}(\boldsymbol{r}t) | i \rangle \right|^2 = \sum_f \langle i | E_\mu^{(-)}(\boldsymbol{r}t) | f \rangle \langle f | E_\mu^{(+)}(\boldsymbol{r}t) | i \rangle \tag{1.15}$$
$$= \langle i | E_\mu^{(-)}(\boldsymbol{r}t) E_\mu^{(+)}(\boldsymbol{r}t) | i \rangle .$$

We may verify immediately from Eq. (1.12) that the rate at which photons are detected in the empty, or vacuum, state vanishes.

The photodetector we have described is the quantum-mechanical analog of what, in classical experiments, has been called a square-law detector. It is important to bear in mind that such a detector for quanta measures the average value of the product $E_\mu^{(-)} E_\mu^{(+)}$, and not that of the square of the real field $E_\mu(\boldsymbol{r}t)$. Indeed, it is easily seen from the foregoing work that the average value of $E_\mu^2(\boldsymbol{r}t)$ does not vanish in the vacuum state;

$$\langle \text{vac} | E_\mu^2(\boldsymbol{r}t) | \text{vac} \rangle > 0 .$$

The electric field in the vacuum undergoes zero-point oscillations which, in the correctly formulated theory, have nothing to do with the detection of photons.

Recording photon intensities with a single detector does not exhaust the measurements we can make upon the field, though it does characterize, in principle, virtually all the classic experiments of optics. A second type of measurement we may make consists of the use of two detectors situated at different points r and r' to detect photon coincidences or, more generally, delayed coincidences. The field matrix element for such transitions takes the form

$$\left\langle f \left| E_\mu^{(+)}(r't') E_\mu^{(+)}(rt) \right| i \right\rangle , \tag{1.16}$$

if both photons are required to be polarized along the μ axis. The total rate at which such transitions occur is proportional to

$$\sum_f \left| \left\langle f \left| E_\mu^{(+)}(r't') E_\mu^{(+)}(rt) \right| i \right\rangle \right|^2 = \left\langle i \left| E_\mu^{(-)}(rt) E_\mu^{(-)}(r't') E_\mu^{(+)}(r't') E_\mu^{(+)}(rt) \right| i \right\rangle . \tag{1.17}$$

Such a total rate is to be interpreted as a probability per unit (time)^2 that one photon is recorded at r at time t and another at r' at time t'. Photon correlation experiments of essentially the type we are describing were performed by Hanbury Brown and Twiss[1] in 1955 and have, subsequently, been performed by others[2].

Whatever may be the practical difficulties of more elaborate experiments, we may at least imagine the possibility of detecting n-fold delayed coincidences of photons for arbitrary n. The total rate per unit (time)^n for such coincidences will be proportional to

$$\left\langle i \left| E_\mu^{(-)}(r_1 t_1) \ldots E_\mu^{(-)}(r_n t_n) E_\mu^{(+)}(r_n t_n) \ldots E_\mu^{(+)}(r_1 t_1) \right| i \right\rangle , \quad n = 1, 2, 3 \ldots \tag{1.18}$$

The entire succession of such expectation values, therefore, possesses a simple physical interpretation.

In closing this survey we add a note on the commutation rules obeyed by the fields $\boldsymbol{E}^{(+)}$ and $\boldsymbol{E}^{(-)}$. It is easy to find these rules from the relation (1.3) for the real field $\boldsymbol{E}$, by decomposing its dependence on the two variables t and t' into positive and negative frequency parts. If the function $D_{\mu\nu}$ has the Fourier transform

$$D_{\mu\nu}(\boldsymbol{r}-\boldsymbol{r}', t-t') = \int_{-\infty}^{\infty} \mathcal{D}_{\mu\nu}(\omega, \boldsymbol{r}-\boldsymbol{r}') e^{-i\omega(t-t')} d\omega , \tag{1.19}$$

we see immediately that the commutator Eq. (1.3) has no part which is of positive frequency in both its t and t' dependences. Neither does it have any part of negative frequency in both its time dependences. It follows that all values of the field $\boldsymbol{E}^{(+)}(rt)$ commute with one another, and so too do those of $\boldsymbol{E}^{(-)}(rt)$, i.e., we have

$$\left[E_\mu^{(+)}(rt), E_\nu^{(+)}(r't') \right] = 0 , \tag{1.20}$$

$$\left[E_\mu^{(-)}(rt), E_\nu^{(-)}(r't') \right] = 0 , \tag{1.21}$$

for all points rt and $r't'$, and all μ and ν. Products of the $\boldsymbol{E}^{(+)}$ operators or products of the $\boldsymbol{E}^{(-)}$ operators such as occur in Eq. (1.18) may, therefore, be freely rearranged, but the operators $\boldsymbol{E}^{(+)}$ and $\boldsymbol{E}^{(-)}$ do not, in general, commute.

1.3 Field Correlations

The electromagnetic field may be regarded as a dynamical system with an infinite number of degrees of freedom. Our knowledge of the condition of such a system is virtually never so complete or so precise in practice as to justify the use of a particular quantum state $|\ \rangle$ in its description. In the most accurate preparation of the state of a field which we can actually accomplish some parameters, usually an indefinitely large number of them, must be regarded as random variables. Since there is no possibility in practice of controlling these parameters, we can only hope ultimately to compare with experiment quantities which are averages over the distributions of the unknown parameters.

Our actual knowledge of the state of the field is specified fully by means of a density operator ρ which is constructed as an average, over the uncontrollable parameters, of an expression bilinear in the state vector. If $|\ \rangle$ is a precisely defined state of the field corresponding to a particular set of random parameters, the density operator is defined as the averaged outer product of state vectors

$$\rho = \{|\ \rangle\langle\ |\}_{\text{av}}. \tag{1.22}$$

The weightings to be used in the averaging are the ones which best describe the actual preparation of the fields. It is clear from the definition that ρ is Hermitian, $\rho^\dagger = \rho$.

The average of an observable O in the quantum state $|\ \rangle$ is the expectation value, $\langle\ |O|\ \rangle$. It is the average of this quantity over the randomly prepared states which we compare with experiment. The average taken in this twofold sense may be written as

$$\{\langle\ |O|\ \rangle\}_{\text{av}} = \text{Tr}\{\rho O\}, \tag{1.23}$$

where the symbol Tr stands for the trace, or the sum of the diagonal matrix elements. Since we require the average of the unit operator to be one, we must have $\text{Tr}\rho = 1$. These considerations show that the average counting rate of an ideal photodetector, which is proportional to Eq. (1.15) in a completely specified quantum state of the field, is more generally proportional to

$$\text{Tr}\{\rho E^{(-)}_\mu(\mathbf{r}t) E^{(+)}_\mu(\mathbf{r}t)\} \tag{1.24}$$

when the state is less completely specified.

It is convenient at this point, as a simplification of notation, to confine our attention to a single vector component of the electric field. We suppose, for the present, that all of our detectors are fitted with polarizers and record only photons polarized parallel to an arbitrary unit vector $\mathbf{e}$. (If $\mathbf{e}$ is chosen as a complex unit vector, $\mathbf{e}^* \cdot \mathbf{e} = 1$, the photons detected may have arbitrary elliptical polarization.) We then introduce the symbols $E^{(+)}$ and $E^{(-)}$ for the projections of the complex fields in the direction $\boldsymbol{e}$ and $\boldsymbol{e}^*$,

$$E^{(+)}(\boldsymbol{r}t) = \boldsymbol{e}^* \cdot \boldsymbol{E}^{(+)}(\boldsymbol{r}t), \tag{1.25}$$

$$E^{(-)}(\boldsymbol{r}t) = \boldsymbol{e} \cdot \boldsymbol{E}^{(-)}(\boldsymbol{r}t). \tag{1.26}$$

We resume a fully general treatment of photon polarizations in Sect. 1.5.

The field average Eq. (1.24) which determines the counting rate of an ideal photodetector is a particular form of a more general type of expression whose properties are of considerable interest. In the more general expression, the fields $E^{(-)}$ and $E^{(+)}$ are evaluated at different space-time points. Statistical averages of the latter type furnish a measure of the correlations of the complex fields at separated positions and times. We shall define such a correlation function, $G^{(1)}$ for the $\boldsymbol{e}$ components of the complex fields as

$$G^{(1)}(\boldsymbol{r}t,\boldsymbol{r}'t') = \text{Tr}\{\rho E^{(-)}(\boldsymbol{r}t) E^{(+)}(\boldsymbol{r}'t')\}. \tag{1.27}$$

Only the values of this function at $\boldsymbol{r} = \boldsymbol{r}'$ and $t = t'$ are needed to predict the counting rate of an ideal photodetector. However, other values of the function become necessary, quite generally, when we use as detectors less ideal systems such as actual photo-ionizable atoms. In actual photodetectors the absorption of photons can not be localized too closely, either in space or in time. Atomic photoionization rates must be written, in general, as double integrals, over a microscopic range, of all the variables in $G^{(1)}(\boldsymbol{r}t,\boldsymbol{r}'t')$. Our interest in the function $G^{(1)}$ extends to widely spaced values of its variables as well. That field correlations may extend over considerable intervals of distance and time is essential to the idea of coherence, which we shall shortly discuss.

As we have noted earlier, our interest in averages of the field operators extends beyond quadratic ones. Just as we generalized the expression for the photon detection rate to define $G^{(1)}$, we may generalize the expression (1.17) for the photon coincidence rate and thereby define a second-order correlation function,

$$G^{(2)}(\boldsymbol{r}_1 t_1 \boldsymbol{r}_2 t_2 \boldsymbol{r}_3 t_3 \boldsymbol{r}_4 t_4) = \text{Tr}\{\rho E^{(-)}(\boldsymbol{r}_1 t_1) E^{(-)}(\boldsymbol{r}_2 t_2) E^{(+)}(\boldsymbol{r}_3 t_3) E^{(+)}(\boldsymbol{r}_4 t_4)\}. \tag{1.28}$$

This too is a function whose values, even at widely separated arguments, interest us.

In view of the possibility of discussing n-photon coincidence experiments for arbitrary n it is natural to define an infinite succession of correlation functions $G^{(n)}$. It is convenient in writing these to abbreviate a set of coordinates $(\boldsymbol{r}_j, t_j)$ by a single symbol, x_j. We then define the n-th order correlation function as

$$G^{(n)}(x_1 \ldots x_n, x_{n+1} \ldots x_{2n}) = $$
$$\text{Tr}\{\rho E^{(-)}(x_1) \ldots E^{(-)}(x_n) E^{(+)}(x_{n+1}) \ldots E^{(+)}(x_{2n})\}. \tag{1.29}$$

The correlation functions have a number of simple properties. It is easily verified that interchanging the arguments in $G^{(1)}$ leads to the complex conjugate function

$$G^{(1)}(\boldsymbol{r}'t',\boldsymbol{r}t) = \{G^{(1)}(\boldsymbol{r}t,\boldsymbol{r}'t')\}^*. \tag{1.30}$$

The same type of relation holds for all of the higher order functions

$$G^{(n)}(x_{2n}\ldots x_1) = \{G^{(n)}(x_1\ldots x_{2n})\}^* \, . \tag{1.31}$$

Furthermore, the commutation relations (1.20) and (1.21) show us that $G^{(n)}$ is unchanged by any permutation of its arguments $(x_1\ldots x_n)$, or its arguments $(x_{n+1}\ldots x_{2n})$. The fact that the complex fields $E^{(\pm)}$ individually satisfy the wave equation (1.2) leads to another useful property of the $G^{(n)}$. The n-th-order function satisfies $2n$ different wave equations, one for each of its arguments x_j, $(j = 1,\ldots,2n)$.

A large number of inequalities satisfied by the functions $G^{(n)}$ may be derived from the positive definite character of the density operator ρ. Derivations of several classes of these are presented in the Appendix. We confine ourselves, in this section, to mentioning some of the simpler and more useful inequalities, those which are linear or quadratic in the correlation functions. It is clear from Eq. (1.31) that all of the functions $G^{(n)}(x_1\ldots x_n, x_n\ldots x_1)$ are real. The linear inequalities assert that these functions are positive definite as well. We have then, in particular for $n = 1$, the self-evident relation

$$G^{(1)}(x_1,x_1) \geq 0 \, , \tag{1.32}$$

and for arbitrary n

$$G^{(n)}(x_1\ldots x_n, x_n\ldots x_1) \geq 0 \, . \tag{1.33}$$

These relations simply affirm that the average photon intensity of a field and the average coincidence counting rates are all intrinsically positive.

The simplest of the quadratic inequalities takes the form

$$G^{(1)}(x_1,x_1)\, G^{(1)}(x_2,x_2) \geq \left|G^{(1)}(x_1,x_2)\right|^2 \, . \tag{1.34}$$

Higher order inequalities of this type are given by

$$G^{(n)}(x_1\ldots x_n, x_n\ldots x_1)\, G^{(n)}(x_{n+1}\ldots x_{2n}, x_{2n}\ldots x_{n+1}) \geq$$
$$\left|G^{(n)}(x_1\ldots x_n, x_{n+1}\ldots x_{2n})\right|^2 , \tag{1.35}$$

which holds for arbitrary n. Different forms of these relations are obtained by permuting or equating coordinates. Various other inequalities are proved in the Appendix along with those noted.

It is interesting to note that when the number of quanta present in the field is bounded, the sequence of functions $G^{(n)}$ terminates. If the density operator restricts the number of photons present to be smaller than or equal to some value M, the properties of $E^{(\pm)}$ as annihilation and creation operators show that $G^{(n)} = 0$ for $n > M$.

Classical correlation functions bearing some analogy to $G^{(1)}$ have received a great deal of discussion in recent years, mainly in connection with the theory of noise in radio waves. A detailed application of the classical correlation theory to optics has been made by Wolf[5]. At the core of Wolf's analysis is a single correlation function Γ, defined as an average over an infinite time span of the product of two fields, evaluated at times separated by a fixed interval. The procedure of time averaging restricts the application of such an approach to the treatment of field distributions which are statistically stationary in time.

If we were to restrict the character of our density operator ρ to describe only stationary field distributions (e.g., by choosing ρ to commute with the field Hamiltonian) our function $G^{(1)}(\mathbf{r}t,\mathbf{r}'t')$ would depend only on the difference of the two times, $t-t'$. In that case the function $G^{(1)}$ would, in the classical limit (strong, low-frequency fields), agree numerically² with Wolf's function Γ. It should be clear, however, that the concepts of correlation and ultimately of coherence are quite useful in the discussion of nonstationary field distributions. The correlation functions $G^{(n)}$ which we have defined are ensemble averages rather than time averages and hence remain well-defined in fields of arbitrary time dependence.

1.4
Coherence

The term "coherence" has had long if somewhat varied use in areas of physics concerned with the electromagnetic field. In physical optics the term is used to denote a tendency of two values of the field at distantly separated points or at greatly separated times to take on correlated values. When optical means are used to superpose the fields at such points (e.g., as in Young's two-slit experiment) intensity fringes result. The possibility of producing such fringes in hypothetical superposition experiments epitomizes the optical definition of coherence. The definition has remained a satisfactorily explicit one only as long as optical experiments were confined to measuring field intensities, or more generally quantities quadratic in the field strengths. We have already noted that the photon correlation experiment of Hanbury Brown and Twiss[1], performed in 1955, is of an altogether new type and measures the average of a quartic expression[6]. The study of quantities of fourth and higher powers in the field strengths is the basis of all work in the recently developed area of nonlinear optics. It appears safe to assume that the number of such experiments will increase in the future, and that the concept of coherence should be extended to apply to them.

Another pressing reason for sharpening the meaning of coherence is provided by the recent development of the optical maser. The maser produces light beams of narrow spectral bandwidth which are characterized by field correlations extending over quite long ranges. Such light is inevitably described as coherent, but the sense in which the term is used has not been made adequately clear. If the sense is simply the optical one then, as we shall see, it may scarcely do justice to the potentialities of the device.

---

2) This is true provided Wolf's "disturbance" field V behaves ergodically and is identified with $E^{(+)}$.

The optical definition does not at all distinguish among the many ways in which fields may vary while remaining equally correlated at all pairs of points. That much greater regularities may exist in the field variations of a maser beam than are required by the optical definition of coherence may be seen by comparing the maser beam with the carrier wave of a radio transmitter. The latter type of wave ideally possesses a stability of amplitude which optically coherent fields need not have[7]. Furthermore, the field values of such a wave possess correlations of a much more detailed sort than the optical definition requires. These are properties best expressed in terms of the higher order correlation functions $G^{(n)}$, for $n > 1$.

To discuss coherence in quantitative terms it is convenient to introduce normalized forms of the correlation functions. Corresponding to the first-order function $G^{(1)}$ we define

$$g^{(1)}(\mathbf{r}t,\mathbf{r}'t') = \frac{G^{(1)}(\mathbf{r}t,\mathbf{r}'t')}{\{G^{(1)}(\mathbf{r}t,\mathbf{r}t)\,G^{(1)}(\mathbf{r}'t',\mathbf{r}'t')\}^{1/2}} \,. \tag{1.36}$$

It is immediately seen from Eq. (1.34) that $g^{(1)}$ obeys the inequality

$$\left|g^{(1)}(\mathbf{r}t,\mathbf{r}'t')\right| \leq 1 \,. \tag{1.37}$$

For $\mathbf{r} = \mathbf{r}'$, $t = t'$ we have, of course, $g^{(1)} \equiv 1$.

The normalized forms of the higher order correlation functions are defined as

$$g^{(n)}(x_1 \ldots x_{2n}) = \frac{G^{(n)}(x_1 \ldots x_{2n})}{\prod_{j=1}^{2n}\{G^{(1)}(x_j,x_j)\}^{1/2}} \,. \tag{1.38}$$

These functions, for $n > 1$, are not, in general, restricted in absolute value as is $g^{(1)}$.

We shall try in this paper to give the concept of coherence as precise a definition as is both realizable in physical terms, and useful as well[8]. We, therefore, begin by stating an infinite sequence of conditions on the functions $g^{(n)}$ which are to be satisfied by a fully coherent field. These necessary conditions for coherence are that the normalized correlation functions all have unit absolute magnitude,

$$\left|g^{(n)}(x_1 \ldots x_{2n})\right| = 1 \,, \quad n = 1,2,\ldots \tag{1.39}$$

That there exist at least some states which meet these conditions at all points of space and time is immediately clear from the example of a classical plane wave, $E^{(+)} \propto \exp[i(\mathbf{k}\cdot\mathbf{r} - \omega t)]$. We shall presently show that the class of coherent fields is vastly larger than that of individual plane waves.

The conditions (1.39) on the functions $g^{(n)}$ are stated only as necessary ones and need not be construed as defining coherence completely. We shall shortly, in fact, sharpen the definition somewhat further. It is worth noting at this point, however, that not all of the fields which have been described as "coherent" in the past meet the set of conditions (1.39) even approximately. There may be some virtue, therefore, in constructing a hierarchy of orders of coherence to discuss fields which do not have that property in its fullest sense. We shall state as a condition necessary for first-order coherence that $\left|g^{(1)}(\mathbf{r}t,\mathbf{r}'t')\right| = 1$. More generally, for a field to be characterized by n-th order coherence we shall require $\left|g^{(j)}\right| = 1$ for $j \leq n$. For fields which occur in practice, one can not expect relations such as these to hold exactly for all points in space and time. We shall, therefore, often employ the term n-th order coherence more loosely to mean that the first n coherence conditions are fairly accurately satisfied over appreciable intervals of the variables surrounding all points $x_1 = x_2 = \cdots = x_{2n}$.

The definition of coherence which has been used to date in all studies of physical optics corresponds only to first-order coherence. The most coherent fields which have been generated by optical means prior to the development of the maser, in fact, lack second and higher order coherence. On the other hand, the optical maser, functioning with ideal stability, may produce fields which are coherent to all orders.

The various orders of coherence may, in principle, be distinguished fairly directly in experimental terms. The inequality (1.33), which states that the n-fold coincidence counting rate is positive, requires that $g^{(n)}(x_1 \ldots x_n, x_n \ldots x_1)$ be positive. If the field in question possesses n-th-order coherence, it must, therefore, have

$$g^{(j)}(x_1 \ldots x_j, x_j \ldots x_1) = 1 , \qquad (1.40)$$

for $j \leq n$. It follows from the definitions of the $g^{(j)}$ that the corresponding values of the correlation functions $G^{(j)}$ factorize, i.e.,

$$G^{(j)}(x_1 \ldots x_j, x_j \ldots x_1) = \prod_{i=1}^{j} G^{(1)}(x_i, x_i) , \qquad (1.41)$$

for $j \leq n$. These relations mean, in observational terms, that the rate at which j-fold delayed coincidences are detected by our ideal photon counters, reduces to a product of the detection rates of the individual counters[6]. In photon coincidence experiments of multiplicity up to and including n, the photon counts registered by the individual counters may then be regarded as statistically independent events. No tendency of photon counts to be statistically correlated will be evident in j-fold coincidence experiments for $j \leq n$.

The experiments of Hanbury Brown and Twiss[1] were designed to detect correlations in the fluctuating outputs of two photomultipliers. These detectors were placed in fields made coherent with one another (in the optical sense) through the use of monochromatic, pinhole illumination and a semitransparent mirror. The photocurrents of the two detectors were observed to show a positive correlation for small delay times, rather than independent fluctuations. A similar experiment has been performed by Rebka and Pound[2], using coincidence counting equipment. Their experiment, performed with a more monochromatic beam and better geometrical definition, shows an explicit correlation in the counting probabilities of the two detectors. These observations verify that light beams from ordinary sources such as discharge tubes, when made optimally coherent in the first-order sense, still lack second-order coherence.

1.4 Coherence

The coherence conditions (1.39) can also be stated as a requirement that the functions $|G^{(n)}(x_1 \ldots x_{2n})|$ factorize into a product of $2n$ functions of the same form, each dependent on a single space-time variable,

$$|G^{(n)}(x_1 \ldots x_{2n})| = \prod_{j=1}^{2n} \left\{ G^{(1)}(x_j, x_j) \right\}^{1/2}. \tag{1.42}$$

This statement of the necessary conditions for coherence suggests that it may be convenient to give a stronger definition to coherence by regarding it as a factorization property of the correlation functions.

Let us suppose that there exists a function $\mathcal{E}(x)$, independent of n, such that the correlation functions for all n may be expressed as the products

$$G^{(n)}(x_1 \ldots x_n, x_{n+1} \ldots x_{2n}) = \mathcal{E}^*(x_1) \ldots \mathcal{E}^*(x_n) \mathcal{E}(x_{n+1}) \ldots \mathcal{E}(x_{2n}). \tag{1.43}$$

It is immediately clear that these functions satisfy the conditions (1.39) and (1.42). To show that fields with such correlations exist we need only refer again to the case of a classical plane wave. In fact, any classical field of predetermined (i.e., nonrandom) behavior has correlation functions which fall into this form, and such fields are at times called coherent in communication theory. We shall, therefore, adopt the factorization conditions (1.43) as the definition of a coherent field and turn next to the question of how they may be satisfied in the quantum domain.

If it were possible for the field to be in an eigenstate of the operators $E^{(+)}$ and $E^{(-)}$ the correlation functions for such states would factorize immediately to the desired form. The operators $E^{(+)}(\mathbf{r}t)$ and $E^{(-)}(\mathbf{r}'t')$ do not commute, however, so no state can be an eigenstate of both in the usual sense. Not only are these operators non-Hermitian, but the failure of each to commute with its adjoint shows that $E^{(+)}$ and $E^{(-)}$ are non-normal as well. Operators of this type can not, as a rule, be diagonalized at all, but may nonetheless have eigenstates. In general, we must distinguish between their left and right eigenstates; the two types need not occur in mutually adjoint pairs. The operator $E^{(+)}(\mathbf{r}t)$, in particular, has no left eigenstates, but does have right eigenstates[3] corresponding to complex eigenvalues for the field, which are functions of position and time. We shall suppose that $|\ \rangle$ is a right eigenstate of $E^{(+)}$ and that the equation it satisfies takes the form

$$E^{(+)}(\mathbf{r}t) |\ \rangle = \mathcal{E}(\mathbf{r}t) |\ \rangle, \tag{1.44}$$

in which the function $\mathcal{E}(\mathbf{r}t)$ is to be interpreted as the complex eigenvalue. The Hermitian adjoint of this relation shows us that the conjugate state, $\langle\ |$, is a left eigenstate of $E^{(-)}(\mathbf{r}t)$,

$$\langle\ | E^{(-)}(\mathbf{r}t) = \langle\ | \mathcal{E}^*(\mathbf{r}t). \tag{1.45}$$

---

[3] States of the harmonic oscillator which have an analogous property were introduced in a slightly different but related connection by E. Schrödinger, *Naturwiss.* **14**, 664 (1926). The electromagnetic field, as is well known, may be treated as an assembly of oscillators.

The density operator for such states is simply the projection operator, $\rho = |\ \rangle\langle\ |$. It follows immediately from these relations that the correlation functions $G^{(n)}$ all factorize into the form of Eq. (1.43). In other words, the state of the field defined by Eqns. (1.44) or (1.45) meets our definition precisely and is fully coherent. We shall discuss the properties of such states[4] at length in the paper to follow. For the present it may suffice to say that we can find an eigenstate $|\ \rangle$ which corresponds to the choice, as an eigenvalue, of any function $\mathcal{E}(rt)$ which satisfies certain conditions. One condition, which is clear from Eq. (1.44), is that $\mathcal{E}(rt)$ must satisfy the wave equation. The other, which corresponds to the positive frequency character of $E^{(+)}$, is that $\mathcal{E}(rt)$, when regarded as a function of a complex time variable, be analytic in the lower half-plane. The eigenstates which correspond to different fields $\mathcal{E}(rt)$ are not mutually orthogonal, but nonetheless form a natural basis for the discussion of photon detection problems. We have introduced them here only to demonstrate the possibility of satisfying the coherence conditions in quantum theory. Such quantum states do not exhaust the possibility of describing coherent fields. Statistical mixtures, for example, of the states for which the eigenvalues $\mathcal{E}(rt)$ differ by constant phase factors satisfy the coherence conditions equally well.

The fields which have been described as most coherent in optical contexts have tended to be those of the narrowest spectral bandwidth. If coherent fields in optics have necessarily been chosen as monochromatic ones, it is because that has been virtually the only means of securing appreciably correlated fields from intrinsically chaotic sources. For this reason, perhaps, there has been a natural tendency to associate the concept of coherence with monochromaticity. The association was, in fact, made an implicitly rigid one by earlier discussions[5] of optical (i.e., first-order) coherence which were applicable only to statistically stationary fields. By extending the definition of coherence to nonstationary fields we see that it places no constraint on the frequency spectrum. Coherent fields exist corresponding to eigenvalues $\mathcal{E}(rt)$ with arbitrary spectra. The coherence conditions restrict randomness of the fields rather than their bandwidth.

Having defined full coherence by means of the factorization conditions (1.43), we may now use them in defining the various orders of coherence. We shall speak of m-th-order coherent fields when the conditions (1.43) are satisfied for $n \leq m$, a definition which accords with our earlier conditions on $|g^{(j)}|$.

Photon correlation experiments have shown the importance of distinguishing between the first two orders of coherence. At the other end of the scale, we have shown that there exist, in principle at least, states which are fully coherent. We are entitled to ask, therefore, whether the intermediate orders of coherence will also be useful classifications. In the absence of any experimental information, we can only guess that they may be useful, though perhaps not in the sharp sense in which we we have defined them. One may easily imagine the possibility that, for light sources such as the maser,

---

[4] Some of the properties of these states have already been noted in Refs.[6] and[8].

the correlation functions $G^{(n)}$ show gradually increasing departure from the factored forms (1.43) as n increases, even when the variables $x_1 \ldots x_{2n}$ are not too widely separated. In such contexts the order of coherence can only be defined approximately.[5] Something of the same approximate character must be present in all applications of the definitions we have given. The field correlations we have discussed can extend over great intervals of distance and time, though never infinite ones in practice. Coherence conditions, such as $|g^{(n)}| = 1$, can only be met within a finite range of relative values of the coordinates $x_1 \ldots x_{2n}$. It is only within such ranges, and therefore as an approximation, that we can speak of coherence at all.

1.5 Coherence and Polarization

We have to this point, in the interest of simplicity, dealt only with the projections of the fields along a single (possibly complex) unit vector $\mathbf{e}$. To take fuller account of the vector nature of the fields we must define tensor rather than scalar correlation functions. The first-order function is taken to be

$$G^{(1)}_{\mu\nu}(x,x') = \mathrm{Tr}\left\{\rho E^{(-)}_\mu(x) E^{(+)}_\nu(x')\right\}, \tag{1.46}$$

in which the indices μ and ν label Cartesian components. This function satisfies the symmetry relation

$$G^{(1)}_{\nu\mu}(x',x) = \{G^{(1)}_{\mu\nu}(x,x')\}^*, \tag{1.47}$$

and is shown in the appendix to obey the inequalities,

$$G^{(1)}_{\mu\mu}(x,x) \geq 0 \tag{1.48}$$

and

$$G^{(1)}_{\mu\mu}(x,x)\, G^{(1)}_{\nu\nu}(x',x') \geq \left|G^{(1)}_{\mu\nu}(x,x')\right|^2. \tag{1.49}$$

The photon intensities which can be detected at the space-time point x are found from $G^{(1)}_{\mu\nu}(x,x')$ for $x' = x$. We shall abbreviate this 3×3 matrix as $\mathbf{G}^{(1)}(x)$, and use it as the basis of a brief discussion of polarization correlations in three dimensions, a subject which seems to have received little attention in comparison to plane polarizations. The symmetry relation (1.47) for $x' = x$ shows that the intensity matrix $\mathbf{G}^{(1)}(x)$ is

---

[5] The characterization we have given the n-th-order coherent fields is, in principle, an accurately realizable one, however. States with such properties may be constructed in a variety of ways. The factorization conditions can be met for $j \leq n$, for example, by suitably chosen states in which the number of photons present may take on any value up to n. The correlation functions of order $j > n$ then vanish, as we have noted earlier. The vanishing of these correlation functions for states with bounded numbers of quanta shows, incidentally, that no bound can be placed on the photon number in a fully coherent field.

Hermitian; an argument given in the Appendix shows it to be a positive definite matrix as well. It follows that $\mathbf{G}^{(1)}(x)$ has positive real eigenvalues, $\lambda_p(x)$, $(p = 1, 2, 3)$, which correspond to a set of (generally, complex) eigenvectors. The eigenvectors, which we write as $\mathbf{e}^{(p)}$ satisfy

$$\begin{aligned} \mathbf{G}^{(1)}(x) \cdot \mathbf{e}^{(p)*} &= \lambda_p \mathbf{e}^{(p)*}, \\ \mathbf{e}^{(p)} \cdot \mathbf{G}^{(1)}(x) &= \lambda_p \mathbf{e}^{(p)}. \end{aligned} \quad (1.50)$$

If the three eigenvalues $\lambda_p(x)$ are all different, it is clear that the three eigenvectors must be orthogonal; if not they may be chosen so. If the eigenvectors are normalized to obey the relations

$$\mathbf{e}^{(p)} \cdot \mathbf{e}^{(q)*} = \delta_{pq}, \quad (1.51)$$

their components form the unitary matrix which diagonalizes $\mathbf{G}^{(1)}(x)$. The eigenvectors, or equivalently the unitary matrix, are determined by a set of eight independent real parameters. A tensor product, such as

$$\mathbf{e}^{(p)} \cdot \mathbf{G}^{(1)}(x) \cdot \mathbf{e}^{(q)*} = \lambda_p \delta_{pq}, \quad (1.52)$$

expresses the correlation, at the point x, of the field components in the $\mathbf{e}^{(p)}$ and $\mathbf{e}^{(q)}$ directions. It is clear, then, that there always exist a set of three (complex) orthogonal polarization vectors such that the field components in these directions are statistically uncorrelated. The eigenvalues λ_p correspond to the intensities for these polarizations. For quantitative discussions of polarization it is convenient to define the normalized intensities $I_p = \lambda_p / \sum_q \lambda_q$ ($p = 1, 2, 3$), which sum to unity, $\sum_p I_p = 1$. When the normalized intensities are all equal to $\frac{1}{3}$ we have the case of an isotropic field, as in a hohlraum filled with thermal radiation.

The triad of eigenvectors at a point in an arbitrary field depends, in general, on time as well as position. If the density operator, ρ, represents a stationary ensemble, however, the triad becomes fixed. A particular example which has been studied in minute detail in optics is that of a beam of plane waves[5,9]. In that case, since the fields are transverse, one of the eigenvectors may be chosen as the beam direction and obviously corresponds to the eigenvalue zero. The net polarization of the beam is usually defined as the magnitude of the difference of the normalized intensities, $|I_1 - I_2|$, which correspond to the remaining two eigenvalues.

We next define the higher order correlation functions as

$$G^{(n)}_{\mu_1 \ldots \mu_{2n}}(x_1 \ldots x_n, x_{n+1} \ldots x_{2n}) = \\ \operatorname{Tr} \left\{ \rho E^{(-)}_{\mu_1}(x_1) \ldots E^{(-)}_{\mu_n}(x_n) E^{(+)}_{\mu_{n+1}}(x_{n+1}) \ldots E^{(+)}_{\mu_{2n}}(x_{2n}) \right\}. \quad (1.53)$$

These functions are unchanged by simultaneous permutations of the coordinates $(x_1 \ldots x_n)$ and the indices $(\mu_1 \ldots \mu_n)$; they are likewise invariant under permutations of the $(x_{n+1} \ldots x_{2n})$ and $(\mu_{n+1} \ldots \mu_{2n})$. They satisfy the symmetry relation

$$G^{(n)}_{\mu_{2n} \ldots \mu_1}(x_{2n} \ldots x_1) = \left\{ G^{(n)}_{\mu_1 \ldots \mu_{2n}}(x_1 \ldots x_{2n}) \right\}^* \quad (1.54)$$

1.5 Coherence and Polarization

and are shown, in the Appendix, to obey the inequalities

$$G^{(n)}_{\mu_1\ldots\mu_n\mu_n\ldots\mu_1}(x_1\ldots x_n x_n\ldots x_1) \geq 0 \tag{1.55}$$

and

$$G^{(n)}_{\mu_1\ldots\mu_n\mu_n\ldots\mu_1}(x_1\ldots x_n, x_n\ldots x_1) \times G^{(n)}_{\mu_{n+1}\ldots\mu_{2n}\mu_{2n}\ldots\mu_{n+1}}(x_{n+1}\ldots x_{2n}, x_{2n}\ldots x_{n+1}) \geq \left|G^{(n)}_{\mu_1\ldots\mu_n\mu_{n+1}\ldots\mu_{2n}}(x_1\ldots x_n, x_{n+1}\ldots x_{2n})\right|^2. \tag{1.56}$$

As in our earlier discussion of coherence, it is convenient to make use of the normalized correlation functions

$$g^{(n)}_{\mu_1\ldots\mu_{2n}}(x_1\ldots x_{2n}) = \frac{G^{(n)}_{\mu_1\ldots\mu_{2n}}(x_1\ldots x_{2n})}{\prod_{j=1}^{2n}\left\{G^{(1)}_{\mu_j\mu_j}(x_j,x_j)\right\}^{1/2}} \tag{1.57}$$

The necessary conditions for full coherence are

$$\left|g^{(n)}_{\mu_1\ldots\mu_{2n}}(x_1\ldots x_{2n})\right| = 1, \tag{1.58}$$

which must hold for all components $\mu_1\ldots\mu_{2n}$, as well as all n. It is clear, however, that these conditions do not constitute an adequate definition of coherence, since they are not, in general, invariant under rotations of the coordinate axes. We therefore turn once again to a definition of coherence as a factorization property of the correlation functions.

We define full coherence to hold when the set of correlation functions $G^{(n)}$ may be expressed as products of the components of a vector field $\mathcal{E}_\mu(x)$, ($\mu = 1,2,3$), i.e.,

$$G^{(n)}_{\mu_1\ldots\mu_{2n}}(x_1\ldots x_n, x_{n+1}\ldots x_{2n}) = \prod_{j=1}^{n}\mathcal{E}^*_{\mu_j}(x_j) \prod_{l=n+1}^{2n}\mathcal{E}_{\mu_l}(x_l), \tag{1.59}$$

where it is understood that the vector field $\mathcal{E}_\mu(x)$ is independent of n. It is immediately clear, from the transformation properties of the definition, that a field coherent in one coordinate frame is equally coherent in any rotated frame. Furthermore, all of the normalized correlation functions $g^{(n)}$, which follow from the definition, satisfy the conditions (1.58).

The coherence conditions (1.59) imply that the field is fully polarized in the direction of the vector $\mathcal{E}(x)$ at each point x. The formal way of seeing this is to note that the intensity matrix $G^{(1)}_{\mu\nu}(x,x)$, which we discussed earlier in general terms, reduces for a coherent field to,

$$G^{(1)}_{\mu\nu}(x,x) = \mathcal{E}^*_\mu(x)\mathcal{E}_\nu(x). \tag{1.60}$$

Such a matrix represents an unnormalized projection operator for the direction of $\mathcal{E}(x)$. It obviously has, as an eigenvector in the sense of Eq. (1.50), the vector $\mathcal{E}_\mu(x)$

itself. The corresponding eigenvalue is the full intensity $\sum_\mu |\mathcal{E}_\mu(x)|^2$. The two remaining eigenvalues, which correspond to orthogonal directions, clearly vanish.

It is interesting to note that for coherent fields many of the inequalities stated earlier, e.g., Eqns. (1.34), (1.35), (1.49), (1.56), reduce to statements of equality. This reduction holds quite generally, as is shown in the Appendix, for those inequalities of quadratic and higher degree in the correlation functions.

The arguments by which we exhibit fields satisfying the coherence conditions, are essentially unchanged from the previous section. In particular, as we shall discuss in the next paper, there exist states which are simultaneously right eigenstates of all three components of $E_\mu^{(+)}(\mathbf{r}t)$ and correspond to a set of three complex eigenvalues $\mathcal{E}_\mu(\mathbf{r}t)$. Such states satisfy the coherence conditions (1.59) precisely.

If we have chosen to discuss only the correlations of the electric field in this paper, it is because that field plays the dominant role in all detection mechanisms for photons of lower frequency than x rays. It is not difficult to construct correlation functions which involve the magnetic field as well as the electric field, and perhaps these too will someday prove useful. One method is to use the relativistic field tensor, $F_{\mu\nu}$, in precisely the way we have used the field E_μ. The field tensor may be written as a 4×4 antisymmetric matrix, made up of the components of both tie electric and magnetic fields. The n-th-order correlation function for the complex components of those fields would have $4n$ four-valued indices. Coherence may then be defined as a requirement that the correlation functions all be separable into the the products of 4×4 antisymmetric fields, just as Eq. (1.59) requires a separation into products of three-vector field components. The advantage of such a definition is to make it clear that coherence is a relativistically invariant concept; that a field which is coherent in any one Lorentz frame is coherent in any other. Fields which are coherent in this relativistic sense are automatically coherent in the more limited senses we have described earlier.

Appendix

In this section we derive a number of inequalities obeyed by the correlation functions defined in the paper. Fundamentally, these relations are all consequences of a single inequality

$$\text{Tr}\left\{\rho A^\dagger A\right\} > 0, \qquad (1.\text{A}1)$$

which holds for arbitrary choice of the operator A. To prove this inequality, we note that the density operator ρ is Hermitian and can always be diagonalized, i.e., we can find a set of basis states such that the matrix representation of ρ is

$$\langle k|\rho|l\rangle = \delta_{kl} p_k. \qquad (1.\text{A}2)$$

The numbers p_k may be interpreted as probabilities associated with the states $|k\rangle$. They are, therefore, non-negative, $p_k \geq 0$; which is to say that ρ is a positive definite operator. The normalization condition on the density operator, $\mathrm{Tr}\,\rho = \sum p_k = 1$, shows that not all the p_k vanish. The trace Eq. (1.A1) may be reduced, in the representation defined by Eq. (1.A2), to the form

$$\mathrm{Tr}\left\{\rho A^\dagger A\right\} = \sum_k p_k \langle k|A^\dagger A|k\rangle \,. \tag{1.A3}$$

The diagonal matrix elements on the right of Eq. (1.A3) are all non-negative since they may be expressed as a sum of squared absolute values,

$$\begin{aligned}\langle k|A^\dagger A|k\rangle &= \sum_l \langle k|A^\dagger|l\rangle \langle l|A|k\rangle \\ &= \sum_l |\langle l|A|k\rangle|^2 \,.\end{aligned} \tag{1.A4}$$

This statement completes the proof of Eq. (1.A1), since the trace is invariant under unitary transformations of the basis states.

The trace which occurs in the inequality (1.A1) has the same basic structure as all of the correlation functions $G^{(n)}$. Various inequalities relating the correlation functions follow, more or less directly, from different choices of the operator A. If, for example, we choose A to be $E^{(+)}(x)$, as defined by Eq. (1.25), we find the inequality (1.32),

$$G^{(1)}(x,x) \geq 0 \,. \tag{1.A5}$$

If we choose A to be the n-fold product $E^{(+)}(x_1)\ldots E^{(+)}(x_n)$ we find the inequality (1.33),

$$G^{(n)}(x_1\ldots x_n, x_n\ldots x_1) \geq 0 \,. \tag{1.A6}$$

The proofs are no different if the components of the three-dimensional field are used in place of $E^{(+)}$, i.e., if a component index μ_j is associated with each coordinate x_j. Hence, we have also derived Eqns. (1.48) and (1.55).

The remaining inequalities are of second and higher degree in the correlation functions. Those obeyed by the first-order function, $G^{(1)}$, may be found as follows: We choose at random a set of m space-time points $x_1\ldots x_m$, and consider as the operator A,

$$A = \sum_{j=1}^m \lambda_j E^{(+)}(x_j) \,, \tag{1.A7}$$

where the superposition coefficients $\lambda_1 \ldots \lambda_m$ are an arbitrary set of complex numbers. When we substitute Eq. (1.A7) into the basic inequality, Eq. (1.A1), we find

$$\sum_{i,j} \lambda_i^* \lambda_j G^{(1)}(x_i,x_j) \geq 0 \,. \tag{1.A8}$$

In other words, the set of correlation functions $G^{(1)}(x_i,x_j)$ for $i,j = 1,\ldots,m$ forms the matrix of coefficients of a positive definite quadratic form. It follows, in particular, that the determinant of the matrix is non-negative,

$$\det[G^{(1)}(x_i,x_j)] \geq 0 \quad i,j = 1,\ldots,m\,. \tag{1.A9}$$

For $m = 1$ this inequality is simply Eq. (1.A5). For $m = 2$ it becomes the one noted in the text as Eq. (1.34),

$$G^{(1)}(x_1,x_1)\,G^{(1)}(x_2,x_2) \geq \left|G^{(1)}(x_1,x_2)\right|^2\,. \tag{1.A10}$$

For larger values of m the inequalities are perhaps best left in the form Eq. (1.A9). When tensor components are introduced, we have only to replace the coordinate x_j in the proofs by the combination of x_j and a tensor index μ_j. The relation (1.49) thereby follows from the form of Eq. (1.A10). If, in particular for $m = 3$, we choose the three coordinates to be the same and the tensor indices all different, i.e., we choose

$$A = \sum_{\nu=1}^{3} \lambda_\nu E_\nu^{(+)}(x)\,, \tag{1.A11}$$

we find that the 3×3 matrix $G^{(1)}_{\mu\nu}(x,x)$ is positive definite, a property used in the text in the discussion of polarizations.

Since the succession of inequalities which follows from Eq. (1.A1) is endless, we only mention the quadratic ones for the higher order functions. To find these, we choose a set of $2n$ coordinates at random and let A be any operator of the form

$$A = \lambda_1 E^{(+)}(x_1)\ldots E^{(+)}(x_n) + \lambda_2 E^{(+)}(x_{n+1})\ldots E^{(+)}(x_{2n})\,. \tag{1.A12}$$

The positive definiteness of the quadratic form which results from substituting this expression in Eq. (1.A1) shows that the inequality (1.35) must hold. When vector indices are attached to the operators $E^{(+)}$, the same proof leads to Eq. (1.56).

We have noted in the text that, for the particular case of coherent fields, the inequalities of second degree in the correlation functions reduce to equalities. The reason for the reduction lies in the way the correlation functions factorize. The factorization causes all of the second and higher order determinants involved in the statement of positive definiteness conditions [e.g., Eq. (1.A9)] to vanish.

References

1 R. Hanbury Brown, R. Q. Twiss, *Nature* **177**, 27 (1956); *Proc. Roy. Soc. (London) A* **242**, 300 (1957); *Proc. Roy. Soc. (London) A* **243**, 291 (1957).

2 G. A. Rebka, R. V. Pound, *Nature* **180**, 1035 (1957).

3 E. M. Purcell, *Nature* **178**, 1449 (1956).

4 See, for example, P. A. M. Dirac, *The Principles of Quantum Mechanics*, 3rd ed., pp. 239–242; Oxford University Press, New York 1947.

5 M. Born, E. Wolf, *Principles of Optics*, Chap. X; Pergamon Press, London, 1959. An extensive bibliography is given there.

6 R. J. Glauber, *Phys. Rev. Lett.* **10**, 84 (1963). The particular field referred to as incoherent in that note may have first-order coherence if it is monochromatic, but not second- or higher order coherence.

7 This point has been noted with particular clarity by M. J. E. Golay, *Proc. IRE* **49**, 959 (1961); also **50**, 223 (1962).

8 A brief account of this work was presented by R. J. Glauber, in *Proc. 3rd Int. Conf. on Quantum Electronics*, Vol. I, p. 111; Dunod, Paris 1964.

9 Most of these studies have been confined to stationary, quasimonochromatic beams. See, for example, G. B. Parrent, Jr., P. Roman, *Nuovo Cimento* **15**, 370 (1960).

2
Optical Coherence and Photon Statistics[1]

2.1
Introduction

The field of optics, after seeming to have reached a sort of maturity, is beginning to undergo some rapid and revolutionary changes. These changes are connected with things which we have, as a matter of principle, known about for many years, but the extent to which we could put our knowledge into practice has, until just a few years ago, been extremely limited. Thus the electromagnetic character of light waves has been familiar knowledge since the last century. A vast body of theory and technique concerning the generation of electromagnetic waves has been built up during these years, but virtually all of it has dealt with radio frequency fields. Light waves of course, are of the same electromagnetic character as radio waves. But because the only ways we had of generating them in the past were extremely clumsy (in a sense we shall presently discuss at some length) there has been very little occasion until recently to apply the insights of radio-frequency theory in optics. A simple physical reason, as we shall see, lies at the bottom of this: all of the traditional types of optical sources possess a certain chaotic quality in common. They are what a radio engineer would refer to as noise generators, and all of the delicate and ingenious techniques of optics are exercises in the constructive use of noise. The invention of the optical maser lias removed this barrier with almost a single stroke. It allows us to presume that we will some day be able to control fields oscillating at optical or higher frequencies with the same sort of precision and versatility that have become familiar in radio frequency technology.

Another recent change is the development of detectors which respond strongly to individual quanta of light. These have permitted us to explore the corpuscular character of optical fields. All of the traditional optical experiments have not only dealt with extremely crude sources, but have paid very little attention to the detection of individual light quanta. The detectors used were topically sensitive only to substantial numbers of photons and were quite slow in action so that we measured only intensities

---

[1] Copyright 2006. From *Quantum Optics and Electronics* by C. deWitt, A. Blandin, C. Cohen-Tannoudji (Eds.), Gordon & Breach, New York 1965, pp. 65–185. Reproduced by permission of Routledge/Taylor and Francis Group, LLC.

Quantum Theory of Optical Coherence. Selected Papers and Lectures. Roy J. Glauber.
Copyright © 2007 WILEY-VCH Verlag GmbH & Co. KGaA, Weinheim
ISBN: 978-3-527-40687-6

which had been averaged over relatively long periods of time. The new light detectors enable us to ask more subtle questions than just ones about average intensities; we can for example, ask questions about the counting of pairs of quanta, and can make measurements of the probability that the quanta are present at an arbitrary pair of space points, at an arbitrary pair of limes.

If the instrumentation in optics has made long strides in the direction of dealing with photons, it is worth mentioning that the instrumentation in the radio frequency field is leading in that direction as well. The energies of radio frequency photons are extremely small, much smaller than the thermal fluctuation energy kT (T = noise temperature $\sim$ room temperature for most amplifiers). There has consequently not been much need in radio frequency technology to date to pay attention to the corpuscular structure of the field. The recent invention, however, of low noise amplifiers, such as the microwave maser, has lowered the noise temperature of the detecting device to such a degree that with further progress it seems not impossible that individual photons maybe detected. So, even in the microwave region, there is now a certain amount of attention being paid to the corpuscular structure of light.

It is interesting, in any case, to investigate the corpuscular nature of electromagnetic fields, because it will set the ultimate limitation to the possibility of transmitting information by means of fields. We will not discuss information theory in this article, but we will have some things to cay which are related to noise theory. Noise theory is the classical form of the theory of fluctuations of the electromagnetic field and is quite naturally related to the theory of quantum fluctuations of the field. All of these subjects fall under a general heading which we might call photon statistics. Coherence theory too, is properly speaking, a rather small area of the same general subject. Its purpose is simply to formulate some useful ways of classifying the statistical behaviour of fields.

The problem to which we shall address ourselves in this article is the construction of a fairly rigorous and general treatment of the problems of photon statistics. There is no need, in doing it, to make any material distinction between radio frequency and optical fields (or between these and X-ray fields for that matter). A part of the formalism, that which has to do with the definition of coherence, is suggested in fact as a way of unifying the rather different concepts of coherence, which have characterized these areas in the past.

We have already remarked that optical experiments have only rarely dealt with individual photons. Much the same observation can be made for optical theory as well. If the photon has to such a remarkable degree remained a stranger to optical theory some justification for that fact surely lies in the great success of the simple, wave models in the analysis of optical experiments. Such models are usually spoken of as being classical in character since they proceed typically from some kind of analogy to classical electromagnetic theory and pay as little attention to the corpuscular character of the radiation as the experimental arrangement will permit.

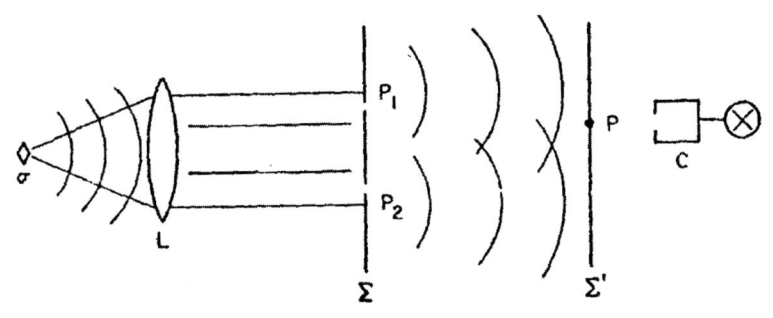

Figure 1

In these approaches one talks typically about some kind of "optical disturbance function" which is assumed to obey the wave equation and perhaps certain boundary conditions as well. The function may represent the components of tho electric vector or possibly other field quantities such as the vector potential, or the magnetic field. In many applications in fact one does not need to be very specific about what it really does or does not represent.

Let us consider the Young interferometer (Fig. 1) in order to illustrate the elementary approaches we are discussing. A plane, quasi-monochromatic wave coming from a point source σ impinges on the screen Σ with two parallel slits at the positions P_1 and P_2.

The two waves emerging from the slits give rise to an interference pattern on the screen Σ', which we can often see with the unaided eye. The simplest way of predicting the form of the interference pattern is to ignore the vector character of the electromagnetic field and introduce a scalar field φ which is presumed to describe the "optical disturbance". We then try to find a function φ which satisfies the wave equation together with a set of boundary conditions which we take to represent the effect of the screen Σ. That problem, as you remember, is in general a good deal too difficult to be solved exactly, and it is customary to make a number of simplifying approximations such as dealing very crudely with the boundary conditions, and making use of Huygens' principle. By these familiar methods we reach a simple evaluation of the field distribution φ on the screen Σ'.

Of course, if we are to predict the form of the interference pattern, we must at some stage face the question of attaching a physical interpretation to the fields. The most familiar approach is to regard $\varphi(r,t)$ as a real field and to identify it, perhaps, with one of the components of the electric field vector. The experimental fringe pattern is then predicted quite accurately, as we all know, if the light intensity on the screen is identified with φ^2, the square of our optical field. The identification possesses the justification, from the standpoint of classical theory, that the Poynting vector, which tells us the energy flux, is indeed quadratic in the field strength. In spite of this evident support the identification is not a unique one, however; it pays too little attention to the way in which the light is detected.

Let us suppose that the light intensity is measured by using a photon counter at the position of the screen. We then ask how we may predict the response of the counter as it is used to probe the pattern. Although the use of the wave equation to find the field amplitude φ did not introduce any distinctions between the classical and the quantum theoretical approaches to the diffraction problem, the use of a photon counter as a detector does introduce a distinction. The photon counter is an intrinsically quantum mechanical instrument. Its output is only predictable in terms of statistical averages even when the state of the field is specified precisely. If we are to predict this average response we must be rather more specific than we have thus far been about the field which the counter sees and we must treat the detection mechanism in a fully quantum mechanical way. What we find when we do these things is that the counter may be more accurately thought of as responding to a complex field φ^+ rather than the real field φ, and as having an output proportional, not to φ^2, but to $|\varphi^+|^2$, (The distinction is not a trivial one physically, since in a monochromatic field φ^2 oscillates rapidly in magnitude while $|\varphi^+|^2$ remains constant.) Once this answer is known it can be used as a crude rule for bypassing the explicit discussion of the detection mechanism in applications to other detection problems.

The use of such rules as a means of avoiding the explicit use of quantum mechanics has several times been called the "semi-classical approach". While approaches of this type clearly need a rule of some sort to bridge the gap between their descriptions of the wave and particle behaviors of photons they may remain perfectly correct approaches in a quantum mechanical sense as long as the rule has been chosen correctly. The fact that a mistaken form of this rule lias been used repeatedly in "semi-classical" discussions is a good indication that the fully quantum mechanical discussion is not entirely beside the point.

One of the properties of the "semi-classical" approaches that makes them elementary is that they deal with ordinary numbers and functions. They make no use of the apparatus of non-commuting operators which, it may appear, ought to be part of any formal quantum mechanical description of the field. Later in this article we shall show that for a certain class of fields there need be no error in a statistical description of the field which is based upon such ordinary functions as we find by solving the wave equation. It is possible to describe these fields fully by means which are rather similar to those used in the classical theory of noise. Where such a description is available it means that there need be nothing incorrect about the so-called "classical" or "semi-classical" approaches except their names, which then become totally misleading. It has recently been claimed that the class of states of the field for which the simple statistical description we have mentioned is available includes all states of the field, and that consequently the quantum theory and the "classical" theory will always yield equivalent results. We shall have to return to this point later in the article when we are better equipped to discuss it, but for the present we may remark that this claim seems to be based more upon wishful thinking than upon accurate mathematics. The quantum theory still offers the only complete and logically consistent basis for discussing field phenomena.

The general subject we shall be discussing, to give it its most imposing name, is quantum electrodynamics. It is an extremely well developed subject. Although it has long been clear that classical electrodynamics is the limit of quantum electrodynamics for $h \to 0$, there have never been any very powerful methods available for discussing electrodynamical problems near the classical limit.

All of quantum electrodynamics has historically been developed in terms of the stationary states $|n\rangle$ of the field Hamiltonian $\mathcal{H}$. These correspond to the presence of an integer number n of quanta, i.e. they obey the equation

$$\mathcal{H}|n\rangle = \left(n + \tfrac{1}{2}\right)\hbar\omega |n\rangle . \tag{2.1}$$

The n-quantum states form a complete set which has usually been regarded as the "natural" basis for the development of all states of the field. To the extent that virtually all electrodynamical calculations have been done by means of expansions in powers of the field strengths, the numbers of photons which have been dealt with in the calculations have usually been very small integers. The classical limit of quantum electrodynamics, on the other hand, is one in which the quantum numbers are typically quite large. Not only are they large but they are typically quite uncertain. If, for example, a harmonic oscillator is vibrating in a state with a relatively well defined phase, it is necessary that it not only be in a state with a large quantum number, but that the quantum number of the state also be quite uncertain, $(\Delta n \Delta \varphi \gtrsim 1)$. When we must deal with quantum states of the electromagnetic field for which the phase of the field is well defined, they can likewise only be states in which the occupation number n is intrinsically rather indefinite. In such cases the description of expectation values in terms of the n-quantum states becomes rather awkward and untransparent.

One of the mathematical tools we shall use in this article is a set of quantum states rather better suited to the description of amplitude and phase variables than the n-quantum states. The use of these states makes the relationship of the classical and quantum mechanical forms of electrodynamics considerably clearer than it has been before.

2.1.1
Classical Theory

It may help to underscore the close connection between the quantum theory we shall develop and the classical theory if we begin by discussing the classical theory alone for a while. We shall describe the classical field in terms of the familiar field vectors, the electric field $\boldsymbol{E}(\boldsymbol{r},t)$ and the magnetic field $\boldsymbol{B}(\boldsymbol{r},t)$. We will take these to obey the source-free Maxwell equations

$$\nabla \cdot \boldsymbol{E} = 0, \qquad \nabla \times \boldsymbol{E} = -\frac{1}{c}\frac{\partial \boldsymbol{B}}{\partial t},$$
$$\nabla \cdot \boldsymbol{B} = 0, \qquad \nabla \times \boldsymbol{B} = \frac{1}{c}\frac{\partial \boldsymbol{E}}{\partial t}, \tag{2.2}$$

by assuming that whatever source has radiated the fields has ceased to radiate further. Since our detectors are usually sensitive to electric rather than magnetic fields, we shall confine ourselves to a discussion of the field $\boldsymbol{E}(\boldsymbol{r},t)$. One of the first things which is done in many classical calculations is to use a Fourier series or integral to expand the time dependence of the field and in that way to separate the field into two complex terms:

$$\boldsymbol{E}(\boldsymbol{r},t) = E^{(+)}(\boldsymbol{r},t) + E^{(-)}(\boldsymbol{r},t) \, . \tag{2.3}$$

The first of these terms, which we shall call the positive frequency part, $E^{(+)}$, contains all the amplitudes which vary as $e^{-i\omega t}$ for $\omega > 0$. The other, the negative frequency part, contains all amplitudes which vary as $e^{i\omega t}$. These terms are complex conjugates of one another

$$E^{(-)} = E^{(+)*} \tag{2.4}$$

and contain equivalent physical information. Either one or the other is frequently used in classical calculations and called either the complex field strength or the complex signal. The use of these complex fields in classical contexts is usually regarded as a mathematical convenience rather than a physical necessity since classical measuring devices tend to respond only to the real field, $E = 2 \operatorname{Re} E^{(\pm)}$.

Quantum mechanical detectors, as we have noted, behave rather differently from classical ones, and for the discussion of these the separation of the field into its positive and negative frequency parts takes on a much deeper significance than it does for classical detectors. As we shall later see, an ideal photon counter (one which has zero size and is equally sensitive to all frequencies) measures the product $E^{(-)}(\boldsymbol{r},t)E^{(+)}(\boldsymbol{r},t) = |E^{(+)}(\boldsymbol{r},t)|^2$. That, at least, is what the detector would measure if we were capable of preparing fields with precisely fixed field strengths. But of course we are never capable of controlling the motions of the charges in our sources with very great precision. In practice all fields are radiated by sources whose behavior is subject to considerable statistical uncertainty. The fields are then correspondingly uncertain and what we require is a way of describing this uncertainty in mathematical terms.

It is more convenient, in describing the randomness of the fields, to deal with a discrete set of variables than to deal with the whole continuum at once. We shall therefore only attempt to describe the field lying inside a certain volume of space within which we can expand it in terms of a discrete set of orthogonal mode functions. We shall take the set of vector mode functions $\{\boldsymbol{u}_k(\boldsymbol{r})\}$ to obey the wave equations

$$\left(\nabla^2 + \frac{\omega_k^2}{c^2}\right)\boldsymbol{u}_k(\boldsymbol{r}) = 0 \, , \tag{2.5}$$

which define a set of frequencies $\{\omega_k\}$ when they are satisfied together with the constraint

$$\nabla \cdot \boldsymbol{u}_k(\boldsymbol{r}) = 0 \tag{2.6}$$

and a suitable set of boundary conditions. These functions may be assumed to form an orthonormal set

$$\int \boldsymbol{u}_k^*(\boldsymbol{r}) \cdot \boldsymbol{u}_{k'}(\boldsymbol{r}) \, d\boldsymbol{r} = \delta_{kk'} \, , \tag{2.7}$$

which is complete within the volume being studied. They may then be used to express the electric field vector in the form

$$\boldsymbol{E}(\boldsymbol{r},t) = \sum_k C_k \boldsymbol{u}_k(\boldsymbol{r}) \, e^{-i\omega_k t} + \sum_k C_k^* \boldsymbol{u}_k^*(\boldsymbol{r}) \, e^{i\omega_k t} \, . \tag{2.8}$$

The two sums on the right are then evidently $E^{(+)}$ and $E^{(-)}$, respectively.

When the expansion in orthogonal modes is used the field is evidently specified completely by the set of complex Fourier amplitudes $\{C_k\}$. To describe random fields we must regard these numbers as random variables in general. Usually the most we can state about these coefficients can be expressed through a probability distribution $p(\{C_k\}) = p(C_1, C_2, C_3, \ldots)$. Then, if we measure some function of $\boldsymbol{E}$ or of $E^{(\pm)}$, the most we can hope to predict is its mean value, i.e., if we measure $F(E^{(+)})$ we can only hope to find the average

$$\left\langle F(E^{(+)}) \right\rangle = \int p(\{C_k\}) \, F\left[E^{(+)}(\{C_k\})\right] \prod_k d^2 C_k \, , \tag{2.9}$$

where the differential element of area is given by $d^2 C_k = d(\operatorname{Re} C_k) \, d(\operatorname{Im} C_k)$.

It is important to remember that this average is an ensemble average. To measure it we must in principle repeat the experiment many times by using the same procedure for preparing the field over and over again. That may not be a very convenient procedure to carry out experimentally but it is the only one which represents the precise meaning of our calculation. The fields we are discussing may vary with time in arbitrary ways. As an example we might take the field generated by a radio transmitter sending some arbitrarily chosen message. There is therefore no possibility in general of replacing the ensemble averages by time averages. The theory of non-stationary statistical phenomena can only be developed in terms of ensemble averages.

The solution of problems in statistical thermodynamics has accustomed us to thinking of statistical fluctuations about the ensemble average as being very small. We are thus usually willing to forget about the need in principle to make an ensemble of thermodynamic measurements and are content to compare just a single measurement with the predicted ensemble average. While the justification of such shortcuts may be excellent in thermodynamic contexts, it is not always so good in statistical optics. Thus when we speak later of the interference patterns produced by superposing light from independent sources we shall find that individual measurements yield results wholly unlike their ensemble averages. The distinction between particular measurements and their averages may thus be quite essential.

2.2
Interference Experiments

One of the classic experiments which exhibits the coherence properties of light is the Young experiment (Fig. 2).

The field present at P at time t may be approximated by a certain linear superposition of the fields present at the two pinholes at earlier times:

$$E^{(+)}(r,t) = \lambda_1 E^{(+)}(r_1,t_1) + \lambda_2 E^{(+)}(r_2,t_2), \qquad (2.10)$$

where the times are given by $t_{1,2} = t - S_{1,2}/c$. The coefficients λ_1, λ_2 depend on the geometry of the arrangement, but are taken to be independent of the properties of the field.

We shall assume, to begin the discussion, that a photodetector placed at P measures the squared absolute value of some component of the complex field strength. (At a later point we shall discuss the validity of the assumption in some detail.) If we write the measured field component as $E^{(+)}(r,t)$, we then have

$$\left|E^{(+)}(r,t)\right|^2 = E^{(-)}(r,t)E^{(+)}(r,t)$$

$$= |\lambda_1|^2 E^{(-)}(r_1,t_1)E^{(+)}(r_1,t_1) + |\lambda_2|^2 E^{(-)}(r_2,t_2)E^{(+)}(r_2,t_2)$$

$$+ 2\operatorname{Re}\left\{\lambda_1^* \lambda_2 E^{(-)}(r_1,t_1)E^{(+)}(r_2,t_2)\right\}. \qquad (2.11)$$

Now since our preparation of the source rarely fixes the Fourier coefficients C_k very precisely we must in principle perform the experiment repeatedly and then average in order to find a non-random result. The only thing we can really predict is the ensemble average of $\left|E^{(+)}(r,t)\right|^2$ taken over the set of random coefficients $\{C_k\}$,

$$\left\langle \left|E^{(+)}(r,t)\right|^2 \right\rangle = |\lambda_1|^2 \left\langle \left|E^{(+)}(r_1,t_1)\right|^2 \right\rangle + |\lambda_2|^2 \left\langle \left|E^{(+)}(r_2,t_2)\right|^2 \right\rangle +$$

$$2\operatorname{Re}\lambda_1^* \lambda_2 \left\langle E^{(-)}(r_1,t_1)E^{(+)}(r_2,t_2) \right\rangle. \qquad (2.12)$$

If we introduce the first order correlation function

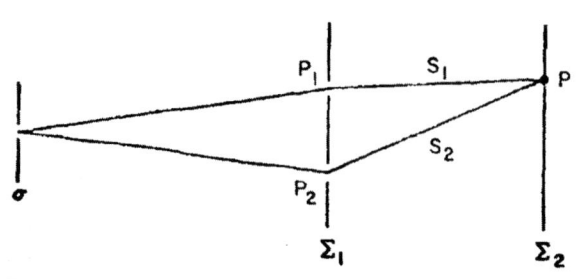

$$G^{(1)}(rt,r't') = \left\langle E^{(-)}(rt)E^{(+)}(r't') \right\rangle, \qquad (2.13)$$

Figure 2

we can rewrite Eq. (2.12) in the following way

$$\left\langle |E^{(+)}(\boldsymbol{r},t)|^2 \right\rangle = |\lambda_1|^2 G^{(1)}(\boldsymbol{r}_1 t_1, \boldsymbol{r}_1 t_1) + |\lambda_2|^2 G^{(1)}(\boldsymbol{r}_2 t_2, \boldsymbol{r}_2 t_2) +$$
$$2\,\mathrm{Re}\left\{\lambda_1^* \lambda_2 G^{(1)}(\boldsymbol{r}_1 t_1, \boldsymbol{r}_2 t_2)\right\}. \quad (2.14)$$

We have omitted consideration of vector and tensor indices of the fields and correlation functions, respectively, since the vector properties of the field are not too important in this experiment. We would have to take careful account of them if somehow a rotation of the plane of polarization were induced behind one pinhole, or if the polarization were in any way made to play a more active role.

A particular case which occurs almost universally in classic optics is that in which the incident field is stationary. The term "stationary" does not mean that nothing is happening. On the contrary, the field is ordinarily oscillating quite rapidly. It means that our knowledge about the field does not change with time. More formally, we associate stationarity with invariance of the statistical description of the beam under displacements of the time variable. The correlation function $G^{(1)}$ for such fields can therefore only depend on the difference $t - t'$

$$G^{(1)}(t,t') = G^{(1)}(t-t') \quad \text{(stationary field)}. \quad (2.15)$$

(Note that by discussing only a single type of correlation function we are stating a necessary condition for stationarity, but not a sufficient one. All average properties of a stationary field must be unchanged by time displacements.) When random classical fields are represented by means of stationary stochastic processes the models used usually have the ergodic property. That property means that the function $G^{(1)}(t-t')$ which is defined as an ensemble average, has the same value as the time averaged correlation function $\Gamma^{(1)}(t-t')$,

$$G^{(1)}(\boldsymbol{r}_1, \boldsymbol{r}_2, \tau) = \Gamma^{(1)}(\boldsymbol{r}_1, \boldsymbol{r}_2, \tau) = \lim_{T \to \infty} \frac{1}{T} \int_0^T E^{(-)}(\boldsymbol{r}_1, t_1 + \tau) E^{(+)}(\boldsymbol{r}_2, t_1)\, dt_1 \,. \quad (2.16)$$

The properties of the time-averaged correlation functions $\Gamma^{(1)}$ for classical fields have been discussed in detail in Chapter X of the text of Born and Wolf[1].

It may be of some help in the sections that follow to have some more concrete applications of interference experiments in mind. Let us take a brief look at one of the fundamental techniques of interferometry by considering a case in which the field incident upon a detector is a superposition of two plane waves. We assume that the propagation vectors of the two plane waves are only very slightly different. This might be the case for example for monochromatically filtered light from the two members of a double star. If we assume that the frequencies of both waves are equal we may write

$$E^{(+)}(\boldsymbol{r},t) = A\,e^{i(\boldsymbol{k}\cdot\boldsymbol{r} - \omega t)} + A\,e^{i(\boldsymbol{k}'\cdot\boldsymbol{r} - \omega t)}. \quad (2.17)$$

2 Optical Coherence and Photon Statistics

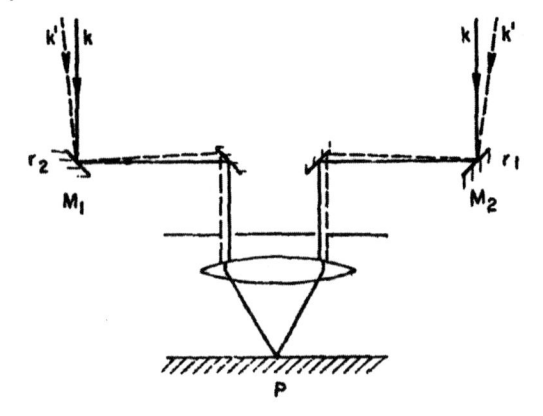

Figure 3

The question we now ask is: what kind of measurement can be performed to determine that we are receiving radiation from two sources and not just one?

Before answering the question let us specify the statistical character of the coefficients A and B. They are, of course, particular examples of the coefficients C_k previously introduced. We will assume A and B to be distributed independently of one another. This means that the probability function $p(A,B)$ factorizes,

$$p(A,B) = p_1(A)\, p_2(B)\,. \tag{2.18}$$

We will assume further more as properties of the distributions p_1 and p_2, that the phases of the complex amplitudes A and B are individually random. We then have $\langle A \rangle = \langle B \rangle = 0$. More generally the mean values of various powers of the amplitudes and their complex conjugates such as $\langle AB^* \rangle$, $\langle |A|^2 A^* B \rangle$, etc. will vanish. Averages in which the amplitudes are paired with their complex conjugates however, take on positive values,

$$\langle |A|^{2n} \rangle \neq 0\,, \qquad \langle |B|^{2n} \rangle \neq 0\,, \qquad n = 1,2,\dots \tag{2.19}$$

A famous device invented to answer the question we have asked is the Michelson stellar interferometer (Fig. 3).

The field at the point P and time t is, in effect, the sum of the two fields impinging on the mirrors M_1, M_2 at the same instant t' (if the optical paths $M_1 P$ and $M_2 P$ are equal). Each of these two fields is of the form Eq. (2.17) evaluated at the points r_1 and r_2, respectively. The average intensity at P will therefore be

$$\left\langle E^{(-)}(\mathbf{r},t) E^{(+)}(\mathbf{r},t) \right\rangle =$$
$$2\,\mathrm{Re}\left\{ \langle |A|^2 + |B|^2 \rangle + \langle |A|^2 \rangle e^{-i\mathbf{k}\cdot(\mathbf{r}_1 - \mathbf{r}_2)} + \langle |B|^2 \rangle e^{-i\mathbf{k}'\cdot(\mathbf{r}_1 - \mathbf{r}_2)} \right\}, \tag{2.20}$$

where we have used $\langle AB^* \rangle = \langle A \rangle \langle B^* \rangle = 0$ in reaching this expression.

If we introduce the correlation function Eq. (2.13),

$$G^{(1)}(r_1 t', r_2 t') = \left\langle E^{(-)}(r_1, t') E^{(+)}(r_2, t') \right\rangle$$
$$= \left\langle |A|^2 \right\rangle e^{-i k \cdot (r_1 - r_2)} + \left\langle |B|^2 \right\rangle e^{-i k' \cdot (r_1 - r_2)},$$
(2.21)

then the intensity may be written as

$$\left\langle E^{(-)}(r,t) E^{(+)}(r,t) \right\rangle = 2\,\mathrm{Re}\left\{ \left\langle |A|^2 + |B|^2 \right\rangle + G^{(1)}(r_1 t', r_2 t') \right\}.$$
(2.22)

The correlation function which describes the interference effect is time independent, because of the stationary character of the field we are treating.

We see from Eq. (2.21) that the correlation function contains two spatially oscillating terms. The way in which these terms reinforce or cancel one another will depend on the displacement $r_1 - r_2$. If $\langle |A|^2 \rangle = \langle |B|^2 \rangle$, Eq. (2.20) yields

$$\left\langle E^{(-)}(r,t) E^{(+)}(r,t) \right\rangle =$$
$$4 \left\langle |A|^2 \right\rangle \left\{ 1 + \cos[\tfrac{1}{2}(k+k') \cdot (r_1 - r_2)] \times \cos[\tfrac{1}{2}(k-k') \cdot (r_1 - r_2)] \right\}.$$
(2.23)

The interference intensity which we see at the point r will be part of a pattern of parallel fringes which we see at the focus of the telescope. Although we have not attempted to describe the fringe pattern in detail, the expression (2.23) for the intensity does indicate one of the characteristic properties of the pattern, that it will vanish altogether when the displacement $r_1 - r_2$ is adjusted so that

$$\cos[\tfrac{1}{2}(k-k') \cdot (r_1 - r_2)]$$

passes through the value zero. By observing the fringes we know that we are dealing with two sources rather than one, and by finding the values of $r_1 - r_2$ at which the fringes disappear we determine their angular separation. The Michelson interferometer has indeed been used to measure the angular separations of double stars, and for measuring angular diameters of stars as well. Only a few stellar diameters have been measured in this way, however, because of the difficulties inherent in working with a large interferometer. An unusually great mechanical stability is clearly required of the apparatus. Furthermore random variations of the index of refraction along the optical path can wash out the pattern.

Instruments quite similar to the Michelson stellar interferometer have been used in radio-astronomy to determine the angular size of celestial radio sources. They consist of two separated antennas supplying signals to a common detector system. In the case of these instruments, as well, it is technically difficult to increase the separation of the antennas without introducing random phase differences in the path between the antennas and the detector. To overcome this difficulty Hanbury Brown and Twiss have devised another form of radio interferometer (Fig. 4).

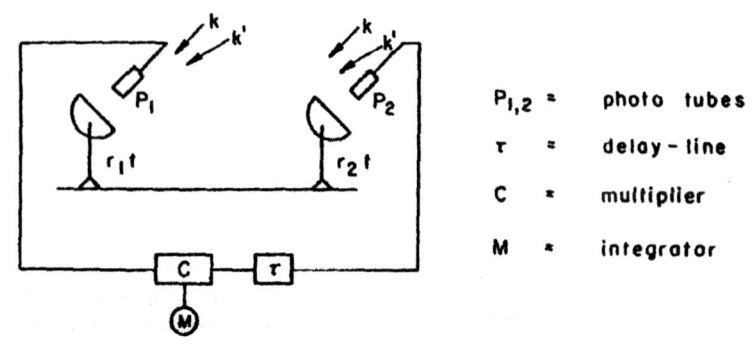

Figure 4

P₁,₂ = photo tubes
τ = delay-line
C = multiplier
M = integrator

The signals at the antennas are detected individually and then the detector outputs, which are of much lower frequency, are transmitted to a central correlating device where they are multiplied together and the product is averaged. The angular size of the source is obtained from measurements of the way in which the correlation of the intensity fluctuations of the signals varies with the separation of the antennas. An equivalent arrangement may be used with visible light.

The essence of the trick used by Hanbury Brown and Twiss was to detect the signals first and by taking away the high frequency components of the incoming radiation, to transmit to the central observation point just a measure of the fluctuations of the intensities arriving at the receivers. Since the detector signals are of relatively low frequency they are easy to transmit faithfully over distances large compared to the limiting dimensions of Michelson interferometers. This experiment is quite different in nature from the interferometer experiment we described earlier because it deals with the average of the product of two random intensities rather than with a single intensity.

It is easy to see that in the average of the product of the two signals there is an interference term, which permits us to resolve the two incoming waves. First we note that a square-law detector placed at P_1 gives a response proportional to

$$\left|E^{(+)}(\mathbf{r}_1,t)\right|^2 = |A|^2 + |B|^2 + AB^*\, e^{i(\mathbf{k}-\mathbf{k}')\cdot\mathbf{r}_1} + A^*B\, e^{-i(\mathbf{k}-\mathbf{k}')\cdot\mathbf{r}_1} \ . \tag{2.24}$$

This output no longer contains the rapid oscillations of the incoming wave. An average of this detected signal, however, would have no interference term (since $\langle AB^*\rangle = 0$). What Hanbury Brown and Twiss did is multiply together the two detected signals and then, and only then, to measure the statistical average. The average of the product of two intensities of the form of Eq. (2.24) is

$$\left\langle |E^{(+)}(\mathbf{r}_1,t)|^2 |E^{(+)}(\mathbf{r}_2,t)|^2\right\rangle = \left\langle (|A|^2+|B|^2)^2\right\rangle + \\ 2\left\langle |A|^2|B|^2\right\rangle \cos[(\mathbf{k}-\mathbf{k}')\cdot(\mathbf{r}_1-\mathbf{r}_2)] \ , \tag{2.25}$$

where we have used the fact that $\langle |A|^2 A^*B\rangle = 0$, etc. The cosine term clearly represents an interference effect. We can use it to resolve the two sources by observing its behavior as $r_1 - r_2$ is varied. It is important to note that the interference effect has been found by considering the average of a quantity quartic in the field amplitudes. In the case of Michelson's interferometer we deal only with expressions quadratic in the field amplitudes.

Although we have discussed the interferometer experiments in terms of ensemble averages, it is clear that they are not ordinarily performed in this way, but rather as time averages. The calculation of time averages, however, is typically at least a little more difficult than the calculation of ensemble averages (and often it is incomparably more difficult). To consider the interferometer measurements as time averages we should have to note that the two plane waves are not, in general, perfectly monochromatic. It follows then that the coefficients A and B, which we were content earlier to evaluate only at a particular instant of time, actually vary with time. To proceed further we should have to adopt models to represent $A(t)$ and $B(t)$ as stochastic functions of time. As we shall see presently, there are extremely persuasive reasons, when we are dealing with natural light sources, to take these models to be Gaussian stochastic processes. Then, since such processes have the ergodic property, we are justified in identifying time averages with ensemble averages.

2.3
Introduction of Quantum Theory

When we describe the electromagnetic field in quantum mechanical terms we must think of the field vectors $\boldsymbol{E}$ and $\boldsymbol{B}$ as operators which satisfy the Maxwell equations. The states, $|\ \rangle$, on which these operators act and their adjoints, $\langle\ |$, contain the information which specifies the field. When measurements are made of the physical quantity which correspond to an operator O, we can not expect in general to find the same results repeatedly. What we find instead is that the measured values fluctuate about the average value given by the product $\langle\ |O|\ \rangle$. The fluctuation is only absent if the state, $|\ \rangle$, happens to be an eigenstate of O, i.e., if we have

$$O|\ \rangle = O'|\ \rangle, \qquad (2.26)$$

where O' is an ordinary number rather than an operator. In that case it is convenient to use Dirac's convention and let the eigenvalue O' be a label for the state by writing the latter as $|O'\rangle$.

As in classical electromagnetic theory, it is convenient to separate the field operator, $\boldsymbol{E}(\boldsymbol{r},t)$, which is naturally Hermitian, into the sum of its positive frequency and negative frequency parts:

$$\boldsymbol{E}(\boldsymbol{r},t) = E^{(+)}(\boldsymbol{r},t) + E^{(-)}(\boldsymbol{r},t). \qquad (2.27)$$

These parts, as we have already noted classically, represent complex rather than real fields. The operators $E^{(\pm)}$ are therefore not Hermitian, but they are Hermitian adjoints of one another

$$E^{(-)}(r,t) = \left\{E^{(+)}(r,t)\right\}^{\dagger}. \tag{2.28}$$

While the fields $E(+)$ and $E^{(-)}$ play essentially indistinguishable roles in classical theory, they tend to play quite dissimilar roles in the quantum theory. The operator $E^{(+)}$ describes the annihilation of a photon while $E^{(-)}$ describes the creation of one. This identification of the operators is virtually the only fact we shall have to borrow from more formal developments of quantum field theory.

We must think fundamentally of all electric field measurements as being made on the Hermitian operator $E(r,t)$ given by Eq. (2.27). In the classical limit it is usually true that the complex fields $E^{(+)}$ and $E^{(-)}$ make contributions of equal magnitude to our measurements. From a quantum mechanical standpoint that is because quantum energies are so small in the classical limit ($\hbar\omega \to 0$), that test charges emit quanta as readily as they absorb them. In the quantum domain, on the other hand, we must expect that the fields $E^{(+)}$ and $E^{(-)}$ will make contributions of altogether different magnitudes to the quantities we measure, such as transition amplitudes.

If we are using atomic systems in their ground states as probes of the electric field for example, then the atoms have no energy to emit photons and can only absorb them. In this case, which corresponds in principle to that of a typical photodetector, only the annihilation operator $E^{(+)}$ figures significantly in determining the transition amplitudes. More exactly, if we do a calculation of the transition amplitude using first order perturbation theory, we easily find that the creation operator $E^{(-)}$ contributes only an extremely small amplitude which varies so rapidly with time that it leads to no observable effect at all. The creation operator can only contribute materially if the detector contains excited atoms. (Thermal energies are a great deal too small to furnish atoms excited to optical energies, but at microwave frequencies it may be necessary to take thermally excited atoms into account.)

In the third and higher orders of perturbation theory, the creation operator can indeed play a tiny role in an absorption experiment. The effect in question is a radiative correction to the first order absorption probability which all estimates indicate will be quite small. We see, therefore, that it is fairly accurate to say that a typical photodetector detects the field $E^{(+)}$ rather than the field E. Although this statement is clearly an approximate one rather than a rigorous one it is none the less important since it furnishes us a reason for formulating the theory is terms of a set of non-Hermitian operators. The formulation, as we shall see, allows in turn a great deal of insight into the way the theory passes to the classical limit.

To gain some further insights into the kinds of quantities measured in photon counting experiments, let us examine the role played by the field operator in the calculation of the appropriate transition probabilities. In the next section we shall indicate how these transition probabilities are calculated in some detail by taking due account of the atomic nature of the detector. Let us for the moment, however, ignore the detailed dynamics of the detector and assume simply that it is an ideally selective device, one which is sensitive to the field $\boldsymbol{E}^{(+)}(\boldsymbol{r}t)$ at a single point of space $\boldsymbol{r}$ at each instant of time t. We may take the transition probability of the detector for absorbing a photon from the field at position $\boldsymbol{r}$ and time t to be proportional to

$$w_{i \to f} = \left| \langle f | \boldsymbol{E}^{(+)}(\boldsymbol{r}t) | i \rangle \right|^2, \tag{2.29}$$

where $|i\rangle$ is the initial state of the field before the detection process, and braf is a final state in which the field could be found after the process. In fact we never measure the final state of the field. The only thing we do measure is the total counting rate. To calculate the total rate we have to sum Eq. (2.29) over all the final states of the field that can be reached from $|i\rangle$ by an absorption process. We can, however, extend the sum over a complete set of final states since the states which cannot be reached (e.g., states $|f\rangle$ which differ from $|i\rangle$ by two or more photons) simply will not contribute to the result since they are orthogonal to the state $\boldsymbol{E}^{(+)}(\boldsymbol{r}t)|i\rangle$.

When the final state summation is carried out the counting rate becomes, in effect,

$$w = \sum_f \left| \langle f | \boldsymbol{E}^{(+)}(\boldsymbol{r},t) | i \rangle \right|^2 = \langle i | \boldsymbol{E}^{(-)}(\boldsymbol{r},t) \boldsymbol{E}^{(+)}(\boldsymbol{r},t) | i \rangle, \tag{2.30}$$

where the completeness relation $\sum |f\rangle \langle f| = 1$ has been used. The counting rate w is proportional to the probability per unit time that an ideal photocounter, placed at $\boldsymbol{r}$, absorbs a photon from the field at time t. It is, according to Eq. (2.30), given by the expectation value of the positive definite Hermitian operator $\boldsymbol{E}^{(-)}(\boldsymbol{r},t) \boldsymbol{E}^{(+)}(\boldsymbol{r},t)$, taken in the state $|i\rangle$ which the field was in prior to the measurement. Equation (2.30) shows explicitly that the photocounter is not sensitive to the square of the real field (as has been assumed in many "semi-classical" calculations), but rather to an operator which corresponds to the squared absolute magnitude the complex field-strength.

We have thus far supposed that we know the state $|i\rangle$ of the field. That does not mean, of course, that we can predict the result of a single measurement made with our counter. If we repeat the measurement another result will quite likely turn out, and Eq. (2.30) gives us only the mean value of many repeated measurements. So quantum mechanics forces us to talk about ensemble averages even if we know the state of the field precisely.

In practice, of course, we almost never know the state $|i\rangle$ very precisely. Radiation sources are usually complicated systems with many degrees of freedom, so the states $|i\rangle$ depend, as a rule, on many uncontrollable parameters. Since we have no possibility of knowing the exact state of a field, we must resort to a statistical description. This description summarizes our knowledge of the field, by averaging over the unknown parameters. The predictions that we make by using this description must therefore, in principle, be compared experimentally with ensemble averages. With this understanding we may write the counting rate as an ensemble average of Eq. (2.30) over all random variables involved in the state $|i\rangle$,

$$w = \left\{ \langle i | \boldsymbol{E}^{(-)}(\boldsymbol{r},t) \boldsymbol{E}^{(+)}(\boldsymbol{r},t) | i \rangle \right\}_{\text{av. over } i}. \tag{2.31}$$

If we introduce the density operator $\rho = \{|i\rangle\langle i|\}_{\text{av. over }i}$, we may write this average as

$$w = \text{Tr}\left\{\rho \boldsymbol{E}^{(-)}(\boldsymbol{r},t)\boldsymbol{E}^{(+)}(\boldsymbol{r},t)\right\}, \qquad (2.32)$$

where Tr stands for the trace of the operator which follows. The density operator is the average of the projection operators on the initial field states. It is obviously Hermitian, $\rho^\dagger = \rho$. Furthermore, it also has the property of positive definiteness, $\langle j|\rho|j\rangle \geq 0$ for any state $|j\rangle$. It is worth emphasizing that a two-fold averaging process is implied by Eq. (2.32). That we must average the measurements made upon a pure state is an intrinsic requirement of quantum mechanics which has no classical analog. The ensemble average over initial states, on the other hand, is analogous to the averaging over the set of random coefficients $\{C_k\}$ which we described in the classical theory.

Equation (2.32) gives the counting rate of a single ideal photodetector in terms of the quantum mechanical correlation function

$$G^{(1)}(x,x') = \text{Tr}\left\{\rho \boldsymbol{E}^{(-)}(x)\boldsymbol{E}^{(+)}(x')\right\}, \qquad x \equiv \{\boldsymbol{r},t\}, \qquad (2.33)$$

which is analogous to the correlation function introduced to describe classical interference experiments. To describe more sophisticated experiments, e.g., the coincidence experiment of Hanbury Brown and Twiss, it is useful to define a more general set of correlation functions

$$G^{(n)}(x_1 \ldots x_n, x_{n+1} \ldots x_{2n}) = $$
$$\text{Tr}\left\{\rho \boldsymbol{E}^{(-)}(x_1) \ldots \boldsymbol{E}^{(-)}(x_n) \boldsymbol{E}^{(+)}(x_{n+1}) \ldots \boldsymbol{E}^{(+)}(x_{2n})\right\}. \qquad (2.34)$$

The function $G^{(n)}$ will be referred to as the n-th order correlation function. The analytical properties of this set of functions and their relation to experimental measurements will be discussed later.

We could, of course, have chosen to define a somewhat larger class of correlation functions than the $G^{(n)}$ by dealing with averages such as $\text{Tr}\{\rho \boldsymbol{E}^{(-)}\boldsymbol{E}^{(+)}\boldsymbol{E}^{(+)}\}$, which contain unequal numbers of creation and annihilation operators. If we have chosen not to set down any special notation for such averages it is because they are not of the types which are measured in typical photon counting experiments. Such averages may, in principle, be measured in other kinds of experiments but they will always vanish in stationary states of the field and, much more generally, whenever the absolute phases of the fields are random. Random absolute phases are, of course, rather characteristic of optical and other extremely high frequency fields.

2.4
The One-Atom Photon Detector

Let us now consider the photodetection process in somewhat more detail. We shall imagine, for the present, that our photon counter is a rather idealized type of device

2.4 The One-Atom Photon Detector

which has as its sensitive element a single atom which is free to undergo photoabsorption transitions such as the photoelectric effect. We assume that the atom is shielded from the radiation field we are investigating by a shutter of some sort which opens at time t_0 and closes again at time t. Our problem will be to calculate the probability that a photoabsorption process takes place during this interval and that it is recorded by our apparatus.

The detector will be assumed to be far enough from the radiation source so that the field behaves as a free field. The Hamiltonian of the system (field + detector) can then be written as

$$\mathcal{H} = \mathcal{H}_0 + \mathcal{H}_1 \; ; \quad \mathcal{H}_0 = \mathcal{H}_{0,\mathrm{at}} + \mathcal{H}_{0,\mathrm{F}} ,$$

where $\mathcal{H}_0$ is the sum of Hamiltonians of the free field and the atom. The interaction term $\mathcal{H}_1$ is time independent in the Schrödinger picture. In the interaction representation, however, it becomes time dependent. If we make use of the electric dipole approximation, which is quite accurate at optical frequencies, we can write the time dependent interaction Hamiltonian as

$$\mathcal{H}_1 = e^{i\mathcal{H}_0 t/\hbar} \, \mathcal{H}_1 \, e^{-i\mathcal{H}_0 t/\hbar} = -e\sum_\gamma \boldsymbol{q}_\gamma(t) \cdot \boldsymbol{E}(\boldsymbol{r},t) . \tag{2.35}$$

In this expression $\boldsymbol{r}$ represents the position of the atomic nucleus and $\boldsymbol{q}_\gamma$ the position operator of the γ-th electron relative to the nucleus. The time dependence of the field $\boldsymbol{E}(\boldsymbol{r},t)$ which occurs in Eq. (2.35) is that of the free field uninfluenced by the presence of the atom.

The Schrödinger equation of the combined system of field and atom in the interaction representation is

$$i\hbar \frac{\partial}{\partial t} |t\rangle = \mathcal{H}_1(t) |t\rangle . \tag{2.36}$$

Its solution can be written in the general form

$$|t\rangle = U(t,t_0) |t_0\rangle ,$$

where $U(t,t_0)$ is the unitary time development operator which describes the way in which the initial state changes under the influence of the perturbation. In the first order of perturbation theory the solution has the well-known form

$$|t\rangle = \left\{ 1 + \frac{1}{i\hbar} \int_{t_0}^{t} \mathcal{H}_1(t') \, dt' \right\} |t_0\rangle . \tag{2.37}$$

Let us suppose now that the system is initially in the state $|gi\rangle = |g\rangle|i\rangle$, where $|i\rangle$ is some known state of the field, and $|g\rangle$ is the ground state of the atom. We ask now for the probability that the system at time t is in a specified state $|af\rangle = |a\rangle|f\rangle$, where $|a\rangle$

is an excited state of the atom and $|f\rangle$ is the final state of the field. This probability is given by the squared absolute value of the matrix element

$$\langle af|U(t,t_0)|gi\rangle = \frac{1}{i\hbar}\int_{t_0}^{t}\langle af|\mathcal{H}_1(t')|gi\rangle\,dt'. \qquad (2.38)$$

(The zeroth order term in $U(t,t_0)$ of Eq. (2.37) does not contribute because of the orthogonality of the electron states $|a\rangle$ and $|g\rangle$.) By substituting the interaction operator from Eq. (2.35) we can separate the matrix element into two parts, a matrix element for the atom and one for the field:

$$\langle af|U(t,t_0)|gi\rangle = \frac{ie}{\hbar}\sum_{\gamma}\int_{t_0}^{t}\langle a|q_\gamma(t')|g\rangle\langle f|E(rt')|i\rangle\,dt'. \qquad (2.39)$$

To evaluate the atomic matrix element we recall that

$$q_\gamma(t') = e^{i\mathcal{H}_0 t'/\hbar}q_\gamma(0)e^{-i\mathcal{H}_0 t'/\hbar} = e^{i\mathcal{H}_{0,\mathrm{at}}t'/\hbar}q_\gamma(0)e^{-i\mathcal{H}_{0,\mathrm{at}}t'/\hbar}.$$

The latter relation holds because the field Hamiltonian $\mathcal{H}_{0,F}$ commutes with the atomic Hamiltonian $\mathcal{H}_{0,\mathrm{at}}$ and with the electron coordinate $q_\gamma(0)$ as well. We may write the matrix element as

$$\left\langle a\left|\sum_{\gamma}q_\gamma(t')\right|g\right\rangle = M_{ag}\,e^{i\omega_{ag}t'}$$

with

$$M_{ag} = \left\langle a\left|\sum_{\gamma}q_\gamma(0)\right|g\right\rangle \quad\text{and}\quad \hbar\omega_{ag} = E_a - E_g.$$

The matrix element M_{ag} occurs simply as a time independent coefficient in the transition amplitude

$$\langle af|U(t,t_0)|gi\rangle = \frac{ie}{\hbar}\int_{t_0}^{t}e^{i\omega_{ag}t'}M_{ag}\langle f|E(r,t')|i\rangle\,dt'. \qquad (2.40)$$

We can now replace $E(r,t')$ in this expression by the sum of the two operators $E^{(+)}(r,t)$ and $E^{(-)}(r,t)$. The emission operator $E^{(-)}(r,t)$ contains only negative frequencies, i.e., exponential time dependences of the form $e^{i\omega t}$ for $\omega > 0$. The time integrals of these terms clearly oscillate rapidly with increasing t. They are furthermore quite small in amplitude compared with the terms contributed by the annihilation operator $E^{(+)}(r,t)$. What we are describing, in fact, is the way in which the transitions are restricted by the conservation of energy. In order to find that the atomic transitions conserve the energy of the field quanta with an accuracy $\Delta E = \hbar\Delta\omega$, we must leave our shutter open for a length of time $t - t_0 \gg 1/\Delta\omega$. In practice we always have $\Delta\omega \ll \omega_{ag}$, i.e., the shutter is open for a great many periods of oscillation and then the contribution of the emission term $E^{(-)}(r,t)$ is entirely negligible. (We are assuming that the detector is at a relatively low temperature, as we have remarked in the preceding section.)

We must next sum the squared modulus of the amplitude Eq. (2.40) over all final states $|f\rangle$ of the field, since no observations are ordinarily made of those states. One of the virtues of working with the expression (2.40) for the amplitude is that in the final state summation we can sum over *all* the states of a complete set; those final states which cannot be reached by the field for physical reasons are present in the sum but contribute nothing, either because the matrix elements leading to them vanish identically, or because the time integrals of the matrix elements vanish. Thus the constraint represented by the conservation of energy, for example, is actually implicit in the structure of the time integrals in the sum of the squared amplitudes,

$$\sum_f |\langle af|U(t,t_0)|gi\rangle|^2 = \left(\frac{e}{\hbar}\right)^2 \int_{t_0}^t \int_{t_0}^t dt' dt'' e^{i\omega_{ag}(t''-t')} \sum_{\mu,\nu} M^*_{ag,\mu} M_{ag,\nu}$$
$$\langle i|E^{(-)}_\mu(\mathbf{r},t') E^{(+)}_\nu(\mathbf{r},t'')|i\rangle, \quad (2.41)$$

which has been derived by using the relation

$$\langle f|E^{(+)}|i\rangle^* = \langle i|E^{(-)}|f\rangle$$

and the completeness relation $\sum_f |f\rangle\langle f| = 1$.

We have already discussed the need to average such expression as Eq. (2.41) over an ensemble of initial states $|i\rangle$ since the initial state is rarely known accurately in practice. We then find for the transition probability the expression

$$p_{g\to a}(t) = \left\{\sum |\langle af|U(t,t_0)|gi\rangle|^2\right\}_{\text{av. over } i}$$
$$= \left(\frac{e}{\hbar}\right)^2 \sum_{\mu,\nu} \int_{t_0}^t \int_{t_0}^t dt' dt'' e^{i\omega_{ag}(t''-t')} M^*_{ag,\mu} M_{ag,\nu}$$
$$\times \text{Tr}\left\{\rho E^{(-)}_\mu(\mathbf{r},t') E^{(+)}_\nu(\mathbf{r},t'')\right\} \quad (2.42)$$
$$= \left(\frac{e}{\hbar}\right)^2 \sum_{\mu,\nu} \int_{t_0}^t \int_{t_0}^t dt' dt'' e^{i\omega_{ag}(t''-t')} M^*_{ag,\mu} M_{ag,\nu} G^{(1)}_{\mu\nu}(\mathbf{r}t',\mathbf{r}t'').$$

The definitions of the density operator ρ of the field and of the first order correlation function $G^{(1)}$ have been given in the preceding section.

The foregoing discussion has assumed that the atom makes a transition to a specified final state $|a\rangle$. Counters employing discrete final states have received a certain amount of discussion recently. Bloembergen and Weber, for instance, have proposed using a scheme illustrated by Fig. 5.

When the atom is excited to the state a by an incident field of frequency ω it is then raised to a higher level b by a pumping field at frequency ω_p. The emission of a photon with the sum frequency $\omega_s = \omega + \omega_p$ indicates the absorption of a photon from the incident field.

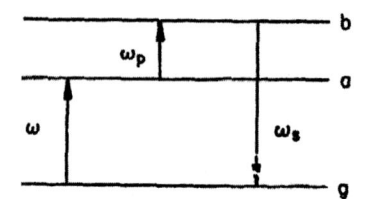

Figure 5

In the detectors used to date, however, the final states $|a\rangle$ of the atoms form an extremely dense set, or a continuum; the atoms are simply ionized, for instance. Since a counter of photoelectrons has only a limited ability to select among final atomic states (e.g., the counting of photoelectrons places only weak restrictions on their momenta), we have to sum the probability given by Eq. (2.42) over at least part of the continuum of final states $|a\rangle$. But not all ejected electrons can really be counted. Often they are ejected in directions for which the counter is insensitive or they are stopped by matter. The device might furthermore be built so as to introduce some explicit selection according to energies before detecting photoelectrons.

We shall not discuss the actual means used for detecting the photoelectrons in any detail here. Instead we shall assume simply that the probability that an electron ejected by photoabsorption is really registered is given by some function $R(a)$. The way in which this function varies with the final state $|a\rangle$ of the electron–ion system will depend, in general, on the geometrical and physical properties of the actual counting device. If we now sum the probabilities given by Eq. (2.42) over the final states $|a\rangle$ using the probability $R(a)$ as a weight, we find for the probability of detecting a photon absorption in our one-atom detector

$$p^{(1)}(t) = \sum_a R(a) p_{g \to a}(t)$$

$$= \left(\frac{e}{\hbar}\right)^2 \sum_{\mu,\nu} \int_{t_0}^{t} \int_{t_0}^{t} dt' \, dt'' \sum_a R(a) M_{ag,\mu}^* M_{ag,\nu} \, e^{i\omega_{ag}(t''-t')} G_{\mu\nu}^{(1)}(\mathbf{r}t', \mathbf{r}t'') \, . \tag{2.43}$$

We now separate the sum over the final states into two parts, a sum over the final electron energies and one over all other variables such as momentum directions, spin, etc. We do this by introducing the sensitivity function,

$$s_{\nu\mu}(\omega) = 2\pi \left(\frac{e}{\hbar}\right)^2 \sum_a R(a) M_{ag,\nu} M_{ag,\mu}^* \delta(\omega - \omega_{ag}) \, , \tag{2.44}$$

which contains contributions only from transitions with a fixed energy transfer, $\hbar\omega$. (Note that $s_{\nu\mu}(\omega)$, although it is written as a sum of delta functions, is actually a well-behaved function for the case we are considering since the sum over states $|a\rangle$ is really an integration over states with a continuum of energies.)

By making use of the sensitivity function and of the properties of the delta function it contains we may write the counting probability in Eq. (2.43) in the form

$$p^{(1)}(t) = \frac{1}{2\pi} \int_{t_0}^{t} dt' \int_{t_0}^{t} dt'' \int_{-\infty}^{\infty} d\omega \sum_{\mu,\nu} s_{\nu\mu}(\omega) e^{i\omega(t''-t')} G_{\mu\nu}^{(1)}(\mathbf{r}t', \mathbf{r}t''). \qquad (2.45)$$

Since $s_{\nu\mu}(\omega) = 0$ for $\omega < 0$ we have extended the integral over the variable ω from $-\infty$ to $+\infty$. If we define the Fourier transform of the sensitivity function by

$$S_{\nu\mu}(t) = \frac{1}{2\pi} \int_{-\infty}^{\infty} s_{\nu\mu}(\omega) e^{i\omega t} d\omega, \qquad (2.46)$$

we finally obtain

$$p^{(1)}(t) = \int_{t_0}^{t} dt' \int_{t_0}^{t} dt'' \sum_{\mu,\nu} S_{\nu\mu}(t''-t') G_{\mu\nu}^{(1)}(\mathbf{r}t', \mathbf{r}t''). \qquad (2.47)$$

Equation (2.47) represents the total transition probability when our shutter is open from time t_0 to t. To obtain the rate at which transitions occur we must differentiate with respect to t.

In general there is nothing very localizable in time about the absorption process. It is not possible to say that the photon has been absorbed in a particular interval of time small compared to the total period during which the shutter has been open. This becomes quite clear if we assume that the sensitivity $s_{\nu\mu}(\omega)$ is sharply peaked with a small width $\Delta\omega$. Then $S_{\nu\mu}(t''-t')$ takes on nonvanishing values for $|t''-t'| \leq 1/\Delta\omega$, which maybe an arbitrarily long interval of time for small $\Delta\omega$. The degree of non-locality in time which enters the integral in Eq. (2.47) is, roughly speaking, just the reciprocal $1/\Delta\omega$ of the bandwidth of our device. If the bandwidth is narrow the counter measures an average of values of $G^{(1)}(\mathbf{r}t', \mathbf{r}t'')$ with t' quite different from t''. In optical experiments a narrowband sensitivity is usually reached by putting narrowband light filters in front of broadband counters, i.e., by "filtering" the correlation function $G^{(1)}$ rather than by discriminating between photoelectrons. Broadband counters are therefore, in this sense, somewhat more basic than narrowband ones.

In the limiting case of extremely broadband detection the detection process becomes approximately local in time. We have already made some mention in the preceding section of an ideal photodetector. Such a detector, we shall assume, has a sensitivity function $s_{\nu\mu}(\omega)$ which is constant for all frequencies. To gain a quick insight into the meaning of this assumption we note that when the sensitivity function is a constant, $s_{\nu\mu}$, independent of frequency, Eq. (2.46) reduces to

$$S_{\nu\mu}(t) = s_{\nu\mu}\delta(t). \qquad (2.48)$$

The photon absorption process then becomes, in effect, localized in time, and the transition probability given by Eq. (2.47) reduces to

$$p^{(1)}(t) = \sum_{\mu,\nu} s_{\nu\mu} \int_{t_0}^{t} G^{(1)}(\mathbf{r}t', \mathbf{r}t') dt'. \qquad (2.49)$$

Now the assumption that $s_{\nu\mu}(\omega)$ is independent of frequency would be quite a difficult one to meet in practice for $\omega > 0$. When we take negative values of ω into account it becomes, strictly speaking, an impossible condition to meet since $s_{\nu\mu}(\omega) = 0$ for $\omega < 0$. But in fact neither of these troubles stands in the way of our constructing actual devices which approximate the behavior of ideal detectors arbitrarily well, as long as we agree to use them on radiation fields of restricted frequency bandwidth. Once we assume that the field excitations have finite bandwidth all we really require of our detector is that its sensitivity be constant over the excited frequency band. The detector then functions in an ideal way no matter how much the sensitivity varies outside the excited band.

To show that we need only be concerned to have the sensitivity remain constant over the band which is actually excited, we shall examine Eq. (2.45) for the transition probability in a little more detail. Let us begin by imagining that the time interval $t - t_0$ is exceedingly great, e.g., we let $t \to \infty$ and $t_0 \to -\infty$. Then if we let $K_{\mu\nu}(\omega)$ be the Fourier integral

$$K_{\mu\nu}(\omega) = \int_{-\infty}^{\infty} dt' \int_{-\infty}^{\infty} dt'' e^{i\omega(t''-t')} G_{\mu\nu}^{(1)}(\mathbf{r}t', \mathbf{r}t''), \qquad (2.50)$$

it is clear that $K_{\mu\nu}$ vanishes for frequencies ω lying outside the excited band. (e.g., the diagonal elements $K_{\mu\mu}(\omega)$ are simply proportional to the power spectra of the three field components.) We may then make use of $K_{\mu\nu}(\omega)$ to rewrite Eq. (2.45) as

$$p^{(1)}(t) = \frac{1}{2\pi} \int_{-\infty}^{\infty} \sum_{\mu\nu} s_{\nu\mu}(\omega) K_{\mu\nu}(\omega) \, d\omega. \qquad (2.51)$$

Now as long as $s_{\nu\mu}(\omega)$ takes on the constant value $s_{\nu\mu}$ over the excited band (and no matter how it behaves elsewhere) we may write Eq. (2.51) as

$$p^{(1)}(t) = \sum_{\mu\nu} s_{\nu\mu} \left(\frac{1}{2\pi}\right) \int_{-\infty}^{\infty} K_{\mu\nu}(\omega) \, d\omega$$
$$= \sum_{\mu\nu} s_{\nu\mu} \int_{-\infty}^{\infty} G_{\mu\nu}^{(1)}(\mathbf{r}t', \mathbf{r}t') \, dt', \qquad (2.52)$$

and the latter of these expressions again shows the locality in time of the photon absorption process which we noted earlier in Eq. (2.49), i.e., the two arguments of the correlation function in the integrand are the same.

In order to derive the foregoing result we imagined that the time interval $t - t_0$ was allowed to become infinite. To see the influence of the fact that the time interval has a finite length, let us define a time-dependent step function

$$\eta(t') = \begin{cases} 0 & \text{for } t' < t_0, \\ 1 & \text{for } t_0 < t' < t, \\ 0 & \text{for } t' > t. \end{cases}$$

Then the limits of the time integrations in Eq. (2.45), for example, may be extended from $-\infty$ to ∞ if we first multiply the correlation function in the integrand by $\eta(t')\eta(t'')$. This extension of the limits of the time integrations means that we may use once more an argument of the type which led to Eq. (2.52). But the difference is that the function $K_{\mu\nu}(\omega)$ must now be regarded as the Fourier transform

$$K_{\mu\nu}(\omega) = \int_{-\infty}^{\infty} dt' \int_{-\infty}^{\infty} dt'' \, e^{i\omega(t''-t)} \eta(t') G_{\mu\nu}^{(1)}(\mathbf{r}t',\mathbf{r}t'') \eta(t'') \,. \tag{2.53}$$

The bandwidth of this function will in general be different from that of the radiation present but the difference will only be significant if the period during which the shutter is open is extremely brief.

Let us suppose the bandwidth of the radiation present, i.e., of the function $G^{(1)}$, is $\delta\omega$. The bandwidth associated with the functions η is of order $(t-t_0)^{-1}$. The frequency width characteristic of $K_{\mu\nu}(\omega)$ is presumably of the magnitude of the larger of these two widths. Then if we assume that the sensitivity function of our detector only varies appreciably over an interval $\Delta\omega$, we shall secure an expression for the transition probability which reduces to the form of Eq. (2.49) as long as $e l\omega$ satisfies the two conditions

$$\Delta\omega \gg \delta\omega \quad \text{and} \quad \Delta\omega \gg (t-t_0)^{-1} \,.$$

The second of these conditions sets a lower bound $1/\Delta\omega$ to the length of time our shutter can be open if we want the behavior of our counter to remain ideal.

If we differentiate Eq. (2.49) with respect to time we find that the rate of increase of the transition probability, i.e., the counting rate, is

$$w^{(1)}(t) = \frac{dp^{(1)}(t)}{dt} = \sum_{\nu\mu} s_{\nu\mu} G_{\nu\mu}^{(1)}(\mathbf{r}t,\mathbf{r}t) \,. \tag{2.54}$$

Having carried the tensor indices of the sensitivity and correlation functions far enough to illustrate their role in determining the transition probabilities we shall now eliminate them by imagining the field to possess a specified polarization $\hat{e}$. This can be accomplished in practice, of course, by putting a polarization filter in front of the counter. With the notation

$$E^{(-)}(\mathbf{r},t) = \hat{e} \cdot \mathbf{E}^{(-)}(\mathbf{r},t) \,,$$
$$E^{(+)}(\mathbf{r},t) = \hat{e}^* \cdot \mathbf{E}^{(+)}(\mathbf{r},t) \,,$$
$$G^{(1)}(\mathbf{r}t,\mathbf{r}'t') = \text{Tr}\left\{\rho E^{(-)}(\mathbf{r},t) E^{(+)}(\mathbf{r}'t')\right\} \,, \tag{2.55}$$
$$s = \sum_{\mu\nu} \hat{e}_\nu s_{\nu\mu} \hat{e}_\mu^* \,,$$

Eq. (2.54) may be rewritten as

$$w^{(1)}(t) = s G^{(1)}(\mathbf{r}t,\mathbf{r}t) \,. \tag{2.56}$$

We have thus justified the assumption, made in the course of the simplified discussions given earlier, that an ideal photon counter can be constructed to respond, in effect, to the field at a given instant of time. Its counting rate is proportional to the first order correlation function evaluated at a single point and a single time.

In deriving the foregoing results we have employed the electric dipole approximation. The use of that approximation has been much more a matter of convenience titan one of necessity. We could as well have retained the general coupling between the momentum of the atomic electrons and the vector potential. We would then have made use of correlation functions for the vector potential rather than for the electric field. The only difference in the calculations would then be a matter of taking account of the finite size of the atom. Instead of having atomic matrix elements simply occurring as constant factors in the transition probabilities we would have to integrate products of the atomic wave functions and the correlation functions The transition probabilities, in other words, would be integrals which involve the correlation functions for finite spatial as well as temporal intervals. Fortunately these unilluminating complications are not too necessary quantitatively at optical and lower frequencies.

2.5
The *n*-Atom Photon Detector

The photon counter we have thus far discussed has as its sensitive element only a single atom. Since that is hardly a very realistic picture of an actual detector, we must generalize our arguments to deal with detectors containing arbitrarily many atoms which may undergo photoabsorption processes. We shall carry out this generalization in two stages. In the present section we consider detectors which consist of a relatively modest number of atoms and show how these can be used to investigate the higher order correlation properties of the fields. We shall postpone until the last section a full discussion of the statistical properties of actual photon counting experiments, since it will be useful to discuss the coherence properties of fields first.

The one-atom detector, as we have seen, furnishes us with measurements of the first-order correlation function of the field, $G^{(1)}$. There exist, however, more general correlation properties of fields; some of these are related, for example, to experiments in which we measure coincidences of photon absorption processes taking place at different points in space and time. Such an experiment has been performed for example by Hanbury Brown and Twiss, and we shall discuss it in some detail in the later sections.

Let us suppose that n similar atoms are placed at different positions $r_1, r_2, \ldots, r_n$ in the field. These atoms, we assume, form the sensitive element of a species of compound detector. A shutter in front of all of the atoms will be opened during the time interval from t_0 to t. We ask for the probability that each of the atoms has absorbed a photon from the field during that time interval. Though this problem is still rather artificial in nature, its solution will be an essential part of the general discussion of photon counting we shall undertake later.

2.5 The n-Atom Photon Detector

The process in question involves the absorption of n photons, and therefore, to calculate its probability, we must, strictly speaking, apply n-th order perturbation theory. Needless to say, a number of simplifications are available to us in doing this.

In order to solve the Schrödinger equation in the interaction representation

$$i\hbar \frac{\partial}{\partial t} |t\rangle = \mathcal{H}_I(t) |t\rangle ,\qquad (2.57)$$

we have already introduced the unitary time development operator $U(t,t_0)$ which transforms the states according to the scheme

$$|t\rangle = U(t,t_0) |t_0\rangle .$$

A formal solution for $U(t,t_0)$ may be written in the form

$$U(t,t_0) = \left\{ \exp\left[-\frac{i}{\hbar} \int_{t_0}^{t} \mathcal{H}_I(t') \, dt' \right] \right\}_+ ,\qquad (2.58a)$$

$$= \sum_{n=0}^{\infty} \frac{1}{n!} \left(\frac{-i}{\hbar} \right)^n \int_{t_0}^{t} \cdots \int_{t_0}^{t} \{\mathcal{H}_I(t_1) \ldots \mathcal{H}_I(t_n)\}_+ \prod_{p=1}^{n} dt_p ,\qquad (2.58b)$$

where the bracket symbol $\{\ \}_+$ stands for a time ordering operation to be carried out on all the operators inside the bracket. It requires that the products of operators be rearranged so that their time arguments increase from right to left. The representations (2.58) for the solution are perhaps most easily derived by writing the Schrödinger equation (2.57) as an integral equation and solving the integral equation by means of a power series.

The interaction Hamiltonian $\mathcal{H}_I(t)$ for the n atoms interacting with the field is given by

$$\mathcal{H}_I(t) = \sum_{j=1}^{n} \mathcal{H}_{I,j}(t) ,\qquad (2.59)$$

where $\mathcal{H}_{I,j}(t)$ represents the coupling of the j-th atom to the field. The individual coupling terms take the form

$$\mathcal{H}_{I,j}(t) = -e \sum_{\gamma} q_{\gamma}(t) \cdot \mathbf{E}(\mathbf{r}_j,t) ,\qquad (2.60)$$

which we have already discussed. We shall assume, for simplicity, that the atoms are dynamically independent of one another, i.e., that their zeroth order Hamiltonians are separable and commute.

The n-fold absorption process is described, to lowest order, by the n-th order term $U^{(n)}(t,t_0)$ of $U(t,t_0)$, i.e., the n-th order term of the series in Eq. (2.58b). By inserting the Hamiltonian given by Eq. (2.59) into Eq. (2.58b), we obtain for $U^{(n)}(t,t_0)$ an expression containing n^n terms, which represent all of the ways in which n atoms can participate in an *n*-th order process. Many of these terms, however, have nothing to do with the process we are considering, since we require each atom to participate by absorbing a photon once and only once. Terms involving repetitions of the Hamiltonian for a given atom describe processes other than those we are interested in. The only terms which do contribute are those in which each of the $\mathcal{H}_{1,j}$ appears only once. There are $n!$ such terms, and all of them contribute equally since the bracket $\{\ \}_+$ is a symmetric function of the operators it contains. Therefore, the part of $U^{(n)}(t,t_0)$ we must consider reduces to

$$\left(\frac{-i}{\hbar}\right)^n \int_{t_0}^t \cdots \int_{t_0}^t \{\mathcal{H}_{1,1}(t_1)\ldots\mathcal{H}_{1,n}(t_n)\}_+ \prod_{p=1}^n dt_p . \tag{2.61}$$

Since none of the n atoms can omit a photon (each of them is in the ground state initially), only the positive frequency part of the electric field operator in each $\mathcal{H}_{1,j}$ will contribute to the transition amplitude. When the electric field operator in Eq. (2.60)) is replaced by $\boldsymbol{E}^{(+)}(\boldsymbol{r}_j,t)$ we shall write the resulting interaction Hamiltonian as $\mathcal{H}_{1,j}^{(+)}$. The operators $\mathcal{H}_{1,j}^{(+)}$ commute with each other since the atoms are dynamically independent and the fields $\boldsymbol{E}^{(+)}(\boldsymbol{r}_j,t)$ commute. We can therefore drop the ordering bracket $\{\ \}_+$ in the expression (2.61), and write the desired part of $U^{(n)}(t,t_0)$ as an n-fold product of single integrals

$$\left(\frac{-i}{\hbar}\right)^n \prod_{j=1}^n \int_{t_0}^t \mathcal{H}_{1,j}^{(+)}(t') dt' . \tag{2.62}$$

The result is a simple one. The operator which induces the transitions which interest us is simply a product of the operators which induce the individual absorption processes. This does not mean, however, that the matrix of the transition operator factorizes.

In evaluating the matrix element of the operator Eq. (2.62) between two states of the entire system we must note that the individual atoms which are all in the *same ground state* initially may make transitions to final states $|a\rangle$, which are *different* for different atoms. If we indicate these initial and final states for the atoms with $|\{g\}\rangle$ and $|\{a_j\}\rangle$, and use the symbols $|i,\{g\}\rangle$ and $|f,\{a_j\}\rangle$ for the initial and final states of the entire system, then the matrix element of Eq. (2.62) or of $U^{(n)}(t,t_0)$ takes the form

$$\langle f,\{a_j\}|U^{(n)}(t,t_0)|i,\{g\}\rangle = \left(\frac{ie}{\hbar}\right)^n \int_{t_0}^t \cdots \int_{t_0}^t \exp\left[i\sum_j \omega_{a_jg}t_j\right] \times$$
$$\langle f|\boldsymbol{E}^{(+)}(\boldsymbol{r}_n t_n)\ldots\boldsymbol{E}^{(+)}(\boldsymbol{r}_1 t_1)|i\rangle \prod_{j=1}^n M_{a_jg} \prod_{p=1}^n dt_p , \tag{2.63}$$

where we have introduced notation for the atomic matrix elements and frequencies analogous to that of the preceding section, and have eliminated tensor indices by assuming the field to have a unique polarization as in Eq. (2.55).

2.5 The n-Atom Photon Detector

We must next carry out upon the amplitude Eq. (2.63) the now familiar procedures of squaring, summing over final states of the field and averaging over initial states of the field. The expression we derive in that way is a transition probability for each of the atoms to reach a specified final state $|a_j\rangle$. Since each of these final states is in general part of a continuum we must sum the probability we have derived over all the relevant final atomic states. We shall again assume that our counting device does not record all of these final states with equal likelihood, but is characterized by a certain probability $R(a_j)$ that any particular photoabsorption process is recorded. For simplicity we shall take this recording probability to be the same function for each of the n atoms of the detector. We may then carry out the final state summations for the atoms by introducing the same sensitivity functions we discussed in Eqns. (2.44) and (2.46) of the preceding section. When these simple sums and averages are all carried out we find for the n-fold counting probability

$$p^{(n)}(t) = \int_{t_0}^{t} \cdots \int_{t_0}^{t} \prod_{j=1}^{n} S(t_j'' - t_j') G^{(n)}(\mathbf{r}_1 t_1' \ldots \mathbf{r}_n t_n', \mathbf{r}_n t_n'' \ldots \mathbf{r}_1 t_1'') \prod_{j=1}^{n} dt_j' dt_j''. \quad (2.64)$$

In this expression $G^{(n)}$ is the n-th order correlation function for the field defined by

$$G^{(n)}(x_1 \ldots x_{2n}) = \text{Tr}\left\{ \rho \mathbf{E}^{(-)}(x_1) \ldots \mathbf{E}^{(-)}(x_n) \mathbf{E}^{(+)}(x_{n+1}) \ldots \mathbf{E}^{(+)}(x_{2n}) \right\}$$

with $x_j = \{\mathbf{r}_j, t\}$.

For broadband detectors Eq. (2.64) reduces to the simpler form

$$p^{(n)}(t) = s^n \int_{t_0}^{t} \cdots \int_{t_0}^{t} G^{(n)}(\mathbf{r}_1 t_1' \ldots \mathbf{r}_n t_n', \mathbf{r}_n t_n' \ldots \mathbf{r}_1 t_1') \prod_{j=1}^{n} dt_j'. \quad (2.65)$$

An ideal n-atom counter thus measures a time integral of the n-th order correlation function.

We have thus far considered the n atoms which undergo photoabsorption to be part of a single detector. But a detector constructed in this way is not very different, really, from a set of n detectors of the one-atom variety we discussed in the last Section. If we regard the n atoms as the sensitive elements of a set of n independent detectors, then the n-fold photoabsorption process we have been discussing furnishes the basis of a primitive technique for n-fold coincidence counting of photons.

The technique may be refined a little if we imagine that there is a separate shutter in front of each one-atom detector. Then we may assume that all the shutters open at the same time t_0 but that the time at which each of them is closed maybe varied arbitrarily. Let us suppose that the time at which the j-th shutter is closed is t_j. Then the j-th atom only sees the field from time t_0 to t_j. The effect of closing the shutter may be simulated by assuming that the atom is decoupled from the field at time t_j. For this purpose we may introduce the step function

$$\theta(t) = \begin{cases} 0 & \text{for } t < 0, \\ 1 & \text{for } t > 0, \end{cases} \quad (2.66)$$

and write an effective interaction Hamiltonian (i.e., one which takes account of the closing of all the shutters) as

$$\mathcal{H}_I(t) = \sum_{j=1}^{n} \theta(t_j - t) \mathcal{H}_{I,j}(t) \,. \tag{2.67}$$

The calculation of the probability that a photoabsorption takes place in each detector is essentially the same with the effective Hamiltonian (2.67) as the calculation we have described earlier. The only real difference, besides the one of interpretation, is that the answer for the total detection probability is now an n-fold time integral in which the upper limits of integration are the times t_j. For the broadband case the answer is, for example

$$p^{(n)}(t_1 \ldots t_n) = s^n \int_{t_0}^{t_1} dt'_1 \ldots \int_{t_0}^{t_n} dt'_n \, G^{(n)}(\mathbf{r}_1 t'_1 \ldots \mathbf{r}_n t'_n, \mathbf{r}_n t'_n \ldots \mathbf{r}_1 t'_1) \,. \tag{2.68}$$

The times $t_1 \ldots t_n$ may be varied independently. An n-fold delayed coincidence rate, i.e., a counting rate per (unit time)^n, may therefore be defined as

$$\begin{aligned} w^{(n)}(t_1 \ldots t_n) &= \frac{\partial^n}{\partial t_1 \ldots \partial t_n} p^{(n)}(t_1 \ldots t_n) \\ &= s^n G^{(n)}(\mathbf{r}_1 t_1 \ldots \mathbf{r}_n t_n, \mathbf{r}_n t_n \ldots \mathbf{r}_1 t_1) \,. \end{aligned} \tag{2.69}$$

This result verifies the statement we made earlier that coincidence experiments performed with ideal detectors furnish measurements of the higher order correlation functions.

It may be worth emphasizing that the kinds of measurement processes we have been describing differ both in method and in spirit from those that are customarily discussed in the formal quantum mechanical theory of measurement. The formal theory of measurement has been useful in establishing the physical interpretation of quantum mechanical expressions. But because there are few areas in which exact statements meeting the required assumptions of the theory can be made, the applications of the formal theory have been quite restricted to date.

The kinds of field measurements we have discussed are, by contrast, explicitly approximate in character. We have only calculated the transition probabilities to the lowest order in which the transitions occur. While this approximation would not bo too difficult to remedy for individual atomic transitions, the higher order effects in multi-atom detectors would be found to have quite a complicated mathematical structure. It is implicit in the approximation we have used that the electromagnetic influences (as well as other influences) of one atom on another are ignored. That can be seen, for example, from the fact that the $\mathbf{E}^{(+)}$ operators which occur in the correlation function $G^{(n)}$ all commute. The transition rate Eq. (2.69), for example, does not depend on the ordering of the times $t_1 \ldots t_n$ even though the points $\mathbf{r}_j t_j$ may have time-like relationships to one another and electromagnetic disturbances can indeed pass from one point to another.

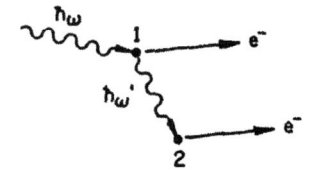

Figure 6

While the atoms may influence one another electromagnetically in ways not described by our lowest-order results, those influences are typically extremely small and are sometimes of a kind that can be eliminated experimentally. To take a specific example, let us suppose, that instead of a simple photoabsorption process in atom 1, we have a type of Raman effect which produces another photon as well as a photoelectron (Fig. 6). The emitted photon may then be absorbed by atom 2, producing a second photoelectron. Not only does this type of process have an extremely small cross section, but it may be eliminated entirely by choosing detector atoms with ionization potentials greater than $\frac{1}{2}\hbar\omega$.

We have mentioned the electromagnetic influences of the atoms upon one another just to underscore the fact that we have not been describing an exact theory of measurement. It may none the less be an extremely useful and accurate theory.

2.6 Properties of the Correlation Functions

The n-th order correlation function was defined as the expectation value

$$G^{(n)}(x_1\ldots x_{2n}) = \mathrm{Tr}\left\{\rho \boldsymbol{E}^{(-)}(x_1)\ldots \boldsymbol{E}^{(-)}(x_n)\boldsymbol{E}^{(+)}(x_{n+1})\ldots \boldsymbol{E}^{(+)}(x_{2n})\right\}. \quad (2.70)$$

The averaging process we carry out to evaluate this expression is the quantum mechanical analog of the classical procedure introduced in the first section. There we spoke of averages over a set of random Fourier coefficients. The resemblance between the two approaches is not yet a very persuasive one, but it will become more so as we proceed.

As a first property of the correlation functions we note that when we have an upper bound on the number of photons present in the field then the functions $G^{(n)}$ vanish identically for all orders higher than a fixed order M. To state the property more explicitly, if $|n\rangle$ is an n-quantum state and the density operator is written in the form

$$\rho = \sum_{m,n} c_{mn} |n\rangle \langle m|, \quad (2.71)$$

then if we have $c_{mn} = 0$ whenever $n > M$ or $m > M$, it follows from the nature of the annihilation operators $\boldsymbol{E}^{(+)}$, that

$$\boldsymbol{E}^{(+)}(x_1)\ldots \boldsymbol{E}^{(+)}(x_p)\rho = 0 \quad (2.72)$$

for $p > M$. Furthermore, the conjugate relation

$$\rho E^{(-)}(x_1)\ldots E^{(-)}(x_p) = 0 \tag{2.73}$$

also holds for $p > M$. Thus it follows that

$$G^{(p)} \equiv 0 \tag{2.74}$$

for $p > M$.

This property of the correlation functions must be regarded as a rather strange one when viewed from the standpoint of classical theory. There the correlation functions are essentially sums of moments of the probability distribution for the Fourier coefficients, and it would be quite difficult to imagine a case for which the moments higher than a certain order vanish identically. We have, in fact, constructed states which have no classical analog by imposing an upper bound on the number of photons present. However, that should not be surprising since in the limit $\hbar \to 0$ these are states whose total energy goes to zero.

A further property of the correlation functions can be derived from the general statement

$$\mathrm{Tr}(A^\dagger) = (\mathrm{Tr} A)^*, \tag{2.75}$$

which holds for all linear operators A. Applying this identity to the correlation function (2.70), we find

$$\begin{aligned}
\left[G^{(n)}(x_1\ldots x_{2n})\right]^* &= \mathrm{Tr}\left\{E^{(-)}(x_{2n})\ldots E^{(-)}(x_{n+1})E^{(+)}(x_n)\ldots E^{(+)}(x_1)\rho^\dagger\right\} \\
&= \mathrm{Tr}\left\{\rho E^{(-)}(x_{2n})\ldots E^{(-)}(x_{n+1})E^{(+)}(x_n)\ldots E^{(+)}(x_1)\right\} \\
&= G^{(n)}(x_{2n}\ldots x_1).
\end{aligned} \tag{2.76}$$

Here we have made use of the Hermitian character of ρ and of the invariance of the trace of a product of operators under a cyclic permutation.

As a consequence of the commutation properties of the $E^{(+)}$ and $E^{(-)}$ we can freely permute the arguments $(x_1\ldots x_n)$ and $(x_{n+1}\ldots x_{2n})$ without altering the value of $G^{(n)}(x_1\ldots x_n, x_{n+1}\ldots x_{2n})$. We cannot, however, interchange any of the first n arguments with any of the remaining n, unless suitable terms are added, since the corresponding operators do not commute.

A number of interesting inequalities can be derived from the general statement

$$\mathrm{Tr}\left\{\rho A^\dagger A\right\} \geq 0. \tag{2.77}$$

This relation, which follows from the positive definite character of the operator in the brackets, holds for any linear operator A. To prove the inequality we note that ρ is

Hermitian and therefore can be diagonalized. Thus, in some representation it has the form

$$\langle k|\rho|m\rangle = \delta_{km} p_k . \tag{2.78}$$

It follows immediately from the definition of the density operator that

$$p_k = \langle k|\rho|k\rangle = \{\langle k|i\rangle\langle i|k\rangle\}_{\text{av}} = \left\{|\langle i|k\rangle|^2\right\}_{\text{av}} \geq 0 . \tag{2.79}$$

(Furthermore, since $\text{Tr}\,\rho = \sum_k p_k = 1$, not all the p_k vanish.) Now a simple application of the completeness relation gives

$$\begin{aligned}
\text{Tr}\{\rho A^\dagger A\} &= \sum_k p_k \langle k|A^\dagger A|k\rangle \\
&= \sum_k p_k \sum_m \langle k|A^\dagger|m\rangle\langle m|A|k\rangle \\
&= \sum_k p_k \sum_m |\langle m|A|k\rangle|^2 \geq 0 .
\end{aligned} \tag{2.80}$$

Of course this value for the trace is independent of the particular representation used. Hence the proof of the inequality (2.77) is completed.

A number of results may be derived from the general inequality (2.77) by means of various substitutions. For example the choice $A = E^{(+)}(x)$ gives at once

$$G^{(1)}(x,x) \geq 0 . \tag{2.81}$$

Similarly the substitution $A = E^{(+)}(x_1)\ldots E^{(+)}(x_n)$ give us

$$G^{(n)}(x_1 \ldots x_n, x_n \ldots x_1) \geq 0 . \tag{2.82}$$

These two relations are also evident from the physical meaning of the "diagonal" forms of the $G^{(n)}$. The forms are interpretable as photon intensities and coincidence rates respectively, and are thus intrinsically positive.

These results and all of our later ones can be generalized immediately to deal with vector fields $E_\mu^{(+)}(x)$ rather than the scalar field $E^{(+)}(x)$. We need only associate a vector index μ_j with each coordinate x_j. We can thus consider x_j as a shorthand for the set of variables $\{r_j, t_j, \mu_j\}$ instead of simply $\{r_j, t_j\}$.

Another possible choice for the operator A is

$$A = \sum_{j=1}^n \lambda_j E^{(+)}(x_j) , \tag{2.83}$$

where the λ_j are a set of arbitrary complex numbers. For this case Eq. (2.77) gives us

$$\sum_{ij} \lambda_i^* \lambda_j G^{(1)}(x_i, x_j) \geq 0 . \tag{2.84}$$

Thus the set of correlation functions $G^{(1)}(x_i,x_j)$ forms a matrix of coefficients for a positive definite quadratic form. Such a matrix has, of course, a positive determinant,

$$\det\left[G^{(1)}(x_i,x_j)\right] \geq 0. \tag{2.85}$$

For $n=1$ this is simply the relation Eq. (2.81). For $n=2$ we find

$$G^{(1)}(x_1,x_1)\,G^{(1)}(x_2,x_2) \geq \left|G^{(1)}(x_1,x_2)\right|^2, \tag{2.86}$$

which is a simple generalization of the Schwarz inequality.

By proceeding along the same line we can derive an infinite sequence of inequalities. We shall confine ourselves however, to mentioning the quadratic ones for the higher order correlation functions. If we write

$$A = \lambda_1 \boldsymbol{E}^{(+)}(x_1)\ldots\boldsymbol{E}^{(+)}(x_n) + \lambda_2 \boldsymbol{E}^{(+)}(x_{n+1})\ldots\boldsymbol{E}^{(+)}(x_{2n}), \tag{2.87}$$

then the positive-definiteness of the related quadratic form requires that we have

$$G^{(n)}(x_1\ldots x_n, x_n\ldots x_1)\,G^{(n)}(x_{n+1}\ldots x_{2n}, x_{2n}\ldots x_{n+1})$$
$$\geq \left|G^{(n)}(x_1\ldots x_n, x_{n+1}\ldots x_{2n})\right|^2. \tag{2.88}$$

2.6.1
Space and Time Dependence of the Correlation Functions

We note that the operators $\boldsymbol{E}^{(\pm)}(\boldsymbol{r},t)$ occurring in the correlation functions obey the Maxwell equations and furthermore satisfy whatever boundary conditions we ordinarily require of the electric field vector (e.g., periodic boundary conditions or the conditions for conducting walls). As a result the functions $G^{(n)}(x_1\ldots x_{2n})$ obey $2n$ wave equations and $2n$ sets of boundary conditions, one for each of the space-time variables.

Let us now consider the structure of the functions $G^{(n)}$ in stationary fields. The best way to define stationarity in quantum mechanics is to require that the density operator commute with the Hamiltonian. This criterion is equivalent to the statement that ρ is independent of time in the Schrödinger picture. (In the Heisenberg picture, however, the density operator for isolated systems is always time-independent.) If we use this definition and the familiar interpretation of the Hamiltonian as an infinitesimal time-displacement operator we may write

$$\mathrm{Tr}\left\{\rho\boldsymbol{E}^{(-)}(x_1)\ldots\boldsymbol{E}^{(+)}(x_{2n})\right\} = \mathrm{Tr}\left\{e^{i\mathcal{H}\tau/\hbar}\rho\boldsymbol{E}^{(-)}(x_1)\ldots\boldsymbol{E}^{(+)}(x_{2n})e^{-i\mathcal{H}\tau/\hbar}\right\}$$

$$= \mathrm{Tr}\left\{e^{i\mathcal{H}\tau/\hbar}\rho e^{-i\mathcal{H}\tau/\hbar}\,e^{i\mathcal{H}\tau/\hbar}\boldsymbol{E}^{(-)}(x_1)e^{-i\mathcal{H}\tau/\hbar}\ldots e^{i\mathcal{H}\tau/\hbar}\boldsymbol{E}^{(+)}(x_{2n})e^{-i\mathcal{H}\tau/\hbar}\right\}$$

$$= \mathrm{Tr}\left\{\rho\boldsymbol{E}^{(-)}(\boldsymbol{r}_1,t_1+\tau)\ldots\boldsymbol{E}^{(+)}(\boldsymbol{r}_{2n},t_{2n}+\tau)\right\},$$

2.6 Properties of the Correlation Functions

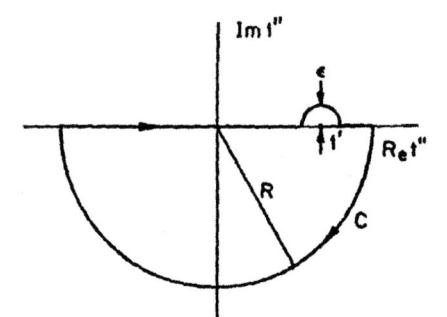

Figure 7

where τ is an arbitrary time parameter. We have thus shown that for stationary fields the correlation functions obey the identity

$$G^{(n)}(r_1 t_1 \ldots r_{2n} t_{2n}) = G^{(n)}(r_1 t_1 + \tau, \ldots r_{2n} t_{2n} + \tau), \qquad (2.89)$$

i.e. they are not changed by a common time displacement of all the arguments. As a result, the $G^{(n)}$ may be thought of as depending only on $(2n-1)$ time differences. The same sort of argument can also be constructed for dealing with spatial displacements. When the density operator commutes with the components of the momentum of the field, the correlation functions are invariant under displacement of the spatial coordinates in the corresponding directions.

One further mathematical property of the correlation functions is a consequence of the way in which the functions are constructed from the positive and negative frequency parts of the fields. The function $G^{(n)}(t_1 \ldots t_n, t_{n+1} \ldots t_{2n})$ has a time dependence which, according to our convention, contains only positive frequencies for the variables $t_{n+1} \ldots t_{2n}$ and only negative frequencies for $t_1 \ldots t_n$. Thus, for example, if we ignore the spatial dependences we may write

$$G^{(1)}(t,t') = \sum_{kk'} c_{kk'} e^{i\omega_{k'} t'} e^{-i\omega_k t} \qquad (2.90)$$

with ω_k and $\omega_{k'} > 0$.

Now if we consider $G^{(n)}(t,t')$ as a function of two complex time variables, t and t', it is clearly an analytic function of t' in the half plane $\mathrm{Im}\, t' \le 0$, and an analytic function of t in the half-plane $\mathrm{Im}\, t \ge 0$.

We can therefore use the Cauchy theorem of complex function theory to construct identities such as

$$G^{(1)}(t,t') = \frac{1}{2\pi i} \int_C -\frac{G^{(1)}(t,t'')}{t'' - t'} dt'', \qquad (2.91)$$

where C is the contour in the complex t'' plane which is shown in Fig. 7.

Now from the boundedness of the coefficients $c_{kk'}$ in Eq. (2.90) we may see that the semi-circular part of the contour in the lower half plane gives no contribution in the limit as the radius R goes to infinity. Furthermore we note that the contribution of the infinitesimal semi-circular contour in the upper half-plane is just $-\pi i$ times the residue at the pole. In this way we find

$$G^{(1)}(t,t') = \frac{i}{\pi} P \int_{-\infty}^{\infty} \frac{G^{(1)}(t,t'')}{t''-t} dt'', \qquad (2.92)$$

where the integration is performed along the real axis and the symbol P denotes the Cauchy principal value. When we take the real and imaginary parts of Eq. (2.92), we obtain the pair of relations

$$\operatorname{Im} G^{(1)}(t,t') = \frac{1}{\pi} P \int_{-\infty}^{\infty} \frac{\operatorname{Re} G^{(1)}(t,t'')}{t''-t'} dt'', \qquad (2.93)$$

$$\operatorname{Re} G^{(1)}(t,t') = -\frac{1}{\pi} P \int_{-\infty}^{\infty} \frac{\operatorname{Im} G^{(1)}(t,t'')}{t''-t'} dt''. \qquad (2.94)$$

These relations enable us in principle to calculate the imaginary part of the correlation functions once we know the real part and vice versa.

Hilbert transform relationships of this type have received a considerable amount of attention in physics and electrical engineering in connection with the requirement that linearly responding systems behave causally. The relations such as (2.93) and (2.94) which are obeyed by the correlation functions, however, have nothing to do with causality. They are simply consequences of the way in which the functions have been defined.

2.7
Diffraction and Interference

From a mathematical standpoint, the quantum mechanical treatment of diffraction problems need not differ too greatly from the classical treatment. The field operators are required in general to obey the same linear differential equations and boundary conditions as the classical fields. The problem of constructing such operators may be reduced to the problem of finding a suitable set of mode functions in which to expand them (i.e., a set of mode functions which satisfies the wave equation together with suitable boundary conditions on any surfaces present). To find these modes we naturally resort to the familiar methods of the classical theory of boundary value problems. The solution for the mode functions is not a quantum dynamical problem at all. On the other hand, the fact that it is a well-explored "classical" problem does not mean, as we all know, that it is necessarily a simple one.

Let us return, for example, to the discussion of Young's experiment, illustrated in Fig. 2 (p. 30). When we said that the field at points on the screen Σ_2 is simply a linear combination of the fields at the two pinholes P_1 and P_3, evaluated at appropriate times, we were not solving the diffraction problem exactly, but making a number of physical approximations. One approximation, for example, was an implicit neglect of the fact that transmission of light through the pinholes has a slightly dispersive character. (This effect can be quite small if the bandwidth of the incident radiation is not too broad.) Approximations such as these are essentially classical in character. They are present simply because we have not taken the trouble to solve the classical diffraction problem more precisely.

With this understanding we can now discuss Young's experiment in fully quantum mechanical terms. The positive frequency part of the field $\boldsymbol{E}^{(+)}(\boldsymbol{r},t)$ when evaluated on the screen Σ_2 will be given, just as in the classical theory, by a linear combination of the fields $\boldsymbol{E}^{(+)}$ evaluated at the pinholes and having the form of Eq. (2.10). The only difference is that the fields $\boldsymbol{E}^{(+)}$ are now operators. If we assume that the two pinholes are not only quite tiny compared with their separation but equal in size then we shall have $\lambda_1 = \lambda_2$ in Eq. (2.10) and we may let the constant λ stand for both coefficients. Now if our observations of the interference pattern on the screen Σ_2 are made with an ideal photon detector, the counting rate of the detector will be proportional to $G^{(1)}(rt,rt)$. In other words, the intensity observed will be proportional to

$$I = \text{Tr}\left\{\rho \boldsymbol{E}^{(-)}(\boldsymbol{r}t)\boldsymbol{E}^{(+)}(\boldsymbol{r}t)\right\}$$

$$= \text{Tr}\left\{\rho |\lambda|^2 \left[\boldsymbol{E}^{(-)}(x_1) + \boldsymbol{E}^{(-)}(x_2)\right]\left[\boldsymbol{E}^{(+)}(x_1) + \boldsymbol{E}^{(+)}(x_2)\right]\right\}, \quad (2.95)$$

where we have again let x_j stand for the point $(\boldsymbol{r}_j, t_j)$. This intensity may be expressed in terms of first order correlation functions by expanding the product in Eq. (2.95). We then find

$$I = |\lambda|^2 \left\{G^{(1)}(x_1,x_1) + G^{(1)}(x_2,x_2) + 2\,\text{Re}\,G^{(1)}(x_1,x_2)\right\}. \quad (2.96)$$

The first two terms on the right side of this equation are the intensities which would be contributed by either pinhole in the absence of the other. These are, according to the assumptions we have made, rather slowly varying functions of x_1 and x_2. The third term on the right side of Eq. (2.96) is the interference term, as we have already noted in the classical discussion. The correlation function for $x_1 \neq x_2$ in general takes on complex values. If we write it as

$$G^{(1)}(x_1,x_2) = |G^{(1)}(x_1,x_2)| e^{i\varphi(x_1,x_2)},$$

then the intensity becomes

$$I = |\lambda|^2 \left\{G^{(1)}(x_1,x_1) + G^{(1)}(x_2,x_2) + 2|G^{(1)}(x_1,x_2)|\cos\varphi(x_1,x_2)\right\}, \quad (2.97)$$

and we see in the oscillation of the cosine term the origin of the familiar interference fringes.

2.7.1
Some General Remarks on Interference

The discussion we have given of Young's experiment is so closely related to the usual classical analysis that it may not be too clear in what way the interference phenomenon is a quantum mechanical one. A few general remarks about the quantum mechanical interpretation of interferences may therefore be in order. Interference phenomena characteristically occur in quantum mechanics whenever the probability amplitude for reaching a given final state from a given initial one is the sum of two or more partial amplitudes which have well defined phase relations. The individual partial amplitudes are usually contributed by alternative ways in which the system can evolve from its initial state to the final one.

The Young experiment furnishes a simple illustration of these generalities. We may consider as the initial state of the system one in which a wavepacket representing a single incident photon lies to the left of the first screen σ (Fig. 2) which has the single pinhole. We assume that initially all atoms of our photodetector are in the ground state. The final state of the system will be taken to be one in which the photon has been absorbed and one of the atoms of the counter has been correspondingly excited. The amplitude for reaching this final state is the sum of two amplitudes, each associated with the passage of the photon through one of the two pinholes in the screen Σ_1.

It is interesting to note that the existence of the interference effect is linked quite essentially with our inability to tell which of the possible paths the photon actually takes. Niels Bohr has shown, in a famous argument, that any attempt to determine which of the two paths the photon has followed will wipe out the interference fringes. One way of making such an attempt, for example, is by trying to measure the recoil of the screen Σ_1 when it deflects the photon. The photon may transfer either of two different recoil momenta to the screen (if it excites the counter). However, if we are to make sufficiently accurate measurements of the momentum of the screen we must be prepared to accept an uncertainty in its position which will mean that no fringes appear when the experiment is performed repeatedly.

This lesson is one which can be generalized to apply to all of the quantum mechanical situations we have described earlier. The different paths by which a system may evolve will contribute amplitudes with well-defined phase relations only as long as we have no way of telling which path the system takes. When we make observations to determine the path we characteristically alter the system by making the phases of the partial amplitudes random relative to one another, i.e., we wipe out any interference of the amplitudes on the average.

The alternative paths we have been speaking of are evolutionary paths or histories. For single particle systems such histories may often be identified with spatial trajectories, but for systems with many particles or variable numbers of particles the concept is a much more general one. It is important to emphasize that the quantities which interfere in quantum mechanics are amplitudes associated with particular histories, since the terminology which has been used has often invited confusion on this score.

An example of a statement which is often quoted and easily misinterpreted is made by Dirac in the first chapter of his classic text, *The Principles of Quantum Mechanics* (3rd edition, p. 9; Clarendon Press, Oxford 1947). There Dirac points out that the interference of the two component beams of the Michelson Interferometer cannot be interpreted as taking place because the photons of one beam sometimes annihilate photons from the other and sometimes combine to produce four photons. "This would contradict the conservation of energy. The new theory, which connects the wave functions with probabilities for one photon, gets over the difficulty by making each photon go partly into each of the two components. Each photon then interferes only with itself. Interference between two different photons never occurs." These remarks were only intended to refer to an experimental situation generically similar to that of Young's experiment, one in which the interference pattern is revealed by detecting single photons. To attempt to apply Dirac's remarks as a general doctrine for dealing with other types of interference experiments may lead to contradictions, as we shall presently see.

2.7.2
First-Order Coherence

The word "coherence" is used not only in optics, but in a variety of quantum mechanical and communication theoretical contexts as well. We shall not attempt to construct an encyclopedia of these usages here. We shall try instead to give the term a precise meaning when applied to electromagnetic fields. The meaning we shall adept is in fact one which links several of these conventional usages together.

The familiar concept of optical coherence is associated with the possibility of producing interference fringes when two fields are superposed. Let us return to the expression Eq. (2.97) for the intensity observed in Young's experiment. It is clear that no fringes will be observed if the correlation function $G^{(1)}(x_1,x_2)$ vanishes, and we may describe that condition by saying that the fields at x_1 and x_2 are incoherent.

It is only natural, on the other hand, to associate the highest degree of coherence with a field which exhibits the strongest possible interference fringes. Now, in the last section, we have derived a general inequality, (Eq. (2.86)), which states

$$|G^{(1)}(x_1,x_2)| \leq \left\{G^{(1)}(x_1,x_1)\, G^{(1)}(x_2,x_2)\right\}^{1/2}.$$

When we keep the intensities $G^{(1)}(x_1,x_2)$ and $G^{(1)}(x_2,x_2)$ fixed, the strongest contrast of the fringe intensities which is possible corresponds to using the equality sign in this relation. Thus we have established the necessary condition for coherence

$$|G^{(1)}(x_1,x_2)| = \left\{G^{(1)}(x_1,x_1)\, G^{(1)}(x_2,x_2)\right\}^{1/2}. \tag{2.98}$$

If we introduce the normalized correlation function

$$g^{(1)}(x_1,x_2) = \frac{G^{(1)}(x_1,x_2)}{\left\{G^{(1)}(x_1,x_1)\, G^{(1)}(x_2,x_2)\right\}^{1/2}}, \tag{2.99}$$

the condition (2.98) becomes

$$\left|g^{(1)}(x_1,x_2)\right| = 1 \tag{2.100}$$

or, in other words,

$$\left|g^{(1)}(x_1,x_2)\right| = e^{i\varphi(x_1,x_2)} \, .$$

Substitution in Eq. (2.97) now gives for the intensity in Young's experiment

$$|\lambda|^{-2} I = G^{(1)}(x_1,x_1) + G^{(1)}(x_2,x_2)$$
$$+ 2\left\{G^{(1)}(x_1,x_1) G^{(1)}(x_2,x_2)\right\}^{1/2} \cos \varphi(x_1,x_2)$$
$$= \left|\left\{G^{(1)}(x_1,x_1)\right\}^{1/2} e^{i\varphi(x_1,x_2)} + \left\{G^{(1)}(x_2,x_2)\right\}^{1/2}\right|^2 . \tag{2.101}$$

This intensity varies between the limits

$$I_{\min} = \left(\left\{G^{(1)}(x_1,x_1)\right\}^{1/2} - \left\{G^{(1)}(x_2,x_2)\right\}^{1/2}\right)^2 \tag{2.102}$$

and

$$I_{\max} = \left(\left\{G^{(1)}(x_1,x_1)\right\}^{1/2} + \left\{G^{(1)}(x_2,x_2)\right\}^{1/2}\right)^2 . \tag{2.103}$$

The parameter which is usually called the visibility of the fringes is given by

$$v = \frac{I_{\max} - I_{\min}}{I_{\max} + I_{\min}} = \frac{2\left\{G^{(1)}(x_1,x_1) G^{(1)}(x_2,x_2)\right\}^{1/2}}{G^{(1)}(x_1,x_1) + G^{(1)}(x_2,x_2)} . \tag{2.104}$$

If the fields incident on the two pinholes have equal intensity, i.e., if $G^{(1)}(x_1,x_1) = G^{(1)}(x_2,x_2)$, then the intensity varies between zero and $4G^{(1)}(x_1,x_2)$ and the visibility is $v = 1$.

The condition Eq. (2.98) is only a condition on the fields at two space-time points x_1 and x_2. When it is satisfied we might speak of the fields at these two points as being coherent with one another. That would correspond to the usage adopted by Born and Wolf in their discussion of classical fields on the basis of time-averaged correlation functions.

In quantum mechanics one characteristically thinks of the entire field as a dynamical system. It will be rather more convenient, therefore, for many analytical and statistical purposes to think of coherence as an idealized property of whole fields. That property can be described in terms of the condition Eq. (2.98), but an equivalent and mathematically more useful description can be given in terms of the requirement that the first order correlation function factorize. Let us suppose that the correlation function $G^{(1)}(x_1,x_2)$ separates into a product of two functions $A(x_1)$ and $B(x_2)$. Then from

$$G^{(1)}(x_1,x_2) = A(x_1) B(x_2) \tag{2.105}$$

we conclude via the symmetry relation, Eq. (2.76), that the functions A and B obey the identity

$$A(x_2) B(x_1) = A^*(x_1) B^*(x_2)$$

or

$$\frac{A(x_2)}{B^*(x_2)} = \frac{A^*(x_1)}{B(x_1)} . \tag{2.106}$$

Since in the latter relation a function of x_1 is equated to one of x_2 both functions must be constant. Furthermore the constant, let us call it μ, must be real as we can see by equating x_1 and x_2. We thus have

$$A(x) = \mu B^*(x) , \tag{2.107}$$

and from the fact that $G^{(1)}(x,x)$ is positive it becomes evident that μ is positive. Hence, if we define the function

$$\mathcal{E}(x) = \sqrt{\mu} B(x) , \tag{2.108}$$

we see that the first order correlation function falls into the form

$$G^{(1)}(x_1,x_2) = \mathcal{E}^*(x_1) \mathcal{E}(x_2) . \tag{2.109}$$

This explicit construction of the factorized form of the correlation function shows that, when factorization does take place, the function $\mathcal{E}(x)$ is almost uniquely determined. The only ambiguity which remains is that of a constant multiplicative phase factor.

We shall find it most convenient to use the factorization property Eq. (2.109) as our definition of optical coherence or first-order coherence of the field. It is immediately evident that this condition implies the conditions (2.98) and (2.100) on the absolute values of the correlation functions. In fact, it is also true that the latter conditions, if they hold at all points in the field, imply in turn the factorization condition (2.109). We shall demonstrate that shortly and thereby show that the two ways of discussing coherence are equivalent. But first let us discuss some examples of coherent fields.

The most elementary example of a field for which $G^{(1)}$ factorizes is any classical field for which the Fourier coefficients C_k are precisely determined, i.e., any field for which the probability distribution $P(\{C_k\})$ reduces to a product of delta functions. In that case the function $\mathcal{E}(x)$ is simply the classical field $\mathbf{E}^{(+)}(x)$ itself. We perceive here a first hint of the close association which exists between coherence and noiselessness, an association which we shall presently explore further. The absence of randomness or noise in the specification of the Fourier coefficients of a field has long been the criterion used by communication engineers for speaking of a "coherent" signal.

To see another illustration of coherence let us note that one of the possible ways of performing Young's experiment, though perhaps not the most practical one, is to begin with a single photon wavepacket incident upon the first pinhole. Then if we repeat the experiment many times, duplicating the wavepacket precisely in each repetition, we should expect to see the familiar interference fringes in the statistical distribution of photons received on the final screen. That pure states for single photons are always capable of giving rise to fringes, in this statistical sense, may be seen by examining the first-order correlation function. Let us suppose that the field is in some pure single-photon state which we denote by $|1 \text{ phot.}\rangle$. Then the density operator for the field is

$$\rho = |1 \text{ phot.}\rangle \langle 1 \text{ phot.}|, \qquad (2.110)$$

and the first order correlation function reduces to

$$G^{(1)}(x_1, x_2) = \left\langle 1 \text{ phot.} \left| \boldsymbol{E}^{(-)}(x_1) \boldsymbol{E}^{(+)}(x_2) \right| 1 \text{ phot.} \right\rangle. \qquad (2.111)$$

Now since $\boldsymbol{E}^{(+)}$ is a photon annihilation operator, the state $\boldsymbol{E}^{(+)}(x_2)|1 \text{ phot.}\rangle$ can only be a multiple of the vacuum state which we denote as $|0\rangle$. It is therefore possible to insert the projection operator upon the vacuum state, $|0\rangle\langle 0|$, between the $\boldsymbol{E}^{(-)}$ and $\boldsymbol{E}^{(+)}$ operators in Eq. (2.111) without altering the value of the correlation function. When we do that we find

$$G^{(1)}(x_1, x_2) = \left\langle 1 \text{ phot.} \left| \boldsymbol{E}^{(-)}(x_1) \right| 0 \right\rangle \left\langle 0 \left| \boldsymbol{E}^{(+)}(x_2) \right| 1 \text{ phot.} \right\rangle, \qquad (2.112)$$

which is exactly the factorized form required by Eq. (2.109). Hence any pure state in which the field is occupied by a single photon possesses first order coherence. (In this way the optical definition of coherence makes contact with some of the ways in which the term is used quantum mechanically in connection with pure states.)

We have, of course, only proved that a pure one-photon state is coherent. If, for example, we repeat our hypothetical one-photon interference experiment without duplicating the same wavepacket each time, i.e., if we consider a mixture of pure states, then we can not expect in general to observe intensity fringes of maximum contrast. Certain particular mixtures of one photon states may, however, preserve the factorization property Eq. (2.109) of the correlation function and thereby preserve the coherence property. Hence we must not think of pure states as the only ones which bring about coherence.

To give an example, let us suppose that only one mode of the field is excited, say the k-th. Then, since the other modes all remain in their ground states, it is easily seen, that we may ignore them altogether in calculating the correlation function. Now if the density operator for the k-th mode assumes the general form

$$\rho = \sum_{n,m} c_{nm} |n\rangle \langle m|, \qquad (2.113)$$

where $|n\rangle$ is the n-th quantum state for the mode, we may write the first-order correlation function as

$$G^{(1)}(\mathbf{r}_1 t_1, \mathbf{r}_2 t_2) = \tfrac{1}{2}\hbar\omega \sum_{n,m} c_{nm} \left\langle m \left| a_k^\dagger a_k \right| n \right\rangle u_k^*(\mathbf{r}_1) u_k(\mathbf{r}_2) e^{i\omega_k(t_1 - t_2)}$$
$$= C^2 u_k^*(\mathbf{r}_1) e^{i\omega_k t_1} u_k(\mathbf{r}_2) e^{-i\omega_k t_2}, \qquad (2.114)$$

where in the first of these expressions we have anticipated some of the notation of Eq. (2.152) and in the second we have used the definition

$$C^2 = \tfrac{1}{2}\hbar\omega_k \sum_n n c_{nn}. \qquad (2.115)$$

It is clear from the possibility of writing

$$\mathcal{E}(\mathbf{r},t) = C u_k(\mathbf{r}) e^{-i\omega_k t} \qquad (2.116)$$

that the correlation function Eq. (2.114) falls into the factorized form Eq. (2.109). Hence the excitation of a single mode, whether it is in a pure state or an arbitrary mixture, leads to fields with first-order coherence.

Although we have been able to give some simple examples of fields which possess first order coherence, it is worth pointing out that the factorization condition (2.109) is quite a restrictive one. It is, for example, not satisfied by pure states of the field in general as one may easily verify by calculating the correlation function for a state in which two or more photons are present and occupy different modes. Initial states such as these may lead to fringes in Young's experiment but the fringes will not, as a rule, satisfy the condition of maximum contrast. While the coherence condition is a restrictive one, we shall show presently that there exists a much broader class of states which satisfy it than those we have considered thus far.

Let us note particularly that no statement has been made requiring that coherent fields be monochromatic. The fields which satisfy the factorization condition (2.109), or for which interference fringes of maximum (instantaneous) contrast occur, can have arbitrary time dependences. The functions $\mathcal{E}(\mathbf{r},t)$ which determine the correlation functions of these fields may consequently have arbitrary Fourier spectra. What seems perhaps curious about these statements is that the experimental effort to produce nearly coherent beams of light has chiefly been a struggle to produce highly monochromatic ones. The reason for this connection has been that all of the effort has involved the use of stationary light sources. Such sources lead to fields for which the first order correlation function depends only on the difference of two times,

$$G^{(1)}(t_1, t_2) = G^{(1)}(t_1 - t_2). \qquad (2.117)$$

If such fields are to be coherent the correlation function must factorize to the form

$$G^{(1)}(t_1 - t_2) = \mathcal{E}^*(t_1) \mathcal{E}(t_2), \tag{2.118}$$

but this is a functional equation which has only exponential solutions. Since the dependence of $G^{(1)}$ on the variable t_2, can only contain positive frequencies we must have $\mathcal{E}(t) \propto e^{-i\omega t}$ for some $\omega > 0$. In other words, a coherent field which is stationary can only be monochromatic.

After giving so precise a definition to first order coherence we must add that it is a rather idealized condition, as is nearly any condition one places upon quantum mechanical states. We must not expect correlation functions for actual fields to obey the factorization condition (2.109) over unlimited ranges of the variables x_1 and x_2. In practice we define coherence lengths and times to describe the ranges of the spatial and temporal variables over which the factorization holds to a good approximation.

2.7.3
Fringe Contrast and Factorization

In the foregoing section we have defined coherence, mainly for reasons of mathematical convenience, in terms of a factorization property of the correlation function. That factorization property, we then showed, implies the condition (2.98) on the absolute value of the correlation function, i.e., the condition that the fringes show maximum contrast. Now it is possible to show that the latter condition, provided it holds for all space-lime points, also implies the factorization property. The proof we present is taken from a forthcoming paper by U. Titulaer and the author.

When the relation

$$\left|G^{(1)}(x_1, x_2)\right|^2 = G^{(1)}(x_1, x_1) G^{(1)}(x_2, x_2) \tag{2.119}$$

holds it places severe constraints upon the density operator for the field. These constraints may be found by first noting that Eq. (2.119) implies the existence of operators A such that

$$\mathrm{Tr}(\rho A^\dagger A) = 0. \tag{2.120}$$

To exhibit such operators A we choose an arbitrary space-time point x_0 at which the intensity of the field is non-vanishing, $G^{(1)}(x_0, x_0) \neq 0$, and write

$$A = E^{(+)}(x) - \frac{G^{(1)}(x_0, x)}{G^{(1)}(x_0, x_0)} E^{(+)}(x_0). \tag{2.121}$$

It then follows that

$$\mathrm{Tr}(\rho A^\dagger A) = G^{(1)}(x, x_0) - \frac{\left|G^{(1)}(x_0, x)\right|^2}{G^{(1)}(x_0, x_0)} = 0 \tag{2.122}$$

for all points x. Now the density operator ρ can be written as an average of products of the state vectors of the system having the form

$$\rho = \sum_i p_i |i\rangle \langle i|, \qquad (2.123)$$

where the probabilities p_i are all positive. The vanishing of the trace given by Eq. (2.120) means that

$$\sum_i p_i \langle i|A^\dagger A|i\rangle = 0. \qquad (2.124)$$

Since all the terms entering the sum are intrinsically positive, we may conclude that

$$\langle i|A^\dagger A|i\rangle = 0 \qquad (2.125)$$

for all states $|i\rangle$ for which $p_i \neq 0$. But this relation implies in turn that these states $|i\rangle$ are eigenstates of A with eigenvalue zero

$$A|i\rangle = 0. \qquad (2.126)$$

What we have shown is that the vanishing of the trace (2.120) implies the pair of operator relations

$$A\rho = \rho A^\dagger = 0. \qquad (2.127)$$

Since these relations hold when the operator A takes on the value given by Eq. (2.121), the density operator must obey the pair of identities

$$E^{(+)}(x)\rho = \frac{G^{(1)}(x_0,x)}{G^{(1)}(x_0,x_0)} E^{(+)}(x_0)\rho \qquad (2.128)$$

$$\rho E^{(-)}(x) = \frac{G^{(1)}(x,x_0)}{G^{(1)}(x_0,x_0)} \rho E^{(-)}(x_0). \qquad (2.129)$$

These identities may now be used to shift the arguments of correlation functions to a common reference point x_0. If we let $x = x_2$ in the first of these identities and $x = x_1$ in the second of them we may then use them to construct the relation

$$\mathrm{Tr}\left\{\rho E^{(-)}(x_1) E^{(+)}(x_2)\right\} = \frac{G^{(1)}(x_1,x_0)}{G^{(1)}(x_0,x_0)} \mathrm{Tr}\left\{\rho E^{(-)}(x_0) E^{(+)}(x_0)\right\} \frac{G^{(1)}(x_0,x_2)}{G^{(1)}(x_0,x_0)},$$

which can also be written as the functional identity

$$G^{(1)}(x_1,x_2) = \frac{G^{(1)}(x_1,x_0) G^{(1)}(x_0,x_2)}{G^{(1)}(x_0,x_0)}.$$

Now we have only to define the function $\mathcal{E}(x)$ as

$$\mathcal{E}(x) = \frac{G^{(1)}(x_0,x)}{\{G^{(1)}(x_0,x_0)\}^{1/2}} \tag{2.130}$$

in order to see that the first order correlation function takes on the factorized form

$$G^{(1)}(x_1,x_2) = \mathcal{E}^*(x_1)\,\mathcal{E}(x_2) \,. \tag{2.131}$$

There is no need to repeat this demonstration in order to deal with the tensor structure of the correlation functions for fields which are not fully polarized. All we need to do is to consider each coordinate x as specifying a tensor index as well as a position and time.

2.8
Interpretation of Intensity Interferometer Experiments

In the preceding Section we have discussed Young's experiment at some length as an example typical of the interference experiments which are based upon the measurement of a first order correlation function. While all of the older interference experiments share this character, we have discussed in Sect. 2.2 some more recent experiments which are of a fundamentally different type. These are the intensity interferometry experiments of Hanbury Brown, and Twiss which measure, in effect, the second order correlation function of the incident field.

We have given a simple classical discussion of the way in which the correlation fringes appear in the intensity interferometer when the field is produced by a pair of sources with small angular separation. It is interesting, therefore, to investigate the quantum mechanical origin of these same fringes. If we remember that the intensity interferometer functions by first detecting the incident fields in each of two receivers, we see immediately that pairs of photons must be involved in the interference effect, i.e., nothing is recorded at all unless different photons are incident on each of the two detectors at more or loss the same time. It is at precisely this point that one is confronted by a serious dilemma if he attaches too great a generality to Dirac's statement that "interference between two different photons never occurs."

The general discussion of interference which we gave in the last Section should make it clear that no such dilemma need exist. The things which should be regarded as interfering are not, strictly speaking, the photons, but alternative "histories" of the system as a whole. Let us imagine that the initial state of the system is one in which two (generally overlapping) single-photon wavepackets are present in the field and the atoms of the two detectors (represented by photon counters) are in the ground state. We may take the final state of the system to be one in which both photons have been absorbed and one atom in each of the counters is correspondingly excited. If we label the photons 1 and 2, and the two counters a and b, we see that there are two alternative

2.8 Interpretation of Intensity Interferometer Experiments

Figure 8

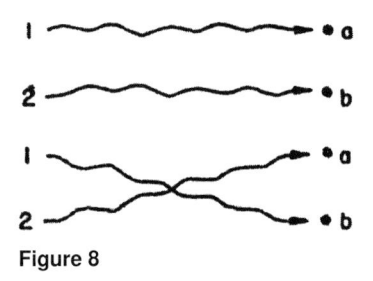

ways in which the final state may be reached. Either photon 1 is absorbed by counter a and 2 by b, or 1 is absorbed by b and 2 by a (Fig. 8).

If the packets had altogether different average propagation vectors these alternative histories would be distinguishable by means of careful measurements made in the counters. But the circumstances in which the fringes are observable are precisely those in which the packets have nearly the same average propagation vectors (e.g., packets with the same frequencies, small angular separation of the sources). In other words the fringes appear once again just when the alternative histories of the system become indistinguishable. Since the amplitudes for the two histories interfere, it becomes meaningless to ask which counter absorbed which photon.

2.8.1
Higher Order Coherence and Photon Coincidences

We recall from our classical discussions of Sect. 2.2 that the intensity interferometer measures the second order correlation function of the incident field. Radiation fields generated by natural sources tend to have a chaotic quality which allows us to construct these correlation functions from a knowledge of the first order functions. However, no such constructions are available in general for dealing with radiation from man-made sources such as the laser or radio transmitters. The fields generated by these sources can have much higher regularity than is ever possible for natural sources. It will be useful, therefore, to sharpen the concept of coherence by defining higher order analogs of optical coherence.

We begin once more by stating conditions on the absolute values of the correlation functions. For full coherence we shall require that the normalized form of the n-th order correlation function,

$$g^{(n)}(x_1 \ldots x_{2n}) = \frac{G^{(n)}(x_1 \ldots x_{2n})}{\prod_{j=1}^{2n} \{G^{(1)}(x_j x_j)\}^{1/2}}, \quad (2.132)$$

have modulus unity for all n and all combinations of arguments x. If the functions have unit modulus only for $n \leq M$ we shall speak of M-th order coherence.

The concept of M-th order coherence has a simple interpretation in terms of n-fold (delayed) coincidence experiments. We know that $G^{(n)}(x_1 \ldots x_n, x_n \ldots x_1)$ is an average coincidence rate for n ideal photodetectors registering at the points $x_1 \ldots x_n$. Since this value of the function is real and positive the condition that $g^{(n)}$ have unit modulus for $n \leq M$ implies that

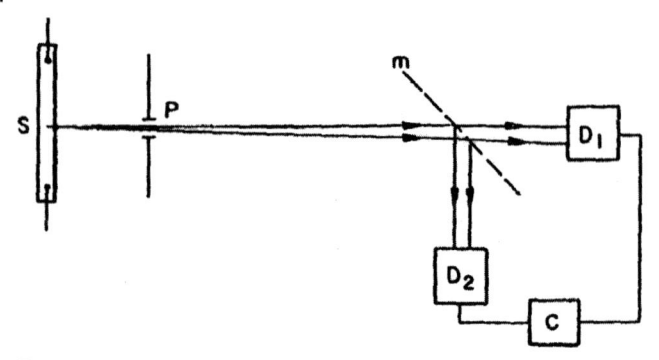

$$g^{(n)}(x_1 \ldots x_n, x_n \ldots x_1) = 1$$

for $n \leq M$. Hence for fields with M-th order coherence, it is clear from the definition of $g^{(n)}$ that we have

$$G^{(n)}(x_1 \ldots x_n, x_n \ldots x_1) = \prod_{j=1}^{n} G^{(1)}(x_j x_j) \qquad (2.133)$$

for $n \leq M$.

Expressed in experimental terms, this means that the n-fold coincidence rate is just the product of the counting rates which would be measured by each counter individually in the absence of the others. Thus there is no tendency toward statistical correlation of the photon counts. In a field with coherence of order $M \geq n$ the n photon counters register in a statistically independent way.

Several investigations of light beams using coincidence counting of photons or equivalent experimental procedures have in fact been carried out during the last few years. The first of these to detect a tendency toward statistical correlation of the arrival times of photons was performed (in addition to the other experiments we have mentioned) by Hanbury Brown, and Twiss[2], In the experiment light from a source S (Fig. 9) passes through a pinhole P and then reaches a half-silvered mirror m, which splits it into two beams. Detectors D_1 and D_2 are placed symmetrically with respect to the mirror. Their photocurrents are multiplied together by the correlator C whose average output is the quantity measured. We may consider the half-silvered mirror m as a device, which permits us, in effect, to place two different photodetectors at essentially the same position in the beam.

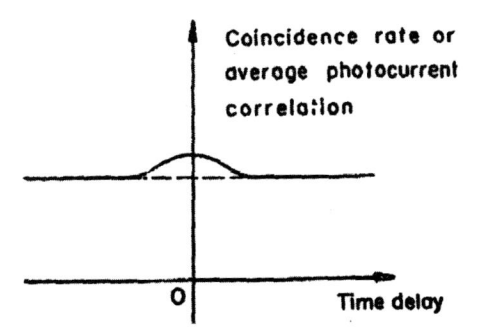

2.8 Interpretation of Intensity Interferometer Experiments

Figure 10

Shortly after the original experiment had been performed another version of it with a slightly more direct interpretation was performed by Rebka and Pound[3]. In the latter experiment D_1 and D_2 are counters of individual photons, and C is a device for registering delayed coincidences. The experiment measures the average coincidence rate as a function of delay time while the counters remain fixed in their symmetrical positions relative to the mirror. Now, even if the photon beams incident on the two counters were statistically independent of one another, there would be a certain background counting rate of accidental coincidences. This rate would, however, be independent of any time delay. Thus any observed dependence of the coincidence rate on the time delay indicates a lack of statistical independence.

The result of the experiments is indicated in Fig. 10. If the responses of the counters were statistically independent the coincidence rate would be independent of time delay. The observation of a small "bump" in the experimental curve indicates that the photons have a distinct tendency to arrive in pairs. Although the effect was at first difficult to observe it is, as we shall show, not necessarily a small one at all. The small magnitude of the observed "bump" and its particular shape in these experiments were determined almost entirely by the relatively slow response times of the counters.

Let us note that, if the counters are placed symmetrically with respect to the mirror, the fields which are incident upon them are essentially identical, apart from a constant multiplicative factor. It follows then that if r_1 and r_2 are mirror-image points in the two detectors we have

$$\left|g^{(1)}(r_1 t, r_2 t)\right| = 1 , \qquad (2.134)$$

i.e., the fields which fall on the two detectors have essentially perfect first order coherence. The observation of a positive correlation in the coincidence rate demonstrates, on the other hand, that the fields are not coherent in the second order sense. We shall show presently that this result is a characteristic one for all experiments performed with natural light sources. These have a random character which destroys second order coherence.

2.8.2
Further Discussion of Higher Order Coherence

Let us return now to the definition of higher order coherence. We have, by analogy with first order coherence, defined M-th order coherence in terms of the succession of conditions

$$|G^{(n)}(x_1 \ldots x_{2n})|^2 = \prod_{j=1}^{2n} G^{(1)}(x_j, x_j) \tag{2.135}$$

on the absolute values of the correlation functions for $n \leq M$. Just as in the first order case we found it convenient to express the coherence condition in an alternative way, as a factorization property of the correlation function, we shall find it even more convenient here to do much the same thing. We shall therefore state as an alternative definition the requirement that there exist a single complex function $\mathcal{E}(x)$ such that

$$G^{(1)}(x_1 \ldots x_{2n}) = \prod_{j=1}^{n} \mathcal{E}^*(x_j) \prod_{j=n+1}^{2n} \mathcal{E}(x_j) \tag{2.136}$$

for all $n \leq M$. If this factorization holds for all n we shall speak of full coherence.

If we note that the definition (2.136) contains the statement

$$G^{(1)}(x,x) = |\mathcal{E}(x)|^2 , \tag{2.137}$$

then we see immediately that it requires that the correlation functions obey the absolute value conditions Eq. (2.135).

It is possible, on the other hand, to show that the absolute value conditions also imply the factorization properties. To do that we note that M-th order coherence always requires first order coherence. We may therefore make use of the identities which were shown in the last Section to be consequences of first order coherence. In particular, since the operators $\boldsymbol{E}^{(-)}(x_j)$ for $j = 1, \ldots n$ all commute with one another, as do the operators $\boldsymbol{E}^{(+)}(x_j)$ for $j = n+1, \ldots 2n$, we can use each of the two identities Eq. (2.128) and Eq. (2.129) n times in order to shift all of the arguments of the n-th order correlation function to a particular reference point x_0. More specifically, we write

$$\text{Tr}\left\{\rho \boldsymbol{E}^{(-)}(x_1) \ldots \boldsymbol{E}^{(-)}(x_n) \boldsymbol{E}^{(+)}(x_{n+1}) \ldots \boldsymbol{E}^{(+)}(x_{2n})\right\} = \prod_{j=1}^{n} \frac{G^{(1)}(x_j, x_0)}{G^{(1)}(x_0, x_0)}$$

$$\times \text{Tr}\left\{\rho \boldsymbol{E}^{(-)}(x_0) \ldots \boldsymbol{E}^{(-)}(x_0) \boldsymbol{E}^{(+)}(x_0) \ldots \boldsymbol{E}^{(+)}(x_0)\right\} \prod_{j=n+1}^{2n} \frac{G^{(1)}(x_0, x_j)}{G^{(1)}(x_0, x_0)} ,$$

which is the identity

$$G^{(n)}(x_1 \ldots x_{2n}) = \frac{G^{(n)}(x_0 \ldots x_0)}{\{G^{(1)}(x_0, x_0)\}^n} = \frac{\prod_{j=1}^{n} G^{(1)}(x_j, x_0) \prod_{j=n+1}^{2n} G^{(1)}(x_0, x_j)}{\{G^{(1)}(x_0, x_0)\}^n} .$$

If we introduce the function $\mathcal{E}(x)$ which is defined by Eq. (2.131), and make use of the normalized form of the correlation function, we may write the latter identity in the form

$$G^{(n)}(x_1\ldots x_{2n}) = g^{(n)}(x_0\ldots x_0) \prod_{j=1}^{n} \mathcal{E}^*(x_j) \prod_{j=n+1}^{2n} \mathcal{E}(x_j) \,. \tag{2.138}$$

Now, as we have shown earlier, the functions $\mathcal{E}(x)$ can only depend on the choice of the arbitrary reference point x_0 through a constant phase factor. Since that phase factor cancels out of the product which occurs in Eq. (2.138), it follows that for fields with first order coherence the functions $g^{(n)}(x_0\ldots x_0)$ are independent of x_0. In other words, the condition of first order coherence alone is sufficient to bring all of the higher order correlation functions into a factorized form, although not exactly the form, in general, which is required for higher order coherence. The difference is that Eq. (2.138) contains the constant factors $g^{(n)}(x_0\ldots x_0)$ which should be unity if higher order coherence is to hold. Now the higher order coherence conditions Eq. (2.135) do require these coefficients to have unit absolute value for $n \leq M$. Then, since the $g^{(n)}(x_0\ldots x_0)$ must be real and positive, they must be equal to one.

Hence the conditions Eq. (2.135) do indeed imply the factorization condition Eq. (2.136).

2.8.3
Treatment of Arbitrary Polarizations

From a mathematical standpoint, very little need be added to our earlier discussions in order to treat fields with arbitrary polarization properties rather than the fully polarized fields we have been discussing. All we need do, as we have already noted, in order to deal with the general tensor character of the correlation functions, is to think of every coordinate in the formulae we have derived as specifying a tensor index as well as a position and time.

Thus the relations (2.76) for $n = 1$ and (2.86), for example, may be generalized to read

$$\left\{ G^{(1)}_{\mu\nu}(x_1,x_2) \right\}^* = G^{(1)}_{\nu\mu}(x_2,x_1) \tag{2.139}$$

and

$$\left| G^{(1)}_{\mu\nu}(x_1,x_2) \right|^2 \leq G^{(1)}_{\mu\mu}(x_1,x_1) G^{(1)}_{\nu\nu}(x_2,x_2) \,. \tag{2.140}$$

It may be worth noting that all information about the state of polarization of the field is contained in the correlation tensor $G^{(1)}_{\mu\nu}(x,x)$. Let us denote this tensor by $\mathcal{G}_{\mu\nu}$. We see immediately that $\mathcal{G}_{\mu\nu}$ is a Hermitian matrix, $\mathcal{G}^*_{\mu\nu} = \mathcal{G}_{\nu\mu}$. If we substitute $A = \sum_{\nu=1}^{3} \lambda_\nu E^{(+)}_\nu(x)$ in the general inequality $\text{Tr}\{\rho A^\dagger A\} \geq 0$ we find

$$\sum_{\mu,\nu=1}^{3} \lambda^*_\mu \lambda_\nu \mathcal{G}_{\mu\nu} \geq 0 \,. \tag{2.141}$$

Thus $G_{\mu\nu}$ is also positive definite. Because of its Hermitian character $G_{\mu\nu}$ can be diagonalized, that is to say there exist three real and positive eigenvalues λ_p and three (generally complex) eigenvectors $\boldsymbol{e}^{(p)}$, such that

$$G \cdot \boldsymbol{e}^{(p)*} = \lambda_p \boldsymbol{e}^{(p)*} \,; \qquad \boldsymbol{e}^{(p)} \cdot G = \lambda_p \boldsymbol{e}^{(p)} \,. \tag{2.142}$$

Note that both the λ_p and the $\boldsymbol{e}^{(p)}$ depend in general on the space-time point x, that occurs in the definition of G.

The $\boldsymbol{e}^{(p)}$ are either found to be mutually orthogonal if the λ's have no degeneracy, or they can be chosen orthogonal if the λ's are degenerate. Hence we may assume

$$\boldsymbol{e}^{(p)} \cdot \boldsymbol{e}^{(q)*} = \delta_{pq} \,. \tag{2.143}$$

Since the tensor product

$$\boldsymbol{e}^{(p)} \cdot G \cdot \boldsymbol{e}^{(q)*} = \lambda_p \delta_{pq} \tag{2.144}$$

expresses the correlation of the field components in the directions of $\boldsymbol{e}^{(p)}$ and $\boldsymbol{e}^{(q)}$ there are three "directions" (i.e., complex directions) in which the field components are mutually uncorrelated. Any field may thus be regarded as a superposition of three orthogonally polarized fields whose amplitudes are (instantaneously) uncorrelated.

The eigenvalues $\lambda^{(p)}$ are the intensities corresponding to the three polarizations. The total intensity is given by

$$\text{Tr}\, G = \sum_p \lambda_p \,. \tag{2.145}$$

A set of normalized intensities can be defined as

$$I_p = \frac{\lambda_p}{\sum_{j=1}^{3} \lambda_j} \qquad (p = 1, 2, 3) \,.$$

These numbers can be interpreted as specifying the degree of polarization of the field. In an isotropic radiation field we must have $I_p = \frac{1}{3}$, $(p = 1, 2, 3)$. If the field is stationary i.e., $[\rho, H] = 0$ then G is time independent and the λ_p and I_p and $\boldsymbol{e}^{(p)}$ become fixed at any spatial position r.

If we are considering a beam with a single direction of propagation $\boldsymbol{k}$, then clearly $\boldsymbol{k} \cdot G = G \cdot \boldsymbol{k} = 0$ (since light is a transverse wave). Hence $\boldsymbol{k}$ is an eigenvector of G corresponding to the eigenvalue $\lambda = 0$. Then there are two remaining eigenvalues λ_p, $p = 1, 2$. The net polarization of the beam is usually defined as $|I_1 - I_2| = |\lambda_1 - \lambda_2|/(\lambda_1 + \lambda_2)$. The two polarizations $\boldsymbol{e}^{(p)}$ for $p = 1, 2$ clearly lie in the plane perpendicular to $\boldsymbol{k}$.

The higher order correlation tensors are defined by

$$G^{(n)}_{\mu_1\ldots\mu_{2n}}(x_1\ldots x_{2n}) = \text{Tr}\left\{\rho \boldsymbol{E}^{(-)}_{\mu_1}(x_1)\ldots \boldsymbol{E}^{(-)}_{\mu_n}(x_n) \boldsymbol{E}^{(+)}_{\mu_{n+1}}(x_{n+1})\ldots \boldsymbol{E}^{(+)}_{\mu_{2n}}(x_{2n})\right\} \,. \tag{2.146}$$

The coherence condition, Eq. (2.136), may evidently be restated for fields of arbitrary polarization by requiring that there exist a vector function $\mathcal{E}_\mu(x)$ such that

$$G_{\mu_1\ldots\mu_{2n}}(x_1\ldots x_{2n}) = \prod_{j=1}^{n} \mathcal{E}_{\mu_j}^*(x_j) \prod_{j=n+1}^{2n} \mathcal{E}_{\mu_j}(x_j) \tag{2.147}$$

for $n \leq M$.

As a last remark on polarizations we note that first order coherence implies full polarization of the field, i.e., if we have

$$G_{\mu\nu}^{(1)}(x,x) = \mathcal{G}_{\mu\nu} = \mathcal{E}_\mu^*(x)\mathcal{E}_\nu(x), \tag{2.148}$$

then clearly the vector $\mathcal{E}_\mu(x)$ itself is an eigenvector. The corresponding intensity is $\sum_{\mu=1}^{3} |\mathcal{E}_\mu(x)|^2$, which is the full intensity of the field present.

Let us try to construct states in which the fields have full coherence, that is to say, states in which all the correlation functions $G^{(n)}$ factorize according to Eqns. (2.136) or (2.147). If there existed simultaneous eigenstates of the operators $\boldsymbol{E}^{(+)}$ and $\boldsymbol{E}^{(-)}$, such eigenstates would clearly bring about the desired factorization. However, since $\boldsymbol{E}^{(+)}$ and $\boldsymbol{E}^{(-)}$ do not commute (and have a commutator which is a c-number) it is clear that no such eigenstates exist. We may reduce our demand to a more plausible level by noting that in the correlation functions the field operators always occur in normal order. Therefore, it is sufficient to secure coherence if the state of the field is simply an eigenstate of $\boldsymbol{E}^{(+)}$ in the restricted sense

$$\boldsymbol{E}_\mu^{(+)}(x)|\ \rangle = \mathcal{E}_\mu(x)|\ \rangle. \tag{2.149}$$

This is true because the adjoint relation is

$$\langle\ |\boldsymbol{E}_\mu^{(-)}(x) = \mathcal{E}_\mu^*(x)\langle\ |, \tag{2.150}$$

and together the two relations lead to the desired factorization of the correlation functions.

Since the operator $\boldsymbol{E}^{(+)}$ is neither Hermitian nor normal (i.e., it does not commute with its Hermitian adjoint), there is no *a priori* reason why eigenstates of this form should exist. Indeed it is easily shown that the similar relation

$$\langle\ |\boldsymbol{E}^{(+)}(x) = \mathcal{E}(x)\langle\ | \tag{2.151}$$

can have no normalizable solution at all. The simplest way to show that Eq. (2.149) has solutions is to construct them.

If any solution of Eq. (2.149) is to exist then it is clear that the function $\mathcal{E}_\mu(x)$ must satisfy the same wave equation and boundary conditions as the operator $\boldsymbol{E}_\mu^{(+)}(x)$. The latter has the Fourier expansion

$$\begin{aligned}\boldsymbol{E}^{(+)}(\boldsymbol{r},t) &= -\frac{1}{c}\frac{\partial A^{(+)}}{\partial t} \\ &= i\sum_k \left(\frac{\hbar\omega_k}{2}\right)^{1/2} a_k u_k(\boldsymbol{r})e^{-i\omega_k t}.\end{aligned} \tag{2.152}$$

Here the time independent operators a_k are described completely by means of their commutation relations

$$[a_k, a_{k'}] = [a_k^\dagger, a_{k'}^\dagger] = 0,$$
$$[a_k, a_{k'}^\dagger] = \delta_{kk'}.$$
(2.153)

For $\mathcal{E}(\mathbf{r},t)$ we must have a corresponding expansion

$$\mathcal{E}(\mathbf{r},t) = i \sum_k \left(\frac{\hbar \omega_k}{2}\right)^{1/2} \alpha_k u_k(\mathbf{r}) e^{-i\omega_k t},$$
(2.154)

where the coefficients α_k are a set of numbers which can take on arbitrary complex values.

Now if we substitute the expansions (2.152) and (2.154) in the equation which determines the eigenstates, we see that the coefficients of each mode function must separately be equal. Hence the eigenstate must satisfy the conditions

$$a_k |\ \rangle = \alpha_k |\ \rangle$$
(2.155)

for all modes k.

The coefficients α_k correspond in a simple way to the classical Fourier coefficients C_k which we introduced in the first section. More specifically if we compare Eqns. (2.8) and (2.154) we see that the correspondence is

$$C_k = i \left(\frac{\hbar \omega_k}{2}\right)^{1/2} \alpha_k.$$
(2.156)

This relation shows that to describe classical fields we shall have to deal with parameters α_k of large modulus, i.e., if we let $\hbar \to 0$ then α_k increases as $\hbar^{-1/2}$.

To construct the desired eigenstate we can begin with the construction of a state $|\alpha_k\rangle_k$ for the single mode k, such that

$$a_k |\alpha_k\rangle_k = \alpha_k |\alpha_k\rangle_k.$$
(2.157)

The state for the entire system is then given by the direct product

$$|\ \rangle = \prod_k |\alpha_k\rangle_k.$$
(2.158)

We shall call these states the *coherent* states. From the fact that they remain the same, up to a numerical factor, when we apply an annihilation operator a_k, it follows immediately that they cannot be eigenstates of the photon number operator.

The sense in which states of the type of Eq. (2.158) are coherent includes, of course, optical coherence (they secure factorization of the first order coherence function). But it also includes a sense used in communication theory which we have mentioned earlier. There a coherent signal is a pure signal, one that has no noise. A classical signal of this type is ideally one with a precisely defined set of Fourier coefficients C_k. But this is exactly the kind of field we are talking about in the more general quantum mechanical context. Our precise specification of the Fourier coefficients α_k means, as we shall see, that we are as close as possible to having no noise in the signal. It can not mean, however, that there is no noise at all. Unpredictably fluctuating fields are present even in the vacuum. Our detectors detect individual photons, and photons tend to arrive randomly. Even when we specify the field as accurately as we can, we can only make predictions about the response of our counter in statistical terms; there will be some inevitable noise, and the coherent states of the field only tend to reduce that noise to a minimum.

2.9
Coherent and Incoherent States of the Radiation Field[2]

2.9.1
Introduction

Few problems of physics have received more attention in the past than those posed by the dual wave-particle properties of light. The story of the solution of these problems is a familiar one. It has culminated in the development of a remarkably versatile quantum theory of the electromagnetic field. Yet, for reasons which are partly mathematical and partly, perhaps, the accident of history, very little of the insight of quantum electrodynamics has been brought to bear on the problems of optics. The statistical properties of photon beams, for example, have been discussed to date almost exclusively in classical or semiclassical terms. Such discussions may indeed be informative, but they inevitably leave open serious questions of self-consistency, and risk overlooking quantum phenomena which have no classical analogs. The wave–particle duality, which should be central to any correct treatment of photon statistics, does not survive the transition to the classical limit. The need for a more consistent theory has led us to begin the development of a fully quantum-mechanical approach to the problems of photon statistics. We have quoted several of the results of this work in a recent note[4], and shall devote much of the present paper to explaining the background of the material reported there.

Most of the mathematical development of quantum electrodynamics to date has been carried out through the use of a particular set of quantum states for the field. These are the stationary states of the non-interacting field, which corresponds to the

---

[2] Although included in the publication mentioned in footnote 1 on p. 23, the original publication of the following section was R. J. Glauber, *Phys. Rev.* **131**, 2766–2788 (1966). Because of their different origin, Sections 2.9.1 through 2.9.2 overlap with the previous sections of Chapter 2, and readers who have followed Chapter 2 this far should have no difficulty in beginning this part at Sect. 2.9.3 (Coherent States of a Single Mode), and following its presentation through Sect. 2.9.9 (Density Operators for the Field). Following the reprint of the *Phys. Rev.* paper, the lecture notes begin again with Sect. 2.10. That section resumes the story near the end of Sect. 2.9.9, which it is intended to amplify.

presence of a precisely defined number of photons. The need to use these states has seemed almost axiomatic inasmuch as nearly all quantum electrodynamical calculations have been carried out by means of perturbation theory. It is characteristic of electrodynamical perturbation theory that in each successive order of approximation it describes processes which either increase or decrease the number of photons present by one. Calculations performed by such methods have only rarely been able to deal with more than a few photons at a time. The description of the light beams which occur in optics, on the other hand, may require that we deal with states in which the number of photons present is large and intrinsically uncertain. It has long been clear that the use of the usual set of photon states as a basis offers at best only an awkward way of approaching such problems.

We have found that the use of a rather different set of states, one which arises in a natural way in the discussion of correlation and coherence[5,6] properties of fields, offers much more penetrating insights into the role played by photons in the description of light beams. These states, which we have called coherent ones, are of a type that has long been used to illustrate the time-dependent behavior of harmonic oscillators. Since they lack the convenient property of forming an orthogonal set, very little attention has been paid them as a set of basis states for the description of fields. We shall show that these states, though not orthogonal, do form a complete set and that any state of the field may be represented simply and uniquely in terms of them. By suitably extending the methods used to express arbitrary states in terms of the coherent states, we may express arbitrary operators in terms of products of the corresponding state vectors. It is particularly convenient to express the density operator for the field in an expansion of this type. Such expansions have the property that whenever the field possesses a classical limit, they render that limit evident while at the same time preserving an intrinsically quantum-mechanical description of the field.

The earlier sections of the paper are devoted to a detailed introduction of the coherent states and a survey of some of their properties. We then undertake in Sections 2.9.4 and 2.9.5 the expansion of arbitrary states and operators in terms of the coherent states. Section 2.9.6 is devoted to a discussion of the particular properties of density operators and the way these properties are represented in the new scheme. The application of the formalism to physical problems is begun in Sect. 2.9.7, where we introduce a particular form for the density operator which seems especially suited to the treatment of radiation by macroscopic sources. This form for the density operator leads to a particularly simple way of describing the superposition of radiation fields. A form of the density operator which corresponds to a very commonly occurring form of incoherence is then discussed in Sect. 2.9.8 and shown to be closely related to the density operator for blackbody radiation. In Sect. 2.9.9 the results established earlier for the treatment of single modes of the radiation field are generalized to treat the entire field. The photon fields generated by arbitrary distributions of classical currents are shown to have an especially simple description in terms of coherent states. Finally, in Sect. 2.9.10 the methods of the preceding sections are illustrated in a discussion of certain forms of coherent and incoherent fields and of their spectra and correlation functions.

2.9.2
Field-Theoretical Background

We have, in an earlier paper[6], discussed the separation of the electric field operator $E(rt)$ into its positive-frequency part $E^{(+)}(rt)$ and its negative-frequency part $E^{(-)}(rt)$. These individual fields were then used to define a succession of correlation functions $G^{(n)}$, the simplest of which takes the form

$$G^{(1)}_{\mu\nu}(rt,r't') = \mathrm{Tr}\left\{\rho E^{(-)}_\mu(rt) E^{(+)}_\nu(r't')\right\}, \qquad (2.159)$$

where ρ is the density operator which describes the field and the symbol Tr stands for the trace. We noted, in discussing these functions, that there exist quantum-mechanical states which are eigenstates of the positive-and negative-frequency parts of the fields in the senses indicated by the relations

$$E^{(+)}_\mu(rt)|\ \rangle = \mathcal{E}_\mu(rt)|\ \rangle, \qquad (2.160)$$

$$\langle\ |E^{(-)}_\mu(rt) = \mathcal{E}^*_\mu(rt)\langle\ |. \qquad (2.161)$$

in which the function $\mathcal{E}_\mu(rt)$ plays the role of an eigenvalue. It is possible, as we shall note, to find eigenstates $|\ \rangle$ which correspond to arbitrary choices of the eigenvalue function $\mathcal{E}_\mu(rt)$, provided they obey the Maxwell equations satisfied by the field operator $E_\mu(rt)$ and contain only positive frequency terms in their Fourier resolutions.

The importance of the eigenstates defined by Eqns. (2.160) and (2.161) is indicated by the fact that they cause the correlation functions to factorize. If the field is in an eigenstate of this type we have $\rho = |\ \rangle\langle\ |$, and the first-order correlation function therefore reduces to

$$G^{(1)}_{\mu\nu}(rt,r't') = \mathcal{E}^*_\mu(rt)\,\mathcal{E}_\nu(r't'). \qquad (2.162)$$

An analogous separation into a product of $2n$ factors takes place in the n-th-order correlation function. The existence of such factorized forms for the correlation functions is the condition we have used to define fully coherent fields. The eigenstates $|\ \rangle$, which we have therefore called the coherent states, have many properties which it will be interesting to study in detail. For this purpose, it will be useful to introduce some of the more directly related elements of quantum electrodynamics.

The electric and magnetic field operators $E(rt)$ and $B(rt)$ may be derived from the operator $A(rt)$, which represents the vector potential, via the relations

$$= -\frac{1}{c}\frac{\partial A}{\partial t}, \qquad B = \nabla \times A. \qquad (2.163)$$

We shall find it convenient, in discussing the quantum states of the field, to describe the field by means of a discrete succession of dynamical variables rather than a continuum of them. For this reason we assume that the field we are discussing is confined within a spatial volume of finite size, and expand the vector potential within that volume in an appropriate set of vector mode functions. The amplitudes associated with these oscillation modes then form a discrete set of variables whose dynamical behavior is easily discussed.

The most convenient choice of a set of mode functions, $\boldsymbol{u}_k(\boldsymbol{r})$, is usually determined by physical considerations which have little direct bearing on our present work. In particular, we need not specify the nature of the boundary conditions for the volume under study; they may be either the periodic boundary conditions which lead to traveling wave modes, or the conditions appropriate to reflecting surfaces which lead to standing waves. If the volume contains no refracting materials, the mode function $\boldsymbol{u}_k(\boldsymbol{r})$, which corresponds to frequency ω_k, may be taken to satisfy the wave equation

$$\nabla^2 \boldsymbol{u}_k + \frac{\omega_k^2}{c^2} \boldsymbol{u}_k = 0 \tag{2.164}$$

at interior points. More generally, whatever the form of the wave equation or the boundary conditions may be, we shall assume that the mode functions form a complete set which satisfies the orthonormality condition

$$\int \boldsymbol{u}_k^*(\boldsymbol{r}) \cdot \boldsymbol{u}_l(\boldsymbol{r}) \, \mathrm{d}\boldsymbol{r} = \delta_{kl} , \tag{2.165}$$

and the transversality condition

$$\nabla \cdot \boldsymbol{u}_k(\boldsymbol{r}) \neq 0 . \tag{2.166}$$

The plane-wave mode functions appropriate to a cubical volume of side L may be written as

$$\boldsymbol{u}_k(\boldsymbol{r}) = L^{-3/2} \hat{\boldsymbol{e}}^{(\lambda)} \exp(i \boldsymbol{k} \cdot \boldsymbol{r}) , \tag{2.167}$$

where $\hat{\boldsymbol{e}}^{(\lambda)}$ is a unit polarization vector. This example illustrates the way in which the mode index k may represent an abbreviation for several discrete variables, i.e., in this case the polarization index ($\lambda = 1, 2$) and the three Cartesian components of the propagation vector $\boldsymbol{k}$. The polarization vector $\hat{\boldsymbol{e}}^{(\lambda)}$ is required to be perpendicular to $\boldsymbol{k}$ by the condition Eq. (2.166), and the permissible values of $\boldsymbol{k}$ are determined in a familiar way by means of periodic boundary conditions.

The expansion we shall use for the vector potential takes the form

$$\boldsymbol{A}(\boldsymbol{r}t) = c \sum_k \left(\frac{\hbar}{2\omega_k}\right)^{1/2} \times \left[a_k \boldsymbol{u}_k(\boldsymbol{r}) \, \mathrm{e}^{-i\omega_k t} + a_k^\dagger \boldsymbol{u}_k^*(\boldsymbol{r}) \, \mathrm{e}^{i\omega_k t}\right] , \tag{2.168}$$

in which the normalization factors have been chosen to render dimensionless the pair of complex-conjugate amplitudes a_k and $a_k^\dagger$. In the classical form of electromagnetic

2.9 Coherent and Incoherent States of the Radiation Field

theory these Fourier amplitudes are complex numbers which may be chosen arbitrarily but remain constant in time when no charges or currents are present. In quantum electrodynamics, on the other hand, these amplitudes must be regarded as mutually adjoint operators. The amplitude operators, as we have defined them, will likewise remain constant when no field sources are active in the system studied.

The dynamical behavior of the field amplitudes is governed by the electromagnetic Hamiltonian which, in rationalized units, takes the form

$$H = \frac{1}{2} \int (\mathbf{E}^2 + \mathbf{B}^2) \, d\mathbf{r} \, . \tag{2.169}$$

With the use of Eqns. (2.165) and (2.166) and of a suitable set of boundary conditions on the mode functions, the Hamiltonian may be reduced to the form

$$H = \frac{1}{2} \sum_k \hbar \omega_k (a_k^\dagger a_k + a_k a_k^\dagger) \, . \tag{2.170}$$

This expression is the source of a well-known and extremely fruitful analogy between the mode amplitudes of the field and the coordinates of an assembly of one-dimensional harmonic oscillators. The quantum mechanical properties of the amplitude operators a_k and $a_k^\dagger$ may be described completely by adopting for them the commutation relations familiar from the example of independent harmonic oscillators:

$$[a_k, a_{k'}] = [a_k^\dagger, a_{k'}^\dagger] = 0 \, , \tag{2.171a}$$

$$[a_k, a_{k'}^\dagger] = \delta_{kk'} \, . \tag{2.171b}$$

Having thus separated the dynamical variables of the different modes, we are now free to discuss the quantum states of the modes independently of one another. Our knowledge of the state of each mode may be described by a state vector $|\ \rangle_k$ in a Hilbert space appropriate to that mode. The states of the entire field are then defined in the product space of the Hilbert spaces for all of the modes.

To discuss the quantum states of the individual modes we need only be familiar with the most elementary aspects of the treatment of a single harmonic oscillator. The Hamiltonian $\frac{1}{2}\omega_k(a_k^\dagger a_k + a_k a_k^\dagger)$ has eigenvalues $\hbar\omega_k(n_k + \frac{1}{2})$, where n_k is an integer ($n_k = 0, 1, 2, \ldots$). The state vector for the ground state of the oscillator will be written as $|0\rangle_k$. It is defined by the condition

$$a_k |0\rangle_k = 0 \, . \tag{2.172}$$

The state vectors for the excited states of the oscillator may be obtained by applying integral powers of the operator $a_k^\dagger$ to $|0\rangle_k$. These states are written in normalized form as

$$|n_k\rangle_k = \frac{(a_k^\dagger)^{n_k}}{(n_k!)^{1/2}} |0\rangle_k \, , \qquad n_k = 0, 1, 2 \ldots \tag{2.173}$$

The way in which the operators a_k and $a_k^\dagger$ act upon these states is indicated by the relations

$$a_k |n_k\rangle_k = n_k^{1/2} |n_k - 1\rangle_k \,, \tag{2.174}$$

$$a_k^\dagger |n_k\rangle = (n_k + 1)^{1/2} |n_k + 1\rangle_k \,, \tag{2.175}$$

$$a_k^\dagger a_k |n_k\rangle = n_k |n_k\rangle \,. \tag{2.176}$$

With these preliminaries completed we are now ready to discuss the coherent states of the field in greater detail. The expansion Eq. (2.168) for the vector potential exhibits its positive frequency part as the sum containing the photon annihilation operators a_k and its negative frequency part as that involving the creation operators $a_k^\dagger$. The positive frequency part of the electric field operator is thus given, according to Eq. (2.168), by

$$\boldsymbol{E}^{(+)}(\boldsymbol{r}t) = i \sum_k (\tfrac{1}{2}\hbar\omega_k)^{1/2} a_k \boldsymbol{u}_k(\boldsymbol{r}) e^{-i\omega_k t} \,. \tag{2.177}$$

The eigenvalue functions $\boldsymbol{\mathcal{E}}(\boldsymbol{r}t)$ defined by Eq. (2.160) must clearly satisfy the Maxwell equations, just as the operator $\boldsymbol{E}^{(+)}(\boldsymbol{r}t)$ does. They therefore possess an expansion in normal modes similar to Eq. (2.177). In other words we may introduce a set of c-number Fourier coefficients α_k which permit us to write the eigenvalue function as

$$\boldsymbol{\mathcal{E}}(\boldsymbol{r}t) = i \sum_k (\tfrac{1}{2}\hbar\omega_k)^{1/2} \alpha_k \boldsymbol{u}_k(\boldsymbol{r}) e^{-i\omega_k t} \,. \tag{2.178}$$

Since the mode functions $\boldsymbol{u}_k(\boldsymbol{r})$ form an orthogonal set, it then follows that the eigenstate $|\ \rangle$ for the field obeys the infinite succession of relations

$$a_k | \ \rangle = \alpha_k | \ \rangle \,, \tag{2.179}$$

for all modes k. To find the states which satisfy these relations we seek states, $|\alpha_k\rangle_k$, of the individual modes which individually obey the relations

$$a_k |\alpha_k\rangle_k = \alpha_k |\alpha_k\rangle_k \,. \tag{2.180}$$

The coherent states $|\ \rangle$ of the field, considered as a whole, are then seen to be direct products of the individual states $|\alpha_k\rangle$,

$$|\ \rangle = \prod_k |\alpha_k\rangle_k \,. \tag{2.181}$$

2.9.3
Coherent States of a Single Mode

The next few sections will be devoted to discussing the description of a single mode oscillator. We may therefore simplify the notation a bit by dropping the mode index

2.9 Coherent and Incoherent States of the Radiation Field

k as a subscript to the state vector and to the amplitude parameters and operators. To find the oscillator state $|\alpha\rangle$ which satisfies

$$a|\alpha\rangle = \alpha|\alpha\rangle , \qquad (2.182)$$

we begin by taking the scalar product of both sides of the equation with the n-th excited state, $\langle n|$. By using the Hermitian adjoint form of the relation Eq. (2.175), we find the recursion relation

$$(n+1)^{1/2} \langle n+1|\alpha\rangle = \alpha \langle n|\alpha\rangle \qquad (2.183)$$

for the scalar products $\langle n|\alpha\rangle$. We immediately find from the recursion relation that

$$\langle n|\alpha\rangle = \frac{\alpha^n}{(n!)^{1/2}} \langle 0|\alpha\rangle . \qquad (2.184)$$

These scalar products are the expansion coefficients of the state $|\alpha\rangle$ in terms of the complete orthonormal set $\langle n|$ ($n = 0, 1, \ldots$). We thus have

$$|\alpha\rangle = \sum_n |n\rangle \langle n|\alpha\rangle$$

$$= \langle 0|\alpha\rangle \sum_n \frac{\alpha^n}{(n!)^{1/2}} |n\rangle . \qquad (2.185)$$

The squared length of the vector $|\alpha\rangle$ is thus

$$\langle \alpha|\alpha\rangle = |\langle 0|\alpha\rangle|^2 \sum_n \frac{|\alpha|^{2n}}{n!}$$

$$= |\langle 0|\alpha\rangle|^2 e^{|\alpha|^2} . \qquad (2.186)$$

If the state $|\alpha\rangle$ is normalized so that $\langle \alpha|\alpha\rangle = 1$ we may evidently define its phase by choosing

$$\langle 0|\alpha\rangle = e^{-|\alpha|^2/2} . \qquad (2.187)$$

The coherent states of the oscillator therefore take the forms

$$|\alpha\rangle = e^{-|\alpha|^2/2} \sum_n \frac{\alpha^n}{(n!)^{1/2}} |n\rangle \qquad (2.188)$$

and

$$\langle \alpha| = e^{-|\alpha|^2/2} \sum_n \frac{(\alpha^*)^n}{(n!)^{1/2}} \langle n| . \qquad (2.189)$$

These forms show that the average occupation number of the n-th state is given by a Poisson distribution with mean value $|\alpha|^2$,

$$|\langle n|\alpha\rangle|^2 = \frac{|\alpha|^{2n}}{n!} e^{-|\alpha|^2} . \qquad (2.190)$$

They also show that the coherent state $|\alpha\rangle$ corresponding to $\alpha = 0$ is the unique ground state of the oscillator, i.e., the state $|n\rangle$ for $n = 0$.

An alternative approach to the coherent states will also prove quite useful in the work to follow. For this purpose we assume that there exists a unitary operator D which acts as a displacement operator upon the amplitudes $a^\dagger$ and a. We let D be a function of a complex parameter β, and require that it displace the amplitude operators according to the scheme

$$D^{-1}(\beta) a D(\beta) = a + \beta, \tag{2.191}$$

$$D^{-1}(\beta) a^\dagger D(\beta) = a^\dagger + \beta^*. \tag{2.192}$$

Then if $|\alpha\rangle$ obeys Eq. (2.182), it follows that $D^{-1}(\beta)|\alpha\rangle$ is an eigenstate of a corresponding to the eigenvalue $\alpha - \beta$,

$$a D^{-1}(\beta)|\alpha\rangle = (\alpha - \beta) D^{-1}(\beta)|\alpha\rangle. \tag{2.193}$$

In particular, if we choose $\beta = \alpha$, we find

$$a D^{-1}(\alpha)|\alpha\rangle = 0.$$

Since the ground state of the oscillator is uniquely defined by the relation Eq. (2.172), it follows that $D^{-1}(\alpha)|\alpha\rangle$ is just the ground state, $|0\rangle$. The coherent states, in other words, are just displaced forms of the ground state of the oscillator,

$$|\alpha\rangle = D(\alpha)|0\rangle. \tag{2.194}$$

To find an explicit form for the displacement operator $D(\alpha)$, we begin by considering infinitesimal displacements in the neighborhood of $D(0) = 1$. For arbitrary displacements $d\alpha$, we see easily from the commutation rules Eq. (2.171) that $D(d\alpha)$ may be chosen to have the form

$$D(d\alpha) = 1 + a^\dagger d\alpha - a d\alpha^*, \tag{2.195}$$

which holds to first order in $d\alpha$. To formulate a simple differential equation obeyed by the unknown operator we consider increments of α of the form $d\alpha = \alpha d\lambda$ where λ is a real parameter. Then if we assume the operators D to possess the group multiplication property

$$D(\alpha(\lambda + d\lambda)) = D(\alpha d\lambda) D(\alpha\lambda), \tag{2.196}$$

we find the differential equation

$$\frac{d}{d\lambda} D(\alpha\lambda) = (\alpha a^\dagger - \alpha^* a) D(\alpha\lambda), \tag{2.197}$$

whose solution, evaluated for $\lambda = 1$, is the unitary operator

$$D(\alpha) = e^{\alpha a^\dagger - \alpha^* a}. \tag{2.198}$$

2.9 Coherent and Incoherent States of the Radiation Field

The coherent states $|\alpha\rangle$ may therefore be written in the form

$$|\alpha\rangle = e^{\alpha a^\dagger - \alpha^* a}|0\rangle \tag{2.199}$$

which is correctly normalized since $D(\alpha)$ is unitary.

It is interesting to discuss the relationship between the two forms we have derived for the coherent states. For this purpose we invoke a simple theorem on the multiplication of exponential functions of operators. If $\mathcal{A}$ and $\mathcal{B}$ are any two operators, whose commutator $[\mathcal{A}, \mathcal{B}]$ commutes with each of them,

$$[[\mathcal{A}, \mathcal{B}], \mathcal{A}] = [[\mathcal{A}, \mathcal{B}], \mathcal{B}] = 0, \tag{2.200}$$

it may be shown[7] that

$$\exp(\mathcal{A})\exp(\mathcal{B}) = \exp\{\mathcal{A} + \mathcal{B} + \tfrac{1}{2}[\mathcal{A}, \mathcal{B}]\}. \tag{2.201}$$

If we write $\mathcal{A} = a^\dagger$ and $\mathcal{B} = a$, this theorem permits us to resolve the exponential $D(\alpha)$ given by Eq. (2.198) into the product

$$D(\alpha) = e^{-|\alpha|^2/2} e^{\alpha a^\dagger} e^{-\alpha^* a}. \tag{2.202}$$

Products of this type, which have been ordered so that the annihilation operators all stand to the right of the creation operators, will be said to be in normal form. Their convenience is indicated by the fact that the exponential $\exp[-\alpha^* a]$, when applied to the ground state $|0\rangle$, reduces in effect to unity, i.e., we have

$$e^{-\alpha^* a}|0\rangle = |0\rangle, \tag{2.203}$$

since the exponential may be expanded in series and the definition Eq. (2.172) of the ground state applied. It follows then that the coherent states may be written as

$$|\alpha\rangle = D(\alpha)|0\rangle$$
$$= e^{-|\alpha|^2/2} e^{\alpha a^\dagger}|0\rangle \tag{2.204}$$
$$= e^{-|\alpha|^2/2} \sum_n \frac{(\alpha a^\dagger)^n}{n!}|0\rangle. \tag{2.205}$$

Since the excited states of the oscillator are given by $|n\rangle = (n!)^{-1/2}(a^\dagger)^n|0\rangle$, we have once again derived the expression

$$|\alpha\rangle = e^{-|\alpha|^2/2} \sum_n \frac{\alpha^n}{n!}|n\rangle.$$

It may help in visualizing the coherent states if we discuss the form they take in coordinate space and in momentum space. We therefore introduce a pair of Hermitian operators q and p to represent, respectively, the coordinate of the mode oscillator and its momentum. These operators, which must satisfy the canonical commutation relation, $[q, p] = i\hbar$, may be defined for our purposes by the familiar expressions

$$q = \left(\frac{\hbar}{2\omega}\right)^{1/2} (a^\dagger + a), \tag{2.206a}$$

$$p = i\left(\frac{\hbar\omega}{2}\right)^{1/2} (a^\dagger - a). \tag{2.206b}$$

To find the expectation value of q and p in the coherent states we need only use Eq. (2.182), which defines these states, and its corresponding Hermitian adjoint form. We have then

$$\langle \alpha | q | \alpha \rangle = \left(\frac{2\hbar}{\omega}\right)^{1/2} \operatorname{Re} \alpha, \tag{2.207a}$$

$$\langle \alpha | p | \alpha \rangle = (2\hbar\omega)^{1/2} \operatorname{Im} \alpha, \tag{2.207b}$$

where $\operatorname{Re} \alpha$ and $\operatorname{Im} \alpha$ stand for the real and imaginary parts of α.

To find the wave functions for the coherent states, we write the defining Eq. (2.182) in the form

$$(2\hbar\omega)^{-1/2}(\omega q + i p)|\alpha\rangle = \alpha |\alpha\rangle, \tag{2.208}$$

and take the scalar product of both members with the conjugate state $\langle q'|$, which corresponds to the eigenvalue q' for q. Since the momentum may be represented by a derivative operator, i.e., $\langle q'|p = -i\hbar(d/dq')\langle q'|$, we find that the coordinate space wave function, $\langle q'|\alpha\rangle$, obeys the differential equation

$$\frac{d}{dq'}\langle q'|\alpha\rangle = -s\left(\frac{\omega}{2\hbar}\right)^{1/2}\left\{\left(\frac{\omega}{2\hbar}\right)^{1/2} q' - \alpha\right\}\langle q'|\alpha\rangle. \tag{2.209}$$

The equation may be integrated immediately to yield a solution for the wave function which, in normalized form, is

$$\langle q'|\alpha\rangle = \left(\frac{\omega}{\pi\hbar}\right)^{1/4} \exp\left\{-\left[\left(\frac{\omega}{2\hbar}\right)^{1/2} q' - \alpha\right]^2\right\}. \tag{2.210}$$

An analogous argument furnishes the momentum space wave function. If we take the scalar product of Eq. (2.208) with a momentum eigenstate $\langle p'|$, and use the relation $\langle p'|q = i\hbar(\partial/\partial p')\langle p'|$, we reach a differential equation whose normalized solution is

$$\langle p'|\alpha\rangle = \frac{1}{(\pi\hbar\omega)^{1/4}} \exp\left\{-\left[\left(\frac{1}{(2\hbar\omega)^{1/2}}\right) p' + i\alpha\right]^2\right\}. \tag{2.211}$$

Both of these wave functions are simply displaced forms of the ground-state wave function of the oscillator. The parameters $(\hbar/\omega)^{1/2}$ and $(\hbar\omega)^{1/2}$ correspond to the amplitudes of the zero-point fluctuations of the coordinate and momentum, respectively, for an oscillator of unit mass. The fact that the wave functions for the coherent states have this elementary structure should be no surprise in view of the way they are generated in Eq. (2.194), by means of displacements in the complex α plane.

The time-independent states $|\alpha\rangle$ which we have been describing are those characteristic of the Heisenberg picture of quantum mechanics. The Schrödinger picture, alternatively, would make use of the time-dependent states $\exp(-iHt/\hbar)|\alpha\rangle$. If we omit the zero-point energy $\frac{1}{2}\hbar\omega$ from the oscillator Hamiltonian and write $H = \hbar\omega a^\dagger a$, it is then clear from the expansion (2.188) for $|\alpha\rangle$ that the corresponding Schrödinger state takes the same form with α replaced by $\alpha e^{-i\omega t}$. We may thus write the Schrödinger state as $|\alpha e^{-i\omega t}\rangle$. With the substitution of $\alpha e^{-i\omega t}$ for α in Eqns. (2.207a) and (2.207b), we see that the expectation values of the coordinate and momentum carry out a simple harmonic motion with coordinate amplitude $(2\hbar/\omega)^{1/2}|\alpha|$. The same substitutions in the wave functions (2.210) and (2.211) show that the Gaussian probability densities characteristic of the ground state of the oscillator are simply carried back and forth in the same motion as the expectation values. Such wavepackets are, of course, quite familiar; they were introduced to quantum mechanics at a very early stage by Schrödinger[8], and have often been used to illustrate the way in which the behavior of the oscillator approaches the classical limit.

Another connection in which the wavepackets (2.210) and (2.211) have been discussed in the past has to do with the particular way in which they localize the coordinate q' and the momentum p'. wavepackets can, of course, be found which localize either variable more sharply, but only at the expense of the localization of the other. There is a sense in which the wavepackets (2.210) and (2.211) furnish a unique compromise; they minimize the product of the uncertainties of the variables q' and p'. If we represent expectation values by means of the angular brackets $\langle\ \rangle$ and define the variances

$$(\Delta q)^2 = \langle q^2 \rangle - \langle q \rangle^2 , \qquad (2.212a)$$

$$(\Delta p)^2 = \langle p^2 \rangle - \langle p \rangle^2 , \qquad (2.212b)$$

we find, for the wave functions (2.210) and (2.211), that the product of the variances is

$$(\Delta p)^2 (\Delta q)^2 = \frac{1}{4}\hbar^2 .$$

According to the uncertainty principle, this is the minimum value such a product can have[9]. There thus exists a particular sense in which the description of an oscillator by means of the wave functions (2.210) and (2.211) represents as close an approach to classical localization as is possible.

The uses we shall make of the coherent states in quantum electrodynamics will not, in fact, require the explicit introduction of coordinate or momentum variables. We have reviewed the familiar representations of the coherent states in terms of these variables in the hope that they may be of some help in understanding the various applications of the states which we shall shortly undertake.

One property of the states $|\alpha\rangle$ which is made clear by the wave-function representations is that two such states are not, in general, orthogonal to one another. If we consider, for example, the wave functions $\langle q'|\alpha\rangle$ and $\langle q'|\alpha'\rangle$ for values of α' close to α, it is evident that the functions are similar in form and overlap one another appreciably. For values of α' quite different from α, however, the overlap is at most quite small. We may therefore expect that the scalar product $\langle \alpha|\alpha'\rangle$, which is unity for $\alpha' = \alpha$, will tend to decrease in absolute magnitude as α' and α recede from one another in the complex plane. The scalar product may, in fact, be calculated more simply than by using wave functions if we employ the representations Eq. (2.188) and (2.189). We then find

$$\langle \alpha|\beta\rangle = e^{-|\alpha|^2/2 - |\beta|^2/2} \sum_{n,m} \frac{(\alpha^*)^n \beta^m}{(n!\,m!)^{1/2}} \langle n|m\rangle,$$

which, in view of the orthonormality of the $|n\rangle$ states, reduces to

$$\langle \alpha|\beta\rangle = \exp\left\{\alpha^*\beta - \tfrac{1}{2}|\alpha|^2 - \tfrac{1}{2}|\beta|^2\right\}. \tag{2.213}$$

The absolute magnitude of the scalar product is given by

$$|\langle \alpha|\beta\rangle|^2 = \exp\left\{-|\alpha - \beta|^2\right\}, \tag{2.214}$$

which shows that the coherent states tend to become approximately orthogonal for values of α and β which are sufficiently different. The fact that these states are not even approximately orthogonal for $|\alpha - \beta|$ of order unity may be regarded as an expression of the overlap caused by the presence of the displaced zero-point fluctuations.

Since the coherent states do not form an orthogonal set, they appear to have received little attention as a possible system of basis vectors for the expansion of arbitrary states[10]. We shall show in the following section that such expansions can be carried out conveniently and uniquely and that they possess exceedingly useful properties. In later sections we shall, by generalizing the procedure to deal with bilinear combinations of states $|\alpha\rangle$ and $\langle\beta|$, develop analogous expansions for operators[4] as well.

2.9.4
Expansion of Arbitrary States in Terms of Coherent States

While orthogonality is a convenient property for a set of basis states it is not a necessary one. The essential property of such a set is that it be complete. The set of coherent states $|\alpha\rangle$ for a mode oscillator can be shown without difficulty to form a complete set. To give a proof we need only demonstrate that the unit operator may be expressed as a suitable sum or an integral, over the complex α plane, of projection operators of the form $|\alpha\rangle\langle\alpha|$. In order to describe such integrals we introduce the differential element of area in the α plane

$$d^2\alpha = d(\mathrm{Re}\,\alpha)\,(d\,\mathrm{Im}\,\alpha) \tag{2.215}$$

(i.e., $d^2\alpha$ is real). If we write $\alpha = |\alpha|\,e^{i\vartheta}$, we may easily prove the integral identity

$$\int (\alpha^*)^n \alpha^m e^{-|\alpha|^2} d^2\alpha = \int_0^\infty |\alpha|^{n+m+1} e^{-|\alpha|^2} d|\alpha| \int_0^{2\pi} e^{i(m-n)\vartheta} d\vartheta$$
$$= \pi n!\,\delta_{nm}\,, \tag{2.216}$$

in which the integration is carried out, as indicated, over the entire area of the complex plane. With the aid of this identity and the expansions (2.188) and (2.189) for the coherent states, we may immediately show

$$\int |\alpha\rangle\langle\alpha|\,d^2\alpha = \pi \sum_n |n\rangle\langle n|\,.$$

Since the n-quantum states are known to form a complete orthonormal set, the indicated sum over n is simply the unit operator. We have thus shown[4]

$$\frac{1}{\pi}\int |\alpha\rangle\langle\alpha|\,d^2\alpha = 1\,, \tag{2.217}$$

which is a completeness relation for the coherent states of precisely the type desired.

An arbitrary state of an oscillator must possess an expansion in terms of the n-quantum states of the form

$$|\,\rangle = \sum_n c_n |n\rangle\,,$$
$$= \sum_n c_n \frac{(a^\dagger)^n}{(n!)^{1/2}}|0\rangle\,, \tag{2.218}$$

where $\sum |c_n|^2 = 1$. The series which occurs in Eq. (2.218) may be used to define a function f of a complex variable z,

$$f(z) = \sum_n c_n \frac{z^n}{(n!)^{1/2}}\,. \tag{2.219}$$

It is clear from the normalization condition on the c_n that this series converges for all finite z, and thus represents a function which is analytic throughout the finite complex plane. We shall speak of the functions $f(z)$ for which $\sum |c_n|^2 = 1$ as the set of normalized entire functions. There is evidently a one-to-one correspondence which exists between such entire functions and the states of the oscillator. One way of approaching the description of the oscillator is to regard the functions $f(z)$ themselves as the elements of a Hilbert space. The properties of this space and of expansions carried out in it have been studied in some detail by Segal[11] and Bargmann[12]. The method we shall use for expanding arbitrary states in terms of the coherent states has been developed as a simple generalization of the usual method for carrying out changes of basis states in quantum mechanics. It is evidently equivalent, however, to one of the expansions stated by Bargmann.

If we designate the arbitrary state which corresponds to the function $f(z)$ by $|f\rangle$, then we may rewrite Eq. (2.218) as

$$|f\rangle = f(a^\dagger)|0\rangle . \tag{2.220}$$

To secure the expansion of $|f\rangle$ in terms of the states $|\alpha\rangle$, we multiply $|f\rangle$ by the representation (2.217) of the unit operator. We then find

$$|f\rangle = \frac{1}{\pi} \int |\alpha\rangle \langle \alpha | f(a^\dagger) | 0 \rangle \, d^2\alpha ,$$

which reduces, since $\langle \alpha | f(a^\dagger) = \langle \alpha | f(\alpha^*)$, to

$$|f\rangle = \frac{1}{\pi} \int |\alpha\rangle f(\alpha^*) e^{-|\alpha|^2/2} \, d^2\alpha , \tag{2.221}$$

which is an expansion of the desired type.

It is worth noting that the expansion (2.221) can easily be inverted to furnish an explicit form for the function $f(\alpha^*)$ which corresponds to any vector $|f\rangle$. For this purpose we take tie scalar product of both sides of Eq. (2.221) with the coherent state $\langle \beta |$, and then, using Eq. (2.213), evaluate the scalar product $\langle \beta | \alpha \rangle$ to find

$$\langle \beta | f \rangle = \frac{1}{\pi} e^{-|\beta|^2/2} \int e^{\beta^* \alpha - |\alpha|^2} f(\alpha^*) \, d^2\alpha . \tag{2.222}$$

Since $f(\alpha^*)$ may be expanded in a convergent power series we note the relation

$$\frac{1}{\pi} \int e^{\beta^* \alpha - |\alpha|^2} (\alpha^*)^n \, d^2\alpha = (\beta^*)^n , \tag{2.223}$$

from which we may derive the more general identity

$$\frac{1}{\pi} \int e^{\beta^* \alpha - |\alpha|^2} f(\alpha^*) \, d^2\alpha = f(\beta^*) . \tag{2.224}$$

On substituting the latter identity in Eq. (2.222) we find

$$f(\beta^*) = e^{|\beta|^2/2} \langle \beta | f \rangle . \tag{2.225}$$

There is thus a unique correspondence between functions $f(\alpha^*)$ which play the role of expansion amplitudes in Eq. (2.221) and the vectors $|f\rangle$ which describe the state of the oscillator.

An expansion analogous to Eq. (2.221) also exists for the adjoint state vectors. If we let $g(\alpha^*)$ be an entire function of α^* we may construct for the state $\langle g |$ the expansion

$$\langle g | = \frac{1}{\pi} \int [g(\beta^*)]^* \langle \beta | e^{-|\beta|^2/2} \, d^2\beta . \tag{2.226}$$

The scalar product of the two states $\langle g|$ and $|f\rangle$ may then be expressed as

$$\langle g|f\rangle = \pi^{-2} \int [g(\beta^*)]^* f(\alpha^*) \exp\{\beta^*\alpha - |\alpha|^2 - |\beta|^2\} \, d^2\alpha \, d^2\beta \, .$$

The identity (2.224) permits us to carry out the integration over the variable α to find

$$\langle g|f\rangle = \frac{1}{\pi} \int [g(\beta^*)]^* f(\beta^*) e^{-|\beta|^2} \, d^2\beta \, . \tag{2.227}$$

This expression for the scalar product of two vectors is, in essence, the starting point used by Bargmann in his discussion[13] of the Hilbert space of functions $f(z)$.

It may be worth noting, for its mathematical interest, that the coherent states $|\alpha\rangle$ are not linearly independent of one another, as the members of a complete orthogonal set would be. Thus, for example, the expansion (2.221) may be used to express any given coherent state linearly in terms of all of the others, i.e., in view of Eqns. (2.225) and (2.213) we may write

$$|\alpha\rangle = \frac{1}{\pi} \int |\beta\rangle \, e^{\beta^*\alpha - |\alpha|^2/2 - |\beta|^2/2} \, d^2\beta \, . \tag{2.228}$$

There exist many other types of linear dependence among the states $|\alpha\rangle$. We may, for example, note the identity

$$\int |\alpha\rangle \, \alpha^n \, e^{-|\alpha|^2/2} \, d^2\alpha = 0 \, , \tag{2.229}$$

which holds for all integral $n > 0$. It is clear from the latter result that if we admitted as expansion coefficients in Eq. (2.221) more general functions than $f(\alpha^*)$, say functions $F(\alpha, \alpha^*)$, there would be many additional ways of expanding any state in terms of coherent states. The constraint implicit in Eq. (2.221), that the expansion function must depend analytically upon the variable α^* is what renders the expansion unique. The virtue of an expansion scheme in which the coefficients are uniquely determined is evident. It becomes possible, by inverting the expansion as in Eq. (2.225), to construct an explicit solution for the expansion coefficient of any state, no matter what representation it was expressed in initially.

2.9.5
Expansion of Operators in Terms of Coherent State Vectors

Our knowledge of the condition of an oscillator mode is rarely explicit enough in practice to permit the specification of its quantum state. Instead, we must describe it in terms of a mixture of states which is expressed by means of a density operator. The same reasons that lead us to express arbitrary states in terms of the coherent states, therefore, suggest that we develop an expansion for the density operator in terms of these states as well. We shall begin by considering in the present section a rather more general class of operators and then specialize to the case of the density operator in the section which follows.

A general quantum mechanical operator T may be expressed in terms of its matrix elements connecting states with fixed numbers of quanta as

$$T = \sum_{n,m} |n\rangle T_{nm} \langle m| , \qquad (2.230)$$

$$= \sum T_{nm} (n!\, m!)^{-1/2} (a^\dagger)^n |0\rangle \langle 0| a^m . \qquad (2.231)$$

If we use this expression for T to calculate the matrix element which connects the two coherent states $\langle \alpha|$ and $\langle \beta|$ we find

$$\langle \alpha|T|\beta\rangle = \sum_{n,m} T_{nm} (n!\, m!)^{-1/2} (\alpha^*)^n \beta^m \langle \alpha|0\rangle \langle 0|\beta\rangle . \qquad (2.232)$$

It is evidently convenient to define a function $\mathcal{T}(\alpha^*,\beta)$ as

$$\mathcal{T}(\alpha^*,\beta) = \sum_{n,m} T_{nm} (n!\, m!)^{-1/2} (\alpha^*)^n \beta^m . \qquad (2.233)$$

The operators which occur in quantum mechanics are often unbounded ones such as those of Eqns. (2.174)–(2.176). Those operators and the others we are apt to encounter have the property that the magnitudes of the matrix elements T_{nm} are dominated by an expression of the form $Mn^j m^k$ for some fixed positive values of M, j, and k. It then follows that the double series Eq. (2.233) converges throughout the finite α^* and β planes and represents an entire function of both variables.

To secure the expansion of the operator T in terms of the coherent states, we may use the representation Eq. (2.217) of the unit operator to write

$$T = \frac{1}{\pi^2} \int |\alpha\rangle \langle \alpha|T|\beta\rangle \langle \beta| \, d^2\alpha \, d^2\beta , \qquad (2.234)$$

$$= \frac{1}{\pi^2} \int |\alpha\rangle \mathcal{T}(\alpha^*,\beta) \langle \beta| \langle \alpha|0\rangle \langle 0|\beta\rangle \, d^2\alpha \, d^2\beta ,$$

$$= \frac{1}{\pi^2} \int |\alpha\rangle \mathcal{T}(\alpha^*,\beta) \langle \beta| \exp\{-\tfrac{1}{2}|\alpha|^2 - \tfrac{1}{2}|\beta|^2\} \, d^2\alpha \, d^2\beta . \qquad (2.235)$$

The inversion of this expansion, or the solution for $\mathcal{T}(\alpha^*,\beta)$, is accomplished by the same method we used to invert Eq. (2.221) and secure the amplitude function Eq. (2.225). The result of the inversion is

$$\mathcal{T}(\alpha^*,\beta) = \langle \alpha|T|\beta\rangle \exp\{\tfrac{1}{2}|\alpha|^2 + \tfrac{1}{2}|\beta|^2\} . \qquad (2.236)$$

We see, thus, that the expansion of operators, as well as of arbitrary quantum states, in terms of the coherent states is a unique one.

The law of operator multiplication is easily expressed in terms of the functions $\mathcal{T}$. If $T = T_1 T_2$ and $\mathcal{T}_1$ and $\mathcal{T}_2$ are the functions appropriate to the latter two operators, we note that

$$\langle \alpha|T|\beta\rangle = \langle \alpha|T_1 T_2|\beta\rangle$$

$$= \frac{1}{\pi} \int \langle \alpha|T_1|\gamma\rangle \langle \gamma|T_2|\beta\rangle \, d^2\gamma . \qquad (2.237)$$

The function $\mathcal{T}$ which represents the product is therefore given by

$$\mathcal{T}(\alpha^*,\beta) = \frac{1}{\pi}\int \mathcal{T}_1(\alpha^*,\gamma)\,\mathcal{T}_2(\gamma^*,\beta)\,e^{-|\gamma|^2}\,d^2\gamma\,. \tag{2.238}$$

The expansion function for the operator $T^\dagger$, the Hermitian adjoint of T, is obtained by substituting T^*_{mn} for T_{nm} in Eq. (2.233). It is given by $[\mathcal{T}(\beta^*,\alpha)]^*$. If the operator T is Hermitian the function $\mathcal{T}$ must satisfy the identity

$$\mathcal{T}(\alpha^*,\beta) = [\mathcal{T}(\beta^*,\alpha)]^*\,, \tag{2.239}$$

since the expansions of T and $T^\dagger$ are unique.

The functions $\mathcal{T}(\alpha^*,\beta)$ which represent normal products of the operators $a^\dagger$ and a such as $(a^\dagger)^n a^m$ are immediately seen from Eqns. (2.236) and (2.213) to be

$$\mathcal{T}(\alpha^*,\beta) = (\alpha^*)^n \beta^m \exp[\alpha^*\beta]\,. \tag{2.240}$$

In particular, the unit operator corresponds to $n = m = 0$.

It may be worth noting at this point that many of the foregoing formulas can be abbreviated somewhat by adopting a normalization different from the conventional one for the coherent states. If we introduce the symbol $\|\alpha\rangle$ for the states normalized in the new way and define these as

$$\|\alpha\rangle = |\alpha\rangle\,e^{|\alpha|^2/2}\,, \tag{2.241}$$

then we may write the scalar product of two such states as $\langle\alpha\|\beta\rangle$. We see from Eq. (2.213) that this scalar product is

$$\langle\alpha\|\beta\rangle = \exp[\alpha^*\beta]\,. \tag{2.242}$$

We may next, following Bargmann[12], introduce an element of measure $d\mu(\alpha)$ which is defined as

$$d\mu(\alpha) = \frac{1}{\pi}e^{-|\alpha|^2}\,d^2\alpha\,. \tag{2.243}$$

With these alterations, all of the Gaussian functions, and factors of π, in the preceding formulas become absorbed, as it were, into the notation. The Eqns. (2.235) and (2.236), for example, reduce to the briefer forms

$$T = \int \|\alpha\rangle\,\mathcal{T}(\alpha^*,\beta)\,\langle\beta\|\,d\mu(\alpha)\,d\mu(\beta) \tag{2.244}$$

and

$$\mathcal{T}(\alpha^*,\beta) = \langle\alpha\|T\|\beta\rangle\,. \tag{2.245}$$

A more significant property of the states $\|\alpha\rangle$ is that they are given by the expansion

$$\|\alpha\rangle = \sum_n \frac{\alpha^n}{(n!)^{1/2}}\,|n\rangle \tag{2.246}$$

and thus obey the relation

$$a^\dagger \|\alpha\rangle = \frac{\partial}{\partial \alpha}\|\alpha\rangle \ . \tag{2.247}$$

While the properties of the alternatively normalized states $\|\alpha\rangle$ are worth bearing in mind, we have chosen not to adopt this normalization in the present paper in order to retain the more conventional interpretation of scalar products as probability amplitudes. The advantage afforded by the relation (2.247) is not a great one since all of the operators we shall have to deal with are either already in normally ordered form, or easily so ordered.

2.9.6
General Properties of the Density Operator

The formalism we have developed in the two preceding sections has been intended to provide a background for the expression of the density operator of a mode in terms of the vectors that represent coherent states. Viewed in mathematical terms, the use of the coherent state vectors in this way leads to considerable simplification in the calculation of statistical averages. The fact that these states are eigenstates of the field operators $E^{(\pm)}(rt)$ means that normally ordered products of the field operators, when they are to be averaged, may be replaced by the products of their eigenvalues, i.e., treated not as operators, but as numbers. The field correlation functions such as $G^{(1)}$ given by Eq. (2.159) are averages of just such operator products. Their evaluation may be carried out quite conveniently through use of the representations we shall discuss.

Any density operator ρ may, according to the methods of the preceding section, be represented in a unique way by means of a function of two complex variables, $R(\alpha^a *, \beta)$, which is analytic throughout the finite α^* and β planes. The function R is given explicitly, by means of Eq. (2.236), as

$$R(\alpha^*, \beta) = \langle \alpha|\rho|\beta\rangle \exp\left[\tfrac{1}{2}|\alpha|^2 + \tfrac{1}{2}|\beta|^2\right] \ . \tag{2.248}$$

If we happen to know the matrix representation of ρ in the basis formed by the n-quantum states, the function R is evidently given by

$$R(\alpha^*, \beta) = \sum_{n,m} \langle n|\rho|m\rangle (n!\,m!)^{-1/2}(\alpha^*)^n \beta^m \ . \tag{2.249}$$

If we do not know the matrix elements $\langle n|\rho|m\rangle$ they may be found quite simply from a knowledge of $R(\alpha^*, \beta)$. One method for finding them is to consider $R(\alpha^*, \beta)$ as a generating function and identify its Taylor series with the series Eq. (2.249). A second method is to note that if we multiply Eq. (2.249) by $\alpha^i(\beta^*)^i \exp[-(|\alpha|^2+|\beta|^2)]$ and integrate over the α and β planes, then all terms save that for $n = i$ and $m = j$ vanish in the sum on the right and we have

$$\langle i|\rho|j\rangle = \frac{1}{\pi^2} \int R(\alpha^*, \beta)(i!\,j!)^{-1/2} \alpha^i (\beta^*)^j e^{-(|\alpha|^2+|\beta|^2)}\, d^2\alpha\, d^2\beta \ . \tag{2.250}$$

2.9 Coherent and Incoherent States of the Radiation Field

Given the knowledge of $R(\alpha^*,\beta)$, we may write the density operator as

$$\rho = \frac{1}{\pi^2}\int |\alpha\rangle R(\alpha^*,\beta)\langle\beta| e^{-\frac{1}{2}(|\alpha|^2+|\beta|^2)} d^2\alpha\, d^2\beta. \tag{2.251}$$

The statistical average of an operator T is given by the trace of the product ρT. If we calculate this average by using the representation (2.251) for ρ we must note that the trace of the expression $|\alpha\rangle\langle\beta|T$, regarded as an operator, is the matrix element $\langle\beta|T|\alpha\rangle$. Then, if we express the matrix element in terms of the function $\mathcal{T}(\alpha^*,\beta)$ defined by Eq. (2.236) we find

$$\text{Tr}\{\rho T\} = \frac{1}{\pi^2}\int R(\alpha^*,\beta)\mathcal{T}(\beta^*,\alpha) e^{-|\alpha|^2-|\beta|^2} d^2\alpha\, d^2\beta. \tag{2.252}$$

If T is any operator of the form $(a^\dagger)^n a^m$, its representation $\mathcal{T}(\beta^*,\alpha)$ is given by Eq. (2.240). In particular for $n = m = 0$, we have the unit operator $T = 1$ which is represented by $\mathcal{T}(\beta^*,\alpha) = \exp[\beta^*\alpha]$. Hence, the trace of ρ itself, which must be normalized to unity, is

$$\text{Tr}\,\rho = 1$$
$$= \frac{1}{\pi^2}\int R(\alpha^*,\beta)\exp\left[\beta^*\alpha - |\alpha|^2 - |\beta|^2\right] d^2\alpha\, d^2\beta.$$

Since $R(\alpha^*,\beta)$ is an entire function of α^*, we may use Eq. (2.224) to carry out the integration over the α plane. In this way we see that the normalization condition on R is

$$\frac{1}{\pi}\int R(\beta^*,\beta) e^{-|\beta|^2} d^2\beta = 1. \tag{2.253}$$

The density operator is Hermitian and hence has real eigenvalues. These eigenvalues may be interpreted as probabilities and so must be positive numbers. Since ρ is thus a positive definite operator, its expectation value in any state, e.g., the state $|f\rangle$ defined by Eq. (2.220), must be non-negative,

$$\langle f|\rho|f\rangle \geq 0. \tag{2.254}$$

If, for example, we choose the state $|f\rangle$ to be a coherent state $|\alpha\rangle$ we find that the function R, which is given by Eq. (2.248), satisfies the inequality

$$R(\alpha^*,\alpha) \geq 0. \tag{2.255}$$

If we let the state $|f\rangle$ be specified as in Eq. (2.221) by an entire function $f(\alpha^*)$, then we find from the inequality (2.254) the more general condition for positive definiteness

$$\int [f(\alpha^*)]^* f(\beta^*) R(\alpha^*,\beta) e^{-|\alpha|^2-|\beta|^2} d^2\alpha\, d^2\beta \geq 0, \tag{2.256}$$

which must hold for all entire functions f.

In many types of physical experiments, particularly those dealing with fields which oscillate at extremely high frequencies, we cannot be said to have any *a priori* knowledge of the time-dependent parameters. The predictions we make in such circumstances are unchanged by displacements in time. They may be derived from a density operator which is stationary, that is, one which commutes with the Hamiltonian operator or, more simply, with $a^\dagger a$. The necessary and sufficient condition that a function $R(\alpha^*, \beta)$ correspond to a stationary density operator is that it depend only on the product of its two variables, $\alpha^* \beta$. There must, in other words, exist an analytic function S such that

$$R(\alpha^*, \beta) = S(\alpha^* \beta). \tag{2.257}$$

That this condition is a sufficient one is clear from the invariance of R under the multiplication of both α and β by a phase factor, $e^{i\varphi}$. The condition may be derived as a necessary one directly from the vanishing of the commutator of ρ with $a^\dagger a$. An alternative and perhaps simpler way of seeing the result depends on noting that a stationary ρ can only be a function of the Hamiltonian for the mode, or of $a^\dagger a$. It is therefore diagonal in the basis formed by the n-quantum states, i.e., $\langle n|\rho|m\rangle = \delta_{nm}\langle n|\rho|n\rangle$. Examination of the series expansion Eq. (2.249) for R then shows that it then takes the form of Eq. (2.257).

2.9.7
The *P* Representation of the Density Operator

In the preceding sections we have demonstrated the generality of the use of the coherent states as a basis. Not all fields require for their description density operators of quite so general a form. Indeed for a broad class of radiation fields which includes, as we shall see, virtually all of those studied in optics, it becomes possible to reduce the density operator to a considerably simpler form. This form is one which brings to light many similarities between quantum electrodynamical calculations and the corresponding classical ones. Its use offers deep insights into the reasons why some of the fundamental laws of optics, such as those for superposition of fields and calculation of the resulting intensities, are the same as in classical theory, even when very few quanta are involved. We shall continue, for the present, to limit consideration to a single mode of the field.

One type of oscillator state which interests us particularly is, of course, a coherent state. The density operator for a pure state $|\alpha\rangle$ is just the projection operator

$$\rho = |\alpha\rangle\langle\alpha|. \tag{2.258}$$

The unique representation of this operator as a function $R(\beta^*, \gamma)$ is easily shown, from Eq. (2.248), to be

$$R(\beta^*, \gamma) = \exp\left[\beta^* \alpha + \gamma \alpha^* - |\alpha|^2\right]. \tag{2.259}$$

Other functions $R(\beta^*,\gamma)$, which satisfy the analyticity requirements necessary for the representations of density operators, may be constructed by forming linear combinations of exponentials such as Eq. (2.259) for various values of the complex parameter α. The functions R, which we form in this way, represent statistical mixtures of the coherent states. The most general such function R may be written as

$$R(\beta^*,\gamma) = \int P(\alpha) \exp\left[\beta^*\alpha + \gamma\alpha^* - |\alpha|^2\right] d^2\alpha, \tag{2.260}$$

where $P(\alpha)$ is a weight function defined at all points of the complex α plane. Since $R(\beta^*,\gamma)$ must satisfy the Hermiticity condition, Eq. (2.239), we require that the weight function be real-valued, i.e., $[P(\alpha)]^* = P(\alpha)$. The function $P(\alpha)$ need not be subject to any regularity conditions, but its singularities must be integrable ones.[3] It is convenient to allow $P(\alpha)$ to have delta function singularities so that we may think of a pure coherent state as represented by a special case of Eq. (2.260). A real-valued two-dimensional delta function which is suited to this purpose may be defined as

$$\delta^{(2)}(\alpha) = \delta(\operatorname{Re}\alpha)\,\delta(\operatorname{Im}\alpha). \tag{2.261}$$

The pure coherent state $|\beta\rangle$ is then evidently described by

$$P(\alpha) = \delta^{(2)}(\alpha - \beta), \tag{2.262}$$

and the ground state of the oscillator is specified by setting $\beta = 0$.

The density operator ρ which corresponds to Eq. (2.260) is just a superposition of the projection operators Eq. (2.258),

$$\rho = \int P(\alpha) |\alpha\rangle\langle\alpha| \, d^2\alpha. \tag{2.263}$$

It is the kind of operator we might naturally be led to if we were given knowledge that the oscillator is in a coherent state, but one which corresponds to an unknown eigenvalue α. The function $P(\alpha)$ might then be thought of as playing a role analogous to a probability density for the distribution of values of α over the complex plane[14]. Such an interpretation may, as we shall see, be justified at times. In general, however, it is not possible to interpret the function $P(\alpha)$ as a probability distribution in any precise way since the projection operators $|\alpha\rangle\langle\alpha|$ with which it is associated are not orthogonal to one another for different values of α. There is an approximate sense, as we have noted in connection with Eq. (2.214), in which two states $|\alpha\rangle$ and $|\alpha'\rangle$ may be said to become orthogonal to one another for $|\alpha - \alpha'| \gg 1$, i.e., when their wavepackets Eq. (2.210) and those of the form Eq. (2.211) do not appreciably overlap. When the function $P(\alpha)$ tends to vary little over such large ranges of the parameter α, the non-orthogonality of the coherent states will make little difference, and $P(\alpha)$

---

[3] If the singularities of $P(\alpha)$ are of types stronger than those of delta functions, e.g., derivatives of delta functions, the field represented will have no classical analog.

will then be interpretable approximately as a probability density. The functions $P(\alpha)$ which vary this slowly will, in general, be associated with strong fields, ones which may be described approximately in classical terms.

We shall call the expression Eq. (2.263) for the density operator the *P* representation in order to distinguish it from the more general form based on the functions *R* discussed earlier. The normalization property of the density operator requires that $P(\alpha)$ obey the normalization condition

$$\operatorname{Tr}\rho = \int P(\alpha)\, d^2\alpha = 1 . \qquad (2.264)$$

It is interesting to examine the conditions that the positive definiteness of ρ places upon $P(\alpha)$. If we apply the condition Eq. (2.256) to the function $R(\beta^*,\gamma)$ given by Eq. (2.260) we find

$$\int [f(\beta^*)]^* f(\gamma^*) P(\alpha) \exp\left[\beta^*\alpha + \gamma\alpha^* - |\alpha|^2 - |\beta|^2 - |\gamma|^2\right] d^2\alpha\, d^2\beta\, d^2\gamma \geq 0 . \qquad (2.265)$$

The γ integration may be carried out via Eq. (2.224) and the β integration by means of its complex conjugate. We then have the condition that

$$\int |f(\alpha^*)|^2 P(\alpha)\, e^{-|\alpha|^2} d^2\alpha \geq 0 \qquad (2.266)$$

must hold for all entire functions $f(\alpha^*)$. In particular, the choice $f(\alpha^*) = \exp\left[\beta^*\alpha - \frac{1}{2}|\beta|^2\right]$ leads to the simple condition

$$\int P(\alpha)\, e^{-|\alpha-\beta|^2} d^2\alpha \geq 0 , \qquad (2.267)$$

which must hold for all complex values of β. It corresponds to the requirement $\langle \beta|\rho|\beta\rangle \geq 0$. These conditions are immediately satisfied if $P(\alpha)$ is positive valued as it would be, were it a probability density. They are not strong enough, however, to exclude the possibility that $P(\alpha)$ takes on negative values over some suitably restricted regions of the plane.[4] This result serves to underscore the fact that the weight function P (a) cannot, in general, be interpreted as a probability density.[5] If a density operator is specified by means of the P representation, its matrix elements connecting the *n*-quantum states are given by

$$\langle n|\rho|m\rangle = \int P(\alpha) \langle n|\alpha\rangle \langle \alpha|m\rangle\, d^2\alpha . \qquad (2.268)$$

---

[4] An example of a weight function $P(\alpha)$ which takes on negative values but leads to a positive definite density operator is given by the form

$$P(\alpha) = \frac{1+\lambda}{\pi n} \exp\left[-\frac{|\alpha|^2}{n}\right] - \lambda \delta^{(2)}(\alpha)$$

for $n > 0$ and $0 < \lambda < n^{-1}$. The matrix representation of the corresponding density operator, which is given by Eq. (2.269), is seen to be diagonal and to have only positive eigenvalues.

[5] A familiar example of a function which plays a role analogous to that of a probability density, but may take on negative values in quantum-mechanical contexts is the Wigner distribution function, E. P. Wigner, *Phys. Rev.* **40**, 749 (1932).

When Eqns. (2.184) and (2.187) are used to evaluate the scalar products in the integrand we find

$$\langle n|\rho|m\rangle = (n!\,m!)^{-1/2} \int P(\alpha)\alpha^n(\alpha^*)^m\, e^{-|\alpha|^2}\, d^2\alpha . \qquad (2.269)$$

This form for the density matrix indicates a fundamental property of the fields which are most naturally described by means of the P representation. If $P(\alpha)$ is a weight function with singularities no stronger than those of delta function type, it will, in general, possess nonvanishing complex moments of arbitrarily high order. [The unique exception is the choice $P(\alpha) = \delta^{(2)}(\alpha)$ which corresponds to the ground state of the mode.] It follows then that the diagonal matrix elements $\langle n|\rho|n\rangle$, which represent the probabilities for the presence of n photons in the mode, take on nonvanishing values for arbitrarily large n. There is thus no upper bound to the number of photons present when the function P is well behaved in the sense we have noted.[6]

Stationary density operators correspond in the P representation to functions $P(\alpha)$ which depend only on $|\alpha|$. This correspondence is made clear by Eq. (2.259) which shows that such $P(\alpha)$ lead to functions $R(\beta^*,\gamma)$ which are unaltered by a common phase change of β and γ. It is seen equally well through Eq. (2.269) which shows that $\langle n|\rho|m\rangle$ reduces to diagonal form when the weight function $P(\alpha)$ is circularly symmetric.

Some indication of the importance, in practical terms, of the P representation for the density operator can be found by considering the way in which photon fields produced by different sources become superposed. Since we are only discussing the behavior of one mode of the field for the present, we are only dealing with a fragment of the full problem, but all the modes may eventually be treated similarly. We shall illustrate the superposition law by assuming there are two different transient radiation sources coupled to the field mode and that they may be switched on and off separately. The first source will be assumed, when it is turned on alone at time t_1, to excite the mode from its ground state $|0\rangle$ to the coherent state $|\alpha_1\rangle$. If we assume that the source has ceased radiating by a time t_2, the state of the field remains $|\alpha_1\rangle$ for all later times. We may alternatively consider the case in which the first source remains inactive and the second one is switched on at time t_2. The second source will then be assumed to bring the mode from its ground state to the coherent state $|\alpha_2\rangle$. We now ask what state the mode will be brought to if the two sources are allowed to act in succession, the first at t_1 and the second at t_2.

The answer for this simple case may be seen without performing any detailed calculations by making use of the unitary displacement operators described in Sect. 2.9.3.

---

[6] Density operators for fields in which the number of photons present possesses an upper bound N are represented by functions $R(\beta^*,\gamma)$ which are polynomials of n-th degree in β^* and in γ. It is evident from the behavior of such polynomials for large $|\beta|$ and $|\gamma|$ that any weight function $P(\alpha)$ which corresponds to $R(\beta^*,\gamma)$ through Eq. (2.259) would have to have singularities much stronger than those of a delta function. Such fields are probably represented more conveniently by means of the R function.

The action of the first source is represented by the unitary operator $D(\alpha_1)$ which displaces the oscillator state from the ground state to the coherent state $|\alpha_1\rangle = D(\alpha_1)|0\rangle$. The action of the second source is evidently represented by the displacement operator $D(\alpha_2)$, so that when it is turned on after the first source, it brings the oscillator to the superposed state

$$|\ \rangle = D(\alpha_2)D(\alpha_1)|0\rangle \ . \tag{2.270}$$

Since the displacement operators are of the exponential form (2.198), their multiplication law is given by Eq. (2.201). We thus find

$$D(\alpha_2)D(\alpha_1) = D(\alpha_1 + \alpha_2) \exp\left[\tfrac{1}{2}(\alpha_2 \alpha_1^* - \alpha_2^* \alpha_1)\right] \ . \tag{2.271}$$

The exponential which has been separated from the D operators in this relation has a purely imaginary argument and, hence, corresponds to a phase factor. The superposed state, Eq. (2.270), in other words, is just the coherent state $|\alpha_1 + \alpha_2\rangle$ multiplied by a phase factor. The phase factor has no influence upon the density operator for the superposed state, which is

$$\rho = |\alpha_1 + \alpha_2\rangle \langle \alpha_1 + \alpha_2| \ . \tag{2.272}$$

To vary the way in which the sources are turned on in the imaginary experiment we have described, e.g., to turn the two sources on at other times or in the reverse order, would only alter the final state through a phase factor and would thus lead to the same final density operator. The amplitudes of successive coherent excitations of the mode add as complex numbers in quantum theory, just as they do in classical theory.

Let us suppose next that the sources in the same experiment are somewhat less ideal and that, instead of exciting the mode to pure coherent states, they excite it to conditions described by mixtures of coherent states of the form (2.263). The first source acting alone, we assume, brings the field to a condition described by the density operator

$$\rho_1 = \int P_1(\alpha_1) |\alpha_1\rangle \langle \alpha_1| \, d^2\alpha_1 \ . \tag{2.273}$$

The condition produced by the second source, when it acts alone, is assumed to be represented by

$$\rho = \int P_2(\alpha_2) |\alpha_2\rangle \langle \alpha_2| \, d^2\alpha_2$$

$$= \int P_2(\alpha_2) D(\alpha_2) |0\rangle \langle 0| D^{-1}(\alpha_2) \, d^2\alpha_2 \ .$$

If the second source is turned on after the first, it brings the field to a condition described by the density operator

$$\rho = \int P_2(\alpha_2) D(\alpha_2) \rho_1 D^{-1}(\alpha_2) \, d^2\alpha_2$$

$$= \int P_2(\alpha_2) P_1(\alpha_1) |\alpha_1 + \alpha_2\rangle \langle \alpha_1 + \alpha_2| \, d^2\alpha_1 \, d^2\alpha_2 \ . \tag{2.274}$$

The latter density operator may be written in the general form

$$\rho = \int P(\alpha) |\alpha\rangle \langle \alpha| \, d^2\alpha \,,$$

if we define the weight function $P(\alpha)$ for the superposed excitations to be

$$P(\alpha) = \int \delta^{(2)}(\alpha - \alpha_1 - \alpha_2) P_1(\alpha_1) P_2(\alpha_2) \, d^2\alpha_1 \, d^2\alpha_2 \qquad (2.275)$$

$$= \int P_1(\alpha - \alpha') P_2(\alpha') \, d^2\alpha' \,. \qquad (2.276)$$

We see immediately from Eq. (2.275) that P is correctly normalized if P_1 and P_2 are. The simple convolution law for combining the weight functions is one of the unique features of the description of fields by means of the P representation. It is quite analogous to the law we would use in classical theory to describe the probability distribution of the sum of two uncertain Fourier amplitudes for a mode.

The convolution theorem can often be used to separate fields into component fields with simpler properties. Suppose we have a field described by a weight function $P(\alpha)$ which has a mean value of a given by

$$\bar{\alpha} = \int \alpha P(\alpha) \, d^2\alpha \,. \qquad (2.277)$$

It is clear from Eq. (2.276) that any such field may be regarded as the sum of a pure coherent field which corresponds to the weight function $\delta^{(2)}(\alpha - \bar{\alpha})$ and an additional field represented by $P(\alpha + \bar{\alpha})$ for which the mean value of α vanishes. Fields with vanishing mean values of α will be referred to as unphased fields.

The use of the P representation of the density operator, where it is not too singular, leads to simplifications in the calculation of statistical averages which go somewhat beyond those discussed in the last section. Thus, for example, the statistical average of any normally ordered product of the creation and annihilation operators, such as $(a^\dagger)^n a^m$, reduces to a simple average of $(\alpha^*)^n \alpha^m$ taken with respect to the weight function $P(\alpha)$, i.e., we have

$$\mathrm{Tr}\left\{\rho(a^\dagger)^n a^m\right\} = \int P(\alpha) \left\langle \alpha \left| (a^\dagger)^n a^m \right| \alpha \right\rangle d^2\alpha ,$$

$$= \int P(\alpha)(\alpha^*)^n \alpha^m \, d^2\alpha \,. \qquad (2.278)$$

This identity means, in practice, that many quantum-mechanical calculations can be carried out by means which are analogous to those already familiar from classical theory.

The mean number of photons which are present in a mode is the most elementary measure of the intensity of its excitation. The operator which represents the number of photons present is seen from Eq. (2.176) to be $a^\dagger a$. The average photon number, written as $\langle n \rangle$, is therefore given by

$$\langle n \rangle = \mathrm{Tr}\left\{\rho a^\dagger a\right\} \,. \qquad (2.279)$$

According to Eq. (2.278), with its two exponents set equal to unity, we have

$$\langle n \rangle = \int P(\alpha) |\alpha|^2 \, d^2\alpha, \qquad (2.280)$$

i.e., the average photon number is just the mean squared absolute value of the amplitude α. When two fields described by distributions P_1 and P_2 are superposed, the resulting intensities are found from rules of the form which have always been used in classical electromagnetic theory. For unphased fields the intensities add "incoherently"; for coherent states the amplitudes add "coherently".

The use of the P representation of the density operator in describing fields brings many of the results of quantum electrodynamics into forms similar to those of classical theory. While these similarities make applications of the correspondence principle particularly clear, they must not be interpreted as indicating that classical theory is any sort of adequate substitute for the quantum theory. The weight functions $P(\alpha)$ which occur in quantum theoretical applications are not accurately interpretable as probability distributions, nor are they derivable as a rule from classical treatments of the radiation sources. They depend upon Planck's constant, in general, in ways that are unfathomable by classical or semiclassical analysis.

Since a number of calculations having to do with photon statistics have been carried out in the past by essentially classical methods, it may be helpful to discuss the relation between the P representation and the classical theory a bit further. It is worth noting in particular that the definition we have given the amplitude α as an eigenvalue of the annihilation operator is an intrinsically quantum-mechanical one. If we wish to represent a given classical field amplitude for the mode as an eigenvalue, then we see from Eq. (2.178) that the appropriate value of α has a magnitude which is proportional to $\hbar^{-1/2}$. In the dimensionless terms in which α is defined, the classical description of the mode only applies to the region $|\alpha| \gg 1$ of the complex α plane, i.e., to amplitudes of oscillation which are large compared with the range of the zero-point fluctuations present in the wavepacket Eq. (2.210) and (2.211). Classical theory can therefore, in principle, only furnish us with the grossest sort of information about the weight function $P(\alpha)$. When the weight function extends appreciably into the classical regions of the plane, classical theory can only be relied upon, crudely speaking, to tell us average values of the function $P(\alpha)$ over areas whose dimensions, $|\Delta\alpha|$, are of order unity or larger. From Eq. (2.267) we see that such average values will always be positive; in the classical limit they may always be interpreted as probabilities.

2.9.8
The Gaussian Density Operator

The Gaussian function is a venerable statistical distribution, familiar from countless occurrences in classical statistics. We shall indicate in this section that it has its place in quantum field theory as well, where it furnishes the natural description of the most commonly occurring type of incoherence[4].

2.9 Coherent and Incoherent States of the Radiation Field

Let us assume that the field mode we are studying is coupled to a number of sources which are essentially similar but are statistically independent of one another in their behavior. Such sources might, in practice, simply be several hypothetical subdivisions of one large source. If we may represent the contribution of each source (numbered $j = 1, \ldots N$) to the excitation of the mode by means of a weight function $p(\alpha_j)$, we may then construct the weight function $P(\alpha)$ which describes the superposed fields by means of the generalized form of the convolution theorem

$$P(\alpha) = \int \delta^{(2)}\left(\alpha - \sum_{j=1}^{N} \alpha_j\right) \prod_{j=1}^{N} p(\alpha_j) \, d^2\alpha_j \, . \tag{2.281}$$

Since the weight functions which appear in this expression are all real valued, it is sometimes convenient to think of the amplitudes α in their arguments not as complex numbers, but as two-dimensional real vectors $\boldsymbol{\alpha}$ (i.e., $\alpha_x = \operatorname{Re}\alpha$, $\alpha_y = \operatorname{Im}\alpha$). Then if λ is an arbitrary complex number represented by the vector $\boldsymbol{\lambda}$, we may use a two-dimensional scalar product for the abbreviation

$$\operatorname{Re}\lambda \operatorname{Re}\alpha + \operatorname{Im}\lambda \operatorname{Im}\alpha = \boldsymbol{\alpha} \cdot \boldsymbol{\lambda} \, . \tag{2.282}$$

Using this notation, we may define the two-dimensional Fourier transform of the weight function $p(\boldsymbol{\alpha})$ as

$$\xi(\boldsymbol{\alpha}) = \int \exp(i\boldsymbol{\lambda} \cdot \boldsymbol{\alpha}) \, p(\boldsymbol{\alpha}) \, d^2\alpha \, . \tag{2.283}$$

The superposition law Eq. (2.281) then shows that the Fourier transform of the weight function $P(\alpha)$ is given by

$$\Xi(\boldsymbol{\lambda}) = \int \exp(i\boldsymbol{\lambda} \cdot \boldsymbol{\alpha}) \, P(\boldsymbol{\alpha}) \, d^2\alpha \, ,$$
$$= [\xi(\boldsymbol{\lambda})]^N \, . \tag{2.284}$$

If the individual sources are stationary ones their weight function $p(\boldsymbol{\alpha})$ depends only on $|\alpha|$. The transform $\xi(\boldsymbol{\lambda})$ may then be approximated for small values of $|\boldsymbol{\lambda}|$ by

$$\xi(\boldsymbol{\lambda}) = 1 - \tfrac{1}{4}\lambda^2 \int |\alpha|^2 p(\boldsymbol{\alpha}) \, d^2\alpha \, ,$$
$$= 1 - \tfrac{1}{4}\lambda^2 \langle |\alpha|^2 \rangle \, . \tag{2.285}$$

For values of $|\boldsymbol{\lambda}|$ which are smaller still (i.e., $|\boldsymbol{\lambda}|^2 < N^{-1/2} \langle |\alpha|^2 \rangle^{-1}$, the transform Ξ for the superposed field may be approximated by

$$\Xi(\boldsymbol{\lambda}) \approx \exp\left[-\tfrac{1}{4}\lambda^2 N \langle |\alpha|^2 \rangle\right] \, . \tag{2.286}$$

Since the weight function $p(\boldsymbol{\alpha})$ may take on negative values it is necessary at this point to verify that the second moment $\langle |\alpha|^2 \rangle$ is positive. That it is indeed positive is indicated by Eqns. (2.279) and (2.280) which show that $\langle|\alpha|^2\rangle$ is the mean number of photons which would be radiated by each source in the absence of the others. For large values of N the transform $\Xi(\lambda)$ therefore decreases rapidly as $|\lambda|$ increases. Since the function becomes vanishingly small for $|\lambda|$ lying outside the range of approximation noted earlier, we may use Eq. (2.286) more generally as an asymptotic approximation to $\Xi(\lambda)$ for large N. When we calculate the transform of this asymptotic expression for $\Xi(\lambda)$ we find

$$P(\boldsymbol{\alpha}) = (2\pi)^{-2} \int \exp\left[-i\boldsymbol{\alpha}\cdot\boldsymbol{\lambda}\right] \Xi(\boldsymbol{\lambda}) \, d^2\lambda ,$$

$$= \frac{1}{\pi N \langle|\alpha|^2\rangle} \exp\left[-\frac{\alpha^2}{N\langle|\alpha|^2\rangle}\right]. \tag{2.287}$$

The mean value of $|\alpha|^2$ for such a weight function is evidently $N\langle|\alpha|^2\rangle$, but by the general theorem expressed in Eq. (2.280), this mean value is just the average of the total number of quanta present in the mode. If we write the latter average as $\langle n \rangle$, and resume the use of the complex notation for the variable α, the weight function Eq. (2.287) may be written as

$$P(\alpha) = \frac{1}{\pi \langle n \rangle} e^{-|\alpha|^2/\langle n \rangle} . \tag{2.288}$$

The weight function $P(\alpha)$ is positive everywhere and takes the same form as the probability distribution for the total displacement which results from a random walk in the complex plane. However, because the coherent states $|\alpha\rangle$ are not an orthogonal set, $P(\alpha)$ can only be accurately interpreted as a probability distribution for $\langle n \rangle \gg 1$. We may note that it is not ultimately necessary, in order to derive Eq. (2.288), to assume that the weight functions corresponding to the individual sources are all the same. All that is required to carry out the proof is that the moments of the individual functions be of comparable magnitudes. The mean squared value of $|\alpha|$ is then given more generally by $\sum_j \langle|\alpha_j|^2\rangle$ rather than the value in Eq. (2.287), but this value is still the mean number of quanta in the mode, as indicated in Eq. (2.288).

It should be clear from the conditions of the derivation that the Gaussian distribution $P(\alpha)$ for the excitation of a mode possesses extremely wide applicability. The random or chaotic sort of excitation it describes is presumably characteristic of most of the familiar types of noncoherent macroscopic light sources, such as gas discharges, incandescent radiators, etc.

The Gaussian density operator

$$\rho = \frac{1}{\pi \langle n \rangle} \int e^{-|\alpha|^2/\langle n \rangle} |\alpha\rangle \langle \alpha| \, d^2\alpha \tag{2.289}$$

may be seen to take on a very simple form as well in the basis which specifies the photon numbers. To find this form we substitute in Eq. (2.289) the expansions Eqns. (2.188) and (2.189) for the coherent states and note the identity

$$\pi^{-1}(l!m!)^{-1/2} \int \exp\left[-C|\alpha|^2\right] \alpha^l (\alpha^*)^m \, d^2\alpha = \delta_{lm} C^{-(m+1)} ,$$

which holds for $C > 0$. If we write $C = (1 + \langle n \rangle)/\langle n \rangle$ we then find

$$\rho = \frac{1}{1 + \langle n \rangle} \sum_m \left\{ \frac{\langle n \rangle}{1 + \langle n \rangle} \right\}^m |m\rangle \langle m| . \tag{2.290}$$

In other words, the number of quanta in the mode is distributed according to the powers of the parameter $\langle n \rangle / (1 + \langle n \rangle)$. The Planck distribution for blackbody radiation furnishes an illustration of a density operator which has long been known to take the form of Eq. (2.290). The thermal excitation which leads to the black-body distribution is an ideal example of the random type we have described earlier, and so it should not be surprising that this distribution is one of the class we have derived. It is worth noting, in particular, that while the Planck distribution is characteristic of thermal equilibrium, no such limitation is implicit in the general form of the density operator Eq. (2.289). It will apply whenever the excitation has an appropriately random quality, no matter how far the radiator is from thermal equilibrium.

The Gaussian distribution function $\exp[-|\alpha|^2/\langle n \rangle]$ is phrased in terms which are explicitly quantum mechanical. In the limit which would represent a classical field both $|\alpha|^2$ and the average quantum number $\langle n \rangle$ become infinite as $\hbar^{-1}$, but their quotient, which is the argument of the Gaussian function, remains well defined. The form which the distribution takes in the classical limit is a familiar one. Historically, one of the origins of the random walk problem is to be found in the discussion of a classical harmonic oscillator which is subject to random excitations[15]. Such oscillators have complex amplitudes which are described under quite general conditions by a Gaussian distribution. If we were armed with this knowledge, and lacked the quantum-mechanical analysis given earlier, we might be tempted to assume that a Gaussian distribution derived in this way from classical theory can describe the photon distribution. To demonstrate the fallacy of this view we must examine more closely the nature of the parameter $\langle n \rangle$ which is, after all, the only physical constant involved in the distribution. We may take, as a simple illustration, the case of thermal excitation corresponding to temperature T. Then the mean photon number is given by $\langle n \rangle = [\exp(\hbar \omega / \kappa T) - 1]^{-1}$, where κ is Boltzmann's constant, and the distribution $P(\alpha)$ takes the form

$$P(\alpha) = \frac{1}{\pi} \left[e^{\hbar \omega / \kappa T} - 1 \right] \exp \left[-(e^{\hbar \omega / \kappa T} - 1)|\alpha|^2 \right] . \tag{2.291}$$

To reach the classical analog of this distribution we would assume that the classical field energy in the mode, $H = \frac{1}{2} \int (\mathbf{E}^2 + \mathbf{B}^2) \, d\mathbf{r}$, is distributed with a probability proportional to $\exp[-H/\kappa T]$. The distribution for the amplitude α that results is

$$P_{\mathrm{cl}}(\alpha) = \left(\frac{\hbar \omega}{\pi \kappa T} \right) \exp \left[-\frac{\hbar \omega |\alpha|^2}{\kappa T} \right] , \tag{2.292}$$

which is seen to be a first approximation in powers of $\hbar$ to the correct distribution. (Again, we must remember that the quantity $\hbar |\alpha|^2$ is to be construed as a classical parameter.) The distribution $P_{cl}(\alpha)$ only extends into the classical region of the plane, $|\alpha| \gg 1$, for low-frequency modes, that is, only for $(\hbar\omega/\kappa T) \ll 1$ are the modes sufficiently excited to be accurately described by classical theory. For higher frequencies the two distributions differ greatly in nature even though both are Gaussian. The classical distribution retains much too large a radius in the α plane as $\hbar\omega$ increases beyond κT, rather than narrowing extremely rapidly as the correct distribution does.[7] That error, in fact, epitomizes the ultraviolet catastrophe of the classical radiation theory. The example we have discussed is, of course, an elementary one, but it should serve to illustrate some of the points noted in the preceding section regarding the limitations of the classical distribution function.

The expression for the thermal density operator of an oscillator in terms of coherent quantum states appears to offer new and instructive approaches to many familiar problems. It permits us, for example, to derive the thermal averages of exponential functions of the operators a and $a^\dagger$ in an elementary way. The thermal average of the operator $D(\beta)$ defined by Eq. (2.198) is an illustration. It is given by

$$\text{Tr}\{\rho D(\beta)\} = \frac{1}{\pi\langle n\rangle} \int e^{-|\alpha|^2/\langle n\rangle} \langle \alpha | D(\beta) | \alpha\rangle \, d^2\alpha \, . \tag{2.293}$$

The expectation value in the integrand is, in this case

$$\begin{aligned}
\langle \alpha | D(\beta) | \alpha\rangle &= \langle 0 | D^{-1}(\alpha) D(\beta) D(\alpha) | 0\rangle \, , \\
&= \exp[\beta\alpha^* - \beta^*\alpha] \langle 0 | D(\beta) | 0\rangle \, , \\
&= \exp[\beta\alpha^* - \beta^*\alpha] \langle 0 | \beta\rangle \, , \\
&= \exp\left[\beta\alpha^* - \beta^*\alpha - \tfrac{1}{2}|\beta|^2\right] \, ,
\end{aligned} \tag{2.294}$$

where the properties of $D(\alpha)$ as a displacement operator have been used in the intermediate steps. When the integration indicated in Eq. (2.293) is carried out, we find

$$\text{Tr}\{\rho D(\beta)\} = \exp\left[-|\beta|^2(\langle n\rangle + \tfrac{1}{2})\right] \, , \tag{2.295}$$

which is a frequently used corollary of Bloch's theorem on the distribution function of an oscillator coordinate[16].

2.9.9
Density Operators for the Field

The developments introduced in Sections 2.9.3 through 2.9.8 have all concerned the description of the quantum state of a single mode of the electromagnetic field. We may describe the field as a whole by constructing analogous methods to deal with all

---

[7] For frequencies in the middle of the visible spectrum and temperatures under 3000 K the quantum mechanical distribution Eq. (2.291) will have a radius which corresponds to $|\alpha|^2 \ll 10^{-3}$, i.e., the distribution is far from classical in nature. Comparable radii characterize the distributions for nonthermal incoherent sources.

its modes at once. For this purpose we introduce a basic set of coherent states for the entire field and write them as

$$|\{\alpha_k\}\rangle \equiv \prod_k |\alpha_k\rangle_k , \qquad (2.296)$$

where the notation $\{\alpha_k\}$, which will be used in several other connections, stands for the set of all amplitudes α_k. It is clear then, from the arguments of Sect. 2.9.4, that any state of the field determines uniquely a function $f(\{\alpha_k^*\})$ which is an entire function of each of the variables α_k^*. If the Hilbert space vector which represents the state is known and designated as $|f\rangle$, the function f is given by

$$f(\{\alpha_k^*\}) = \langle\{\alpha_k\}|f\rangle \exp\left[\tfrac{1}{2}\sum_k |\alpha_k|^2\right] , \qquad (2.297)$$

which is the direct generalization of Eq. (2.225). The expansion for the state $|f\rangle$ in terms of coherent states is then

$$|f\rangle = \int |\{\alpha_k\}\rangle f(\{\alpha_k^*\}) \prod_k \pi^{-1} e^{-\tfrac{1}{2}|\alpha_k|^2} d^2\alpha_k , \qquad (2.298)$$

which generalizes Eq. (2.221).

All of the operators which occur in field theory possess expansions in terms of the vectors $|\{\alpha_k\}\rangle$ and their adjoints. To construct such representations is simply a matter of generalizing the formulas of Sect. 2.9.5 to deal with an infinite set of amplitude variables. We therefore proceed directly to a discussion of the density operator. For any density operator ρ we may define a function $R(\{\alpha_k^*\},\{\beta_k\})$ which is an entire function of each of the variables α_k^* and β_k for all modes k. This function, as may be seen from Eq. (2.248), is given by

$$R(\{\alpha_k^*\},\{\beta_k\}) = \langle\{\alpha_k\}|\rho|\{\beta_k\}\rangle \times \exp\left[\tfrac{1}{2}\sum (|\alpha_k|^2+|\beta_k|^2)\right] . \qquad (2.299)$$

The corresponding representation of the density operator is

$$\rho = \int |\{\alpha_k\}\rangle R(\{\alpha_k^*\},\{\beta_k\}) \langle\{\beta_k\}| \prod_k \pi^{-2} e^{-(|\alpha_k|^2+|\beta_k|^2)/2} d^2\alpha_k d^2\beta_k . \qquad (2.300)$$

If the set of integers $\{n_k\}$ is used to specify the familiar stationary states which have n_k photons in the k-th mode, we may regard R as a generating function for the matrix elements of ρ connecting these states, i.e., as a generalization of Eq. (2.249) we have

$$R(\{\alpha_k^*\},\{\beta_k\}) = \sum_{\{n_k\},\{m_k\}} \langle\{n_k\}|\rho|\{m_k\}\rangle \prod_k (n_k! m_k!)^{-1/2} (\alpha_k^*)^{n_k} \beta_k^{m_k} . \qquad (2.301)$$

The matrix elements of ρ in the stationary basis are then given by

$$\langle\{n_k\}|\rho|\{m_k\}\rangle = \int R(\{\alpha_k^*\},\{\beta_k\}) \prod_k \pi^{-2} (n_k! m_k!)^{-1/2} \alpha_k^{n_k} (\beta_k^*)^{m_k}$$
$$\times e^{-(|\alpha_k|^2+|\beta_k|^2)/2} d^2\alpha_k d^2\beta_k . \qquad (2.302)$$

The normalization condition on R is clearly

$$\int R(\{\beta_k^*\},\{\beta_k\}) \prod_k \pi^{-1} e^{-|\beta_k|^2} d^2\beta_k = 1 . \tag{2.303}$$

The positive definiteness condition, Eq. (2.256), may also be generalized in an evident way to deal with the full set of amplitude variables.

It may help as a simple illustration of the foregoing formulae to consider the representation of a single-photon wavepacket. The state which is empty of all photons is the one for which the amplitudes α_k all vanish. If we write that state as $|\text{vac}\rangle$, then we may write the most general one-photon state as $\sum_k q(k) a_k^\dagger |\text{vac}\rangle$, where the function $q(k)$ plays the role of a packet amplitude. The function f which represents this state is then

$$f(\{\alpha_k^*\}) = \sum_k q(k) \alpha_k^* , \tag{2.304}$$

and the corresponding function R which determines the density operator is

$$R(\{\alpha_k^*\},\{\beta_k\}) = \sum_k q(k)\alpha_k^* \sum_{k'} q^*(k')\beta_{k'} . \tag{2.305}$$

The normalization condition (2.303) corresponds to the requirement $\sum |q(k)|^2 = 1$. Since the state we have considered is a pure one, the function R factorizes into the product of two functions, one having the form of f and the other of its complex conjugate. If the packet amplitudes $q(k)$ were in some degree unpredictable, as they usually are, the packet could no longer be represented by a pure state. The function R would then be an average taken over the distribution of the amplitudes $q(k)$ and hence would lose its factorizable form in general. Whenever an upper bound exists for the number of photons present, i.e., the number of photons is required to be less than or equal to some integer N, we will find that R is a polynomial of at most N-th degree in the variables $\{\alpha_k^*\}$ and of the same degree in the $\{\beta_k\}$.

There will, of course, exist many types of excitation for which the photon numbers are unbounded. Among these are the ones which are more conveniently described by means of a generalized P distribution, i.e., the excitations for which there exists a reasonably well-behaved real-valued function $P(\{\alpha_k\})$ such that

$$R(\{\beta_k^*\},\{\gamma_k\}) = \int P(\{\alpha_k\}) \exp\left[\sum_k (\beta_k^* \alpha_k + \gamma_k \alpha_k^* - |\alpha_k|^2)\right] \prod_k d^2\alpha_k . \tag{2.306}$$

When R possesses a representation of this type the density operator (2.300) may be reduced by means of Eq. (2.228) and its complex conjugate to the simple form

$$\rho = \int P(\{\alpha_k\}) |\{\alpha_k\}\rangle \langle\{\alpha_k\}| \prod_k d^2\alpha_k , \tag{2.307}$$

which is the many-mode form of the P representation given by Eq. (2.263). The function P must satisfy the positive definiteness condition

$$\int |f(\{\alpha_k^*\})|^2 \, P(\{\alpha_k\}) \prod_k e^{-|\alpha_k|^2} \, d^2\alpha_k \geq 0 \tag{2.308}$$

for all possible choices of entire functions $f(\{\alpha_k^*\})$. The matrix elements of the density operator in the representation based on the n-photon states are

$$\langle \{nk\} | \rho | \{m_k\} \rangle = \int P(\{\alpha_k\}) \prod (n_k! m_k!)^{-1/2} \alpha_k^{n_k} (\alpha_k^*)^{m_k} e^{-|\alpha_k|^2} \, d^2\alpha_k. \tag{2.309}$$

Stationary density operators, i.e., ones which commute with the Hamiltonian correspond to functions $P(\{\alpha_k\})$ which depend on the amplitude variables only through their magnitudes $\{|\alpha_k|\}$.

The superposition of two fields is described by forming the convolution integral of their distribution functions, much as in the case of a single mode. Thus, if two fields, described by $P_1(\{\beta_k\})$ and $P_2(\{\gamma_k\})$, respectively, are superposed, the resulting field has a distribution function

$$P(\{\alpha_k\}) = \int \prod_k \delta^{(2)}(\alpha_k - \beta_k - \gamma_k) P_1(\{\beta_k\}) \, P_2(\{\gamma_k\}) \prod_k d^2\beta_k \, d^2\gamma_k. \tag{2.310}$$

For fields which are represented by means of the density operator Eq. (2.307) all of the averages of normally ordered operator products can be calculated by means of formulas which, as in the case of a single mode, greatly resemble those of classical theory. Thus, the parameters $\{\alpha_k\}$ play much the same role in these calculations as the random Fourier amplitudes of the field do in the familiar classical theory of microwave noise[17]. Furthermore, the weight function $P(\{\alpha_k\})$ plays a role similar to that of the probability distribution for the Fourier amplitudes. Although this resemblance is extremely convenient in calculations, and offers immediate insight into the application of the correspondence principle, we must not lose sight of the fact that the function $P(\{\alpha_k\})$ is, in general, an explicitly quantum-mechanical structure. It may assume negative values, and is not accurately interpretable as a probability distribution except in the classical limit of strongly excited or low frequency fields.

In the foregoing discussions we have freely assumed that the density operator which describes the field is known and that it may, therefore, be expressed either in the representation of Eq. (2.300) or in the P representation of Eq. (2.307). For certain types of incoherent sources which we have discussed in Sect. 2.9.8 and will mention again in Sect. 2.9.10, the explicit construction of these density operators is not at all difficult. But to find accurate density operators for other types of sources, including the recently developed coherent ones, will require a good deal of physical insight. The general problem of treating quantum mechanically the interaction of a many-atom source both with the radiation field and with an excitation mechanism of some sort promises to be a complicated one. It will have to be approached, no doubt, through greatly simplified models.

Since very little is known about the density operator for radiation fields, some insight may be gained by examining the form it takes on in one of the few completely soluble problems of quantum electrodynamics. We shall study the photon field radiated by an electric current distribution which is essentially classical in nature, one that does not suffer any noticeable reaction from the process of radiation. We may then represent the radiating current by a prescribed vector function of space and time $j(r,t)$. The Hamiltonian which describes the coupling of the quantized electromagnetic field to the current distribution takes the form

$$H_1(t) = -\frac{1}{c} \int j(r,t) \cdot A(r,t) \, dr . \qquad (2.311)$$

The introduction of an explicitly time-dependent interaction of this type means that the state vector for the field, $|\ \rangle$, which previously was fixed (corresponding to the Heisenberg picture) will begin to change with time in accordance with the Schrödinger equation

$$i\hbar \frac{\partial}{\partial t} |\ \rangle = H_1(t) |\ \rangle , \qquad (2.312)$$

which is the one appropriate to the interaction representation. The solution of this equation is easily found[18]. If we assume that the initial state of the field at time $t = -\infty$ is one empty of all photons, then the state of the field at time t may be written in the form

$$|t\rangle = \exp\left\{ \frac{i}{\hbar c} \int_{-\infty}^{t} dt' \int j(r,t') \cdot A(r,t') \, dr + i\varphi(t) \right\} |\text{vac}\rangle . \qquad (2.313)$$

The function $\varphi(t)$ which occurs in the exponent is a real-valued c-number phase function. It is easily evaluated, but cancels out of the product $|t\rangle\langle t|$ and so has no bearing on the construction of the density operator. The exponential operator which occurs in Eq. (2.313) may be expressed quite simply in terms of the displacement operators we discussed in Sect. 2.9.3. For this purpose we define a displacement operator D_k for the k-th mode as

$$D_k(\beta_k) = \exp\left[\beta_k a_k^\dagger - \beta_k^* a_k \right] . \qquad (2.314)$$

Then it is clear from the expansion Eq. (2.168)) for the vector potential that we may write

$$\exp\left\{ \frac{i}{\hbar c} \int_{-\infty}^{t} dt' \int j(r,t') \cdot A(r,t') \, dr \right\} = \prod_k D_k [\alpha_k(t)] , \qquad (2.315)$$

where the time-dependent amplitudes $\alpha_k(t)$ are given by

$$\alpha_k(t) = \frac{i}{(2\hbar\omega)^{1/2}} \int_{-\infty}^{t} dt' \int dr\, u_k^*(r) \cdot j(r,t') e^{i\omega t'} . \qquad (2.316)$$

2.9 Coherent and Incoherent States of the Radiation Field

The density operator at time t may therefore be written as

$$|t\rangle\langle t| = \prod_k D_k[\alpha_k(t)] |\text{vac}\rangle\langle\text{vac}| \prod_k D_k^{-1}[\alpha_k(t)] \tag{2.317}$$

$$= |\{\alpha_k(t)\}\rangle\langle\{\alpha_k(t)\}| . \tag{2.318}$$

The radiation by any prescribed current distribution, in other words, always leads to a pure coherent state.

It is only a slight generalization of the model we have just considered to imagine that the current distribution $j(r,t)$ is not wholly predictable. In that case the amplitudes $\alpha_k(t)$ defined by Eq. (2.316) become random variables which possess collectively a probability distribution function which we may write as $p(\{\alpha_k\},t)$. The density operator for the field radiated by such a random current then becomes

$$\rho(t) = \int p(\{\alpha_k\},t) |\{\alpha_k\}\rangle\langle\{\alpha_k\}| \prod_k d^2\alpha_k . \tag{2.319}$$

We see that the density operator for a field radiated by a random current which suffers no recoil in the radiation process always takes the form of the P representation of Eq. (2.307). The weight function in this case does admit interpretation as a probability distribution, but it has a classical structure associated directly with the properties of the radiating current rather than with particular (nonorthogonal) states of the field. The assumption we have made in defining the model, that the current suffers negligible reaction, is a strong one but is fairly well fulfilled in radiating systems operated at radio or microwave frequencies. The fields produced by such systems should be accurately described by density operators of the form Eq. (2.319).

2.9.10
Correlation and Coherence Properties of the Field

Any eigenvalue function $\mathcal{E}(rt)$ which satisfies the appropriate field equations and contains only positive frequency terms determines a set of mode amplitudes $\{\alpha_k\}$ uniquely through the expansion Eq. (2.178). This set of mode amplitudes then determines a coherent state of the field, $|\{\alpha_k\}\rangle$, such that

$$E^{(+)}(rt)|\{\alpha_k\}\rangle = \mathcal{E}(rt)|\{\alpha_k\}\rangle . \tag{2.320}$$

To discuss the general form which the field correlation functions take in such states it is convenient to abbreviate a set of coordinates (r_j,t_j) by a single symbol x_j. The n-th-order correlation function is then defined as[6]

$$G^{(n)}_{\mu_1\ldots\mu_{2n}}(x_1\ldots x_{2n}) = \text{Tr}\left\{\rho E^{(-)}_{\mu_1}(x_1)\ldots E^{(-)}_{\mu_n}(x_n) E^{(+)}_{\mu_{n+1}}(x_{n+1})\ldots E^{(+)}_{\mu_{2n}}(x_{2n})\right\} . \tag{2.321}$$

The density operator for the coherent state defined by Eq. (2.320) is the projection operator

$$\rho = |\{\alpha_k\}\rangle\langle\{\alpha_k\}| . \tag{2.322}$$

For this operator it follows from Eq. (2.320) and its Hermitian adjoint that the correlation functions reduce to the factorized form

$$G^{(n)}_{\mu_1\ldots\mu_{2n}}(x_1\ldots x_{2n}) = \prod_{j=1}^{n} \mathcal{E}^*_{\mu_j}(x_j) \prod_{l=n+1}^{2n} \mathcal{E}^*_{\mu_l}(x_l) \,. \tag{2.323}$$

In other words, the field which corresponds to the state $|\{\alpha_k\}\rangle$ satisfies the conditions for full coherence according to the definition[6] given earlier.

It is worth noting that the state $|\{\alpha_k\}\rangle$ is not the only one which leads to the set of correlation functions Eq. (2.323)). Indeed, let us consider a state which corresponds not to the amplitudes $\{\alpha_k\}$, but to a set $\{e^{i\varphi}\alpha_k\}$ which differs by a common phase factor (i.e., φ is real and independent of k). Then the corresponding eigenvalue function becomes $e^{i\varphi}\mathcal{E}(rt)$, but such a change leaves the correlation functions (2.323) unaltered. It is clear from this invariance property of the correlation functions that certain mixtures of the coherent states also lead to the same set of functions. Thus, if $|\{\alpha^*_k\}\rangle$ is the state defined by Eq. (2.320), and $L(\varphi)$ is any real-valued function of φ normalized in the sense

$$\int_0^{2\pi} L(\varphi)\,d\varphi = 1\,, \tag{2.324}$$

we see that the density operator

$$\rho = \int_0^{2\pi} L(\varphi) \left|\{e^{i\varphi}\alpha_k\}\right\rangle \left\langle\{e^{i\varphi}\alpha_k\}\right| d\varphi \tag{2.325}$$

leads for all choices of $L(\varphi)$ to the set of correlation functions (2.323). Such a density operator is, of course, a special case of the general form (2.307), one which corresponds to an over-all uncertainty in the phase of the $\{\alpha_k\}$. The particular choice $L(\varphi) = (2\pi)^{-1}$, which corresponds to complete ignorance of the phase, represents the usual state of our knowledge about high-frequency fields. We have been careful, therefore, to define coherence in terms of a set of correlation functions which are independent of the over-all phase.

Since nonstationary fields of many sorts can be represented by means of eigenvalue functions, it becomes a simple matter to construct corresponding quantum states. As an illustration we may consider the example of an amplitude-modulated plane wave. For this purpose we make use of the particular set of mode functions defined by Eq. (2.167). Then if the carrier wave has frequency ω and the modulation is periodic and has frequency $\zeta\omega$ where $0 < \zeta < 1$, we may write an appropriate eigenvalue function as

$$\mathcal{E}(rt) = i\left(\frac{\hbar\omega}{2L^3}\right)^{1/2} \hat{e}^{(\lambda)}\alpha_k\{1 + M\cos[\zeta(\mathbf{k}\cdot\mathbf{r} - \omega t) - \delta]\}\, e^{i(\mathbf{k}\cdot\mathbf{r} - \omega t)}\,. \tag{2.326}$$

When this expression is expanded in plane-wave modes it has only three nonvanishing amplitude coefficients. These are α_k itself and the two sideband amplitudes

$$\alpha_{k(1-\zeta)} = \tfrac{1}{2}M(1-\zeta)^{-1/2}e^{i\delta}\alpha_k,$$
$$\alpha_{k(1+\zeta)} = \tfrac{1}{2}M(1+\zeta)^{-1/2}e^{-i\delta}\alpha_k. \tag{2.327}$$

The coherent state which corresponds to the modulated wave may be constructed immediately from the knowledge of these amplitudes. In practice, of course, we will not often know the phase of α_k, and so the wave should be represented not by a single coherent state, but by a mixture of the form Eq. (2.325). Representations of other forms of modulated waves may be constructed similarly. Incoherent fields, or the broad class of fields for which the correlation functions do not factorize, must be described by means of density operators which are more general in their structure than those of Eqns. (2.322) or (2.325). To illustrate the form taken by the correlation functions for such cases we may suppose the field to be described by the P representation of the density operator. Then the first-order correlation function is given by

$$G^{(1)}_{\mu\nu}(\mathbf{r}t,\mathbf{r}'t') = \int P(\{\alpha_k\}) \sum_{k,k'} \tfrac{1}{2}\hbar(\omega\omega')^{1/2} u^*_{k\mu}(\mathbf{r}) e_{k'\nu}(\mathbf{r}') \alpha^*_k \alpha_{k'} e^{i(\omega t - \omega' t')} \prod_l d^2\alpha_l. \tag{2.328}$$

Fields for which the P representation is inconveniently singular may, as we have noted earlier, always be described by means of analytic functions $R(\{\alpha^*_k\},\{\beta_k\})$ and corresponding density operators of the form Eq. (2.300). When that form of density operator is used to evaluate the first-order correlation function we find

$$G^{(1)}_{\mu\nu}(\mathbf{r}t,\mathbf{r}'t') = \int R(\{\alpha^*_k\},\{\beta_k\}) \sum_{k',k''} \tfrac{1}{2}\hbar(\omega'\omega'')^{1/2} u^*_{k'\mu}(\mathbf{r}) u_{k''\nu}(\mathbf{r}') \beta^*_{k'}\alpha_{k''}$$
$$\times e^{i(\omega' t - \omega'' t')} \prod_l e^{\beta^*_l \alpha_l} d\mu(\alpha_l) d\mu(\beta_l), \tag{2.329}$$

where the differentials $d\mu(\alpha_l)$ and $d\mu(\beta_l)$ are those defined by Eq. (2.243). The higher order correlation functions are given by integrals analogous to Eqns. (2.328) and (2.329). Their integrands contain polynomials of the $2n$-th degree in the amplitude variables α_k and β^*_k in place of the quadratic forms which occur in the first-order functions.

The energy spectrum of a radiation field is easily derived from a knowledge of its first-order correlation function. If we return for a moment to the expansion Eq. (2.177) for the positive-frequency field operator, and write the negative-frequency field as its Hermitian adjoint, we see that these operators obey the identity

$$2\int \mathbf{E}^{(-)}(\mathbf{r}t)\cdot \mathbf{E}^{(+)}(\mathbf{r}t')\, d\mathbf{r} = \sum_k \hbar\omega a^\dagger_k a_k \exp[i\omega(t-t')]. \tag{2.330}$$

If we take the statistical average of both sides of this equation we may write the resulting relation as

$$\sum_\mu \int G^{(1)}_{\mu\mu}(rt,rt')\,dr = \tfrac{1}{2}\sum_k \hbar\omega \langle n_k\rangle \exp[i\omega(t-t')], \qquad (2.331)$$

where $\langle n_k\rangle$ is the average number of photons in the k-th mode. The Fourier representation of the volume integral of $\sum_\mu G^{(1)}_{\mu\mu}$ therefore identifies the energy spectrum $\hbar\omega \langle n_k\rangle$ quite generally.

For fields which may be represented by stationary density operators, it becomes still simpler to extract the energy spectrum from the correlation function. For such fields the weight function $P(\{\alpha_k\})$ depends only on the absolute values of the α_k, so that we have

$$\int P(\{\alpha_k\}) \alpha^*_{k'} \alpha_{k''} \prod_l d^2\alpha_l = \langle |\alpha_{k'}|^2\rangle \delta_{k'k''}$$

$$= \langle n_{k'}\rangle \delta_{k'k''}. \qquad (2.332)$$

By using Eq. (2.328) to evaluate the correlation function, and specializing to the case of plane-wave modes, we then find

$$\sum_\mu G^{(1)}_{\mu\mu}(rt,rt') = \tfrac{1}{2} L^{-3} \sum_{k,\lambda} \hbar\omega \langle n_{k,\lambda}\rangle e^{i\omega(t-t')}, \qquad (2.333)$$

in which we have explicitly indicated the role of the polarization index λ. If the volume which contains the field is sufficiently large in comparison to the wavelengths of the excited modes, the sum over the modes in Eq. (2.333) may be expressed as an integral over k space ($\sum_k \to \int L^3 (2\pi)^{-3}\,dk$). By defining an energy spectrum for the quanta present (i.e., an energy per unit interval of ω) as

$$w(\omega) = (2\pi)^{-3} \hbar k^3 \sum_\lambda \int \langle n_{k,\lambda}\rangle\, d\Omega_k, \qquad (2.334)$$

where $d\Omega_k$ is an element of solid angle in k space, we may then rewrite Eq. (2.333) in the form

$$\sum_\mu G^{(1)}_{\mu\mu}(rt,rt') = \tfrac{1}{2} \int_0^\infty w(\omega) e^{i\omega(t-t')}\, d\omega. \qquad (2.335)$$

With the understanding that $w(\omega) = 0$ for $\omega < 0$, we may extend the integral over ω from $-\infty$ to ∞. It is then clear that the relation (2.335) may be inverted to express the energy spectrum as the Fourier transform of the time-dependent correlation function,

$$w(\omega) = \frac{1}{\pi} \int_{-\infty}^\infty \sum_\mu G^{(1)}_{\mu\mu}(r0,rt) e^{i\omega t}\, dt. \qquad (2.336)$$

A pair of relations analogous to Eqns. (2.335) and (2.336), and together called the Wiener–Khintchine theorem, has long been of use in the classical theory of random fields.[8] The relations we have derived are, in a sense, the natural quantum mechanical generalization of the Wiener–Khintchine theorem. All we have assumed is that the field is describable by a stationary form of the P representation of the density operator. The proof need not, in fact, rest upon the use of the P representation since we can construct a corresponding statement in terms of the more general representation Eq. (2.300).

Stationary fields, according to Eq. (2.257), are represented by entire functions $R = S(\{\alpha_k^* \beta_k\})$, i.e., functions which depend only on the set of products $\alpha_k^* \beta_k$. For such fields, then, the integral over the α and β planes which is required in Eq. (2.329) takes the form

$$\langle \beta_{k'}^* \alpha_{k''} \rangle = \int S(\{\alpha_k^* \beta_k\}) \beta_{k'}^* \alpha_{k''} \prod_l e^{\beta_l^* \alpha_l} \, d\mu(\alpha_l) \, d\mu(\beta_l) \,. \tag{2.337}$$

Since the range of integration of each of the α and β variables covers the entire complex plane, this integral cannot be altered if we change the signs of any of the variables. If, however, we replace the particular variables $\alpha_{k''}$ and $\beta_{k''}$ by $-\alpha_{k''}$ and $\beta_{k''}$ the integral is seen to reverse in sign, unless we have

$$\langle \beta_{k'}^* \alpha_{k''} \rangle = \delta_{k'k''} \langle \beta_{k'}^* \alpha_{k'} \rangle \,. \tag{2.338}$$

The average $\langle \beta_k^* \alpha_k \rangle$, we may note from Eqns. (2.240) and (2.252), is just the mean number of quanta in the k-th mode,

$$\langle \beta_k^* \alpha_k \rangle = \mathrm{Tr}\left\{\rho a_k^\dagger a_k\right\} = \langle n_k \rangle \,. \tag{2.339}$$

We have thus shown that the general expression (2.329) for the first-order correlation function always satisfies Eq. (2.333) when the field is described by a stationary density operator. The derivation of the equations relating the energy spectrum to the time-dependent correlation function then proceeds as before.

The simplest and most universal example of an incoherent field is the type generated by superposing the outputs of stationary sources. We have shown in some detail in Sect. 2.9.8 that as the number of sources which contribute to the excitation of a single mode increases, the density operator for the mode takes on a Gaussian form in the P representation. It is not difficult to derive an analogous result for the case of sources which excite many modes at once. We shall suppose that the sources ($j = 1 \ldots N$) are essentially identical, and that their contributions to the excitation are described by a weight function $p(\{\alpha_{jk}\})$. The weight function $P(\{\alpha_k\})$ for the superposed fields is then given by the convolution theorem as

$$P(\{\alpha_k\}) = \int \prod_k \delta^{(2)}\left(\alpha_k - \sum_{j=1}^N \alpha_{jk}\right) \prod_{j=1}^N p(\{\alpha_{jk}\}) \prod_k d^2\alpha_{jk} \,. \tag{2.340}$$

---

[8] The Wiener–Khintchine theorem is usually expressed in terms of cosine transforms since it deals with a real-valued correlation function for the classical field $\mathbf{E}$, rather than a complex one for the fields $\mathbf{E}^{(\pm)}$. The complex correlation functions are considerably more convenient to use for quantum mechanical purposes, as is shown in Ref.[6].

Since the individual sources are assumed to be stationary, the function $p(\{\alpha_{jk}\})$ will only depend on the variables α_{jk} through their absolute magnitudes, $|\alpha_{jk}|$. The derivation which leads from Eq. (2.340) to a Gaussian asymptotic form for $P(\{\alpha_k\})$ is so closely parallel to that of Eqns. (2.281)–(2.288) that there is no need to write it out in detail. The argument makes use of second-order moments of the function p which may, with the same type of vector notation used previously, be written as

$$\langle \alpha_k \alpha_{k'} \rangle = \int \alpha_k \alpha_{k'} p(\{\alpha_k\}) \prod_l d^2 \alpha_l . \tag{2.341}$$

The stationary character of the function p implies that such moments vanish for $k \neq k'$. With this observation, we may retrace our earlier steps to show that the many-dimensional Fourier transform of P takes the form of a product of Gaussians, one for each mode and each similar in form to that of Eq. (2.286). It then follows immediately that the weight function P for the field as a whole is given by a product of Gaussian factors each of the form of Eq. (2.288). We thus have

$$P(\{\alpha_k\}) = \prod_k \frac{1}{\pi \langle n_k \rangle} e^{-|\alpha_k|^2 / \langle n_k \rangle} , \tag{2.342}$$

where $\langle n_k \rangle$ is the average number of photons present in the k-th mode when the fields are fully superposed. One of the striking features of this weight function is its factorized form. It is interesting to remember, therefore, that no assumption of factorizability has been made regarding the weight functions p which describe the individual sources. These sources may, indeed, be ones for which the various mode amplitudes are strongly coupled in magnitude. It is the stationary property of the sources which leads, because of the vanishing of the moments Eq. (2.341) for $k \neq k'$, to the factorized form for the weight function Eq. (2.342).

The density operator which corresponds to the Gaussian weight function Eq. (2.342) evidently describes an ideally random sort of excitation of the field modes. We may reasonably surmise that it applies, at least as a good approximation, to all of the familiar sorts of incoherent sources in laboratory use. It is clear, in particular, from the arguments of Sect. 2.9.7 that the Gaussian weight function describes thermal sources correctly. The substitution of the Planck distribution $\langle n_k \rangle = [\exp(\hbar \omega_k / \kappa T) - 1]^{-1}$ into Eq. (2.342) leads to the density operator for the entire thermal radiation field. To the extent that the Gaussian weight function Eq. (2.342) may describe radiation by a great variety of incoherent sources there will be certain deep-seated similarities in the photon fields generated by all of them. One may, for example, think of these sources all as resembling thermal ones and differing from them only in the spectral distributions of their outputs. As a way of illustrating these similarities we might imagine passing blackbody radiation through a filter which is designed to give the spectral distribution of the emerging light a particular line profile. We may choose this artificial line profile to be the same as that of some true emission line radiated, say, by a discharge tube. We then ask whether measurements carried out upon the photon field can distinguish the true emission-line source from the artificial one. If the radiation by the discharge tube is described, as we presume, by a Gaussian weight function, it is clear that the two sources will be indistinguishable from the standpoint of any photon counting experiments. They are equivalent sorts of narrow-band, quantum-mechanical noise generators.

It is a simple matter to find the correlation functions for the incoherent fields[5] described by the Gaussian weight function Eq. (2.342). If we substitute this weight function into the expansion Eq. (2.328) for the first-order correlation function we find

$$G^{(1)}_{\mu\nu}(rt,r't') = \tfrac{1}{2}\sum_k \hbar\omega u^*_{k\mu}(r)\, u_{k\nu}(r')\, \langle n_k \rangle\, e^{i\omega(t-t')}. \tag{2.343}$$

When the mode functions $u_k(r)$ are the plane waves of Eq. (2.167), and the volume of the system is sufficiently large, we may write the correlation function as the integral

$$G^{(1)}_{\mu\nu}(rt,r't') = \frac{\hbar c}{2(2\pi)^3}\int\sum_\lambda e^{(\lambda)*}_\mu e^{(\lambda)}_\nu\, \langle n_{k,\lambda}\rangle\, k\, \exp\{-i[k\cdot(r-r')-\omega(t-t')]\}\, dk, \tag{2.344}$$

in which the index λ again labels polarizations. To find the second-order correlation function defined by Eq. (2.321) we may write it likewise as an expansion in terms of mode functions. The only new moments of the weight function which we need to know are those given by $\langle |\alpha_k|^4\rangle = 2\langle |\alpha_k|^2\rangle^2 = 2\langle n_k\rangle^2$. We then find that the second-order correlation function may be expressed in terms of the first-order function as

$$G^{(2)}_{\mu_1\mu_2\mu_3\mu_4}(x_1 x_2, x_3 x_4) = G^{(1)}_{\mu_1\mu_3}(x_1,x_3)G^{(1)}_{\mu_2\mu_4}(x_2,x_4) + G^{(1)}_{\mu_1\mu_4}(x_1,x_4)G^{(1)}_{\mu_2\mu_3}(x_2,x_3). \tag{2.345}$$

It is easily shown that all of the higher order correlation functions as well reduce to sums of products of the first-order function. The n-th-order correlation function may be written as

$$G^{(n)}_{\mu_1\ldots\mu_{2n}}(x_1\ldots x_n, x_{n+1}\ldots x_{2n}) = \sum_{\mathcal{P}}\prod_{j=1}^{n} G^{(1)}_{\mu_j\nu_j}(x_j,y_j), \tag{2.346}$$

where the indices ν_j and the coordinates y_j for $j=1\ldots n$ are a permutation of the two sets $\mu_{n+1}\ldots\mu_{2n}$ and $x_{n+1}\ldots x_{2n}$, and the sum is carried out over all of the $n!$ permutations. One of the family resemblances which links all fields represented by the weight function Eq. (2.342) is that their properties may be fully described through knowledge of the first-order correlation function.

The fields which have traditionally been called coherent ones in optical terminology are easily described in terms of the first-order correlation function given by Eq. (2.344). Since the light in such fields is accurately collimated and nearly monochromatic, the mean occupation number $\langle n_{k,\lambda}\rangle$ vanishes outside a small volume of k-space. The criterion for accurate coherence is ordinarily that the dimensions of this volume be extremely small in comparison to the magnitude of k. It is easily verified, if the field is fully polarized, and the two points (r,t) and (r',t') are not too distantly separated, that the correlation function Eq. (2.344) falls approximately into the factorized form of Eq. (2.162). That is to say, fields of the type we have described approximately fulfill the condition for first-order coherence[6]. It is easily seen, however, from the structure of the higher order correlation functions that these fields can never have second or higher order coherence. In fact, if we evaluate the function $G^{(n)}$ given by Eq. (2.346) for the particular case in which all of the coordinates are set equal, $x_1 = \cdots = x_{2n} = x$, and all of the indices as well, $\mu_1 = \cdots = \mu_{2n} = \mu$, we find the result

$$G^{(n)}_{\mu\ldots\mu}(x\ldots x, x\ldots x) = n! \left[G^{(1)}_{\mu\mu}(x,x)\right]^n. \tag{2.347}$$

The presence of the coefficient $n!$ in this expression is incompatible with the factorization condition Eq. (2.323) for the correlation functions of order n greater than one. The absence of second or higher order coherence is thus a general feature of stationary fields described by the Gaussian weight function Eq. (2.342). There exists, in other words, a fundamental sense in which these fields remain incoherent no matter how monochromatic or accurately collimated they are. We need hardly add that other types of fields such as those generated by radio transmitters or masers may possess arbitrarily high orders of coherence.

During the completion of the present paper a note by Sudarshan[14] has appeared which deals with some of the problems of photon statistics that have been treated here.[9] Sudarshan has observed the existence of what we have called the P representation of the density operator and has stated its connection with the representation based on the n-quantum states. To that extent, his work agrees with ours in Sections 2.9.7 and 2.9.9. He has, however, made a number of statements which appear to attach an altogether different interpretation to the P representation. In particular, he regards its existence as demonstrating the "complete equivalence" of the classical and quantum mechanical approaches to photon statistics. He states further that there is a "one-to-one correspondence" between the weight functions P and the probability distributions for the field amplitudes of classical theory.

The relation between the P representation and classical theory has already been discussed at some length in Sections 2.9.7–2.9.9. We have shown there that the weight function $P(\alpha)$ is, in general, an intrinsically quantum-mechanical structure and not derivable from classical arguments. In the limit $\hbar \to 0$, which corresponds to large amplitudes of excitation for the modes, the weight functions $P(\alpha)$ may approach classical probability functions as asymptotic forms. Since infinitely many quantum states of the field may approach the same asymptotic form, it is clear that the correspondence between the weight functions $P(\alpha)$ and classical probability distributions is not at all one-to-one.

---

[9] In an accompanying note, L. Mandel and E. Wolf [*Phys. Rev. Lett.* **10**, 276 (1963)] warmly defend the classical approach to photon problems. Some of the possibilities and fundamental limitations of this approach should be evident from our earlier work. We may mention that the "implication" they draw from Ref.[4] and disagree with cannot be validly inferred from any reading of that paper.

2.10
Radiation by a Predetermined Charge–Current Distribution

Not many problems of quantum electrodynamics are in any sense exactly soluble. But there does exist one simple, completely soluble problem which has considerable physical meaning. That is the problem of finding the photon field radiated by an electric current distribution which is essentially classical in nature. By "classical" in this case we mean that we may represent the current by a prescribed vector function of space and time, $j(rt)$.

Such a model clearly can not represent the process of radiation by an individual atom, since the atomic current is affected by radiation recoil in essentially unpredictable ways. The model may, however, be an excellent approximation for dealing with radiation by aggregates of atoms which are large enough to show statistically predictable behavior for the total current vector. Note that in saying this we are not at all ignoring the reaction of the radiation process back upon the current. All we require is that whatever the reaction is, it be predictable at least in principle (as the radiation resistance of an antenna is, for example. It seems likely that this model, when allowance is made for statistical uncertainties in the current distribution, will accurately account for the photon fields generated by most macroscopic sources.

The interaction Hamiltonian which describes the coupling of the quantized electromagnetic field to the current distribution takes the form

$$\mathcal{H}_I = \frac{1}{c} \int j(r,t) \cdot A(r,t) \, dr . \tag{2.348}$$

The state vector of the field changes with time in the interaction representation, obeying the Schrödinger equation

$$i\hbar \frac{\partial}{\partial t} |t\rangle = \mathcal{H}_I |t\rangle . \tag{2.349}$$

Now let us, as an abbreviation, introduce the operator $B(t)$ which is defined as

$$B(t) = \frac{1}{\hbar c} \int j(r,t) \cdot A(r,t) \, dr . \tag{2.350}$$

The operator $B(t)$ is simply a linear combination of values of the vector potential, and hence obeys a commutation relation of the same general type as the vector potential. In general $[B(t), B(t')]$ will be different from zero, but it is always an ordinary number.

Now the Schrödinger equation, Eq. (2.349), can be rewritten as

$$\frac{d}{dt} |t\rangle = B(t) |t\rangle . \tag{2.351}$$

Because of the operator character of $B(t)$ the solution of this equation is *not*

$$\exp\left\{ \int_{t_0}^{t} B(t') \, dt' \right\} |t_0\rangle \tag{2.352}$$

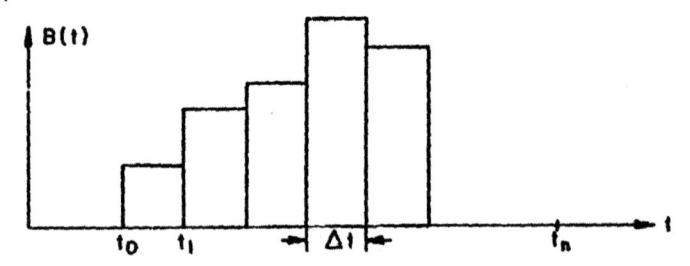

[Figure 11: Graph showing B(t) as a step function versus t, with sub-intervals marked at t_0, t_1, ..., t_n and width Δt.]

Figure 11

as it would be if $B(t)$ were an ordinary number. However, because of the simple commutation relation obeyed by the B's this expression will turn out not to be quite as wrong as we might perhaps expect.

We know that the state $|t\rangle$ at time t can be expressed by means of a unitary operator, $U(t,t_0)$, applied to the state $|t_0\rangle$ as time t_0, i.e.,

$$|t\rangle = U(t,t_0)|t_0\rangle . \tag{2.353}$$

The equations which determine $U(t,t_0)$ are evidently

$$\frac{d}{dt}U(t,t_0) = B(t)U(t,t_0) \tag{2.354}$$

and the initial condition $U(t,t_0) = 1$.

In order to solve for the operator U let us begin by dividing the time interval $(t_0 - t)$ into sub-intervals of length Δt extending between the times $t_j = t_0 + j\Delta t$, where j is an integral. We may then reach the solution of Eq. (2.354) through a simple limiting process. We assume that the operator $B(t)$ is constant in value during each of the sub-intervals of time and allow its value to change at the times t_j. A rather fanciful picture of this variation is shown in the "graph" of the operator B versus time given by Fig. 11.

Since the operator B is constant in each of the sub-intervals, we can easily integrate the differential equation (2.354) for the individual sub-intervals. If $B(t)$ takes on the value B_j in the interval from t_{j-1} to t_j then we evidently have

$$U(t_j, t_{j-1}) = e^{B_j \Delta t} . \tag{2.355}$$

Hence the transformation operator which corresponds to a succession of sub-intervals must be

$$U(t_n, t_0) = e^{B_n \Delta t} e^{B_{n-1} \Delta t} \ldots e^{B_1 \Delta t} . \tag{2.356}$$

Now we can use the familiar theorem for multiplication of exponentials, Eq. (2.201), to evaluate the product. For $n = 2$, for example, we have

$$U(t_2, t_0) = e^{B_2 \Delta t} e^{B_1 \Delta t} = \exp\left\{(B_1 + B_2)\Delta t + \tfrac{1}{2}[B_2, B_1](\Delta t)^2\right\} . \tag{2.357}$$

2.10 Radiation by a Predetermined Charge–Current Distribution

The repetition of similar multiplications clearly leads to

$$U(t_n,t_0) = \exp\left\{\sum_{j=1}^{n} B_j \Delta t + \tfrac{1}{2}\sum_{j>k}[B_j,B_k](\Delta t)^2\right\}, \qquad (2.358)$$

which is an exact solution as long as $B(t)$ has the discontinuous time variation we have assumed.

We may consider the case in which the operator $B(t)$ varies continuously with time to be the limit in which $\Delta t \to 0$, i.e., we assume $t_n = t$ remains fixed and let $n \to \infty$. In that limit Eq. (2.358) becomes the general solution

$$U(t,t_0) = \exp\left\{\int_{t_0}^{t} B(t')\,dt' + \tfrac{1}{2}\int_{t_0}^{t} dt' \int_{t_0}^{t} dt''\,[B(t'),B(t'')]\right\}. \qquad (2.359)$$

If we compare this solution with the expression (2.352), which was reached by naively ignoring the operator character of $B(t)$, we may see that the difference lies only in the addition of the term

$$\tfrac{1}{2}\int_{t_0}^{t} dt' \int_{t_0}^{t} dt''\,[B(t'),B(t'')] \qquad (2.360)$$

to the exponent. The commutator in this integral is an ordinary number and, in fact, a purely imaginary one. Hence the solution Eq. (2.359) only differs from Eq. (2.352) by a time-dependent phase factor. If we let $i\varphi(t)$ represent the integral Eq. (2.360), then we may write the transformation operator as

$$\begin{aligned} U(t,t_0) &= \exp\left\{\int_{t_0}^{t} B(t')\,dt' + i\varphi(t)\right\} \\ &= \exp\left\{\frac{i}{\hbar c}\int_{t_0}^{t} \boldsymbol{j}(\boldsymbol{r},t')\cdot\boldsymbol{A}(\boldsymbol{r},t')\,dt'\,d\boldsymbol{r} + i\varphi(t)\right\}. \end{aligned} \qquad (2.361)$$

Although the phase function $\varphi(t)$ is not altogether lacking in physical interest, (there is information contained in it, for example, on the interaction energy of the current and field) it does not have any influence on the calculation of density operators for the field, i.e., if the density operator has the initial value $\rho(t_0)$, then its value at time t is

$$\rho(t) = U(t,t_0)\rho(t_0)U^{\dagger}(t,t_0), \qquad (2.362)$$

and we see immediately that the phase factor cancels.

If in particular the initial state is the vacuum state

$$|t_0\rangle = |\text{vac}\rangle, \qquad (2.363)$$

then at time t we have

$$e^{-i\varphi(t)}|t\rangle = \exp\left\{\frac{i}{\hbar c}\int_{t_0}^{t} \boldsymbol{j}(\boldsymbol{r},t')\cdot\boldsymbol{A}(\boldsymbol{r},t')\,d\boldsymbol{r}\,dt'\right\}|\text{vac}\rangle. \qquad (2.364)$$

Now if we introduce the expansion of the operator A in normal modes, Eq. (2.168), we see that the unitary operator which is applied to the vacuum state on the right side of Eq. (2.364) is simply a product of displacement operators which take the form

$$D_k(\alpha_k) = \exp\left[\alpha_k a_k^\dagger - \alpha_k^* a_k\right] . \tag{2.365}$$

More precisely, if we define the set of time-dependent amplitudes

$$\alpha_k(t) = \frac{i}{(2\hbar\omega_k)^{1/2}} \int_{t_0}^{t} \boldsymbol{j}(\boldsymbol{r},t') \cdot \boldsymbol{u}_k^*(\boldsymbol{r},t') e^{-i\omega_k t'} \, d\boldsymbol{r} \, dt' , \tag{2.366}$$

then Eq. (2.364) may be rewritten as

$$e^{-i\varphi(t)} |t\rangle = \prod_k D_k(\alpha_k(t)) |\text{vac}\rangle . \tag{2.367}$$

It is clear from this result that a prescribed current distribution, radiating into the vacuum, always brings the field to a coherent state

$$e^{-i\varphi(t)} |t\rangle = |\{\alpha_k(t)\}\rangle . \tag{2.368}$$

More generally, if the field is initially in an arbitrary coherent state its state remains coherent under the influence of the current distribution. The solution to the radiation problem we have found takes accurate account of the quantum mechanical properties of the field. It is related, however, in a simple way to the solution of the corresponding classical problem. The amplitudes $\alpha_k(t)$ are simply related to the time-dependent mode amplitudes for the classically radiated field through Eq. (2.153).

The density operator at time t which corresponds to the coherent state Eq. (2.368) is simply

$$\rho(t) = |\{\alpha_k(t)\}\rangle \langle\{\alpha_k(t)\}| , \tag{2.369}$$

which may be written in the P representation as

$$\rho(t) = \int P(\{\beta_k\}) |\{\beta_k\}\rangle \langle\{\beta_k\}| \prod_k d^2\beta_k , \tag{2.370}$$

by making use of the P function

$$P(\{\beta_k\}) = \prod_k \delta^{(2)}(\beta_k - \alpha_k(t)) . \tag{2.371}$$

The calculations we have carried out have dealt with a predetermined current distribution, i.e., one which behaves in a way which is *in principle* predictable. But in practice, of course, we may lack the information necessary to make such predictions and may have to resort to a statistical description of the behavior of the current. In that case, since we do not know the current $\boldsymbol{j}(\boldsymbol{r},t)$ at any given time, it becomes impossible to make an exact specification of the set of amplitudes $\alpha_k(t)$ through Eq. (2.366). The best we can do is to state that the coefficients α_k have a certain probability distribution $p(\{\alpha_k\},t)$ at time t whose dispersion corresponds to whatever randomness is present. Then it is clear that the density operator can be written in the form

$$\rho = \int p(\{\alpha_k\},t) |\{\alpha_k\}\rangle \langle\{\alpha_k\}| \prod_k d^2\alpha_k , \qquad (2.372)$$

which is a fairly general form for the P representation, but one in which the function P is obviously always positive.

Density operators having the general form of Eq. (2.372) with $p(\{\alpha_k\},t)$ positive may arise from a variety of sources (e.g., thermal radiators, discharge tubes, etc.). Hence it is interesting to note that our arguments indicate that we can always construct for these cases some sort of random classical current distribution which will lead to the same field, i.e., the same density operator.

2.11
Phase-Space Distributions for the Field

In classical mechanics we can specify the state of a system by giving the instantaneous values of all coordinates and momenta. The evolution of the system then follows uniquely from the equations of motion. It can be visualized by considering the n coordinates and n momenta of the system as the coordinates of a point in a $2n$-dimensional space, the phase space. The point which represents a system in this space moves along a uniquely determined trajectory. This picture is easily adapted to the uses of classical statistical mechanics. There, since we are characteristically uncertain of the initial coordinates and momenta of the system, we can speak only of probability distributions $P_{cl}(p'_1 \ldots p'_n, q'_1 \ldots q'_n)$ for these variables. Instead of following the motion of a single point through the phase space, we must follow the motion of a whole "cloud" of them representing an ensemble of similarly prepared systems. The expectation value of any function of the p'_i and q'_i can then be calculated by means of an integral, involving the probability P_{cl} as a weight function.

There has been, since the earliest days of quantum mechanics, a prevailing temptation to use the same sort of phase space picture for the description of quantum mechanical uncertainties. We shall not attempt to discuss these representations here in much generality since our interests are confined to the electromagnetic field. From a dynamical standpoint, the oscillations of each mode of the field are those of a harmonic oscillator. It will be quite sufficient, for the present discussion, to confine our attention to a single mode. In that case, the classical phase space has only two dimensions, corresponding to the variables p' and q'. The phase point for a mode with energy E moves classically along the ellipse $p'^2 + \omega^2 q'^2 = 2E$. (The mass parameter is set equal to unity.)

A coherent state of the mode will exist corresponding to any complex eigenvalue we specify for the operator

$$a = (2\hbar\omega)^{-1/2}(\omega q + ip). \tag{2.373}$$

The amplitude α corresponding to the state $|\alpha\rangle$ may be written as

$$\alpha = (2\hbar\omega)^{-1/2}(\omega q' + ip'), \tag{2.374}$$

where q' and p' are real numbers. Now we have shown in Sect. 2.9.3 that the state $|\alpha\rangle$ may be described by a wavepacket which has minimum uncertainty and the mean coordinate q' and the mean momentum p'.

Furthermore if we use the Schrödinger picture and follow the motion of the state with time, we know that the state remains coherent at all times, and that its time-dependent amplitude is simply $\alpha e^{-i\omega t}$. The motion of the amplitude vector in the complex α plane takes place on the circle $|\alpha|$ = constant, which simply represents an ellipse of the type noted earlier in the p', q' plane.

It is clear that the complex α plane is simply a species of two-dimensional phase space. One therefore inevitably feels a great temptation to think of the coherent state wavepackets in terms of probability "clouds" whose centers move on circular paths. Such an image, however, is an intrinsically classical one. In quantum mechanics the observables p and q are not simultaneously measurable (with more than limited accuracy), and therefore a certain lack of meaning, or at best an arbitrariness of meaning characterizes any attempt to speak of a joint probability distribution for the variables p' and q'. We can, of course, speak of the distribution of either variable in precisely defined terms, but these are *alternative* descriptions of the oscillator rather than a way of dealing with p' and q' simultaneously.

The P representation of the density operator, which we introduced in Sect. 2.9.7, can often be regarded as defining something at least comparable to a phase space distribution. The complex α plane on which the P function is defined, is indeed a species of phase space. Furthermore as we have noted in the paper, the P function has a number of properties in common with probability distributions. However, as we have also seen, the function may take on negative values, and behave in singular ways which are altogether unlike those of a probability density. There is nothing inconsistent about such strange behavior because the function is not accessible to measurement as a joint probability distribution.

From the standpoint of similarity to classical theory, the function $P(\alpha)$ is simply one of a class of functions which possess, by definition, some of the properties of a phase space distribution and then inevitably lack others. We will discuss some other examples of such functions, which are perhaps best called quasiprobability densities, later in this Section, and show their relation to the P representation. First, however, let us turn to the question of how generally applicable the P representation is.

2.11.1
The P Representation and the Moment Problem

Although it is clear, from the examples given in Sect. 2.9, that the P representation of the density operator is capable of representing a fairly broad variety of fields, no effort was made there to characterize that class of fields. Sudarshan has stated in a brief note[19], however, that a "diagonal" representation of the density operator in terms of the coherent states may be used to represent an arbitrary field. He has given an explicit construction for the weight function of such a representation as an infinite sum of arbitrarily high-order derivatives of a delta function. He has said that, as a consequence of this construction, "the description of statistical states of a quantum mechanical system... is completely equivalent to the description in terms of classical probability distributions".

The way in which Sudarshan's construction for the function $P(\alpha)$ may be reached is as follows: we consider the matrix elements of the density operator in the n-quantum state representation as known and note that, according to Eq. (2.269), these matrix elements are the complex moments,

$$\langle n|\rho|m\rangle = (n!\,m!)^{-1/2} \int P(\alpha)\,(\alpha^*)^m \alpha^n \, d^2\alpha$$

of the weight function $P(\alpha)$. We then consider this sequence of equations for all n and m to define a species of two-dimensional moment problem, i.e., we seek a function $P(\alpha)$ which has the correct matrix of moments. The general problem when stated thus becomes a notoriously difficult one, and one which need not, for arbitrary matrix elements $\langle n|\rho|m\rangle$, have a reasonable solution of any sort. Sudarshan's solution corresponds to taking advantage of some remarkable properties of the delta function and its derivatives which are perhaps most easily illustrated in a one-dimensional context.

Let us suppose that we are given the problem of finding a function $f(x)$ on the interval $-\infty < x < \infty$ which has a specified set of moments M_n, i.e., we have

$$\int_{-\infty}^{\infty} f(x) x^n \, dx = M_n, \qquad n = 0, 1, 2\ldots \tag{2.375}$$

If we write the j-th derivative of the delta function as

$$\delta^{(j)}(x) = \frac{d^j}{dx^j} \delta(x), \tag{2.376}$$

then we observe that its moments are given by

$$\int_{-\infty}^{\infty} x^k \delta^{(j)}(x) \, dx = (-1)^j j!\, \delta_{jk}. \tag{2.377}$$

In other words, each derivative of the delta function has one and only one non-vanishing moment. It would seem then that we can construct a "solution" of the general moment problem simply by writing

$$f(x) = \sum_{n=0}^{\infty} \frac{(-1)^n}{n!} M_n \delta^{(n)}(x). \tag{2.378}$$

The test of such a "solution" is ultimately whether or not it means anything.

Mathematicians have long noted that the delta function and its derivatives are not, strictly speaking, functions at all. More recently they have provided us with the theory of distributions (or generalized functions) as a means of dealing with these structures in more meaningful and rigorous terms.

Equations (2.376) and (2.377) assume a well-defined meaning in terms of distribution theory, but the theory shows that there is in general no useful meaning which can be attached to an infinite sum such as Eq. (2.378).

The "solution" exhibited by Sudarshan for the two-dimensional moment problem takes the explicit form

$$P_s(\alpha) = \sum_{n=0}^{\infty} \sum_{m=0}^{\infty} \frac{(n!\,m!)^{1/2}}{(n+m)!} \langle n|\rho|m\rangle \frac{1}{2\pi|\alpha|} e^{|\alpha|^2 - i(n-m)\theta} \left\{ \left(-\frac{\partial}{\partial |\alpha|}\right)^{n+m} \delta(|\alpha|) \right\}, \tag{2.379}$$

where we have written $\alpha = |\alpha|e^{i\theta}$. Recently Holliday and Sage[20] have shown, by considering a simple example explicitly, that this expression cannot be construed as a generalized function of any sort. The example was that of the thermal density operator, and for it they showed that when the series Eq. (2.379) is multiplied by an extremely well-behaved test function (which vanishes outside a circle of finite radius in the α plane) and the product is then integrated, the integral diverges. More recently, Cahill has shown that whenever there is no upper bound to the number of quanta present, the series Eq. (2.379) will fail to be interpretable as a distribution (or a generalized function).

While these results indicate that Sudarshan's proposed representation is not, in general, meaningful, they leave open the larger question of the generality of the P representation. They allow the possibility, in other words, that there might exist other constructions of the P representation which are meaningful for all states of the field. Recently, however, D. Kastler and the author[22] have demonstrated that the P representation lacks the generality necessary to represent all states. They have shown in particular that there exist quantum states of the field (or which it is not possible to find functions $P(\alpha)$ which are distributions. That means that all general results derived by using the P representation must be qualified by the assumption that the representation exists.

2.11.2
A Positive-Definite "Phase Space Density"

We will now consider some other examples of quasiprobability functions, with different types of behavior and different degrees of usefulness. The first of these is the diagonal element, $\langle\alpha|\rho|\alpha\rangle$, of the density operator. It is clear that $\langle\alpha|\rho|\alpha\rangle$ is nonnegative and that it is a well-defined function of α for all ρ. It is therefore a good deal closer to being a phase space density in its behavior than is $P(\alpha)$.

2.11 Phase-Space Distributions for the Field

From the general expression for $R(\alpha^*, \beta)$ given by Eq. (2.248),

$$R(\alpha^*, \beta) = \pi \langle \alpha | \rho | \beta \rangle \exp\left\{\tfrac{1}{2}(|\alpha|^2 + |\beta|^2)\right\},$$

we have

$$\langle \alpha | \rho | \alpha \rangle = R(\alpha^*, \alpha) e^{-|\alpha|^2}. \tag{2.380}$$

Hence, according to Eq. (2.253), the normalization condition on $\langle \alpha | \rho | \alpha \rangle$ is

$$\frac{1}{\pi}\int \langle \alpha | \rho | \alpha \rangle \, d^2\alpha = \frac{1}{\pi}\int R(\alpha^*, \alpha) e^{-|\alpha|^2} d^2\alpha = 1. \tag{2.381}$$

If the P representation exists for the density operator ρ and has a weight function $P(\beta)$, we clearly have

$$\langle \alpha | \rho | \alpha \rangle = \int P(\beta) |\langle \alpha | \beta \rangle|^2 \, d^2\beta \tag{2.382}$$

$$= \int P(\beta) e^{-|\alpha - \beta|^2} d^2\beta.$$

The function we are considering is simply a Gaussian convolution of the P function.

We can use the function $\langle \alpha | \rho | \alpha \rangle$ to calculate averages of products of operators which are in antinormal order in much the same way as products in normal order are averaged by means of the P representation. Let us consider, for example, the average

$$\mathrm{Tr}\left\{\rho J(a) K(a^\dagger)\right\},$$

where J and K can be any functions of the annihilation and creation operators, respectively.

We can write this average as

$$\mathrm{Tr}\left\{K(a^\dagger)\rho J(a)\right\} = \frac{1}{\pi}\int \mathrm{Tr}\left\{|\alpha\rangle\langle\alpha| K(a^\dagger)\rho J(a)\right\} d^2\alpha$$

$$= \frac{1}{\pi}\int d^2\alpha \left\langle \alpha \left| K(a^\dagger)\rho J(a) \right| \alpha \right\rangle = \frac{1}{\pi}\int \langle \alpha | \rho | \alpha \rangle \, K(\alpha^*) J(\alpha) \, d^2\alpha. \tag{2.383}$$

Unfortunately we are not too often interested in evaluating the expectation values of antinormally ordered products of field operators. When the full set of modes of the field is considered such expectation values tend to contain divergent contributions from the vacuum fluctuations.

The function $\langle \alpha | \rho | \alpha \rangle$ takes an interesting form for the n-th excited state of the oscillator. For these states we have

$$\rho_n = |n\rangle \langle n| = \frac{1}{n!}(a^\dagger)^n |0\rangle \langle 0| a^n, \tag{2.384}$$

and therefore the result

$$\langle \alpha | \rho | \alpha \rangle = \frac{1}{n!}|\langle \alpha | n \rangle|^2 = \frac{|\alpha|^{2n}}{n!} e^{-|\alpha|^2}. \tag{2.385}$$

This is an extremely well-behaved function, especially when we compare it with the analogous expression in the P representation, which contains the $2n$-th derivative of a delta function. The function $x^n e^{-x}$ has a maximum at $x = n$ and is quite sharply peaked there for large values of n. If we want to express the result Eq. (2.385) as a distribution in phase space we can substitute the expression Eq. (2.374) for α and write

$$\langle \alpha | \rho | \alpha \rangle = \frac{1}{n!} \frac{(p'^2 + \omega^2 q'^2)^n}{(2\hbar\omega)^n} \exp\left\{ -\frac{p'^2 + \omega^2 q'^2}{2\hbar\omega} \right\}. \tag{2.386}$$

This function evidently has its maximum value on the ellipse $(1/2)(p'^2 + \omega^2 q'^2) = n\hbar\omega$, that is to say on the classical orbit in phase space. It drops to zero on either side of the classical orbit while remaining positive everywhere.

Another example for which we can easily illustrate this "phase space density" is that of the Gaussian density operator. For that case we have

$$\begin{aligned} R(\alpha^*, \beta) &= \int P(\gamma) \exp\{\alpha^* \gamma + \beta \gamma^* - |\gamma|^2\} \, d^2\gamma \\ &= \frac{1}{\pi \langle n \rangle} \int \exp\left\{ -\frac{|\gamma|^2}{\langle n \rangle} + \alpha^* \gamma + \beta \gamma^* - |\gamma|^2 \right\} d^2\gamma \tag{2.387} \\ &= \frac{1}{\pi \langle n \rangle} \int \exp\left\{ -|\gamma|^2 \frac{1 + \langle n \rangle}{\langle n \rangle} + \alpha^* \gamma + \beta \gamma^* \right\} d^2\gamma. \end{aligned}$$

We can now make the substitution $\xi = \gamma \left\{ \frac{1 + \langle n \rangle}{\langle n \rangle} \right\}^{1/2}$, which reduces the integral to the standard form

$$\begin{aligned} R(\alpha^*, \beta) &= \frac{1}{\pi(1 + \langle n \rangle)} \int \exp\left\{ -|\xi|^2 + \left[\frac{\langle n \rangle}{1 + \langle n \rangle} \right]^{1/2} (\alpha^* \xi + \beta \xi^*) \right\} d^2\xi \\ &= \frac{1}{\pi(1 + \langle n \rangle)} \int \exp\left\{ \frac{\langle n \rangle}{1 + \langle n \rangle} \alpha^* \beta \right\}. \tag{2.388} \end{aligned}$$

Hence we find

$$\begin{aligned} \langle \alpha | \rho | \alpha \rangle &= R(\alpha^*, \alpha) e^{-|\alpha|^2} \\ &= \frac{1}{\pi(1 + \langle n \rangle)} \exp\left\{ -\frac{|\alpha|^2}{1 + \langle n \rangle} \right\}. \tag{2.389} \end{aligned}$$

If $\langle n \rangle$ goes to zero this expression becomes the Gaussian function $(1/\pi) \exp(-|\alpha|^2)$. In the same case the weight function $P(\alpha)$ would be a delta function at the origin. If $\langle n \rangle$ goes to infinity we have

$$\langle \alpha | \rho | \alpha \rangle \cong \frac{1}{\pi \langle n \rangle} e^{-|\alpha|^2 / \langle n \rangle} \cong P(\alpha). \tag{2.390}$$

In this limit $\langle \alpha | \rho | \alpha \rangle$ becomes equal to the P-distribution. That is so because the limit of large $\langle n \rangle$ is just the classical limit. There $P(\alpha)$ does indeed become interpretable as a classical phase space density, and the distinction between normally and antinormally ordered operators also vanishes, as a consequence of the correspondence principle.

2.11.3
Wigner's "Phase Space Density"

The Wigner distribution can be considered as the grandfather of all our quasiprobability functions. It exists and is well-behaved for all quantum states but seems to take on negative values without hesitation. We shall follow the approach used by Moyal[23] to define the Wigner distribution.

We begin by discussing a species of characteristic function which is defined as

$$X(\mu,\nu) = \left\langle e^{i(\mu p + \nu q)} \right\rangle, \qquad (2.391)$$

where p and q are operators. By using our theorem for the decomposition of exponentials, Eq. (2.201), we may write this expression as

$$X(\mu,\nu) = \mathrm{Tr}\left\{ \rho e^{i\mu p/2} e^{i\nu q} e^{i\mu p/2} \right\}. \qquad (2.392)$$

If we restrict consideration to a pure state, use the coordinate representation, and recall the interpretation of exponential functions of the momentum as coordinate displacement operators, we may rewrite Eq. (2.392) as

$$X(\mu,\nu) = \int \psi^*\left(q'' - \frac{\mu\hbar}{2}\right) e^{i\nu q''} \psi\left(q'' + \frac{\mu\hbar}{2}\right) dq'', \qquad (2.393)$$

where $\psi(q')$ is the wave function of the pure state. The Wigner function is then the Fourier transform of this characteristic function

$$\begin{aligned}
W(p',q') &\equiv \frac{1}{(2\pi)^2} \int \exp\{-i(\mu p' + \nu q')\} X(\mu,\nu) \, d\mu \, d\nu \\
&= \frac{1}{(2\pi)^2} \int \exp\{-i(\mu p' + \nu q')\} \int \psi^*\left(q'' - \frac{\mu\hbar}{2}\right) e^{i\nu q''} \\
&\qquad \times \psi\left(q'' + \frac{\mu\hbar}{2}\right) dq'' \, d\mu \, d\nu \\
&= \frac{1}{2\pi} \int e^{-i\mu p'} \int \psi^*\left(q'' - \frac{\mu\hbar}{2}\right) \delta(q' - q'') \psi\left(q'' + \frac{\mu\hbar}{2}\right) dq'' \, d\mu \\
&= \frac{1}{2\pi} \int \psi^*\left(q' - \frac{\mu\hbar}{2}\right) e^{-i\mu p'} \psi\left(q' + \frac{\mu\hbar}{2}\right) d\mu. \qquad (2.394)
\end{aligned}$$

If we substitute $y = -\mu\hbar$ in the latter expression we derive the form of the distribution originally stated by Wigner,

$$W(p',q') = \frac{1}{2\pi\hbar} \int \psi^*\left(q' + \tfrac{1}{2}y\right) e^{ip'y/\hbar} \psi^*\left(q' - \tfrac{1}{2}y\right) dy. \qquad (2.395)$$

It is obvious that whenever we have a wave function we can derive a Wigner distribution from it. Thus the distribution always exists, but it is not necessarily positive. When we have a mixture of states we must of course take a suitably weighted average of Eq. (2.395) over all the states which occur. The normalization condition for $W(q',p')$ is

$$\int W(q',p')\,dp'\,dq' = \int \delta(\mu)\,\delta(\nu)\,X(\mu,\nu)\,d\mu\,d\nu$$
$$= X(0,0)$$
$$= 1. \tag{2.396}$$

To compare the Wigner distribution with the others we have discussed, it is useful to express it in terms of the creation and annihilation operators $a^\dagger$ and a. Then if we define a complex Fourier transform variable

$$\lambda = -\mu\left\{\frac{\hbar\omega}{2}\right\}^{1/2} + i\nu\left\{\frac{\hbar}{2\omega}\right\}^{1/2}, \tag{2.397}$$

we may write the operator which occurs in the exponent of the characteristic function as

$$-i(\mu p + \nu q) = \lambda a^\dagger - \lambda^* a, \tag{2.398}$$

and the characteristic function itself becomes

$$X(\mu,\nu) = \left\langle e^{\lambda a^\dagger - \lambda^* a}\right\rangle$$
$$= \mathrm{Tr}\left\{\rho e^{\lambda a^\dagger} e^{-\lambda^* a}\right\} e^{-|\lambda|^2/2}$$
$$= \mathrm{Tr}\left\{\rho e^{-\lambda^* a} e^{\lambda a^\dagger}\right\} e^{|\lambda|^2/2}. \tag{2.399}$$

We can now use the normally ordered form to express the Wigner function in terms of the P representation. If we assume that the density operator possesses a P representation, the characteristic function is given by

$$X(\mu,\nu) = \int P(\beta)\left\langle \beta\left|e^{\lambda a^\dagger} e^{-\lambda^* a}\right|\beta\right\rangle e^{-|\lambda|^2/2}\,d^2\beta$$
$$= \int P(\beta)\exp\left\{\lambda\beta^* - \lambda^*\beta - \tfrac{1}{2}|\lambda|^2\right\}\,d^2\beta. \tag{2.400}$$

In calculating the Fourier transform of X, i.e., the Wigner function, it is convenient to use a linear combination of α and α^* in the exponent rather than a combination of the classical variables q' and p'. We therefore write

$$i(\mu p' + \nu q') = \lambda \alpha^* - \lambda^* \alpha \tag{2.401}$$

and

$$d\mu\,d\nu = \frac{2}{\hbar}d^2\lambda. \tag{2.402}$$

2.11 Phase-Space Distributions for the Field

Then the Fourier transform becomes

$$W(q',p') = \frac{1}{(2\pi)^2} \int \text{Tr}\left\{\rho e^{\lambda(a^\dagger-\alpha^*)} e^{-\lambda^*(a-\alpha)}\right\} e^{-|\lambda|^2/2} \frac{2}{\hbar} d^2\lambda$$

$$= \frac{1}{2\pi\hbar} \int P(\beta) \exp\left\{\lambda(\beta^*-\alpha^*) - \lambda^*(\beta-\alpha) - \tfrac{1}{2}|\lambda|^2\right\} d^2\lambda\, d^2\beta. \quad (2.403)$$

We can reduce this integral to a standard form by the substitution $\xi = \lambda/\sqrt{2}$ which leads to

$$W(q',p') = \frac{1}{\pi\hbar} \int P(\beta) \exp\left\{\sqrt{2}\xi(\beta^*-\alpha^*) - \sqrt{2}\xi^*(\beta-\alpha) - |\xi|^2\right\} d^2\xi\, d^2\beta$$

$$= \frac{1}{\pi\hbar} \int P(\beta) \exp\left\{-2|\beta-\alpha|^2\right\} d^2\beta. \quad (2.404)$$

It is sometimes convenient to think of the Wigner function more directly as a function of the complex variable α, and to change its normalization accordingly. We therefore recall that

$$d^2\alpha = \frac{1}{\{2\hbar\omega\}^{1/2}} \left\{\frac{\omega}{2\hbar}\right\}^{1/2} dp'\, dq' = \frac{dp'\, dq'}{2\hbar} \quad (2.405)$$

and define the function

$$W(\alpha) = 2\hbar W(p',q') \quad (2.406)$$

so that

$$\int W(\alpha)\, d^2\alpha = 1. \quad (2.407)$$

The Wigner function of complex argument is then given by

$$W(\alpha) = \frac{2}{\pi} \int P(\beta) e^{-2|\beta-\alpha|^2} d^2\beta. \quad (2.408)$$

When we compare this expression with the one derived in the preceding section,

$$\langle \alpha|\rho|\alpha\rangle = \int P(\beta) e^{-|\beta-\alpha|^2} d^2\beta, \quad (2.409)$$

we see that both of these expressions are simply Gaussian convolutions of the P distribution (when the latter exists). The quality which the Wigner distribution shares with the P-distribution, of becoming negative in places, would seem to be due to the fact that the averaging process expressed by Eq. (2.408) takes place over a radius which is $(\sqrt{2})^{-1}$ times smaller than that expressed by Eq. (2.409).

As an example, let us evaluate the Wigner distribution for a field described by a Gaussian density operator. For this case we have, according to Eq. (2.408),

$$W(\alpha) = \frac{2}{\pi^2 \langle n\rangle} \int \exp\left\{-\frac{|\beta|^2}{\langle n\rangle} - 2|\beta-\alpha|^2\right\} d^2\beta$$

$$= \frac{2e^{-2|\alpha|^2}}{\pi^2 \langle n\rangle} \int \exp\left\{-|\beta|^2\left(2 + \frac{1}{\langle n\rangle}\right) + 2(\beta^*\alpha + \alpha^*\beta)\right\} d^2\beta. \quad (2.410)$$

We now use the substitution $\gamma = \left\{ \frac{1+2\langle n \rangle}{\langle n \rangle} \right\}^{1/2} \beta$ to reduce the integral to the standard form

$$W(\alpha) = \frac{2 e^{-2|\alpha|^2}}{\pi^2 (2\langle n \rangle + 1)} \int \exp\left\{ -|\gamma|^2 + 2 \left[\frac{\langle n \rangle}{2\langle n \rangle + 1} \right]^{1/2} (\gamma^* \alpha + \alpha^* \gamma) \right\} d^2\gamma$$

$$= \frac{2}{\pi^2 (2\langle n \rangle + 1)} \int \exp\left\{ -\left| \gamma - 2 \left[\frac{\langle n \rangle}{2\langle n \rangle + 1} \right]^{1/2} \alpha \right|^2 \right\} d^2\gamma$$

$$\times \exp\left\{ \left[\frac{4\langle n \rangle}{2\langle n \rangle + 1} - 2 \right] |\alpha|^2 \right\} . \quad (2.411)$$

The latter integral leads immediately to the result

$$W(\alpha) = \frac{2}{\pi (2\langle n \rangle + 1)} \exp\left\{ -\frac{2}{2\langle n \rangle + 1} |\alpha|^2 \right\} . \quad (2.412)$$

Thus, the Wigner distribution also has the Gaussian shape. We consider again the two limiting cases $\langle n \rangle = 0$, for which

$$W(\alpha) = \frac{2}{\pi} e^{-2|\alpha|^2} , \quad (2.413)$$

and $\langle n \rangle \to \infty$, for which

$$W(\alpha) \cong \frac{1}{\pi \langle n \rangle} e^{-|\alpha|^2/\langle n \rangle} = P(\alpha) . \quad (2.414)$$

The latter result is the one we anticipate for the correspondence limit.

The simple Gaussian form given by Eq. (2.412) may be used to derive the complete set of Wigner distributions for the n-quantum states. This is possible because the function Eq. (2.412) may be regarded as a generating function for the Wigner distributions. Let us consider, for a moment, the general case of a density operator which may be written in the form

$$\rho = (1-x) \sum_{n=0}^{\infty} x^n |n\rangle \langle n| , \quad (2.415)$$

where x is an arbitrary parameter. If we let $W_n(\alpha)$ be the Wigner function for the n-th quantum state, then, as a consequence of the linearity of W in ρ, we must have

$$W(\alpha) = (1-x) \sum_{n=0}^{\infty} x^n W_n(\alpha) . \quad (2.416)$$

Now if we make the identification $x = \langle n \rangle / (1 + \langle n \rangle)$, it becomes clear from Eq. (2.290) that ρ given by Eq. (2.415) is simply the Gaussian density operator. We

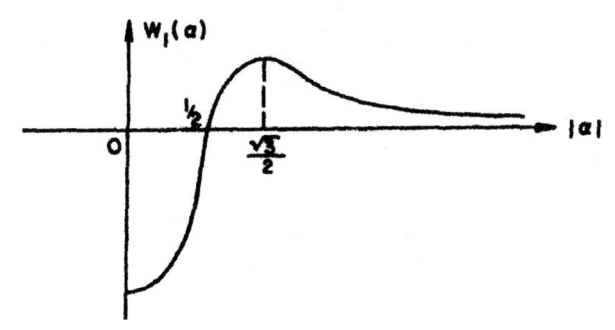

Figure 12

can therefore write Eq. (2.412) alternatively in terms of the variable x as

$$W(\alpha) = \frac{2(1-x)}{\pi(1+x)} \exp\left\{-2\left(\frac{1-x}{1+x}\right)|\alpha|^2\right\}$$
$$= \frac{2(1-x)}{\pi(1+x)} \exp\left\{\frac{x}{1+x}4|\alpha|^2\right\} \exp\{-2|\alpha|^2\}. \qquad (2.417)$$

This rather complicated exponential is just the generating function for the Laguerre polynomials L_n. In more familiar notation the generating function reads as

$$\frac{\exp\left\{-\frac{\rho u}{1-u}\right\}}{1-u} = \sum_{n=0}^{\infty} L_n(\rho) \frac{u^n}{n!}. \qquad (2.418)$$

Hence Eq. (2.417) yields the expansion

$$W(\alpha) = (1-x)\frac{2}{\pi} \sum_{n=0}^{\infty} x^n \frac{(-1)^n}{n!} L_n(4|\alpha|^2) e^{-2|\alpha|^2}. \qquad (2.419)$$

The Wigner function for the n-th excited state of the oscillator may thus be identified as

$$W_n(\alpha) = \frac{2}{\pi} \frac{(-1)^n}{n!} L_n(4|\alpha|^2) e^{-2|\alpha|^2}. \qquad (2.420)$$

These functions have quite a wiggly behavior in the complex phase plane. The n-th function has nodes on n concentric circles.

For the first two states we have, more explicitly,

$$W_0(\alpha) = \frac{2}{\pi} e^{-2|\alpha|^2} = \frac{2}{\pi} \exp\left\{-\frac{p'^2 + \omega^2 q'^2}{2\hbar\omega}\right\}, \qquad (2.421)$$

$$W_1(\alpha) = \frac{2}{\pi}(4|\alpha|^2 - 1) e^{-2|\alpha|^2}. \qquad (2.422)$$

The function $W_1(\alpha)$ is sketched in Fig. 12. Its maximum lies at the radius $\alpha = \sqrt{3}/2$.

Each of the functions we have considered (the P function, the function $\langle\alpha|\rho|\alpha\rangle$, and the Wigner distribution) has its particular advantages. It should, however, be clear from the preceding discussions that we can construct numerous other such functions, each with virtues of its own. An element of arbitrariness underlies all such discussions of phase space distributions.

Note added in proof [of the original publication]: In a recent preprint, Klauder, McKenna, and Currie confirm the conclusion that no useful weight function P need exist for arbitrary density operators. To minimize this difficulty they express matrix elements of the density operator through a limiting procedure involving an infinite sequence of operators expressed as P representations. This procedure, however, does not preserve the most useful property of the P representation, the reduction of statistical averages to simple integrals over the complex α plane.

2.12
Correlation Functions and Quasiprobability Distributions

In this Section and in the ones which follow we shall begin to discuss applications of our formalism in somewhat more concrete terms. As a first step in that direction it will be useful to amplify several of the points which are stated rather briefly in Sect. 2.9.10.

Let us suppose that the electromagnetic field is in a pure coherent state which we denote by $|\{\alpha_k\}\rangle$. Then $|\{\alpha_k\}\rangle$ is an eigenstate of the operator $\boldsymbol{E}^{(+)}$,

$$\boldsymbol{E}^{(+)}(\boldsymbol{r}t)|\{\alpha_k\}\rangle = \mathcal{E}(\boldsymbol{r}t\{\alpha\})|\{\alpha_k\}\rangle \ . \tag{2.423}$$

and the corresponding eigenvalue function $\mathcal{E}$ is a linear form in the variables $\{\alpha_k\}$, i.e., we have

$$\mathcal{E}(\boldsymbol{r}t\{\alpha_k\}) = \mathrm{i}\sum_k \left(\frac{\hbar\omega}{2}\right)^{1/2} \boldsymbol{u}_k(\boldsymbol{r})\,\mathrm{e}^{\mathrm{i}\omega_k t}\alpha_k \ . \tag{2.424}$$

The corresponding field is fully coherent since the correlation functions for all orders n fall into the factorized form

$$G^{(n)}_{\mu_1\ldots\mu_{2n}}(x_1\ldots x_{2n}) = \prod_{j=1}^{n} \mathcal{E}^*_{\mu_j}(x_j\{\alpha_k\}) \prod_{j=n+1}^{2n} \mathcal{E}_{\mu_j}(x_j\{\alpha_k\}) \ . \tag{2.425}$$

We have already noted that the term "coherence" is used frequently in the discussion of quantum mechanical problems of all sorts. Since the term is usually meant to imply that interference phenomena can take place, many of its uses are to be found in discussions of pure quantum mechanical states. Pure states, however, by no means exhaust the possibilities of securing interference. For most quantum mechanical systems there exist certain statistical mixtures of states which preserve essentially the same interference phenomena as are found for pure states. It is easy to exhibit these mixtures for the case of the electromagnetic field and to show that they may correspond to fields which are fully coherent in the sense of Eq. (2.425).

Instead of considering the field which corresponds to the set of amplitudes $\{\alpha_k\}$, let us consider the field corresponding to a set $\{\alpha'_k\}$ which we obtain by multiplying each of the coefficients α_k by a phase factor, $e^{i\phi}$, which ia the same for all modes. If we have

$$\alpha'_k = e^{i\phi} \alpha_k, \qquad (2.426)$$

then, since the eigenvalue function, $\mathcal{E}$, is linear, we must have

$$\mathcal{E}_\mu(\mathbf{r}t\{\alpha'_k\}) = e^{i\phi} \mathcal{E}_\mu(\mathbf{r}t\{\alpha_k\}). \qquad (2.427)$$

Because the phase factors cancel when we construct the correlation functions, it is clear that the altered state of the field leads to the same set of correlation functions (2.425) as the original state. This invariance property, which is implicit in our definition of the correlation functions, means that we secure the same correlation functions not only for pure states corresponding to different values of the phase ϕ, but for arbitrary mixtures of such states as well.

Let us suppose that $\mathcal{L}(\phi)$ is a function which satisfies the normalization condition

$$\int_0^{2\pi} \mathcal{L}(\phi)\,d\phi = 1. \qquad (2.428)$$

Then we may construct a density operator

$$\rho = \int_0^{2\pi} \mathcal{L}(\phi) \left|\{\alpha_k e^{i\phi}\}\right\rangle \left\langle\{\alpha_k e^{i\phi}\}\right| d\phi, \qquad (2.429)$$

which represents mixtures of states with different values of the overall phase ϕ. (Note that $\mathcal{L}(\phi)$ must also satisfy a positive definiteness condition analogous to Eq. (2.266).) All such mixtures, i.e. all choices of $\mathcal{L}(\phi)$, lead to precisely the set of correlation functions (2.425); hence all such mixed states correspond to fully coherent fields.

It is most important, from a practical standpoint, that our definitions permit these mixed states to correspond to coherent fields. Our *a priori* knowledge of the state of high frequency fields usually contains no information about the overall phase ϕ. An ensemble of experiments performed with such fields must then be described by using a density operator of the form Eq. (2.429) with the special choice

$$\mathcal{L}(\phi) = \frac{1}{2\pi}, \qquad (2.430)$$

which represents our total ignorance of the phase. The indefinite character of this phase does not influence any of the interference intensities we have discussed thus far. It must therefore have no bearing on the coherence properties of a field. Our definition of coherence would hardly be very useful physically if it did not allow the appropriate mixed states as well as pure ones to be coherent.

2.12.1
First Order Correlation Functions for Stationary Fields

Virtually all of the famous experiments of optics may be described in terms of the first order correlation function for stationary light beams. Let us begin the evaluation of such a correlation function by using the normal mode expansion for the field operators to write it in the form

$$G^{(1)}_{\mu\nu}(\mathbf{r}t,\mathbf{r}'t') = \tfrac{1}{2}\sum_{k,k'} \hbar(\omega_k\omega_{k'})^{1/2} \operatorname{Tr}\{\rho a^\dagger_k a_{k'}\} u^*_{k\mu}(\mathbf{r}) u_{k'\nu}(\mathbf{r}') e^{i(\omega_k t - \omega_{k'} t')}. \qquad (2.431)$$

To evaluate the statistical averages $\operatorname{Tr}\{\rho a^\dagger_k a_{k'}\}$ we first note that these will always vanish when the modes k and k' are non-degenerate. We may prove that they vanish in this case by recalling that for stationary fields ρ commutes with the field Hamiltonian $\mathcal{H}_0$. Thus we have, for example,

$$\rho = e^{-i\mathcal{H}_0 t/\hbar} \rho\, e^{i\mathcal{H}_0 t/\hbar} \qquad (2.432)$$

for all values of the parameter t. If we substitute the latter form for the operator into the expression for the desired trace we find

$$\begin{aligned}
\operatorname{Tr}\{\rho a^\dagger_k a_{k'}\} &= \operatorname{Tr}\left\{\rho\, e^{i\mathcal{H}_0 t/\hbar} a^\dagger_k a_{k'}\, e^{-i\mathcal{H}_0 t/\hbar}\right\} \\
&= \operatorname{Tr}\left\{\rho a^\dagger_k a_{k'}\right\} e^{i(\omega_k - \omega_{k'})t}. \qquad (2.433)
\end{aligned}$$

Since the trace is independent of the parameter t, it must vanish whenever $\omega_k \neq \omega_{k'}$.

For the case of two different but degenerate modes, k and k', on the other hand, the quantity $\operatorname{Tr}(\rho a^\dagger_k a_{k'})$ need not vanish. More generally, if there are N degenerate modes the corresponding averages $\operatorname{Tr}(\rho a^\dagger_k a_{k'})$ can be regarded as forming the elements of an $N \times N$ Hermitian matrix which is not, in general, diagonal. It is always possible to diagonalize this matrix, however, by means of a linear transformation which amounts simply to a redefinition of the set of degenerate mode functions. For any stationary state of the field represented by a density operator ρ, in other words, there will exist some particular choice of mode functions $\mathbf{u}_k(\mathbf{r})$ such that the matrix reduces to diagonal form, i.e. we have

$$\operatorname{Tr}\left\{\rho a^\dagger_k a_{k'}\right\} = \langle n_k \rangle \delta_{kk'}, \qquad (2.434)$$

where $\langle n_k \rangle$ is the mean occupation number of the k-th mode.

The convenience of working with particular choices of degenerate mode functions is easily illustrated by means of the polarization properties of light beams. For any plane wave state of a beam there are two degenerate polarization modes which are orthogonal. If we were to choose a pair of plane polarization states as a basis, and were to describe a circularly polarized beam, for example, the quantities $\operatorname{Tr}(\rho a^\dagger_k a_{k'})$ would form a 2×2 matrix with four non-vanishing components. It is no surprise then that a more convenient choice of mode functions for that case consists of the two orthogonal circular polarizations. That choice reduces the matrix to one with only a single non-vanishing component.

Let us now return to our calculation of the first order correlation function for stationary fields. We see from Eqns. (2.434) and (2.431) that with a suitable choice of basis functions it is always possible to write the correlation function as an expansion of the form

$$G^{(1)}_{\mu\nu}(rt,r't') = \tfrac{1}{2}\sum_k \hbar\omega_k \langle n_k\rangle u^*_{k\mu}(r)u_{k\nu}(r')\, e^{i\omega_k(t-t')}\,, \tag{2.435}$$

which is determined simply by the set of average occupation numbers $\langle n_k\rangle$. An expansion of this type which is often useful is based on the set of plane wave modes of a large cubical volume of side L. These modes, whose functions $u_k(r)$ are given by Eq. (2.167), are so densely distributed in the space of the propagation vector k, when the volume of the system is large, that the sum over the states required in Eq. (2.435) may be replaced by the integral $(L/2\pi)^3 \int dk\ldots$ The expansion of the correlation function is then

$$G^{(1)}(rt,r't') = \frac{\hbar c}{2(2\pi)^3}\int \sum_{\lambda=1,2} \hat{e}^{(\lambda)*}_\mu \hat{e}^{(\lambda)}_\nu \langle n_{k,\lambda}\rangle k$$
$$\times \exp\{-i[k\cdot(r-r')-\omega_k(t-t')]\}\, dk\,, \tag{2.436}$$

where λ is an index which labels the polarizations associated with the propagation vector.

Let us suppose that the field consists of a well collimated light beam which is nearly monochromatic and is fully polarized. Then the mean occupation number $\langle n_{k,\lambda}\rangle$ will only take on non-vanishing values within a very small cell of k-space and, say, for $\lambda = 1$. Under these circumstances, if the magnitudes of $|r-r'|$ and $c|t-t'|$ remain small in comparison to the reciprocal dimensions of the volume in which $\langle n_{k,\lambda}\rangle$ differs from zero, it becomes possible to approximate the integral in Eq. (2.436) by neglecting the variation of the exponential in the integrand. If k_0 and ω_0 are the mean propagation vector and frequency of the beam we have

$$G^{(1)}(rt,r't') \approx \frac{\hbar c}{2(2\pi)^3} N \hat{e}^{(1)*}_\mu \hat{e}^{(1)}_\nu\, e^{-i[(k_0(r-r')-\omega_0(t-t')]}\,, \tag{2.437}$$

where

$$N = \int \langle n_{k,1}\rangle\, dk\,. \tag{2.438}$$

The light beam we have described is of just the sort most often used in interference experiments. It is also the kind most often referred to as "coherent" in the traditional terminology of optics. Now it is evident that by defining the field

$$\mathcal{E}(r,t) = \left(\frac{\hbar c}{2(2\pi)^3}\right) N^{1/2} \hat{e}^{(1)}\, e^{i(k_0\cdot r-\omega_0 t)} \tag{2.439}$$

we may write the expression Eq. (2.437) for the correlation function in the factorized form

$$G^{(1)}_{\mu\nu}(\boldsymbol{r}t,\boldsymbol{r}'t') \approx \mathcal{E}^*_\mu(\boldsymbol{r}t)\,\mathcal{E}_\nu(\boldsymbol{r}'t')\,. \tag{2.440}$$

Hence the field in question does indeed satisfy the condition for first order coherence. It is worth emphasizing, however, that the factorization in Eq. (2.440) is an approximate one which tends to be most accurate for points $\boldsymbol{r}',t'$ near $\boldsymbol{r},t$. The imperfect collimation and monochromaticity of the beam define finite ranges of the variables $\boldsymbol{r}-\boldsymbol{r}'$ and $t-t'$, i.e. coherence distances and a coherence time, within which the factorization condition is obeyed. These ranges can, in principle, be made arbitrarily large by improving the quality of the beam.

This example illustrates the sense In which the coherence conditions must usually be regarded as idealizations. Given the practical sorts of field sources at our disposal, we cannot expect that the field correlations they generate will obey the coherence conditions over infinite ranges of the coordinate variables (even though in the case of laser fields these conditions may be known to hold over tens of thousands of miles).

2.12.2
Correlation Functions for Chaotic Fields

A particularly important class of stationary fields, which arises whenever the source is essentially chaotic in nature, is one in which the weight function in the P representation is a product of Gaussian factors, one for each mode. The density operator is then specified by

$$P(\{\alpha_k\}) = \prod_k \frac{1}{\pi \langle n_k \rangle}\, e^{-|\alpha_k|^2/\langle n_k \rangle}\,, \tag{2.441}$$

and it follows that all of the statistical properties of the field are determined by the set of average occupation numbers $\langle n_k \rangle$. The knowledge of this same set of numbers, on the other hand, is equivalent, according to Eq. (2.435), to specifying the first order correlation function for the field. There thus exists a fundamental sense in which the first order correlation function furnishes all the information we need for the description of fields specified by Gaussian weight functions. We may demonstrate this simplifying property more explicitly by showing that all of the higher order correlation functions for such fields can be expressed as sums of products of first order correlation functions.

In order to prove this theorem we shall construct a species of generating functional for the set of all correlation functions of the field. The essential tool for doing this is the operation of functional differentiation. If $F[\zeta(x)]$ is a functional of $\zeta(x)$, i.e. a function of the set of values of $\zeta(x)$ for all x, then we define its functional derivative with respect to $\zeta(x_0)$ to be

$$\frac{\delta F}{\delta \zeta(x_0)} = \lim_{\epsilon \to 0} \frac{1}{\epsilon}\left\{ F\left[\zeta(x) + \epsilon \delta^{(4)}(x-x_0)\right] - F[\zeta(x)] \right\}\,, \tag{2.442}$$

where $\delta^{(4)}$ is a four-dimensional (space-time) delta function. As an illustration, if we apply this definition to an integral operator of the form

$$F = \int \zeta(x) \boldsymbol{E}^{(-)}(x) \, d^4x, \tag{2.443}$$

we find

$$\frac{\delta F}{\delta \zeta(x)} = \int \delta^{(4)}(x - x_0) \boldsymbol{E}^{(-)}(x) \, d^4x = \boldsymbol{E}^{(-)}(x_0). \tag{2.444}$$

Now, let us define the generating functional

$$\Xi[\zeta(x), \eta(x)] = \mathrm{Tr}\left\{ \rho \, e^{\int \zeta(x) \boldsymbol{E}^{(-)}(x) \, d^4x} \, e^{\int \eta(x') \boldsymbol{E}^{(+)}(x') \, d^4x'} \right\}. \tag{2.445}$$

which depends upon two independent functions $\zeta(x)$ and $\eta(x)$ and is the trace of a normally ordered product. Then we easily see that the functional derivatives of this expression, evaluated for $\zeta(x) = \eta(x) = 0$, are the correlation functions of the field; i.e. we have

$$\left. \frac{\delta^2}{\delta \zeta(x_1) \delta \eta(x_2)} \Xi \right|_{\zeta=\eta=0} = \mathrm{Tr}\left\{ \rho \boldsymbol{E}^{(-)}(x_1) \boldsymbol{E}^{(+)}(x_2) \right\} = G^{(1)}(x_1, x_2), \tag{2.446}$$

and more generally

$$\left. \frac{\delta^{2n}}{\delta \zeta(x_1) \ldots \delta \zeta(x_n) \delta \eta(x_{n+1}) \ldots \eta(x_{2n})} \Xi \right|_{\zeta=\eta=0} = G^{(n)}(x_1 \ldots x_n, x_{n+1} \ldots x_{2n}). \tag{2.447}$$

(The tensor indices which have been suppressed in these expressions may be restored by considering each coordinate x to specify a component index as well as a position and a time, e.g., the function $\zeta(x)$ is actually a set of four functions $\zeta_\mu(\boldsymbol{r},t)$ for $\mu = 1, \ldots, 4$, etc.)

It is convenient, at this point, to introduce the abbreviation

$$\boldsymbol{e}(x,k) = i \left(\frac{\hbar \omega_k}{2} \right)^{1/2} \boldsymbol{u}_k(\boldsymbol{r}) e^{-i \omega_k t}, \tag{2.448}$$

which permits us to write the expansion of the operator $\boldsymbol{E}^{(+)}$ in terms of the mode functions as

$$\boldsymbol{E}^{(+)}(x) = \sum_k \boldsymbol{e}(x,k) a_k. \tag{2.449}$$

Then when we use the P representation for the density operator with the Gaussian weight function Eq. (2.441), the generating functional Eq. (2.445) may be written as

$$\Xi = \int \exp\left\{ -\sum_k \frac{|\alpha_k|^2}{\langle n_k \rangle} \right\} \exp\left\{ \sum_k \int \zeta(x) \boldsymbol{e}^*(xk) \alpha_k^* \, d^4x \right\}$$
$$\times \exp\left\{ \sum_k \int \eta(x') \boldsymbol{e}(x'k) \alpha_k \, d^4x' \right\} \prod \frac{d^2 \alpha_k}{\pi \langle n_k \rangle}. \tag{2.450}$$

This multiple integral factors into a product of integrals, one for each mode k. If we introduce the pair of complex parameters

$$\beta_k = \int \zeta(x) \boldsymbol{e}^*(x,k) \, \mathrm{d}^4 x \,,$$
$$\gamma_k = \int \eta(x') \boldsymbol{e}(x',k) \, \mathrm{d}^4 x' \,, \tag{2.451}$$

the integral factor for the k-th mode takes the familiar form

$$\int \exp\left\{ -\frac{|\alpha_k|^2}{\langle n_k \rangle} + \beta_k \alpha_k^* + \gamma_k \alpha_k \right\} \frac{\mathrm{d}^2 \alpha_k}{\pi \langle n_k \rangle} = \exp\{\beta_k \gamma_k \langle n_k \rangle\} \,. \tag{2.452}$$

Hence the generating functional is given by the product

$$\Xi = \sum_k \exp\{\beta_k \gamma_k \langle n_k \rangle\}$$
$$= \exp\left\{ \int \zeta(x) \sum_k \boldsymbol{e}^*(x,k) \boldsymbol{e}(x',k) \langle n_k \rangle \eta(x') \, \mathrm{d}^4 x \, \mathrm{d}^4 x' \right\} \,. \tag{2.453}$$

Now, according to Eqns. (2.435) and (2.448), the first order correlation function for the field is given by the expansion

$$G^{(1)}(x,x') = \sum_k \boldsymbol{e}^*(x,k) \boldsymbol{e}(x',k) \langle n_k \rangle \,, \tag{2.454}$$

which is just the sum which occurs in the exponential function of Eq. (2.453). Hence the generating functional for the correlation functions of all orders may be expressed in terms of the first order correlation function as

$$\Xi[\zeta(x), \eta(x')] = \exp\left\{ \int \zeta(x) G^{(1)}(x,x') \eta(x') \, \mathrm{d}^4 x \, \mathrm{d}^4 x' \right\} \,. \tag{2.455}$$

We may now derive explicit expressions for the higher order correlation functions by evaluating the appropriate functional derivatives. In particular the n-th derivative with respect to ζ may be written as

$$\frac{\delta^n}{\delta \zeta(x_1) \dots \delta \zeta(x_n)} \Xi = \left\{ \prod_{j=1}^n \int G^{(1)}(x_j, x') \eta(x') \, \mathrm{d}^4 x' \right\} \Xi \,. \tag{2.456}$$

To evaluate the n-th order correlation function we must next differentiate n times with respect to the function η. Since $\zeta(x)$ is finally to be set equal to zero it is easy to see that all of the terms which come from differentiating the factor Ξ on the right side of

Eq. (2.456) with respect to η will finally vanish. Hence we have simply

$$\frac{\delta^{2n}}{\delta\zeta(x_1)\ldots\delta\zeta(x_n)\delta\eta(x_{n+1})\ldots\delta\eta(x_{2n})}\Xi\bigg|_{\zeta=0,\,\eta=0}$$

$$=\frac{\delta^n}{\delta\eta(x_{n+1})\ldots\delta\eta(x_{2n})}\prod_{j=1}^{n}\int G^{(1)}(x_j,x')\eta(x')\,\mathrm{d}^4x'$$

$$=\sum_{P}\prod_{j=1}^{n}G^{(1)}(x_j,x_{P(n+j)})\,; \qquad (2.457)$$

i.e., the derivative is a sum taken over the $n!$ possible ways of permuting the set of coordinates $x_{n+1}\ldots x_{2n}$. Since the derivative we have evaluated, according to Eq. (2.447), is the n-th order correlation function, we have finally

$$G^{(n)}(x_1\ldots x_n, x_{n+1}\ldots x_{2n}) = \sum_{P}\prod_{j=1}^{n}G^{(1)}(x_j,x_{P(n+j)})\,. \qquad (2.458)$$

The n-th order correlation function for Gaussian fields is just a symmetrical sum of products of first order correlation functions.

To illustrate this result for the second order correlation function we may write

$$G^{(2)}(x_1x_2, x_3x_4) = G^{(1)}(x_1x_3)G^{(1)}(x_2x_4) + G^{(1)}(x_1x_4)G^{(1)}(x_2x_3)\,. \qquad (2.459)$$

Now if the field in question possesses first order coherence, we may write the first order correlation function in the factorized form of Eq. (2.109). The two terms of Eq. (2.459) are then equal and we find

$$G^{(2)}(x_1x_2, x_3x_4) = 2\mathcal{E}^*(x_1)\mathcal{E}^*(x_2)\mathcal{E}(x_3)\mathcal{E}(x_4)\,. \qquad (2.460)$$

The second order correlation function factorizes, but because of the presence of the factor of 2, it does so in a way which precludes the possibility that the field has second or higher order coherence. The n-th order correlation function for such fields in evidently given by

$$G^{(n)}(x_1\ldots x_{2n}) = n!\prod_{j=1}^{n}\mathcal{E}^*(x_j)\prod_{j=n+1}^{2n}\mathcal{E}(x_j)\,. \qquad (2.461)$$

2.12.3
Quasiprobability Distribution for the Field Amplitude

Whenever the density operator for the field may be specified by means of the P representation the function $P(\{\alpha_k\})$ plays a role analogous to that of a probability density for the individual mode amplitudes α_k. Of course when we make measurements upon a light beam, we are typically measuring not the individual amplitudes α_k, but the average values of various functions of the complex field strength eigenvalue, $\mathcal{E}(rt)$, which is a particular linear sum of the mode amplitudes,

$$\mathcal{E}(x,\{\alpha_k\}) = \sum_k e(x,k)\alpha_k . \tag{2.462}$$

To describe the fullest variety of such measurements which we can make at a single space-time point $x = (\mathbf{r},t)$, it is convenient to derive from $P(\{\alpha_k\})$ a species of reduced quasiprobability distribution for the complex field amplitude $\mathcal{E}(x,\{\alpha_k\})$. This distribution function for the field amplitude will be quite useful in discussing the origin of the photon correlation effect discovered by Hanbury Brown, and Twiss.

To illustrate the kinds of averages we frequently want to discuss, let us note that the average intensity of the field at the point x is

$$G^{(1)}(x,x) = \int P(\{\alpha_k\}) |\mathcal{E}(x\{\alpha_k\})|^2 \prod_k d^2\alpha_k , \tag{2.463}$$

and the average coincidence rate for the limiting case in which the two counters are placed at the same point and are sensitive at the same time is

$$G^{(2)}(xx,xx) = \int P(\{\alpha_k\}) |\mathcal{E}(x\{\alpha_k\})|^4 \prod_k d^2\alpha_k . \tag{2.464}$$

These are examples of a general class of averages which take the form

$$\int P(\{\alpha_k\}) F(\mathcal{E}(x\{\alpha_k\})) \prod_k d^2\alpha_k \tag{2.465}$$

for suitably determined functions F. It is convenient now to separate the multidimensional integration over the complex amplitude parameters α_k into two steps, the integration over the subspace of the α_k parameters in which the linear combination

$$\mathcal{E}(x\{\alpha_k\}) = \sum_k e(x,k)\alpha_k$$

remains constant, and then the further integration over the values this sum may take on. The first of these integrations is accomplished by defining the function

$$W(\mathcal{E},x) = \int P(\{\alpha_k\}) \delta^{(2)}\left[\mathcal{E} - \sum_k e(x,k)\alpha_k\right] \prod_k d^2\alpha_k . \tag{2.466}$$

We may then write the complete integral Eq. (2.465) in the form

$$\int P(\{\alpha_k\}) F[\mathcal{E}(x\{\alpha_k\})] \prod_k d^2\alpha_k = \iint P(\{\alpha_k\}) \delta^{(2)}\left[\mathcal{E} - \sum_k e(x,k)\alpha_k\right]$$
$$\times F(\mathcal{E}) \prod_k d^2\alpha_k\, d^2\mathcal{E}$$
$$= \int W(\mathcal{E},x) F(\mathcal{E})\, d^2\mathcal{E} , \tag{2.467}$$

where $d^2\mathcal{E} = d(\text{Re}\,\mathcal{E})\,d(\text{Im}\,\mathcal{E})$ is a real element of area in the complex field amplitude plane. The function $W(\mathcal{E},x)$ defined by Eq. (2.466) evidently plays a role analogous to that of a probability distribution for the complex field amplitude at the space-time point x. Of course, since the function P from which it is derived is only a quasiprobability distribution, and is subject to all the restrictions mentioned in the last Section, the same limitations will apply to the physical interpretation of the function $W(\mathcal{E},x)$. It too can take on negative values, for example.

The function W furnishes a particularly simple description of fields which consist of many independently excited modes. Since the total field amplitude $\mathcal{E}$ is then the sum of a large number of independently distributed complex amplitudes proportional to the α_k, the distribution of the amplitude $\mathcal{E}$ will correspond to that of the endpoint of a many-step random walk in the complex plane. This distribution tends to take on a Gaussian form when the number of contributing modes is large, no matter how the mode amplitudes may be distributed individually. From a mathematical standpoint this argument differs hardly at all from the discussion of the central limit theorem given in Sect. 2.9.8; i.e., the starting point, Eq. (2.466), becomes similar in structure to Eq. (2.281) when the function $P(\{\alpha_k\})$ is assumed to factorize into a product $\prod_k P_k(\alpha_k)$. As a slight generalization of the discussion given there we may let the individual mode excitations be non-stationary in character and have mean amplitudes

$$\int P_k(\alpha_k)\alpha_k\,d^2\alpha_k = \langle\alpha_k\rangle\,. \tag{2.468}$$

Then by applying the central limit theorem, we find

$$W(\mathcal{E},x) = \frac{1}{\pi\sum_k |e(x,k)|^2\{\langle|\alpha_k|^2\rangle - |\langle\alpha_k\rangle|^2\}} \times \exp\left\{-\frac{|\mathcal{E} - \sum_k e(x,k)\langle\alpha_k\rangle|^2}{\sum_k |e(x,k)|^2\{\langle|\alpha_k|^2\rangle - |\langle\alpha_k\rangle|^2\}}\right\}. \tag{2.469}$$

If the mean amplitudes $\langle\alpha_k\rangle$ vanish, as they do for example in the case of stationary fields, we have

$$W(\mathcal{E},x) = \frac{1}{\pi\sum_k |e(x,k)|^2 \langle n_k\rangle}\exp\left\{-\frac{|\mathcal{E}|^2}{\sum_k |e(x,k)|^2 \langle n_k\rangle}\right\}$$

$$= \frac{1}{\pi G^{(1)}(x,x)}\exp\left\{-\frac{|\mathcal{E}|^2}{G^{(1)}(x,x)}\right\}. \tag{2.470}$$

To illustrate the use of this expression for $W(\mathcal{E},x)$, let us calculate the n-th order correlation function with all arguments equal. By letting $F(\mathcal{E}) = |\mathcal{E}|^{2n}$ in Eq. (2.467) we find

$$G^{(n)}(x\ldots x) = \int W(\mathcal{E},x)|\mathcal{E}|^{2n}\,d^2\mathcal{E}\,. \tag{2.471}$$

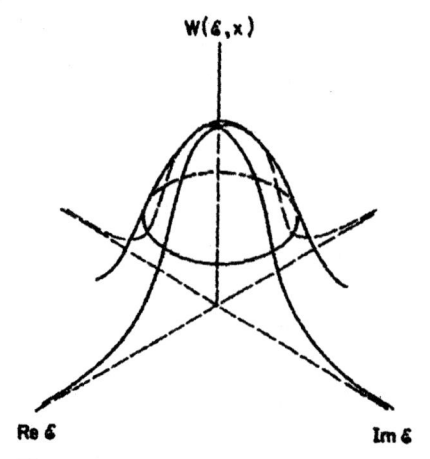

Figure 13

For the Gaussian form of W given by Eq. (2.470), the latter integral is simply

$$G^{(n)}(x\ldots x) = n!\left\{G^{(1)}(x,x)\right\}^n. \tag{2.472}$$

An important class of fields which obey the separability conditions we have assumed in deriving these results is that specified by the Gaussian density operators discussed earlier. For these fields, in fact, Eq. (2.472) follows directly from Eq. (2.458). But since we have not had to assume that the functions $P_k(\alpha_k)$ are individually Gaussian in form to derive Eqns. (2.470) and (2.472), these results evidently hold true for a considerably broader variety of field excitations.

A sketch of the Gaussian distribution function $W(\mathcal{E},x)$ is given in Fig. 13. Since this function plays a role akin to that of a probability distribution for the complex field amplitude $\mathcal{E}$, it is evident that the absolute magnitude of the field undergoes a considerable amount of fluctuation. Thus, while the most probable value of the field amplitude is $\mathcal{E} = 0$, the amplitude will occasionally stray out into the regions of the complex plane which represent the "tail" of the Gaussian and correspond to arbitrarily strong fields. The relation Eq. (2.472) between values of the correlation functions may also be stated as the relation

$$\langle |\mathcal{E}|^{2n}\rangle = n!\left\{\langle |\mathcal{E}|^2\rangle\right\}^n \tag{2.473}$$

between average moments $\langle |\mathcal{E}|^j\rangle$ of the function W. The extremely rapid increase with n of the ratio $\langle |\mathcal{E}|^{2n}\rangle / \{\langle |\mathcal{E}|^2\rangle\}^n$, which the Gaussian distribution shows, is due to its "long-tailed" character.

Although the Gaussian form for the function $W(\mathcal{E},x)$ will presumably apply almost universally to the radiation from natural or essentially chaotic sources, altogether different distributions may be required to describe the radiation from certain man-made sources. In fact the avoidance of fields which have the extremely random

2.12 Correlation Functions and Quasiprobability Distributions | 143

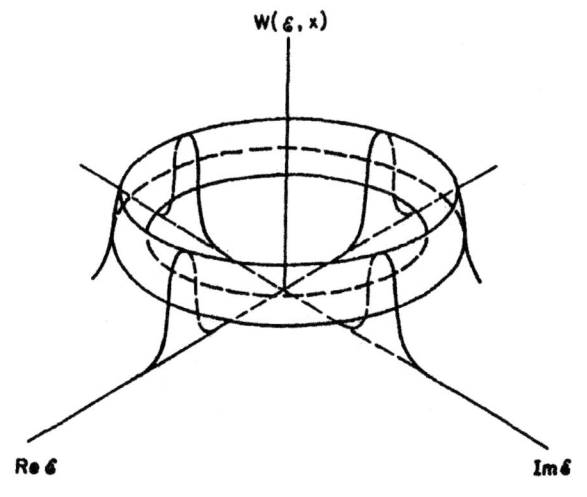

Figure 14

or noisy character of the Gaussian form of $W(\mathcal{E},x)$ has been one of the major goals of radio-frequency technology. One of its earliest accomplishments was the development of oscillators which generate fields of extremely stable modulus, e.g., broadcast carrier waves. These oscillators are nonlinear devices and the contributions of the various mode amplitudes to the total field are not at all independently distributed as in the Gaussian case. For a stationary field generated by such an oscillator we might find the function $W(\mathcal{E},x)$ to assume a form similar to that shown in Fig. 14; i.e., the modulus of the field, $|\mathcal{E}|$, has only a very small probability for taking on values either appreciably smaller or larger than its root-mean-square value, $\{\langle|\mathcal{E}|^2\rangle\}^{1/2}$.

The shape of the function $W(\mathcal{E},x)$ furnishes an elementary insight into the origin of the photon correlation effect which was discovered by Hanbury Brown and Twiss by means of the experiment described in Sect. 2.8. Let us consider the two-fold coincidence counting rate for photons when the two detectors D_1 and D_2 of Fig. 9 occupy precisely symmetrical positions relative to the half-silvered mirror m, and when the detectors are adjusted so that they register coincidences with no time delay. Since the arrangement is one in which the counters, in effect, occupy the same position and are sensitive at the same time, the coincidence rate is given by a correlation function of the form

$$G^{(2)}(xx,xx) = \langle|\mathcal{E}(x)|^4\rangle \,. \tag{2.474}$$

Now, according to Eqns. (2.472) or (2.473), for all chaotic light sources we should find

$$\langle|\mathcal{E}(x)|^4\rangle = 2\langle|\mathcal{E}(x)|^2\rangle^2$$
$$= 2\left\{G^{(1)}(x,x)\right\}^2 \,. \tag{2.475}$$

The amount by which $G^{(2)}(xx,xx)$ exceeds $\{G^{(1)}(x,x)\}^2$ is a measure of the non-random tendency of the photons to be recorded as simultaneously arriving pairs; i.e., it is a measure of the height of the "bump" on the coincidence rate curve shown in Fig. 10. Since the coincidence rate for zero time delay is twice the background or accidental coincidence rate, the correlation effect is not a small one. (The original observations of the effect were made difficult by the relatively long response times of the counting systems compared with the time interval over which the correlation persists.)

To see the nature of the photon correlation effect for other types of distributions $W(\mathcal{E},x)$, let us note that it is proportional to

$$G^{(2)}(xx,xx) - \{G^{(1)}(x,x)\}^2 = \langle|\mathcal{E}(x)|^4\rangle - \langle|\mathcal{E}(x)|^2\rangle^2$$
$$= \int W(\mathcal{E},x)\{|\mathcal{E}|^2 - \langle|\mathcal{E}|^2\rangle\}^2 \, d^2\mathcal{E}. \quad (2.476)$$

One of the curious quantum mechanical properties of this expression is that, although it resembles a statistical variance for the quantity $|\mathcal{E}|^2$, it may actually take on negative as well as positive values. That is true since $W(\mathcal{E},x)$ as we have noted, is not strictly speaking a probability distribution. It is not difficult to find states of the field for which W takes on negative values at least locally avid for which the average Eq. (2.476) is consequently negative. When the field is in such states photon coincidences will be recorded with less than the random background rate by the Hanbury Brown–Twiss detection apparatus, an effect which is the reverse of the one observed for natural radiation sources.

Whenever the field is generated by an essentially classical source, i.e., one with predetermined behavior, it will be possible, as we have seen in Sect. 2.7, to construct a P representation for the density operator with a non-negative weight function $P(\{\alpha_k\})$. Then the function $W(\mathcal{E},x)$ defined by Eq. (2.466) will likewise take on no negative values. We may thus state that for all classically generatable fields, the Hanbury Brown–Twiss correlation is positive,

$$G^{(2)}(xx,xx) - \{G^{(1)}(x,x)\}^2 \geq 0. \quad (2.477)$$

If the correlation effect is to vanish for fields of this type we must evidently have

$$W(\mathcal{E},x)\{|\mathcal{E}|^2 - \langle|\mathcal{E}|^2\rangle\}^2 = 0 \quad (2.478)$$

for all $\mathcal{E}$. The function $W(\mathcal{E},x)$ can therefore only take on non-vanishing values at points lying on the circle $|\mathcal{E}|^2 = \langle|\mathcal{E}|^2\rangle$. If the function $W(\mathcal{E},x)$, in other words, is of a form which allows no amplitude modulation of the field, the correlation effect will vanish and conversely. In fact in that limit we have more generally

$$G^{(n)}(x\ldots x) = \langle|\mathcal{E}|^{2n}\rangle = \langle|\mathcal{E}|^2\rangle^n = \{G^{(1)}(x,x)\}^n \quad (2.479)$$

and all *n*-fold coincidence experiments show an absence of any tendency toward statistical correlations.

A number of the published discussions of the Hanbury Brown–Twiss effect explain it as being caused by the fact that photons are Bose particles and consequently have a certain tendency to cluster. That such explanations are far from complete is made evident by the fact that the quantum mechanical form of the effect may have either sign; it may constitute an anticorrelation or "repulsion", rather than a positive correlation or "clumping". Furthermore the fact that classical fields have only a positive correlation effect is a clear demonstration that the average quantities one evaluates by means of the correlation functions (even where the *P* representation exists) are not always equivalent in quantum theory and classical theory. The variety of fields encountered in the quantum theory is simply much larger than that allowed by classical theory.

It should be evident that the measurement of the photon correlation effect, at least at zero delay time, simply furnishes a measure of the amount of random amplitude modulation present in fields with positive $W(\mathcal{E},x)$. The effect should be nearly absent from the field generated by a well stabilized oscillator. In particular since a gas laser operating well above its threshold is presumably quite a stable oscillator, any Hanbury Brown–Twiss correlation found in its beam should be quite small in magnitude.

The fact that a photon correlation experiment, or its analog in the radio-frequency region, an intensity correlation experiment, can furnish a simple way of telling whether a radiation field comes from a natural source or a man-made one could have some interesting if rather far-fetched astronomical consequences. If intelligent beings elsewhere in the galaxy want to communicate with us, it seems reasonable to suppose that they would use amplitude-stabilized oscillators of some sort as radiators. In that case their signals, as we have seen, would have an unmistakable character even when no message was being transmitted. In fact the unmodulated signal could be easier to distinguish from background noise than the modulated one.

2.12.4
Quasiprobability Distribution for the Field Amplitudes at Two Space-Time Points

A number of the correlation functions and other expectation values which interest us depend on the fields at two different space-time points x_1 and x_2. These averages may be expressed, when the *P* representation exists, in the general form

$$\int P(\{\alpha_k\}) F[\mathcal{E}(x_1\{\alpha_k\}), \mathcal{E}(x_2\{\alpha_k\})] \prod_k d^2\alpha_k , \qquad (2.480)$$

where the function *F* is suitably defined for each case. Two familiar examples of such averages are the first order correlation function $G^{(1)}(x_1,x_2)$, for which we would choose

$$F = \mathcal{E}^*(x_1\{\alpha_k\}) \mathcal{E}(x_2\{\alpha_k\}) , \qquad (2.481)$$

and the delayed coincidence counting rate, $G^{(2)}(x_1 x_2, x_2 x_1)$, for which we would choose

$$F = |\mathcal{E}(x_1\{\alpha_k\})|^2 \, |\mathcal{E}(x_2\{\alpha_k\})|^2 \,. \tag{2.482}$$

Now, if we define a species of distribution function $W(\mathcal{E}_1 x_1, \mathcal{E}_2 x_2)$, for the complex field amplitudes at the two points by means of the relation

$$W(\mathcal{E}_1 x_1, \mathcal{E}_2 x_2) = \int P(\{\alpha_k\}) \, \delta^{(2)}[\mathcal{E}_1 - \mathcal{E}(x_1\{\alpha_k\})] \, \delta^{(2)}[(\mathcal{E}_2 - \mathcal{E}(x_2\{\alpha_k\}))] \prod_k d^2 \alpha_k \,, \tag{2.483}$$

then an average quantity of the form Eq. (2.480) is given by the integral

$$\int W(\mathcal{E}_1 x_1, \mathcal{E}_2 x_2) \, F(\mathcal{E}_1, \mathcal{E}_2) \, d^2 \mathcal{E}_1 \, d^2 \mathcal{E}_2 \,. \tag{2.484}$$

The function $W(\mathcal{E}_1 x_1, \mathcal{E}_2 x_2)$, more specifically, is a quasiprobability distribution which plays the same role in averaging functions of two space-time variables as the function $W(\mathcal{E},x)$, which we discussed earlier, plays in the calculation of averages for a single space-time point. We may, in fact obtain $W(\mathcal{E},x)$ from the two-point function by integrating over either of the field variables,

$$W(\mathcal{E},x) = \int W(\mathcal{E}x, \mathcal{E}'x') \, d^2 \mathcal{E}'$$

$$= \int W(\mathcal{E}'x', \mathcal{E}x) \, d^2 \mathcal{E}' \,. \tag{2.485}$$

When the function $P(\{\alpha_k\})$ factorizes into a product of independent weight functions, one for each mode, and when the number of excited modes is large, it is easy to show, again by techniques similar to those used in Sect. 2.9.8, that $W(\mathcal{E}_1 x_1, \mathcal{E}_2 x_2)$ assumes a Gaussian form in the two complex amplitude variables $\mathcal{E}_1$ and $\mathcal{E}_2$. To carry out the derivation we simply show that the double Fourier transform of $W(\mathcal{E}_1 x_1, \mathcal{E}_2 x_2)$ with respect to the amplitude variables $\mathcal{E}_1$ and $\mathcal{E}_2$ is asymptotically Gaussian in form when the number of excited modes becomes infinite. Inversion of the transform then yields a result which, for the case of stationary fields, can be written as

$$W(\mathcal{E}_1 x_1, \mathcal{E}_2 x_2) = \frac{1}{\pi G^{(1)}(x_1 x_1) \, G^{(1)}(x_2 x_2) \left[1 - |g^{(1)}(x_1 x_2)|^2\right]}$$

$$\times \exp\left\{ -\frac{\dfrac{|\mathcal{E}_1|^2}{G^{(1)}(x_1 x_1)} + \dfrac{|\mathcal{E}_2|^2}{G^{(1)}(x_2 x_2)} - 2\operatorname{Re}\dfrac{\mathcal{E}_1 \mathcal{E}_2^* g^{(1)}(x_1 x_2)}{\left[G^{(1)}(x_1 x_1) \, G^{(1)}(x_2 x_2)\right]^{1/2}}}{1 - |g^{(1)}(x_1 x_2)|^2} \right\}, \tag{2.486}$$

where $g^{(1)}$ the normalized form of the first order correlation function defined by

2.12 Correlation Functions and Quasiprobability Distributions

Eq. (2.99). As a simple check of this result it is easy to verify that the average of the function Eq. (2.481) is

$$\left\{G^{(1)}(x_1 x_1) G^{(1)}(x_2 x_2)\right\}^{1/2} g^{(1)}(x_1 x_2) = G^{(1)}(x_1 x_2) \tag{2.487}$$

as required, and that the average of the function Eq. (2.482) is indeed

$$G^{(1)}(x_1 x_1) G^{(1)}(x_2 x_2) + \left|G^{(1)}(x_1 x_2)\right|^2 = G^{(2)}(x_1 x_2, x_2 x_1). \tag{2.488}$$

The function $W(\mathcal{E}_1 x_1, \mathcal{E}_2 x_2)$ plays a role in the theory which is analogous to that of a probability density for a compound event, i.e., finding the field $\mathcal{E}_1$ at $x_1 = (\mathbf{r}_1, t_1)$ and $\mathcal{E}_2$ at $x_2 = (\mathbf{r}_2, t_2)$. In probability theory it is often of interest, in dealing with such compound events, to imagine that the first part of the event has already taken place and to calculate the probability that the compound event is then completed. We may define an analog of such a conditioned probability function by means of the relation

$$W(\mathcal{E}_1 x_1 | \mathcal{E}_2 x_2) = \frac{W(\mathcal{E}_1 x_1, \mathcal{E}_2 x_2)}{W(\mathcal{E}_1, x_1)}, \tag{2.489}$$

where $W(\mathcal{E}_1, x_1)$ is the function defined by Eq. (2.466). The function $W(\mathcal{E}_1 x_1 | \mathcal{E}_2 x_2)$ is analogous to a probability density for the field amplitude to have values in the neighborhood of $\mathcal{E}_2$ at $x_2 = (\mathbf{r}_2, t_2)$, given that it had the value $\mathcal{E}_1$ at $x_1 = (\mathbf{r}_1, t_1)$. We shall call the function the conditioned quasiprobability density; it is, strictly speaking, only measurable as a probability density in the classical or strong field limit.

When we calculate the ratio of the functions given by Eqns. (2.486) and (2.470) we find the result

$$W(\mathcal{E}_1 x_1 | \mathcal{E}_2 x_2) = \frac{1}{\pi} \frac{1}{G^{(1)}(x_2 x_2) \left[1 - \left|g^{(1)}(x_1 x_2)\right|^2\right]}$$

$$\times \exp\left\{-\frac{\left|\dfrac{\mathcal{E}_2}{\{G^{(1)}(x_2 x_2)\}^{1/2}} - \dfrac{\mathcal{E}_1}{\{G^{(1)}(x_1 x_1)\}^{1/2}} g^{(1)}(x_1 x_2)\right|^2}{1 - \left|g^{(1)}(x_1 x_2)\right|^2}\right\} \tag{2.490}$$

for the conditioned quasiprobability distribution. The field $\mathcal{E}_2$ in other words, has a Gaussian distribution about the mean value

$$\langle \mathcal{E}_2 \rangle = \mathcal{E}_1 \frac{g^{(1)}(x_1 x_2)}{G^{(1)}(x_1 x_1)} \tag{2.491}$$

with a dispersion proportional to $G^{(2)}(x_2 x_2)\{1 - |g^{(1)}(x_1 x_2)|^2\}$, which vanishes for x_2 near x_1 and tends to approach $G^{(2)}(x_2 x_2)$ as x_2 recedes from x_1. We shall examine these expressions more closely once we have illustrated the evaluation of the correlation functions on which they depend.

2.13
Elementary Models of Light Beams

Since our results to this point have all been stated in fairly general terms, it may be of help to discuss an illustrative example or two. Let us consider, as a particularly simple example, a stationary light beam which may be thought of as a plane wave progressing along the positive y axis. We shall allow the beam to have an arbitrary frequency bandwidth, but shall take it to have a specific polarization $\hat{e}$. The first order correlation function for the beam may then be evaluated as a sum over plane wave mode functions by means of Eq. (2.435). The index which labels the mode functions in this case may be taken to k_y, the y-component of the propagation vector. (The other components vanish.) Since the values of k_y are densely distributed, when the size L of the quantization volume is large, the sum over k_y is equivalent to a one-dimensional integration

$$\sum_{k_y} \to \frac{L}{2\pi} \int dk_y \ldots$$

When the mode functions given by Eq. (2.167) are substituted in Eq. (2.435) and the sum is replaced by an integral, we find

$$G^{(1)}(y_1 t_1, y_2 t_2) = \frac{\hbar c}{4\pi} \int_0^\infty \frac{\langle n_k \rangle}{L^2} k \exp\{-i[k_y(y_1 - y_2) - \omega_k(t_1 - t_2)]\} \, dk_y, \quad (2.492)$$

where $G^{(1)}$ is understood to be a correlation function for the field components in the direction $\hat{e}$, as in Eq. (2.55). Since the beam contains no backward traveling waves (which would be represented by negative values of k_y), we may write the integral equally well as one over the frequency variable $\omega_k = ck_y$. Then if we introduce the parameter

$$s = t_1 - t_2 + \frac{1}{c}(y_1 - y_2) \quad (2.493)$$

to express the space-time interval which occurs as an argument, we may write

$$G^{(1)}(y_1 t_1, y_2 t_2) = \frac{1}{4\pi c} \int_0^\infty \frac{\langle n_k \rangle \hbar \omega_k}{L^2} e^{i\omega_k s} \, d\omega_k. \quad (2.494)$$

The expression $\langle n_k \rangle \hbar \omega_k$, which occurs in the integrand of Eq. (2.494), is the average energy of excitation of the k-th mode. Let us assume, as an example, that our beam has a spectral profile of the Lorentz form by writing

$$\frac{\langle n_k \rangle \hbar \omega_k}{cL^2} = \frac{2\gamma}{(\omega - \omega_0)^2 + \gamma^2} U. \quad (2.495)$$

Here ω_0 is the central frequency, γ is the half-width at half height, and the constant U is a measure of the intensity of the beam. Since the frequency ω_0 is typically much larger than γ, only a very small numerical error is made in the integration over the spectral profile if the lower limit $\omega = 0$ in Eq. (2.494) is replaced by $\omega = -\infty$. By making this approximation and letting $\omega' = \omega - \omega_0$ find

$$G^{(1)}(y_1 t_1, y_2 t_2) = \frac{\gamma}{2\pi} U e^{i\omega_0 s} \int_{-\infty}^{\infty} \frac{e^{i\omega' s}}{\omega'^2 + \gamma^2} d\omega' . \tag{2.496}$$

The singularities of the function

$$\frac{1}{\omega'^2 + \gamma^2} = \frac{1}{2i\gamma} \left\{ \frac{1}{\omega' - i\gamma} - \frac{1}{\omega' + i\gamma} \right\} \tag{2.497}$$

are a pair of simple poles lying at $\pm i\gamma$ in the complex ω' plane. The integral in Eq. (2.496) can be written as a contour integral around a closed path in the ω' plane in either of two simple ways, depending on the sign of the variable s. For $s > 0$ the contour may be closed by means of an infinite semicircle in the upper half plane (Im $\omega' > 0$); for $s < 0$ it may be closed by a semicircle in the lower half plane. Since the integrals along both semicircles vanish, we find by applying the residue theorem

$$\int_{-\infty}^{\infty} \frac{1}{2i\gamma} \left\{ \frac{1}{\omega - i\gamma} - \frac{1}{\omega + i\gamma} \right\} e^{i\omega s} d\omega = 2\pi i \begin{cases} \frac{1}{2i\gamma} e^{-\gamma s}, & s > 0, \\ \frac{1}{2i\gamma} e^{\gamma s}, & s < 0. \end{cases} \tag{2.498}$$

The first order correlation function, according to Eq. (2.496), is therefore given by

$$G^{(1)}(y_1 t_1, y_2 t_2) = \tfrac{1}{2} U e^{i\omega_0 s - \gamma |s|} . \tag{2.499}$$

The intensity of the field is found by letting $y_1 = y_2$ and $t_1 = t_2$. For these values of the coordinates, which correspond to $s = 0$, we have

$$G^{(1)}(y_1 t_1, y_1 t_1) = \tfrac{1}{2} U . \tag{2.500}$$

This is the average of the squared magnitude of the complex field $\boldsymbol{E}^{(+)}$. It is easy to see, if we recall the formulae of elementary electrodynamics, that the parameter U is equal to the average total of the electric and magnetic energy densities for the field.

The correlation function given by Eq. (2.499) shows that our light beam exhibits approximate first order coherence when its frequency band width γ is sufficiently small. Thus, when we have

$$\frac{1}{\gamma} \gg |s| = \left| t_1 - t_2 - \frac{1}{c}(y_1 - y_2) \right| , \tag{2.501}$$

the factor $e^{-\gamma|s|}$ in Eq. (2.499) may be approximated by unity, and the remainder of the expression for the correlation function may be written in the appropriate factorized form. As an alternative way of discussing first order coherence we note that the normalized form of the correlation function is

$$g^{(1)}(y_1t_1,y_2t_2) = \frac{G^{(1)}(y_1t_1,y_2t_2)}{\left\{G^{(1)}(y_1t_1,y_1t_1)\,G^{(1)}(y_2t_2,y_2t_2)\right\}^{1/2}}$$
$$= \exp\left[i\omega_0 s - \gamma|s|\right]. \tag{2.502}$$

This function indeed has absolute magnitude close to unity as long as $\gamma|s|$ is sufficiently small.

A good deal of attention has been directed experimentally to the problem of developing light sources with narrow line width. In the best of these sources of the ordinary gas discharge or chaotic variety γ is of the order of 10^9 cycles per second. In ordinary laboratory sources it is often of order 10^{11} cycles per second or larger. The corresponding coherence ranges are 30 cm and 0.3 cm, respectively.

Although we have been discussing the way in which monochromaticity may imply coherence, it may be worth recalling that it is not a necessary condition even for first order coherence. The coherence condition only becomes linked to a requirement of monochromaticity when we restrict our consideration to stationary fields, as we noted in connection with Eq. (2.118). For the case of stationary laser beams, the range of first order coherence is determined by the spectral bandwidth just as for ordinary sources. For the case of gas lasers it is possible to reduce the band width γ to values of the order of 10^3 cycles per second without too much difficulty, and it seems possible to achieve frequency stabilization to within about 10 cycles per second over brief intervals. The coherence ranges corresponding to these band widths are 300 km and 30 000 km, respectively.

Before we can calculate the second and higher order correlation functions for our light beam, we must specify its statistical nature somewhat further. It is at this point that the descriptions of beams generated by natural sources and those generated by coherent sources become qualitatively different. Let us assume that our source is of the usual chaotic variety. Then the higher order correlation functions may all be expressed as sums of products of first order correlation functions, as we have seen in Eq. (2.458). The spectral density function of our plane wave beam, in other words, completely determines the statistical properties of the field. In particular the delayed coincidence rate for counting pairs of photons is given by

$$G^{(2)}(y_1t_1,y_2t_2,y_2t_2,y_1t_1) = G^{(1)}(y_1t_1,y_1t_1)\,G^{(1)}(y_2t_2,y_2t_2) + |G^{(1)}(y_1t_1,y_2t_2)|^2$$
$$= G^{(1)}(y_1t_1,y_1t_1)\,G^{(1)}(y_2t_2,y_2t_2)$$
$$\quad + \left\{1 + |g^{(1)}(y_1t_1,y_2t_2)|^2\right\}$$
$$= \left(\frac{1}{2}U\right)^2 \left\{1 + e^{-2\gamma|s|}\right\}. \tag{2.503}$$

The presence of the term $e^{-2\gamma|s|}$ in this expression shows that the beam can never possess second order coherence. Furthermore when we plot the coincidence rate against

2.13 Elementary Models of Light Beams

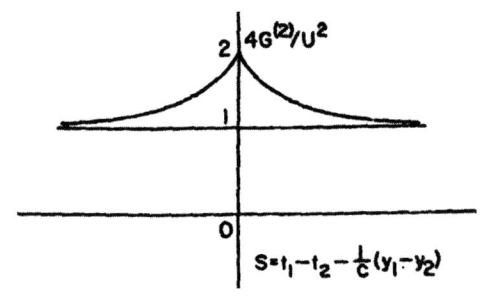

Figure 15

s as in Fig. 15 we see that that term constitutes the "bump" on the Hanbury Brown–Twiss correlation curve, i.e., the deviation of the curve from the accidental or background coincidence rate. The experimental curve shown earlier in Fig. 10 corresponds to a curve of the form shown here after the resolution properties of the counter system have been folded in.

We have noted in the last Section that the origin of the correlation effect lies in the random amplitude modulation of our light beam. Thus the factors of $n!$ by which the n-fold coincidence rate (at zero time delay) exceeds the random coincidence rate are easily explained in terms of the moments of the Gaussian amplitude distribution $W(\mathcal{E},x)$ given by Eq. (2.470). To understand the behavior of the correlation effect for non-vanishing time delays, and to see, for example, why the effect disappears for $|s| \gg 1/2\gamma$, we may make use of the quasiprobability distributions defined for pairs of values of the field amplitude in the last Section. When we substitute the values given by Eqns. (2.499) and (2.502) for the correlation functions into the expression (2.487) for the conditioned quasiprobability function $W(\mathcal{E}_1 x_1 | \mathcal{E}_2 x_2)$, we find

$$W(\mathcal{E}_1 y_1 t_1, \mathcal{E}_2 y_2 t_2) = \frac{1}{\frac{1}{2}\pi U (1 - e^{-2\gamma|s|})} \exp\left\{ -\frac{|\mathcal{E}_2 - \mathcal{E}_1 e^{i\omega_0 s - \gamma|s|}|^2}{\frac{1}{2} U (1 - e^{-2\gamma|s|})} \right\}. \quad (2.504)$$

This function is to be interpreted as the distribution of values of the field amplitude $\mathcal{E}_2$ at $y_2 t_2$, when the amplitude is known to take on the value $\mathcal{E}_1$ at $y_1 t_1$. When the parameter s vanishes, the mean radius of the Gaussian peak of this expression vanishes and the distribution reduces to the delta function $\delta^{(2)}(\mathcal{E}_2 - \mathcal{E}_1)$. As $|s|$ increases from zero, the mean value of $\mathcal{E}_2$, which is given by $\mathcal{E}_1 e^{i\omega_0 s - \gamma|s|}$, describes an exponential spiral in the complex $\mathcal{E}_2$ plane while relaxing to the value zero. The spiral which corresponds to $s < 0$ is shown in exaggerated form in Fig. 16. At the same time the mean squared radius of the Gaussian peak of the distribution increases to the asymptotic value $(1/2)U$. For values of $|s|$ much greater than $1/\gamma$ the conditioned distribution Eq. (2.504) relaxes to a form centered on the origin, which is simply the unconditioned distribution $W(\mathcal{E}_2, y_2 t_2)$ given by Eq. (2.470). The time $1/\gamma$ is a relaxation time for the field amplitude distributions. Our knowledge of $\mathcal{E}_1$ ceases to have much influence on the distribution of $\mathcal{E}_2$ for $|s| > 1/\gamma$. It is not surprising then that for intervals for

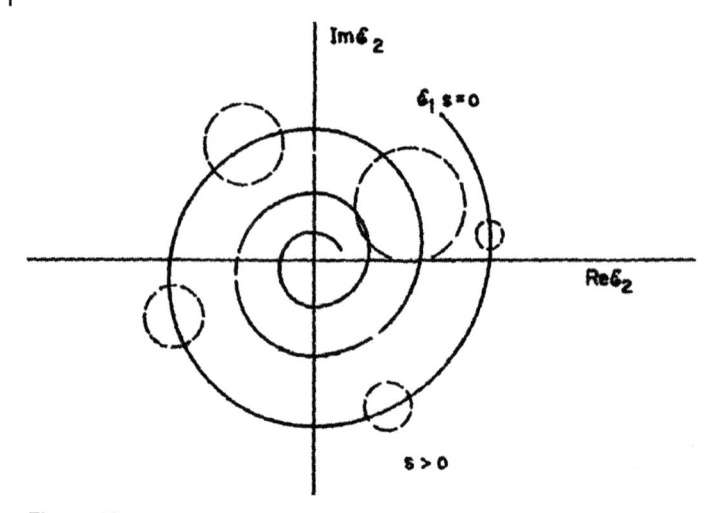

Figure 16

which $|s| \gg 1/\gamma$ the two-photon coincidence rate, which is given by

$$G^{(2)}(y_1t_1y_2t_2,y_2t_2y_1t_1) = \int W(\mathcal{E}_1y_1t_1,\mathcal{E}_2y_2t_2)|\mathcal{E}_1|^2|\mathcal{E}_2|^2\,d^2\mathcal{E}_1\,d^2\mathcal{E}_2$$

$$= \int W(\mathcal{E}_1y_1t_1)W(\mathcal{E}_1y_1t_1|\mathcal{E}_2y_2t_2)|\mathcal{E}_1|^2|\mathcal{E}_2|^2\,d^2\mathcal{E}_1\,d^2\mathcal{E}_2, \quad (2.505)$$

reduces to the factorized form

$$G^{(2)}(y_1t_1y_2t_2,y_2t_2y_1t_1) = G^{(1)}(y_1t_1,y_1t_1)G^{(1)}(y_2t_2,y_2t_2).$$

The tendency toward photon coincidences is wiped out, in other words, when the interval $s = t_1 - t_2 - c^{-1}(y_1 - y_2)$ becomes large because the field amplitudes $\mathcal{E}(y_1t_1)$ and $\mathcal{E}(y_2t_2)$ cease to be statistically correlated.

To see how the full time dependence of the coincidence rate emerges from the integral Eq. (2.505), we note that when the conditioned distribution function is given by Eq. (2.504), the average value of $|\mathcal{E}_2|^2$ when $\mathcal{E}_1$ is fixed is

$$\int W(\mathcal{E}_1y_1t_1|\mathcal{E}_2y_2t_2)|\mathcal{E}_2|^2\,d^2\mathcal{E}_2 = |\mathcal{E}_1|^2 e^{-2\gamma|s|} + \tfrac{1}{2}U(1-e^{-2\gamma|s|}). \quad (2.506)$$

When this expression is multiplied by $|\mathcal{E}_1|^2$ and averaged, as in Eq. (2.505), over the Gaussian form for $W(\mathcal{E}_1y_1t_1)$, we find

$$G^{(2)}(y_1t_1,y_2t_2,y_2t_2,y_1t_1) = \left(\tfrac{1}{2}U\right)^2\left\{2e^{-2\gamma|s|} + 1 - e^{-2\gamma|s|}\right\}$$

$$= \left(\tfrac{1}{2}U\right)^2\left\{1 + e^{-2\gamma|s|}\right\}, \quad (2.507)$$

which verifies the value of the coincidence rate found earlier in Eq. (2.503).

The values we have derived for the correlation functions have all been based on the assumption that the energy spectrum of our light beam has the Lorentz shape. The corresponding results are easily derived for other spectra for which the Fourier transform of the energy distribution is known. Other simple, smooth representations of the profile of a spectrum line, for example, lead to results which are qualitatively similar to those for the Lorentz line.

Since the photon correlation effect extends over delay times of the order of the inverse band width, γ, it might appear that this time can be stretched out by a factor of a million or more by using the extremely monochromatic light of the laser rather than light from natural sources. The error in such reasoning lies in the fact that the statistical properties of the laser beam are quite different from those of the chaotically generated beams we have been discussing. Lasers, when they are operating most monochromatically, generate beams with very little amplitude modulation, and for these, as we have seen in the last Section, there would be virtually no photon correlation effect at all.

2.13.1
Model for Ideal Laser Fields

For fields generated by chaotic sources, knowledge simply of the average occupation numbers $\langle n_k \rangle$ is sufficient to determine the density operator ρ, and from it all of the statistical properties of the field. However if our source is not chaotic in nature, we cannot expect that there will exist any self-evident way of finding the density operator for the field it generates without analyzing the mechanism by which it radiates in some detail. The only reliable method we have of constructing density operators, in general, is to devise theoretical models of the system under study and to integrate corresponding Schrödinger equation, or equivalently to solve the equation of motion for the density operator. These assignments are formidable ones for the case of the laser oscillator and have not been carried out to date in quantum mechanical terms. The greatest part of the difficulty lies in the mathematical complications associated with the nonlinearity of the device. The nonlinearity plays an essential role in stabilizing the field generated by the laser. It seems unlikely, therefore, that we shall have a quantum mechanically consistent picture of the frequency bandwidth of the laser or of the fluctuations of Its output until further progress is made with these problems.

If we are willing to overlook the noise and band width problems for the moment, and to confine our discussion to the case of an ideally monochromatic laser, then it is not difficult to find a representation for the density operator of the beam it generates. The radiation field is coupled within the laser to the electric dipole vectors of all of the atoms of the active medium. These atoms have a polarization which oscillates with the field and at the same time radiates energy into it. If we view the active medium as a whole, we see that it has an oscillating polarization density of macroscopic proportions, i.e., all neighboring atoms contribute similarly to the total polarization density. If we remember that the time derivative of a polarization density is, in effect, a current distribution, then we may think of the field as being radiated by the oscillating current distribution. When the laser is operating well above its threshold there is nothing weak about this current distribution; it is essentially of classical magnitude. Furthermore, if the laser has the ideal stability we have assumed, the current simply oscillates steadily in a perfectly predictable way. We may, in other words, to an excellent approximation, describe the bound current in the active medium as a *c*-number current density.

The general problem of finding the fields radiated by prescribed current distributions has been solved in Sect. 2.10. The most important property of the solution is that radiation by a known current distribution always brings the field to a coherent state (assuming that no other radiation was present initially). If the current oscillates with a single frequency, only the field modes with precisely that frequency will be excited. If we assume, for simplicity, that the geometry of our system favors the excitation of only one mode of the field, then the density operator for the field may be written in the form

$$\rho = |\alpha\rangle\langle\alpha| \, , \tag{2.508}$$

where $|\alpha\rangle$ is a coherent state for the excited mode, and the amplitude α is given by an integral of the form Eq. (2.367) taken over the bound current distribution.

Let us write the complex field eigenvalue which corresponds to the amplitude α as

$$\mathcal{E}(rt) = i\left(\frac{\hbar\omega}{2}\right)^{1/2} u(r)\,e^{-i\omega t}\alpha \, . \tag{2.509}$$

Then, since the density operator Eq. (2.508) corresponds to a pure coherent state, the correlation functions of all orders will factorize to the form of Eq. (2.136), i.e., the beam will possess full coherence. It follows then that the *n*-fold delayed coincidence rates will factorize to the form

$$G^{(2)}(x_1\ldots x_n, x_n\ldots x_1) = \prod_{j=1}^{n} G^{(1)}(x_j, x_j) \, , \tag{2.510}$$

and no photon coincidence correlations of any order will be detectable in the ideal laser beam.

The argument which led to the density operator Eq. (2.508) for the laser beam assumed that the oscillating current distribution is known precisely, i.e., that we know its phase of oscillation as well as its amplitude. In practice our knowledge about quantities which oscillate at extremely high frequencies rarely includes any information about their absolute phase. (This is due more to the absence of a suitable clock to use as a reference standard than it is to any difficulty of principle in defining or measuring the phase of essentially classical quantities such as the bound current in the laser.) When we lack any knowledge of the phase of oscillation of the current, the density operator should be written in an appropriately specialized form of Eq. (2.372). It is clear that this form is simply the expression Eq. (2.508) for the density operator averaged over the phase of the complex amplitude α, i.e.,

$$\rho = \int_0^{2\pi} \left||\alpha|e^{i\theta}\right\rangle \left\langle|\alpha|e^{i\theta}\right| \frac{d\theta}{2\pi}$$

$$= \int \frac{1}{2\pi|\alpha|} \delta(|\beta| - |\alpha|) |\beta\rangle\langle\beta| \, d^2\beta. \tag{2.511}$$

These forms of the density operator depend on a only through its absolute value, and hence represent stationary fields. They represent mixed rather than pure states of the field, but as we have noted in the last Section, mixtures corresponding to averaging an overall phase variable do not alter the coherence properties of the field. It is easy to verify that the correlation functions which are derived from the density operator Eq. (2.511) are identical to those which follow from Eq. (2.508).

The explicit construction of the density operator for an ideal laser beam shows that no photon correlations are to be detected in such a beam. The reason for the absence of such correlations is evident from the analysis of the last Section. The quasiprobability function $W(\mathcal{E}, x)$ which corresponds to the stationary density operator Eq. (2.511) is immediately seen from Eqns. (2.466) and (2.509) to be

$$W(\mathcal{E}, x) = \frac{1}{2\pi \left(\frac{\hbar\omega}{2}\right)^{1/2} |\mathbf{u}(\mathbf{r})\alpha|} \delta\left(|\mathcal{E}| - \left(\frac{\hbar\omega}{2}\right)^{1/2} |\mathbf{u}(\mathbf{r})\alpha|\right). \tag{2.512}$$

This function vanishes everywhere in the complex $\mathcal{E}$ plane except on a circle where the delta function is singular. It describes a field which undergoes no amplitude modulation at all, and that is the basic reason for the absence of photon correlations in an ideal laser beam.

It is also possible, by making use of the correspondence principle, to see the origin of this property of coherently radiated beams more directly. We shall simplify our picture of the laser by regarding it simply as an oscillating charge distribution which radiates much as an antenna does. The charge, we assume, has only a single mode of vibration whose amplitude is, in effect, that of a harmonic oscillator. Since the electric polarization of this oscillator assumes macroscopic proportions we must regard the oscillator coordinate as an essentially classical quantity; i.e., the oscillator is typically in highly excited quantum states which have enormous quantum numbers.

When the oscillator is decoupled from whatever mechanism has excited it and allowed to radiate spontaneously, its amplitude of vibration will decrease quite slowly in relation to the oscillation period. Since the behavior of the oscillator is essentially classical, the current due to its moving charge distribution is quite predictable. As we have noted earlier, the radiation by such a current brings the field to a coherent state. If, on the other hand, we look at the oscillator from a quantum mechanical standpoint, we may think of it as making transitions downward in energy, step by step, passing through states with quantum numbers n, $n-1$, $n-2$... where $n \gg 1$. The length of time the oscillator spends in each of these states is distributed exponentially and, since n is so large, the average lifetimes of the states do not vary significantly from one state to the next. Each transition is accompanied by the emission of a photon. We are therefore not surprised to find that when the photons are detected by a counter, the intervals between their successive arrival times are exponentially distributed. This exponential distribution of time intervals indicates the absence of any tendency toward pair or higher order correlations. It is the characteristic distribution for the intervals between totally uncorrelated events which happen at a fixed average rate. It is clear that where two or more counters are used there will be no time-dependent correlations of their outputs.

2.13.2
Model of a Laser Field With Finite Bandwidth

An actual laser beam, in contrast to the ideal variety we have just discussed, will never be precisely monochromatic. Its frequency is bound to vary more or less randomly over a narrow range due to disturbances which have their origin both inside and outside the laser itself. We shall construct a simple model of a laser field with finite frequency bandwidth by assuming that the mechanism which disturbs the laser is essentially stochastic in nature.

Let us assume, for simplicity, that the laser excites only a single mode of the electromagnetic field which has frequency ω_0. Then the field Hamiltonian for that mode is

$$H_0 = \hbar \omega_0 a^\dagger a$$

and, in the absence of any perturbing influences, the time-dependent operators $a(t)$ and $a^\dagger(t)$ are given in terms of the time-independent ones, a and $a^\dagger$, by

$$\begin{aligned} a(t) &= a\,e^{-i\omega_0 t} \\ a^\dagger(t) &= a^\dagger\,e^{i\omega_0 t} \end{aligned} \tag{2.513}$$

The completely harmonic behavior of the oscillating field will be perturbed by various interactions of the field with other systems. We shall assume that the effect of these interactions can be represented by the addition of a term to the field Hamiltonian which depends on one or more random functions of time, $f(t)$. If we write this stochastic addition to the Hamiltonian as $H_f(t)$, the total field Hamiltonian becomes

$$H = H_0 + H_f(t) \,. \tag{2.514}$$

To see the influence of the stochastic term most clearly we shall solve the Schrödinger equation in the interaction representation. The interaction Hamiltonian is then

$$H'_f(t) = e^{iH_0 t/\hbar} H_f(t)\, e^{-iH_0 t/\hbar} \,. \tag{2.515}$$

2.13 Elementary Models of Light Beams

We define the unitary operator $U_f(t,t')$ as the solution of the Schrödinger equation

$$i\hbar \frac{\partial}{\partial t} U_f(t,t') = H'_f(t) U_f(t,t') \tag{2.516}$$

which obeys the initial condition

$$U_f(t',t') = 1 . \tag{2.517}$$

Then, if we write the state vector of the field at time t as $|t\rangle$, we see that it evolves according to the transformation

$$|t\rangle = U_f(t,t') |t'\rangle .$$

The equation of motion for the density operator in the interaction representation, which we shall write as $\rho_i(t)$, is

$$i\hbar \frac{\partial}{\partial t} \rho_i(t) = [H'_f(t), \rho_i(t)] . \tag{2.518}$$

The solution for the time development of the density operator may be written in terms of the unitary operator U_f as

$$\rho_i(t) = U_f(t,t') \rho_i(t') U_f^{-1}(t,t') . \tag{2.519}$$

The expressions for the field correlation functions which we have discussed earlier in this chapter have all been constructed according to Heisenberg picture of quantum mechanics in which the state vectors and the density operator are independent of time. When these vary with time, as in the interaction representation, the expectation values we require must be constructed somewhat differently. The required expressions can be found by starting with the form the expectation values take in the Heisenberg representation and carrying out the unitary transformation to the interaction representation.

Let us consider two arbitrary operators which take the time-dependent forms $L(t)$ and $M(t)$ in the Heisenberg representation. An example of the kind of statistical average which is used in the construction of the correlation functions is the averaged product which may be written as $\langle L(t)M(t')\rangle_f$. The subscript on the average means that it is computed for a particular behavior of the random function $f(t)$ on which the stochastic Hamiltonian depends. The average, when evaluated in the Heisenberg representation, is clearly

$$\langle L(t)M(t')\rangle_f = \text{Tr}\{L(t)M(t')\rho\} , \tag{2.520}$$

where ρ is the time-independent Heisenberg density operator.

One of the ways of defining the Heisenberg representation (which is unitarily equivalent to all other ways) is to let the fixed Heisenberg state vector for the system be identical to the state vector in the interaction representation at a particular time t_0. Then the relation

$$|t\rangle = U_f(t,t_0) |t_0\rangle \tag{2.521}$$

expresses the unitary transformation from Heisenberg states $|t_0\rangle$ to states $|t\rangle$ in the interaction representation. The corresponding transformations of the operators L, M and ρ are

$$L_i(t) = U_f(t,t_0)\, L(t)\, U_f^{-1}(t,t_0),$$

$$M_i(t) = U_f(t,t_0)\, M(t)\, U_f^{-1}(t,t_0), \qquad (2.522)$$

$$\rho_i(t) = U_f(t,t_0)\, \rho\, U_f^{-1}(t,t_0),$$

where the subscripts i denote the forms of the operators in the interaction representation. When the inverted forms of these relations are used to express the operators in Eq. (2.520) we find

$$\langle L(t) M(t')\rangle_f = \mathrm{Tr}\left\{U_f^{-1}(t,t_0) L_i(t) U_f(t,t_0) U_f^{-1}(t',t_0) M_i(t') \rho_i(t') U_f(t',t_0)\right\}. \qquad (2.523)$$

Since the time displacement operator U_f obeys the multiplication law

$$U_f(t,t') U_f(t',t_0) = U_f(t,t_0), \qquad (2.524)$$

the expression for the average may be reduced to the form

$$\langle L(t) M(t')\rangle_f = \mathrm{Tr}\left\{L_i(t)\, U_f(t,t')\, M_i(t')\, \rho_i(t')\, U_f^{-1}(t,t')\right\}. \qquad (2.525)$$

The occurrences of the operator U_f in this expression evidently take into account the effect of the disturbance of the field during the interval from t' to t. The disturbance, we are assuming, is a random one and the average Eq. (2.525) has been evaluated for some particular way in which it may behave, i.e., it is evaluated for a particular random function $f(t)$. Before the average can be compared with experiments it must again be averaged over a suitable ensemble of random functions $f(t)$. The latter averaging process is simplified by our use of the interaction representation.

Since the products LM which interest us are in normally ordered form it will be extremely convenient to make use of the P representation for the density operator. We shall therefore only consider the class of stochastic Hamiltonians which preserve the possibility of expressing the density operator by means of the P representation. We assume, in other words, that $\rho_i(t)$ may be written in the form

$$\rho_i(t) = \int P(\alpha,t) |\alpha\rangle\langle\alpha|\, d^2\alpha \qquad (2.526)$$

at all times t.

If the density operator at time t' corresponds to the pure coherent state $|\alpha\rangle$, i.e.

$$\rho_i(t') = |\alpha\rangle\langle\alpha|, \qquad (2.527)$$

then, according to Eq. (2.519), at time t it will be

$$\rho_i(t) = U_f(t,t') \rho_i(t') U_f^{-1}(t,t')$$
$$= U_f(t,t') |\alpha\rangle \langle \alpha| U_f^{-1}(t,t') . \quad (2.528)$$

Now, according to Eq. (2.526), this operator too will have a P representation for which we may introduce the special notation

$$\rho_i(t) = \int P(\alpha t'|\beta t) |\beta\rangle \langle \beta| \, d^2\beta . \quad (2.529)$$

The function $(\alpha t'|\beta t)$ is evidently a conditioned quasiprobability function. It corresponds in the classical limit to a probability distribution for the complex amplitude β at time t, when we are given the knowledge that it had (or will have) the value α at time t'.

To illustrate the use of these relations in evaluating statistical averages, let us consider the average of the product $a^\dagger(t)a(t')$ which occurs in the first order correlation function. If we substitute $L(t) = a^\dagger(t)$ and $M(t) = a(t)$ into Eq. (2.525) we find, by using Eq. (2.513)

$$\left\langle a^\dagger(t)a(t') \right\rangle_t = \text{Tr}\left\{ a^\dagger e^{i\omega_0 t} U_f(t,t') a e^{-i\omega_0 t'} \rho_i(t') U_f^{-1}(t,t') \right\} . \quad (2.530)$$

Next we make use of Eq. (2.526) for the density operator, and the fact that $|\alpha\rangle$ is an eigenstate of a to write

$$\left\langle a^\dagger(t)a(t') \right\rangle_t = \text{Tr}\left\{ U_f(t,t') \int P(\alpha t') \alpha |\alpha\rangle \langle \alpha| \, d^2\alpha \, U_f^{-1}(t,t') a^\dagger \right\} e^{i\omega_0(t-t')} . \quad (2.531)$$

The unitary transformation inside the brackets may now be carried out by using Eq. (2.529) to represent the density operator indicated in Eq. (2.528). We then have

$$\left\langle a^\dagger(t)a(t') \right\rangle_t = \text{Tr}\left\{ \int P(\alpha t') \alpha P(\alpha t'|\beta t) |\beta\rangle \langle \beta| \beta^* \, d^2\alpha \, d^2\beta \right\} e^{i\omega_0(t-t')}$$
$$= \int P(\alpha t') P(\alpha t'|\beta t) \alpha \beta^* \, d^2\alpha \, d^2\beta \, e^{i\omega_0(t-t')} . \quad (2.532)$$

The latter expression for the average bears a close resemblance to forms which occur in the classical theory of continuous Markoff processes. We must now remember that the average we have constructed corresponds to some particular behavior of the random Hamiltonian. The quantity to be compared with experiment is not any one such value, but the average of all such values taken over a suitable ensemble of random functions $f(t)$. We may write this average as

$$\left\langle a^\dagger(t)a(t') \right\rangle = \int \left\langle P(\alpha t') P(\alpha t'|\beta t) \right\rangle_{\text{av. over f}} \alpha \beta^* \, d^2\alpha \, d^2\beta \, e^{i\omega_0(t-t')} . \quad (2.533)$$

The foregoing equations furnish us with a fairly general framework for discussing the influence of random disturbances on the oscillations of the field. We shall now use this formalism in constructing a simple model of a laser beam of finite bandwidth.

Surely the simplest way to give the oscillating mode of the field a finite frequency bandwidth is to assume that its frequency is a random function of time. We may do this by writing the total field Hamiltonian of Eq. (2.514) as

$$H = \hbar[\omega_0 + f(t)] a^\dagger a , \qquad (2.534)$$

where $f(t)$ is a random function of some sort whose ensemble average, $\langle f(t) \rangle$, vanishes.

Since the random Hamiltonian is evidently

$$H_f(t) = \hbar f(t) a^\dagger a , \qquad (2.535)$$

and it commutes with $H_0 = \hbar \omega a^\dagger a$, the interaction Hamiltonian according to Eq. (2.515) is simply H_f itself.

The Schrödinger equation (2.516) then takes the form

$$i\frac{\partial}{\partial t} U_f(t,t') = f(t) a^\dagger a U_f(t,t') . \qquad (2.536)$$

Its solution is simply an exponential function which may be written in the form

$$U_f(t,t') = e^{-i a^\dagger a \phi(tt')} \qquad (2.537)$$

where ϕ is defined by

$$\phi(tt') = \int_{t'}^{t} f(t'') dt'' . \qquad (2.538)$$

To see the effect of the transformation U_f on the states of the field, let us suppose that the field is in the coherent state $|\alpha\rangle$ at time t'. Then at time t the state will be

$$\begin{aligned}
|t\rangle &= U_f(t,t') |\alpha\rangle \\
&= e^{-i a^\dagger a \phi(tt')} |\alpha\rangle \\
&= e^{-|\alpha|^2/2} e^{-i a^\dagger a \phi(tt')} \sum_{n=0}^{\infty} \frac{\alpha^n}{(n!)^{1/2}} |n\rangle \\
&= e^{-|\alpha|^2/2} \sum \frac{\alpha^n}{(n!)^{1/2}} e^{-in\phi(tt')} |n\rangle \\
&= \left| \alpha e^{-i\phi(tt')} \right\rangle .
\end{aligned} \qquad (2.539)$$

The particular random Hamiltonian we have assumed just transforms one coherent state into another for which the amplitude parameter differs from the original one by a phase factor. There is evidently no amplitude modulation in this model at all.

2.13 Elementary Models of Light Beams

When we use Eq. (2.536) to construct the density operator represented by Eqns. (2.528) and (2.529) we find

$$\left|\alpha e^{-i\phi(tt')}\right\rangle\left\langle\alpha e^{-i\phi(tt')}\right| = \int P(\alpha t'|\beta t)\,|\beta\rangle\,\langle\beta|\,d^2\beta, \tag{2.540}$$

from which we see that we may take the conditioned quasiprobability density to be simply the delta function

$$P(\alpha t'|\beta t) = \delta^{(2)}\left(\beta - \alpha e^{-i\phi(tt')}\right). \tag{2.541}$$

If we introduce the phases of the amplitudes α and β via the definitions

$$\begin{aligned}\alpha &= |\alpha|\,e^{i\theta_0}, \\ \beta &= |\beta|\,e^{i\theta},\end{aligned} \tag{2.542}$$

then the two-dimensional delta function Eq. (2.541) can be written in terms of a product of two one-dimensional ones as

$$P(\alpha t'|\beta t) = \frac{1}{|\alpha|}\,\delta(|\beta|-|\alpha|)\,\delta(\theta-\theta_0+\phi(tt')). \tag{2.543}$$

This function describes the evolution of the state of the field from the coherent state $|\alpha\rangle$ at time t', when we are given any particular random function $f(t)$. To find the state at time t which is typical of the set of possible random functions, we must average Eq. (2.543) over the ensemble of functions $f(t)$. We may write this average as

$$P_{\mathrm{av}}(\alpha t'|\beta t) = \frac{1}{|\alpha|}\,\delta(|\beta|-|\alpha|)\,\langle\delta(\theta-\theta_0+\phi(tt'))\rangle_{\mathrm{av.\ over}\ f}. \tag{2.544}$$

Now, if we recall that the function $\delta(\theta)$ has the Fourier series expansion

$$\delta(\theta) = \frac{1}{2\pi}\sum_{m=-\infty}^{\infty} e^{im\theta}, \tag{2.545}$$

we see that the averaged delta function in Eq. (2.544) may be written as

$$\delta(\theta) = \frac{1}{2\pi}\sum_{m=-\infty}^{\infty} e^{im(\theta-\theta_0)}\left\langle\exp\left\{im\int_{t'}^{t} f(t'')\,dt''\right\}\right\rangle_{\mathrm{av.\ over}\ f}. \tag{2.546}$$

We must clearly specify some of the properties of the random functions $f(t)$ before the exponential functions in Eq. (2.546) can be averaged over them.

The different physical processes which may perturb the frequency of our field oscillator require in general that we discuss various kinds of random functions $f(t)$. For the present, however, we shall only consider one of the simpler types of random functions. We shall assume that $f(t)$ is a stationary Gaussian stochastic process, i.e., that at any time t the ensemble of values of $f(t)$ has a fixed Gaussian distribution. Then it is not difficult to show that the averaged exponentials in Eq. (2.546) are given by

$$\left\langle \exp\left\{ im \int_{t'}^{t} f(t'') \, dt'' \right\} \right\rangle_{\substack{\text{av.}\\\text{over } f}} = \exp\left\{ -\tfrac{1}{2} m^2 \int_{t'}^{t}\int_{t'}^{t} \langle f(t'')f(t''')\rangle \, dt'' \, dt''' \right\}, \tag{2.547}$$

where the ensemble average $\langle f(t'')f(t''')\rangle$ is simply the autocorrelation function of the random process $f(t)$.

Let us assume, simply as an illustration, that the function $f(t)$ fluctuates so rapidly that its autocorrelation function can be taken to have the form

$$\langle f(t'')f(t''')\rangle = 2\zeta\, \delta(t'' - t'''), \tag{2.548}$$

where ζ is a positive constant. Then the averaged exponential in Eq. (2.547) reduces to

$$\left\langle \exp\left\{ im \int_{t'}^{t} f(t'') \, dt'' \right\} \right\rangle_{\substack{\text{av.}\\\text{over } f}} = \exp\left\{ -m^2 \zeta |t - t'| \right\}, \tag{2.549}$$

and the averaged delta function in Eq. (2.546) becomes

$$\langle \delta(\theta - \theta_0 + \phi(tt')) \rangle_{\substack{\text{av.}\\\text{over } f}} = \frac{1}{2\pi} \sum_{m=-\infty}^{\infty} e^{im(\theta - \theta_0) - m^2 \zeta |t - t'|}. \tag{2.550}$$

It is interesting to note that this function is simply the Green's function of the partial differential equation for the diffusion of heat on a circular ring, i.e., it satisfies the equation

$$\left(\frac{\partial}{\partial t} - \zeta \frac{\partial^2}{\partial \theta^2} \right) \langle \delta(\theta - \theta_0 + \phi) \rangle_{\text{av}} = 0$$

for $t > t'$ and reduces to $\delta(\theta - \theta_0)$ for $t = t'$. It is clear then that the conditioned quasiprobability function Eq. (2.544), which we may write as

$$P_{\text{av}}(\alpha t' | \beta t) = \frac{1}{2\pi |\alpha|} \delta(|\beta| - |\alpha|) \sum_{m=-\infty}^{\infty} e^{im(\theta - \theta_0) - m^2 \zeta |t - t'|} \tag{2.551}$$

describes a kind of random phase modulation in which the phase variable $\theta = \arg \beta$ "diffuses" away from its initial value, θ_0.

The reciprocal of the diffusion constant ζ defines a relaxation time for the phase variable. For time intervals $t - t'$ which greatly exceed $1/\zeta$ the distribution Eq. (2.551) reduces to a constant, circularly symmetric form; the phase θ becomes completely random.

2.13 Elementary Models of Light Beams

Let us now return to the question of evaluating the first order correlation function for the field. According to Eq. (2.533) we may construct the function as soon as we have evaluated the average

$$\langle P(\alpha,t') P(\alpha t'|\beta t) \rangle_{\text{av. over } f} . \tag{2.552}$$

We shall assume that we have no knowledge of the initial phase of oscillation of the field. Since the random perturbation of the field only shifts its phase, the phase remains uniformly distributed at all times; i.e., we never know more about the phase than we did initially. The density operator which represents the field is therefore stationary. The function $P(\alpha,t)$ in Eq. (2.526) depends on α only through its absolute value and is independent of t, and of the behavior of the function $f(t)$ as well. In this most frequently occurring case, the function $P(\alpha,t')$ may be written as $P(|\alpha|)$ and removed from the averaging brackets in the expression (2.552). That expression then reduces to the form

$$P(|\alpha|) P_{\text{av}}(\alpha t'|\beta t) , \tag{2.553}$$

where the second factor is given by Eq. (2.551). Now it is evident from Eq. (2.551) that

$$\int P_{\text{av}}(\alpha t'|\beta t)\beta^* \, \mathrm{d}^2\beta = \frac{1}{2\pi|\alpha|} \int_0^\infty \delta(|\beta|-|\alpha|) |\beta|^2 \, \mathrm{d}|\beta|$$

$$\times \int_0^{2\pi} \mathrm{e}^{-\mathrm{i}\theta} \sum \mathrm{e}^{\mathrm{i}m(\theta-\theta_0) - m^2\zeta|t-t'|} \, \mathrm{d}\theta$$

$$= |\alpha| \mathrm{e}^{-\mathrm{i}\theta_0 - \zeta|t-t'|} = \alpha^* \mathrm{e}^{-\zeta|t-t'|} . \tag{2.554}$$

On substituting the expression (2.553) into the correlation function Eq. (2.533) and making use of the integral just evaluated we find

$$\langle a^\dagger(t) a(t') \rangle = \int P(|\alpha|) |\alpha|^2 \, \mathrm{d}^2\alpha \, \mathrm{e}^{\mathrm{i}\omega_0(t-t') - \zeta|t-t'|}$$

$$= \langle |\alpha|^2 \rangle \, \mathrm{e}^{\mathrm{i}\omega_0(t-t') - \zeta|t-t'|} , \tag{2.555}$$

where the symbol $\langle |\alpha|^2 \rangle$ has been used for the mean squared amplitude of excitation, or equivalently the average number of photons in the mode.

If we assume that the mode function $\boldsymbol{u}(\boldsymbol{r})$ for the field does not change as a result of the perturbation, then the full space-time dependence of the first order correlation function may be found by multiplying the expression Eq. (2.555) by a product of the form $\boldsymbol{u}^*(\boldsymbol{r})\boldsymbol{u}(\boldsymbol{r}')$. According to Eq. (2.336), which is a quantum mechanical form of the Wiener–Khintchine theorem, the energy spectrum of the field will be proportional to the Fourier transform of the correlation function Eq. (2.555). When we calculate the transform we find

$$\int_{-\infty}^\infty \langle a^\dagger(0) a(t') \rangle \, \mathrm{e}^{\mathrm{i}\omega t'} \, \mathrm{d}t' = \langle |\alpha|^2 \rangle \int_{-\infty}^\infty \mathrm{e}^{\mathrm{i}(\omega-\omega_0)t' - \zeta|t'|} \, \mathrm{d}t'$$

$$= \langle |\alpha|^2 \rangle \frac{2\zeta}{(\omega-\omega_0)^2 + \zeta^2} . \tag{2.556}$$

Our phase diffusion model thus has an energy spectrum of Lorentzian shape, the diffusion constant ζ is its half-width.

From a spectroscopic standpoint, the field we are describing could not be distinguished from the chaotically generated field of Lorentzian line shape which we discussed earlier, if we happened to have $\zeta = \gamma$. The fundamentally different nature of these two fields is best expressed by means of their higher order correlation functions. These functions may be evaluated for the phase diffusion model through simple extensions of the methods we have developed, but we shall not do so here. One fairly obvious result, however, is worth mentioning. Since the random phase modulation we have described carries no amplitude modulation with it, it will not introduce any photon coincidence correlations.

There are a number of ways in which the simple phase diffusion model which we have presented as an illustration can be generalized and made more realistic. We may easily remove, for example, the assumption that the stochastic process $f(t)$ has a vanishingly small relaxation time. Furthermore we may consider other types of stochastic processes than Gaussian ones. Finally, we may consider other forms of the random Hamiltonian than Eq. (2.535) and attempt in that way to account for some of the effects of random amplitude modulation as well as phase modulation.

2.14
Interference of Independent Light Beams

One of the questions having to do with coherence which has given rise to much discussion and a certain amount of confusion recently is that of interference between independent light beams. That such interference phenomena can exist should come as no great surprise; they have been observed long ago with radio waves of fixed frequency. If we have had to wait until recently[24] to see such phenomena at optical frequencies, the delay has been wholly due to instrumental difficulties.

The problems which have arisen in the discussion of these interference phenomena concern the precise way in which they should be understood and described. It would be quite difficult to say how much of the misunderstanding we have mentioned is simply semantic in nature and how much is more deeply conceptual. There is, for example, nothing intrinsically quantum mechanical about the interference of independent beams. Yet the fact that altogether different sets of quanta must somehow interfere with one another seems to have contributed greatly to the confusion. We shall not recount the history of this subject here but shall only discuss a few of the simplest possible examples of the interference phenomenon.

The simplest sort of experimental arrangement we can have is essentially that illustrated in Fig. 17. Two independent laser sources (or possibly other types of sources), L_1 and L_2 project their beams in directions which are nearly parallel, but slightly convergent. The beams fall upon overlapping areas of a screen Σ. If the light intensities are high enough, or we have sufficient time available to record over a long period, we

2.14 Interference of Independent Light Beams

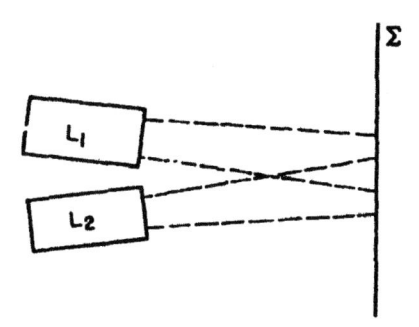

Figure 17

may let our detector be a photographic film in the plane Σ. If the conditions do not favor photography, on the other hand, we might use a mosaic of photon counters in the plane Σ. In either case we will look for interference fringes in the area of overlap of the beams.

Let us assume that the way in which each light source excites the field can be described in the P representation by means of functions $P_1(\{\alpha_{1k}\})$ and $P_2(\{\alpha_{2k}\})$. The single P function which describes the superposed fields is then given, according to Eq. (2.275)) or (2.310), by

$$P(\{\alpha_k\}) = \int P_1(\{\alpha_{1k}\}) P_2(\{\alpha_{2k}\}) \prod_k \delta^{(2)}(\alpha_k - \alpha_{1k} - \alpha_{2k}) d^2\alpha_{1k} d^2\alpha_{2k} . \quad (2.557)$$

The average intensity of the superposed fields at any space-time point x is given by the first order correlation function

$$G^{(1)}(x,x) = \int P(\{\alpha_k\}) |\mathcal{E}(x,\{\alpha_k\})|^2 \prod_k d^2\alpha_k$$

$$= \int P_1(\{\alpha_{1k}\}) P_2(\{\alpha_{2k}\}) |\mathcal{E}(x,\{\alpha_{1k}+\alpha_{2k}\})|^2 \prod_k d^2\alpha_{1k} d^2\alpha_{2k} . \quad (2.558)$$

In reaching the second of these expressions we have made use of Eq. (2.557) and have carried out the integrations over the variables $\{\alpha_k\}$. Now let us note that the eigenvalue field $\mathcal{E}(x,\{\alpha_k\})$ depends linearly upon the amplitudes α_k so that we have

$$\mathcal{E}(x,\{\alpha_{1k}+\alpha_{2k}\}) = \mathcal{E}(x,\{\alpha_{1k}\}) + \mathcal{E}(x,\{\alpha_{2k}\}) , \quad (2.559)$$

a statement which corresponds to the classical superposition principle. If we substitute this relation in Eq. (2.558), and let the symbols $\{G^{(1)}(x,x)\}_j$ with $j = 1,2$, be the intensities which would be produced by either source in the absence of the other, then we may write the total intensity as

$$G^{(1)}(x,x) = \{G^{(1)}(x,x)\}_1 + \{G^{(1)}(x,x)\}_2$$
$$+ 2\operatorname{Re}\left\{\int P_1(\{\alpha_{1k}\})\, \mathcal{E}^*(x\{\alpha_{1k}\}) \prod_k d^2\alpha_{1k} \int P_2(\{\alpha_{2k}\})\, \mathcal{E}^*(x\{\alpha_{2k}\}) \prod_k d^2\alpha_{2k}\right\}. \quad (2.560)$$

The third term of this sum is evidently an interference term. We must next ask when it contributes to the observed intensities and when it does not.

We have noted in Sect. 2.9.7 that any light beam described in the P representation can be regarded as the superposition of two fields, one of which corresponds to a pure coherent state and the other of which is of the unphased form, i.e., it has vanishing expectation value for the complex field strength. When each of the fields generated by the two sources is analyzed in this way, it becomes clear that the unphased components of the fields will not contribute to the interference term in Eq. (2.560). The interference term will, in fact, vanish completely unless the field generated by each of the two sources has a non-zero coherent component.

The most elementary kind of example in which the interference term is different from zero is one in which the two sources acting separately bring the field to coherent states represented by

$$P_1(\{\alpha_{1k}\}) = \prod_k \delta^{(2)}(\alpha_{1k} - \beta_{1k}),$$
$$P_2(\{\alpha_{2k}\}) = \prod_k \delta^{(2)}(\alpha_{2k} - \beta_{2k}). \quad (2.561)$$

Then the interference term of Eq. (2.560) reduces to

$$2\operatorname{Re}\{\mathcal{E}^*(x,\{\beta_{1k}\})\, \mathcal{E}(x,\{\beta_{2k}\})\}. \quad (2.562)$$

The analysis of this term may be simplified by assuming that the two sources are ideal lasers which are similar in construction and that each excites only a single plane wave mode. The two plane wave modes are then not identical since their propagation vectors are not quite parallel, but they have the came frequency. Under these conditions it is easy to see that the interference term Eq. (2.562) describes stationary intensity fringes which are seen on the screen in the area in which the two beams overlap. The fringes are perpendicular to the plane which contains the two propagation vectors and may be made narrow or broad by making the angle between the beams large or small.

Let us suppose that the single mode excited by source 1 has amplitude β_1 and that excited by source 2 has amplitude β_2. Then, since the plane wave mode functions are intrinsically complex, it is clear that the position of the fringe system on the screen Σ (i.e., its displacement in the direction perpendicular to the fringes) will depend on the phase difference of the complex amplitudes β_1 and β_2. If the geometry of the

2.14 Interference of Independent Light Beams

experiment is sufficiently well determined, then by observing the fringe system we may measure the phase difference.

No difficulty of principle stands in the way of our actually carrying out experiments of the type we have just described with two laser beams. But in practice we never have the complete knowledge of the excitation amplitudes which we assumed, for example, in constructing Eqns. (2.561) and (2.562). As we have remarked many times earlier, we are almost always lacking knowledge of overall phase parameters. As long as this is so we do not know the phases of oscillation of our lasers, and the only way we can honestly represent the density operators for the modes they excite is by means of the functions

$$P_j(\alpha_j) = \frac{1}{2\pi|\beta_j|}\delta(|\alpha_j| - |\beta_j|) \tag{2.563}$$

for $j = 1,2$. These functions represent the stationary density operators which are obtained, as in Eq. (2.511), by averaging the coherent states over phase. But the P functions (2.563) are of the unphased variety; they correspond to vanishing averaged complex fields. When the descriptions of our two sources are stationary, in other words, the interference term in Eq. (2.560) vanishes identically.

If this result is taken to mean that there are no fringes to be seen on the screen, then our ignorance of the phase parameters has somehow wiped out a large-scale physical phenomenon. To bring the paradox of such a conclusion into sharper focus it is possible to argue that each of our laser sources is essentially classical in nature and really has a well defined phase of oscillation. Consequently the fringes should be visible on the screen both to people who do and who don't know the phases alike.

To see that we have not really encountered any fundamental dilemma we must recall that density operators are constructed for the purpose of describing *ensembles* of quantum mechanical experiments. The need to repeat experiments upon many similarly prepared systems arises for reasons which are quite basic to quantum mechanics. The quantities measured in general fluctuate unpredictably from one system to another, even when all the systems are prepared in precisely the same quantum state. When the quantum state itself is random there is still a further reason for carrying out experiments on a large number of systems and averaging their results.

The two P functions given by Eq. (2.561) represent, for example, pure states of the field. In any single experiment carried out with two sources for which all the excitation amplitudes and phases are known, we would probably detect a more-or-less noisy form of the interference pattern we have been discussing. The interference pattern would assume the smooth form given by Eq. (2.562) only after we had averaged over many experiments performed with identically prepared sources.

Now when we have no knowledge of the phases of oscillation of our two laser sources, our formalism describes an ensemble of experiments in which the phases are allowed to be completely random. It is true that the contribution of the interference effect to the average intensity for this ensemble vanishes. But one can not conclude from the vanishing of the ensemble average that the fringes do not show up in the individual experiments. This experiment is one in which the members of the ensemble are individually quite unlike their ensemble average. Each of the experiments will exhibit a stationary fringe pattern on the screen, just as when the oscillation phases are known. But since the phases are random, the displacement of the pattern will vary randomly from one experiment to the next. It is the averaging over the random displacement which wipes away the fringes in the ensemble average.

A question we might now ask is how we can use the density operator formalism at all to make statistical statements about the fringe pattern. When the sources are stationary it has appeared to tell us nothing but that the ensemble average of the interference intensity vanishes at every point on the screen. Let us imagine that we are performing the experiment with a pair of lasers chosen from our random phase ensemble. To determine that there is indeed an interference pattern on the screen we must measure the intensity at a considerable number of points on the screen. We do not prepare the system anew for each of these measurements; they are carried out for a single preparation of the lasers. Now just the first of the intensity measurements at a known point on the screen goes a long way toward determining the phase difference of the two lasers. It determines a linear combination of the sine and cosine of the phase difference of the amplitudes β_1 and β_2 which restricts the phase difference to either of two discrete values. Measurement of the intensity at another point then determines the phase difference.

Once we have used intensity measurements at a couple of points to determine the phase difference we can predict the appearance of the rest of the interference pattern in an ensemble average sense. Of course the ensemble in this case is no longer the one we began with, though it still remains a stationary one. Our initial intensity measurements furnish us with information which requires that we reduce the size of our initial ensemble by retaining only those experiments in which the phase difference is found to be nearly the same. This reduced ensemble will be described by a stationary density operator since a phase factor common to the amplitudes β_1 and β_2 of a pair of degenerate modes remains completely random. Let us suppose that we find the phase difference of the two beams to be

$$\arg \beta_1 - \arg \beta_2 = \theta . \tag{2.564}$$

Then the selection process by which we reduce the ensemble to one appropriate to experiments for fixed θ can be represented by inserting a factor

$$\frac{1}{2\pi} \delta(\arg \alpha_1 - \arg \alpha_2 - \theta) \tag{2.565}$$

into the integrand of the P function (2.557). Once we have located the fringe pattern by experimentally determining its unpredictable position, we have no difficulty in constructing a stationary density operator which predicts the average intensities in the pattern.

The idea of reducing the size of our ensemble to reflect the acquisition of knowledge about a system should not be too unfamiliar. In any multi-step game of chance,

2.14 Interference of Independent Light Beams

for example, the odds for winning, which one hopes are even initially, change as one completes each move. The initial odds are calculated by using the complete ensemble of possible games, but the odds calculated at the later states use only the reduced ensembles appropriate to the information which was revealed by the earlier moves.

Another sense, though a rather different one, in which the use of the stationary density operator furnishes information about the randomly placed interference pattern may be seen by discussing the second order correlation function. It is easy to show that the two-fold coincidence counting rate

$$G^{(2)}(\mathbf{r}t\,\mathbf{r}'t',\mathbf{r}'t'\,\mathbf{r}t) = \int P(\{\alpha_k\})|\mathcal{E}(\mathbf{r}t\{\alpha_k\})|^2 |\mathcal{E}(\mathbf{r}'t'\{\alpha_k\})|^2 \prod_k d^2\alpha_k \quad (2.566)$$

contains a term which oscillates as a function of the positions $\mathbf{r}$ and $\mathbf{r}'$ on the screen. This type of interference term may be derived by means of essentially the same argument as we used in discussing the intensity interference experiments in Sect. 2.2. The oscillation of the intensity correlation function must evidently reflect oscillation of the intensity itself. Furthermore since the unknown phase angles of β_1 and β_2 cancel out of the second order correlation function nothing need be known about them to calculate it.

However a simple measurement of the intensity of a random fringe pattern (e.g., by examining a photograph) is not the same as a measurement of $G^{(2)}$, and there is no simple way of concluding in general from a knowledge of $G^{(2)}$ what the intensity pattern of the random fringe system should be. Thus, while $G^{(2)}$ and the other even order correlation functions are useful in their own right, they offer no alternative way of discussing the fringe intensities. If we want the intensities we must derive them from the density operators for appropriately reduced ensembles.

We have assumed to this point that our light sources are ideal noise-free lasers. We now ask what happens when the random modulation of the devices is taken into account. Since the most important of the parameters in determining the two-beam interference pattern is the phase of oscillation of the laser, we can secure a good idea of what goes on by using the phase diffusion model to represent the laser beams. According to that model, the phase of a laser beam wanders appreciably over time-intervals long compared to a relaxation time $1/\zeta$, and remains relatively fixed over time intervals which are much shorter in length.

When the two laser beams are represented by such models, the light intensities we record on the screen will depend on the length of time we require to make our measurements. If the intensities are sufficiently great that we can record them in a time short compared to $1/\zeta$, then the two beams will retain nearly the initial values of their phases while the measurements are being made. A randomly situated fringe pattern of the sort we have already discussed should then show up. But a similar measurement made, say, half a relaxation time later would reveal a differently placed set of fringes, corresponding to the fluctuation that had taken place in the phase difference of the two beams.

If we could follow the fringe intensity as a function of time, we should see the parallel fringe system execute a sort of random wandering back and forth on the screen. If we were to try recording the intensities on the screen by integrating these over a period much longer than the relaxation time we would find that the fringe structure is washed out and only a uniform intensity remains.

Laser sources are convenient ones for such two-beam experiments, because they are intense, and monochromatic enough to have relatively long relaxation times. It is also quite possible, in principle, to carry out such experiments with beams from ordinary chaotic sources. The random amplitude modulation of these beams will mean that the fringes fluctuate greatly in contrast as well as in position. The relaxation time for these variations will be the inverse frequency bandwidth of the sources. If such fringes have not been photographed to date, it is because exposure times shorter than 10^{-10} s would be necessary.

2.15
Photon Counting Experiments

The number of photons which a counter records in any interval of time fluctuates randomly. In a simple type of counting experiment we might imagine that the counter is exposed to the field for a fixed interval of time t. Then, by repeating the experiment many times, we should find a distribution function for the number of counts received in that interval. Although the average number of counts is frequently all that we require, the way in which the number fluctuates about its average value can be fully understood only when we know the distribution function and its moments. In this Section we shall discuss ways of predicting the distribution function and the relation between the form of the distribution and the coherence of the field.

Let us first recall some of the results we established in Sect. 2.5. We calculated there the probability that in an interval of time from t_0 to t all n atoms of a hypothetical n-atom photodetector undergo photoabsorption transitions which are registered as photon counts. When we eliminate the tensor indices by assuming the field to be fully polarized, this probability is given by Eq. (2.64), i.e., we have

$$p^{(n)}(t) = \int_0^{t_0} \cdots \int_0^{t_0} \prod_{j=1}^n S(t_j'' - t_j') \, G^{(n)}(\mathbf{r}_1 t_1' \ldots \mathbf{r}_n t_n', \mathbf{r}_n t_n'' \ldots \mathbf{r}_1 t_1'') \prod_{j=1}^n dt_j' \, dt_j'',$$

(2.567)

where the sensitivity function S is defined by Eqns. (2.46) and (2.44), and we have set $t_0 = 0$. If our detector happens to be of the broadband variety, we may use Eq. (2.48) to reduce the number of time integrations in this integral from $2n$ to n, but this reduction is not a necessary one for the arguments to follow.

We must now consider a more realistic model of a counter which contains an enormous number of atoms, say $N \approx 10^{20}$, which are capable of detecting photons by undergoing photoabsorption processes. Needless to say, it will virtually never happen that all N of these atoms do undergo absorption processes in any finite interval of time. The total number of photoabsorptions is much smaller as a rule, and we shall try to use Eq. (2.567) to find its distribution law.

The total number of photocounts recorded in any interval of time may be regarded as a sum of random variables, one for each atom of the detector. To do this, let us introduce the random variable z_j for the j-th atom, which takes on the values

$$z_j = \begin{cases} 0 \text{ if no photoabsorption process is recorded for the } j\text{-th atom,} \\ 1 \text{ if a photoabsorption process is recorded for the } j\text{-th atom.} \end{cases} \quad (2.568)$$

Then the random variable which represents the total number of counts will be

$$C = \sum_{j=1}^{n} z_j . \quad (2.569)$$

Associated with each final state of the system, i.e., any set of values $z_1 \ldots z_N$, there is a probability function $\mathcal{P}(z_1 \ldots z_N, t)$. The statistical average of any function of the z_j's is then found by averaging the function over the probability distribution. For example, the average number of counts is given by

$$\langle C \rangle = \sum_{z_j=0,1} \sum_{J=1}^{N} z_j \mathcal{P}(z_1 \ldots z_N, t) , \quad (2.570)$$

where the final summation is over the values 0 and 1 for the entire set of variables z_j. We shall write such sums in the future as sums over $\{z_j\}$. We next introduce the reduced probability function for the j-th atom which we define as

$$p_j(z_j, t) = \sum_{z_k, k \neq j} \mathcal{P}(z_1 \ldots z_N, t) . \quad (2.571)$$

The average number of counts may be written in terms of the reduced probabilities p_j as

$$\langle C \rangle = \sum_{\{z_k\}} \sum_{j=1}^{N} z_j p_j(z_j, t)$$

$$= \sum_{j=1}^{N} p_j(1, t) . \quad (2.572)$$

The probability $p_j(1,t)$ which occurs in the latter expression is clearly equal to the one-atom transition probability $p^{(1)}(t)$ evaluated for the j-th atom. That probability is given by Eq. (2.567) for $n = 1$, with $\mathbf{r}_1 = \mathbf{r}_j$, and we shall write it as $p_j^{(1)}(t)$. The average number of counts is thus

$$\langle C \rangle = \sum_{j=1}^{N} p_j^{(1)}(t) . \quad (2.573)$$

We shall now introduce a generating function which will enable us to solve simultaneously for the unknown distribution of photocounts and for its moments. We could, of course, find the moments directly by generalizing the way in which $\langle C \rangle$ was obtained, but the present method has the advantage of enabling us to obtain all the quantities of interest from a single function. The generating function we choose is

$$Q(\lambda,t) = \langle (1-\lambda)^C \rangle, \qquad (2.574)$$

where C is the random integer given by Eq. (2.569), the brackets indicate an ensemble average, and the variable λ is intended simply to be a useful parameter. If we write Q as a sum over the integer values which C may take on we have an expansion of the form

$$Q(\lambda,t) = \sum_{m=0}^{N} (1-\lambda)^m p(m,t), \qquad (2.575)$$

where $p(m,t)$ is the probability that the counter has recorded m photocounts at the time t. It is clear that if $Q(\lambda,t)$ is known $p(m,t)$ can be obtained by differentiation,

$$p(m,t) = \frac{(-1)^m}{m!} \left[\frac{d^n}{d\lambda^n} Q(\lambda,t) \right]_{\lambda=1}, \qquad (2.576)$$

since Eq. (2.575) may be regarded as a Taylor expansion for Q about $\lambda = 1$.

If, on the other hand, we expand $Q(\lambda,t)$ in a power series about $\lambda = 0$ we have

$$Q(\lambda,t) = \sum_{n=0}^{N} \frac{\lambda^n}{n!} \left[\frac{d^n}{d\lambda^n} Q(\lambda,t) \right]_{\lambda=0}. \qquad (2.577)$$

The derivatives which occur in this expansion are given by

$$(-1)^n \left[\frac{d^n}{d\lambda^n} Q(\lambda,t) \right]_{\lambda=0} = \left\langle \frac{C!}{(C-n)!} \right\rangle$$
$$= \langle C(C-1)\ldots(C-n+1) \rangle. \qquad (2.578)$$

The averages on the right of this equation are known as factorial moments. They are simple linear combinations of the ordinary moments $\langle C^n \rangle$ of the distribution of photocounts. It is clear from these relations that a knowledge of the generating function enables us to find both the probability distribution and its moments. We must next show how it is possible to evaluate the generating function in terms of the photoabsorption probabilities $p^{(n)}(t)$.

First let us note that $Q(\lambda,t)$ can be written as

$$Q(\lambda,t) = \sum_{\{z_k\}} \mathcal{P}(z_1 \ldots z_N, t)(1-\lambda)^{\sum_{j=1}^{N} z_j}$$
$$= \sum_{\{z_k\}} \mathcal{P}(z_1 \ldots z_N, t) \prod_{j=1}^{N} (1-\lambda)^{z_j}. \qquad (2.579)$$

2.15 Photon Counting Experiments

The latter form, however, may be simplified by using the identity

$$(1 - \lambda)^{z_j} = 1 - z_j \lambda, \tag{2.580}$$

which holds because z_j takes on only the values zero and one. With this simplification, Eq. (2.579) becomes

$$Q(\lambda, t) = \sum_{\{z_j\}} P(z_1 \ldots z_N, t) \prod_{j=1}^{N} (1 - \lambda z_j). \tag{2.581}$$

When the N-fold product in this expression is expanded in powers of λ, we have

$$Q(\lambda, t) = \sum_{n=0}^{N} (-\lambda)^n \sum_{\{z_j\} \ n\text{-fold} \atop \text{combinations}} \sum z_{j_1} \ldots z_{j_n} P(z_1 \ldots z_N, t), \tag{2.582}$$

where the first sum is taken over all the ways of choosing n atoms from the set of N.

If we now define the n-fold joint probability that atoms $j_i \ldots j_n$ all undergo photoabsorption processes as

$$p_{j_1 \ldots j_n}^{(n)}(t) = \sum_{\{z_k\}} z_{j_1} \ldots z_{j_n} P(z_1 \ldots z_N, t), \tag{2.583}$$

then we may write the generating function in the form

$$Q(\lambda, t) = \sum_{n=0}^{N} (-\lambda)^n \sum_{n\text{-fold} \atop \text{combinations}} p_{j_1 \ldots j_n}^{(n)}(t). \tag{2.584}$$

Now the number $p_{j_1 \ldots j_n}^{(n)}(t)$ has been defined as the probability that each of a particular set of n atoms absorbs a photon, regardless of what all the other atoms do. This probability is simply the expression $p^{(n)}(t)$ given by Eq. (2.567) and evaluated for the particular atoms $j_1 \ldots j_n$. Hence we know all the terms of Eq. (2.584) and the problem is simply to sum them. What we shall do, in fact, is to turn the sums over atoms into volume integrations.

Since the probabilities $p^{(n)}(t)$ are only large for values of n which are extremely small in comparison with N, we may approximate the sums over n-fold combinations by writing

$$\sum_{n\text{-fold} \atop \text{combinations}} \approx \frac{1}{n!} \sum_{j_1=1}^{N} \sum_{j_2=1}^{N} \cdots \sum_{j_n=1}^{N}. \tag{2.585}$$

Then the sums over the individual atoms may be carried out as spatial integrations by letting the number of atoms per unit volume be $\sigma(r)$ and writing

$$\sum_{j_1=1}^{N} \cdots = \int d\mathbf{r}_1 \, \sigma(\mathbf{r}_1) \ldots \tag{2.586}$$

2 Optical Coherence and Photon Statistics

We are, in effect, dealing with the limit $N \to \infty$. When the probabilities given by Eq. (2.567) are substituted in the expression Eq. (2.584) for the generating function and the sum over combinations of atoms is transformed as we have indicated, we find

$$Q(\lambda,t) = \sum_{n=1}^{\infty} \frac{(-\lambda)^n}{n!} \int_{t_0}^{t} \cdots \int_{t_0}^{t} \int_{\text{Vol. of detector}} \cdots \int_{\text{Vol. of detector}}$$

$$\times G^{(n)}(r'_1 t'_1 \ldots r'_n t'_n, r''_n t''_n \ldots r''_1 t''_1) \prod_{j=1}^{n} \sigma(r'_j) S(t''_j - t'_j) \, dr'_j \, dt'_j \, dt''_j \, . \quad (2.587)$$

To abbreviate this expression a bit, let us define the function

$$V(x',x'') = \sigma(r') \delta(r' - r'') S(t'' - t') , \quad (2.588)$$

where x indicates both the position r and the time t. Then the expression for the generating function reduces to

$$Q(\lambda,t) = \sum_{n=1}^{\infty} \frac{(-\lambda)^n}{n!} \int \cdots \int G^{(n)}(x'_1 \ldots x'_n, x''_n \ldots x''_1) \prod_{j=1}^{n} V(x'_j x''_j) \, d^4 x'_j \, d^4 x''_j \, .$$

$$(2.589)$$

Since this is a power series expansion about $\lambda = 0$, the factorial moments must be given, according to Eqns. (2.105) and (2.106), by

$$\left\langle \frac{C!}{(C-n)!} \right\rangle = \int \cdots \int G^{(n)}(x'_1 \ldots x'_n, x''_n \ldots x''_1) \prod_{j=1}^{n} V(x'_j x''_j) \, d^4 x'_j \, d^4 x''_j , \quad (2.590)$$

where the integrations are carried out over the sensitive volume of the counter and the time interval from 0 to t.

As an illustration of the usefulness of these results, let us consider the case of a fully coherent field. For such a field we have the factorization

$$G^{(n)}(x'_1 \ldots x'_n, x''_n \ldots x''_1) = \prod_{j=1}^{n} G^{(1)}(x'_j, x''_j) , \quad (2.591)$$

so that the series for $Q(\lambda,t)$ may be summed to the form

$$Q(\lambda,t) = e^{-\lambda \iint G^{(1)}(x',x'') V(x'x'') \, d^4 x' \, d^4 x''} \, . \quad (2.592)$$

But from Eq. (2.590) we see that the average number of counts is just

$$\langle C \rangle = \iint G^{(1)}(x',x'') V(x'x'') \, d^4 x' \, d^4 x'' ; \quad (2.593)$$

so that the generating function may be written as

$$Q(\lambda,t) = e^{-\lambda \langle C \rangle} \, . \quad (2.594)$$

Now by using Eq. (2.578) we derive the factorial moments

$$\left\langle \frac{C!}{(C-n)!} \right\rangle = \langle C \rangle^n , \qquad (2.595)$$

and by using Eq. (2.576) we find that the probability distribution is

$$p(m,t) = \frac{\langle C \rangle^m}{m!} e^{-\langle C \rangle} , \qquad (2.596)$$

i.e., when the field is fully coherent we always have a Poisson distribution for the number of counts.

When the field does not possess full coherence we can nevertheless use the coherent states as a basis for describing it. To illustrate the form the statistical calculations take, we shall use the P representation for the density operator of the field. The R representation, which applies more generally, can also be used similarly. In the P representation $G^{(n)}$ is given by the integral

$$G^{(n)}(x_1 \ldots x_{2n}) = \int P(\{\alpha_k\}) \prod_{j=1}^{n} \mathcal{E}^*(x_j\{\alpha_k\}) \prod_{j=n+1}^{2n} \mathcal{E}(x_j\{\alpha_k\}) \prod_{k} d^2\alpha_k . \qquad (2.597)$$

When this expression is substituted into the series Eq. (2.589) we find that the series may be summed to the closed form

$$Q(\lambda,t) = \int P(\{\alpha_k\}) e^{-\lambda \Omega(\{\alpha_k\})} \prod_{k} d^2\alpha_k , \qquad (2.598)$$

where

$$\Omega(\{\alpha_k\}) = \int \mathcal{E}^*(x'\{\alpha_k\}) \mathcal{E}(x''\{\alpha_k\}) V(x',x'') d^4x' d^4x'' . \qquad (2.599)$$

Furthermore we see from Eq. (2.578) that the factorial moments are

$$\left\langle \frac{C!}{(C-n)!} \right\rangle = \int P(\{\alpha_k\}) \Omega^n(\{\alpha_k\}) \prod_{k} d^2\alpha_k , \qquad (2.600)$$

and from Eq. (2.576) that the probability distribution is given by

$$p(m,t) = \int P(\{\alpha_k\}) \frac{\Omega^m(\{\alpha_k\})}{m!} e^{-\Omega(\{\alpha_k\})} \prod_{k} d^2\alpha_k . \qquad (2.601)$$

The probability of counting m photons is evidently a species of average over the corresponding probabilities for an ensemble of Poisson distributions. We hardly need emphasize that the averaging process is not a classical one and that the quasiprobability function P may assume negative values.

As a further illustration of the methods we are discussing let us consider the general case of a chaotically generated field. The density operators of such fields may be represented by means of the Gaussian function

$$P(\{\alpha_k\}) = \prod_{k} \frac{1}{\pi \langle n_k \rangle} e^{-|\alpha_k|^2/\langle n_k \rangle} . \qquad (2.602)$$

Then, since the function Ω is a quadratic form in the variables α_k, it will be possible to evaluate the integral Eq. (2.598) for the generating function in full generality.

Before we do this, however, let us introduce some useful notation. We may express the function $\mathcal{E}(x,\{\alpha_k\})$ as a linear form in the variables α_k by using the normal mode expansion

$$\mathcal{E}(x,\{\alpha_k\}) = \sum_k e(x,k)\alpha_k , \tag{2.603}$$

where the functions e are given by Eq. (2.448). If we then define the matrix

$$B_{k'k''} = \int e^*(x'k')V(x'x'')e(x''k'') \, \mathrm{d}^4x' \, \mathrm{d}^4x'' , \tag{2.604}$$

we may write the quadratic form Ω as

$$\Omega(\{\alpha_k\}) = \sum_{k'k''} \alpha_{k'}^* B_{k'k''} \alpha_{k''} . \tag{2.605}$$

When this expression for Ω and the Gaussian form for P are substituted in Eq. (2.598) we find that the generating function is given by

$$Q(\lambda,t) = \int \cdots \int \exp\left\{ -\sum_k \frac{|\alpha_k|^2}{\langle n_k \rangle} - \lambda \sum_{k'k''} \alpha_{k'}^* B_{k'k''} \alpha_{k''} \right\} \prod_k \frac{\mathrm{d}^2\alpha_k}{\pi \langle n_k \rangle} .$$

If we then introduce the variables

$$\beta_k = \frac{\alpha_k}{\langle n_k \rangle^{1/2}} \tag{2.606}$$

and define the matrix

$$M_{k'k''} = \langle n_{k'} \rangle^{1/2} B_{k'k''} \langle n_{k''} \rangle^{1/2} , \tag{2.607}$$

the integral for the generating function may be simplified to the form

$$Q(\lambda,t) = \int \cdots \int \exp\left\{ -\sum_k |\beta_k|^2 - \lambda \sum_{k'k''} \beta_{k'}^* M_{k'k''} \beta_{k''} \right\} \prod_k \frac{\mathrm{d}^2\beta_k}{\pi} . \tag{2.608}$$

Now we can consider the set of numbers β_k as forming the components of a complex vector β. Then if we let M represent the matrix whose components are given by Eq. (2.607), we may write the exponent in the integrand of Eq. (2.608) as the product

$$-\beta^\dagger(1+\lambda M)\beta .$$

Since the matrix M is Hermitian it may be diagonalized by carrying out a unitary transformation upon the vector β. Then if we let the eigenvalues of M be $\mathcal{M}_i$, and let the transformed complex coordinates be γ_i, the integral for the generating function reduces to the elementary form

$$Q(\lambda,t) = \int \cdots \int \exp\left\{-\sum_i (1+\lambda \mathcal{M}_i)|\gamma_i|^2\right\} \prod_i \frac{d^2\gamma_i}{\pi}$$

$$= \frac{1}{\prod_i (1+\lambda \mathcal{M}_i)} \qquad (2.609a)$$

$$= \frac{1}{\det(1+\lambda M)} . \qquad (2.609b)$$

It is worth noting that the matrix M must be positive definite, since the quadratic form Ω defined by Eqns. (2.599) or (2.605) is the average number of photons counted in a particular coherent field. Hence the eigenvalues $\mathcal{M}_i$ are positive, and the singularities of the generating function lie on the negative real axis of the variable λ. Since Q is analytic in the half-plane $\mathrm{Re}\,\lambda \geq 0$, we see that if we are given Q as a power series expansion about either of the points $\lambda = 0$ or $\lambda = 1$, the series expansion about the other of the points may be evaluated, in principle by analytic continuation. This argument shows that the procedure we have been using, of evaluating the generating function by means of its expansion about $\lambda = 0$, actually leads to a unique answer for the probability distribution.

Since the matrix M is in general of infinite rank, neither of the expressions (2.609) is easy to evaluate directly. Let us note, however, that $\det(1+\lambda M)$ may be written as

$$\prod_i (1+\lambda \mathcal{M}_i) = \exp\left\{\sum_i \log(1+\lambda \mathcal{M}_i)\right\} .$$

Now for $|\lambda| < (\mathcal{M}_{\max})^{-1}$, where $\mathcal{M}_{\max}$ is the largest of the eigenvalues $\mathcal{M}_i$, we may expand the logarithm in the exponent in a convergent power series. In this way we see that

$$\det(1+\lambda M) = \exp\left\{\sum_i (\lambda \mathcal{M}_i - \tfrac{1}{2}\lambda^2 \mathcal{M}_i^2 + \ldots)\right\}$$

$$= \exp\left\{\mathrm{Tr}(\lambda M - \tfrac{1}{2}\lambda^2 M^2 + \ldots)\right\}$$

$$= \exp\{\mathrm{Tr}\,\log(1+\lambda M)\} , \qquad (2.610)$$

where Tr, as always, stands for the trace. By making use of this identity we can express the generating function as

$$Q(\lambda,t) = e^{-\mathrm{Tr}\,\log(1+\lambda M)} . \qquad (2.611)$$

If we expand the logarithm in powers of λ, we may write this function in the form

$$Q(\lambda,t) = \exp\left\{\sum_{r=1}^{\infty} \frac{(-\lambda)^r}{r} I_r\right\} , \qquad (2.612)$$

where I_r is defined by

$$I_r = \text{Tr}\{M^r\} . \tag{2.613}$$

If we recall the definition of the matrix M given by Eqns. (2.607) and (2.604), then we see that for $r = 1$ we have

$$I_1 = \iint \sum_k e^*(x'k) e(x''k) \langle n_k \rangle V(x'x'') \, d^4x' \, d^4x'' .$$

The sum over k in the integrand, according to Eq. (2.454), is simply the first order correlation function. The integral thus reduces to

$$I_1 = \iint G^{(1)}(x',x'') V(x'x'') \, d^4x' \, d^4x'' . \tag{2.614}$$

If we compare Eq. (2.612) with Eqns. (2.449) and (2.450) we see that this $r = 1$ term is of the same form as the exponent of the generating function for the case of a pure coherent field. The lack of coherence for the Gaussian case is reflected by the presence in the exponent of the additional terms with $r \geq 2$. By making further use of the matrix M we can show that the general expression for I_r is the cyclic integral

$$I_r = \int \prod_{j=1}^{r} G^{(1)}(x'_j, x''_{j+1}) V(x'_j x''_j) \, d^4x'_j \, d^4x''_j , \tag{2.615}$$

in which the coordinate x''_{r+1} is to be interpreted as x''_1. For the case of broadband detectors the definitions (2.588) and (2.48) allow us to simplify this integral to the form

$$I_r = s^r \int_0^t \cdots \int_0^t \prod_j dt'_j \int \cdots \int \prod_j G^{(1)}(r'_j t'_j, r'_{j+1} t'_{j+1}) \sigma(r'_j) \, dr'_j . \tag{2.616}$$

To discuss the evaluation of these integrals let us suppose that our counting experiment has particularly simple geometry. We shall assume that our field consists of plane waves traveling in the positive y direction, so that the first order correlation function is given by Eq. (2.492). This function naturally depends only on the y coordinates of its spatial arguments. We next assume that the sensitive region of the counter, i.e., its photocathode, is a very thin layer of atoms lying in a plane perpendicular to the y axis. The function $\sigma(r)$, in other words, is essentially a delta function of the y coordinate. With these assumptions, which experiments often approximate quite closely in practice, the spatial integrations in Eq. (2.616) become trivial. The functions $G^{(1)}$ are independent of their position variables for all of the points for which $\sigma(r)$ differs from zero.

The time integrals in Eq. (2.616) are considerably less trivial, but we may discuss the forms they take for short times and for long times. If the time t is much smaller than the inverse frequency bandwidth of the radiation present, the functions $G^{(1)}$ will hardly vary at all in the interval from 0 to t. For such times the integral I_r must simply be proportional to t^r. If we write I_1 as wt, where w is a proportionality constant, then the elementary character of the spatial integrations shows that the general result must be

$$I_r = (wt)^r. \qquad (2.617)$$

When this result is substituted in Eq. (2.612), we find that the generating function for small values of t is

$$Q(\lambda,t) = \exp\{-\log(1+\lambda wt)\}$$
$$= \frac{1}{1+\lambda wt}. \qquad (2.618)$$

The probability distribution for the number of counts is then given, according to Eq. (2.576), by

$$p(m,t) = \frac{(wt)^m}{(1+wt)^{m+1}}. \qquad (2.619)$$

The distribution for short times is thus given by a power law not unlike the Planck distribution. The mean number of counts is wt, so that w is simply the average counting rate.

For times t which considerably exceed the inverse bandwidth of the radiation field, it is also possible to simplify the integrals I_r. In this case, however, their values depend sensitively on the spectral distribution of the energy present in the field. Let us therefore assume, as an example, that the frequency spectrum has the Lorentz form

$$\langle n_k \rangle \hbar \omega_k = \frac{\text{constant}}{(\omega - \omega_0)^2 + \gamma^2}. \qquad (2.620)$$

The time dependence of the first order correlation function is then given by Eq. (2.499). When this function is substituted into the integral Eq. (2.616), we see that, because of the cyclical structure of the integrand, all of the I_r will increase linearly with time for $t \gg \gamma^{-1}$. We may again define the average counting rate, w, by writing the integral I_1 as wt. Then it is not difficult to show that the full set of integrals I_r may be written in the form

$$I_r = \frac{t(2\gamma w)^r}{2(r-1)!} \left(-\frac{1}{2\gamma} \frac{d}{d\gamma}\right)^{r-1} \frac{1}{\gamma} \qquad (2.621)$$

for $t \gg \gamma^{-1}$.

With these values for the I_r it is possible to sum the series in the exponent of Eq. (2.612) in closed form. When this is done we find that the generating function is

$$Q(\lambda,t) = \exp\left\{-\left[(\gamma^2 + 2\gamma w\lambda)^{1/2} - \gamma\right]t\right\}. \qquad (2.622)$$

When the counting rate w is small compared to the frequency bandwidth, i.e., $w \ll \gamma$, then the expression in the exponent may be expanded, and we find that in the lowest approximation the generating function reduces to

$$Q(\lambda,t) = e^{-\lambda wt} . \qquad (2.623)$$

This function, as we have seen, leads to a Poisson distribution. It is the distribution we would find if there were no tendency for the photons to arrive in correlated bunches, or for the field amplitude to fluctuate randomly.

To discuss the distribution and moments which follow from the generating function Eq. (2.571), it is useful to introduce the set of inverse polynomials

$$\begin{aligned} s_0(\xi) &= s_1(\xi) = 1 , \\ s_2(\xi) &= 1 + \frac{1}{\xi} , \\ s_3(\xi) &= 1 + \frac{3}{\xi} + \frac{3}{\xi^2} , \\ s_4(\xi) &= 1 + \frac{6}{\xi} + \frac{15}{\xi^2} + \frac{15}{\xi^3} . \end{aligned} \qquad (2.624)$$

The further members of the sequence are given by the recursion formula

$$s_{n+1}(\xi) = -s_n(\xi) + \left(1 + \frac{n}{\xi}\right) s_n(\xi) . \qquad (2.625)$$

These polynomials are quite familiar in the theory of Bessel functions. They may also be calculated from the expression

$$s_n(\xi) = e^\xi \left(\frac{2\xi}{\pi}\right)^{1/2} K_{n-1/2}(\xi) , \qquad (2.626)$$

where $K_{n-1/2}$ is a modified Hankel function of half-integral order.

If we now expand the generating function Eq. (2.622) in a power series about $\lambda = 1$ and examine its coefficients we find that the probability of receiving m counts in time t is

$$p(m,t) = \frac{1}{m!} \left(\frac{\gamma wt}{\Gamma}\right)^m s_m(\Gamma t) e^{-(\Gamma-\gamma)t} , \qquad (2.627)$$

where we have written

$$\Gamma = (\gamma^2 - 2w\gamma)^{1/2} . \qquad (2.628)$$

The distribution Eq. (2.627) has the same mean value, wt, as the Poisson distribution which follows from the generating function Eq. (2.623). Its variance, however, is always larger than that of the Poisson distribution because of the photon clumping effect.

The power series expansion of the generating function Eq. (2.622) about $\lambda = 0$ is

$$Q(\lambda, t) = \sum_{n=0}^{\infty} \frac{(-\lambda wt)^n}{n!} s_n(\gamma t) . \qquad (2.629)$$

We conclude from this expansion that the factorial moments of the distribution Eq. (2.627) are given by

$$\left\langle \frac{C!}{(C-n)!} \right\rangle = (wt)^n s_n(\gamma t)$$

$$= \langle C(C-1) \ldots (C-n+1) \rangle . \qquad (2.630)$$

For a Poisson distribution these moments would be simply $(wt)^n$. The first two of the moments (2.630) are

$$\langle C \rangle = wt , \qquad (2.631)$$

$$\langle C(C-1) \rangle = (wt)^2 \left(1 + \frac{1}{\gamma t}\right) . \qquad (2.632)$$

The variance of the number of counts is thus

$$\langle C^2 \rangle - \langle C \rangle^2 = \langle C \rangle \left\{ 1 + \frac{\langle C \rangle}{\gamma t} \right\} . \qquad (2.633)$$

The term $\langle C \rangle / \gamma t$ is the addition to the variance which is due to the fact that the photon arrival times are not statistically independent of one another.

References

1 M. Born, E. Wolf, *Principles of Optics*, Chap. X; Pergamon Press, London 1959.

2 R. Hanbury Brown, R. Q. Twiss, *Nature* **177**, 27 (1956); *Proc. Roy. Soc. (London) A* **242**, 300 (1957); *Proc. Roy. Soc. (London) A* **243**, 291 (1957).

3 G. A. Rebka, R. V. Pound, *Nature* **180**, 1035 (1957).

4 R. J. Glauber, *Phys. Rev. Lett.* **10**, 84 (1963).

5 R. J. Glauber, in *Proc. 3rd Int. Conf. on Quantum Electronics*, Vol. I, p. 111; Dunod, Paris 1964.

6 R. J. Glauber, *Phys. Rev.* **130**, 2529 (1963); reprinted as Chapter 1 in this volume.

7 A. Messiah, *Quantum Mechanics*, Vol. I, p. 442; North-Holland, Amsterdam 1961.

8 E. Schrödinger, *Naturwiss.* **14**, 664 (1926). For a more recent treatment see L. I. Schiff, *Quantum Mechanics*, 2nd ed., p. 67; McGraw-Hill, New York 1955.

9 W. Heisenberg, *The Physical Principles of the Quantum Theory*, pp. 16–19; University of Chicago Press, Chicago 1930. Reprinted by Dover Publications, New York, 1930.

10 Uses of these states as generating functions for the n-quantum states have, however, been made by J. Schwinger, *Phys. Rev.* **91**, 728 (1953).

11 E. Segal, *Illinois J. Math.* **6**, 520 (1962).

12 V. Bargmann, *Commun. Pure and Appl. Math.* **14**, 187 (1961); *Proc. Natl. Acad. Sci. U. S.* **48**, 199 (1962).

13 Some of Bargmann's arguments are summarized by S. Schweber, *J. Math. Phys.* **3**, 831 (1962), who has used them in connection with the formulation of quantum mechanics in terms of Feynman amplitudes. We are indebted to Dr. S. Bergmann for calling this reference to our attention.

14 The existence of this form for the density operator has also been observed by E. C. G. Sudarshan, *Phys. Rev. Lett.* **10**, 277 (1963). His note is discussed briefly at the end of Sect. 2.9.10.

15 Lord Rayleigh, *The Theory of Sound*, 2nd ed., Vol. I, p. 35; MacMillan, London 1894. *Scientific Papers*, Vol. I, p. 491, Vol. IV, p. 370; Cambridge University Press, Cambridge 1899–1920.

16 F. Bloch, *Z. Phys.* **74**, 295 (1932).

17 J. Lawson and G. E. Uhlenbeck, *Threshold Noise Signals*, pp. 33–56; McGraw-Hill, New York 1950.

18 R. J. Glauber, *Phys. Rev.* **84**, 395 (1951).

19 E. C. G. Sudarshan, *Phys. Rev. Lett.* **10**, 277 (1963).

20 D. Holliday, M. L. Sage (to be published).

21 K. Cahill (private communication).

22 D. Kastler, R. J. Glauber (to be published).

23 J. E. Moyal, *Proc. Camb. Phil. Soc.* **46**, 99 (1948).

24 G. Magyar, L. Mandel, *Nature* **198**, 255 (1963).

3
Correlation Functions for Coherent Fields[1]

3.1
Introduction

To secure a complete description of the coherence properties of an electromagnetic field it is useful to distinguish between various orders of coherence. A field which possesses m-th-order coherence, for example, will exhibit particularly simple properties in measurements which detect average m-th powers of field intensities or m-fold products of them. From a historic standpoint nearly all of the measurements which could be carried out in physical optics were, until recently, just measurements of quantities proportional to the average light intensity. It is natural therefore that the most familiar meaning of optical coherence, the one which describes the intensity fringes seen in a multitude of optical experiments, corresponds only to the case of first-order coherence, $m = 1$.

Averages of nonlinear functions of the intensity, on the other hand, are measured either implicitly or explicitly by a number of techniques which have recently been introduced in optical experiments. Individual moments of the intensity distribution may be measured, for example, in photon-coincidence-counting experiments, or by making use of nonlinear media, while implicit measurements of the full set of moments may be made by determining sufficiently accurately the statistical distribution of photons detected by a single counter. The higher order coherence properties of the field furnish a natural basis for describing the results of such experiments.

The precise definition of the different orders of coherence is best stated in terms of a set of quantum-mechanical correlation functions for the electromagnetic field. These functions are defined as ensemble averages of normally ordered products of equal numbers of photon creation and annihilation operators. Their definition is presented in further detail in Sect. 3.2 together with a review of some of their elementary properties.

The most convenient definition of m-th-order coherence, from a mathematical standpoint, takes it to correspond to a simple factorization property of the correlation functions of order up to and including m. In Ref.[1], where this definition was introduced, it was also noted that a necessary condition for m-th-order coherence can

---

[1] Reprinted with permission from U. M. Titulaer, R. J. Glauber, *Phys. Rev.* **140**, 676–682 (1965). Copyright 2006 by the American Physical Society.

Quantum Theory of Optical Coherence. Selected Papers and Lectures. Roy J. Glauber.
Copyright © 2007 WILEY-VCH Verlag GmbH & Co. KGaA, Weinheim
ISBN: 978-3-527-40687-6

be stated in terms of the absolute values of the correlation functions of order $\leq m$. For the case of first-order coherence the latter condition corresponds to the requirement of maximal contrast in the interference fringes which could be formed by superposing the fields occurring at different space-time points. In Sections 3.3 and 3.4 we shall show that this set of conditions is also sufficient to secure m-th-order coherence. It may thus be used as an equivalent definition of m-th-order coherence.

We show further that for a field which has precise first-order coherence, all of the higher order correlation functions reduce to a factorized form. The factorized correlation functions differ from the ones which would be required for coherence to all orders only by a set of real multiplicative constants. These constants furnish an extremely convenient description of the higher order coherence properties of a field which has first-order coherence. They are measurable through their proportionality to the probability per unit [time]^m for detecting m photons in coincidence by means of m ideal photodetectors. The values the constants take on are illustrated for the cases of several special fields.

In Sect. 3.5 we derive a number of more specific results for a class of quantum fields which are generated by radiation sources whose behavior may be described as predetermined. These fields are characterized by the fact that their density operator possesses a positive definite weight function $P(\{\alpha_j\})$ in the representation which is diagonal in the eigenstates of the annihilation operators[2]. For this class of fields we show that the set of coefficients we have mentioned earlier forms a monotonically increasing sequence, and we derive some rigorous inequalities governing its rate of increase. We derive in addition certain upper and lower bounds for the higher order correlation functions. As a last result we show that for this class of fields the combination of first- and second-order coherence is sufficient to assure coherence to all orders.

The results of Sect. 3.5 are derived only for fields with positive definite P functions. Those are, in fact, precisely the quantum fields which may be described in a natural way as possessing classical analogs. For the case of each of the relations derived in Sect. 3.5 we give examples which show that these relations do not hold for quantum fields of more general type. The analysis thus serves to illustrate the physical meaning of the fact that a much greater variety of states is available to quantized fields than to classical ones.

3.2
Correlation Functions and Coherence Conditions

We begin the description of field correlations by introducing the familiar operator representation for the quantized electric-field vector

$$E(r,t) = i\sum_k (\tfrac{1}{2}\hbar\omega_k)^{1/2} \left\{ a_k u_k(r) e^{-i\omega_k t} - a_k^\dagger u_k^*(r) e^{i\omega_k t} \right\}. \tag{3.1}$$

In this expression the index k labels the normal modes of the field. Their number is denumerable if we think of the field as enclosed in a finite volume. The functions $\boldsymbol{u}_k(\boldsymbol{r})$ form an orthonormal set of vector mode functions and the ω_k are the corresponding frequencies. The operators a_k and $a_k^\dagger$ are the photon annihilation and creation operators for the k-th mode. They obey the familiar commutation relations

$$[a_k, a_{k'}] = [a_k^\dagger, a_{k'}^\dagger] = 0,$$
$$[a_k, a_{k'}^\dagger] = \delta_{kk'}.$$
(3.2)

The field operator Eq. (3.1) consists of a positive-frequency part

$$\boldsymbol{E}^{(+)}(\boldsymbol{r},t) = i \sum_k (\tfrac{1}{2}\hbar\omega_k)^{1/2} a_k \boldsymbol{u}_k(\boldsymbol{r}) e^{-i\omega_k t}$$
(3.3)

and a negative-frequency part $\boldsymbol{E}^{(-)}(\boldsymbol{r},t)$ which is its Hermitian adjoint. These complex field operators describe the absorption and emission, respectively, of a single photon at the point $\boldsymbol{r}$ and time t. Their commutators are easily found from those of the a_k and $a_k^\dagger$.

The state of the field is described by a state vector, denoted by $|i\rangle$, or more generally by a density operator $\rho = \{|i\rangle\langle i|\}_{\text{av}}$ where the average is taken over an ensemble appropriate to the way in which the system is prepared. We can express the expectation value of an operator O as $\{\langle i|O|i\rangle\}_{\text{av}} = \text{Tr}\{\rho O\}$. The operator ρ is Hermitian; it is positive definite and its trace is equal to unity.

In terms of these quantities the m-th-order correlation function for the electromagnetic field is defined as

$$G^{(n)}_{\mu_1\ldots\mu_n,\mu_{n+1}\ldots\mu_{2n}}(\boldsymbol{r}_1 t_1, \ldots \boldsymbol{r}_n t_n, \boldsymbol{r}_{n+1} t_{n+1} \ldots \boldsymbol{r}_{2n} t_{2n})$$
$$= \text{Tr}\left\{ \rho E^{(-)}_{\mu_1}(\boldsymbol{r}_1,t_1) \ldots E^{(-)}_{\mu_n}(\boldsymbol{r}_n,t_n) E^{(+)}_{\mu_{n+1}}(\boldsymbol{r}_{n+1},t_{n+1}) \ldots E^{(+)}_{\mu_{2n}}(\boldsymbol{r}_{2n},t_{2n}) \right\}.$$
(3.4)

To write this and other tensor functions of several space-time variables more compactly we shall introduce a simple abbreviation. We let the variable x_j stand for the combination of the arguments $\boldsymbol{r}_j, t_j$ and the polarization index μ_j. Then the correlation function Eq. (3.4) becomes simply $G^{(n)}(x_1 \ldots x_n, x_{n+1} \ldots x_{2n})$. More generally, any function in which p variables x occur is to be interpreted as possessing p vector indices when written out more explicitly.

The detailed ways in which the correlation functions may be measured have been discussed elsewhere[1,3]. For our present purposes it is sufficient to note that $G^{(1)}(x,x)$ is proportional to the average counting rate of an ideal photodetector recording photons with a specific polarization at a specific space-time point. The functions $G^{(n)}(x_1 \ldots x_n, x_n \ldots x_1)$ are related in a similar way to the delayed m-fold coincidence rate in an experiment with n ideal counters[1,3]. The function $G^{(1)}$ with arguments $x_2 \neq x_1$ furnishes a basis for the description of different kinds of interference experiments, and its higher order analogs can be used to describe combined interference and coincidence experiments.

A field is said to have m-th-order coherence[1] if there exists a single function $\mathcal{E}(x)$ such that for all arguments x_j and for all $n \leq m$ the correlation functions factorize according to the scheme

$$G^{(n)}(x_1 \ldots x_n, x_{n+1} \ldots x_{2n}) = \prod_{j=1}^{n} \mathcal{E}^*(x_j) \mathcal{E}(x_{j+n}) \,. \tag{3.5}$$

It follows immediately from this definition that the absolute value of $G^{(n)}$ obeys the relation

$$\left|G^{(n)}(x_1 \ldots x_n, x_{n+1} \ldots x_{2n})\right|^2 = \prod_{j=1}^{2n} G^{(1)}(x_j, x_j) \,. \tag{3.6}$$

That conditions of this type are necessary for coherence can also be understood from the physical meaning of the functions involved. For $n = 1$, Eq. (3.6) contains the statement that the interference patterns obtained by superposing the fields from two different points, have the greatest possible contrast. The higher order conditions Eq. (3.6) relate similarly the quantities which are measured in combined interference and coincidence experiments, and are thus more directly accessible to experimental test than the factorization conditions Eq. (3.5). It is therefore interesting to examine whether the conditions Eq. (3.6) for all $n \leq m$ are also sufficient for m-th-order coherence, i.e., whether they imply the mathematically more useful statement Eq. (3.5). In the next section we shall prove that this is indeed the case.

Similar problems have been considered by Parrent[4] and by Mandel and Wolf[5] for a classical theory of coherence. They treated only the first-order scalar case and discussed a correlation function defined as a time average rather than an instantaneous ensemble average. Their definitions thus limit their treatments to fields which are statistically stationary in time. The physical meanings of their definitions of coherence, which they restrict to quasimonochromatic fields, are substantially different from ours. We shall not go into these approaches here, but we may note that they also led to a factorization theorem for the first-order correlation function, which is similar in part to the one we find for $G^{(1)}$.

3.3
Correlation Functions as Scalar Products

The correlation functions defined in the last section are of the general form $\text{Tr}\{\rho A^\dagger B\}$, where A and B are certain products of annihilation operators. Since ρ is a positive definite Hermitian operator, we can show that the form $\text{Tr}\{\rho A^\dagger B\}$, which we denote for brevity by (A,B) fulfills the familiar axioms of a scalar product

$$(A, \lambda B + \mu C) = \lambda(A,B) + \mu(A,C) \,, \tag{3.7a}$$
$$(A,B) = (B,A)^* \,, \tag{3.7b}$$
$$(A,A) \geq 0 \,. \tag{3.7c}$$

From these axioms, one easily derives[6] a generalization of Schwarz's inequality, which states that for any two operators A and B, we have

$$(A,A)(B,B) \geq |(A,B)|^2 . \tag{3.8}$$

The only difference between Eqns. (3.7) and the usual definition of a scalar product is that we allow the possibility that $(A,A) = 0$ for $A \neq 0$. As a consequence of Eq. (3.8), we see that, for an operator A with $(A,A) = 0$ and an arbitrary operator B, we have $(A,B) = (B,A) = 0$, or, more explicitly,

$$\text{Tr}\left\{\rho A^\dagger B\right\} = \text{Tr}\left\{\rho B^\dagger A\right\} = 0 .$$

If, in particular, the space in which the operators are defined has a denumerable basis $|q_n\rangle$, $n = 1, 2, 3 \ldots$, we can choose for B the operator $|q_n\rangle\langle q_m|$. Then it follows that the nm matrix element of $A\rho$ and of $\rho A^\dagger$ must vanish for all n and m, a condition which implies the operator relations

$$A\rho = \rho A^\dagger = 0 . \tag{3.9}$$

This result can be applied to the case in which the two members of the inequality (3.8) become equal. If A and B are operators which satisfy the relation

$$(A,A)(B,B) = |(A,B)|^2 \tag{3.10}$$

and $(B,B) \neq 0$, we clearly have

$$\left(A - \frac{(B,A)}{(B,B)}B,\ A - \frac{(B,A)}{(B,B)}B\right) = 0 .$$

Application of Eq. (3.9) then shows that the density operator obeys the identities

$$\left[A - \frac{(B,A)}{(B,B)}B\right]\rho = \rho\left[A^\dagger - \frac{(A,B)}{(B,B)}B^\dagger\right] = 0 . \tag{3.11}$$

If, for example, we let $A = E^{(+)}(x_1)$ and $B = E^{(+)}(x_2)$, we see that the scalar product (A,B) is simply the first-order correlation function $G^{(1)}(x_1,x_2)$. The Schwarz inequality (3.8) for this case is

$$G^{(1)}(x_1,x_1)\, G^{(1)}(x_2,x_2) \geq |G^{(1)}(x_1,x_2)|^2 . \tag{3.12}$$

The condition that the two members of this inequality be equal,

$$G^{(1)}(x_1,x_1)\, G^{(1)}(x_2,x_2) = |G^{(1)}(x_1,x_2)|^2 , \tag{3.13}$$

is the condition for maximum fringe contrast noted in Eq. (3.6), for $m = 1$. The restriction that it imposes on the density operator may be found by noting that Eq. (3.13) takes the form of Eq. (3.10). Hence, if we choose x_0 to be a coordinate for which

$$\text{Tr}\left\{\rho E^{(-)}(x_0)\, E^{(+)}(x_0)\right\} = G^{(1)}(x_0,x_0) \neq 0 \tag{3.14}$$

and let $x_2 = x_0$, we find from Eq. (3.11) the relations

$$E^{(+)}(x_1)\rho = \frac{G^{(1)}(x_0,x_1)}{G^{(1)}(x_0,x_0)} E^{(+)}(x_0)\rho, \tag{3.15a}$$

$$\rho E^{(-)}(x_1) = \frac{G^{(1)}(x_1,x_0)}{G^{(1)}(x_0,x_0)} \rho E^{(-)}(x_0). \tag{3.15b}$$

These relations must hold as identities for all x_1; they imply rigorous restrictions on the density operator which will be discussed further in a forthcoming paper[7]. For the present, however, we will confine ourselves to discussing the restrictions they imply upon the form of the correlation functions.

If we apply the identities Eq. (3.15) to the definition of $G^{(1)}(x_1,x_2)$ we find the relation

$$\mathrm{Tr}\left\{\rho E^{(-)}(x_1) E^{(+)}(x_2)\right\} = \frac{G^{(1)}(x_1,x_0) G^{(1)}(x_0,x_2)}{\left[G^{(1)}(x_0,x_0)\right]^2} \mathrm{Tr}\left\{\rho E^{(-)}(x_0) E^{(+)}(x_0)\right\}.$$

The correlation function, in other words, obeys the functional equation

$$G^{(1)}(x_1,x_2) = \frac{G^{(1)}(x_1,x_0) G^{(1)}(x_0,x_2)}{G^{(1)}(x_0,x_0)} \tag{3.16}$$

for all x_1 and x_2 and all x_0 for which $G^{(1)}(x_0,x_0) \neq 0$.

Let us now define the complex function $\mathcal{E}(x)$ as

$$\mathcal{E}(x) = G^{(1)}(x_0,x) \left[G^{(1)}(x_0,x_0)\right]^{-1/2}. \tag{3.17}$$

With this definition we can formulate our result as follows: There exists a complex function $\mathcal{E}(x)$ such that for all x_1 and x_2 we have

$$G^{(1)}(x_1,x_2) = \mathcal{E}^*(x_1) \mathcal{E}(x_2), \tag{3.18}$$

which is exactly the condition Eq. (3.5) for first-order coherence. Since it is obvious that this factorization condition in turn implies the condition Eq. (3.13) we see that the two conditions are equivalent.

The definition Eq. (3.17) of $\mathcal{E}(x)$ would seem to imply a dependence of the way in which the form Eq. (3.18) factorizes on the arbitrary choice of the reference point x_0. We shall show, however, that if condition (3.18) holds for all pairs of arguments this dependence can only he a trivial one. Let us suppose that $G^{(1)}(x_1,x_2)$ has a second factorized form in which $\mathcal{E}(x)$ is replaced by $\mathcal{E}'(x)$. Then we have the relations (valid for all x_1 and x_2)

$$\mathcal{E}'^*(x_1) \mathcal{E}'(x_2) = \mathcal{E}^*(x_1) \mathcal{E}(x_2)$$

or

$$\mathcal{E}'^*(x_1)/\mathcal{E}^*(x_1) = \mathcal{E}(x_2)/\mathcal{E}'(x_2).$$

From the second form of this identity we see that there must exist a constant λ such that $\mathcal{E}'(x) = \lambda \mathcal{E}(x)$. Furthermore, from the first form of the identity we see that $|\lambda|^2 = 1$. For first-order coherent fields a change of the point x_0 can thus only correspond to multiplying $\mathcal{E}(x)$ by a phase factor. Such a phase factor clearly cancels in the calculation of any correlation function.

3.4
Application to Higher Order Correlation Functions

Although Eq. (3.13) is a condition imposed only on the first-order correlation function, its consequences include a remarkable sequence of identities which must be obeyed by the higher order correlation functions as well. These identities allow us to reduce the $G^{(n)}$ for arbitrary arguments to a standard form. To derive those identities we begin by letting x_0 again be a point for which $G^{(1)}(x_0, x_0) \neq 0$. We then observe that the operators $E^{(+)}(x_j)$ all commute with one another, as do the $E^{(-)}(x_j)$. If we now apply each of the identities (3.15a) and (3.15b) n times to the definition of the n-th-order correlation function given by Eq. (3.4), we find

$$\mathrm{Tr}\left\{\rho E^{(-)}(x_1)\ldots E^{(-)}(x_n) E^{(+)}(x_{n+1})\ldots E^{(+)}(x_{2n})\right\} =$$

$$\prod_{j=1}^{n} \frac{G^{(1)}(x_j, x_0)\, G^{(1)}(x_0, x_{j+n})}{G^{(1)}(x_0, x_0)\, G^{(1)}(x_0, x_0)}\, \mathrm{Tr}\left\{\rho \left[E^{(-)}(x_0)\right]^n \left[E^{(+)}(x_0)\right]^n\right\}.$$

By using the definition Eq. (3.17) of the field $\mathcal{E}(x)$ we can reformulate this result as

$$G^{(n)}(x_1 \ldots x_n, x_{n+1} \ldots x_{2n}) = g_n \prod_j \mathcal{E}^*(x_j)\, \mathcal{E}(x_{j+n}) \tag{3.19}$$

with

$$g_n = G^{(n)}(x_0 \ldots x_0, x_0 \ldots x_0)\left[G^{(1)}(x_0, x_0)\right]^{-n}. \tag{3.20}$$

The quantities g_n may be regarded as constants in Eq. (3.19) since they are independent of $x_1 \ldots x_{2n}$. It is easy to show that the g_n cannot actually depend on the choice of x_0 either. To see this we need only note that $G^{(n)}$ is independent of x_0 and recall that in the last section we showed that products of the fields $\mathcal{E}(x)$ such as the one occurring in Eq. (3.19) are independent of x_0 as well. The g_n are simply a set of constants determined only by the state of the field; it is evident from Eq. (3.20) that they are real and non-negative.

We have shown that for any field which possesses first-order coherence, the higher order correlation functions must factorize into the forms given by Eq. (3.19). These forms are quite similar in structure to those which define higher-order coherence [cf. Eq. (3.5)], and differ from them only through the inclusion of the factors g_n. It is clear, from the assumption of first-order coherence, e.g., from Eq. (3.18), that $g_1 = 1$, but the

g_n for $n \neq 1$ only assume the value unity for special choices of fields. If, for example, the field is one for which the conditions (3.6) on the absolute value of $G^{(n)}$ are fulfilled for $n \leq m$, then we see that $|g_n|^2 = 1$ for $n \leq m$. Since the g_n are real and positive we must in fact have $g_n = 1$ for $n \leq m$. The conditions (3.6), in other words, are sufficient to require that the correlation functions fall into precisely the form needed for m-th-order coherence. The case of full coherence corresponds, of course, to $g_n = 1$ for all n.

It is possible at present to generate electromagnetic fields which possess the property of first-order coherence to an excellent approximation, i.e., fields for which the factorization condition (3.18) holds for quite large space-time separations of x_1 relative to x_2. Since for such fields the constants g_n should play an important role in the description of higher order coherence properties, we shall make a number of comments on the values they may take on.

From our definition of g_n and the discussions in Refs.[1,3] it is clear that g_n is proportional to the probability per unit [time]^n of detecting n photons with an idealized photodetector. If the density operator is such that the number of photons in the field cannot exceed some value M, then, of course, $g_n = 0$ for $n \geq M$. In particular, if exactly M photons are present in a single mode of the field, we easily find from the commutation rules of the mode amplitudes that

$$g_n = M! \left[M^n (M-n)! \right]^{-1} \quad \text{for} \quad n \leq M. \tag{3.21}$$

A particularly important case for which the coefficients g_n are known is formed by the fields generated by chaotic sources, a class which includes thermal ones. For such fields, it is shown in Ref.[2] that

$$g_n = n!. \tag{3.22}$$

In the next section we shall demonstrate that for a category of fields which may be thought of as generated by classical sources, the coefficients g_n always form a monotonically increasing sequence.

We conclude this section with an observation about photon coincidence experiments such as those performed by Hanbury Brown and Twiss. When the separation of the two counters used in such experiments is quite small compared to the coherence length of the field (and when their relative delay time is small compared to the coherence time) the number $g_2 - 1$ furnishes a measure[1,3] of the difference between the observed and the accidental coincidence rates. The quantity analogous to $g_2 - 1$ in classical electromagnetic theory is proportional to the variance of the energy density of the field and is intrinsically positive. From the preceding discussion, however, it is clear that $g_2 - 1$ is by no means necessarily positive [e.g., for the case Eq. (3.21) it is not]. The Hanbury Brown–Twiss effect assumes the particular form observed for natural light sources because of the particular statistical mixtures of quantum states which such sources tend to produce. Artificial sources could in principle produce fields with coincidence rates smaller than the product of the individual counting rates.

3.5
Fields With Positive-Definite *P* Functions

In this section we derive some relations obeyed by the correlation functions for a special class of fields corresponding to sources with predetermined behavior. In order to give a mathematical characterization of these fields we first consider a particular set of states, the coherent states $|\{\alpha_k\}\rangle$, which have been discussed in Ref.[2]. These are simultaneous right eigenstates of the full set of annihilation operators and have complex eigenvalues α_k corresponding to the a_k. Thus they are also right eigenstates of the positive frequency part of the field at any point x. The latter property is expressed by the relations

$$E^{(+)}(x)|\{\alpha_k\}\rangle = \mathcal{E}(x,\{\alpha_k\})|\{\alpha_k\}\rangle, \qquad (3.23a)$$

$$\langle\{\alpha_k\}|E^{(-)}(x) = \mathcal{E}^*(x,\{\alpha_k\})\langle\{\alpha_k\}| \qquad (3.23b)$$

with

$$\mathcal{E}_\mu(r,t,\{\alpha_k\}) = i\sum_k \left(\tfrac{1}{2}\hbar\omega_k\right)^{1/2} \alpha_k u_{k\mu}(r) e^{-i\omega_k t}.$$

The latter equation illustrates our convention regarding the meaning of the variable x.

We now consider fields which can be described fay means of a density operator of the form

$$\rho = \int P(\{\alpha_k\})|\{\alpha_k\}\rangle\langle\{\alpha_k\}| \prod_k d^2\alpha_k \qquad (3.24)$$

with a positive definite weight function $P(\{\alpha_k\})$ which has, at most, δ-type singularities. In this integral the differential $d^2\alpha_k$ stands for $d\,\text{Re}(\alpha_k)\,d\,\text{Im}(\alpha_k)$. Since all our integrals will be taken over the full set of amplitude variables $\{\alpha_k\}$ we shall simply write α in the remainder of this section instead of $\{\alpha_k\}$ and $d^2\alpha$ instead of $\prod_k d^2\alpha_k$. The function $P(\alpha)$ has to be normalized so that $\int P(\alpha)\,d^2\alpha = 1$. Density operators of the type Eq. (3.24) can be shown[2] to furnish a description appropriate for any field generated by a prescribed charge-current distribution. They also describe the light emitted by a completely chaotic source, a model appropriate for virtually all known natural light sources.

The representation Eq. (3.24) has the pre-eminent advantage of permitting us to express the correlation functions as integrals over $P(\alpha)$:

$$G^{(n)}(x_1\ldots x_{2n}) = \int P(\alpha) \prod_{j=1}^n \mathcal{E}^*(x_j,\alpha)\,\mathcal{E}(x_{j+n},\alpha)\,d^2\alpha. \qquad (3.25)$$

This expression for $G^{(n)}$ as an integral, rather than the more general form of scalar product considered in Sect. 3.3, makes it possible to derive a number of inequalities which need not hold for more general quantum mechanical fields.

One such inequality can be derived by considering real-valued functions $A(\alpha)$ and $B(\alpha)$ which satisfy the relation

$$[A(\alpha) - A(\beta)][B(\alpha) - B(\beta)] \geq 0 \tag{3.26}$$

for all α and β. For such functions one can easily prove the following generalization of the Tchebycheff inequality [using positive definiteness and normalization of $P(\alpha)$][8]

$$\int P(\alpha) A(\alpha) B(\alpha) \, d^2\alpha \geq \int P(\alpha) A(\alpha) \, d^2\alpha \int P(\beta) B(\beta) \, d^2\beta . \tag{3.27}$$

Condition (3.26) may be fulfilled by choosing $A(\alpha)$ and $B(\alpha)$ each to be a power of the same positive function $|\mathcal{E}(x,\alpha)|^2$. The integral of the n-th power of this expression, taken over $P(\alpha)$, is just equal to the correlation, function $G^{(n)}(x\ldots x)$, with all of its $2n$ arguments set equal. Now if we let $A(\alpha) = |\mathcal{E}(x,\alpha)|^{2m}$ and $B(\alpha) = |\mathcal{E}(x,\alpha)|^{2(n-m)}$ where $n \geq m \geq 0$, the inequality (3.27) implies the relation

$$G^{(n)}(x\ldots x) \geq G^{(m)}(x\ldots x) G^{(n-m)}(x\ldots x) . \tag{3.28}$$

If both members are divided by $[G^{(1)}(x,x)]^n$, we find the inequality

$$g_n \geq g_m \, g_{n-m} \tag{3.29}$$

for the coefficients g_n. If we let $m = 1$, in particular, and recall that $g_1 = 1$, we see that

$$g_n \geq g_{n-1} , \tag{3.30}$$

or the constants g_n form a monotonically increasing sequence. We note in particular $g_2 \geq 1$, which means that, for the class of fields under consideration, the coincidence rate in a Hanbury Brown–Twiss experiment always exceeds or equals the product of the individual counting rates as the separation of the detectors approaches zero.

Another sequence of inequalities can be derived from the Schwarz inequality expressed in the form which applies to integrals of complex functions,

$$\int P(\alpha) |A(\alpha)|^2 \, d^2\alpha \int P(\beta) |B(\beta)|^2 \, d^2\beta \geq \left| \int P(\alpha) A^*(\alpha) B(\alpha) \, d^2\alpha \right|^2 . \tag{3.31}$$

If we now substitute $A(\alpha) = [\mathcal{E}(x,\alpha)]^{n-m}[G^{(1)}(x,x)]^{-n}$ and $B(\alpha) = [\mathcal{E}^*(x,\alpha)]^m \times [\mathcal{E}(x,\alpha)]^n$ and recall the definition of g_n, we see that

$$g_{n-m} \, g_{n+m} \geq g_n^2 . \tag{3.32}$$

This inequality shows that

$$\tfrac{1}{2}(\ln g_{n+m} + \ln g_{n-m}) \geq \ln g_n , \tag{3.33}$$

or $\ln g_n$ is a convex function[2] of n. The convexity property permits us to show that for $n \geq l \geq 0$ and $m \geq 0$ we have[2]

$$l \ln g_{n+m} + m \ln g_{n-l} \geq (l+m) \ln g_n . \tag{3.34}$$

If we let $l = n - 1$ in this relation and write $p = n + m$, then by recalling that $g_1 = 1$ we find the inequality

$$g_p \geq g_n^{(p-1)/(n-1)} \quad \text{for} \quad p \geq n . \tag{3.35}$$

This relation sets lower limits to the rapidity with which the g_p increase. It shows furthermore that, for any field which does not possess second-order coherence, i.e., for which $g_2 > 1$, the g_p increase without bound as $p \to \infty$.

We turn next to the consideration of some bounds on the values of the correlation functions of the form $G^{(n)}(x_1 \ldots x_n, x_n \ldots x_1)$, i.e., on the values of the n-fold coincidence counting rates. For this purpose we can use the inequality

$$\prod_{j=1}^{n} \int P(\alpha) \, |\mathcal{E}(x_j,\alpha)|^{2n} \, d^2\alpha \geq \left[\int P(\alpha) \prod_{j=1}^{n} |\mathcal{E}(x_j,\alpha)|^2 \, d^2\alpha \right]^n ,$$

which is a simple consequence of the Hölder inequality,[3] as stated for integrals. When expressed in terms of correlation functions this relation becomes

$$\prod_{j=1}^{n} G^{(n)}(x_j \ldots x_j) \geq \left[G^{(n)}(x_1 \ldots x_n, x_n \ldots x_1) \right]^n . \tag{3.36}$$

For fields with first-order coherence the inequality reduces to an equality.

A slightly more general inequality can be derived when we combine Eq. (3.36) with the inequality

$$G^{(n)}(x_1 \ldots x_n, x_n \ldots x_1) \, G^{(n)}(x_{n+1} \ldots x_{2n}, x_{2n} \ldots x_{n+1}) \geq \left| G^{(n)}(x_1 \ldots x_{2n}) \right|^2$$

which has been derived from the Schwarz inequality in Ref.[1]. In this way we find

$$\prod_{j=1}^{2n} G^{(n)}(x_j \ldots x_j) \geq \left| G^{(n)}(x_1 \ldots x_{2n}) \right|^2 . \tag{3.37}$$

This relation leads, for example, to a simple property of linearly polarized fields which are invariant under space-time translations. For these fields the $G^{(n)}(x_j \ldots x_j)$ are independent of x_j as long as the vector index it specifies corresponds to the direction of polarization. The inequality (3.37) then implies for all x_j

$$G^{(n)}(x_j \ldots x_j) \geq \left| G^{(n)}(x_1 \ldots x_{2n}) \right| , \tag{3.38}$$

---

[2] Ref.[8], p. 70.
[3] Ref.[8], p. 140.

i.e., the absolute value of $G^{(n)}$ reaches its maximum when the $2n$ arguments are all equal.

For correlation functions of even order we can also state a lower limit for the coincidence rate. This may be done by substituting in the Schwarz inequality, Eq. (3.31), the functions

$$A(\alpha) = 1$$

and

$$B(\alpha) = \prod_{j=1}^{n} \mathcal{E}^*(x_j, \alpha)\, \mathcal{E}(x_{j+n}, \alpha) \,.$$

We then find

$$G^{(2n)}(x_1 \ldots x_{2n}, x_{2n} \ldots x_1) \geq \left| G^{(n)}(x_1 \ldots x_n, x_{n+1} \ldots x_{2n}) \right|^2 . \tag{3.39}$$

An interesting property of the fields under consideration is that the combination of first- and second-order coherence implies coherence to all orders. To show this we note that as a consequence of first-order coherence the correlation functions all factorize and we need only consider the coefficients g_n which characterize the field. Second-order coherence implies $g_2 = 1$, which in turn implies that $G^{(2)}(x_0, x_0, x_0, x_0) = [G^{(1)}(x_0, x_0)]^2$. The latter relation is the statement

$$\int P(\alpha)\, |\mathcal{E}(x_0, \alpha)|^4 \, d^2\alpha - [G^{(1)}(x_0, x_0)]^2$$
$$= \int P(\alpha) \left\{ |\mathcal{E}(x_0, \alpha)|^4 - [G^{(1)}(x_0, x_0)]^2 \right\} d^2\alpha$$
$$= 0,$$

which may also be written in the form

$$\int P(\alpha) \left\{ |\mathcal{E}(x_0, \alpha)|^2 - G^{(1)}(x_0, x_0) \right\}^2 d^2\alpha = 0. \tag{3.40}$$

If we now define the scalar product of two functions $f(\alpha)$ and $g(\alpha)$ to be

$$(f(\alpha), g(\alpha)) = \int P(\alpha)\, f^*(\alpha)\, g(\alpha)\, d^2\alpha \,,$$

then we see that Eq. (3.40) states that the norm of the function

$$f(\alpha) = |\mathcal{E}(x_0, \alpha)|^2 - G^{(1)}(x_0, x_0)$$

vanishes. This is a situation we have already encountered in Sect. 3.3. Application of the Schwarz inequality, Eq. (3.8), shows that the scalar product of any other function $g(\alpha)$ with it vanishes. We have shown, in other words, that

$$\int P(\alpha)\, g^*(\alpha) \left\{ |\mathcal{E}(x_0, \alpha)|^2 - G^{(1)}(x_0, x_0) \right\} d^2\alpha = 0,$$

or that $|\mathcal{E}(x_0,\alpha)|^2$ can be replaced by $G^{(1)}(x_0,x_0)$ in all integrals taken over the weight function $P(\alpha)$. In particular evaluation of the correlation function $G^{(n)}(x_0\ldots x_0)$ in this way yields

$$\begin{aligned}G^{(n)}(x_0\ldots x_0) &= [G^{(1)}(x_0,x_0)]^n \int P(\alpha)\,d^2\alpha \\ &= [G^{(1)}(x_0,x_0)]^n\,.\end{aligned} \tag{3.41}$$

By referring to the definition of the g_n, Eq. (3.20), we see that this relation implies $g_n = 1$ for all n and thus full coherence of the field in question.

As a final remark we wish to emphasize once more that the results we have derived in this section are only valid for fields which have a positive definite $P(\alpha)$ and need not hold for more general types of fields. It is not difficult, in fact, to find among the more general fields explicit counterexamples for each of the relations proved. We note first that the inequalities (3.30) and (3.32)–(3.35) fail to hold for fields with a finite number of photons present; this can be seen for example from Eq. (3.21) when only a single mode is occupied. For any n-photon state, the inequality (3.39) is clearly disobeyed, since the left-hand side is always zero while the right-hand side is not. The inequalities (3.36)–(3.38) cannot hold for a field that is constructed as the superposition of n one-photon wavepackets which nave no spatial overlap. Once again, for this case, the left-hand sides of the inequalities vanish while the right-hand sides can be made different from zero. Finally, if we consider the state $2^{-1/2}|\text{vac}\rangle + 2^{-1/2}|2_k\rangle$, where $|2_k\rangle$ is a state with just two photons present in a particular mode k, then this state is easily shown to have first-order coherence, and further to have g_2 equal to unity, but nevertheless to be far from fully coherent since all the g_n for $n > 2$ are equal to zero.

References

1 R. J. Glauber, *Phys. Rev.* **130**, 2529 (1963); reprinted as Chapter 1 in this volume.

2 R. J. Glauber, *Phys. Rev.* **131**, 2766 (1963); reprinted as Sect. 2.9 in this volume..

3 R. J. Glauber, *Quantum Optics and Electronics*, Les Houches 1964, edited by C. deWitt, A. Blandin, C. Cohen-Tannoudji; Gordon & Breach, New York 1965, p. 63; reprinted as Chapter 2 in this volume.. Some analogous results have also been obtained by P. L. Kelley, W. H. Kleiner, *Phys. Rev. A* **136**, 316 (1964).

4 G. B. Parrent, *J. Opt. Soc. Am.* **49**, 787 (1959). For some related results see also C. L. Mehta, E. Wolf, A. P. Balachandran, to be published.

5 L. Mandel, E. Wolf, *J. Opt. Soc. Am.* **51**, 815 (1961).

6 See, e.g., R. Courant, D. Hilbert, *Methods of Mathematical Physics*, Interscience, New York 1953, Vol. I, p. 2. For the case that A and B are products of photon annihilation operators, (the case to which we will apply the theorem) a proof of Eq. (3.8) is contained in the Appendix of Ref.[1].

7 U. M. Titulaer, R. J. Glauber, to be published.

8 G. H. Hardy, J. E. Littlewood, G. Polya, *Inequalities*, Cambridge University Press, Cambridge 1952, 2nd ed., pp. 43, 168.

4
Density Operators for Coherent Fields[1]

4.1
Introduction

The statistical information which describes the state of the quantized electromagnetic field is implicitly contained in its density operator. We have shown in earlier papers[1–3] how the knowledge of the density operator enables us, for example, to evaluate all of the correlation functions for the field vectors. An important property of the correlation functions is that they furnish a concise description of the coherence properties of the field; when the correlation functions obey appropriate factorization conditions the fields are described as coherent. Such restrictions on the form of the correlation functions, it is clear, may also be regarded as constraints placed upon the density operator of the field. In the present paper, we shall study in some detail the constraints which the coherence conditions impose upon the density operator, and discuss some explicit features of the fields whose density operators satisfy them.

It is convenient at this point to recall the definitions of the correlation functions, and of the coherence properties which they describe. The first-order correlation function for a radiation field described by the density operator ρ is given by

$$G^{(1)}_{\mu\nu}(rt,r't') = \text{Tr}\left\{\rho E^{(-)}_\mu(rt) E^{(+)}_\nu(r't')\right\}, \qquad (4.1)$$

in which the operators $E^{(+)}_\mu(rt)$ and $E^{(-)}_\nu(rt)$ are the positive and negative frequency parts, respectively, of the operators for the vector components of the electric field at position r and time t. The correlation function of n-th order is defined as the average of a $2n$-fold product of components of the field strength. We may write it as

$$G^{(n)}(x_1 \ldots x_n, x_{n+1} \ldots x_{2n})$$
$$= \text{Tr}\left\{\rho E^{(-)}(x_1)\ldots E^{(-)}(x_n) E^{(+)}(x_{n+1})\ldots E^{(+)}(x_{2n})\right\}, \qquad (4.2)$$

where each variable x_j is an abbreviation for the space and time coordinates r_j and t_j and a vector index μ_j as well. A field possesses m-th order coherence if the first m of

---

[1] Reprinted with permission from U. M. Titulaer, R. J. Glauber, *Phys. Rev.* **145**, 1041–1050 (1966). Copyright 2006 by the American Physical Society.

Quantum Theory of Optical Coherence. Selected Papers and Lectures. Roy J. Glauber.
Copyright © 2007 WILEY-VCH Verlag GmbH & Co. KGaA, Weinheim
ISBN: 978-3-527-40687-6

its correlation functions can be written in the factorized form

$$G^{(n)}(x_1\ldots x_n, x_{n+1}\ldots x_{2n}) = \prod_{j=1}^{n} \mathcal{E}^*(x_j)\,\mathcal{E}(x_{j+n})\,, \qquad (4.3)$$

where the function $\mathcal{E}(x)$ is a complex solution of the wave equation. In a previous paper[4] we have studied the restrictions which this succession of conditions imposes on the structure of the correlation functions. We shall show that an analysis closely related to the one we have used in discussing the correlation functions may be used to derive a number of properties of the density operators.

The density operators of fields which possess first-order coherence are discussed in Sect. 4.2. There exists a sense, as we shall show, in which first-order coherence corresponds to the restriction that only one variety of photons be present in the field. This variety may correspond to the excitation of an arbitrary superposition of the various monochromatic modes of the field. If we use this superposition to define a new mode of the field, which is, in general, a nonmonochromatic one, we find that the density operator for a first-order coherent field reduces to a form which has only this single mode excited.

The reduction of the density operator to a single-mode form makes it possible to specify the operator by means of a single matrix B_{mn}, which is its occupation number representation. A diagonal element of this matrix, B_{nn}, is the probability for the presence of n photons in the one excited mode. We show in Sect. 4.3 that there exists a simple connection between these diagonal elements and the set of real constants g_n which were introduced in Chapter 3 as a means of describing higher order coherence properties of the field. Each of the two sets of constants, $\{g_n\}$ and $\{B_{nn}\}$, can easily be evaluated if the other is given. We show, in particular, that for any fully coherent field, i.e., one which is characterized by $g_n = 1$ for all n, the probabilities B_{nn} correspond to a Poisson distribution for the number of photons present in the field. The coherence conditions, on the other hand, do not impose any restrictions on the off-diagonal elements of the matrix B_{mn}. There exists accordingly a broad variety of states which fulfill the coherence conditions precisely.

A well-known example of a fully coherent field is one which is in an eigenstate of the annihilation operators $a^\dagger$ for all the modes of the field[2,5]. Such an eigenstate, denoted by $|\{\alpha_k\}\rangle$, is characterized by the relation

$$a_k|\{\alpha_k\}\rangle = \alpha_k|\{\alpha_k\}\rangle\,. \qquad (4.4)$$

This state can be expressed in terms of the eigenstates of the number operator by means of the expression[5]

$$|\{\alpha_k\}\rangle = \prod_k \left[\sum_{n=0}^{\infty} e^{-|\alpha_k|^2/2}\frac{(\alpha_k)^n}{(n!)^{1/2}}|n\rangle_k\right]\,. \qquad (4.5)$$

Although these states possess a number of unique physical properties, they are not the only ones which possess full coherence. The wider class of states which satisfy the full set of coherence conditions is discussed in Sect. 4.3.

In Sect. 4.4 we describe some of the ways in which the eigenstates of the annihilation operators are distinguished from the other states which exhibit full coherence. Their first characterization, which follows immediately from their nature as eigenstates of the annihilation operators, is that they are the only ones in which the variance of any annihilation operator vanishes. The characterization proves to be useful in deriving a number of other unique properties of these states. We find, for example, that the eigenstates of the annihilation operator are the only ones for which the density operator for the field reduces to a product of density operators for the individual modes, regardless of which system of orthogonal modes we use. They are, furthermore, the only states with first-order coherence in which the function $\mathcal{E}(x)$ in the expression (4.3) can be taken to be the expectation value of $E^{(+)}(x)$.

4.2
Evaluation of the Density Operator

In this section we begin the systematic study of the restrictions imposed upon the density operator by the requirement of first-order coherence. Before doing so, however, we shall find it instructive to examine some particular examples of fields which exhibit first-order coherence. These examples will later be useful as illustrations of the general theory. We consider them here since they suggest some of its important features.

One of the simplest examples of a state which has first-order coherence is an arbitrary pure single photon state, which we may denote by $|1 \text{ phot}\rangle$. For this field the first-order correlation function $G^{(1)}(x_1, x_2)$ takes the form

$$G^{(1)}(x_1, x_2) = \langle 1 \text{ phot} | E^{(-)}(x_1) E^{(+)}(x_2) | 1 \text{ phot} \rangle . \tag{4.6}$$

Since $E^{(+)}(x)$ is a photon annihilation operator, $E^{(+)}(x)|1 \text{ phot}\rangle$ can only be a multiple of the vacuum state, $|\text{vac}\rangle$. We may therefore insert the projection operator on the vacuum state in Eq. (4.6) and write

$$G^{(1)}(x_1, x_2) = \langle 1 \text{ phot} | E^{(-)}(x_1) | \text{vac} \rangle \langle \text{vac} | E^{(+)}(x_2) | 1 \text{ phot} \rangle .$$

This means that for a single-photon field $G^{(1)}(x_1, x_2)$ fulfills the requirement for first-order coherence,

$$G^{(1)}(x_1, x_2) = \mathcal{E}^*(x_1) \mathcal{E}(x_2) , \tag{4.7}$$

when we choose the function $\mathcal{E}(x)$ to be

$$\mathcal{E}(x) = \langle \text{vac} | E^{(+)}(x) | 1 \text{ phot} \rangle . \tag{4.8}$$

We note that no restrictions whatever are placed upon the spectral properties of the state $|1 \text{ phot}\rangle$; any pure one-photon wavepacket will do, whatever its frequency distribution may be.

A second, and more familiar type of coherent field is one which is monochromatic and completely polarized. Examples of such fields are ones in which only a single mode, say the k-th, is excited. To show that such a field fulfills the requirement for first-order coherence we make use of the mode expansion for the positive frequency part of the field,

$$E^{(+)}(r,t) = i \sum_k \left(\tfrac{1}{2}\hbar\omega_k\right)^{1/2} u_k(r) e^{-i\omega_k t} a_k . \tag{4.9}$$

The summation is carried out over a complete set of modes, characterized by the orthonormal set of vector mode functions $u_k(r)$ and the corresponding frequencies ω_k. The operator a_k is an annihilation operator for a photon in the k-th mode.

If the k-th mode is the only one excited, the density operator for the field must obey the relations $a_l \rho = \rho a_l^\dagger = 0$ unless $l = k$. These relations imply that the first-order correlation function takes the form

$$G^{(1)}(x_1,x_2) = \mathrm{Tr}\left\{\rho E^{(-)}(x_1) E^{(+)}(x_2)\right\}$$
$$= \left(\tfrac{1}{2}\hbar\omega_k\right) u_k^*(r_1) u_k(r_2) e^{i\omega_k(t_1-t_2)} \mathrm{Tr}\left\{\rho a_k^\dagger a_k\right\} . \tag{4.10}$$

Since the trace that appears here is independent of x_1 and x_2, we can obviously construct a field $\mathcal{E}(x)$ in terms of which the correlation function factorizes according to the coherence condition Eq. (4.7). A field confined to a single mode can still take a variety of quantum mechanical forms; its density operator can be written as

$$\rho = \sum_{m_k,n_k} c_{m_k n_k} \frac{(a_k^\dagger)^{m_k}}{(m_k!)^{1/2}} |\mathrm{vac}\rangle\langle\mathrm{vac}| \frac{(a_k)^{n_k}}{(n_k!)^{1/2}} . \tag{4.11}$$

Such density operators may describe pure states as well as mixtures.

To exhibit more general forms of coherent fields, let us note that the field operator $E^{(+)}(r,t)$ may be expanded in terms of other solutions of the wave equation than the monochromatic mode functions $u_k(r) e^{-i\omega_k t}$. We may, for example, define a different set of solutions $v_l(rt)$ of the wave equation as the set of linear combinations

$$v_l(rt) = i \sum_k \gamma_{lk} \left(\tfrac{1}{2}\hbar\omega_k\right)^{1/2} u_k(r) e^{-i\omega_k t} , \tag{4.12}$$

where the coefficients γ_{lk} are the elements of a unitary matrix, i.e., they obey the unitarity relation

$$\sum_l \gamma_{lk}^* \gamma_{lm} = \delta_{lm} .$$

If we let b_l be the annihilation operator for a photon in the mode $v_l(rt)$, then we require that the positive frequency field operator have the expansion

$$E^{(+)}(r,t) = \sum_l v_l(rt) b_l . \tag{4.13}$$

Comparison of this expansion with Eq. (4.9) shows that we must then have

$$\sum_l b_l \gamma_{lk} = a_k \,.$$

By inverting this relation and its Hermitian adjoint, we see that b_l and $b_l^\dagger$ are given by the linear combinations

$$b_l = \sum_k \gamma_{lk}^* a_k \,, \qquad (4.14\text{a})$$

$$b_l^\dagger = \sum_k \gamma_{lk} a_k^\dagger \,. \qquad (4.14\text{b})$$

These operators are seen to satisfy the commutation relations

$$[b_l, b_m] = [b_l^\dagger, b_m^\dagger] = 0 \,,$$
$$[b_l, b_m^\dagger] = \delta_{lm} \,, \qquad (4.15)$$

which are of the same form as those of the operators a and $a^\dagger$. The transformation (4.14) which is induced by Eq. (4.12) is therefore a canonical one.[2]

Let us now suppose that only one of the modes corresponding to the functions v_l is occupied, say the one for $l = l_0$. Then the density operator obeys the relations

$$b_l \rho = \rho b_l^\dagger = 0$$

for $l \neq l_0$. By using these relations in conjunction with the expression (4.13) and its adjoint for the complex field operators, we find that the first-order correlation function takes the form

$$G^{(1)}(\mathbf{r}t, \mathbf{r}'t') = v_{l_0}^*(\mathbf{r}t)\, v_{l_0}(\mathbf{r}'t')\, \text{Tr}\left\{\rho b_{l_0}^\dagger b_{l_0}\right\} \,. \qquad (4.16)$$

This form satisfies the requirements for first-order coherence just as the form in Eq. (4.10) did.

Since the fields we shall consider below are only excited in the single mode associated with a particular function $v_{l_0}(\mathbf{r}t)$, we may simplify the notation a bit by dropping the indices l_0. It is evident that by defining the transformation coefficients suitably the function v for the single mode which is excited may be taken to be any function of the form

$$v(\mathbf{r}t) = i \sum_k \gamma_k \left(\tfrac{1}{2}\hbar\omega_k\right)^{1/2} u_k(\mathbf{r})\, e^{-i\omega_k t} \,, \qquad (4.17)$$

where the complex coefficients γ_k obey the normalization condition

$$\sum_k |\gamma_k|^2 = 1 \,. \qquad (4.18)$$

---

[2] It may be noted that, although the transformation is canonical, the mode functions $v_l(\mathbf{r},t)$ are not in general an orthonormal set, because of the frequency-dependent normalization factors in Eq. (4.12).

The corresponding annihilation and creation operators are then

$$b = \sum_k \gamma_k^* a_k, \tag{4.19a}$$

$$b^\dagger = \sum_k \gamma_k a_k^\dagger. \tag{4.19b}$$

An n photon state of the mode (4.17) may evidently be expressed in the form $(n!)^{-1/2}(b^\dagger)^n |\text{vac}\rangle$. It follows then that the most general density operator for which this mode alone is excited may be written in the form

$$\rho = \sum_{m,n} B_{mn} \frac{(b^\dagger)^n}{(n!)^{1/2}} |\text{vac}\rangle \langle \text{vac}| \frac{b^m}{(m!)^{1/2}}, \tag{4.20}$$

where the coefficients B_{mn}, like the $c_{m_k n_k}$ in Eq. (4.11), may be chosen in a great variety of ways.

By generalizing our definition of a mode to include nonmonochromatic solutions of the wave equation we have derived the density operators Eq. (4.20), which describe a broad class of fields possessing first-order coherence. We shall now show that the density operators Eq. (4.20) form the most general class possessing this property; the density operator for any first-order coherent field can be written in the form Eq. (4.20).

To demonstrate this theorem we make use of an identity, satisfied by the density operators of all first-order coherent fields, as shown in our previous paper. This is the relation Eq. (3.15a), which may be written as

$$E_\mu^{(+)}(r,t)\rho = \frac{G_{\mu_0\mu}^{(1)}(r_0 t_0, rt)}{G_{\mu_0\mu_0}^{(1)}(r_0 t_0, r_0 t_0)} E_{\mu_0}^{(+)}(r_0 t_0)\rho,$$

where r_0, t_0 is an arbitrary point for which the correlation function in the denominator does not vanish for an appropriate choice of μ_0. When both sides of this equation are multiplied by

$$-i\left(\tfrac{1}{2}\hbar\omega_k\right)^{-1/2} u_{k\mu}^*(r)\, e^{i\omega_k t},$$

summed over μ, and integrated over r the resulting equation may be written in the form

$$a_k \rho = \beta_k E^{(+)}(x_0)\rho, \tag{4.21}$$

where the constant β_k is given by

$$\beta_k = \sum_\mu \int \frac{u_{k\mu}^*(r)\, e^{i\omega_k t}}{i\left(\tfrac{1}{2}\hbar\omega_k\right)^{1/2}} \frac{G_{\mu_0\mu}^{(1)}(r_0 t_0, rt)}{G_{\mu_0\mu_0}^{(1)}(r_0 t_0, r_0 t_0)}\, d\mathbf{r}. \tag{4.22}$$

It is easily seen from the fact that $G^{(1)}$ satisfies the wave equation that the dependence on t in this equation cancels exactly, and the coefficient β_k is independent of time. We show further that $\sum_k |\beta_k|^2$ is finite as long as the mean total number of photons $\langle N \rangle$ in the field is finite[3]; this follows directly from the expression

$$\langle N \rangle = \mathrm{Tr}\left\{\sum_k a_k^\dagger a_k \rho\right\} = \sum_k |\beta_k|^2 G^{(1)}(x_0,x_0) < \infty.$$

With these preliminaries we are able to evaluate the density operator ρ in the occupation number representation,

$$\rho = \sum_{\{n_k\},\{m_l\}} |\{n_k\}\rangle \langle\{n_k\}|\rho|\{m_l\}\rangle \langle\{m_l\}|.$$

If we insert the definitions of the states $|\{n_k\}\rangle$ in this equation we have

$$\rho = \sum_{\{n_k\},\{m_l\}} \prod_k \frac{(a_k^\dagger)^{n_k}}{(n_k!)^{1/2}} |\mathrm{vac}\rangle \langle\mathrm{vac}| \prod_k \frac{(a_k)^{n_k}}{(n_k!)^{1/2}} \rho$$
$$\times \prod_l \frac{(a_l^\dagger)^{m_l}}{(m_l!)^{1/2}} |\mathrm{vac}\rangle \langle\mathrm{vac}| \prod_l \frac{(a_l)^{m_l}}{(m_l!)^{1/2}}.$$

We may now make use of relation (4.21) and its adjoint and of the commutation properties of creation and annihilation operators to rewrite this expression for ρ in the form

$$\rho = \sum_{\{n_k\},\{m_l\}} \prod_k \frac{(\beta_k a_k^\dagger)^{n_k}}{n_k!} |\mathrm{vac}\rangle \langle\mathrm{vac}| \left[E^{(+)}(x_0)\right]^{\sum_k n_k} \rho$$
$$\times \left[E^{(-)}(x_0)\right]^{\sum_l m_l} |\mathrm{vac}\rangle \langle\mathrm{vac}| \prod_l \frac{(\beta_l^* a_l)^{m_l}}{m_l!}. \quad (4.23)$$

We next define $n = \sum_k n_k$, and $m = \sum_l m_l$ and use the multinomial theorem to carry out the partial summations in which n and m remain fixed. In this way we obtain

$$\rho = \sum_{n,m} \frac{(\sum_k \beta_k a_k^\dagger)^n}{n!} |\mathrm{vac}\rangle \langle\mathrm{vac}| \frac{(\sum_l \beta_l^* a_l)^m}{m!}$$
$$\times \left\langle \mathrm{vac} \left| \left[E^{(+)}(x_0)\right]^n \rho \left[E^{(-)}(x_0)\right]^m \right| \mathrm{vac}\right\rangle. \quad (4.24)$$

This expression is seen to have exactly the form of Eq. (4.20) when the coefficients γ_k in the definitions (4.19) of b and $b^\dagger$ are identified as

$$\gamma_k = \beta_k \left[\sum_k |\beta_k|^2\right]^{-1/2}. \quad (4.25)$$

---

[3] The assumption that the total number of photons is finite is a natural one in the present context. We have assumed that the modes of the field are a denumerable set, as is the case for the field in any finite enclosure. For such a field there exists a minimum frequency, and infrared divergences do not occur.

The coefficients β_k are given by Eq. (4.22) and the boundedness of $\sum_k |\beta_k|^2$ was shown for any field with an average occupation number which is bounded. When this identification is made, we see that the coefficients B_{nm} of Eq. (4.20) are given by

$$B_{nm} = (n!\,m!)^{-\frac{1}{2}} \left[\sum_k |\beta_k|^2\right]^{\frac{-(n+m)}{2}} \left\langle \text{vac} \left| \left[E^{(+)}(x_0)\right]^n \rho \left[E^{(-)}(x_0)\right]^m \right| \text{vac} \right\rangle . \tag{4.26}$$

It is clear from this expression for the coefficients B_{nm} that they form a Hermitian matrix. From the definition Eq. (4.19a) of b we find the relations $b|\text{vac}\rangle = 0$ and $[b, b^\dagger] = 1$. They enable us to evaluate the trace of the density operator in terms of the B_{nm} as

$$\begin{aligned}\text{Tr}\,\rho &= \sum_{n,m} B_{nm} (n!\,m!)^{-1/2} \left\langle \text{vac} \left| b^m (b^\dagger)^n \right| \text{vac} \right\rangle \\ &= \sum_{n,m} B_{nm} (n!\,m!)^{-1/2} n!\, \delta_{nm} \\ &= \sum_n B_{nn} = 1 .\end{aligned}$$

The matrix B_{nm} therefore has unit trace. Its diagonal element B_{mm} is easily seen to be the probability that m photons occupy the mode. Our main result in this section can be summarized quite concisely. The density operator for the most general type of field possessing first order coherence is a simple generalization of the density operator Eq. (4.11) for a field with a single monochromatic mode excited. To achieve full generality it is only necessary to replace the creation operator $a_k^\dagger$ in the density operator Eq. (4.11) by the more general creation operator $b^\dagger$ which creates a photon in a particular superposition of modes. If we think of this superposition as specifying a particular type of photon wavepacket, then we see that the field may be regarded as consisting entirely of photons of that type.[4]

We have proved both that a field specified by the density operator Eq. (4.20) has first-order coherence and that conversely every field with first-order coherence must have a density operator of the form Eq. (4.20). This means that we have obtained a third way of characterizing first-order coherent fields. The first two are the factorization condition Eq. (4.6) and the maximum fringe visibility condition. The latter condition and its equivalence with the factorization condition are discussed in Sect. 3.3.

---

[4] It is perhaps worth noting that this situation strongly resembles the one encountered in the study of superfluids. The superfluid component has long-range correlations which may be expressed by means of a factorization of its correlation functions. This component consists, furthermore, entirely of particles in a small number of similar quantum states. The fact that these states may be nonstationary and quite arbitrary in form was stressed recently by P. C. Hohenberg and P. C. Martin, *Ann. Phys. New York* **34**, 291 (1965).

4.3
Fully Coherent Fields

To discuss the higher order coherence properties of a field with first-order coherence we consider the sequence of numbers

$$g_n = \frac{G^{(n)}(x_0\ldots x_0)}{[G^{(1)}(x_0,x_0)]^n}. \tag{4.27}$$

In Chapter 3 we showed that these numbers completely describe the higher order coherence properties of a field which has first-order coherence. Such a field is specified, as shown in Sect. 4.2, by a density operator of the type Eq. (4.20). The correlation functions with all arguments set equal are therefore given by

$$G^{(n)}(x_0\ldots x_0) = \sum_m B_{mm} \left\langle \mathrm{vac} \left| \frac{b^m}{(m!)^{1/2}} \left[E^{(-)}(x_0)\right]^n \left[E^{(+)}(x_0)\right]^n \frac{(b^\dagger)^m}{(m!)^{1/2}} \right| \mathrm{vac} \right\rangle. \tag{4.28}$$

To evaluate this expression we make use of the commutator

$$\left[b^m,\,[E^{(-)}(x_0)]^n\right] = \left[\frac{m!}{(m-n)!}\right] b^{m-n} \left[b,\,E^{(-)}(x_0)\right]^n \tag{4.29}$$

for $m \geq n$. The terms of the sum (4.28) for $m < n$ are easily seen to vanish.

The relation (4.29) together with its adjoint enables us to write $G^{(n)}(x_0\ldots x_0)$ as

$$G^{(n)}(x_0\ldots x_0) = \left|\left[b,\,E^{(-)}(x_0)\right]\right|^{2n} \sum_{m=n}^\infty B_{mm} \frac{1}{m!} \left[\frac{m!}{(m-n)!}\right]^2$$

$$\times \left\langle \mathrm{vac} \left| b^{m-n}(b^\dagger)^{m-n} \right| \mathrm{vac} \right\rangle$$

$$= \left|\left[b,\,E^{(-)}(x_0)\right]\right|^{2n} \sum_{m=n}^\infty B_{mm} \frac{m!}{(m-n)!}. \tag{4.30}$$

In particular for $n = 1$, we have

$$G^{(1)}(x_0,x_0) = \left|\left[b,\,E^{(-)}(x_0)\right]\right|^2 \sum_{m=1}^\infty B_{mm}\, m\,. \tag{4.31}$$

If we introduce the notation

$$\langle f(m) \rangle \equiv \sum_m f(m) B_{mm}, \tag{4.32}$$

we can express the coefficient g_n as

$$g_n = \frac{G^{(n)}(x_0\ldots x_0)}{[G^{(1)}(x_0,x_0)]^n} = \frac{\langle m!/(m-n)! \rangle}{\langle m \rangle^n}. \tag{4.33}$$

A fully coherent field, we have shown in Chapter 3, is characterized by $g_n = 1$ for all n. This condition can be written as

$$\left\langle \frac{m!}{(m-n)!} \right\rangle = \langle m \rangle^n . \tag{4.34}$$

In other words, the factorial moments of the photon number distribution reduce to powers of the mean photon number. This characterization of the field leads uniquely to a Poisson distribution for the number of photons present in it. The distribution is most easily derived by making use of a generating function method. A number of details of this method which we shall not examine here are discussed in Sect. 2.15.

If we define a generating function $F(\lambda)$ as

$$F(\lambda) = \sum_{m=0}^{\infty} \lambda^m B_{mm} , \tag{4.35}$$

then we see immediately that

$$B_{mm} = \left[\frac{1}{m!} \left(\frac{d}{d\lambda} \right)^m F(\lambda) \right]_{\lambda=0} . \tag{4.36}$$

On the other hand, the factorial moments can also be expressed in terms of $F(\lambda)$ as

$$\left\langle \frac{m!}{(m-n)!} \right\rangle = \sum_{m=n}^{\infty} m(m-1)\ldots(m-n+1) B_{mm}$$

$$= \left[\left(\frac{d}{d\lambda} \right)^n F(\lambda) \right]_{\lambda=1} . \tag{4.37}$$

Since Eq. (4.34) tells us the derivatives of the generating function at $\lambda = 1$, we may construct the function explicitly by means of a Taylor series expansion about $\lambda = 1$. In this way, we find

$$F(\lambda) = \sum_{n=0}^{\infty} \frac{(\lambda-1)^n}{n!} \langle m \rangle^n = e^{(\lambda-1)\langle m \rangle} .$$

By carrying out the differentiations indicated in Eq. (4.36) we see that the B_{nn} are given by

$$B_{nn} = \frac{\langle m \rangle^n}{n!} e^{-\langle m \rangle} , \tag{4.38}$$

which is exactly a Poisson distribution.

The converse of this theorem is also true. If the density operator for a field takes the form of Eq. (4.20), and if the diagonal matrix elements B_{nn} form a Poisson distribution, then the field is fully coherent. To prove this we need only note that under the conditions stated the relations (4.34) will all hold, or the g_n, in other words, will all equal unity.

It is interesting to observe that once a field possesses first-order coherence, and the density operator consequently falls into the form of Eq. (4.20), the higher order coherence conditions only place constraints on the diagonal elements of the matrix B_{mn}. The great freedom which remains available in the choice of the off-diagonal elements means that a considerable variety of mixed states as well as pure ones is capable of satisfying the conditions for full coherence. (The definition of full coherence has intentionally been chosen to admit appropriate types of mixtures as well as pure states, since virtually no optical experiments deal with pure states.)

The eigenstates of the photon annihilation operators are obvious examples of pure states which are fully coherent. We may use the theorems we have proved to derive a more general class of pure states which possess the same property. To do this we note that for any pure state, $|\ \rangle$, the density operator takes the form $\rho = |\ \rangle\langle\ |$. Its matrix elements in the occupation-number representation must therefore take the form

$$\langle\{n_k\}|\rho|\{m_k\}\rangle = r^*(\{n_k\})\, r(\{m_k\})\,, \tag{4.39}$$

where $r(\{m_k\})$ is a suitably determined function of the occupation numbers. We see then that if a density operator of the form Eq. (4.20) is to represent a pure state, the matrix B_{mn} must factorize according to the scheme

$$B_{nm} = C_n^* C_m\,, \tag{4.40}$$

in which the set of coefficients C_n obey the normalization condition

$$\sum_n |C_n|^2 = 1\,.$$

The density operator which results from substituting Eq. (4.40) into Eq. (4.20) may be regarded as a projection operator on the state

$$e^{i\phi} \sum_n \frac{C_n}{(n!)^{1/2}} (b^\dagger)^n |\text{vac}\rangle\,, \tag{4.41}$$

where ϕ is an arbitrary phase angle. This is the most general pure state which possesses first-order coherence. The conditions for full coherence impose the further requirement that the $B_{nn} = |C_n|^2$ be given by Eq. (4.38). By satisfying this requirement we see that the most general pure state which has full coherence may be written in the form

$$\sum_n \frac{\langle m\rangle^{n/2}}{n!} e^{-\langle m\rangle/2 + i\theta_n} (b^\dagger)^n |\text{vac}\rangle\,, \tag{4.42}$$

where the phases θ_n may be chosen arbitrarily. Comparison of this expression with that for an eigenstate of a photon annihilation operator, Eq. (4.4), shows that it is an eigenstate of the operator b if and only if the θ_m obey a linear relationship of the form

$$\theta_m = m\theta + \phi \quad (\text{mod}\, 2\pi) \tag{4.43}$$

for some θ and ϕ.

By making use of Eq. (3.18) of Ref.[5], we can easily show that an eigenstate of b, denoted by $|\beta\rangle$, is also an eigenstate of each of the annihilation operators a_k. We use the definition of b, Eq. (4.19a), and the commutativity of all $a_k^\dagger$ and $a_{k'}$ for $k \neq k'$ to write

$$|\beta\rangle = \exp\left[\beta b^\dagger - \beta^* b\right] |\text{vac}\rangle$$

$$= \exp\left[\sum_k (\beta \gamma_k a_k^\dagger - \beta^* \gamma_k^* a_k)\right] |\text{vac}\rangle$$

$$= \prod_k \exp\left[\beta \gamma_k a_k^\dagger - \beta^* \gamma_k^* a_k\right] |\text{vac}\rangle$$

$$= |\{\beta \gamma_k\}\rangle . \tag{4.44}$$

Here $|\{\beta \gamma_k\}\rangle$ is the eigenstate of the operators a_k with eigenvalues $\beta \gamma_k$. The constants γ_k are defined in Eq. (4.25).

The difference between an eigenstate of the annihilation operators[5] and the more general type of fully coherent state, described by Eq. (4.42) does not become manifest in experiments, which can be described completely in terms of the correlation functions $G^{(n)}$, since these functions are the same for both. Examples of such experiments are measurements of the field intensity, photon coincidence rates and the distribution functions of the number of photons counted. The quantities these experiments measure are independent of the absolute phase of the field. A typical example of a quantity which does depend on the absolute phase is an average value of the product of unequal powers of the creation and annihilation operators $b^\dagger$ and b. If ρ is the projection operator on the state (4.42), such a quantity is given by

$$\text{Tr}\left\{\rho (b^\dagger)^l b^n\right\} = \langle m \rangle^{(l+n)/2} \sum_p \frac{\langle m \rangle^p}{p!} e^{-\langle m \rangle} e^{i(\theta_{p+n} - \theta_{p+l})} . \tag{4.45}$$

It is interesting to compare this average value with the one we obtain if ρ corresponds to a mixture of eigenstates of b with the same amplitude, $|\beta| = \langle m \rangle^{1/2}$, but different phases,

$$\rho = \sum_j p_j \left|\langle m \rangle^{1/2} e^{i\theta_j}\right\rangle \left\langle \langle m \rangle^{1/2} e^{i\theta_j}\right| , \tag{4.46}$$

where the p_j are non-negative numbers and have the normalization $\sum_j p_j = 1$. For this density operator we find, in place of Eq. (4.45),

$$\text{Tr}\left\{\rho (b^\dagger)^l b^n\right\} = \langle m \rangle^{(l+n)/2} \sum_j p_j e^{i(n-l)\theta_j} . \tag{4.47}$$

---

5) In Ref.[5] and elsewhere the eigenstates of the annihilation operator have been referred to as coherent states. We have used their lengthier and more specific designation in the present paper, however, in order to avoid confusion in discussing broader classes of states which satisfy the coherence conditions. Since the eigenstates of the annihilation operators are the only easily generated members of this set, and mathematically the most useful ones, their designation simply as coherent states will remain convenient in most other contexts.

Comparison of this result with Eq. (4.45) shows us that for each l and n we can construct a set of coefficients p_j and θ_j such that the result Eq. (4.45) is reproduced by a ρ of the form Eq. (4.46). The average expressed in Eq. (4.45), in other words, is equivalent to one in which the phase of the field is allowed to vary while the modulus remains fixed. For different l and n, however, we must use different sets of p_j and θ_j.

The density operator Eq. (4.46) corresponds in the classical limit to an ensemble of fields with fixed modulus and a phase randomly distributed among certain discrete values. A state of the field of the form given by Eq. (4.42) can evidently be replaced, within the context of Eq. (4.45), by such a random-phase ensemble. While this observation is of help in picturing the physical character of the states Eq. (4.42), we must remember that the equivalent ensembles will be different for each pair of integers l and n in Eq. (4.45).

The subtle correlations which may be present in a quantum mechanical state cannot be reproduced faithfully by means of a single classical ensemble, as can be seen by considering a state (4.42) for which

$$\theta_n = n\theta + (-1)^n \phi.$$

This state may show a large phase uncertainty in the expression for $\text{Tr}\{\rho b\}$, but none at all in that for $\text{Tr}\{\rho b^2\}$. States of such thoroughly unclassical nature[6] will, of course, not be produced by everyday light sources. However, in a complete quantum mechanical discussion we cannot exclude them from consideration.

4.4
Unique Properties of the Annihilation Operator Eigenstates

The eigenstates of the annihilation operators can be distinguished from the more general class of states which exhibit full coherence in a variety of ways. We shall discuss several of the respects in which these states are unique in the present section.

It is well known that the eigenstates of the annihilation operators are ones for which the product of the uncertainties of the coordinate and of the momentum of each field oscillator is a minimum. This minimum property does not characterize the eigenstates uniquely, however, since each oscillator has a continuum of states with the same value of the uncertainty product. A more useful way of characterizing the eigenstates of the annihilation operators is through the discussion of a rather different type of uncertainty. For this purpose we note that for any density operator ρ, and any set of complex numbers $\{\alpha_k\}$ we have the positive-definiteness inequalities[7]

$$\text{Tr}\left\{\rho(a_k^\dagger - \alpha_k^*)(a_k - \alpha_k)\right\} \geq 0, \tag{4.48}$$

---

[6] We may note, incidentally, that this state is an example of one which cannot be represented by means of the P representation defined in Ref.[5]. It is a particular case of the example stated in Ref.[3], Eq. (23).

[7] See, e.g., Eq. (3.7c) in Chapter 3 or the Appendix of Chapter 1.

which hold for all k. If the field is in a pure eigenstate of the annihilation operators corresponding to the eigenvalues $\{\alpha_k\}$, i.e., we have $\rho = |\{\alpha_k\}\rangle \langle\{\alpha_k\}|$ then it is clear that the averages given by Eq. (4.48) vanish for all k. Since the numbers $\{\alpha_k\}$ are in that case the mean values of the mode amplitudes, $a_k = \text{Tr}\{\rho a_k\}$, the eigenstates of the annihilation operators may be said to minimize the uncertainties (i.e., the variances) of the complex field amplitudes. [There exists some ambiguity in defining the variance of a non-Hermitian operator. The expression (4.48) is chosen so that the zero-point oscillations do not contribute to the variance.]

To show that the minimum property we have just exhibited characterizes the eigenstates uniquely, let us assume that for some sets of amplitudes $\{\alpha_k\}$ we have

$$\text{Tr}\left\{\rho(a_k^\dagger - \alpha_k^*)(a_k - \alpha_k)\right\} = 0 \tag{4.49}$$

for all k, and solve for the density operator ρ. As a first step we may make use of the theorem proved in Sect. 2.2 which shows that the identities Eq. (4.49) imply

$$(a_k - \alpha_k)\rho = \rho(a_k^\dagger - \alpha_k^*) = 0 \tag{4.50}$$

for all k. It is clear from these relations that α_k is the mean value of a_k and α_k^* is that of the conjugate operator. By using the brackets $\langle\ \rangle$ to denote statistical averages, we may therefore write Eq. (4.49) in the alternative form

$$\left\langle a_k^\dagger a_k \right\rangle - |\langle a_k \rangle|^2 = 0. \tag{4.51}$$

Let us consider the case of a field in which only the k-th mode is excited. For brevity we shall temporarily drop the index k. The density operator for this field can be written in the occupation number representation as

$$\rho = \sum_{n,m} |n\rangle \langle n|\rho|m\rangle \langle m| . \tag{4.52}$$

We may now use the definitions of the states $|n\rangle$ and the identities Eq. (4.50) to bring the density operator to the form

$$\rho = \sum_{n,m} |n\rangle \left\langle \text{vac} \left| \frac{a^n}{(n!)^{1/2}} \rho \frac{(a^\dagger)^m}{(m!)^{1/2}} \right| \text{vac} \right\rangle \langle m|$$

$$= \langle \text{vac}|\rho|\text{vac}\rangle \sum_n \frac{\alpha^n}{(n!)^{1/2}} |n\rangle \sum_m \langle m| \frac{(\alpha^*)^m}{(m!)^{1/2}} . \tag{4.53}$$

Since $\text{Tr}\rho = 1$ we must have

$$\langle \text{vac}|\rho|\text{vac}\rangle = e^{-|\alpha|^2} . \tag{4.54}$$

We see then that ρ is a projection operator on the eigenstate $|\alpha\rangle$ given by Eq. (4.5). Hence the necessary and sufficient condition that Eq. (4.49) holds for a single-mode excitation is that the mode be in the eigenstate of the annihilation operator having eigenvalue α.

The generalization of this result to the case in which arbitrarily many modes are excited is immediate. By expanding the density operator in the occupation number representation for the full set of modes and again using the identities Eq. (4.50) we may see that ρ is the projection operator on the state $|\{\alpha_k\}\rangle$ which is an eigenstate of all the operators a_k.

The theorem we have just proved applies to non-monochromatic modes as well as monochromatic ones. Thus if the operators b_l are the set defined by Eq. (4.14a) and there exists a set of complex numbers β_l such that

$$\operatorname{Tr}\left\{\rho(b_l^\dagger - \beta_l^*)(b_l - \beta_l)\right\} = 0 \tag{4.55}$$

for all l, then the β_l must be the mean values of the b_l,

$$\beta_l = \operatorname{Tr}\{\rho b_l\}, \tag{4.56}$$

and ρ must be the projection operator on the state

$$|\{\beta_l\}\rangle = \exp\left(-\frac{1}{2}\sum_l |\beta_l|^2\right) \sum_n \left[\frac{\left(\sum_l \beta_l b_l^\dagger\right)^n}{n!}\right] |\text{vac}\rangle$$

$$= \exp\left[\sum_l (\beta_l b_l^\dagger - \beta_l^* b_l)\right] |\text{vac}\rangle, \tag{4.57}$$

which is an eigenstate of the $\{b_l\}$. By using Eqns. (4.14a) and (4.14b) to expand b_l in terms of the sets of operators $\{a_k\}$ and $\{a_k^\dagger\}$, respectively, we see that the state $|\{\beta_l\}\rangle$ may be written as

$$|\{\beta_l\}\rangle \prod_k \exp\left[\sum_l (\beta_l \gamma_{lk} a_k^\dagger - \beta_l^* \gamma_{lk}^* a_k)\right] |\text{vac}\rangle. \tag{4.58}$$

We have shown in Eq. (4.44) that such a state is an eigenstate of the $\{a_k\}$. It may be written alternatively in the form $|\{\alpha_k\}\rangle$, where the eigenvalues α_k are given by

$$\alpha_k = \sum_l \beta_l \gamma_{lk}. \tag{4.59}$$

To obtain further insight into the meaning of the inequality (4.48), let us consider a single mode for which the annihilation operator is a, and introduce a coordinate and a momentum operator for it by writing

$$q = \left(\frac{\hbar}{2\omega}\right)^{1/2} (a^\dagger + a), \tag{4.60a}$$

$$p = i\left(\frac{\hbar\omega}{2}\right)^{1/2} (a^\dagger - a). \tag{4.60b}$$

If we then let the constant α be

$$\alpha = \langle a \rangle = (2\hbar\omega)^{-1/2}\{\omega\langle q \rangle + i\langle p \rangle\}, \tag{4.61}$$

where the brackets $\langle\ \rangle$ stand for statistical averages, we find that the inequality (4.48) may be written in the form

$$\left(\frac{\omega}{2\hbar}\right)\langle(q-\langle q\rangle)^2\rangle + \frac{1}{2\hbar\omega}\langle(p-\langle p\rangle)^2\rangle - \tfrac{1}{2} \geq 0. \tag{4.62}$$

The eigenstates of the annihilation operator are the states which reduce this particular linear combination of the variances of q and p to its minimum value. The linear combination is a familiar one. For the case $\langle q \rangle = \langle p \rangle = 0$, the inequality asserts simply that the energy of any state of the oscillator exceeds that of its ground state, $\tfrac{1}{2}\hbar\omega$. It is clear, furthermore, that for $\langle q \rangle$ and $\langle p \rangle$ different from zero, the state which minimizes the sum, i.e., the eigenstate of the annihilation operator, may be found simply by displacing the ground state in coordinate and momentum space. The latter property of the eigenstates of a has already been noted in Ref.[5]. The inequality (4.62) may indeed be shown to be a consequence of the uncertainty principle. If we apply first the arithmetic mean-geometric mean inequality and then the uncertainty principle we find

$$\tfrac{1}{2}\omega^2\langle(q-\langle q\rangle)^2\rangle + \tfrac{1}{2}\langle(p-\langle p\rangle)^2\rangle \geq \omega\{\langle(q-\langle q\rangle)^2\rangle\langle(p-\langle p\rangle)^2\rangle\}^{1/2} \geq \tfrac{1}{2}\hbar\omega.$$

It is interesting to apply the inequality we have been discussing to the set of states which correspond to fully coherent fields. If we let b be the annihilation operator defined by Eq. (4.19a) and let β be an arbitrary complex number, then the inequality becomes

$$\text{Tr}\left\{\rho(b^\dagger - \beta^*)(b-\beta)\right\} \geq 0. \tag{4.63}$$

If, in particular, we let β be the average value of b, $\beta = \text{Tr}\{\rho b\} = \langle b \rangle$, then Eq. (4.63) reduces to the Schwarz inequality

$$\langle b^\dagger b \rangle \geq |\langle b \rangle|^2. \tag{4.64}$$

As we have noted earlier, all states which possess full coherence have Poisson distributions for their occupation numbers; the distributions which correspond, for example, to the states specified by Eq. (4.42) are all identical and have the mean occupation number $\langle b^\dagger b \rangle = \langle m \rangle$. The mean value of the operator b is therefore constrained in such states by the inequality

$$\langle m \rangle \geq |\langle b \rangle|^2, \tag{4.65}$$

and the upper bound of $|\langle b \rangle|$ is only attained when the state reduces to an eigenstate of b.

Our discussion at the end of Sect. 4.3 of the physical nature of the states Eq. (4.42) indicates that the difference between the two members of the inequality (4.65) arises largely from the phase uncertainty of the operator b. The phase uncertainty of the field evidently tends to be minimized, within the class of fully coherent states, by the eigenstate of the annihilation operator.[8]

One of the interesting features of the eigenstates of the annihilation operators is evident in Eq. (4.44). When a field with only the b mode occupied is in an eigenstate of b, then its state vector, when written in the representation associated with the a_k modes, factorizes into a product of eigenstates of the $\{a_k\}$. The density operator for the field, in other words, takes the product form $\rho = \prod_k \rho_k$, and measurements made on the individual modes yield statistically independent results. We shall show that this condition of statistical independence is actually sufficient to single out the eigenstates of the annihilation operator from among all states of the field which possess coherence of any order. To do this let us suppose that the density operator ρ describes a first-order coherent field. Then, as we have noted earlier, only a single mode of the field is excited, and we label it the b_{l_0}-mode. We shall assume that none of the orthogonal set of modes we label with the amplitudes $\{a_k\}$ is identical to the b_{l_0}-mode. These statements imply, as we shall show, that if ρ can be expressed in the factorized form $\prod_k \rho_k$, it is a projection operator on an eigenstate of the annihilation operator b_{l_0}. The factorization, in other words, is a sufficient condition for the field to be a pure eigenstate as well as a necessary one.

To begin the proof of this theorem we note that the b_{l_0}-mode may be regarded as a member of an orthogonal set of modes for which the b_l are given as linear combinations of the $\{a_k\}$ by Eqns. (4.14a) and (4.14b). These relations are

$$b_l = \sum \gamma_{lk}^* a_k, \tag{4.66a}$$

$$b_l^\dagger = \sum \gamma_{lk} a_k^\dagger, \tag{4.66b}$$

where the coefficients γ_{lk} form a unitary matrix. The b_l-modes with $l \neq l_0$ are not excited. We therefore have

$$b_l^\dagger \rho = \rho b_l = 0 \quad \text{for} \quad l \neq l_0, \tag{4.67}$$

a statement which implies that the variance of b_l also vanishes;

$$\text{Tr}\left\{\rho(b_l^\dagger - \langle b_l^\dagger \rangle)(b_l - \langle b_l \rangle)\right\} = 0 \quad \text{for} \quad l \neq l_0. \tag{4.68}$$

---

[8] Calculations verifying that the eigenstates of the annihilation operators possess relatively well-defined phases have been carried out by P. Carruthers and M. M. Nieto [Phys. Rev. Lett. **14**, 387 (1965)]. The phase uncertainties are small in the sense that they approach the lower bound set by the commutation relations for $\langle m \rangle \gg 1$, and are not far from this limit for smaller values of $\langle m \rangle$.

This variance can also be written as

$$\mathrm{Tr}\left\{\rho \sum_k \gamma_{lk}(a_k^\dagger - \alpha_k^*) \sum_{k'} \gamma_{lk'}^*(a_{k'} - \alpha_{k'})\right\} = 0, \quad (4.69)$$

where $\alpha_k = \langle a_k \rangle$ is the expectation value of a_k.

If ρ is assumed to factorize into the form $\prod_k \rho_k$ it is clear that the terms with $k \neq k'$ vanish in Eq. (4.69). The vanishing of the variance Eq. (4.68) therefore implies the condition

$$\sum_k |\gamma_{lk}|^2 \mathrm{Tr}\left\{\rho_k(a_k^\dagger - \alpha_k^*)(a_k - \alpha_k)\right\} = 0 \quad \text{for} \quad l \neq l_0. \quad (4.70)$$

Since, according to Eq. (4.48), all the terms of this sum are non-negative, we must have

$$\mathrm{Tr}\left\{\rho_k(a_k^\dagger - \alpha_k^*)(a_k - \alpha_k)\right\} = 0 \quad (4.71)$$

for all k, providing there exists at least one value of $l \neq l_0$ such that $\gamma_{lk} \neq 0$. If, on the other hand, no such value of l exists, i.e., we have $\gamma_{lk} = 0$ for $l \neq l_0$, then the relation which is inverse to Eq. (4.66a),

$$a_k = \sum_l b_l \gamma_{lk}, \quad (4.72)$$

reduces to the form

$$a_k = b_{l_0} \gamma_{l_0 k}, \quad (4.73)$$

and the coefficient $\gamma_{l_0 k}$ is simply a phase factor. This statement contradicts our explicit assumption that the excited mode is not one of the set associated with the amplitudes a_k.

The relation Eq. (4.71) has been shown earlier in this section to imply that ρ_k is the projection operator on an eigenstate of the annihilation operator a_k. The over-all density operator ρ is therefore the projection operator on a simultaneous eigenstate of all the a_k. This result completes the proof of our theorem.[9]

It is perhaps worth noting that the general class of fully coherent fields only possesses a factorization property much weaker in form than the one we have demonstrated for the eigenstates of the annihilation operator. The density operators for fully

---

[9] For the particular case in which the excited mode is a superposition of two other modes a related result has been noted by Y. Aharonov, D. Falkoff, E. Lemer, and H. Pendleton, *Ann. Phys. New York*, to be published; a partial report is contained in the conference proceedings mentioned in Ref.[3]. These authors give a physical interpretation of the result in terms of a comparison between quantum mechanical and classical features of measurement processes. The demonstration we have given provides a rigorous basis for their theorem in that it overcomes the reservations they point out at the end of Section II of their paper. Furthermore, it extends their theorem to deal with arbitrary numbers of superposed modes.

coherent fields are given by Eq. (4.20) with the diagonal coefficients B_{nn} specified by Eq. (4.38). The diagonal elements of the density operator in the n-quantum-state representation are just the probabilities for having a specified number of quanta in each field mode. It is easily shown, by calculating the diagonal element that the joint probability distribution for the set of occupation numbers $\{n_k\}$ is

$$p(\{n_k\}) = \langle\{n_k\}|\rho|\{n_k\}\rangle$$
$$= \prod_k \frac{\langle m_k \rangle^{n_k}}{n_k!} e^{-\langle m_k \rangle}, \qquad (4.74)$$

where

$$\langle m_k \rangle = |\gamma_k|^2 \langle m \rangle, \qquad (4.75)$$

and γ_k is given by Eq. (4.25). The joint probability distribution for the occupation numbers of the modes reduces to a product of independent Poisson distributions. The factorization property of the density operator, however, need not extend to its off-diagonal matrix elements. The various modes will not, in general, contribute in a statistically independent way to physical processes which depend on such matrix elements.

We conclude this section with another corollary of the theorem proved at the beginning of this section. We note that the positive frequency part of the field at $x = (\mathbf{r}, t, \mu)$ is an operator of the form Eq. (4.19a) with $\gamma_k^* = i(\frac{1}{2}\hbar\omega_k)^{1/2} u_{k\mu}(\mathbf{r}) e^{-i\omega_k t}$. Vanishing of the variance of $E^{(+)}(x)$ therefore implies vanishing of the variance of a_k for all k such that $u_{k\mu}(\mathbf{r}) \neq 0$. If we require the variance of $E^{(+)}(x)$ to vanish for all x, then the variance of a_k vanishes for all k, and the field is in a pure eigenstate of the annihilation operators.

On the other hand, the variance of $E^{(+)}(x)$ can be written as

$$\mathrm{Tr}\left\{\rho E^{(+)}(x)\right\} - \left\langle E^{(+)}(x)\right\rangle^* \left\langle E^{(+)}(x)\right\rangle = G^{(1)}(x,x) - \left\langle E^{(+)}(x)\right\rangle^* \left\langle E^{(+)}(x)\right\rangle. \qquad (4.76)$$

Here we have used the definition Eq. (4.1) of $G^{(1)}$, and have written $\langle E^{(+)}(x)\rangle$ for the expectation value of $E^{(+)}(x)$. If we compare the condition for the vanishing of the variance,

$$G^{(1)}(x,x) = \left\langle E^{(+)}(x)\right\rangle^* \left\langle E^{(+)}(x)\right\rangle, \qquad (4.77)$$

with the condition (4.7) for first-order coherence, we see that the field $\mathcal{E}(x)$ in terms of which $G^{(1)}$ factorizes, must be equal to $\langle E^{(+)}(x)\rangle$, apart from a phase factor. It is obvious, furthermore, that Eq. (4.77) is obeyed by any eigenstate of the annihilation operators. Thus we may formulate the following theorem.

A field which obeys the first-order coherence condition Eq. (4.7) is in a pure eigenstate of the annihilation operators if and only if the function $\mathcal{E}(x)$ in this condition may be taken to be the expectation value of $E^{(+)}(x)$.

This result may be formulated somewhat differently if we introduce a larger set of correlation functions defined by

$$G^{(n,m)}(x_1 \ldots x_n, x_{n+1} \ldots x_{n+m}) =$$
$$\text{Tr}\left\{ \rho E^{(-)}(x_1) \ldots E^{(-)}(x_n) E^{(+)}(x_{n+1}) \ldots E^{(+)}(x_{n+m}) \right\} . \quad (4.78)$$

It is seen immediately that if the field is in an eigenstate of the annihilation operators we have

$$G^{(n,m)}(x_1 \ldots x_{n+m}) = \prod_{j=1}^{n} \mathcal{E}^*(x_j) \prod_{j=n+1}^{n+m} \mathcal{E}(x_j) \quad (4.79)$$

for all n and m. The function $\mathcal{E}(x)$ may be identified as

$$\mathcal{E}(x) = G^{(0,1)}(x) = \left\langle E^{(+)}(x) \right\rangle . \quad (4.80)$$

Factorization of the $G^{(n,m)}$ as indicated in Eq. (4.79) implies, in turn, that the field is a pure eigenstate of the annihilation operators. This result is in fact implied, according to the theorem proved earlier, by just the two conditions Eq. (4.79) for (n,m) equal to $(0,1)$ and $(1,1)$.

References

1 R. J. Glauber, *Phys. Rev.* **130**, 2529 (1963); reprinted as Chapter 1 in this volume.

2 R. J. Glauber, *Quantum Optics and Electronics*, Les Houches 1964, edited by C. deWitt, A. Blandin, C. Cohen-Tannoudji; Gordon & Breach, New York 1965, p. 63. Reprinted as Chapter 2 in this volume.

3 R. J. Glauber, in *Proceedings of the Physics of Quantum Electronics Conference*, San Juan, Puerto Rico 1965, edited by P. Kelley, B. Lax, P. E. Tannenwald; McGraw-Hill, New York 1966, p. 788.

4 U. M. Titulaer, R. J. Glauber, *Phys. Rev.* B **140**, 676 (1965); reprinted as Chapter 3 in this volume. The main results of that paper, and some of the present paper as well, are also described by U. M. Titulaer and R. J. Glauber, Ref.[3], p. 812.

5 R. J. Glauber, *Phys. Rev.* **131**, 2766 (1963); reprinted as Sect. 2.9 in this volume. See in particular Sect. 2.9.3 and Eq. (2.296).

5
Classical Behavior of Systems of Quantum Oscillators[1]

There are many contexts in which the quantum mechanical behavior of a harmonic oscillator or of a set of coupled oscillators is found to be quite similar to the classical behavior of the same systems. A considerable number of instances in which this similarity exists may be explained by means of an elementary theorem which we shall prove in the present note.

It has recently been observed that a number of problems involving harmonic oscillator degrees of freedom may be simplified by making systematic use of the coherent states of the oscillators in describing their motion. For a single oscillator these states are the right eigenstates[1] of the annihilation operator a. For any complex number α there exists a coherent state $|\alpha\rangle$ of the oscillator with the property $a|\alpha\rangle = \alpha|\alpha\rangle$. Such states have the same uncertainty of position and momentum as the ground state of the oscillator; they may be regarded simply as forms of the ground state which have been displaced both in coordinate and momentum space. While the states $|\alpha\rangle$ do not form an orthogonal set, they are complete (in fact overcomplete). Simple means are available, notwithstanding the non-orthogonality of the states, for expanding any state in terms of them.

There exists a well-defined sense in which the coherent states of an oscillator are as nearly classical in character as it is possible for a quantum state to be. If $(\delta q)^2$ and $(\delta p)^2$ are the mean variances of the coordinate and momentum of an oscillator respectively, and ω and m its angular frequency and mass, the coherent states are uniquely determined by the condition[2] that $(2m)^{-1}\{(\delta p)^2 + m^2\omega^2(\delta q)^2\}$ assume its absolute minimum value[2], which is $\frac{1}{2}\hbar\omega$. It was first noted by Schrödinger[3] that if a single free oscillator is initially in such a state then its Gaussian wavepacket undergoes no spreading with time, and the center of the packet moves both in coordinate and momentum space precisely as a classical oscillator. This observation corresponds to the statement that an initially coherent state of the oscillator remains coherent at all later times. We ask, therefore, whether more general oscillator systems do not show analogous behavior in preserving the coherent character of their quantum states as a function of time.

---

[1] Reprinted from R. J. Glauber *Phys. Lett.* **21**, 650–652 (1996), copyright 2006, with permission from Elsevier.

[2] An additional sense in which the coherent states have minimum uncertainty is that they minimize the product $(\delta q)^2(\delta p)^2$. This condition, however, does not characterize them uniquely.

Quantum Theory of Optical Coherence. Selected Papers and Lectures. Roy J. Glauber.
Copyright © 2007 WILEY-VCH Verlag GmbH & Co. KGaA, Weinheim
ISBN: 978-3-527-40687-6

We consider systems of oscillators for which the Heisenberg equations of motion can be written in the form

$$\dot{a}_j(t) = F_j(\{a_k(t)\}, t), \qquad j, k = 1 \ldots n, \tag{5.1}$$

where the functions F_j may, as indicated, depend explicitly on time. For such systems we shall show that if the Schrödinger state is initially a coherent state it remains a coherent state at all times. To prove the theorem we define the Heisenberg and Schrödinger pictures of the motion of the system so that they coincide at time $t = 0$. The initial Schrödinger state of the system, which we take to be the coherent state $|\{\alpha_k\}\rangle$, is therefore the constant Heisenberg state vector. This state is an eigenstate of the initial values $a_k(0) \equiv a_k$ of the amplitude operators,

$$a_k |\{\alpha_k\}\rangle = \alpha_k |\{\alpha_k\}\rangle . \tag{5.2}$$

We now form the variances of the complex amplitude operators, which are defined as

$$V_j(t) = \left\langle \{a_j^\dagger(t) - \langle a_j^\dagger(t)\rangle\}\{a_j(t) - \langle a_j(t)\rangle\} \right\rangle$$
$$= \langle a_j^\dagger(t) a_j(t)\rangle - \langle a_j^\dagger(t)\rangle \langle a_j(t)\rangle , \tag{5.3}$$

where the state for which the expectation values are taken is given by Eq. (5.2). It is clear from Eq. (5.2) that the initial values of the variances $V_j(t)$ vanish. If we evaluate the time derivative of $V_j(t)$ by making use of Eq. (5.1) and its Hermitian adjoint, we see that the initial values of the time derivatives also vanish. More generally, the n-th time derivative of $V_j(t)$ may be constructed by making repeated use of the equations of motion, and it is clear from Eq. (5.2) that the initial values of all of these derivatives vanish as well. Since $V_j(t)$ vanishes at $t = 0$ together with its derivatives of all orders we infer that $V_j(t)$ vanishes at times t different from zero.

Let us write the state of the system at time t in the Schrödinger picture as $|t\rangle$. Then the variances $V_j(t)$ are given in the Schrödinger picture by

$$V_j(t) = \left\langle t \left| \{a_j^\dagger - \langle a_j^\dagger(t)\rangle\}\{a_j - \langle a_j(t)\rangle\} \right| t \right\rangle . \tag{5.4}$$

These variances are non-negative, according to the Schwarz inequality, which tells us that the values $V_j(t) = 0$ can only be attained if

$$a_j |t\rangle = \langle a_j(t)\rangle |t\rangle , \qquad j = 1 \ldots n . \tag{5.5}$$

The state $|t\rangle$, in other words, is always an eigenstate of all the operators a_j. It may be written as the coherent state $|\{\alpha_k(t)\}\rangle$, where the eigenvalues of the complex amplitude operators are just

$$\alpha_j(t) = \langle a_j(t)\rangle , \tag{5.6}$$

i.e. the time-dependent eigenvalues are just the expectation values of the amplitudes.

The same result may be seen somewhat more directly if we solve the differential Eqns. (5.1) and express the solutions in the form

$$a_j(t) = L_j(\{a_k\}, t) .\tag{5.7}$$

Then it is clear that the state $|\{\alpha_k\}\rangle$ remains an eigenstate of $a_j(t)$ for all times t, i.e. we have

$$a_j(t)|\{\alpha_k\}\rangle = L_j(\{\alpha_k\}, t)|\{\alpha_k\}\rangle$$
$$= \langle a_j(t)\rangle |\{\alpha_k\}\rangle .\tag{5.8}$$

Now the time dependent operators $a_j(t)$ may be written in terms of their initial values as

$$a_j(t) = U^{-1}(t) a_j U(t) ,\tag{5.9}$$

where $U(t)$ is the unitary operator which relates the Heisenberg and Schrödinger pictures. It follows then that

$$a_j U(t)|\{\alpha_k\}\rangle = \langle a_j(t)\rangle U(t)|\{\alpha_k\}\rangle .\tag{5.10}$$

But $U(t)|\{\alpha_k\}\rangle$ is the Schrödinger state of the system at time t and we see once again that it is coherent in character for all t. This property of the state permits us to solve for it in a particularly simple way. By writing the state as $|\{\alpha_k(t)\}\rangle$ and making use of Eqns. (5.6) and (5.7) we see that the complex amplitudes $\alpha_k(t)$ satisfy the set of c-number equations of motion

$$\dot{\alpha}_j(t) = F_j(\{\alpha_k(t)\}t) ,\tag{5.11}$$

which show that the motion of the system is nearly classical in character. The wavepackets which represent all of the oscillators in configuration or momentum space retain their minimum uncertainty character at all times and move along trajectories which are precisely those of classical theory.

A simple example of a system exhibiting this behavior is an oscillator whose motion is forced by an arbitrary external field[4]. Other examples are the phase diffusion model of a laser beam[5] and the parametric frequency converter[6]. More generally, any coupling of the oscillators for which the interaction Hamiltonian takes the form

$$H' = \sum_{jk} f_{jk}(t) a_j^\dagger a_k + \sum_j \left\{ g_j(t) a_j^\dagger + g_j^*(t) a_j \right\} ,\tag{5.12}$$

where $f_{jk} = f_{kj}^*$ and the f_{jk} and g_j are arbitrary functions of time, leads to states which remain coherent at all times[8].

The theorem we have demonstrated does not extend in general to cases in which the equations of motion express the time derivatives $\dot{a}_j(t)$ in terms of the adjoint operators $\{a_k^\dagger(t)\}$ as well as the $\{\alpha_k(t)\}$. When the equations of motion take this more general form, as they do for example in the case of the parametric amplifier[7], the behavior of the system is typically less susceptible to classical description.

References

1 R. J. Glauber, *Proc. 3rd Intern. Congr. on Quantum Electronics, Paris 1963*, Vol. I, p. 111; Dunod, Paris 1964. *Phys. Rev.* **131**, 2766 (1963); reprinted as Sect. 2.9 in this volume.

2 U. M. Titulaer, R. J. Glauber, *Phys. Rev.*, to be published.

3 E. Schrödinger, *Naturwiss.* **14**, 664 (1926).

4 P. Carruthers, M. M. Nieto, *Am. J. Phys.* **33**, 537 (1965).

5 R. J. Glauber, *1964 Summer School for Theoretical Physics, Les Houches, France*, p. 165, Gordon & Breach, New York 1965; reprinted as Chapter 2 in this volume..

6 W. H. Louisell, *Radiation and noise in quantum electronics*, p. 274; McGraw-Hill, New York 1964.

7 The case of $g_j = 0$ and f_{jk} time independent has been discussed by C. W. Helstrom, *J. Math. Phys.*, to be published.

6
Quantum Theory of Parametric Amplification I[1]

6.1
Introduction

The fundamental process which has become known as parametric amplification in electronic contexts plays a central role in several physical phenomena of interest. These include the coherent Raman and Brillouin effects and the frequency splitting of light beams in nonlinear media. The most familiar form of the parametric amplifier is designed to amplify an oscillating signal by means of a particular coupling of the mode in which it appears to a second mode of oscillation, the idler mode. The coupling parameter is made to oscillate with time in a way which gives rise to a steady increase of the energy in both the signal and idler modes. The physical processes we have indicated as depending upon parametric amplification may be described in parallel terms. In the coherent Raman effect, for example, the presence of a monochromatic light wave in a Raman active medium gives rise to parametric coupling between an optical vibrational mode and a mode of the radiation field which represents the scattered (Stokes) wave. In the case of Brillouin scattering a similar form of coupling holds, with the vibrational mode oscillating at an acoustic rather than an optical frequency. The frequency splitting of light beams is an example of parametric amplification in which both of the coupled modes are electromagnetic. An intense light wave in a nonlinear dielectric medium couples pairs of electromagnetic field modes whose frequencies sum to the frequency of the original wave. It is worth noting that in this example and in the case of the Raman effect, the modes of oscillation in which the fields are amplified are initially free of excitation or very nearly so. The amplification process, in other words, not only intensifies pre-existing fields, but creates fields as well.

It is this property which indicates most clearly that the theory of the amplification process must be constructed in quantum-mechanical terms. An initially unexcited field mode can receive quanta only by means of spontaneous emission processes, which bring about what may be described as amplification of the zero-point oscillations of the mode. The correct treatment of such processes obviously lies outside the scope of

---

[1] Reprinted with permission from B. R. Mollow, R. J. Glauber, *Phys. Rev.* **160**, 1076–1096 (1967). Copyright 2006 by the American Physical Society.

Quantum Theory of Optical Coherence. Selected Papers and Lectures. Roy J. Glauber.
Copyright © 2007 WILEY-VCH Verlag GmbH & Co. KGaA, Weinheim
ISBN: 978-3-527-40687-6

classical theory. Classical analysis can only be applied to the amplification of fields which already contain many quanta, and that condition is met in the optical frequency range only by fields of high intensity.

In the radio-frequency region of the spectrum, signals containing few quanta are quite weak in comparison to the noise levels of the most sensitive detectors. The recent development of extremely low-noise amplifiers, however, has raised the possibility that it may not be long before such weak signals are indeed detectable at the higher microwave frequencies. This possibility has already inspired a lively discussion of the quantum-mechanical theory of amplification[1-16]. A quantum-mechanical model of the parametric amplifier has been proposed within this context by Louisell, Yariv, and Siegman[6]. Their model is an elementary one which may be applied as well to the optical phenomena we have noted. In this paper and the one that follows we shall adopt it as the basis of our analysis.

Previous discussions of the quantum-mechanical parametric amplifier have been based on the equations of motion for the field variables. The solutions to these equations have been used to find various time-dependent expectation values and moments of the field strengths. These data are only a part of the information implicitly contained in the time-dependent density operator for the fields. Since the density operator provides the most complete statistical description available for the system we shall devote most of our analysis to the ways in which it can be found and represented.

In a sense the amplifier we discuss is a device for transforming quantum fields into classical ones. The mathematical methods we use to discuss the amplification process must be well adapted to the treatment of both extremes. We will find it particularly convenient, in this connection, to make use of the set of coherent states in describing the quantum state of the system. In many of the examples we consider, the density operator for the system may be expressed as a species of statistical mixture of pure coherent states which we have called the P representation[17]. This representation, when it is available, can be the source of a good deal of insight, since it describes quantum states in terms not unlike those of classical probability theory. Part of our interest in the present work has therefore been directed toward determining, in a dynamical context, the usefulness and the limitations of this way of representing the density operator.

The kinds of observations which one may imagine making upon the parametric amplifier fall naturally into two classes, those which measure the field of only one mode, either the signal mode or the idler, and those which measure the fields in both. In the present paper, which is the first of two on the amplification process, we shall restrict ourselves to describing the time-dependent behavior of just one of the two interacting modes. This restriction allows us to base our analysis on a reduced form of the density operator, which is simpler in its structure than the full density operator and is somewhat more easily found. In the paper which follows we discuss the complete form of the density operator and correlations in the behavior of the two modes which it describes.

In the next two sections of this paper we outline some of the basic properties of the coherent states and the P representation, and then describe the theoretical model of the parametric amplifier. The expression for the reduced density operator for a single mode of the amplifier system is formulated in Sect. 6.4, and calculations of its value for a variety of initial states of the system are presented in Sections 6.5–6.8. The case in which both modes are initially in coherent states represents an ideally precise specification of the initial fields and therefore is discussed in some detail in Sections 6.5 and 6.6. By superposing solutions of this form we then discuss a broader variety of initial states. A number of examples of the important case in which one of the two modes is initially in a chaotic state, e.g., a thermal state, are discussed in Sections 6.7 and 6.8. The remaining three sections of the paper are devoted to more general questions concerning the properties of the P representation. In Sect. 6.9 we discuss cases in which the mode of interest is in an arbitrary initial state, while the other mode is initially in a chaotic state. The arbitrary initial state need not possess a P representation. We show that as the field is altered by the amplification process, a critical time is reached after which the P representation necessarily exists for the mode of interest. We show that the weight function for the representation must be non-negative beginning at somewhat later times. We thus follow the evolution of the function which, in the classical limit of prolonged amplification, becomes the probability distribution for the mode amplitude. An illustration of this evolution is discussed in Sect. 6.10, and the arguments we have presented are extended to the treatment of somewhat more general initial states in Sect. 6.11.

6.2
The Coherent States and the P Representation

The dynamical behavior of a single mode of the electromagnetic field may be described in terms analogous to those used for a harmonic oscillator. Each field mode is characterized by an annihilation operator a and its adjoint $a^\dagger$; these operators obey the commutation relation

$$[a, a^\dagger] = 1 . \tag{6.1}$$

In the case of the free field, the different modes are dynamically independent. The Hamiltonian for a freely oscillating mode of frequency ω may be written in the form

$$H_0 = \hbar \omega a^\dagger a . \tag{6.2}$$

The stationary states of this Hamiltonian are the n-quantum states

$$|n\rangle = \frac{1}{(n!)^{1/2}} (a^\dagger)^n |0\rangle , \tag{6.3}$$

where $n = 0, 1, 2, \ldots$ and $|0\rangle$ is the ground state of the mode, which is defined to satisfy

$$a|0\rangle = 0 . \tag{6.4}$$

The n-quantum states form a complete set, but one which is not particularly convenient to use when the quantum numbers of the field are large or quite uncertain, as they are in cases in which we have some knowledge of the phase of oscillation of the field. For such cases an alternative set of states, the coherent states, has been found especially useful[17–22]. The coherent state with complex amplitude α is defined to satisfy the eigenvalue equation

$$a|\alpha\rangle = \alpha|\alpha\rangle . \tag{6.5}$$

An explicit expression for $|\alpha\rangle$, which is determined apart from a phase factor by Eq. (6.5), is

$$|\alpha\rangle = e^{-|\alpha|^2/2} \sum_{n=0}^{\infty} \frac{(\alpha a^\dagger)^n}{n!} |0\rangle . \tag{6.6}$$

It should be emphasized that $|\alpha\rangle$ is not an eigenstate of $a^\dagger$, or of the real and imaginary parts of a considered individually. Such properties are excluded by the commutation relation Eq. (6.1).

The configuration-space wave function for a coherent state has the form of a minimum-uncertainty wavepacket; it is simply a displaced[2] form of the wave function for the ground state of the oscillator, which is a coherent state with complex eigenvalue equal to zero. The properties of the vacuum fluctuations present in a coherent state are such that two coherent states with different complex eigenvalues are not orthogonal; it is easily shown that

$$\langle \alpha | \beta \rangle = e^{-|\alpha|^2/2 - |\beta|^2/2 + \alpha^* \beta} , \tag{6.7}$$

and therefore

$$|\langle \alpha | \beta \rangle|^2 = e^{-|\alpha-\beta|^2} . \tag{6.8}$$

Although the coherent states lack orthogonality, they do form a complete set. They may be shown to satisfy the completeness relation[17,24]

$$\frac{1}{\pi} \int |\alpha\rangle\langle\alpha| \, d^2\alpha = 1 , \tag{6.9}$$

where $d^2\alpha \equiv d(\operatorname{Re}\alpha) \, d(\operatorname{Im}\alpha)$. This relation may be used to expand arbitrary state vectors and operators in terms of the coherent state vectors[17].

An operator we shall be particularly interested in expanding in terms of the coherent states is the density operator. This operator provides the most general statistical description of the state of a system; indeed no simpler description is useful, in general, for individual systems which interact with others. A density operator ρ must be Hermitian, must have non-negative eigenvalues, and must satisfy the trace relation

$$\operatorname{Tr}\rho = 1 . \tag{6.10}$$

---

[2] The ground state of the oscillator is displaced in both coordinate and momentum space in general. See, for example, Ref.[17], p. 2771.

The mean value of a dynamical operator F for a system described by the density operator ρ is $\text{Tr}\{\rho F\}$.

There exists a considerable variety of ways of representing the density operator for oscillator systems. The most common, perhaps, is the use of the n-quantum state matrix elements $\langle n|\rho|m\rangle$, which are the basis of the expansion

$$\rho = \sum_{n,m=0}^{\infty} |n\rangle \langle n|\rho|m\rangle \langle m| \ . \tag{6.11}$$

This expansion is simplest when few quanta are present, and when the off-diagonal matrix elements of ρ vanish, as they do for many stationary fields. For nonstationary fields, on the other hand, e.g., fields for which we have some information about the phase of the complex field strength, ρ is expressed more conveniently, as a rule, by means of the expansion

$$\rho = \frac{1}{\pi^2} \int |\alpha\rangle \langle \alpha|\rho|\beta\rangle \langle \beta| \, d^2\alpha \, d^2\beta \ , \tag{6.12}$$

which holds for an arbitrary density operator, according to the completeness relation Eq. (6.9).

We may also note that ρ is uniquely determined by its characteristic function[3] $\chi(\eta)$, which is defined for arbitrary complex η by the expression[4]

$$\chi(\eta) \equiv \text{Tr}\left\{\rho e^{\eta a^\dagger - \eta^* a}\right\} \ . \tag{6.13}$$

A particularly simple way of representing the density operator, the P representation[17,18,21,24,25], corresponds to writing it as a statistical mixture of pure coherent states:

$$\rho = \int P(\alpha) |\alpha\rangle \langle \alpha| \, d^2\alpha \ . \tag{6.14}$$

The P representation, when it is available, leads to great simplifications in the calculation of statistical averages of quantum-mechanical operators. It is especially suitable for making comparisons between quantum and classical theory, and for exhibiting the classical limit of quantum mechanics. The principal drawback of the P representation is that it cannot be used to represent all varieties of quantum states[24]. We shall presently discuss this limitation in greater detail.

The Hermiticity of ρ implies that the function P which appears in the expansion Eq. (6.14) must be real. The relation $\text{Tr}\,\rho = 1$ leads to the requirement

$$\int P(\alpha) \, d^2\alpha = 1 \ . \tag{6.15}$$

---

[3] That χ determines ρ uniquely follows from an expansion theorem due to H. Weyl, *The Theory of Groups and Quantum Mechanics*, Dover Publications, New York 1950, p. 272. See also Ref.[16], Sect. III.

[4] For a discussion of the characteristic function with complex argument, see Sect. 2.11.

For any real non-negative function P(α) satisfying this integral condition, the operator ρ defined by Eq. (6.14) is necessarily a Hermitian positive definite operator with unit trace, and hence is a permissible density operator. The physical constraints imposed on the density operator do not, however, exclude negative values $P(\alpha)$ for or require that it be a very well-behaved function.

The nonorthogonality of the projection operators $|\alpha\rangle\langle\alpha|$, which is a reflection of the noncommutation of the real and imaginary parts of a, makes it impossible in general to interpret the function $P(\alpha)$ as a probability density. In the classical limit, however, which corresponds to fields with $|\alpha| \gg 1$, we often deal with density operators for which the function $P(\alpha)$ varies slowly over unit distances in the complex α plane. In such cases the lack of orthogonality of the states $|\alpha\rangle$ becomes unimportant and the function P may be identified in an asymptotic sense with the classical probability distribution for finding the oscillator with the complex coordinate α.

The normally ordered form of an operator expression is obtained by writing the creation operators to the left of the annihilation operators wherever they appear. The mean value of a normally ordered operator is evaluated in the P representation by an integral which has the form of a classical statistical average, with the function P playing the role of the classical distribution function. By using Eqns. (6.5) and (6.14) we find, for example,

$$\mathrm{Tr}\left\{\rho\,(a^\dagger)^n\,a^m\right\} = \int \langle\alpha|(a^\dagger)^n\,a^m|\alpha\rangle\,P(\alpha)\,\mathrm{d}^2\alpha$$
$$= \int (\alpha^*)^n\,\alpha^m\,P(\alpha)\,\mathrm{d}^2\alpha\,. \tag{6.16}$$

We have noted that the characteristic function uniquely determines the density operator. In order to evaluate the function P in terms of the characteristic function, we first introduce the normally ordered characteristic function[24] $\chi_N(\eta)$. This function is defined by an expression analogous to the definition Eq. (6.13) of the ordinary characteristic function $\chi(\eta)$, but with the exponential written in normally ordered form:

$$\chi_N(\eta) \equiv \mathrm{Tr}\left\{\rho\,e^{\eta a^\dagger}\,e^{-\eta^* a}\right\}\,. \tag{6.17}$$

By making use of the well-known operator identity[26]

$$e^A\,e^B = e^{A+B+[A,B]/2} \tag{6.18}$$

which holds for any two operators A and B satisfying

$$[[A,B],A] = [[A,B],B] = 0\,, \tag{6.19}$$

we find that $\chi_N(\eta)$ and $\chi(\eta)$ are related by

$$\chi_N(\eta) = e^{|\eta|^2/2}\,\chi(\eta)\,. \tag{6.20}$$

If the density operator ρ has a P representation, then $\chi_N(\eta)$ is given by

$$\chi_N(\eta) = \int \langle \alpha | e^{\eta a^\dagger} e^{-\eta^* a} | \alpha \rangle P(\alpha) \, d^2\alpha$$
$$= e^{\eta \alpha^* - \eta^* \alpha} P(\alpha) \, d^2\alpha . \qquad (6.21)$$

If we write η and α in terms of their real and imaginary parts, we find that Eq. (6.21) expresses $\chi_N(\eta)$ as the two-dimensional Fourier transform of $P(\alpha)$. The solution to Eq. (6.21) for $P(\alpha)$, which follows simply from the Fourier inversion theorem, is[24]

$$P(\alpha) = \frac{1}{\pi^2} \int e^{\alpha \eta^* - \alpha^* \eta} \chi_N(\eta) \, d^2\eta . \qquad (6.22)$$

We have shown, then, that if the P representation exists, the function $\chi_N(\eta)$ has the Fourier transform $P(\alpha)$. The converse, which follows simply from the uniqueness of the correspondence between a density operator and its characteristic function, is also true: If we *define* $P(\alpha)$ by Eq. (6.22), where $\chi_N(\eta)$ is defined for an arbitrary density operator ρ by Eq. (6.17), then we may construct ρ by substituting $P(\alpha)$ into Eq. (6.14). Our criterion for the existence of the P representation will therefore be simply the existence of a Fourier transform for the normally ordered characteristic function $\chi_N(\eta)$.

Although the class of states for which the Fourier transform of $\chi_N(\eta)$ exists is a broad one, there exist many well-behaved quantum states for which the function $\chi_N(\eta)$ increases so rapidly as $|\eta| \to \infty$ that no meaning can be attached to the integral Eq. (6.22), even as a tempered distribution[27,28]. In such unmanageably singular cases it seems most reasonable to say that the P representation does not exist.[5] The density operator may always be expressed in terms of the coherent states, on the other hand, by using the more general expansion of Eq. (6.12).

6.3
Model of the Parametric Amplifier

The dynamical elements of the amplifier model proposed by Louisell, Yariv, and Siegman[6] are two modes of oscillation of the electromagnetic field. These play a symmetrical role in the amplification process; it will be somewhat briefer to refer to them as the A and B modes in the work which follows, than to designate one arbitrarily as the signal mode and the other as the idler. It is assumed that the uncoupled

---

[5] An alternative approach is suggested by J. R. Klauder, J. McKenna, and D. G. Currie, *J. Math. Phys.* **6**, 734 (1965), C. L. Mehta and E. C. G. Sudarshan, *Phys. Rev.* B **138**, 274 (1965), and J. R. Klauder, *Phys. Rev. Lett.* **16**, 534 (1966), who represent the density operator as the limit of an infinite sequence of P representations. Statistical averages, they show, may be evaluated for any density operator by carrying out an appropriate limiting procedure in each instance. The usefulness of this approach in physical contexts is not yet clear. Restriction of the functions $P(\alpha)$, on the other hand, to a narrower class than that of tempered distributions has been suggested by R. Bonifacio, L. M. Narducci, and E. Montaldi, *Phys. Rev. Lett.* **16**, 1125 (1966).

A and B modes have the dynamical behavior of harmonic oscillators, which are described by the annihilation operators $a(t)$ and $b(t)$, respectively. These operators and their ad-joints satisfy the canonical commutation relations

$$[a(t),b(t)] = [a(t),b^\dagger(t)] = 0,$$
$$[a(t),a^\dagger(t)] = [b(t),b^\dagger(t)] = 1.$$
(6.23)

The A and B modes are assumed to be coupled by a parameter which oscillates harmonically at a frequency ω equal to the sum of the natural frequencies ω_a and ω_b of the unperturbed oscillators:

$$\omega = \omega_a + \omega_b. \tag{6.24}$$

The Hamiltonian for the two coupled modes is taken to have the form[6]

$$H(t) = \hbar\omega_a a^\dagger(t)a(t) + \hbar\omega_b b^\dagger(t)b(t) - \hbar\kappa[a^\dagger(t)b^\dagger(t)e^{-i\omega t} + a(t)b(t)e^{i\omega t}],$$
(6.25)

in which κ is a coupling constant, and the phase of the externally imposed oscillation of the coupling is chosen to be zero[6] at $t = 0$. This Hamiltonian contains all of the terms which are most essential to the description of a number of physical realizations of the parametric amplifier. It is possible to construct such an amplifier, for example, by using a sample of a lossless nonlinear dielectric substance to couple the modes of a resonant cavity with reflecting walls[6]. By imposing on the dielectric an external field, the "pump" field, which oscillates at a frequency equal to the sum of the frequencies of two particular modes, those modes are made to undergo a closely coupled forced oscillation. The Hamiltonian Eq. (6.25) is intended only to describe the behavior of the two modes which are resonantly coupled in this way; the nonresonant couplings to other modes have been omitted, and the pump field has been assumed strong enough to be represented in classical terms.

The frequency-splitting of light beams may be described by essentially the same Hamiltonian as is given in Eq. (6.25). The pump field in this case is the incident fight beam, which excites, by means of a nonlinear dielectric, the emission of light in each of two modes which meet appropriate resonance conditions. Coherent Raman and Brillouin scattering may be described in a similar way. In these cases one of the two harmonic oscillators described by the Hamiltonian Eq. (6.25) represents a vibrational mode of the medium (an optical mode for the Raman effect and an acoustic mode for Brillouin scattering). The coupling of the vibrational mode with the scattered light wave is provided by the presence of the intense incident fight wave of frequency ω. While these familiar phenomena are all described by the Hamiltonian Eq. (6.25), we

---

[6] Any other initial phase for the coupling may be treated either by redefining the initial time or by performing a canonical transformation which appropriately readjusts the phases of the operators a and b.

should emphasize that we are discussing a model of an amplifier rather than an oscillator. Raman and Brillouin scattering have typically been observed under conditions in which the leakage of scattered light from the medium has tended to quench the amplification process. The scattering medium has functioned as an amplifier only for a brief interval after the appearance of the incident field, and then has continued to function as an oscillator. The effects of leakage and dissipation, on the other hand, are omitted from the Hamiltonian Eq. (6.25), and the amplification process it describes therefore continues indefinitely without quenching.

The Heisenberg equations of motion which follow from the Hamiltonian $H(t)$ are

$$i\hbar \frac{d}{dt} a(t) = [a(t), H(t)],$$
$$i\hbar \frac{d}{dt} b(t) = [b(t), H(t)]. \qquad (6.26)$$

Before discussing these equations further, let us note that the Hamiltonian Eq. (6.25) possesses a simple invariance property. It remains unchanged under the transformation $a(t) \to a(t)\,e^{i\theta}$, $b(t) \to b(t)\,e^{-i\theta}$ (θ real), which is generated by the unitary operator

$$V(\theta,t) = e^{i\theta[a^\dagger(t)\,a(t) - b^\dagger(t)\,b(t)]} \qquad (6.27)$$

via the relations

$$V^{-1}(\theta,t)\,a(t)\,V(\theta,t) = a(t)\,e^{i\theta},$$
$$V^{-1}(\theta,t)\,b(t)\,V(\theta,t) = b(t)\,e^{-i\theta}. \qquad (6.28)$$

Since $H(t)$ is invariant under the group of transformations defined by $V(\theta,t)$, i.e.,

$$V^{-1}(\theta,t)\,H(t)\,V(\theta t) = H(t), \qquad (6.29)$$

it follows that $H(t)$ commutes with the generator of the group:

$$\left[(a^\dagger(t)\,a(t) - b^\dagger(t)\,b(t)),\, H(t)\right] = 0. \qquad (6.30)$$

This equation may also be deduced directly from the explicit form for $H(t)$ and the commutation relations Eq. (6.23). It implies, according to the equations of motion Eq. (6.26), that the generator is a constant of the motion:

$$a^\dagger(t)\,a(t) - b^\dagger(t)\,b(t) = a^\dagger(0)\,a(0) - b^\dagger(0)\,b(0). \qquad (6.31)$$

This relation expresses a conservation law for the difference in the number of quanta in the A and B modes; the law follows directly from the form of the coupling between the modes.

The foregoing equations have been formulated in the Heisenberg picture, which is characterized by the time-dependent operators $a(t)$ and $b(t)$, and by a time-independent state vector $|\ \rangle$ for the system. The Schrödinger picture, on the other hand, is characterized by a time-dependent state vector $|t\rangle$, and by the time-independent operators a and b. We take the two representations to coincide at $t=0$ by writing

$$a(0) \equiv a, \\ b(0) \equiv b, \tag{6.32}$$

and

$$|\ \rangle \equiv |t=0\rangle. \tag{6.33}$$

The time dependence of the Schrödinger state vector is given by

$$i\hbar \frac{d}{dt}|t\rangle = H_S(t)|t\rangle, \tag{6.34}$$

in which the Schrödinger Hamiltonian $H_S(t)$ may be obtained from the Heisenberg Hamiltonian $H(t)$ by making the substitutions $a(t) \to a$, $b(t) \to b$ in Eq. (6.25). The Hamiltonian remains time-dependent in the Schrödinger picture because of the explicit time dependence of the coupling.

The significance of the conservation law Eq. (6.31) in the Schrödinger picture may be expressed as follows: If a Schrödinger state vector $|t\rangle$ is initially an eigenstate of $a^\dagger a - b^\dagger b$ with eigenvalue m, then it remains so at all times; if $|t\rangle$ is expanded in the n-quantum representation, only those terms appear for which the difference in the number of quanta in the A and B modes is equal to m.

The Heisenberg equations of motion for the operators $a(t)$ and $b(t)$ are obtained by substituting Eq. (6.25) for $H(t)$ into Eq. (6.26), and making use of the commutation relations Eq. (6.23); we find

$$i\frac{d}{dt}a(t) = \omega_a a(t) - \kappa b^\dagger(t) e^{-i\omega t}, \\ i\frac{d}{dt}b(t) = \omega_b b(t) - \kappa a^\dagger(t) e^{-i\omega t}. \tag{6.35}$$

Although these are operator equations their linear character means that they are no more difficult to solve than the corresponding linear equations for c-numbers. The solutions to the coupled equations (6.35) may be written in the form

$$a(t) = a c_a(t) + b^\dagger s_a(t), \tag{6.36a}$$
$$b(t) = b c_b(t) + a^\dagger s_b(t), \tag{6.36b}$$

in which the abbreviations $c_a(t)$, $s_a(t)$, $c_b(t)$, and $s_b(t)$ have been introduced for the c-number functions defined by

$$c_a(t) \equiv e^{-i\omega_a t} \cosh \kappa t, \\ s_a(t) \equiv i e^{-i\omega_a t} \sinh \kappa t, \\ c_b(t) \equiv e^{-i\omega_b t} \cosh \kappa t, \\ s_b(t) \equiv i e^{-i\omega_b t} \sinh \kappa t. \tag{6.37}$$

6.3 Model of the Parametric Amplifier

A direct insight into the way the number of quanta present changes with time may be gained by noting that the equations of motion Eq. (6.35) lead to a second-order rate equation involving only the occupation numbers of the A and B modes[30]. By using Eq. (6.35) to evaluate the second derivative of $a^\dagger(t)a(t)$ we find

$$\frac{d^2}{dt^2}\left\{a^\dagger(t)a(t)\right\} = 2\kappa^2\left\{a^\dagger(t)a(t) + b(t)b^\dagger(t)\right\}. \tag{6.38}$$

If we now introduce the notation

$$\begin{aligned} N_a(t) &\equiv a^\dagger(t)a(t), \\ N_b(t) &\equiv b^\dagger(t)b(t) \end{aligned} \tag{6.39}$$

for the occupation number operators for the two modes, and use the commutation relations Eq. (6.23) for the b operators, we may write Eq. (6.38) as

$$\frac{d^2}{dt^2}N_a(t) = 2\kappa^2[N_a(t) + N_b(t) + 1]. \tag{6.40}$$

We now make use of the conservation law Eq. (6.31) to write the difference $N_a(t) - N_b(t)$ as a time-independent operator:

$$N_a(t) - N_b(t) \equiv M. \tag{6.41}$$

By using this equation to eliminate $N_b(t)$ from Eq. (6.40) we obtain a rate equation for the operator $N_a(t)$:

$$\frac{d^2}{dt^2}N_a(t) = 2\kappa^2[2N_a(t) + 1 - M]. \tag{6.42}$$

The solution of this equation for $N_a(t)$ in terms of its initial value $2N_a(0)$ and the initial value of its time derivative $\dot{N}_a(0)$ is

$$N_a(t) = \frac{\dot{N}_a(0)}{2\kappa}\sinh 2\kappa t + [N_a(0) + \tfrac{1}{2}(1-M)]\cosh 2\kappa t - \tfrac{1}{2}(1-M). \tag{6.43}$$

A similar result, with the sign of M reversed, holds for the operator $N_b(t)$. It is clear, therefore, that the number of quanta in both modes tends, at large enough times, to increase exponentially with time.

The operator solutions Eq. (6.36) to the equations of motion may be used to calculate the mean values of products of arbitrary numbers of creation and annihilation operators as a function of time. As the simplest example, we consider first the mean values of $a(t)$ and $b(t)$, which are defined in terms of the Heisenberg density operator ρ by

$$\begin{aligned} \overline{\alpha}(t) &\equiv \operatorname{Tr}\{\rho a(t)\}, \\ \overline{\beta}(t) &\equiv \operatorname{Tr}\{\rho b(t)\}. \end{aligned} \tag{6.44}$$

The solutions for $\bar{\alpha}(t)$ and $\bar{\beta}(t)$ as a function of time take the same form as those for the complex mode amplitudes in the corresponding classical problem:

$$\bar{\alpha}(t) = \alpha_0 c_a(t) + \beta_0^* s_a(t), \qquad (6.45a)$$

$$\bar{\beta}(t) = \beta_0 c_b(t) + \alpha_0^* s_b(t), \qquad (6.45b)$$

where we have written

$$\begin{aligned} \bar{\alpha}(0) &\equiv \alpha_0, \\ \bar{\beta}(0) &\equiv \beta_0. \end{aligned} \qquad (6.46)$$

It is possible to calculate the expectation values and variances of numbers of quanta and complex field strengths for a variety of initial states in a straightforward manner with the aid of the solutions Eq. (6.36) and the commutation relations Eq. (6.23). Let us assume for example that both the A and B modes are initially in pure coherent states. We may write this initial state of the system as $|\alpha_0, \beta_0\rangle$, where α_0 and β_0 are the complex amplitudes of the A and B modes, respectively. For certain types of couplings of harmonic oscillator systems it has been shown that an initially coherent Schrödinger state retains its coherent character at later times[30]. The state vectors for the parametric amplifier modes behave quite differently, however. A simple indication of their behavior may be obtained by evaluating the variance of the complex field strength for the A mode. We easily deduce with the aid of Eq. (6.5), the commutation relations Eq. (6.23), and Eq. (6.36a), that

$$\left\langle \alpha_0, \beta_0 \left| \left(a^\dagger(t) - \bar{\alpha}^*(t)\right)\left(a(t) - \bar{\alpha}(t)\right) \right| \alpha_0, \beta_0 \right\rangle = |s_a(t)|^2 = \sinh^2 \kappa t. \qquad (6.47)$$

In the Schrödinger picture this equation takes the form

$$\left\langle t \left| \left(a^\dagger - \bar{\alpha}^*(t)\right)\left(a - \bar{\alpha}(t)\right) \right| t \right\rangle = \sinh^2 \kappa t, \qquad (6.48)$$

in which the Schrödinger state vector $|t\rangle$ is initially coherent:

$$|t = 0\rangle = |\alpha_0, \beta_0\rangle, \qquad (6.49)$$

and satisfies the Schrödinger equation (6.34).

In either of the two pictures it is evident that the variance of the complex field amplitude grows exponentially with time, much as the average occupation number of the field does. It is clear from this behavior of the variance that the state which evolves from an initially coherent state does not retain its coherent character. The uncertainty of the field, which is initially minimal, grows rapidly with time[12]. The amplification process amplifies the vacuum fluctuations along with the expectation value of the field strength, and the minimum uncertainty character of the initial state is quickly lost.

6.4
Reduced Density Operator for the A Mode

To evaluate statistical averages of time-dependent operators in the Heisenberg picture we must make explicit use of the solutions to the operator equations of motion for the system. The operators $a(t)$ and $b(t)$, for example, must be expressed in terms of their initial values before the statistical average of a function which depends on them can be evaluated in the initial state of the system. The Schrödinger picture, on the other hand, offers a more compact way of evaluating such averages; it combines the dynamical part of the calculation with the statistical part by describing the system in terms of a time-dependent density operator $\rho(t)$. We shall confine our attention in the present paper to discussing the statistical behavior of either one of the two modes of the amplifier system. For this purpose we are led to consider a reduced form of the Schrödinger density operator which is defined in terms of the variables for the mode of interest.

The Heisenberg and Schrödinger pictures of the motion of the system are related by the unitary time translation operator $U(t)$, which is defined by the equations

$$i\hbar \frac{d}{dt} U(t) = H_S(t) U(t) \tag{6.50}$$

and

$$U(0) = 1 . \tag{6.51}$$

The Heisenberg operators $a(t)$ and $b(t)$, for example, are given formally in terms of their initial values a and b by

$$a(t) = U^{-1}(t) a U(t) , \tag{6.52a}$$
$$b(t) = U^{-1}(t) b U(t) . \tag{6.52b}$$

In the Schrödinger picture the density operator, like the state vector, is a time-dependent quantity. The Schrödinger density operator $\rho(t)$ is given in terms of the time-independent Heisenberg density operator ρ by the relation

$$p(t) = U(t) \rho U^{-1}(t) . \tag{6.53}$$

The reduced Schrödinger density operator for the A node is defined by

$$\rho_A(t) = \text{Tr}_B \, \rho(t) , \tag{6.54}$$

where Tr_B means trace with respect to the (initial or time-independent) states of the B mode. The mean value of an operator T_A which refers to the variables of the A mode only is given by

$$\text{Tr} \{\rho(t) T_A\} = \text{Tr}_A \, \text{Tr}_B \{\rho(t) T_A\}$$
$$= \text{Tr}_A \{\rho_A(t) T_A\} . \tag{6.55}$$

The time-dependent form of the normally ordered characteristic function $\chi_N(\eta,t)$ for the A mode is given by

$$\chi_N(\eta,t) = \mathrm{Tr}\left\{\rho(t)\,e^{\eta a^\dagger} e^{-\eta^* a}\right\} \tag{6.56}$$

$$= \mathrm{Tr}_A\left\{\rho_A(t)\,e^{\eta a^\dagger} e^{-\eta^* a}\right\}. \tag{6.57}$$

The time-dependent function $\chi_N(\eta,t)$, according to Eq. (6.57), is defined in terms of the reduced density operator $\rho_A(t)$ by an expression identical to the definition Eq. (6.17) for the normally ordered characteristic function corresponding to an arbitrary (single mode) density operator.

We shall say that a P representation for the A mode exists at time t if the reduced Schrödinger density operator $\rho_A(t)$ can be written in the form

$$\rho_A(t) = \int P(\alpha,t)\,|\alpha\rangle\langle\alpha|\,d^2\alpha. \tag{6.58}$$

From the discussion of Sect. 6.2 it is evident that $\rho_A(t)$ has a P representation if $\chi_N(\eta,t)$ possesses a Fourier transform,

$$P(\alpha,t) = \frac{1}{\pi^2}\int e^{\alpha\eta^* - \alpha^*\eta}\,\chi_N(\eta,t)\,d^2\eta. \tag{6.59}$$

The function $\chi_N(\eta,t)$, which has been defined as the trace of a Schrödinger operator, is invariant under transformation to the Heisenberg picture. By substituting Eq. (6.53) for $\rho(t)$ into Eq. (6.56), and making use of the cyclical symmetry of the traces of products, we obtain

$$\chi_N(\eta,t) = \mathrm{Tr}\left\{\rho\,U^{-1}(t)\,e^{\eta a^\dagger} e^{-\eta^* a}\,U(t)\right\}. \tag{6.60}$$

By making use of Eq. (6.52a) and its adjoint we then find

$$\chi_N(\eta,t) = \mathrm{Tr}\left\{\rho\,e^{\eta a^\dagger(t)} e^{-\eta^* a(t)}\right\}. \tag{6.61}$$

This equation expresses $\chi_N(\eta,t)$ in terms of the initial density operator ρ for the joint system of A and B modes, and the time-dependent operator $a(t)$ and its adjoint. A formal solution for $\chi_N(\eta,t)$ may thus be constructed by substituting the solution Eq. (6.36a) for $a(t)$ into Eq. (6.61). If the function $\chi_N(\eta,t)$ obtained in this way has the Fourier transform $P(\alpha,t)$, then the reduced density operator $\rho_A(t)$ is given by Eq. (6.58).

6.5
Initially Coherent State: *P* Representation for the A Mode

In Sect. 6.3 we considered the case in which the joint system of A and B modes is initially described by the pure coherent state vector $|\alpha_0,\beta_0\rangle$. We showed, by evaluating the variance of the complex field strength for the A mode, that such a state does not remain coherent at later times. In this section we shall consider the case of an initially coherent state in greater detail, and solve for the function $P(\alpha,t)$, which provides a full description of the behavior of the A mode.

Let us assume, then, that the density operator for the joint system of A and B modes is initially given by

$$\rho = |\alpha_0, \beta_0\rangle \langle \alpha_0, \beta_0| \ . \tag{6.62}$$

To evaluate the time-dependent normally ordered characteristic function for the A mode, we first use the operator identity Eq. (6.18) to write

$$e^{\eta a^\dagger(t)} e^{-\eta^* a(t)} = e^{|\eta|^2/2} e^{\eta a^\dagger(t) - \eta^* a(t)} \ . \tag{6.63}$$

By substituting this relation into Eq. (6.61) and using Eq. (6.62) for ρ and Eq. (6.36a) for $a(t)$, we obtain

$$\chi_N(\eta,t) = e^{|\eta|^2/2}$$
$$\times \langle \alpha_0, \beta_0 | \exp\left[\eta\left(a^\dagger c_a^*(t) + b s_a^*(t)\right) - \eta^*\left(a c_a(t) + b^\dagger s_a(t)\right)\right] | \alpha_0, \beta_0 \rangle \ . \tag{6.64}$$

We denote by $s(t)$ and $c(t)$ the moduli of the complex functions $s_{a,b}(t)$ and $c_{a,b}(t)$, respectively:

$$\begin{aligned} s(t) &\equiv \sinh \kappa t \ , \\ c(t) &\equiv \cosh \kappa t \ . \end{aligned} \tag{6.65}$$

By using Eq. (6.18) to write the exponential operator in Eq. (6.64) in normally ordered form, we find

$$\chi_N(\eta,t) = \exp\left[|\eta|^2 \left(\tfrac{1}{2} - \tfrac{1}{2}c^2(t) - \tfrac{1}{2}s^2(t)\right)\right]$$
$$\times \langle \alpha_0, \beta_0 | \exp\left[a^\dagger \eta c_a^*(t) - b^\dagger \eta^* s_a(t)\right] \exp\left[-a \eta^* c_a(t) + b \eta s_a^*(t)\right] | \alpha_0, \beta_0 \rangle$$
$$= \exp\left[-|\eta|^2 s^2(t) + \alpha_0^* \eta c_a^*(t) - \beta_0^* \eta^* s_a(t) - \alpha_0 \eta^* c_a(t) + \beta_0 \eta s_a^*(t)\right] \tag{6.66}$$
$$= \exp\left[-|\eta|^2 s^2(t) + \eta \overline{\alpha}^*(t) - \eta^* \overline{\alpha}(t)\right] \ , \tag{6.67}$$

in which $\overline{\alpha}(t)$, the mean value of $a(t)$, is defined by Eq. (6.45a).

Substituting Eq. (6.67) for $\chi_N(\eta,t)$ into Eq. (6.59), we find that $P(\alpha,t)$ is given by the complex Fourier integral

$$P(\alpha,t) = \frac{1}{\pi^2} \int \exp\left[-|\eta|^2 s^2(t) + \eta(\overline{\alpha}^*(t) - \alpha^*) - \eta^*(\overline{\alpha}(t) - \alpha)\right] d^2\eta \ . \tag{6.68}$$

This integral is easily evaluated with the aid of the useful identity

$$\frac{1}{\pi} \int d^2\eta \, e^{-\mu|\eta|^2 + \lambda \eta + \nu \eta^*} = \frac{1}{\mu} \exp\left(\frac{\lambda \nu}{\mu}\right) \ , \tag{6.69}$$

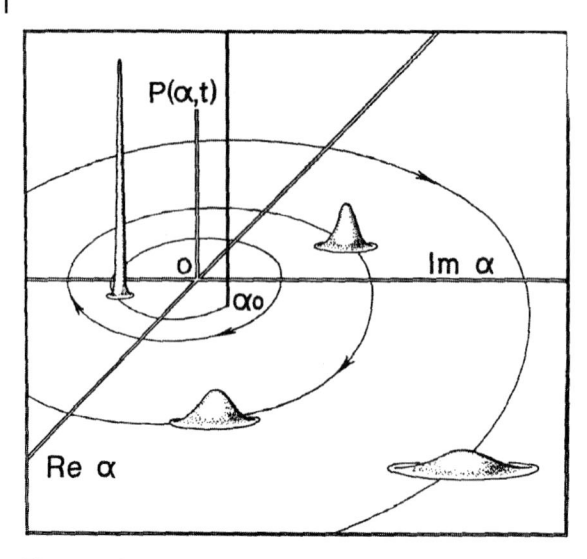

Figure 1 Schematic picture of the way in which the function $P(\alpha,t)$ varies with time when the system is initially in a pure coherent state. At the initial time $t=0$ the function $P(\alpha,t)$ is the delta function $\delta^{(2)}(\alpha-\alpha_0)$. The amplification process leads to a function $P(\alpha,t)$ which is Gaussian in form at later times and has a variance which increases monotonically with time. The mean value of the complex amplitude α describes a spiral trajectory in the complex α plane.

which holds for $\operatorname{Re}\mu > 0$, and for arbitrary complex numbers λ and ν. By making the appropriate identifications of these parameters we find from Eq. (6.68) that $P(\alpha,t)$ takes the form

$$P(\alpha,t) = \frac{1}{\pi s^2(t)} \exp\left[-\frac{|\alpha-\overline{\alpha}(t)|^2}{s^2(t)}\right]. \tag{6.70}$$

The function $P(\alpha,t)$ for the A mode at time t is thus a Gaussian function about the complex mean value $\overline{\alpha}(t)$. The variance of the distribution is $s^2(t) = \sinh^2 \kappa t$, a result which was obtained in Sect. 6.3 from the solution to the equations of motion.

In Ref.[17] it is shown that the P representation provides a natural means of extending to the quantum-mechanical domain the classical concept of the superposition of two independent fields: The P function for the superposition of two fields is the convolution of the P functions for each field considered individually. Since the function $P(\alpha)$ for a coherent state with complex amplitude $\overline{\alpha}$ is the delta function $\delta^{(2)}(\alpha-\overline{\alpha}) \equiv \delta(\operatorname{Re}(\alpha-\overline{\alpha}))\,\delta(\operatorname{Im}(\alpha-\overline{\alpha}))$, we may say that Eq. (6.70) describes the superposition of a coherent state with complex amplitude $\overline{\alpha}(t)$, and a chaotic or Gaussian mixture with variance $\sinh^2 \kappa t$.

In Fig. 1 we have plotted the function $P(\alpha,t)$ at several points along the curve $\overline{\alpha}(t)$, for the case $\beta_0 = 0$, and therefore $\overline{\alpha}(t) = e^{-i\omega_a t}\alpha_0 \cosh \kappa t$. The variance of the distribution, which is initially zero, grows rapidly with time; for large amplification

($\kappa t \gg 1$), both the variance and the square of the mean field strength are proportional to $e^{2\kappa t}$.

The result Eq. (6.70) has been derived for the initially coherent state $|\alpha_0, \beta_0\rangle$. If the joint system of A and B modes is initially described by a P representation,

$$\rho = \int P(\alpha_0, \beta_0) |\alpha_0, \beta_0\rangle \langle \alpha_0, \beta_0| \, d^2\alpha_0 \, d^2\beta_0 , \tag{6.71}$$

then the function P for the A mode at time t is obtained by averaging the right-hand side of Eq. (6.70) with respect to the weight function $P(\alpha_0, \beta_0)$:

$$P(\alpha,t) = \frac{1}{\pi s^2(t)} \int P(\alpha_0, \beta_0) \exp\left[-\frac{|\alpha - \bar{\alpha}(\alpha_0, \beta_0, t)|^2}{s^2(t)}\right] d^2\alpha_0 \, d^2\beta_0 , \tag{6.72}$$

in which the function $\bar{\alpha}(\alpha_0, \beta_0, t)$ is defined by Eq. (6.45a).

The initial state specified by $\alpha_0 = \beta_0 = 0$ in Eq. (6.62) corresponds to the absence of any initial excitation in the system. A field is generated nonetheless; it may be thought of as an amplified form of the zero-point fluctuations of the vacuum. The function $P(\alpha,t)$ for this case is obtained by setting $\bar{\alpha}(t) = 0$ in Eq. (6.70):

$$P(\alpha,t) = \frac{1}{\pi s^2(t)} \exp\left[-\frac{|\alpha|^2}{s^2(t)}\right] . \tag{6.73}$$

This function describes a chaotic mixture with variance $s^2(t)$; since $\bar{\alpha}(t) = 0$ the variance is equal to the mean number of quanta present in the mode,

$$\langle n(t) \rangle = s^2(t) = \sinh^2 \kappa t . \tag{6.74}$$

In Ref.[17] it is shown that a density operator with a Gaussian P function may be written in the n-quantum representation in the form characteristic of thermal equilibrium. The density operator $\rho_A(t)$ is then given by

$$\rho_A(t) = \sum_{n=0}^{\infty} \frac{[\langle n(t) \rangle]^n}{[1 + \langle n(t) \rangle]^{n+1}} |n\rangle \langle n|$$

$$= \sum_{n=0}^{\infty} \frac{[s^2(t)]^n}{[c^2(t)]^{n+1}} |n\rangle \langle n| , \tag{6.75}$$

where $|n\rangle$ represents the n-quantum state of the A mode. The probability of finding n quanta in the A mode after time t is given by the coefficient of $|n\rangle\langle n|$ in this expansion,

$$p_n(t) = \frac{[s^2(t)]^n}{[c^2(t)]^{n+1}} . \tag{6.76}$$

6.6
Initially Coherent State;
Moments, Matrix Elements, and Explicit Representation for $\rho_A(t)$

We have shown that the reduced density operator for the A mode at time t, corresponding to an initially coherent state for the joint system, has the P representation

$$\rho_A(t) = \frac{1}{\pi s^2(t)} \int \exp\left[-\frac{|\alpha - \overline{\alpha}|^2}{s^2(t)}\right] |\alpha\rangle\langle\alpha| \, d^2\alpha . \quad (6.77)$$

In this section we shall derive a number of statistical properties of the mode which follow from this description, and exhibit an explicit form for the density operator in terms of the operators a and $a^\dagger$.

Let us find the mean value of the normally ordered product of an arbitrary number of factors of $a^\dagger$ and a. We may express this average in terms of the function $\chi_N(\eta,t)$, by means of the formula

$$\text{Tr}\left\{\rho_A(t)(a^\dagger)^n a^m\right\} = \left(\frac{\partial}{\partial\eta}\right)^n \left(-\frac{\partial}{\partial\eta^*}\right)^m \chi_N(\eta,t)\Big|_{\eta=\eta^*=0}, \quad (6.78)$$

which may be deduced by differentiating Eq. (6.57) directly. If we substitute Eq. (6.67) for $\chi_N(\eta,t)$ into Eq. (6.78), we find

$$\text{Tr}\left\{\rho_A(t)(a^\dagger)^n a^m\right\} = \left(\frac{\partial}{\partial\eta}\right)^n \left(-\frac{\partial}{\partial\eta^*}\right)^m$$
$$\times \exp\left[\eta\overline{\alpha}^*(t) - \eta^*\overline{\alpha}(t) - \eta\eta^* s^2(t)\right]\Big|_{\eta=\eta^*=0}. \quad (6.79)$$

We show in the Appendix that an expression of this form may be reduced, apart from some simple factors, to an associated Laguerre polynomial. By making use of Eqns. (6.A6) and (6.A7) of the Appendix we obtain for the average in Eq. (6.79) the two equivalent expressions

$$\text{Tr}\left\{\rho_A(t)(a^\dagger)^n a^m\right\} = n! \, [s^2(t)]^n \, [\overline{\alpha}(t)]^{m-n} \, L_n^{(m-n)}\left(-\frac{|\overline{\alpha}(t)|^2}{s^2(t)}\right) \quad (6.80)$$

$$= m! \, [s^2(t)]^m \, [\overline{\alpha}^*(t)]^{n-m} \, L_m^{(n-m)}\left(-\frac{|\overline{\alpha}(t)|^2}{s^2(t)}\right). \quad (6.81)$$

It follows from the easily proved identity

$$(a^\dagger)^n a^n = a^\dagger a(a^\dagger a - 1)\ldots(a^\dagger a - n + 1)$$
$$\equiv \frac{(a^\dagger a)!}{(a^\dagger a - n)!}$$

that the factorial moments of the quantum distribution in the mode, which are obtained by evaluating Eq. (6.80) or Eq. (6.81) at $n = m$, are given by

$$\text{Tr}\left\{\rho_A(t) \frac{(a^\dagger a)!}{(a^\dagger a - n)!}\right\} = n! \, [s^2(t)]^n \, L_n\left(-\frac{|\overline{\alpha}(t)|^2}{s^2(t)}\right). \quad (6.82)$$

6.6 Initially Coherent State; Moments, Matrix Elements, and Explicit Representation for $\rho_A(t)$

For the case of vacuum amplification, i.e., when no quanta are initially present in either mode, the quantity $\mathrm{Tr}\{\rho_A(t)(a^\dagger)^n a^m\}$ is evaluated by letting $\bar{\alpha}(t) \to 0$, in either of the relations Eq. (6.80) or Eq. (6.81). By making use of the identity $L_n(0) = 1$ we find

$$\mathrm{Tr}\{\rho_A(t)(a^\dagger)^n a^m\} = \delta_{nm} n! [s^2(t)]^n . \tag{6.83}$$

These expectation values vanish for $n \neq m$ since the phase of the field is completely uncertain.

The matrix elements of $\rho_A(t)$ in the n-quantum representation are easily evaluated. By multiplying Eq. (6.58) on the left by $\langle m|$ and on the right by $|n\rangle$, we obtain the relation

$$\langle m|\rho_A(t)|n\rangle = \int \langle m|\alpha\rangle \langle \alpha|n\rangle P(\alpha,t) d^2\alpha$$

$$= \int \frac{\alpha^m (\alpha^*)^n}{(m!\,n!)^{1/2}} P(\alpha,t) e^{-|\alpha|^2} d^2\alpha . \tag{6.84}$$

If we define a generating function $R(w,z,t)$ by the equation[17]

$$R(w,z,t) = \int P(\alpha,t) e^{-|\alpha|^2 + w\alpha + z\alpha^*} d^2\alpha , \tag{6.85}$$

we find, on differentiating this expression and comparing the result to Eq. (6.84), that the matrix elements of $\rho_A(t)$ are given by

$$\langle m|\rho_A(t)|n\rangle = \frac{1}{(m!\,n!)^{1/2}} \left(\frac{\partial}{\partial w}\right)^m \left(\frac{\partial}{\partial z}\right)^n R(w,z,t)\Big|_{w=z=0} . \tag{6.86}$$

If we substitute Eq. (6.70) for $P(\alpha,t)$ into Eq. (6.85), we find that the desired generating function is given by

$$R(w,z,t) = \frac{1}{\pi s^2(t)} \int \exp\left[-\frac{|\alpha - \bar{\alpha}(t)|^2}{s^2(t)} - |\alpha|^2 + w\alpha + u\alpha^*\right] d^2\alpha . \tag{6.87}$$

This integral may be evaluated straightforwardly with the aid of the identity Eq. (6.69). We find then

$$R(w,z,t) = \frac{1}{c^2(t)} \exp\left[-\frac{|\bar{\alpha}(t)|^2}{c^2(t)} + w\left(\frac{\bar{\alpha}(t)}{c^2(t)}\right) + z\left(\frac{\bar{\alpha}^*(t)}{c^2(t)}\right) + wz\left(\frac{s^2(t)}{c^2(t)}\right)\right] . \tag{6.88}$$

The derivatives of this expression, which according to Eq. (6.86) correspond to the matrix elements of $\rho_A(t)$, may be expressed in terms of associated Laguerre polynomials much as in the case of the derivatives we discussed in Eq. (6.79). By using

Eqns. (6.A6) and (6.A7) we find for the matrix elements of $\rho_A(t)$ the two equivalent expressions

$$\langle m|\rho_A(t)|n\rangle = \exp\left[-\frac{|\bar{\alpha}(t)|^2}{c^2(t)}\right] \frac{[s^2(t)]^m}{[c^2(t)]^{n+1}} [\bar{\alpha}^*(t)]^{n-m}$$
$$\times \left(\frac{m!}{n!}\right)^{1/2} L_m^{(n-m)}\left(-\frac{|\bar{\alpha}(t)|^2}{s^2(t)c^2(t)}\right) \quad (6.89)$$

$$= \exp\left[-\frac{|\bar{\alpha}(t)|^2}{c^2(t)}\right] \frac{[s^2(t)]^n}{[c^2(t)]^{m+1}} [\bar{\alpha}^*(t)]^{m-n}$$
$$\times \left(\frac{n!}{m!}\right)^{1/2} L_n^{(m-n)}\left(-\frac{|\bar{\alpha}(t)|^2}{s^2(t)c^2(t)}\right). \quad (6.90)$$

The right-hand side of Eq. (6.90) may be obtained from the right-hand side of Eq. (6.89) by taking the complex conjugate of the latter and interchanging n and m; this relation is a reflection of the Hermiticity requirement

$$\langle m|\rho_A(t)|n\rangle = (\langle n|\rho_A(t)|m\rangle)^*. \quad (6.91)$$

The probability of finding n quanta in the A mode at time t, which is obtained by evaluating Eq. (6.89) or Eq. (6.90) for $n = m$ is

$$\langle n|\rho_A(t)|n\rangle = \exp\left[-\frac{|\bar{\alpha}(t)|^2}{c^2(t)}\right] \frac{[s^2(t)]^n}{[c^2(t)]^{n+1}} L_n\left(-\frac{|\bar{\alpha}(t)|^2}{s^2(t)c^2(t)}\right). \quad (6.92)$$

For the case of vacuum amplification, the matrix elements of $\rho_A(t)$ in the n-quantum representation are given by Eq. (6.89) or Eq. (6.90) evaluated at $\bar{\alpha}(t) = 0$:

$$\langle m|\rho_A(t)|n\rangle = \delta_{nm} \frac{[s^2(t)]^n}{[c^2(t)]^{n+1}}, \quad (6.93)$$

which is equivalent to the result already noted in Eq. (6.75).

We need not confine our discussion of the density operator $\rho_A(t)$ to the form it takes in the P representation or to its matrix elements in terms of n-quantum states. It is a simple matter to give an explicit construction of the operator itself. For the case of vacuum amplification, for example, $\rho_A(t)$ may be obtained by expressing Eq. (6.75) formally as follows:

$$\rho_A(t) = \frac{1}{c^2(t)}\left[\frac{s^2(t)}{c^2(t)}\right]^{a^\dagger a} \quad (6.94)$$

$$= \frac{1}{c^2(t)} \exp\left[a^\dagger a \ln\left(\frac{s^2(t)}{c^2(t)}\right)\right]. \quad (6.95)$$

If the initial state of the system is an arbitrary coherent state, it is easy to show that $\rho_A(t)$ is given by Eq. (6.95), with a replaced by $a - \bar{\alpha}(t)$ and $a^\dagger$ by $a^\dagger - \bar{\alpha}^*(t)$:

$$\rho_A(t) = \frac{1}{c^2(t)} \exp\left[[a^\dagger - \bar{\alpha}^*(t)][a - \bar{\alpha}(t)] \ln\left(\frac{s^2(t)}{c^2(t)}\right)\right]. \quad (6.96)$$

This result applies to the initial density operator Eq. (6.62). If the initial density operator for the system has the P representation Eq. (6.71), then the linearity of the dependence of $\rho_A(t)$ on ρ enables us to see that $\rho_A(t)$ is given more generally by

$$\rho_A(t) = \frac{1}{c^2(t)} \int \exp\left[[a^\dagger - \overline{\alpha}^*(\alpha_0,\beta_0,t)] [a - \overline{\alpha}(\alpha_0,\beta_0,t)] \ln\left(\frac{s^2(t)}{c^2(t)}\right) \right]$$
$$\times P(\alpha_0,\beta_0) \, d^2\alpha_0 \, d^2\beta_0, \quad (6.97)$$

where $\overline{\alpha}(\alpha_0,\beta_0,t)$ is defined by Eq. (6.45a).

6.7
Solutions for an Initially Chaotic B Mode

Cases in which one of the modes is in a chaotically mixed state (e.g., a thermal equilibrium distribution) are important from a practical standpoint and relatively simple to treat. In this section we shall consider the behavior of the A mode of the amplifier when the initial state of the B mode is chaotic in character. The simplest such case is the one in which both the A and B modes are initially in chaotic states. Other cases we shall discuss are those in which the A mode is initially in the vacuum state or a coherent state, and cases in which its initial state is specified more generally by a P representation.

If the initial states of the A and B modes are independent chaotic mixtures with mean quantum numbers $\langle n \rangle$ and $\langle m \rangle$, respectively, we may write the initial density operator in the form

$$\rho_c = \int P_c(\alpha_0,\beta_0) |\alpha_0,\beta_0\rangle \langle \alpha_0,\beta_0| \, d^2\alpha_0 \, d^2\beta_0, \quad (6.98)$$

where

$$P_c(\alpha_0,\beta_0) = \frac{1}{\pi^2 \langle n \rangle \langle m \rangle} \exp\left[-\frac{|\alpha_0|^2}{\langle n \rangle} - \frac{|\beta_0|^2}{\langle m \rangle} \right]. \quad (6.99)$$

One way of finding the function $P(\alpha,t)$ is to substitute this Gaussian expression into the formula given by Eq. (6.72) and evaluate the resulting integral. An equivalent and somewhat simpler procedure is to find $P(\alpha,t)$ by beginning once again with the normally ordered characteristic function. That function, it is clear, may be obtained by averaging the result in Eq. (6.66) with respect to the weight function $P_c(\alpha_0,\beta_0)$. We have then

$$\chi_{N,c}(\eta,t) = \int \exp\left[-|\eta|^2 s^2(t) + \alpha_0^* \eta c_a^*(t) - \beta_0^* \eta^* s_a(t) - \alpha_0 \eta^* c_a(t) + \beta_0 \eta s_a^*(t) \right]$$
$$\times P_c(\alpha_0,\beta_0) \, d^2\alpha_0 \, d^2\beta_0. \quad (6.100)$$

By substituting the Gaussian form for $P_c(\alpha_0,\beta_0)$ given by Eq. (6.99) into this expression and performing the integration with the aid of Eq. (6.69), we obtain

$$\chi_{N,c}(\eta,t) = e^{-|\eta|^2 N(t)}, \qquad (6.101)$$

where

$$N(t) = \langle n \rangle c^2(t) + (1 + \langle m \rangle) s^2(t). \qquad (6.102)$$

The function P for the A mode at time t is evaluated as the Fourier transform of $\chi_{N,c}(\eta,t)$:

$$P_c(\alpha,t) = \pi^{-2} \int e^{-|\eta|^2 N(t) + \alpha\eta^* - \alpha^*\eta} d^2\eta$$

$$= \frac{1}{\pi N(t)} \exp\left[-\frac{|\alpha|^2}{N(t)}\right]. \qquad (6.103)$$

The reduced Schrödinger density operator $\rho_{A,c}(t)$ for the A mode thus corresponds to a chaotic mixture with mean quantum number $N(t)$. For $\langle n \rangle = \langle m \rangle = 0$ the joint system is initially in the vacuum state; for this case $N(t) = s^2(t)$, and Eq. (6.103) becomes identical to the result Eq. (6.73) found earlier for the case of vacuum amplification. The effect of the chaotic fields initially present in both the A and B modes is to increase the fluctuations in the field strength of the A mode at time t, from $s^2(t)$ to $N(t)$.

Holliday and Glassgold[16] have discussed an approximate model of laser amplification described by equations which may be cast in a form similar to those for the parametric amplifier. The B mode of oscillation in their model represents formally the effect of the pumping molecules; its state is taken to be a chaotic mixture. They show that if the initial state of the field mode in their model is also taken to be chaotic, then the function $P(\alpha,t)$ which describes it takes the form of Eq. (6.103).

An explicit expression for the density operator $\rho_{A,c}(t)$ may be obtained by replacing $s^2(t)$ in Eq. (6.95) by $N(t)$ (and $c^2(t)$ by $N(t)+1$). Formulas for the moments and n-quantum state matrix elements of $\rho_{A,c}(t)$ are obtained by making the same substitutions in Eqns. (6.83) and (6.93), respectively. The probability of finding l quanta in the A mode at time t, for the initially chaotic mixture Eq. (6.98), is

$$p(l,t|\langle n \rangle, \langle m \rangle) = \frac{[N(t)]^l}{[1+N(t)]^{l+1}}. \qquad (6.104)$$

The choice $\langle n \rangle = 0$ in Eq. (6.99) implies that the initial state of the A mode is the ground state $|0\rangle_A$. The initial density operator for the joint system in this case may be written in the form

$$\rho = |0\rangle_A {}_A\langle 0| \, \rho_{B,\langle m \rangle}, \qquad (6.105)$$

6.7 Solutions for an Initially Chaotic B Mode

in which the chaotic density operator $\rho_{B,\langle m \rangle}$ for the B mode is defined by

$$\rho_{B,\langle m \rangle} = \frac{1}{\pi \langle m \rangle} \int \exp\left[-\frac{|\beta_0|^2}{\langle m \rangle}\right] |\beta_0\rangle \langle \beta_0| \, d^2\beta_0 \, . \tag{6.106}$$

To obtain the function $P(\alpha,t)$ which corresponds to the initial state Eq. (6.105), we substitute $\langle n \rangle = 0$ in Eqns. (6.102) and (6.103). We thereby find

$$P(\alpha,t) = \frac{1}{\pi(1+\langle m \rangle)s^2(t)} \exp\left[-\frac{|\alpha|^2}{(1+\langle m \rangle)s^2(t)}\right] . \tag{6.107}$$

If we begin with the A mode in the coherent state $|\alpha'\rangle$ and the B mode in the chaotic mixture $\rho_{B,\langle m \rangle}$, then the initial density operator for the system is

$$\rho = |\alpha'\rangle\langle\alpha'| \, \rho_{B,\langle m \rangle} \, , \tag{6.108}$$

which has the P representation

$$P(\alpha_0,\beta_0) = \delta^{(2)}(\alpha_0 - \alpha') \frac{1}{\pi \langle m \rangle} \exp\left[-\frac{|\beta_0|^2}{\langle m \rangle}\right] . \tag{6.109}$$

The normally ordered characteristic function for the A mode at time t may be obtained by evaluating the integral in Eq. (6.100), with $P_c(\alpha_0,\beta_0)$ replaced by $P(\alpha_0,\beta_0)$ as defined by Eq. (6.109). We find

$$\chi_N(\eta,t) = \exp\left[\eta \alpha'^* c_a^*(t) - \eta^* \alpha' c_a(t) - |\eta|^2 (1+\langle m \rangle) s^2(t)\right] . \tag{6.110}$$

By substituting this relation into Eq. (6.59) and performing the integration with the aid of the integral identity Eq. (6.69) we find that the function P for the A mode at time t is given by[16]

$$P(\alpha,t) = \frac{1}{\pi(1+\langle m \rangle)s^2(t)} \exp\left[-\frac{|\alpha - \alpha' c_a(t)|^2}{(1+\langle m \rangle)s^2(t)}\right] . \tag{6.111}$$

Thus the effect of choosing the initial state of the A mode to be the coherent state $|\alpha'\rangle$ rather than the ground-state $|0\rangle_A$ is to shift the mean complex amplitude of the distribution from zero to $\alpha' c_a(t)$; the variance of the distribution remains unchanged.

The result Eq. (6.111) may be immediately generalized to the case in which the A mode initially has an arbitrary P representation. If we assume that the initial state of the B mode is the chaotic mixture $\rho_{B,\langle m \rangle}$, then the initial density operator for the system takes the form

$$\rho = \int P(\alpha',0) |\alpha'\rangle\langle\alpha'| \, d^2\alpha' \rho_{B,\langle m \rangle} \, , \tag{6.112}$$

where $P(\alpha',0)$ is the initial P function for the A mode. The value of this function at time t may be obtained by averaging Eq. (6.111) with respect to the weight function $P(\alpha',0)$. We thus have

$$P(\alpha,t) = \frac{1}{\pi(1+\langle m \rangle)s^2(t)} \int \exp\left[-\frac{|\alpha - \alpha' c_a(t)|^2}{(1+\langle m \rangle)s^2(t)}\right] P(\alpha',0) \, d^2\alpha' . \tag{6.113}$$

The effect of amplification upon the function $P(\alpha,t)$ may thus be expressed quite generally by means of a convolution transform with a Gaussian kernel.

It is interesting to compare the asymptotic form of $P(\alpha,t)$ for large times, $\kappa t \gg 1$, with the probability distribution for the amplitude of the field which would emerge from a purely classical linear amplifier taken to have the same amplitude gain factor, $c_a(t) \sim \frac{1}{2}e^{\kappa t - i\omega_a t}$. If the probability distribution for the input amplitude α in the classical linear amplifier were $p(\alpha)$, then the probability distribution for the output amplitude at large times would be

$$P(\alpha,t) \approx 4e^{-2\kappa t} p(2\alpha e^{i\omega_a t - \kappa t}), \qquad (6.114)$$

if the amplification process were noiseless. By comparing this equation with Eq. (6.113) for $\kappa t \gg 1$ we see that the output of the quantum-mechanical parametric amplifier may be represented by thinking of the classical linear amplifier as having an input amplitude distribution

$$p(\alpha) = \frac{1}{\pi(1+\langle m\rangle)} \int \exp\left[-\frac{|\alpha-\alpha'|^2}{1+\langle m\rangle}\right] P(\alpha',0) \, d^2\alpha'. \qquad (6.115)$$

The distribution represented by this convolution integral corresponds to the superposition of two fields. One of these fields is the true input or signal field represented by $P(\alpha',0)$, and the other is a chaotic field with mean quantum number $1+\langle m\rangle$. The effect of noise in the quantum amplifier is thus equivalent to the addition at the input of the classical amplifier of an intensity of Gaussian noise corresponding to $1+\langle m\rangle$ quanta. The $\langle m\rangle$ quanta of noise are contributed by the chaotic nature of the excitation of the B mode. The single quantum of input noise which remains when $\langle m\rangle = 0$ represents the unavoidable quantum noise[6] intrinsic to the amplification process.

6.8
Solution for Initial n-Quantum State of A Mode; B Mode Chaotic

The density operator for the case in which both modes are initially in chaotic mixtures can be used to generate the density operators for cases in which the initial quantum states of the modes are specified more closely. Let us assume, for example, that the A mode is initially in the n-quantum state $|n\rangle_A$. Then if we again assume that the B mode is initially in a chaotic mixture with mean quantum number $\langle m\rangle$, the system is described at $t=0$ by the density operator

$$\rho_{n,c} = |n\rangle_A{}_A\langle n| \rho_{B,\langle m\rangle} \qquad (6.116)$$

where $\rho_{B,\langle m\rangle}$ is defined by Eq. (6.106)

Let us now recall that the initial density operator ρ_c is the product of density operators representing individual chaotic mixtures for each of the two modes:

$$\rho_c = \rho_{A,\langle n\rangle} \rho_{B,\langle m\rangle}. \qquad (6.117)$$

The operator $\rho_{A,\langle n\rangle}$ is defined by an expression similar to Eq. (6.106) but involving the variables of the A mode. In the n-quantum representation $\rho_{A,\langle n\rangle}$ takes the form

$$\rho_{A,\langle n\rangle} = \frac{1}{1+\langle n\rangle} \sum_{n=0}^{\infty} \left[\frac{\langle n\rangle}{1+\langle n\rangle}\right]^n |n\rangle_A \, {}_A\langle n| \,. \tag{6.118}$$

It is convenient at this point to introduce a parameter x defined as

$$x = \frac{\langle n\rangle}{1+\langle n\rangle}, \tag{6.119}$$

which permits us to write the density operator $\rho_{A,\langle n\rangle}$ in the more compact form

$$\rho_{A,\langle n\rangle} = (1-x) \sum_{n=0}^{\infty} x^n |n\rangle_A \, {}_A\langle n| \,. \tag{6.120}$$

If we now construct the operator ρ_c by substituting the form for $\rho_{A,\langle n\rangle}$ given by Eq. (6.120) into Eq. (6.117), and make use of the definition stated in Eq. (6.116), we find the identity

$$\rho_c = (1-x) \sum_{n=0}^{\infty} x^n \rho_{n,c} \,. \tag{6.121}$$

This equation expresses the relationship between the initial density operators ρ_c and $\rho_{n,c}$ in terms of the parameter x. Since the time-dependent density operator is linearly related to its initial value, it follows that the density operators at time t corresponding to the initial values ρ_c and $\rho_{n,c}$, respectively, satisfy a relation identical to Eq. (6.121); the symbol x is just a constant parameter. Quantities linearly related to the density operator, such as the weight function P or the mean values of dynamical operators, obey similar identities. If we know these quantities for the initially chaotic state, we can find them for the state in which the A mode begins with a specified number of quanta by substituting

$$\langle n\rangle = \frac{x}{1-x} \tag{6.122}$$

into the known solutions, and identifying the coefficients in the expansion of these solutions in powers of x.

Let us consider, as an example of some interest, the P function for the A mode at time t, corresponding to the initial density operator $\rho_{n,c}$; we shall denote this function by $P_{n,c}(\alpha,t)$. It follows from Eqns. (6.103) and (6.121) that $P_{n,c}(\alpha,t)$ satisfies the relation

$$\frac{1}{\pi N(x,t)} \exp\left[-\frac{|\alpha|^2}{N(x,t)}\right] = (1-x) \sum_{n=0}^{\infty} x^n P_{n,c}(\alpha,t) \,, \tag{6.123}$$

in which $N(x,t)$ is obtained by substituting Eq. (6.122) for $\langle n\rangle$ into Eq. (6.102):

$$N(x,t) = \left[\frac{x}{1-x}\right] c^2(t) + (1+\langle m\rangle) s^2(t) \,. \tag{6.124}$$

If we insert this expression in Eq. (6.123) and divide by $1-x$, we obtain

$$\frac{1}{\pi\{x[1-\langle m\rangle s^2(t)]+(1+\langle m\rangle)s^2(t)\}}\exp\left[-\frac{|\alpha|^2(1-x)}{x[1-\langle m\rangle s^2(t)]+(1+\langle m\rangle)s^2(t)}\right] \quad (6.125)$$

The function $P_{n,c}(\alpha,t)$ is thus the coefficient of x^n in the power series expansion of the left-hand side of Eq. (6.125). This expression is in fact very closely related to the familiar generating function for the Laguerre polynomials. We may transform it into this generating function by separating from it the factor

$$M(\alpha,t) = \frac{1}{\pi(1+\langle m\rangle)s^2(t)}\exp\left[-\frac{|\alpha|^2}{(1+\langle m\rangle)s^2(t)}\right], \quad (6.126)$$

and introducing the parameter z through the scale transformation

$$z = \mu(t)x, \quad (6.127)$$

where

$$\mu(t) = \frac{\langle m\rangle s^2(t)-1}{(1+\langle m\rangle)s^2(t)}. \quad (6.128)$$

We may then make use of the generating expansion[31]

$$\frac{1}{1-z}e^{-yz/(1-z)} = \sum_{n=0}^{\infty} z^n L_n(y) \quad (6.129)$$

by letting the parameter y be

$$y(\alpha,t) = \frac{|\alpha|^2 c^2(t)}{(1+\langle m\rangle)s^2(t)[1-\langle m\rangle s^2(t)]}. \quad (6.130)$$

If we express Eq. (6.125) in terms of the variable z rather than x, then on equating coefficients of z^n in Eqns. (6.125) and (6.129) we find[16]

$$P_{n,c}(\alpha,t) = M(\alpha,t)[\mu(t)]^n L_n[y(\alpha,t)], \quad (6.131)$$

where the functions $M(\alpha,t)$, $\mu(t)$, and $y(\alpha,t)$ are defined by Eqns. (6.126), (6.128), and (6.130), respectively.

One of the senses in which the function $P_{n,c}(\alpha,t)$ differs from a probability density is immediately evident. The Laguerre polynomial $L_n(y)$ has n real roots which are simple and positive. As long as the function $y(\alpha,t)$ is positive, the function $P_{n,c}(\alpha,0)$ must take on negative values in concentric rings of the α plane which lie between pairs of zeros. As t approaches zero, $P_{n,c}(\alpha,t)$ becomes a highly singular, rapidly oscillating function which vanishes everywhere except within an infinitesimal neighborhood of $\alpha = 0$.

Although the function $P_{n,c}(\alpha,t)$ is not a probability density, we may expect it, in general, to approximate one when the fields we are dealing with are strong enough to be described in classical terms. The linear amplification process we are considering leads by its very nature to fields of arbitrarily great strength when it continues for a sufficiently long time. It is reasonable, therefore, to expect the functions P to approach classical probability densities in the limit of large times.

As far as the sign of $P_{n,c}(\alpha,t)$ is concerned this expectation is easily shown to be justified. For $\langle m \rangle \neq 0$ the function $y(\alpha,t)$ defined by Eq. (6.130) is positive for times smaller than the root of the equation $s(t) = (\langle m \rangle) - 1/2$. For all larger times $y(\alpha,t)$ is negative-valued and $\mu(t)$ is positive-valued; since $L_n(y) \geq 0$ for $y \leq 0$, it follows from Eq. (6.131) that $P_{n,c}(\alpha,t) \geq 0$. Thus for $\langle m \rangle \neq 0$ the amplification process does lead before too long a time to a non-negative function P.

In the particular case $\langle m \rangle = 0$, which corresponds to the B mode beginning in its ground state, the behavior of the function P is somewhat different. In this case Eq. (6.131) reduces to

$$P_{n,c}(\alpha,t) = \frac{1}{\pi s^2(t)} \exp\left[-\frac{|\alpha|^2}{s^2(t)}\right] \left(-\frac{1}{s^2(t)}\right) L_n\left(|\alpha|^2 \frac{c^2(t)}{s^2(t)}\right). \tag{6.132}$$

As the time t increases the annular regions of the α plane in which P is negative do not shrink to zero radius as they do for $\langle m \rangle \neq 0$. Instead they become asymptotically fixed in radius and the function $P_{n,c}(\alpha,t)$ always takes on negative values as well as positive ones. It is clear, however, from the nature of the α dependence of the factor multiplying the Laguerre polynomial in Eq. (6.132), that for large times, i.e., when $s^2(t)$ greatly exceeds the largest root of $L_n(y)$, the negative values of P will be small in magnitude compared with the positive values which occur for $|\alpha| \sim s(t)$.

The answer to the question whether the function P approaches a classical probability density in the limit of large times depends in this case on the physical nature of the quantities we are investigating. If we wish to find the mean value of an operator F for which the expectation value $\langle \alpha | F | \alpha \rangle$ assumes its most significant values for $|\alpha|^2$ smaller than the largest root of L_n, then the nonclassical character of the function P given by Eq. (6.132) will in general be quite significant. If on the other hand we seek the average of an operator F for which $\langle \alpha | F | \alpha \rangle$ assumes its most significant values for large α, e.g., $F = a^\dagger a$, and hence $\langle \alpha | F | \alpha \rangle = |\alpha|^2$, then $P_{n,c}(\alpha,t)$ can be accurately approximated for large t by the positive function

$$P_{n,c}(\alpha,t) \approx \frac{1}{n!\pi s^2(t)} \left[\frac{|\alpha|^2}{s^2(t)}\right]^n \exp\left[-\frac{|\alpha|^2}{s^2(t)}\right], \tag{6.133}$$

which is obtained by approximating the Laguerre polynomial by its dominant term.

As a further example of the use of the generating function technique let us now calculate the probability of finding l quanta in the A mode at time t, given that the system is initially described by the density operator $\rho_{n,c}$ defined by Eq. (6.116). If we write this probability as $p(l,t|\langle n \rangle,\langle m \rangle)$, and recall that $p(l,t|\langle n \rangle,\langle m \rangle)$ represents the probability of finding l quanta in the A mode at time t when the two modes are initially described by ρ_c, then these probabilities must by virtue of Eq. (6.121) satisfy the identity

$$p(l,t\,\langle n\rangle,\langle m\rangle) = (1-x)\sum_{n=0}^{\infty} x^n\, p(l,t|n,\langle m\rangle)\,. \tag{6.134}$$

The probability $p(l,t\,\langle n\rangle,\langle m\rangle)$ may be expressed as a function of x by replacing the quantity $N(t)$ in Eq. (6.104) by the function $N(x,t)$ defined by Eq. (6.124), so that we have

$$p(l,t\,\langle n\rangle,\langle m\rangle) = \frac{[N(x,t)]^l}{[1+N(x,t)]^{l+1}}\,. \tag{6.135}$$

We next substitute the expression Eq. (6.124) for $N(x,t)$ into Eq. (6.135), and equate the result to the summation in Eq. (6.134). To express the identity derived in this way more compactly it is convenient to introduce the functions

$$\mathcal{E}(t) = 1 - \langle m\rangle s^2(t)\,, \tag{6.136}$$

$$\mathcal{F}(t) = (1+\langle m\rangle)s^2(t)\,, \tag{6.137}$$

$$\mathcal{S}(t) = -\langle m\rangle s^2(t)\,, \tag{6.138}$$

and

$$\mathcal{T}(t) = c^2(t) + \langle m\rangle s^2(t)\,. \tag{6.139}$$

Then the identity found by equating the expressions in Eqns. (6.134) and (6.135) may be written in the form

$$\frac{[\mathcal{E}(t)x + \mathcal{F}(t)]^l}{[\mathcal{S}(t)x + \mathcal{T}(t)]^{l+1}} = \sum_{n=0}^{\infty} x^n p(l,t|n,\langle m\rangle)\,. \tag{6.140}$$

The probability $p(l,t|n,\langle m\rangle)$ may be solved for by evaluating the coefficient of x^n in the power series expansion of the rational fraction on the left side of Eq. (6.140).

As a simple illustration, let us consider the case $\langle m\rangle = 0$, for which Eq. (6.140) takes the form

$$\frac{[x + s^2(t)]^l}{[c^2(t)]^{l+1}} = \sum_{n=0}^{\infty} x^n p(l,t|n,0)\,. \tag{6.141}$$

By equating coefficients of x^n, we find

$$p(l,t|n,0) = \binom{l}{n}\frac{[s^2(t)]^{l-n}}{[c^2(t)]^{l+1}} \quad \text{for} \quad l \geq n\,, \tag{6.142}$$

$$= 0 \quad \text{for} \quad l < n\,. \tag{6.143}$$

It is interesting to note that for this case the A mode can never have fewer than n quanta. The reason for this behavior is indicated by the conservation law stated in Eq. (6.31). For the case $\langle m \rangle = 0$ the modes are initially in quantum-number eigenstates, and the operator $a^\dagger a - b^\dagger b$ has the eigenvalue n. The state of the system must retain this eigenvalue for the quantum-number difference at all later times, even though the two modes are no longer in states with well determined quantum numbers. The A mode cannot have fewer than n quanta because in that case the B mode would have to have fewer than zero.

In the general case ($\langle m \rangle \neq 0$) the left-hand side of Eq. (6.140) can be expanded in powers of x without great difficulty. We find that the coefficient of x^n in this expansion may be expressed in terms of the associated Jacobi[32] polynomials $P_i^{(j,k)}(x)$ as follows:

$$p(l,t|n,\langle m\rangle) = \frac{[\mathcal{E}(t)]^l[-\mathcal{S}(t)]^{n-l}}{[\mathcal{T}(t)]^{n+1}} P_l^{(n-l,0)}\left(1 - 2\frac{\mathcal{F}(t)\mathcal{S}(t)}{\mathcal{E}(t)\mathcal{T}(t)}\right) \quad \text{for} \quad n \geq l \tag{6.144}$$

$$= \frac{[\mathcal{E}(t)]^n[\mathcal{F}(t)]^{l-n}}{[\mathcal{T}(t)]^{l+1}} P_l^{(l-n,0)}\left(1 - 2\frac{\mathcal{F}(t)\mathcal{S}(t)}{\mathcal{E}(t)\mathcal{T}(t)}\right) \quad \text{for} \quad n \leq l. \tag{6.145}$$

6.9
General Discussion of Amplification With B Mode Initially Chaotic

Let us now generalize some of the considerations of the preceding sections. We continue to assume that the B mode is initially in the mixed state $\rho_{B,\langle m\rangle}$ defined by Eq. (6.106); the A mode, on the other hand, we take to be described initially by an arbitrary density operator ρ_A. Since the two modes are assumed to be initially independent, the Heisenberg density operator for the joint system is

$$\rho = \rho_A \, \rho_{B,\langle m\rangle}. \tag{6.146}$$

We have remarked that the amplification process leads to arbitrarily strong fields, and that it is therefore reasonable to expect a classical description of the fields to be valid in the limit of large times. As far as the behavior of the A mode is concerned, then, we may expect that the reduced density operator $\rho_A(t)$ will eventually be described by a P representation, and that this P representation will eventually become non-negative. We shall show that for the initial density operator defined by Eq. (6.146), a P representation for the A mode does indeed exist after a certain characteristic time, and does eventually become non-negative, unless $\langle m \rangle = 0$.

We begin by considering the time-dependent form of the ordinary characteristic function for the A mode, which is defined by

$$\chi(\eta,t) = \text{Tr}_A\left\{\rho_A(t)\, e^{\eta a^\dagger - \eta^* a}\right\}. \tag{6.147}$$

The function $\chi(\eta,t)$ like $\chi_N(\eta,t)$, may be expressed in terms of the Heisenberg density operator ρ and the Heisenberg operators $a(t)$ and $a^\dagger(t)$. By steps precisely analogous to those leading to Eq. (6.61) for $\chi_N(\eta,t)$, we find

$$\chi(\eta,t) = \mathrm{Tr}\left\{\rho\, e^{\eta a^\dagger(t) - \eta^* a(t)}\right\}. \tag{6.148}$$

If we substitute Eq. (6.36a) for $a(t)$ into Eq. (6.148), we find

$$\chi(\eta,t) = \mathrm{Tr}\left\{\rho\exp\left[\eta(a^\dagger c_a^*(t) + b s_a^*(t)) - \eta^*(a c_a(t) + b^\dagger s_a(t))\right]\right\}, \tag{6.149}$$

and if we then substitute Eq. (6.146) for ρ into this relation, we obtain the separated expression for the characteristic function

$$\chi(\eta,t) = \mathrm{Tr}_A\left\{\rho_A \exp\left[\eta c_a^*(t) a^\dagger - \eta^* c_a(t) a\right]\right\}$$
$$\times \mathrm{Tr}_B\left\{\rho_{B,\langle m\rangle} \exp\left[-\eta^* s_a(t) b^\dagger + \eta s_a^*(t) b\right]\right\}$$
$$= \chi(\eta c_a^*(t),0)\,\mathrm{Tr}_B\left\{\rho_{B,\langle m\rangle} \exp\left[-\eta^* s_a(t) b^\dagger + \eta s_a^*(t) b\right]\right\}. \tag{6.150}$$

The factor multiplying $\chi(\eta c_a^*(t),0)$ in the latter equation is just the ordinary characteristic function, evaluated at the complex argument $-\eta^* s_a(t)$, for the chaotic density operator $\rho_{B,\langle m\rangle}$. This function may be evaluated by using the expression Eq. (6.106) for $\rho_{B,\langle m\rangle}$ and employing the integral identity Eq. (6.69); we then find

$$\mathrm{Tr}_B\left\{\rho_{B,\langle m\rangle} \exp\left[-\eta^* s_a(t) b^\dagger + \eta s_a^*(t) b\right]\right\} = \exp\left[-|\eta|^2 s^2(t)(\langle m\rangle + \tfrac{1}{2})\right]. \tag{6.151}$$

By substituting this relation into Eq. (6.150) and making use of Eq. (6.20), we find that the normally ordered characteristic function for the A mode at time t is

$$\chi_N(\eta,t) = \exp\left[-|\eta|^2\left\{s^2(t)(\langle m\rangle + \tfrac{1}{2}) - \tfrac{1}{2}\right\}\right] \chi(\eta c_a^*(t),0). \tag{6.152}$$

In Sect. 6.4 it was shown that a P representation for the A mode exists at time t if $\chi_N(\eta,t)$ has a Fourier transform. It can be shown, on the other hand, that the ordinary characteristic function corresponding to an arbitrary density operator necessarily has a well-defined Fourier transform. That transform is in fact the Wigner function[33] written with complex argument. It follows that when the coefficient in square brackets in the exponent of Eq. (6.152) is non-negative, a P representation for the A mode exists. We may define a characteristic time t_0 by the relation

$$s^2(t_0)(\langle m\rangle + \tfrac{1}{2}) - \tfrac{1}{2} = 0,$$

so that

$$\sinh^2 \kappa t_0 = \frac{1}{2\langle m\rangle + 1}. \tag{6.153}$$

6.9 General Discussion of Amplification With B Mode Initially Chaotic

We have shown, then, that a P representation for the A mode must exist for $t \geq t_0$. For many initial states, of course, it will begin to exist at a time prior to $t = t_0$, and for some it will exist at all times $t \geq 0$.

In order to discuss the sign of the function P, we introduce the antinormally ordered characteristic function, which is defined by

$$\chi_{\text{ant.}}(\eta,t) = \text{Tr}_A \left\{ \rho_A(t) e^{-\eta^* a} e^{\eta a^\dagger} \right\} \tag{6.154}$$

$$= e^{|\eta|^2/2} \chi(\eta,t). \tag{6.155}$$

If we insert the expression for the unit operator given by the completeness relation Eq. (6.9) between the exponentials in Eq. (6.154) and make use of Eq. (6.5), we find

$$\chi_{\text{ant.}}(\eta,t) = \int e^{\eta \alpha^* - \eta^* \alpha} \left[\pi^{-1} \langle \alpha | \rho_A(t) | \alpha \rangle \right] d^2\alpha. \tag{6.156}$$

The function $\chi_{\text{ant.}}(\eta,t)$ is thus the Fourier transform of the non-negative function $\pi^{-1} \langle \alpha | \rho_A(t) | \alpha \rangle$. The inverse of Eq. (6.156) is

$$\pi^{-1} \langle \alpha | \rho_A(t) | \alpha \rangle = \pi^{-2} \int e^{\alpha \eta^* - \alpha^* \eta} \chi_{\text{ant.}}(\eta,t) d^2\eta. \tag{6.157}$$

By evaluating Eq. (6.155) at $t = 0$ and with η replaced by $\eta c_a^*(t)$, we find

$$\chi(\eta c_a^*(t),0) = e^{|\eta|^2 c^2(t)/2} \chi_{\text{ant.}}(\eta c_a^*(t),0), \tag{6.158}$$

and if we substitute this relation into Eq. (6.152), we obtain

$$\chi_N(\eta,t) = \exp\left\{-|\eta|^2 \left[\langle m \rangle s^2(t) - 1\right]\right\} \chi_{\text{ant.}}(\eta c_a^*(t),0). \tag{6.159}$$

We now define a second characteristic time t_1 by the relation

$$\sinh^2 \kappa t_1 = \frac{1}{\langle m \rangle} \tag{6.160}$$

which implies that

$$t_1 > t_0,$$

and hence that a P representation for the A mode exists at time t_1. By evaluating Eq. (6.159) at $t = t_1$, we find

$$\chi_N(\eta,t_1) = \chi_{\text{ant.}}(\eta c_a^*(t_1),0). \tag{6.161}$$

If we substitute this expression for $\chi_N(\eta,t_1)$ into Eq. (6.59), we obtain

$$P(\alpha,t_1) = \pi^{-2} \int e^{\alpha \eta^* - \alpha^* \eta} \chi_{\text{ant.}}(\eta c_a^*(t_1),0) d^2\eta. \tag{6.162}$$

It is convenient, before performing the integration in this equation, to change the variable of integration from η to $\eta c_a^*(t_1)$, so that we have

$$P(\alpha,t_1) = \frac{1}{\pi^2 c^2(t_1)} \int \exp\left[\frac{\alpha \eta^*}{c_a(t_1)} - \frac{\alpha^* \eta}{c_a^*(t_1)}\right] \chi_{\text{ant.}}(\eta,0)\, d^2\eta\,. \tag{6.163}$$

By evaluating Eq. (6.157) at $t = 0$ and with α replaced by $\alpha/c_a(t_1)$, and then comparing the result to Eq. (6.163) we deduce that

$$P(\alpha,t_1) = \frac{1}{\pi c^2(t_1)} \langle \alpha[c_a(t_1)]^{-1} | \rho_A(0) | \alpha[c_a(t_1)]^{-1} \rangle\,. \tag{6.164}$$

Since the function $\langle \alpha|\rho|\alpha \rangle$ is intrinsically positive for an arbitrary density operator, it follows that the P function for the A mode is positive definite at $t = t_1$.

It is not difficult to see that the P function for the A mode remains positive definite for $t > t_1$. The proof may be outlined briefly as follows: For $t > t_1$ the coefficient in square brackets in the exponent in Eq. (6.159) is positive, and hence the exponential function in which it appears is the Fourier transform of a Gaussian function which is always real and positive. The Fourier transform of the right-hand side of Eq. (6.159) is therefore the convolution of this Gaussian function with the Fourier transform of $\chi_{\text{ant.}}(\eta c_a^*(t),0)$. We have shown that Fourier transform to be a diagonal matrix element of the density operator and hence to be positive definite. It follows that for $t > t_1$, or $s^2(t) > (\langle m \rangle)^{-1}$, the function $P(\alpha,t)$ is positive definite. It is worth noting that if the initial chaotic distribution in the B mode corresponds to thermal equilibrium, then in the high-temperature limit (or equivalently in the classical limit) the existence and positive-definiteness of the P representation for the A mode are guaranteed after an infinitesimally small time interval.

For the case $\langle m \rangle = 0$, no conclusion about the sign of $P(\alpha,t)$ can be drawn. In Sect. 6.8 it was shown that for $\rho_A = |n\rangle_{A\,A}\langle n|$ and $\langle m \rangle = 0$, the function P for the A mode continues to take on negative values at all times.

6.10
Discussion of *P* Representation: Characteristic Functions Initially Gaussian

The cases we examined in Sect. 6.7 in which the characteristic functions were initially Gaussian in form were ones in which the P representation for the A mode exists at $t = 0$ and at all later times. It is worth noting, therefore, that when the amplifier system begins in states which are described by somewhat more general Gaussian forms for the characteristic functions, the P representation only comes into existence at times $t > 0$. These cases provide simple illustrations of some of the results derived in the preceding section.

Let us begin by discussing some of the properties of a single mode which is described by a Gaussian characteristic function. We define the real variables x and y

6.10 Discussion of P Representation: Characteristic Functions Initially Gaussian

which are proportional to the real and imaginary parts of η as

$$x = 2^{-1/2}(\eta^* + \eta),$$
$$y = i2^{-1/2}(\eta^* - \eta).$$
(6.165)

A simple example of a characteristic function which takes a more general Gaussian form than those considered earlier is

$$\chi(\eta) = \exp\left[-\tfrac{1}{2}(y^2 \mathcal{Z}^2 + x^2 \mathcal{P}^2)\right],$$
(6.166)

where $\mathcal{Z}$ and $\mathcal{P}$ are a pair of real numbers which may be taken to be positive.

If we define the pair of Hermitian operators q and p via the relations

$$q = 2^{-1/2}(a^\dagger + a),$$
$$p = i2^{-1/2}(a^\dagger - a),$$
(6.167)

then according to Eq. (6.13), the general expression for the characteristic function $\chi(\eta)$ may be written as

$$\chi(\eta) = \text{Tr}\left\{\rho e^{i(yq - xp)}\right\}.$$
(6.168)

It is evident that when this function takes the form given in Eq. (6.166), the mean values of q^2 and p^2 are given by

$$\text{Tr}\{\rho q^2\} = -\frac{d^2}{dy^2}\chi(\eta)\bigg|_{\eta=0} = \mathcal{Z}^2,$$
$$\text{Tr}\{\rho p^2\} = -\frac{d^2}{dx^2}\chi(\eta)\bigg|_{\eta=0} = \mathcal{P}^2.$$
(6.169)

Since q and p satisfy the canonical commutation relation $[q, p] = i$, the second moments of these operators must satisfy the inequality

$$\mathcal{Z}^2 \mathcal{P}^2 \geq \tfrac{1}{4},$$
(6.170)

which corresponds to the Heisenberg uncertainty relation.

The normally ordered characteristic function, when it is expressed in terms of the real variables x and y, takes the form

$$\chi_N(\eta) = e^{|\eta|^2/2}\chi(\eta)$$
$$= \exp\left\{-\tfrac{1}{2}\left[y^2(\mathcal{Z}^2 - \tfrac{1}{2}) + x^2(\mathcal{P}^2 - \tfrac{1}{2})\right]\right\}.$$
(6.171)

This function becomes infinite with great rapidity as $|\eta| \to \infty$ unless the coefficients of x^2 and y^2 in the exponential are both negative, i.e., unless

$$\mathcal{Z}^2 \geq \tfrac{1}{2},$$
$$\mathcal{P}^2 \geq \tfrac{1}{2}.$$
(6.172)

If these inequalities are satisfied, the function $\chi_N(\eta)$ possesses a two-dimensional Fourier transform which is the function $P(\alpha)$. If we introduce the real variables q' and p' via the relations

$$\alpha = 2^{-1/2}(q' + ip'),$$
$$\alpha^* = 2^{-1/2}(q' - ip'),$$
(6.173)

then we may write the function $P(\alpha)$ which corresponds to the Fourier integral Eq. (6.22) as

$$P(\alpha) = \pi^{-1}(\mathcal{Z}^2 - \tfrac{1}{2})^{-1/2}(\mathcal{P}^2 - \tfrac{1}{2})^{-1/2} \exp\left\{-\tfrac{1}{2}\left[\frac{q'^2}{\mathcal{Z}^2 - \tfrac{1}{2}} + \frac{p'^2}{\mathcal{P}^2 - \tfrac{1}{2}}\right]\right\}. \quad (6.174)$$

In the limit in which $\mathcal{Z}^2 \to \tfrac{1}{2}$ this function reduces to the one-dimensional δ function

$$P(\alpha) = \left(\frac{2}{\pi}\right)^{1/2} (\mathcal{P}^2 - \tfrac{1}{2})^{-1/2} \delta(q') \exp\left\{-\tfrac{1}{2}\left[\frac{p'^2}{\mathcal{P}^2 - \tfrac{1}{2}}\right]\right\}, \quad (6.175)$$

and a corresponding result holds for $\mathcal{P}^2 \to \tfrac{1}{2}$. When both $\mathcal{Z}^2$ and $\mathcal{P}^2$ approach $\tfrac{1}{2}$, the function $P(\alpha)$ reduces to

$$P(\alpha) = 2\delta(q')\delta(p') = \delta^{(2)}(\alpha), \quad (6.176)$$

which represents the ground state of the mode.

If either of the inequalities (6.172) fails to be satisfied, on the other hand, the function $\chi_N(\eta)$ increases so rapidly as $|\eta| \to \infty$ that no P representation exists. It is not difficult, of course, to find states for which either $\mathcal{Z}^2 = \mathrm{Tr}\{\rho q^2\}$ or $\mathcal{P}^2 = \mathrm{Tr}\{\rho p^2\}$ is less than $\tfrac{1}{2}$, and there is nothing unphysical about them. Indeed the condition $\mathcal{Z}^2 \mathcal{P}^2 = \tfrac{1}{4}$ leads uniquely to a family of minimum uncertainty states, one member of which corresponds to any positive value for $\mathcal{Z}^2$. The only case among these for which the P representation exists corresponds to $\mathcal{Z}^2 = \mathcal{P}^2 = \tfrac{1}{2}$, which specifies the ground state of the mode. In all other cases either $\mathcal{Z}^2$ or $\mathcal{P}^2$ is smaller than $\tfrac{1}{2}$, and the normally ordered characteristic function has no Fourier transform.

The arguments we have given regarding the existence of the P representation are not materially altered if the characteristic function specified by Eq. (6.166) is replaced by more general types of Gaussian functions. If, for example, we have

$$\chi(\eta) = \exp\left[-\tfrac{1}{2}(y^2 \mathcal{Z}^2 + x^2 \mathcal{P}^2) + i(y\bar{q} - x\bar{p})\right] \quad (6.177)$$

for some pair of real numbers $\bar{q}$ and $\bar{p}$, then we see from Eq. (6.168) that the mean values of q and p are just

$$\mathrm{Tr}\{\rho q\} = -i\frac{\partial}{\partial y}\chi(\eta)\bigg|_{\eta=0} = \bar{q},$$
$$\mathrm{Tr}\{\rho p\} = i\frac{\partial}{\partial x}\chi(\eta)\bigg|_{\eta=0} = \bar{p}.$$
(6.178)

6.10 Discussion of P Representation: Characteristic Functions Initially Gaussian

The variances of q and p are given by

$$\Delta q^2 = \text{Tr}\{\rho(q^2 - \bar{q}^2)\}$$

$$= -\frac{\partial^2}{\partial y^2} \ln \chi(\eta)\bigg|_{\eta=0} = Z^2 , \qquad (6.179a)$$

$$\Delta p^2 = \text{Tr}\{\rho(p^2 - \bar{p}^2)\}$$

$$= -\frac{\partial^2}{\partial x^2} \ln \chi(\eta)\bigg|_{\eta=0} = P^2 . \qquad (6.179b)$$

The condition that a Fourier transform exist for the normally ordered characteristic function is once again that both the inequalities (6.172) hold. When they do hold the function $P(\alpha)$ is given by a displaced form of the Gaussian function in Eq. (6.174) obtained by letting $q' \to q' - \bar{q}$ and $p' \to p' - \bar{p}$. Let us define the complex number $\bar{\alpha}$ as

$$\bar{\alpha} = \text{Tr}\{\rho a\} = 2^{-1/2}(\bar{q} - i\bar{p}) . \qquad (6.180)$$

Then in the limit $\Delta q^2 \to \tfrac{1}{2}$, $\Delta p^2 \to \tfrac{1}{2}$, the function $P(\alpha)$ reduces to

$$P(\alpha) = 2\delta(q' - \bar{q})\delta(p' - \bar{p}) = \delta^{(2)}(\alpha - \bar{\alpha}) , \qquad (6.181)$$

which represents the pure coherent state $|\bar{\alpha}\rangle$. This state is but one member of an infinite family of minimum uncertainty states for which $\Delta q^2 \Delta p^2 = Z^2 P^2 = \tfrac{1}{4}$ and $\text{Tr}\{\rho a\} = \bar{\alpha}$. The fact that the other members of the set, for which $Z^2 \neq P^2$, have no P representations in the basis we are using, simply expresses the fact that there is no way of mixing coherent states, which have $\Delta q^2 = \Delta p^2 = \tfrac{1}{2}$, to form states with smaller values of either Δq^2 or Δp^2 and minimum uncertainty $\Delta q^2 \Delta p^2 = \tfrac{1}{4}$.

A theorem which holds for all quantum states is perhaps worth noting at this point: A non-negative P representation can only exist for states for which the variances Δq^2 and Δp^2 obey the inequalities

$$\Delta q^2 \geq \tfrac{1}{2} , \qquad \Delta p^2 \geq \tfrac{1}{2} . \qquad (6.182)$$

Indeed, if we calculate the variance Δq^2 in the P representation we find

$$\Delta q^2 = \text{Tr}\{\rho(q - \bar{q})^2\}$$

$$= \tfrac{1}{2} \text{Tr}\{\rho(a^\dagger + a - \bar{\alpha}^* - \bar{\alpha})^2\}$$

$$= \tfrac{1}{2} + \tfrac{1}{2} \int P(\alpha)|\alpha^* + \alpha - \bar{\alpha}^* - \bar{\alpha}|^2 \, d^2\alpha , \qquad (6.183)$$

where the commutation relation has been used to reach the latter expression. We similarly find

$$\Delta p^2 = \tfrac{1}{2} + \tfrac{1}{2} \int P(\alpha)|\alpha^* - \alpha - \bar{\alpha}^* + \bar{\alpha}|^2 \, d^2\alpha . \qquad (6.184)$$

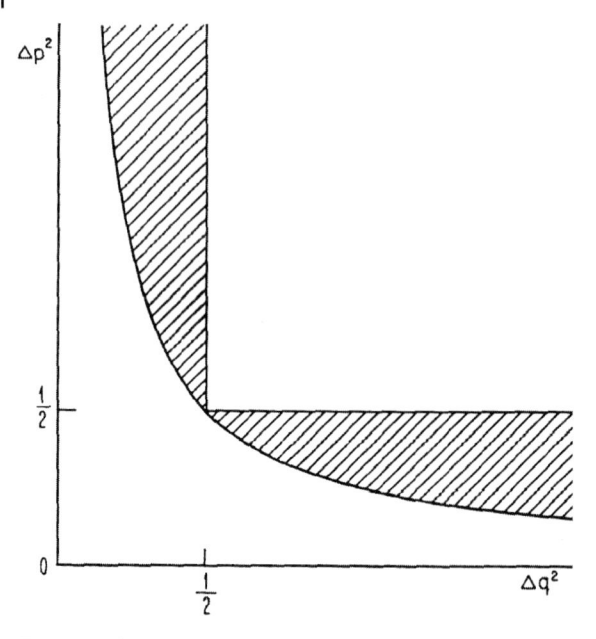

Figure 2 Characterization of quantum states according to their coordinate and momentum uncertainties. The variables Δq^2 and Δp^2 represent the variances of the coordinate and the momentum, respectively, for an arbitrary state of a one-mode system. The hyperbola is defined by the equation $\Delta q^2 \Delta p^2 = \frac{1}{4}$. Points within the shaded region represent states allowed by the uncertainty relation, but for which no function P can exist which takes on only positive values.

It is clear from these expressions that the variances obey the inequalities (6.182) as long as the function $P(\alpha)$ exists and takes on no negative values.

Quantum states of the field exist for all values of Δq^2 and Δp^2 for which $\Delta q^2 \Delta p^2 \geq \frac{1}{4}$. These states are indicated by the points in the $\Delta q^2 \Delta p^2$ plane lying within the hyperbola shown in Fig. 2. The states which correspond to points lying within the shaded regions between the hyperbola and the bounds $\Delta q^2 = \frac{1}{2}$ and $\Delta p^2 = \frac{1}{2}$ are the ones for which no positive-valued function $P(\alpha)$ can exist. In the examples we have considered, which correspond to Gaussian characteristic functions, no function $P(\alpha)$ exists at all in the sense discussed in Sect. 6.2 for the states lying within the shaded regions.

To illustrate the foregoing arguments in a dynamical context let us assume that the A mode of the parametric amplifier is initially in the state specified by the characteristic function Eq. (6.166) and that the B mode is initially in the chaotic mixture $\rho_{B,\langle m \rangle}$ defined by Eq. (6.106). Let us also assume that $\mathcal{Z} < \frac{1}{2}$ (and therefore $\mathcal{P} > \frac{1}{2}$), so that no P representation exists initially for the A mode. We may use Eq. (6.152) to evaluate the normally ordered characteristic function at times $t > 0$. When the characteristic function is not circularly symmetric in the complex η plane, its behavior is most simply described in a frame of reference which rotates uniformly with angular velocity

$-\omega_a$. If we therefore write the argument of χ_N as $\eta\,\mathrm{e}^{-\mathrm{i}\omega_a t}$, we find

$$\chi_N(\eta\,\mathrm{e}^{-\mathrm{i}\omega_a t},t) = \exp\left\{-\tfrac{1}{2}\left[y^2(Z'^2(t)-\tfrac{1}{2})+x^2(P'^2(t)-\tfrac{1}{2})\right]\right\} \qquad (6.185)$$

where

$$\begin{aligned} Z'^2(t) &= Z^2 c^2(t) + (\langle m\rangle + \tfrac{1}{2})s^2(t)\,, \\ P'^2(t) &= P^2 c^2(t) + (\langle m\rangle + \tfrac{1}{2})s^2(t)\,, \end{aligned} \qquad (6.186)$$

and x and y are defined by Eqns. (6.165).

Let us define the Hermitian operators $q(t)$ and $p(t)$ appropriate to a rotating coordinate system by the relation

$$a = 2^{-1/2}[q(t)+\mathrm{i}p(t)]\,\mathrm{e}^{-\mathrm{i}\omega_a t}\,, \qquad (6.187)$$

so that we have

$$\begin{aligned} q(t) &= 2^{-1/2}[a^{\dagger}\,\mathrm{e}^{-\mathrm{i}\omega_a t} + a\,\mathrm{e}^{\mathrm{i}\omega_a t}]\,, \\ p(t) &= \mathrm{i}2^{-1/2}[a^{\dagger}\,\mathrm{e}^{-\mathrm{i}\omega_a t} - a\,\mathrm{e}^{\mathrm{i}\omega_a t}]\,. \end{aligned} \qquad (6.188)$$

Then it is easily shown that the functions $Z'^2(t)$ and $P'^2(t)$ are just the mean-squared values of $q(t)$ and $p(t)$, i.e., we have

$$\begin{aligned} \mathrm{Tr}\{\rho(t)q^2(t)\} &= \tfrac{1}{2}\mathrm{Tr}\left\{\rho\left[a^{\dagger}(t)\,\mathrm{e}^{-\mathrm{i}\omega_a t}+a(t)\,\mathrm{e}^{\mathrm{i}\omega_a t}\right]^2\right\} \\ &= Z'^2(t) \end{aligned} \qquad (6.189)$$

and

$$\begin{aligned} \mathrm{Tr}\{\rho(t)p^2(t)\} &= \tfrac{1}{2}\mathrm{Tr}\left\{-\rho\left[a^{\dagger}(t)\,\mathrm{e}^{-\mathrm{i}\omega_a t}-a(t)\,\mathrm{e}^{\mathrm{i}\omega_a t}\right]^2\right\} \\ &= P'^2(t)\,. \end{aligned} \qquad (6.190)$$

It is clear from Eq. (6.186) that these mean-squared values increase monotonically with time.

The function χ_N given by Eq. (6.185) will possess a two-dimensional Fourier transform only if $Z'^2(t) \geq \tfrac{1}{2}$ and $P'^2(t) \geq \tfrac{1}{2}$. The latter condition is satisfied at all times, since we have assumed that the initial value $P^2 > \tfrac{1}{2}$. The condition on $Z'(t)$ implies that a P representation for the A mode only exists for times t satisfying the inequality

$$\sinh^2 \kappa t \geq \frac{1-2Z^2}{1+2\langle m\rangle + 2Z^2}\,. \qquad (6.191)$$

For $Z^2 \neq 0$, then, a P representation comes into existence for the A mode at a time prior to the time t_0 defined by Eq. (6.153).

For times which satisfy the condition Eq. (6.191)) the function P is given by

$$P(\alpha e^{-i\omega_a t}, t) = \frac{1}{\pi \left[Z'^2(t) - \frac{1}{2}\right]^{1/2} \left[P'^2(t) - \frac{1}{2}\right]^{1/2}}$$
$$\times \exp\left\{-\frac{1}{2}\left[\frac{q'^2}{Z'^2(t) - \frac{1}{2}} + \frac{p'^2}{P'^2(t) - \frac{1}{2}}\right]\right\}, \quad (6.192)$$

where q' and p' are the variables defined by the Eqns. (6.173). At the instant at which the inequality (6.191) is initially satisfied the function P is a one-dimensional δ function similar to that in Eq. (6.175).

6.11
Some General Properties of $P(\alpha, t)$

The results we have derived in the previous sections have corresponded to the choice of particular initial states for the B mode. We shall now derive a general expression for the function $P(\alpha, t)$ which corresponds to the choice of an arbitrary initial density operator for the two-mode system. We illustrate the use of this expression by proving a simple theorem about the existence of the P representation for the case in which the two modes are statistically independent of each other in the initial state.

To treat arbitrary initial states we first introduce the ordinary characteristic function for the joint system of A and B modes, which is defined at $t = 0$ by

$$\chi(\eta, \zeta, 0) = \text{Tr}\left\{\rho e^{\eta a^\dagger + \zeta b^\dagger - \eta^* a - \zeta^* b}\right\}, \quad (6.193)$$

where η and ζ are complex variables. By comparing this expression with Eq. (6.149) we find that the time-dependent characteristic function for the A mode, which we now designate by $\chi_A(\eta, t)$, is related to the initial value of the characteristic function for the joint system by the equation

$$\chi_A(\eta, t) = \chi(\eta c_a^*(t), -\eta^* s_a(t), 0). \quad (6.194)$$

The normally ordered characteristic function $\chi_{N,A}(\eta, t)$ is therefore given by

$$\chi_{N,A}(\eta, t) = e^{|\eta|^2/2} \chi(\eta c_a^*(t), -\eta^* s_a(t), 0). \quad (6.195)$$

If a P representation for the A mode exists at time t, the function $P(\alpha, t)$ is the Fourier transform of $\chi_{N,A}(\eta, t)$, so that we have

$$P(\alpha, t) = \pi^{-2} \int e^{\alpha \eta^* - \alpha^* \eta + |\eta|^2/2} \chi(\eta c_a^*(t), -\eta^* s_a(t), 0) \, d^2\eta. \quad (6.196)$$

The forms of this function derived in the earlier sections correspond to appropriately specialized forms of the characteristic function χ.

Let us now assume that the initial state of the two-mode system is separable, i.e., that the Heisenberg density operator factors into a direct product of density operators for each of the two modes,

$$\rho = \rho_A \rho_B. \tag{6.197}$$

We shall allow ρ_B to represent an arbitrary initial state of the B mode. The density operator ρ_A, on the other hand, is assumed to have the P representation

$$\rho_A = \int P(\alpha,0) |\alpha\rangle \langle\alpha| \, d^2\alpha. \tag{6.198}$$

It follows from Eq. (6.197) that the ordinary characteristic function for the joint system of A and B modes is given at time $t = 0$ by

$$\chi(\eta,\zeta,0) = \text{Tr}_A \left\{ \rho_A e^{\eta a^\dagger - \eta^* a} \right\} \text{Tr}_B \left\{ \rho_B e^{\zeta b^\dagger - \zeta^* b} \right\}$$
$$= \chi_A(\eta,0) \chi_B(\zeta,0)$$
$$= e^{-|\eta|^2/2} \chi_{N,A}(\eta,0) \chi_B(\zeta,0). \tag{6.199}$$

Since ρ_A possesses the P representation Eq. (6.198), the function $\chi_{N,A}(\eta,0)$ has the Fourier transform

$$P(\alpha,0) = \pi^{-2} \int e^{\alpha\eta^* - \alpha^*\eta} \chi_{N,A}(\eta,0) \, d^2\eta. \tag{6.200}$$

If we use the relation Eq. (6.199) for $\chi(\eta,\zeta,0)$ in Eq. (6.195), we obtain

$$\chi_{N,A}(\eta,t) = e^{-|\eta|^2 s^2(t)/2} \chi_{N,A}(\eta c_a^*(t),0) \chi_B(-\eta^* s_a(t),0). \tag{6.201}$$

The absolute magnitude of an ordinary characteristic function such as $\chi_B(\zeta,0)$ possesses a simple upper bound. Since the operator $\exp(\zeta b^\dagger - \zeta^* b)$ is unitary, its expectation value can not exceed unity in modulus. Hence we have

$$|\chi_B(\zeta)| \leq 1. \tag{6.202}$$

It follows then from Eq. (6.201) that

$$|\chi_{N,A}(\eta,t)|^2 \leq e^{-|\eta|^2 s^2(t)} |\chi_{N,A}(\eta c_a^*(t),0)|^2. \tag{6.203}$$

It is easily seen that this inequality implies the continued existence of the P representation for the A mode.

Let us suppose, for example, that the function $P(\alpha,0)$ which describes the initial state of the A mode is quadratically integrable. Then its Fourier transform $\chi_{N,A}(\eta,0)$ must be quadratically integrable, and it follows from the inequality (6.203) that $\chi_{N,A}(\eta,t)$ is quadratically integrable at all times t. The quadratic integrability of $\chi_{N,A}(\eta,t)$ implies that its Fourier transform $P(\alpha,t)$ must exist and remain quadratically integrable at all times.

If, for the sake of greater generality, we take $P(\alpha,0)$ to be a tempered distribution[27], then its Fourier transform $\chi_{N,A}(\eta,0)$ must also be a tempered distribution and so too is the expression for $\chi_{N,A}(\eta,t)$ given by Eq. (6.201). It follows then that $P(\alpha,t)$ must remain a tempered distribution at all times t.

The general formula Eq. (6.196) enables us to discuss the asymptotic form of $P(\alpha,t)$ at large times for a much wider class of initial states than we assumed in deriving the asymptotic relation Eq. (6.114). By changing the variable of integration in Eq. (6.196) from η to $\eta c_a^*(t)$, we find

$$P(\alpha,t) = \frac{1}{\pi^2 c^2(t)} \int \exp\left[\frac{\alpha \eta^*}{c_a(t)} - \frac{\alpha^* \eta}{c_a^*(t)}\right]$$
$$\times \left\{\exp\left[\frac{1}{2}\frac{|\eta|^2}{c^2(t)}\right] \chi(\eta, -i\eta^* \tanh \kappa t, 0)\right\} d^2\eta. \quad (6.204)$$

The function $P(\alpha,t)$ is thus the Fourier transform, evaluated at the complex argument $\alpha/c_a(t)$, of the time-dependent function in curly brackets in Eq. (6.204). As long as the asymptotic value of the Fourier transform is equal to the Fourier transform of the asymptotic value of the function in curly brackets, the function $P(\alpha,t)$ takes the form given by Eq. (6.114) in the limit of large times. The effective input amplitude distribution $p(\alpha)$ is then given by the integral

$$p(\alpha) = \pi^{-2} \int e^{\alpha \eta^* - \alpha^* \eta} \chi(\eta, -i\eta^*, 0) d^2\eta. \quad (6.205)$$

The function given by Eq. (6.115) corresponds to the special case in which the B mode is initially in a chaotic state and the A mode is described by means of a P representation.

Appendix

In Sect. 6.6 use was made of a theorem involving the associated Laguerre polynomials, which we shall now prove. Let us define the set of functions $D_{nm}(\lambda,\mu,\nu)$ as the coefficients in the double power-series expansion

$$e^{\lambda w + \mu z + \nu w z} \equiv \sum_{n,m=0}^{\infty} \frac{w^n z^m}{n! m!} D_{nm}(\lambda,\mu,\nu). \quad (6.A1)$$

Alternatively, we may write

$$D_{nm}(\lambda,\mu,\nu) \equiv \left(\frac{\partial}{\partial w}\right)^n \left(\frac{\partial}{\partial z}\right)^m e^{\lambda w + \mu z + \nu w z}\bigg|_{w,z=0}. \quad (6.A2)$$

The function $D_{nm}(\lambda,\mu,\nu)$ clearly obeys the symmetry relation

$$D_{nm}(\lambda,\mu,\nu) = D_{mn}(\mu,\lambda,\nu). \quad (6.A3)$$

By expanding the exponential first in powers of z, we find

$$e^{\lambda w + \mu z + \nu wz} = \sum_{m=0}^{\infty} \frac{z^m}{m!} (\mu + \nu w)^m e^{\lambda w}$$

$$= \sum_{m=0}^{\infty} \frac{z^m}{m!} \mu^m \left(1 + \frac{\nu w}{\mu}\right)^m e^{\lambda w} . \tag{6.A4}$$

We now make use of the identity[34]

$$\sum_{n=0}^{\infty} L_n^{(m-n)}(x) y^n = (1+y)^m e^{-xy} . \tag{6.A5}$$

If we write $y = \nu w/\mu$ and $x = -\lambda\mu/\nu$, and substitute the resulting form of Eq. (6.A5) into Eq. (6.A4), we deduce, by referring to the definition Eq. (6.A1),

$$D_{nm}(\lambda, \mu, \nu) = n! \, \nu^n \, \mu^{m-n} \, L_n^{(m-n)}(-\lambda\mu/\nu) . \tag{6.A6}$$

By making use of the symmetry relation Eq. (6.A3) we obtain the alternative expression

$$D_{nm}(\lambda, \mu, \nu) = m! \, \nu^m \, \lambda^{n-m} \, L_m^{(n-m)}(-\lambda\mu/\nu) . \tag{6.A7}$$

If we equate the expressions Eq. (6.A6) and Eq. (6.A7), we obtain

$$\frac{t^n}{n!} L_m^{(n-m)}(-t) = \frac{t^m}{m!} L_n^{(m-n)}(-t) , \tag{6.A8}$$

where $t = \lambda\mu/\nu$. This identity also follows for arbitrary t directly from the explicit expression for the associated Laguerre polynomials.

References

1 M. W. Muller, *Phys. Rev.* **106**, 8 (1957).

2 K. Shimoda, H. Takahasi, C. H. Townes, *J. Phys. Soc. Jpn.* **12**, 686 (1957).

3 R. V. Pound, *Ann. Phys. New York* **1**, 24 (1957).

4 M. W. P. Strandberg, *Phys. Rev.* **106**, 617 (1957).

5 R. Serber, C. H. Townes, *Quantum Electronics – A Symposium*, C. H. Townes, Ed., Columbia University Press, New York 1960, p. 233.

6 W. H. Louisell, A. Yariv, A. E. Siegman, *Phys. Rev.* **124**, 1646 (1961).

7 J. Schwinger, *J. Math. Phys.* **2**, 407 (1961).

8 W. H. Wells, *Ann. Phys. New York* **12**, 1 (1961).

9 A. E. Siegman, *Proc. Inst. Radio Engrs.* **49**, 633 (1961).

10 H. A. Haus, J. A. Mullen, *Phys. Rev.* **128**, 2407 (1962).

11 R. Senitzky, *Phys. Rev.* **128**, 2864 (1962).

12 J. P. Gordon, W. H. Louisell, L. R. Walker, *Phys. Rev.* **129**, 481 (1963).

13 J. P. Gordon, L. R. Walker, W. H. Louisell, *Phys. Rev.* **130**, 806 (1963).

14 R. P. Feynman, F. L. Vernon, Jr., *Ann. Phys. New York* **24**, 118 (1963).

15 W. H. Louisell, L. R. Walker, *Phys. Rev. B* **137**, 204 (1965).

16 D. Holliday, A. E. Glassgold, *Phys. Rev. A* **139**, 1717 (1965).

17 R. J. Glauber, *Phys. Rev.* **131**, 2766 (1963); reprinted as Sect. 2.9 in this volume.

18 R. J. Glauber, *Phys. Rev. Lett.* **10**, 84 (1963).

19 R. J. Glauber, *Phys. Rev.* **130**, 2529 (1963); reprinted as Chapter 1 in this volume.

20 R. J. Glauber, in *Quantum Electronics, Proceedings of the Third International Congress, Paris 1963*, N. Bloembergen, P. Grivet, Eds., Columbia University Press, New York 1964, Vol. I, p. 111.

21 R J. Glauber, in *Quantum Optics and Electronics*, C. de Witt A. Blandin, C. Cohen-Tannoudji, Eds., Gordon & Breach, New York 1965; reprinted as Chapter 2 in this volume.

22 P. Carruthers, M. M. Nieto, *Phys. Rev. Lett.* **14**, 387 (1965).

23 J. R. Klauder, *Ann. Phys. New York* **11**, 123 (1960).

24 R. J. Glauber, in *Physics of Quantum Electronics*, P. L. Kelley el al., Eds., McGraw-Hill, New York 1966, p. 788.

25 The same kind of representation is discussed from a different standpoint by E. C. G. Sudarshan, *Phys. Rev. Lett.* **10**, 277 (1963).

26 See, for example, A. Messiah, *Quantum Mechanics*, North-Holland, Amsterdam 1961,, Vol. I, p. 442.

27 L. Schwartz, *Théorie des Distributions*, Hermann et Cie, Paris 1957, Vol. II, Chap. VII.

28 K. E. Cahill, *Phys. Rev. B* **138**, 1566 (1965).

29 T. Von Foerster (private communication).

30 See, for example, R. J. Glauber, *Phys. Lett.* **21**, 650 (1966) (reprinted as Chapter 5 in this volume).

31 *Handbook of Mathematical Functions*, M. Abramovitz, I. A. Stegun, Eds., National Bureau of Standards Applied Mathematics Series 55, US Government Printing and Publishing Office, Washington, DC, 1964, p. 784.

32 See Ref.[31], p. 775.

33 See, for example, J. E. Moyal, *Proc. Cambridge Phil. Soc.* **45**, 99 (1948), or Sect. 2.11.

34 W. Magnus, F. Oberhettinger, *Formulas and Theorems for the Special Functions of Mathematical Physics*, Chelsea Publishing, New York 1949, p. 85.

7
Quantum Theory of Parametric Amplification II[1]

7.1
Introduction

In the preceding Chapter[1] we have begun an analysis of the statistical behavior of the quantum-mechanical parametric amplifier. The model of the amplifier we have discussed is one in which two modes of oscillation, represented by harmonic oscillators, are coupled by means of a time-dependent parameter[2]. The analysis presented in Chapter 6 took the coupling between the modes fully into account, but was concerned with describing the output of only one of the two modes of the system. A complete description of the parametric amplifier, however, must specify the state of both of the interacting modes, since the coupling mechanism leads to correlations between them which are readily detected experimentally. If the two modes of oscillation correspond, for example, to optical-frequency electromagnetic fields, then the correlations of the mode amplitudes can be measured by observing coincidences of photons recorded by two counters, one sensitive to the field of each mode. To be able to predict the results of this type of experiment and others which respond to the state of both modes, we must develop the full statistical description of the two-mode system.

We shall base our discussion of the statistical behavior of the amplifier system on the time-dependent density operator $\rho(t)$ for the field amplitudes of its two modes. As in the case of single-mode systems, one of the more useful ways of expressing the density operator is by means of the P representation[3]. When this representation exists it tends to permit particularly direct comparisons with classical theory, and to allow the computation of certain expectation values by methods similar to those of classical probability theory. The weight function $P(\alpha, \beta, t)$ which appears in the P representation of $\rho(t)$ is in this sense a quantum analog of the classical joint-probability distribution for finding the A and B modes of the system with complex amplitudes α and β, respectively. By finding the time evolution of this function we are able to express the solution for the density operator in a form which exhibits quite directly the correlation between the mode amplitudes.

We shall also discuss the dynamics of the amplifier system in terms of the Wigner function[4,5], a species of quantum-mechanical phase-space distribution which is de-

---

[1] Reprinted with permission from B. R. Mollow, R. J. Glauber, *Phys. Rev.*
160, 1097–1108 (1967). Copyright 2006 by the American Physical Society.

finable for arbitrary density operators. The Wigner function $W(\alpha,\beta,t)$ for the parametric amplifier evolves in time a particularly simple way: It obeys Liouville's equation, and is therefore constant along a classical trajectory.

The time evolution of the function $P(\alpha,\beta,t)$, on the other hand, exhibits more explicitly quantum-mechanical features. We are able to show, for example, that a non-negative function $P(\alpha,\beta,t)$ can exist for the system only during a finite time interval. For times outside that interval the function must either take on negative values or fail to be definable. The simple initial states of the amplifier which we treat indeed lead to functions $P(\alpha,\beta,t)$ which become strongly singular at the end of the interval and then fail to be definable at later times. A general integral representation which we find for $P(\alpha,\beta,t)$ shows that for nearly all initial states of the system the function sooner or later becomes undefinable. Such behavior means simply that the description of the system as being in a "diagonal" mixture of coherent states has become inappropriate.

The physical reason for the breakdown of the P representation lies in the extremely close correlation in the amplitudes of the two modes of the system which is brought about by the amplification process. In a pure coherent state of the amplifier system the field amplitudes of the two modes fluctuate independently of one another, with variances which are the same as those of vacuum fluctuations. It is easily seen, however, from the equations of motion for the system, that as the amplification continues the fields in the two modes become so tightly correlated that they may no longer fluctuate independently to even the small degree allowed by pure coherent states. It is at that point in the time development of the system that it becomes impossible to have a non-negative function $P(\alpha,\beta,t)$, and the singularities occur for the examples we discuss. The density operator itself, on the other hand, is at no time singular. It may always be constructed explicitly in terms of creation and annihilation operators, or characterized by means of its matrix elements or the Wigner function which remain well-behaved at all times.

An arbitrary density operator may be expressed in a particularly simple form in terms of its characteristic function. In Sect. 7.2 we find this function at an arbitrary time t in terms of its form at $t=0$, and thereby obtain a formal solution for the density operator. The solution for the characteristic function is then used in Sect. 7.3 to find the Wigner function in terms of the form it takes initially.

In Sect. 7.4 we derive a decoupled form for the equations of motion of the system which exhibits the time-dependent correlation of the mode amplitudes. The characteristic functions and the functions $W(\alpha,\beta,t)$ and $P(\alpha,\beta,t)$ are then expressed in terms of suitably defined decoupled variables in Sections 7.5 and 7.6. An integral solution is obtained for $P(\alpha,\beta,t)$ and is shown to diverge after some finite time for a broad class of initial states. An example which illustrates this behavior is discussed in Sect. 7.7, and in Sect. 7.8 we present a general analysis of the relationship between the breakdown of the two-mode P representation and the correlation between the mode amplitudes.

7.2
The Two-Mode Characteristic Function

The model[2] of the parametric amplifier on which our analysis is based consists of two quantum-mechanical harmonic oscillator modes, the A and B modes, described by annihilation operators $a(t)$ and $b(t)$, respectively. The Hamiltonian for the system is assumed to have the form

$$H(t) = \hbar\omega_a a^\dagger(t)a(t) + \hbar\omega_b b^\dagger(t)b(t) - \hbar\kappa[a^\dagger(t)b^\dagger(t)e^{-i\omega t} + a(t)b(t)e^{i\omega t}], \quad (7.1)$$

where ω_a and ω_b are the natural frequencies of oscillation of the uncoupled A and B modes, and the frequency ω of the time-varying coupling is assumed to satisfy the resonance condition

$$\omega = \omega_a + \omega_b. \quad (7.2)$$

The solutions to the Heisenberg equations of motion which follow from the Hamiltonian Eq. (7.1) may be written as in Eq. (6.36) in the form

$$a(t) = a c_a(t) + b^\dagger s_a(t), \quad (7.3a)$$
$$b(t) = b c_b(t) + a^\dagger s_b(t), \quad (7.3b)$$

where $a \equiv a(0)$, $b \equiv b(0)$, and the c-number functions $c_a(t)$, $s_a(t)$, $c_b(t)$, and $s_b(t)$ are defined by

$$\begin{aligned}
c_a(t) &\equiv e^{-i\omega_a t} \cosh \kappa t, \\
s_a(t) &\equiv i e^{-i\omega_a t} \sinh \kappa t, \\
c_b(t) &\equiv e^{-i\omega_b t} \cosh \kappa t, \\
s_b(t) &\equiv i e^{-i\omega_b t} \sinh \kappa t.
\end{aligned} \quad (7.4)$$

Our problem is to solve for the density operator which describes the two-mode system at an arbitrary time t, when we are given the density operator at $t = 0$. The density operator may be specified at any time by means of its characteristic function. This function is defined for arbitrary complex η and ζ by

$$\chi(\eta,\zeta,t) = \mathrm{Tr}\left\{\rho(t)\, e^{\eta a^\dagger + \zeta b^\dagger - \eta^* a - \zeta^* b}\right\}, \quad (7.5)$$

in which $\rho(t)$ is the time-dependent Schrödinger density operator for the system, and a and b are the Schrödinger annihilation operators for the A and B modes, respectively. We may express $\chi(\eta,\zeta,t)$ equally well in terms of the Heisenberg density operator ρ and the Heisenberg annihilation operators $a(t)$ and $b(t)$ by means of the canonical transformation of Eqns. (6.52) and (6.53). The trace when written in this form is

$$\chi(\eta,\zeta,t) = \mathrm{Tr}\left\{\rho\, e^{\eta a^\dagger(t) + \zeta b^\dagger(t) - \eta^* a(t) - \zeta^* b(t)}\right\}. \quad (7.6)$$

Since the characteristic function χ determines the density operator uniquely[6,7], we can exhibit a solution to our initial value problem by expressing the function χ at time t in terms of its initial value $\chi(\eta,\zeta,0)$. To construct the solution in this way we begin by substituting Eq. (7.3) for $a(t)$ and $b(t)$ into Eq. (7.6),

$$\chi(\eta,\zeta,t) = \text{Tr}\left\{\rho \exp\left[\eta(a^\dagger c_a^*(t) + b s_a^*(t)) + \zeta(b^\dagger c_b^*(t) + a s_b^*(t)) - \text{H.c.}\right]\right\}, \tag{7.7}$$

where H.c. means Hermitian conjugate. If we now define c-number functions $\eta_0(\eta,\zeta,t)$ and $\zeta_0(\eta,\zeta,t)$ by the equations

$$\begin{aligned}\eta_0(\eta,\zeta,t) &\equiv \eta c_a^*(t) - \zeta^* s_b(t), \\ \zeta_0(\eta,\zeta,t) &\equiv \zeta c_b^*(t) - \eta^* s_a(t),\end{aligned} \tag{7.8}$$

then we find by collecting coefficients of $a^\dagger$, $b^\dagger$, a, and b in Eq. (7.7) that the function $\chi(\eta,\zeta,t)$ obeys the functional identity

$$\begin{aligned}\chi(\eta,\zeta,t) &= \text{Tr}\left\{\rho \exp\left[a^\dagger \eta_0(\eta,\zeta,t) + b^\dagger \zeta_0(\eta,\zeta,t) - \text{H.c.}\right]\right\} \\ &= \chi(\eta_0(\eta,\zeta,t), \zeta_0(\eta,\zeta,t), 0).\end{aligned} \tag{7.9}$$

The characteristic function $\chi(\eta,\zeta,t)$ is thus specified in terms of the form it takes at $t = 0$, and the c-number functions defined by the Eqns. (7.8).

A formal solution for the Schrödinger density operator $\rho(t)$ may now be constructed from our knowledge of the characteristic function. To accomplish this, let us first note that an arbitrary density operator ρ for a single-mode system may be expressed in terms of its characteristic function

$$\chi(\eta) = \text{Tr}\left\{\rho e^{\eta a^\dagger - \eta^* a}\right\} \tag{7.10}$$

by means of the expansion[6,7]

$$\rho = \frac{1}{\pi}\int \chi(\eta) e^{-\eta a^\dagger - \eta^* a} d^2\eta, \tag{7.11}$$

where $d^2\eta = d(\text{Re}\,\eta)\,d(\text{Im}\,\eta)$. This relation is an operator analog of the two-dimensional Fourier transform; it may be verified by multiplying the right side by $\exp(\eta a^\dagger - \eta^* a)$ and evaluating the trace of the resulting expression.

The generalization of the expansion Eq. (7.11) appropriate to the two-mode density operator $\rho(t)$ is

$$\rho(t) = \frac{1}{\pi^2}\int \chi(\eta,\zeta,t) e^{-\eta a^\dagger - \zeta b^\dagger + \eta^* a + \zeta^* b} d^2\eta\, d^2\zeta. \tag{7.12}$$

If we substitute Eq. (7.9) for $\chi(\eta,\zeta,t)$ into this relation, we obtain

$$\rho(t) = \frac{1}{\pi^2}\chi(\eta_0(\eta,\zeta,t), \zeta_0(\eta,\zeta,t), 0) e^{-\eta a^\dagger - \zeta b^\dagger + \eta^* a + \zeta^* b} d^2\eta\, d^2\zeta. \tag{7.13}$$

The integration in this equation may be carried out most conveniently by changing the variables of integration from η and ζ to $\eta_0 = \eta_0(\eta,\zeta,t)$ and $\zeta_0 = \zeta_0(\eta,\zeta,t)$ and making use of the readily proved identity

$$d^2\eta\, d^2\zeta = d^2\eta_0\, d^2\zeta_0 . \tag{7.14}$$

The Eqns. (7.8) are easily inverted; by taking into account the definitions Eq. (7.4) of the coefficients $c_a(t)$, $s_a(t)$, $c_b(t)$, and $s_b(t)$, we find that the solutions for η and ζ take the form

$$\begin{aligned}\eta(\eta_0,\zeta_0,t) &= \eta_0 c_a(t) + \zeta_0^* s_a(t) , \\ \zeta(\eta_0,\zeta_0,t) &= \zeta_0 c_b(t) + \zeta_0^* s_b(t) .\end{aligned} \tag{7.15}$$

When this change of variables is carried out on Eq. (7.13), we find that the density operator takes the form

$$\rho(t) = \frac{1}{\pi^2}\int \chi(\eta_0,\zeta_0,0)\exp\left[-\eta(\eta_0,\zeta_0,t)a^\dagger - \zeta(\eta_0,\zeta_0,t)b^\dagger - \text{H.c.}\right] d^2\eta_0\, d^2\zeta_0 . \tag{7.16}$$

The time-dependent density operator is thus expressed in terms of the initial form of the characteristic function and the time-independent Schrödinger operators $a^\dagger$, $b^\dagger$, a, and b; the time dependence of $\rho(t)$ is contained completely within the functions $\eta(\eta_0,\zeta_0,t)$ and $\zeta(\eta_0,\zeta_0,t)$, which depend only linearly on the variables η_0 and ζ_0.

7.3
The Wigner Function

The Wigner function was originally introduced as a quantum analog of the classical phase-space distribution function[4,6]. It is useful for evaluating the mean values of certain quantum-mechanical operators as integrals of c-number functions carried out over the phase space defined by the eigenvalues of the coordinates and momenta of the system under consideration. Although the Wigner function approaches a classical distribution function in the classical limit, it cannot be unambiguously interpreted as a probability density in quantum-mechanical contexts. The Wigner function is like the weight function of the P representation in that it may take on negative values for suitably restricted values of its arguments. It is unlike the P function, however, in that it exists in a well-defined sense for arbitrary density operators, and is never singular.

In this section we shall solve for the Wigner function for the parametric amplifier in terms of its initial form. We shall show that for arbitrary initial quantum states of the system, the Wigner function satisfies Liouville's equation.

To define the Wigner function, we first note that for a classical oscillator of frequency ω_j, the coordinate q_j and momentum p_j are related to the complex amplitude α_j by means of the expressions

$$q_j = \left(\frac{\hbar}{2\omega_j}\right)^{1/2} (\alpha_j^* + \alpha_j),$$

$$p_j = i\left(\frac{\hbar\omega_j}{2}\right)^{1/2} (\alpha_j^* - \alpha_j). \tag{7.17}$$

These relations may be used to write the Wigner function in terms of the complex arguments α_j rather than the real arguments q_j and p_j. Let us now consider a single-mode system, i.e., a single harmonic oscillator, described by the density operator ρ. The Wigner function in this case may be defined[8,9] as the two-dimensional Fourier transform of the characteristic function $\chi(\eta)$

$$W(\alpha) = \frac{1}{\pi^2} \int e^{\alpha\eta^* - \alpha^*\eta} \chi(\eta)\, d^2\eta. \tag{7.18}$$

We may gain some insight into the meaning of the Wigner function by supposing that the density operator ρ has a P representation

$$\rho = \int P(\alpha) |\alpha\rangle \langle \alpha|\, d^2\alpha, \tag{7.19}$$

and then expressing $W(\alpha)$ in terms of $P(\alpha)$. By combining Eqns. (6.20) and (6.21) we find

$$\chi(\eta) = e^{-|\eta|^2/2} \int e^{\eta\alpha'^* - \eta^*\alpha'} P(\alpha')\, d^2\alpha', \tag{7.20}$$

and by substituting this equation for $\chi(\eta)$ into Eq. (7.18) we then obtain

$$W(\alpha) = \frac{1}{\pi^2} \int e^{-|\eta|^2 + \eta^*(\alpha - \alpha') - \eta(\alpha^* - \alpha'^*)} P(\alpha')\, d^2\eta\, d^2\alpha'. \tag{7.21}$$

If we now perform the integration over η with the aid of the integral identity Eq. (6.69), we find[9]

$$W(\alpha) = \frac{2}{\pi} \int e^{-2|\alpha - \alpha'|^2} P(\alpha')\, d^2\alpha'. \tag{7.22}$$

If a P representation exists, then the Wigner function is the convolution of a Gaussian function with the function $P(\alpha)$. If the latter function varies little over distances in the α plane comparable to unity, then it is clear from Eq. (7.22) that the functions $W(\alpha)$ and $P(\alpha)$ are nearly equal. In this limit, which is the classical one, either function may be identified with the probability density for finding the oscillator with the complex amplitude α.

7.3 The Wigner Function

The Wigner function $W(\alpha,\beta,t)$ for the parametric amplifier is given in terms of the characteristic function $\chi(\eta,\zeta,t)$ by the two-mode generalization of Eq. (7.18)

$$W(\alpha,\beta,t) = \frac{1}{\pi^4} \int e^{\alpha\eta^* + \beta\zeta^* - \text{c.c}} \chi(\eta,\zeta,t) \, d^2\eta \, d^2\zeta \,, \tag{7.23}$$

where c.c. means complex conjugate.

The identity Eq. (7.9) which permits us to express the characteristic function in terms of its initial form permits us to do the same thing with the Wigner function. By substituting Eq. (7.9) for $\chi(\eta,\zeta,t)$ into Eq. (7.23) we find

$$W(\alpha,\beta,t) = \frac{1}{\pi^4} \int e^{\alpha\eta^* + \beta\zeta^* - \text{c.c}} \chi(\eta_0(\eta,\zeta,t), \zeta_0(\eta,\zeta,t), 0) \, d^2\eta \, d^2\zeta \,. \tag{7.24}$$

Let us change the variables of integration in this equation from η and ζ to $\eta_0 = \eta_0(\eta,\zeta,t)$ and $\zeta_0 = \zeta_0(\eta,\zeta,t)$. If we then make use of Eqns. (7.14) and (7.15), we may write the integral as

$$W(\alpha,\beta,t) = \frac{1}{\pi^4} \int \exp\left[\alpha\eta^*(\eta_0,\zeta_0,t) + \beta\zeta^*(\eta_0,\zeta_0,t) - \text{c.c.}\right] \chi(\eta_0,\zeta_0,0) \, d^2\eta_0 \, d^2\zeta_0 \,. \tag{7.25}$$

We now introduce the c-number functions $\alpha_{0c}(\alpha,\beta,t)$ and $\beta_{0c}(\alpha,\beta,t)$, which are defined by

$$\alpha_{0c}(\alpha,\beta,t) \equiv \alpha c_a^*(t) - \beta^* s_b(t) \,, \tag{7.26a}$$
$$\beta_{0c}(\alpha,\beta,t) \equiv \beta c_b^*(t) - \alpha^* a_s(t) \,. \tag{7.26b}$$

It is easily seen that the exponent in Eq. (7.25) can be written in the form

$$\alpha\eta^*(\eta_0,\beta_0,t) + \beta\zeta^*(\eta_0,\beta_0,t) - \text{c.c.} = \alpha_{0c}(\alpha,\beta,t)\eta_0^* + \beta_{0c}(\alpha,\beta,t)\zeta_0^* - \text{c.c.} \tag{7.27}$$

By substituting this identity into Eq. (7.25), we then obtain the relation

$$W(\alpha,\beta,t) = \frac{1}{\pi^4} \int \exp\left[\alpha_{0c}(\alpha,\beta,t)\eta_0^* + \beta_{0c}(\alpha,\beta,t)\zeta_0^* - \text{c.c.}\right]$$
$$\times \chi(\eta_0,\zeta_0,0) \, d^2\eta_0 \, d^2\zeta_0$$
$$= W(\alpha_{0c}(\alpha,\beta,t), \beta_{0c}(\alpha,\beta,t), 0) \,, \tag{7.28}$$

which expresses the Wigner function at time t in terms of the form it takes initially.

To obtain some insight into the meaning of Eq. (7.28), let us write $\alpha_{0c}(\alpha,\beta,t) = \alpha_0$, $\beta_{0c}(\alpha,\beta,t) = \beta_0$, and define functions $\alpha_c(\alpha_0,\beta_0,t)$ and $\beta_c(\alpha_0,\beta_0,t)$ as the solutions to Eq. (7.26) for α and β

$$\alpha_c(\alpha_0,\beta_0,t) = \alpha_0 c_a(t) + \beta_0^* s_a(t) \,, \tag{7.29a}$$
$$\beta_c(\alpha_0,\beta_0,t) = \beta_0 c_b(t) + \alpha_0^* s_b(t) \,. \tag{7.29b}$$

When we write Eq. (7.28) in terms of these functions, it takes the form

$$W(\alpha_c(\alpha_0,\beta_0,t),\beta_c(\alpha_0,\beta_0,t),t) = W(\alpha_0,\beta_0,0) , \qquad (7.30)$$

which is valid for arbitrary complex numbers α_0 and β_0.

The right-hand sides of Eqns. (7.29) are identical to the right-hand sides of Eqns. (6.45) for the mean values of the operators $a(t)$ and $b(t)$. In the latter equations α_0 and β_0 represent the initial values of the average amplitudes. The structure of Eqns. (6.45) is essentially classical in nature; the mean value of the quantum-mechanical operators $a(t)$ and $b(t)$ obey the same equations of motion as the complex amplitudes for the classical parametric amplifier. It follows that the functions $\alpha_c(\alpha_0,\beta_0,t)$ and $\beta_c(\alpha_0,\beta_0,t)$ are just the complex amplitudes evaluated at time t for the A and B modes of a classical parametric amplifier with initial complex amplitudes α_0 and β_0. Conversely, the functions $\alpha_{0c}(\alpha,\beta,t)$ and $\beta_{0c}(\alpha,\beta,t)$ are the initial amplitudes which correspond classically to the amplitudes α and β at time t. Equations (7.30) and (7.28) thus assert that *the Wigner function for the parametric amplifier is constant along a classical trajectory*. This is a property which the Wigner function shares with the classical phase-space distribution function, which it approaches in the classical limit. It should be emphasized, however, that the result is valid for arbitrary initial density operators; in particular, it holds for arbitrarily weak fields (small quantum numbers), and when the Wigner function takes on negative values. The fact that the Wigner function has this property is a consequence of the form taken by the Hamiltonian Eq. (7.1). It may be shown that whenever the Hamiltonian of a system of oscillators is given by a quadratic form in the creation and annihilation operators, the Wigner function is constant along classical trajectories[10–12]. This property does not extend to systems with arbitrary Hamiltonians, as it does in the case of the classical phase-space distribution.

To express Eq. (7.28) or (7.30) in differential form, we first introduce the total time derivative d/dt, which is defined as the ordinary time derivative taken along a classical trajectory. For a system of particles with coordinates q_j and momenta p_j, d/dt is given by

$$\frac{d}{dt} = \frac{\partial}{\partial t} + \sum_j \left(\frac{\partial H}{\partial p_j} \frac{\partial}{\partial q_j} - \frac{\partial H}{\partial q_j} \frac{\partial}{\partial p_j} \right) , \qquad (7.31)$$

where H is the Hamiltonian for the system. If we use Eq. (7.17) to define derivatives with respect to the complex quantities α_j and α_j^* by means of the relations

$$\begin{aligned}\frac{\partial}{\partial \alpha_j} &= \left(\frac{\hbar}{2\omega_j}\right)^{1/2} \frac{\partial}{\partial q_j} - i(\tfrac{1}{2}\hbar\omega_j)^{1/2} \frac{\partial}{\partial p_j} , \\ \frac{\partial}{\partial \alpha_j^*} &= \left(\frac{\hbar}{2\omega_j}\right)^{1/2} \frac{\partial}{\partial q_j} + i(\tfrac{1}{2}\hbar\omega_j)^{1/2} \frac{\partial}{\partial p_j} ,\end{aligned} \qquad (7.32)$$

then by solving these equations for $\partial/\partial q_j$ and $\partial/\partial p_j$ and substituting the results into Eq. (7.31) we find

$$\frac{d}{dt} = \frac{\partial}{\partial t} + \frac{i}{\hbar}\sum_j\left(\frac{\partial H}{\partial\alpha_j}\frac{\partial}{\partial\alpha_j^*} - \frac{\partial H}{\partial\alpha_j^*}\frac{\partial}{\partial\alpha_j}\right). \qquad (7.33)$$

For the parametric amplifier, the index j takes on the values a and b, corresponding to the complex amplitudes $\alpha_a = \alpha$, $\alpha_b = \beta$. The Hamiltonian for the classical parametric amplifier is obtained by replacing the operators $a(t)$ and $b(t)$ in Eq. (7.1) by α and β, respectively. If we substitute the resulting expression for the classical Hamiltonian into Eq. (7.33), we find

$$\frac{d}{dt} = \frac{\partial}{\partial t} + \left[i(\omega_a\alpha^* - \kappa\beta\,e^{i\omega t})\frac{\partial}{\partial\alpha^*} + i(\omega_b\beta^* - \kappa\alpha\,e^{i\omega t})\frac{\partial}{\partial\beta^*} + \text{c.c.}\right]. \qquad (7.34)$$

Since the Wigner function is constant along a classical trajectory, it satisfies Liouville's equation

$$\frac{d}{dt}W(\alpha,\beta,t) = 0. \qquad (7.35)$$

7.4
Decoupled Equations of Motion

The time dependence of the various functions which characterize the density operator $\rho(t)$ may be greatly simplified by carrying out a transformation of variables which decouples the basic equations of motion for the system. We introduce two operators $c^{(+)}(t)$ and $c^{(-)}(t)$, which we define in terms of the Heisenberg operators $a(t)$ and $b(t)$ by the relations

$$c^{(+)}(t) \equiv 2^{-1/2}\left[a(t)\,e^{i\omega_a t} + ib^\dagger(t)\,e^{-i\omega_b t}\right], \qquad (7.36a)$$

$$c^{(-)}(t) \equiv 2^{-1/2}\left[a(t)\,e^{i\omega_a t} - ib^\dagger(t)\,e^{-i\omega_b t}\right]. \qquad (7.36b)$$

By substituting the solutions Eq. (7.3) to the Heisenberg equations of motion for $a(t)$ and $b(t)$ into Eq. (7.36), we observe that the operators $c^{(\pm)}(t)$ have the simple time dependence

$$c^{(+)}(t) = c^{(+)}\,e^{\kappa t}, \qquad (7.37a)$$

$$c^{(-)}(t) = c^{(-)}\,e^{-\kappa t}, \qquad (7.37b)$$

in which the operators $c^{(\pm)}$ are the initial values of $c^{(\pm)}(t)$, and are given in terms of the Schrödinger operators a and b by

$$c^{(\pm)} = 2^{-1/2}(a \pm ib^\dagger). \qquad (7.38)$$

It is clear from Eqns. (7.37) that the operators $c^{(+)}(t)$ and $c^{(-)}(t)$ are decoupled from one another. The transformation Eq. (7.36) evidently defines a species of normal coordinates for the system, but one with exponential rather than oscillatory time dependence. The operators $c^{(+)}(t)$ and $c^{(-)}(0)$ are not canonical; they satisfy the commutation relations

$$[c^{(+)}(t), c^{(+)\dagger}(t)] = [c^{(-)}(t), c^{(-)\dagger}(t)] = [c^{(+)}(t), c^{(-)}(t)] = 0,$$
$$[c^{(+)}(t), c^{(-)\dagger}(t)] = [c^{(-)}(t), c^{(+)\dagger}(t)] = 1. \tag{7.39}$$

The operators $c^{(\pm)}(t)$ are defined by Eqns. (7.36) as linear combinations of the operators

$$a'(t) \equiv e^{i\omega_a t} a(t), \tag{7.40a}$$
$$b'(t) \equiv e^{i\omega_b t} b(t), \tag{7.40b}$$

and their adjoints. The time dependence of these operators is entirely due to the coupling of the modes; they reduce to the Schrödinger operators a and b when the coupling vanishes. It is possible, indeed, to construct a formal solution for $a'(t)$ and $b'(t)$ as a canonical transformation on a and b, in which the transformation depends only on the interaction term of the Hamiltonian Eq. (7.1). To accomplish this, let us first consider the operator $a(t)$, which according to Eq. (6.52a) is given by

$$a(t) = U^{-1}(t) a U(t), \tag{7.41}$$

where $U(t)$ is the unitary time translation operator defined by Eqns. (6.50) and (6.51). If we substitute this form for $a(t)$ into Eq. (7.40a), and then use the identity

$$e^{i\omega_a t} a = e^{-iH_0(0)t/\hbar} a e^{iH_0(0)t/\hbar}, \tag{7.42}$$

where $H_0(0)$ is the initial value of the uncoupled part of the Hamiltonian, we see that we may write Eq. (7.40a) in the form

$$a'(t) = V^{-1}(t) a V(t), \tag{7.43}$$

in which the unitary operator $V(t)$ is defined by

$$V(t) \equiv e^{iH_0(0)t/\hbar} U(t). \tag{7.44}$$

In the same way, we see that $b'(t)$ is given by

$$b'(t) = V^{-1}(t) b V(t). \tag{7.45}$$

The operators $c^{(\pm)}(t)$, which are defined as linear combinations of $a'(t)$ and $b'^{\dagger}(t)$, may be expressed by means of the same canonical transformation as

$$c^{(\pm)}(t) = V^{-1}(t) c^{(\pm)} V(t). \tag{7.46}$$

An explicit expression for $V(t)$ may be derived without difficulty. If we differentiate Eq. (7.44) with respect to time and make use of Eq. (6.50) for the time derivative of $U(t)$, we find

$$i\hbar \frac{d}{dt} V(t) = H_{1,\text{int}}(t) V(t) . \qquad (7.47)$$

In this equation, $H_{1,\text{int}}(t)$ refers to the interaction term in the Hamiltonian, evaluated in the interaction representation, i.e., with $a(t)$ and $b(t)$ replaced by the corresponding operators in the interaction representation, which are $a e^{-i\omega_a t}$ and $b e^{-i\omega_b t}$, respectively. Because of the special form of the interaction we have used, i.e., since the frequency ω is chosen to be equal to $\omega_a + \omega_b$, the operator $H_{1,\text{int}}(t)$ is actually independent of time, and we have

$$H_{1,\text{int}}(t) = H_{1,\text{int}}(0) = -\hbar\kappa(a^\dagger b^\dagger + ab) . \qquad (7.48)$$

The solution to Eq. (7.47) is therefore

$$V(t) = e^{i\kappa(a^\dagger b^\dagger + ab)t} , \qquad (7.49)$$

which obviously reduces to unity when the coupling vanishes.

7.5
Characteristic Functions Expressed in Terms of Decoupled Variables

We have shown in Sect. 7.2 that the characteristic function $\chi(\eta,\zeta,t)$ is expressible at any time t in terms of the form it takes at time $t = 0$. The explicit expression for it may be reduced to a particularly simple form by making use of the decoupled operators $c^{(\pm)}(t)$. To this end we introduce the transformation of variables

$$\sigma^{(+)} \equiv 2^{-1/2} \left(\eta e^{i\omega_a t} - i\zeta^* e^{-i\omega_b t} \right) , \qquad (7.50a)$$

$$\sigma^{(-)} \equiv 2^{-1/2} \left(\eta e^{i\omega_a t} + i\zeta^* e^{-i\omega_b t} \right) . \qquad (7.50b)$$

The functional form of the characteristic function will change when it is expressed in terms of $\sigma^{(+)}$ and $\sigma^{(-)}$ rather than η and ζ. We shall indicate this new functional form by attaching the subscript σ to χ. If we write the solutions to Eqns. (7.50) for η and ζ as $\eta(\sigma^{(+)},\sigma^{(-)},t)$ and $\zeta(\sigma^{(+)},\sigma^{(-)},t)$, respectively, then we may define the function $\chi_\sigma(\sigma^{(+)},\sigma^{(-)},t)$ in terms of $\chi(\eta,\zeta,t)$ by means of the relation

$$\chi_\sigma\left(\sigma^{(+)},\sigma^{(-)},t\right) \equiv \chi\left(\eta(\sigma^{(+)},\sigma^{(-)},t),\zeta(\sigma^{(+)},\sigma^{(-)},t),t\right) . \qquad (7.51)$$

To evaluate $\chi_\sigma(\sigma^{(+)},\sigma^{(-)},t)$ we have only to express the exponent in Eq. (7.6) in terms of $\sigma^{(+)}$ and $\sigma^{(-)}$. If we make use of the identity

$$\eta a^\dagger(t) + \zeta b^\dagger(t) - \text{H.c.} = \sigma^{(+)} c^{(+)\dagger}(t) + \sigma^{(-)} c^{(-)\dagger}(t) - \text{H.c.} , \qquad (7.52)$$

which follows directly from Eqns. (7.50) and (7.36), we find

$$\chi_\sigma\left(\sigma^{(+)},\sigma^{(-)},t\right) = \text{Tr}\left\{\rho \exp\left[\sigma^{(+)}c^{(+)\dagger}(t) + \sigma^{(-)}c^{(-)}(t) - \text{H.c.}\right]\right\}. \quad (7.53)$$

If we now substitute Eq. (7.37) for $c^{(\pm)}(t)$ into Eq. (7.53), we obtain

$$\chi_\sigma\left(\sigma^{(+)},\sigma^{(-)},t\right) = \text{Tr}\left\{\rho \exp\left[\sigma^{(+)}e^{\kappa t}c^{(+)\dagger} + \sigma^{(-)}e^{-\kappa t}c^{(-)\dagger} - \text{H.c.}\right]\right\}$$

$$= \chi_\sigma\left(\sigma^{(+)}e^{\kappa t}, \sigma^{(-)}e^{-\kappa t}, 0\right). \quad (7.54)$$

The function χ_σ at time t is thus obtained from its form at $t = 0$ simply by multiplying one of its arguments by a positive exponential and the other by a negative one.

Let us now turn to the evaluation of the normally ordered characteristic function $\chi_N(\eta,\zeta,t)$. This function is defined in terms of the Schrödinger density operator $\rho(t)$ and the Schrödinger annihilation operators a and b as

$$\chi_N(\eta,\zeta,t) = \text{Tr}\left\{\rho(t)\,e^{\eta a^\dagger + \zeta b^\dagger}\,e^{-\eta^* a - \zeta^* b}\right\}, \quad (7.55)$$

and may be expressed in terms of the Heisenberg density operator ρ and the Heisenberg operators $a(t)$ and $b(t)$ by means of the relation

$$\chi_N(\eta,\zeta,t) = \text{Tr}\left\{\rho\,e^{\eta a^\dagger(t) + \zeta b^\dagger(t)}\,e^{-\eta^* a(t) - \zeta^* b(t)}\right\}. \quad (7.56)$$

It is related to the ordinary characteristic function by

$$\chi_N(\eta,\zeta,t) = e^{|\eta|^2/2 + |\zeta|^2/2}\,\chi(\eta,\zeta,t), \quad (7.57)$$

which is the two-mode generalization of Eq. (6.20).

We shall now express χ_N in terms of the variables $\sigma^{(\pm)}$ defined by Eq. (7.50); as in the case of the ordinary characteristic function, we add the subscript σ to χ_N to indicate the explicit functional form which results. If we use the identity

$$|\eta|^2 + |\zeta|^2 = |\sigma^{(+)}|^2 + |\sigma^{(-)}|^2, \quad (7.58)$$

which follows simply from Eq. (7.50), then by expressing Eq. (7.57) in terms of $\sigma^{(+)}$ and $\sigma^{(-)}$ we find

$$\chi_{N,\sigma}\left(\sigma^{(+)},\sigma^{(-)},t\right) = e^{|\sigma^{(+)}|^2/2 + |\sigma^{(-)}|^2/2}\,\chi_\sigma\left(\sigma^{(+)},\sigma^{(-)},t\right). \quad (7.59)$$

By making use of Eq. (7.54) we may then write

$$\chi_{N,\sigma}\left(\sigma^{(+)},\sigma^{(-)},t\right) = e^{|\sigma^{(+)}|^2/2 + |\sigma^{(-)}|^2/2}\,\chi_\sigma\left(\sigma^{(+)}e^{\kappa t}, \sigma^{(-)}e^{-\kappa t}, 0\right). \quad (7.60)$$

If we evaluate Eq. (7.59) at $t = 0$ and with $\sigma^{(+)}$ and $\sigma^{(-)}$ replaced by $\sigma^{(+)}e^{\kappa t}$ and $\sigma^{(-)}e^{-\kappa t}$, respectively, we find

$$\chi_\sigma\left(\sigma^{(+)}e^{\kappa t}, \sigma^{(-)}e^{-\kappa t}, 0\right) = \exp\left[-\tfrac{1}{2}|\sigma^{(+)}|^2 e^{2\kappa t} - \tfrac{1}{2}|\sigma^{(-)}|^2 e^{-2\kappa t}\right]$$

$$\times \chi_{N\sigma}\left(\sigma^{(+)}e^{\kappa t}, \sigma^{(-)}e^{-\kappa t}, 0\right), \quad (7.61)$$

and if we then substitute this relation into Eq. (7.60) we obtain

$$\chi_{N\sigma}\left(\sigma^{(+)},\sigma^{(-)},t\right) = \exp\left[\tfrac{1}{2}|\sigma^{(+)}|^2\left(1-e^{2\kappa t}\right) + \tfrac{1}{2}|\sigma^{(-)}|^2\left(1-e^{-2\kappa t}\right)\right]$$
$$\times \chi_{N\sigma}\left(\sigma^{(+)}e^{\kappa t},\sigma^{(-)}e^{-\kappa t},0\right), \quad (7.62)$$

which expresses the normally ordered characteristic function at time t in terms of its value at $t = 0$.

7.6
W and P Expressed in Terms of Decoupled Variables

The time dependence of the Wigner function and the function $P(\alpha,\beta,t)$ takes an especially simple form when these functions are expressed in terms of variables whose definition parallels that of the operators $c^{(\pm)}(t)$ given by Eq. (7.36). Let us define the variables $\gamma^{(\pm)}$ in terms of α and β by means of the relations

$$\gamma^{(+)} \equiv 2^{-1/2}\left(\alpha e^{i\omega_a t} + i\beta^* e^{-i\omega_b t}\right), \quad (7.63a)$$

$$\gamma^{(-)} \equiv 2^{-1/2}\left(\alpha e^{i\omega_a t} - i\beta^* e^{-i\omega_b t}\right). \quad (7.63b)$$

In order to express the Fourier transform relationship Eq. (7.23) in terms of the variables $\gamma^{(\pm)}$ for W and $\sigma^{(\pm)}$ for χ, we first note the identities

$$\alpha\eta^* + \beta\zeta^* - \text{c.c.} = \gamma^{(+)}\sigma^{(+)*} + \gamma^{(-)}\sigma^{(-)*} - \text{c.c.} \quad (7.64)$$

and

$$d^2\eta\, d^2\zeta = d^2\sigma^{(+)}\, d^2\sigma^{(-)}, \quad (7.65)$$

which follow from the definitions Eq. (7.50) and (7.63) of $\sigma^{(\pm)}$ and $\gamma^{(\pm)}$, respectively. If we substitute the last two relations into Eq. (7.23), and attach the subscript γ to W to indicate its functional form when it is expressed in terms of $\gamma^{(\pm)}$ we obtain

$$W_\gamma\left(\gamma^{(+)},\gamma^{(-)},t\right) = \frac{1}{\pi^4}\int \exp\left[\gamma^{(+)}\sigma^{(+)*} + \gamma^{(-)}\sigma^{(-)*} - \text{c.c.}\right]$$
$$\times \chi_\sigma\left(\sigma^{(+)},\sigma^{(-)},t\right) d^2\sigma^{(+)}\, d^2\sigma^{(-)}. \quad (7.66)$$

By substituting Eq. (7.54) for $\chi_\sigma(\sigma^{(+)},\sigma^{(-)},t)$ into this equation, and then changing the variables of integration from $\sigma^{(+)},\sigma^{(-)}$ to

$$\xi^{(+)} \equiv \sigma^{(+)}e^{\kappa t},$$
$$\xi^{(-)} \equiv \sigma^{(-)}e^{-\kappa t}, \quad (7.67)$$

we find

$$W_\gamma\left(\gamma^{(+)},\gamma^{(-)},t\right) = \frac{1}{\pi^4}\int \exp\left[\gamma^{(+)}e^{-\kappa t}\xi^{(+)*} + \gamma^{(-)}e^{\kappa t}\xi^{(-)*} - \text{c.c.}\right]$$
$$\times \chi_\sigma\left(\xi^{(+)},\xi^{(-)},0\right)d^2\xi^{(+)}\,d^2\xi^{(-)}$$
$$= W_\gamma\left(\gamma^{(+)}e^{-\kappa t},\gamma^{(-)}e^{\kappa t},0\right). \tag{7.68}$$

When the Wigner function is expressed in terms of the decoupled variables in other words, it be evaluated at any given time in terms of its initial form simply by performing a pair of exponential scale transformations on the arguments $\gamma^{(\pm)}$.

The classical trajectories which we defined in Sect. 7.3 take a particularly simple form when expressed in terms of the variables $\gamma^{(\pm)}$. If we let $\gamma_c^{(\pm)}(t)$ be the variables which correspond through Eq. (7.63) to the classical amplitudes $\alpha_c(t)$ and $\beta_c(t)$ given by Eq. (7.29), then we find

$$\gamma_c^{(+)}(t) = \gamma_c^{(+)}(0)e^{\kappa t}, \tag{7.69a}$$
$$\gamma_c^{(-)}(t) = \gamma_c^{(-)}(0)e^{-\kappa t}. \tag{7.69b}$$

We find from these relations and Eq. (7.68) that the Wigner function, along a classical trajectory defined by particular values of $\gamma^{(\pm)}(0)$, obeys the identity

$$W_\gamma\left(\gamma_c^{(+)}(t),\gamma_c^{(-)}(t),t\right) = W_\gamma\left(\gamma_c^{(+)}(0),\gamma_c^{(-)}(0),0\right), \tag{7.70}$$

which exhibits explicitly the constancy of the function along such a trajectory.

Let us now consider the *P* representation. We shall say that a *P* representation for the joint system of A and B modes exists at time *t* if the Schrödinger density operator $\rho(t)$ can be expressed in the form

$$\rho(t) = \int P(\alpha,\beta,t)|\alpha,\beta\rangle\langle\alpha,\beta|\,d^2\alpha\,d^2\beta. \tag{7.71}$$

A *P* representation exists if the normally ordered characteristic function $\chi_N(\eta,\zeta,t)$ possesses a Fourier transform. The function $P(\alpha,\beta,t)$ is then given by

$$P(\alpha,\beta,t) = \frac{1}{\pi^4}\int e^{\alpha\eta^* + \beta\zeta^* - \alpha^*\eta - \beta^*\zeta}\chi_N(\eta,\zeta,t)\,d^2\eta\,d^2\zeta, \tag{7.72}$$

which is the two-mode generalization of Eq. (6.59).

If we express *P* in terms of the variables $\gamma^{(\pm)}$, and χ_N in terms of the variables $\sigma^{(\pm)}$ defined by Eq. (7.50), then we may show by steps similar to those leading to Eq. (7.66) that Eq. (7.72) falls into the form

$$P_\gamma\left(\gamma^{(+)},\gamma^{(-)},t\right) = \frac{1}{\pi^4}\int \exp\left[\gamma^{(+)}\sigma^{(+)*} + \gamma^{(-)}\sigma^{(-)*} - \text{c.c.}\right]$$
$$\times \chi_{N\sigma}\left(\sigma^{(+)},\sigma^{(-)},t\right)d^2\sigma^{(+)}\,d^2\sigma^{(-)}. \tag{7.73}$$

Here we have again introduced the subscripts γ and σ to indicate the explicit form which the functions P and χ_N take in terms of their new arguments.

The integration in Eq. (7.73) may be carried out most conveniently by changing the variables of integration from $\sigma^{(\pm)}$ to $\xi^{(\pm)}$ as defined by Eq. (7.67). If we define the function $\chi_{N\xi}$ as

$$\chi_{N\xi}\left(\xi^{(+)},\xi^{(-)},t\right) = \chi_{N\sigma}\left(\xi^{(+)}e^{-\kappa t},\xi^{(-)}e^{\kappa t},t\right), \tag{7.74}$$

then Eq. (7.73) takes the form

$$P_\gamma\left(\gamma^{(+)},\gamma^{(-)},t\right) = \frac{1}{\pi^4}\int \exp\left[\gamma^{(+)}e^{-\kappa t}\xi^{(+)*} + \gamma^{(-)}e^{\kappa t}\xi^{(-)*} - \text{c.c.}\right]$$
$$\times \chi_{N\xi}\left(\xi^{(+)},\xi^{(-)},t\right) d^2\xi^{(+)} d^2\xi^{(-)}. \tag{7.75}$$

The function $\chi_{N\xi}(\xi^{(+)},\xi^{(-)},t)$ may be expressed in terms of the form it takes at $t=0$ by using Eq. (7.61) to evaluate the right-hand side of Eq. (7.74). We find

$$\chi_{N\xi}\left(\xi^{(+)},\xi^{(-)},t\right) = \exp\left[\tfrac{1}{2}|\xi^{(+)}|^2\left(e^{-2\kappa t}-1\right) + \tfrac{1}{2}|\xi^{(-)}|^2\left(e^{2\kappa t}-1\right)\right]$$
$$\times \chi_{N\sigma}\left(\xi^{(+)},\xi^{(-)},0\right). \tag{7.76}$$

Equation (7.75) states that $P_\gamma(\gamma^{(+)},\gamma^{(-)},t)$ is the Fourier transform, evaluated at the complex arguments $\gamma^{(+)}e^{-\kappa t}$ and $\gamma^{(-)}e^{\kappa t}$, of the function $\chi_{N\xi}(\xi^{(+)},\xi^{(-)},t)$. We indicated earlier in this section in connection with the function W_γ, that a function remains constant along a classical trajectory if it depends upon the variables $\gamma^{(+)}$, $\gamma^{(-)}$, and t only through the combinations $\gamma^{(+)}e^{-\kappa t}$ and $\gamma^{(-)}e^{\kappa t}$. The function $P_\gamma(\gamma^{(+)},\gamma^{(-)},t)$ on the other hand, has in addition to an implicit time dependence of this kind, an explicit time dependence due to the time dependence of the function $\chi_{N\xi}(\xi^{(+)},\xi^{(-)},t)$ as given by Eq. (7.76). It is evident that the function P, unlike the Wigner function, does not in general remain constant along a classical trajectory.

It is clear from the form of $\chi_{N\xi}(\xi^{(+)},\xi^{(-)},t)$ as given by Eq. (7.76), that the integral representation Eq. (7.75) of the function P need not always converge. Indeed, the exponential which contains the argument $\tfrac{1}{2}|\xi^{(-)}|^2(e^{2\kappa t}-1)$ increases so rapidly as t increases that the integral will always begin to diverge at some finite time unless $\chi_{N\sigma}(\xi^{(+)},\xi^{(-)},0)$ decreases with remarkable rapidity as $|\xi^{(-)}|\to\infty$. A similar requirement must be placed on the dependence of $\chi_{N\sigma}$ on $\xi^{(+)}$ if the integral is not to diverge for times prior to some negative time. When, as is typically the case, $\chi_{N\sigma}$ fails to satisfy either of these conditions of rapid decrease, the integral in Eq. (7.75) diverges for large values of $\pm t$. The integral then converges only within a finite interval of time at most, and no P representation as defined by Eq. (7.72) exists outside it. This result is in marked contrast to the result established in Chapter 6 for the P representation for a single mode of the parametric amplifier. It was shown there that for a broad class of initial states the P representation for a single mode must exist after a certain characteristic time, and in many cases exists at all times.

To state our result for the function P more explicitly, let us return to our original sets of variables, α and β for P, and η and ζ for χ_N. We find then that Eq. (7.75) may be written as

$$P(\alpha,\beta,t) = \frac{1}{\pi^4} \int \exp\left[\alpha_{0c}(\alpha,\beta,t)\eta^* + \beta_{0c}(\alpha,\beta,t)\zeta^* - \text{c.c.} \right.$$
$$\left. + (|\eta|^2 + |\zeta|^2)\, s^2(t) + is(t)c(t)\,(\eta^*\zeta^* - \eta\zeta)\right] \chi_N(\eta,\zeta,0)\, d^2\eta\, d^2\zeta, \quad (7.77)$$

in which $\alpha_{0c}(\alpha,\beta,t)$ and $\beta_{0c}(\alpha,\beta,t)$ are defined by Eq. (7.26) and

$$c(t) \equiv \cosh \kappa t, \qquad s(t) \equiv \sinh \kappa t.$$

We may verify from Eq. (7.77) that $P(\alpha,\beta,t)$ satisfies the differential equation

$$\left[\frac{d}{dt} + i\kappa \left(e^{i\omega t}\frac{\partial^2}{\partial\alpha^* \partial\beta^*} - e^{-i\omega t}\frac{\partial^2}{\partial\alpha\, \partial\beta}\right)\right] P(\alpha,\beta,t) = 0, \quad (7.78)$$

in which the total time derivative d/dt is defined by Eq. (7.34). Equation (7.78) is the analog for the function P of the Liouville equation which is satisfied by the Wigner function.

7.7
Results for Chaotic Initial States

To illustrate the results of the previous section let us consider a particularly simple example. We assume the two modes of the amplifier to be in independent chaotic states initially, so that the state of each mode is specified by its mean initial occupation number. To begin with the simplest case, let us assume that the two mean occupation numbers are equal. Then the initial density operator for the two-mode system has a P representation

$$\rho = \int P(\alpha,\beta,0)\,|\alpha,\beta\rangle\langle\alpha,\beta|\,d^2\alpha\, d^2\beta, \quad (7.79)$$

with the function P given by

$$P(\alpha,\beta,0) = \frac{1}{\pi(\langle n\rangle)^2} \exp\left[-\frac{|\alpha|^2}{\langle n\rangle} - \frac{|\beta|^2}{\langle n\rangle}\right], \quad (7.80)$$

where $\langle n\rangle$ is the mean initial occupation number of both modes.

The normally ordered characteristic function at $t = 0$ is the Fourier transform of $P(\alpha,\beta,0)$

$$\chi_N(\eta,\zeta,0) = \int e^{\eta\alpha^* + \zeta\beta^* - \text{c.c.}}\, P(\alpha,\beta,0)\, d^2\alpha\, d^2\beta. \quad (7.81)$$

7.7 Results for Chaotic Initial States

By substituting Eq. (7.80) for $P(\alpha, \beta, 0)$ into this equation and performing the integrations with the aid of the integral identity Eq. (6.69) we find

$$\chi_N(\eta, \zeta, 0) = \exp\left[-\langle n \rangle \left(|\eta|^2 + |\zeta|^2\right)\right] . \tag{7.82}$$

We may express this result in terms of the variables $\sigma^{(+)}$ and $\sigma^{(-)}$ by making use of the identity Eq. (7.58) to write

$$\chi_{N\sigma}\left(\sigma^{(+)}, \sigma^{(-)}, 0\right) = \exp\left[-\langle n \rangle \left(|\sigma^{(+)}|^2 + |\sigma^{(-)}|^2\right)\right] . \tag{7.83}$$

The time-dependent function $\chi_{N\sigma}(\sigma^{(+)}, \sigma^{(-)}, 0)$ may be obtained from its form at $t = 0$ by making use of the identity Eq. (7.61). The result may be written in the form

$$\chi_{N\sigma}\left(\sigma^{(+)}, \sigma^{(-)}, t\right) = \exp\left[-|\sigma^{(+)}|^2 \lambda(t) + |\sigma^{(-)}|^2 \lambda(-t)\right] , \tag{7.84}$$

where the function $\lambda(t)$ is defined by

$$\lambda(t) \equiv \tfrac{1}{2}\left[(1 + 2\langle n \rangle) e^{2\kappa t} - 1\right] . \tag{7.85}$$

Let us define a time $\tau > 0$ as the root of the equation $\lambda(-t) = 0$, or equivalently by the relation

$$e^{2\kappa\tau} = 1 + 2\langle n \rangle . \tag{7.86}$$

Then the function $\lambda(t)$ is positive for $t > -\tau$ and similarly the function $\lambda(-t)$ is positive for $t < \tau$. For $|t| < \tau$, then, the factors multiplying $|\sigma^{(+)}|^2$ and $|\sigma^{(-)}|^2$ in the argument of the exponential function in Eq. (7.84) are both negative. For these values of t the function $\chi_{N\sigma}(\sigma^{(+)}, \sigma^{(-)}, t)$ therefore has a well-defined Fourier transform over the $\sigma^{(+)}$ and $\sigma^{(-)}$ planes. By substituting Eq. (7.84) for $\chi_{N\sigma}(\sigma^{(+)}, \sigma^{(-)}, t)$ into Eq. (7.73) and performing the integrations with the aid of the identity Eq. (6.69), we find for $|t| < \tau$ the P function

$$P_\gamma(\gamma^{(+)}, \gamma^{(-)}, t) = \frac{1}{\pi^2 \lambda(t) \lambda(-t)} \exp\left[-\frac{|\gamma^{(+)}|^2}{\lambda(t)} - \frac{|\gamma^{(-)}|^2}{\lambda(-t)}\right] . \tag{7.87}$$

For $|t| < \tau$, then, a P representation exists, and the function P, when it is expressed in terms of the variables $\gamma^{(\pm)}$ defined by Eq. (7.63), is the product of two Gaussian functions, with variances $\lambda(t)$ and $\lambda(-t)$ for the variables $\gamma^{(+)}$ and $\gamma^{(-)}$, respectively. For $|t| > \tau$, on the other hand, the function $\chi_{N\sigma}(\sigma^{(+)}, \sigma^{(-)}, t)$ given by Eq. (7.84) becomes infinite as $|\sigma^{(+)}| \to \infty$ or $|\sigma^{(-)}| \to \infty$ and the integral in Eq. (7.73) is seen to be strongly divergent. It follows that for $|t| > \tau$ no P representation as defined by Eq. (7.72) exists for the joint system of A and B modes.

It is instructive to consider the behavior of the P function as $t \to \tau$. Since in this limit $\lambda(-t) \to 0$, the function P_γ given by Eq. (7.87) approaches a delta function in $\gamma^{(-)}$,

$$P_\gamma\left(\gamma^{(+)},\gamma^{(-)},t\right) \to \delta^{(2)}\left(\gamma^{(-)}\right)\frac{1}{\pi\lambda(t)}\exp\left[-\frac{|\gamma^{(+)}|^2}{\lambda(\tau)}\right], \qquad (7.88)$$

and thus the function $P(\alpha,\beta,\tau)$, by virtue of the definition Eq. (7.63b) of $\gamma^{(-)}$, contains the factor

$$\delta^{(2)}\left(\alpha\,e^{i\omega_a\tau} - i\beta^*\,e^{-i\omega_b\tau}\right).$$

The mean value of a normally ordered operator function at $t = \tau$ is therefore given by an integral of classical type over the four-dimensional phase space of the system, in which the only contributions come from the two-dimensional subspace defined by the condition $\alpha\,e^{i\omega_a\tau} = i\beta^*\,e^{-i\omega_b\tau}$. This result reflects in an extreme form one of the characteristic features of the time dependence of the complex mode amplitudes. While they are being amplified in absolute magnitude, the complex amplitudes for the two modes are brought into an exceedingly close, though time dependent, state of correlation. We shall discuss this correlation further as we proceed.

The Wigner function $W_\gamma(\gamma^{(+)},\gamma^{(-)},t)$ corresponding to the same initial density operator may be obtained by evaluating the ordinary characteristic function $\chi_\sigma(\sigma^{(+)},\sigma^{(-)},t)$ and substituting it into the integral in Eq. (7.66). In this way we find

$$W_\gamma(\gamma^{(+)},\gamma^{(-)},t) = \frac{1}{\pi^2\,\nu(t)\,\nu(-t)}\exp\left[-\frac{|\gamma^{(+)}|^2}{\nu(t)} - \frac{|\gamma^{(-)}|^2}{\nu(-t)}\right], \qquad (7.89)$$

where

$$\nu(t) \equiv \tfrac{1}{2}e^{2\kappa(\tau+t)} \qquad (7.90)$$

$$= \lambda(t) + \tfrac{1}{2}. \qquad (7.91)$$

The Wigner function is thus, like the function P, a product of Gaussian functions in the variables $\gamma^{(+)}$ and $\gamma^{(-)}$.

We may obtain some insight into the relationship between the functions P_γ and W_γ by generalizing Eq. (7.22) to the case of a two-mode system. We may then write the function W in terms of P as

$$W(\alpha,\beta,t) = \left(\frac{2}{\pi}\right)^2 \int \exp\left[-2|\alpha-\alpha'|^2 - 2|\beta-\beta'|\right]P(\alpha',\beta',t)\,d^2\alpha'\,d^2\beta'. \qquad (7.92)$$

If we next make use of the identities

$$d^2\alpha\,d^2\beta = d^2\gamma^{(+)}\,d^2\gamma^{(-)} \qquad (7.93)$$

and

$$|\alpha|^2 + |\beta|^2 = |\gamma^{(+)}|^2 + |\gamma^{(-)}|^2, \tag{7.94}$$

which follow directly from the definitions Eq. (7.63) of $\gamma^{(\pm)}$ we find that Eq. (7.92) may be expressed in terms of $\gamma^{(\pm)}$ in the form

$$W_\gamma\left(\gamma^{(+)}, \gamma^{(-)}, t\right) = \left(\frac{2}{\pi}\right)^2 \int \exp\left[-2|\gamma^{(+)} - \gamma'^{(+)}|^2 - 2|\gamma^{(-)} - \gamma'^{(-)}|^2\right]$$

$$\times P_\gamma\left(\gamma'^{(+)}, \gamma'^{(-)}, t\right) d^2\gamma'^{(+)} d^2\gamma'^{(-)}. \tag{7.95}$$

For the state with initial P function Eq. (7.80) we found that a P representation exists for $|t| < \tau$, with P_γ given by Eq. (7.87). If we substitute Eq. (7.87) for P_γ into Eq. (7.95), then by performing the integrations we once again find the result in Eq. (7.89) for W_γ, with $\nu(t)$ given by Eq. (7.91). This derivation of Eq. (7.89) depends on the existence of the P representation, and hence is valid only for $|t| < \tau$. The derivation of the Wigner function given by Eq. (7.89), on the other hand, is valid at all times.

If we examine the time dependence of the Wigner function Eq. (7.89), we see that its form does not change in any dramatic way near the times $t = \pm\tau$. As $t \to \infty$, in fact, the variance of $\gamma^{(-)}$ which it contains goes uniformly to zero. As far as the description implicit in the Wigner function is concerned, the complex amplitudes of the A and B modes tend steadily to become more correlated as $t \to \infty$. In the function P_γ, on the other hand, the expression $\lambda(-t)$, which is the variance of $\gamma^{(-)}$, goes to zero for $t \to \tau$, and P_γ fails to be well-defined thereafter. The convolution transform Eq. (7.95) fails to have a meaningful inverse for $t > \tau$, since $\nu(-t) < \frac{1}{2}$. Evidently the P representation ceases to exist because it is incapable of describing closer field correlations than exist at time $t = \tau$. We shall discuss this point further in the next section.

When the initial mean occupation numbers of the two modes are allowed to take on different values, we may write the initial P function as

$$P(\alpha, \beta, 0) = \frac{1}{\pi^2 \langle n \rangle \langle m \rangle} \exp\left[-\frac{|\alpha|^2}{\langle n \rangle} - \frac{|\beta|^2}{\langle m \rangle}\right], \tag{7.96}$$

where $\langle n \rangle$ and $\langle m \rangle$ are the initial mean quantum numbers for the A and B modes, respectively. By following the obvious generalizations of the steps used earlier in this section we may derive the corresponding time-dependent P function. Let us now define the function $\lambda(t)$ by the equation

$$\lambda(t) \equiv \tfrac{1}{2}\left[(1 + \langle n \rangle + \langle m \rangle)e^{2\kappa t} - 1\right], \tag{7.97}$$

and the time $\tau \geq 0$ as the root of the equation $\lambda(-t) = 0$, so that

$$e^{2\kappa t} = 1 + \langle n \rangle + \langle m \rangle . \tag{7.98}$$

If we then introduce the quantities

$$\Delta = \tfrac{1}{2}(\langle n \rangle - \langle m \rangle) \tag{7.99}$$

and

$$M(t) = \lambda(t)\lambda(-t) - \Delta^2 , \tag{7.100}$$

we find that the function P_γ at time t may be written in the form

$$P_\gamma\left(\gamma^{(+)}, \gamma^{(-)}, t\right) = \pi^{-2} M^{-1}(t) \exp\left\{ -M^{-1}(t)\left[|\gamma^{(+)}|^2 \lambda(-t) + |\gamma^{(-)}|^2 \lambda(t) - \Delta\left(\gamma^{(+)*}\gamma^{(-)} + \gamma^{(+)}\gamma^{(-)*}\right)\right]\right\}. \tag{7.101}$$

The Fourier transform which leads to this result exists only when $M(t) \geq 0$; this is equivalent to the requirement

$$\cosh 2\kappa t \leq 1 + \frac{2\langle n \rangle \langle m \rangle}{1 + \langle n \rangle + \langle m \rangle}, \tag{7.102}$$

which may be shown to imply the weaker conditions $\lambda(t) \geq 0$ and $\lambda(-t) \geq 0$, or $|t| \leq \tau$. For times too large to satisfy the inequality (7.102) no P representation exists.

We may note that if either $\langle n \rangle$ or $\langle m \rangle$ vanishes, a P representation exists only at $t = 0$. In particular, if the initial state is the vacuum ($\langle n \rangle = \langle m \rangle = 0$), then the amplification process immediately destroys the P representation.

It is not difficult to extend our analysis to treat more general forms for the initial P function. Let us suppose, for example, that the system is initially described by the displaced Gaussian function

$$P(\alpha, \beta, 0) = \frac{1}{\pi^2 \langle n \rangle \langle m \rangle} \exp\left[-\frac{|\alpha - \alpha_0|^2}{\langle n \rangle} - \frac{|\beta - \beta_0|^2}{\langle m \rangle}\right], \tag{7.103}$$

where α_0 and β_0 are the mean values of a and b, respectively. Let us define the functions $\bar{\gamma}^{(+)}(t)$ and $\bar{\gamma}^{(-)}(t)$ as the mean values of the operators $c^{(+)}(t)$ and $c^{(-)}(t)$ defined by Eq. (7.36). Then we may show that the function $P_\gamma(\gamma^{(+)}, \gamma^{(-)}, t)$ which corresponds to the state described initially by the P function Eq. (7.103) is just the function given by Eq. (7.101), with $\gamma^{(+)}$ replaced by $\gamma^{(+)} - \bar{\gamma}^{(+)}(t)$, and $\gamma^{(-)}$ by $\gamma^{(-)} - \bar{\gamma}^{(-)}(t)$. The time interval during which a P representation exists is once again the interval defined by the inequality (7.102). If $\langle n \rangle = 0$, the initial state of the A mode is the coherent state $|\alpha_0\rangle$; likewise if $\langle m \rangle = 0$, the initial state of the B mode is the coherent state $|\beta_0\rangle$. In either of these cases the inequality (7.102) is satisfied only for $t = 0$ and a P representation fails to exist for all times other than $t = 0$.

7.8
Correlations of the Mode Amplitudes

We have remarked that the amplification process leads as it continues to an increasingly close correlation between the complex amplitudes of the two modes of the amplifier. That the moduli of the complex amplitudes become effectively equal in the limit of large times is clear from the conservation law Eq. (6.31), which states that the difference $a^\dagger(t)a(t) - b^\dagger(t)b(t)$ is a constant of the motion. Since the magnitudes of both $a^\dagger(t)a(t)$ and $b^\dagger(t)b(t)$ increase exponentially as $t \to \infty$, the relative magnitude of the difference between them becomes exceedingly small.

The complex mode amplitudes become strongly correlated not only in modulus, but in phase as well. The asymptotic form as $t \to \infty$ of the solutions Eq. (7.3) to the Heisenberg equations of motion is

$$a(t) \sim \tfrac{1}{2}(a + ib^\dagger) e^{-i\omega_a t + \kappa t} ,$$
$$b(t) \sim \tfrac{1}{2}(b + ia^\dagger) e^{-i\omega_b t + \kappa t} .$$
(7.104)

By recalling that $\omega = \omega_a + \omega_b$ we see that in the limit of large times the amplitude operators obey the relation

$$a(t) \sim ib^\dagger(t) e^{-i\omega t} .$$
(7.105)

Similarly, if we consider the motion of the system as extending over an infinite period of time prior to $t = 0$, then the equations of motion lead to the correlation

$$a(t) \sim -ib^\dagger(t) e^{-i\omega t}$$
(7.106)

in the limit $t \to \infty$.

To investigate the time-dependent correlation between the modes in greater detail, let us assume that the system is described by a given Heisenberg density operator ρ. Let us next define the c-number functions $\overline{\alpha}(t)$ and $\overline{\beta}(t)$ through Eqns. (6.44) as the mean values of $a(t)$ and $b(t)$, respectively. We then introduce the operators

$$\widehat{a}(t) \equiv a(t) - \overline{\alpha}(t) ,$$
$$\widehat{b}(t) \equiv b(t) - \overline{\beta}(t) ,$$
(7.107)

which describe the quantum fluctuations of the mode amplitudes about their mean values. It is clear that the operators $\widehat{a}(t)$ and $\widehat{b}(t)$ obey the same linear equations of motion as do $a(t)$ and $b(t)$. If we similarly define the operators $\widehat{c}^{(\pm)}(t)$ as the deviations of the operators $c^{(\pm)}(t)$ defined by Eq. (7.36) from their mean values,

$$\widehat{c}^{(\pm)}(t) \equiv c^{(\pm)}(t) - \mathrm{Tr}\left\{\rho c^{(\pm)}(t)\right\} ,$$
(7.108)

then it is clear that the defining Eq. (7.36) and the solutions to the equations of motion (7.37) remain valid when $c^{(\pm)}(t)$, $a(t)$, and $b(t)$ are replaced by $\widehat{c}^{(\pm)}(t)$, $\widehat{a}(t)$, and $\widehat{b}(t)$, respectively.

The fluctuations of the mode amplitudes about their mean values become correlated as $t \to \infty$ much as the mode amplitudes themselves do. To discuss these correlations we introduce the correlation function

$$\mu^{(-)}(t) = \mathrm{Tr}\left\{\rho \left[\hat{a}^{\dagger}(t) e^{-i\omega_a t} - i \hat{b}(t) e^{i\omega_b t}\right] \left[\hat{a}(t) e^{i\omega_a t} - i \hat{b}^{\dagger}(t) e^{-i\omega_b t}\right]\right\} \quad (7.109)$$

$$= 2\,\mathrm{Tr}\left\{\rho\, \hat{c}^{(-)\dagger}(t)\, \hat{c}^{(-)}(t)\right\}. \quad (7.110)$$

To discuss the correlation of the fluctuations as $t \to -\infty$ we likewise form the correlation function

$$\mu^{(+)}(t) = 2\,\mathrm{Tr}\left\{\rho\, \hat{c}^{(+)\dagger}(t)\, \hat{c}^{(+)}(t)\right\}. \quad (7.111)$$

Since ρ is Hermitian and positive definite, both of these functions are real and non-negative

$$\mu^{(-)}(t) \geq 0, \qquad \mu^{(+)}(t) \geq 0. \quad (7.112)$$

The solutions Eq. (7.37) for the operators $c^{(\pm)}(t)$ imply that the correlation functions vary exponentially with time, i.e., we have

$$\mu^{(-)}(t) = \mu^{(-)}(0)\,e^{-2\kappa t}, \qquad \mu^{(+)}(t) = \mu^{(+)}(0)\,e^{2\kappa t}. \quad (7.113)$$

It is clear that the fluctuations of the mode amplitudes tend to become extremely closely correlated as t increases or decreases from zero, no matter how uncorrected they may be initially.

Let us consider the form which the correlation functions $\mu^{(\pm)}(t)$ take in the Schrödinger picture. If we make use in Eq. (7.109) [and in the corresponding equation for $\mu^{(+)}(t)$] of the transformation rules Eq. (6.52) and Eq. (6.53), we find

$$\mu^{(\pm)}(t) = \mathrm{Tr}\left\{\rho(t)\left[(a^{\dagger} - \overline{\alpha}^{*}(t))e^{-i\omega_a t} \mp i(b - \overline{\beta}(t))e^{i\omega_b t}\right]\right.$$
$$\left.\left[(a - \overline{\alpha}(t))e^{i\omega_a t} \pm i(b^{\dagger} - \overline{\beta}^{*}(t))e^{-i\omega_b t}\right]\right\}, \quad (7.114)$$

which expresses the correlation functions in terms of the Schrödinger density operator $\rho(t)$.

The time dependence of the correlation functions expressed by Eq. (7.113) enables us to draw some quite general inferences concerning the time-dependent state of the amplifier system. For large values of $|t|$, for example, the relations Eq. (7.113) tend to specify correlations of the field amplitudes which are even closer than can be realized by any pure coherent state $|\alpha,\beta\rangle$ for the system. Such states specify minimal uncertainties in the fields of the A mode and the B mode individually, but leave small margins equal to those zero-point uncertainties, within which the two fields may fluctuate independently. Even so minute a degree of independence is too great to be consistent with Eq. (7.113). Indeed, if at any time t the density operator takes the form

$\rho(t) = |\alpha, \beta\rangle \langle \alpha, \beta|$, we find by using Eq. (7.114) and the commutation relations for b and $b^\dagger$ that

$$\mu^{(\pm)}(t) = 1 . \tag{7.115}$$

It is clear from Eq. (7.113) that for all times other than this time either $\mu^{(-)}(t)$ or $\mu^{(+)}(t)$ is smaller than unity, and that the field correlations are therefore too close to be described by a coherent state. They are in fact too close to be described within the much more general framework of the P representation, since, as we noted at the end of Sect. 7.7, if the system is initially in any coherent state, no P representation exists at times other than the initial time.

It is appropriate at this point to ask in more general terms what conditions the values of the correlation functions impose on the P representation. We shall show that, if either of the functions $\mu^{(\pm)}(t)$ is smaller than unity, no non-negative P function for the two-mode density operator can exist. A corresponding theorem was proved in Sect. 6.10 for the single mode P function. That function, it was shown, must take on negative values or fail to exist if the variance of either of the variables defined by Eq. (6.167) is less than one half. For the two-mode case let us assume that $\rho(t)$ has a P representation as expressed by Eq. (7.71). Then by making use of Eq. (7.114) and the commutation relations for b and $b^\dagger$ we find

$$\mu^{(\pm)}(t) = 1 + \int \left| \left\{ [\alpha - \overline{\alpha}(t)] e^{i\omega_a t} \pm i[\beta^* - \overline{\beta}^*(t)] e^{-i\omega_b t} \right\} \right|^2 P(\alpha, \beta, t) \, d^2\alpha \, d^2\beta . \tag{7.116}$$

It is clear that if $P(\alpha, \beta, t)$ is to be non-negative then we must have $\mu^{(\pm)}(t) \geq 1$. By virtue of the values Eq. (7.113) for the time-dependent correlation functions we see then that $P(\alpha, \beta, t)$ can remain non-negative only within the time interval specified by

$$\frac{1}{\mu^{(+)}(0)} \leq e^{2\kappa t} \leq \mu^{(-)}(0) . \tag{7.117}$$

We have not of course proved in general that no P representation exists outside the interval defined by the condition Eq. (7.117), though that is indeed the case for the examples considered in the last section. For example in the case of the initially chaotic mixture with equal mean quantum numbers $\langle n \rangle$ for each mode, we have $\mu^{(\pm)}(0) = 1 + 2\langle n \rangle$, and the time interval defined by Eq. (7.117) is thus $-\tau \leq t \leq \tau$, with τ defined by Eq. (7.86); we have seen that for this example a P representation ceases to exist outside the stated interval.

It is worth emphasizing that the condition Eq. (7.117) is merely a necessary one for the existence of a non-negative P function. We should not be surprised, therefore, if a P representation fails to exist inside the indicated interval of time. This is indeed the case for an initially chaotic mixture when the two mean occupation numbers are unequal. In that case the bounds in Eq. (7.117) are specified by $\mu^{(\pm)}(0) = 1 + \langle n \rangle + \langle m \rangle$, but the interval within which the P representation exists is the one given by Eq. (7.102). It is in general a smaller interval.

References

1 B. R. Mollow, R. J. Glauber, *Phys. Rev.* **160**, 1076 (1967); reprinted as Chapter 6 in this volume.

2 W. H. Louisell, A. Yariv, A. E. Siegman, *Phys. Rev.* **124**, 1646 (1961).

3 R. J. Glauber, *Phys. Rev.* **131**, 2766 (1963); reprinted as Sect. 2.9 in this volume.

4 E. Wigner, *Phys. Rev.* **40**, 749 (1932).

5 J. E. Moyal, *Proc. Cambridge Phil. Soc.* **45**, 99 (1948).

6 H. Weyl, *The Theory of Groups and Quantum Mechanics*, Dover Publications, New York 1950, pp. 272–276.

7 A. E. Glassgold, D. Holliday, *Phys. Rev. A* **139**, 1717 (1965), Sect. III.

8 Ref.[5], Sect. 3.

9 R. J. Glauber, in *Quantum Optics and Electronics*, C. de Witt, A. Blandin, C. Cohen-Tannoudji, Eds., Gordon & Breach, New York 1965, Lecture No. 13; reprinted as Sect. 2.11 in this volume.

10 W. H. Wells, *Ann. Phys. New York* **1**(12), 1961 (.)

11 R. P. Feynman, F. L. Vernon, Jr., *Ann. Phys. New York* **24**, 118 (1963).

12 B. R.Mollow, thesis, Harvard University 1966 (unpublished).

8
Photon Statistics[1]

8.1
Introduction

Statistical experiments on photon fields have only recently become a practical reality. To give some examples of what we mean by photon statistics let us think of some simple experiments which can be performed with a light source and a photon counter. We can imagine a shutter of some sort, either mechanical or electrical, to be placed between the source and the counter. We may open the shutter for a specified time and then close it again, observing how many photons are recorded by the counter in the interval. That number is clearly a random variable. Repetitions of the experiment will lead to a statistical distribution for the number of photons counted. A satisfactory theory should furnish us with a prediction of that distribution, its moments and other properties.

There are other kinds of experiments which can be performed with essentially the same apparatus. If the shutter is opened during two different time intervals, for example, we may ask for the joint distribution, or the degree of correlation of the random numbers of photons counted during the pairs of intervals. If we have a second photon counter available we may begin to inquire about spatial as well as temporal correlations. If the two counters are arranged to register in delayed coincidence with one another then we may measure the combined space-time correlations of their counting rates. The prediction of all of these statistical quantities should be the task of a well-formulated theory. But more important perhaps, the theory should be capable of dealing with a much broader variety of light sources than have been familiar in the past.

The light source in the experiment need not be of the usual gas discharge or incandescent type. It might be any of numerous varieties of laser oscillators or it might be one of the new types of amplifiers operating at optical frequencies. We might alternatively have a compound source which emits scattered light, the light from a laser, say, scattered by the density fluctuations of a fluid. All of these sources can emit fields which are rather different in character from one another. The best way of making these

---

[1] Reprinted, with permission from the author, from *Fundamental Problems in Statistical Mechanics II*, E. G. D. Cohen, Ed., North-Holland, Amsterdam 1968, pp. 140–187.

Quantum Theory of Optical Coherence. Selected Papers and Lectures. Roy J. Glauber.
Copyright © 2007 WILEY-VCH Verlag GmbH & Co. KGaA, Weinheim
ISBN: 978-3-527-40687-6

differences evident is through experiments of the kind we have noted, that is to say experiments which reveal a good deal more about the statistical properties of a field than the usual static measurements of its intensity and its spectrum.

It is only within roughly the last ten years that measurements of the sort we are discussing have actually been carried out. The first such experiment, which was performed by Hanbury Brown and Twiss[1], used a gas discharge tube as a source and revealed a distinct tendency for two photon detectors, placed in equivalent positions in the field, to register photons simultaneously. Many of the early experiments on photon statistics were in effect repetitions of this experiment with one or another improvement of method. More recently, as the technique of making statistical measurements upon fields has become refined and as the new coherent light sources have been developed, we have seen an increase in the number of such measurements and in their variety as well.

Since the applications of such experiments are likely to increase further it is worth developing a systematic way of discussing them. Let us begin therefore by noting that there are two rather different kinds of statistical uncertainty which are probed by each of the types of experiments we have mentioned. One kind of uncertainty concerns the state of the radiation source. The source is a macroscopic system which can rarely if ever be prepared in a pure quantum state. The best we can hope to do in general is to describe its uncertain state or the uncertain state of the field to which it leads by the methods of statistical mechanics and in that way to predict the associated fluctuations.

The second source of statistical uncertainty is inherent in the photodetection process. The photon counter is an intrinsically quantum mechanical instrument. Even if an incident field is fully predetermined in its behavior (a condition which can only be approached in the classical or strong field limit), the response of the photon counter to the field is never fully predictable. The random integers it records are numbers which are meaningless in the context of classical electromagnetic theory. The only language which can be used with logical consistency in discussing counting experiments is the language of quantum mechanics[2–6]. It will be evident nonetheless that many features of the quantum mechanical discussion exhibit simple correspondences with classical theory. In a sense what we propose to do is in fact to extend the classical theory of noise in electromagnetic fields down into the quantum domain and in so doing to take account also of the noise which is an unavoidable part of the detection process. It may provide some useful background therefore if we begin by saying a few words about classical noise theory.

8.2
Classical Theory

In order to be able to describe the electromagnetic field in terms of a discrete set of variables let us consider the field inside a volume of finite dimensions. A set of homogeneous boundary conditions which we need not specify for present purposes is assumed to constrain the fields at the surface; they could be periodic boundary conditions for example, or the conditions appropriate to reflecting (i.e. perfectly conducting) walls. For an appropriate set of frequencies ω_k we can find a sequence of vector mode functions $\boldsymbol{u}_k(\boldsymbol{r})$ which satisfy the Helmholtz equation

$$\left(\nabla^2 + \frac{\omega_k^2}{c^2}\right)\boldsymbol{u}_k(\boldsymbol{r}) = 0, \tag{8.1}$$

the transversality condition,

$$\nabla \cdot \boldsymbol{u}_k(\boldsymbol{r}) = 0, \tag{8.2}$$

and the boundary conditions on the electric fields. We can furthermore choose these functions to form a set which is complete and orthonormal,

$$\int \boldsymbol{u}_k^*(\boldsymbol{r}) \cdot \boldsymbol{u}_l(\boldsymbol{r})\, d\boldsymbol{r} = \delta_{kl}. \tag{8.3}$$

The corresponding solutions to the time dependent wave equation take the form $\boldsymbol{u}_k(\boldsymbol{r}) \exp[\pm i\omega_k t]$.

It is often convenient in classical theory to express the field variables as the sum of two complex conjugate terms. We may write the electric field vector, for example, as

$$\boldsymbol{E}(\boldsymbol{r}t) = \boldsymbol{E}^{(+)}(\boldsymbol{r}t) + \boldsymbol{E}^{(-)}(\boldsymbol{r}t), \tag{8.4}$$

where $\boldsymbol{E}^{(+)}$ is the positive frequency part of the field, i.e. the sum of all terms varying with time as $\exp[-i\omega_k t]$ for all $\omega_k > 0$, and $\boldsymbol{E}^{(-)}$ is the negative frequency part of the field,

$$\boldsymbol{E}^{(-)} = \left(\boldsymbol{E}^{(+)}\right)^*. \tag{8.5}$$

Let us now expand a component of the field $\boldsymbol{E}^{(+)}$ in terms of the set of mode functions u_k. To avoid the use of vector indices we can write the appropriate vector components as $E^{(+)}$ and u_k, respectively. If the Maxwell equations are homogeneous, i.e. all sources are turned off after they have radiated, then there must exist a set of complex constants C_k such that

$$E^{(+)}(\boldsymbol{r}t) = \sum_k C_k u_k(\boldsymbol{r}) \exp[-i\omega_k t]. \tag{8.6}$$

The set of all solutions to the Maxwell equations is described by giving all possible complex values to the C_k.

In actual practice we rarely know the set of numbers C_k with more than limited certainty; the fields we examine are usually radiated by systems whose behavior can only be described in statistical terms. The best we can do then is to describe our knowledge of the coefficients C_k by means of a probability distribution $P(\{C_k\})$ defined over the space of the C_k. The real and imaginary parts of each C_k can be varied independently, so that the differential element of the subspace corresponding to one coefficient is

$$d^2 C_k = d(\operatorname{Re} C_k) \, d(\operatorname{Im} C_k), \tag{8.7}$$

and the normalization condition on the probability distribution is

$$\int P(\{C_k\}) \prod_k d^2 C_k = 1. \tag{8.8}$$

When we measure the intensity of a field we are in effect measuring the ensemble average

$$\left\{ |E^{(+)}(\mathbf{r}t)|^2 \right\}_{\text{av}} = \left\{ |E^{(-)}(\mathbf{r}t) E^{(+)}(\mathbf{r}t)| \right\}_{\text{av}}. \tag{8.9}$$

In interference experiments one or another means is often used to superpose fields which correspond to different positions or times before the total field intensity is measured. In such cases the intensity may be expressed directly, as a rule, in terms of a correlation function for the field which we may define as

$$G_{\text{cl}}^{(1)}(\mathbf{r}t, \mathbf{r}'t') = \left\{ E^{(-)}(\mathbf{r}t) E^{(+)}(\mathbf{r}'t') \right\}_{\text{av}}. \tag{8.10}$$

To illustrate the meaning of the ensemble average more explicitly let us note that it is given by the integral

$$G_{\text{cl}}^{(1)}(\mathbf{r}t, \mathbf{r}'t') = \int P(\{C_k\}) E^{(-)}(\mathbf{r}t) E^{(+)}(\mathbf{r}'t') \prod_k d^2 C_k, \tag{8.11}$$

in which the field $E^{(+)}$ is a linear form in the $\{C_k\}$ and $E^{(-)}$ a linear form in the $\{C_k^*\}$.

When the coefficients C_k are regarded as random the function $E^{(+)}(\mathbf{r}t)$ becomes a species of time-dependent stochastic process. The detailed study of such processes is the province of classical noise theory and a vast literature exists on that subject. We shall see presently that a large part of quantum mechanical noise theory can be discussed in closely analogous terms.

8.3
Quantum Theory: Introduction

In quantum mechanics the fields $\mathbf{E}$ and $\mathbf{B}$ must be regarded not as numbers but as operators in the space of states which describe the field. The Hamiltonian operator for the free field is

$$H = \tfrac{1}{2} \int (\mathbf{E}^2 + \mathbf{B}^2) \, d\mathbf{r} \tag{8.12}$$

(in rationalized units). When sources are coupled to the field the vector $\boldsymbol{B}$ in this expression must be interpreted as the transverse part of the electric field, and the source and interaction Hamiltonians must be added to H.

The operator $\boldsymbol{E}$, like its classical counterpart, can be separated into its positive and negative frequency parts

$$\boldsymbol{E} = \boldsymbol{E}^{(+)} + \boldsymbol{E}^{(-)} . \tag{8.13}$$

The requirement that $\boldsymbol{E}$ be Hermitian is secured by requiring $\boldsymbol{E}^{(\pm)}$ to be mutually adjoint,

$$\boldsymbol{E}^{(-)} = \left(\boldsymbol{E}^{(+)}\right)^\dagger . \tag{8.14}$$

Since $\boldsymbol{E}^{(\pm)}$ represent intrinsically complex fields neither is Hermitian. It is perhaps worth remarking at this early point that many of the most familiar features of quantum mechanical calculations depend on the use of Hermitian operators. In order to make the most direct use of the non-Hermitian fields $\boldsymbol{E}^{(\pm)}$ we shall see that it is necessary to introduce some rather interesting and unfamiliar mathematical methods.

If the operator $E^{(+)}$, which represents a positive frequency field component, is expanded in terms of the set of mode functions discussed earlier, then the role of the classical Fourier coefficients $\{C_k\}$ must be taken by a sequence of quantum mechanical amplitude operators. It is conventional to write those amplitude operators as $\{a_k\}$ and to normalize them so that the expansion of $E^{(+)}$ takes the form

$$E^{(+)}(\boldsymbol{r}t) = i\sum_k \left(\tfrac{1}{2}\hbar\omega_k\right)^{1/2} a_k\, u_k(\boldsymbol{r})\, \exp[-i\omega_k t] . \tag{8.15}$$

The canonical commutation relations which the fields must satisfy then reduce to the simple relations

$$\begin{aligned}{[a_k, a_{k'}^\dagger]} &= \delta_{kk'} \\ [a_k, a_{k'}] &= [a_k^\dagger, a_{k'}^\dagger] = 0 \end{aligned} \tag{8.16}$$

between the operators $\{a_k\}$ and their adjoints. The operators a_k and $a_k^\dagger$ which correspond to any mode have exactly the same algebraic properties as the complex amplitude operators which are used to describe the quantum mechanical form of the simple harmonic oscillator. The commutation relations (8.16) thus define the amplitude operators for an infinite set of oscillators, one for each of the field modes.

When the fields in the Hamiltonian (8.12) are expressed in terms of the amplitude operators we find that the Hamiltonian reduces to the form

$$H = \sum_k \hbar\omega_k\, a_k^\dagger a_k + \text{constant} . \tag{8.17}$$

This operator, apart from the additive constant, is simply the Hamiltonian of a set of dynamically independent harmonic oscillators. What we have shown is just the observation with which Dirac began the development of quantum electrodynamics; the free electromagnetic field in a vacuum is equivalent in its dynamical properties to an infinite sequence of harmonic oscillators.

One of the familiar properties of the amplitude operators for harmonic oscillators is the fact that the products $a_k^\dagger a_k$ have as eigenvalues the integers $n_k = 0, 1, 2, \ldots$ It follows then from Eq. (8.17) that each mode has equally spaced energy levels; the integer n_k is evidently just the number of photons occupying the k-th mode.

The eigenstates of the Hamiltonian (8.17) may easily be constructed from oscillator eigenstates. The ground state of the field is defined as the state $|\text{vac}\rangle$ which satisfies the conditions

$$a_k |\text{vac}\rangle = 0 \tag{8.18}$$

for all modes k. It is clearly the state for which all of the n_k vanish; we can regard it simply as a product of ground state vectors of all of the individual mode oscillators.

Any set of integers $\{n_k\}$ which specifies the number of photons in each mode defines an eigenstate of the Hamiltonian. It is easy to show that this eigenstate, which we shall write as $|\{n_k\}\rangle$, can be generated from the vacuum state by writing

$$|\{n_k\}\rangle = \prod_k \frac{(a_k^\dagger)^{n_k}}{n_k!} |\text{vac}\rangle . \tag{8.19}$$

This relation illustrates one of the most important properties of the operators $a_k^\dagger$; they are photon creation operators. It follows from the commutation rules Eq. (8.16) that the adjoint operators a_k are photon annihilation operators. The definition of the vacuum state given by Eq. (8.18), for example, is simply the condition that no photons are available for annihilation.

The state vectors $|\{n_k\}\rangle$ for all values of the integers $\{n_k\}$ form a complete orthogonal set; the space they span is the familiar Fock space of quantum field theory. Although other sets of basis vectors are available in the Fock space virtually all quantum electrodynamical calculations to date have been based on the states $|\{n_k\}\rangle$. Their use is quite natural, for example, in calculations based on time-dependent perturbation theory, since they are after all the stationary states of the free-field Hamiltonian. Stationary states, however, are characteristically ones in which we lack all knowledge of the phase associated with the motion of a system. To deal with any phenomena which involve the phase of oscillation of the electromagnetic field, for example, we must be prepared to deal with superpositions of states $|\{n_k\}\rangle$ having different sets of quantum numbers. In general the more accurate and specific our phase information is, the more different states must be superposed. Nearly all of the applications of quantum electrodynamics to date have been to elementary particle problems and these have rarely required that we deal with more than one or two photons at a time; phase information, for example, tends to play a minimal role in such problems.

The classical limit of quantum electrodynamics represents the opposite extreme. In strong fields the quantum numbers n_k are not only large, they are typically quite uncertain. To use the states $|\{n_k\}\rangle$ in discussing the classical limit would be to obscure precisely what is simple about that limit. What we shall do instead is to make use of a rather different set of basis states, ones which are equally applicable to the discussion of weak and strong fields but which make the classical limit of the theory a great deal clearer. Before introducing these states, however, it will be useful to survey some of the quantities which can be measured in the simplest photon counting experiments and to use them in defining what we mean by the coherence properties of fields.

8.4
Intensity and Coincidence Measurements

Photon counters function by absorbing photons from the field. What they count are, strictly speaking, not photons but atomic photoabsorption processes rendered individually detectable by an amplification mechanism of some sort. In most of the devices used experimentally the fundamental absorption process is the photo-emission of an electron. The amplifying mechanism for the photo-electrons is usually a cascade multiplier. The precise way in which amplification is achieved need not concern us here however, since it plays no direct role in determining what the counter detects.

An ideal photon counter would be one which is quite small in size and has a sensitivity independent of photon frequencies (at least over the spectral range of the incident field). Both requirements can be met rather well, in fact, by the simplest possible model of a counter, a single atom which we observe to see whether and when it emits a photoelectron. Since the atom is quite small compared with the wavelength of visible light most of the transitions it can undergo may be treated by means of the electric dipole approximation. In this approximation the atom is coupled to the field through the interaction Hamiltonian

$$H_\mathrm{I} = -e\sum_{\gamma} \boldsymbol{q}_\gamma \cdot \boldsymbol{E}(\boldsymbol{r},t) , \qquad (8.20)$$

in which $\boldsymbol{q}_\gamma$ is the spatial coordinate of the γ-th electron of the atom relative to its nucleus which is located at $\boldsymbol{r}$.

The intensity of a field is proportional to the rate at which a counter records photons, or in the case of our single-atom counter to the probability per unit time of our observing a photoabsorption process. Let us suppose that the initial state of the field is given by the state vector $|i\rangle$ and that the atom is initially in its ground state. Then it is quite a simple matter to use first order perturbation theory to find the photoabsorption probability. There is no need for us to go through more than one or two of the details of that calculation here since we are only concerned with the way in which the field operators enter it. (The remaining details can be found for example in Sect. 2.4).

The transition amplitude, calculated in lowest order perturbation theory, is proportional simply to the matrix element of H_I between the initial and final states. If the final state of the field is $|f\rangle$ the contribution of the field to the amplitude is a factor $\langle f|E(\mathbf{r}t)|i\rangle$. Let us recall now that the field E is the sum of an annihilation operator $E^{(+)}$ (a linear combination of the $\{a_k\}$) and a creation operator $E^{(-)}$ which is its Hermitian adjoint. The fact that the atom began in its ground state means that it can only undergo transitions to states of higher energy. Since energy is conserved in these transitions (to an excellent approximation) photon emission processes are excluded The effect of energy conservation, in other words, is to insure that only the annihilation operator $E^{(+)}$ contributes to the desired transition amplitude which is thus proportional to $\langle f|E^{(+)}(\mathbf{r}t)|i\rangle$.

The final state of the field usually remains unobserved. The transition probability we seek is therefore proportional to a sum of the squared moduli of the matrix elements taken over a complete set of final states,

$$\sum_f \left|\langle f|E^{(+)}(\mathbf{r}t)|i\rangle\right|^2 = \langle i|E^{(-)}(\mathbf{r}t)E^{(+)}(\mathbf{r}t)|i\rangle. \tag{8.21}$$

The transition rate which we measure is proportional to the expectation value in the initial state of the field of the product $E^{(-)}E^{(+)}$ which we may note is a Hermitian operator with eigenvalue zero in the vacuum state.

The initial state of the field is rarely if ever a pure quantum state. Our knowledge of the many parameters upon which the field depends is usually quite incomplete and some or all of them must be regarded as random. All field measurements must be carried out therefore as averages over the ensemble of ways in which the field can be prepared. (The need to interpret measurements as ensemble averages is of course already intrinsic in our use of a quantum mechanical detector; there are simply more things we must average over because the source behaves randomly as well as the detector.)

The ensemble-averaged intensity is proportional to the expression Eq. (8.21) averaged with the appropriate weight over all possible initial states $|i\rangle$. If we introduce the density operator

$$\rho = \{|i\rangle\langle i|\}_{\text{av}} \tag{8.22}$$

to express our knowledge of the state of the field, we may write the average intensity as

$$\left\{\langle i|E^{(-)}(\mathbf{r}t)E^{(+)}(\mathbf{r}t)|i\rangle\right\}_{\text{av}} = \text{Tr}\left\{\rho E^{(-)}(\mathbf{r}t)E^{(+)}(\mathbf{r}t)\right\}, \tag{8.23}$$

where Tr stands for trace.

The average intensity is the particular value for $\mathbf{r} = \mathbf{r}'$ and $t = t'$ of a more general function

$$G^{(1)}(\mathbf{r}t,\mathbf{r}'t') = \text{Tr}\left\{\rho E^{(-)}(\mathbf{r}t)E^{(+)}(\mathbf{r}'t')\right\}, \tag{8.24}$$

which we shall call the first order correlation function for the field. It is a complex-valued function in general and satisfies wave equations in both the primed and unprimed variables. As we shall indicate in the next section it is in fact this function which we determine when we superpose the fields at two different points before measuring the field intensity.

There are two other details of the calculation of the photoabsorption probabilities which are worth at least mentioning. First, the electric field is of course a vector quantity and so the correlation function $G^{(1)}$ should strictly speaking be defined as a second rank tensor. Adding all of the tensor indices to the functions we shall discuss however would be a real encumbrance of the notation. It will be simpler instead to omit the indices and to understand that each specification of a space-time coordinate r,t carries with it implicitly the specification of a vector index. Alternatively the need for vector indices can be avoided altogether by assuming that our detectors are fitted with polarizing filters so that they are sensitive to only one component of the field.

A second point worth noting is that our rather schematic argument may appear not to have used the assumption that the sensitivity of the counter is independent of frequency. Implicit use of that assumption has in fact been made, however. It is only when the frequency sensitivity is flat that we can say in general that the transition probability at time t depends only on the field operators at time t and no other times. To give the argument more fully would require our going through the time-dependent perturbation calculation in some detail. It is sufficient for present purposes to note that when we do so[5] we find that the probability that an atom has undergone a detected photoabsorption process between times 0 and t can be written as

$$p^{(1)}(t) = s \int_0^t G^{(1)}(rt, r't') \, dt', \qquad (8.25)$$

where s, the sensitivity factor, has been assumed to be frequency-independent. (If, by contrast, the sensitivity varies appreciably with frequency then the transition probability is given by a double time integral which involves the correlation function $G^{(1)}$ at pairs of unequal time arguments.)

Let us suppose now that we use not just one atom as a detector but n atoms whose nuclei are at the positions $r_1 \ldots r_n$. We want to calculate the probability that all n of the atoms undergo detected photoabsorption process, more specifically that atom 1 does so between times 0 and t_1, atom 2 between times 0 and t_2 etc. The calculation must be carried out in n-th order perturbation theory but it differs from carrying out the first order perturbation calculation n times in only one essential respect. The absorption of n photons is regarded as a single compound process. No observations are made of the state of the field in between the individual absorption processes. Hence there is no way of telling in general which atom has absorbed which photon. The amplitudes corresponding to processes in which the photons are permuted contribute coherently in general and interfere with one another.

On the basis of the same assumptions that we made in discussing the single atom it is easy to show (Sect. 2.5) that the compound transition probability can be written as

$$p^{(n)}(t_1 \ldots t_n) = s^n \int_0^{t_1} \ldots \int_0^{t_n} G^{(n)}(r_1 t_1' \ldots r_n t_n', r_n t_n' \ldots r_1 t_1') \prod_j dt_j', \qquad (8.26)$$

where the function $G^{(n)}$ is the n-th order correlation function defined by

$$G^{(n)}(r_1 t_1 \ldots r_n t_n, r_{n+1} t_{n+1} \ldots r_{2n} t_{2n})$$
$$= \mathrm{Tr}\left\{\rho E^{(-)}(r_1 t_1) \ldots E^{(-)}(r_n t_n) E^{(+)}(r_{n+1} t_{n+1}) \ldots E^{(+)}(r_{2n} t_{2n})\right\}. \qquad (8.27)$$

The n atoms may be regarded as n independent counters consisting of a single atom each. In that case their joint counting rate corresponding to the set of times $t_1 \ldots t_n$ is given by

$$w(t_1 \ldots t_n) \equiv \frac{\partial^n}{\partial t_1 \ldots \partial t_n} p^{(n)}(t_1 \ldots t_n)$$
$$= s^n G^{(n)}(r_1 t_1 \ldots r_n t_n, r_n t_n \ldots r_1 t_1). \qquad (8.28)$$

For $n = 1$ the function is simply the intensity noted earlier. Evidently for $n > 1$ the function $G^{(n)}$ with its $2n$ space-time arguments set equal in pairs tells us a time-delayed joint counting rate, one which we could measure in principle by using n-fold coincidence counting techniques. Two-fold coincidence experiments of the type which measure the function $G^{(2)}$ are not especially difficult to perform and a considerable number of them have been carried out in the last few years. Triple and higher order coincidence experiments are considerably more difficult, however, and the functions $G^{(n)}$ for $n \geq 3$ are best evaluated in practice by other means. They have an important bearing as we shall see on the statistical distributions of counted photons.

One of the properties that the correlation functions $G^{(n)}$ all share is that they are the averages of operator products which are ordered in a similar way. This scheme, usually referred to as normal order, is one in which annihilation operators (factors of $E^{(+)}$) precede creation operators (factors of $E^{(-)}$) when we read from right to left. The occurrence of this particular ordering is a direct consequence of the fact that we are discussing experiments in which photons are absorbed. There are other ways in which photons could be detected in principle, even if they would be rather hard to apply in practice. The presence of photons of the appropriate frequency, for example, will tend to stimulate the radiative decay of an initially excited atom. It is easy to see that the correlation functions which would be measured by detectors based in this way on stimulated emission[7] are ones in which the roles of the operators $E^{(+)}$ and $E^{(-)}$ are interchanged, that is to say they would be the expectation values of antinormally ordered products. Still other experiments could be devised to measure the averages of symmetrically ordered products, etc. The expectation values which correspond to other orderings can all be found, of course, from the set of normally ordered ones by applying the commutation rules for the creation and annihilation operators.

8.5 First and Higher Order Coherence

To illustrate the meaning of the correlation function $G^{(n)}$ let us imagine an interference experiment similar to the famous one carried out by Young. We let a beam of light fall on an opaque screen with two tiny identical pinholes one at position r_1 the other at r_2. Some distance behind the screen at a position r we place a photon counter. If the distances from the counter to the two pinholes are s_1 and s_2 respectively then the field which arrives at the counter at time t is a superposition of two fields, a field proportional to the one at pinhole 1 at time $t_1 = t - (s_1/c)$ and another proportional to the field at pinhole 2 at time $t_2 = t - (s_2/c)$. The intensity detected by the counter,

$$I(rt) = G^{(1)}(rt, rt)$$
$$= \text{Tr}\left\{\rho E^{(-)}(rt) E^{(+)}(rt)\right\},$$

is therefore proportional to

$$Tr\left\{\rho \left[E^{(-)}(x_1) + E^{(-)}(x_2)\right]\left[E^{(+)}(x_1) + E^{(+)}(x_2)\right]\right\} = G^{(1)}(x_1, x_1) + G^{(1)}(x_2, x_2) + 2\,\text{Re}\, G^{(1)}(x_1, x_2), \quad (8.29)$$

where we have written x_1 for the coordinates r_1, t_1 and x_2 for r_2, t_2.

It is clear that in the expression (8.29) the terms $G^{(1)}(x_1, x_1)$ and $G^{(1)}(x_2, x_2)$ represent the intensities which would be contributed by each pinhole in the absence of the other. The remaining cross-correlation term is the contribution of interference to the total intensity. If we write the complex function $G^{(1)}$ as $|G^{(1)}|\exp(i\varphi)$ where φ is an appropriate phase function then the interference term is

$$2\,\text{Re}\, G^{(1)}(x_1, x_2) = 2|G^{(1)}(x_1, x_2)|\cos\varphi. \quad (8.30)$$

The oscillations of $\cos\varphi$ correspond to the intensity fringes which are observed as the counter is moved about behind the screen. The degree of contrast of these fringes is clearly governed by the function $|G^{(1)}(x_1, x_2)|$. If the fields incident on the two pinholes were uncorrelated in any way, for example, then we should have $G^{(1)}(x_1, x_2) = 0$ and no fringes present. The two incident fields are usually referred to in that case as being incoherent.

We can define the opposite condition, that is to say coherence, by asking how great the contrast of the fringe pattern can be made. Since $|\cos\varphi|$ never exceeds one the contrast of the fringes is maximized by making the function $|G^{(1)}(x_1, x_2)|$

as large as possible. It is easy to see that this function possesses an upper bound however, since the total intensity given by Eq. (8.29) can never become negative, nor can the intensity in any sort of superposition experiment with arbitrary superposition coefficients. This positive definiteness condition on the intensity shows us directly (see Ref.[3], for example) that the function $G^{(1)}$ obeys the Schwarz inequality

$$G^{(1)}(x_1,x_1)\, G^{(1)}(x_2,x_2) \geq \left|G^{(1)}(x_1,x_2)\right|^2. \tag{8.31}$$

The condition for maximum fringe contrast therefore corresponds to the bound

$$G^{(1)}(x_1,x_1)\, G^{(1)}(x_2,x_2) = \left|G^{(1)}(x_1,x_2)\right|^2. \tag{8.32}$$

Let us now consider the set of fields for which the function $G^{(1)}(x_1,x_2)$ factorizes into the product of a function of x_1 and a function of x_2. We can easily use the general properties of $G^{(1)}$ to show that any such factorized form must be expressible as

$$G^{(1)}(x_1,x_2) = \mathcal{E}^*(x_1)\, \mathcal{E}(x_2), \tag{8.33}$$

where the function $\mathcal{E}(x)$ is a solution of the Maxwell equations containing positive frequency components only. If the correlation function factorizes in this way it is obvious that it satisfies the condition Eq. (8.32) for maximum fringe contrast. Factorization of the function $G^{(1)}$, in other words, implies precise coherence in the sense which has become familiar in optical interference experiments.

Not only does the factorization condition Eq. (8.33) imply maximum fringe contrast, but the converse statement holds as well. By making use of no relations more complicated than the Schwarz inequality it is possible to show[8] that if the maximum fringe contrast condition holds for all x_1 and x_2 then the first order correlation function must fall into the factorized form of Eq. (8.33). The factorization condition on $G^{(1)}$ and the maximum fringe contrast conditions in other words furnish equivalent definitions of fields which are optically coherent. Since this form of coherence constrains only the function $G^{(1)}$ we shall refer to it as first order coherence.

By analogy with the factorization condition for first order coherence we may construct a succession of further factorization conditions which require that the higher order correlation functions take the form

$$G^{(n)}(x_1\ldots x_{2n}) = \mathcal{E}^*(x_1)\ldots\mathcal{E}^*(x_n)\, \mathcal{E}(x_{n+1})\ldots\mathcal{E}(x_{2n}), \tag{8.34}$$

in which the function $\mathcal{E}(x)$ is the same for all values of n. If the functions $G^{(n)}$ satisfy these conditions for $j \leq n$ we shall speak of the field as having n-th order coherence. For $n > 1$ this is of course a much more restrictive definition than first order coherence, which it implies. If the condition Eq. (8.34) holds for all values of n we shall speak of the field as being fully coherent.

That fully coherent fields do exist becomes immediately evident if we consider the classical limit. For any classical field which has completely predetermined behavior, i.e. no statistical uncertainty, the probability density $\mathcal{P}(\{C_k\})$ is a delta function and the correlation functions $G_{cl}^{(n)}$ all factorize into the form of Eq. (8.34). The function $\mathcal{E}(x)$ is then simply the positive frequency part of the field itself. This simple example also illustrates a close relation between coherence and noiselessness; the sense in which we are extending the optical definition of the term "coherence" corresponds closely to the sense in which pure signals are referred to as "coherent" in communication theory. In the next section we shall show that there also exists a considerable variety of quantum mechanical fields which satisfy the conditions for full coherence.

The higher order coherence properties of a field have a direct bearing on the results of coincidence counting experiments performed in it. The n-fold joint counting rate, as we have noted in Eq. (8.28), is given by an n-th order correlation function with its space-time arguments occurring in repeated pairs. If the field possesses coherence of order $m \geq n$ then the joint counting rate obeys the identity

$$G^{(n)}(x_1 \ldots x_n, x_n \ldots x_1) = \prod_{j=1}^{n} G^{(1)}(x_j, x_j) ; \qquad (8.35)$$

that is to say the joint counting rate is equal to the product of the counting rates which would be measured by each of the n counters in the absence of all the others. The responses of the counters in other words, are statistically independent of one another; if the average intensity of the field is independent of time for example, the counters detect no tendency toward any sort of correlation in the arrival times of photons.

Fields which obey the higher order coherence conditions can only be provided by rather special sources in general. We have already noted that when a two-fold coincidence experiment is performed with an ordinary source such as a gas discharge tube the counters detect a distinct tendency for the photons to arrive in correlated pairs. That result means that the field of an ordinary source never has second or higher order coherence. When twofold coincidence experiments are performed in the output beam of a stably operating laser however, the results are rather different; they do indeed show statistical independence in the responses of the counters. It is now a good deal more than a surmise[2] that laser beams can possess coherence of quite a high order; all present theories of the laser indicate that and several experimental results as well.

High order coherence implies the existence of detailed space-time correlations in the behavior of the field amplitude, so it may perhaps be a little surprising that it also implies the absence of photon counting correlations. One of the things the higher order coherence conditions do is to restrict greatly the amount of randomness or noisiness which can be present in the field. As we shall see presently the photon correlations which lead to breakdown of relations such as Eq. (8.35) are an expression of noisiness of the field, precisely the sort of noisiness which higher order coherence eliminates.

A number of simple theorems which characterize fields possessing first and higher order coherence have been proved by Titulaer and Glauber[8,11]. A few of them are perhaps worth mentioning here:

1. The condition for first order coherence, if it holds at all space-time points, goes a long way toward implying the higher order coherence conditions Eq. (8.34). It implies that the higher order correlation functions factorize into the forms

$$G^{(n)}(x_1\ldots x_{2n}) = g_n\, \mathcal{E}^*(x_1)\ldots \mathcal{E}^*(x_n)\, \mathcal{E}(x_{n+1})\ldots \mathcal{E}(x_{2n}), \qquad (8.36)$$

which differ in general form those of Eq. (8.34) by the presence of the constant factor g_n. The numbers of g_n are real and non-negative ($g_1 \equiv 1$). The condition for full coherence is that they all be equal to one.

2. By using linear combinations of the solutions $u_k(r)\exp[-i\omega_k t]$ of the wave equation we can form other sets of mode functions which need not in general be monochromatic. By taking suitable linear combinations of the operators, a_k, we can form annihilation operators appropriate to these generalized field modes. The first order coherence condition is equivalent to the requirement that there be a single generalized mode of the field which contains all of the photons present; all orthogonal modes are empty. Precise first order coherence, in other words, means that all of the photons in the field have identical wavepackets.

3. The higher order coherence conditions put constraints only on the statistical distribution of the number of identical photons present; i.e. on the diagonal elements of the density operator in the representation based on the appropriate choice of a generalized mode. In particular the conditions for full coherence of the field simply require that there be a Poisson distribution of the identical photons present. We shall shortly see an illustration of this theorem.

8.6
The Coherent States

Let us try to construct some of the states of the field which satisfy the conditions for full coherence. We shall begin by seeking pure states since they are likely to be the simplest. For a pure state represented by the state vector $|\ \rangle$ the density operator is

$$\rho = |\ \rangle\langle\ |, \qquad (8.37)$$

and the n-th order correlation function is

$$G^{(n)}(x_1\ldots x_{2n}) = \langle\ |E^{(-)}(x_1)\ldots E^{(-)}(x_n)\, E^{(+)}(x_{n+1})\ldots E^{(+)}(x_{2n})|\ \rangle. \qquad (8.38)$$

We must now ask how we can choose the state $|\ \rangle$ so that $G^{(n)}$ separates into the factorized form

$$G^{(n)}(x_1\ldots x_{2n}) = \mathcal{E}^*(x_1)\ldots \mathcal{E}^*(x_n)\, \mathcal{E}(x_{n+1})\ldots \mathcal{E}(x_{2n}). \qquad (8.39)$$

The fact that the fields $E^{(+)}$ and $E^{(-)}$ do not commute means that we cannot solve the problem by finding states $|\ \rangle$ which are simultaneously eigenstates of $E^{(+)}$ and $E^{(-)}$; no such states exist.

What we can do instead is to exploit the fact that the operator products occurring in the correlation functions are all in normally ordered form. It is quite sufficient, for that reason, to require only that the unknown state satisfy the condition

$$E^{(+)}(rt)|\ \rangle = \mathcal{E}(rt)|\ \rangle , \tag{8.40}$$

since that condition implies the adjoint relation

$$\langle\ |E^{(-)}(rt) = \mathcal{E}^*(rt)\langle\ | , \tag{8.41}$$

and the two relations together imply that the functions $G^{(n)}$ factorize to precisely the form of Eq. (8.39).

The function $\mathcal{E}(r,t)$ must be a positive frequency solution of the Maxwell equations which satisfies the same boundary conditions as the field $E^{(+)}(rt)$. We can thus expand it in terms of the mode functions, just as we expanded $E^{(+)}$ in Eq. (8.15) by writing

$$\mathcal{E}(rt) = i\sum_k \left(\tfrac{1}{2}\hbar\omega_k\right)^{1/2} \alpha_k\, u_k(r)\, \exp[-i\omega_k t] , \tag{8.42}$$

where the α_k are a suitable set of complex numbers. Since the mode functions $u_k(r)$ form an orthogonal set the eigenstate condition Eq. (8.40) implies that we must have

$$a_k|\ \rangle = \alpha_k|\ \rangle \tag{8.43}$$

for all values of the mode index k.

Let us define for the k-th mode alone a state $|\alpha_k\rangle_k$ with the property

$$a_k|\alpha_k\rangle_k = \alpha_k|\alpha_k\rangle_k . \tag{8.44}$$

Once we have found such states for the individual modes we can use their products to form the states

$$|\{\alpha_k\}\rangle = \prod_k |\alpha_k\rangle_k , \tag{8.45}$$

which satisfy Eq. (8.43)) and therefore obey the full set of coherence conditions. We shall refer to these as coherent states of the field.

One problem is now to find the state of each mode which satisfies the condition Eq. (8.44). Since the problem is essentially the same for all modes we can consider a single arbitrary mode and drop the use of its index k. We want then to find the harmonic oscillator state $|\alpha\rangle$ which satisfies the condition

$$a|\alpha\rangle = \alpha|\alpha\rangle . \tag{8.46}$$

One of the simplest ways of solving for the state is to expand it in terms of the n-quantum states. It is clear that the state $|\alpha\rangle$ can not have any precisely defined value of n, in general, since Eq. (8.46) requires that the state remain unchanged when a quantum is removed. To find the appropriate expansion we note the relations

$$a^\dagger |n\rangle = \sqrt{n+1}\,|n+1\rangle ,$$
$$\langle n|a = \sqrt{n+1}\,\langle n+1| ,$$

which follow from the commutation rules (8.16) and the definition of the states $|n\rangle$. By using the latter of these relations we may write Eq. (8.46) in the form

$$\langle n|a|\alpha\rangle = \sqrt{n+1}\,\langle n+1|\alpha\rangle = \alpha \langle n|\alpha\rangle ,$$

which implies

$$\langle n|\alpha\rangle = \frac{\alpha}{\sqrt{n}} \langle n-1|\alpha\rangle$$

and therefore

$$\langle n|\alpha\rangle = \frac{\alpha^n}{\sqrt{n!}} \langle 0|\alpha\rangle . \tag{8.47}$$

It follows that the expansion of the state $|\alpha\rangle$ takes the form

$$|\alpha\rangle = \sum_n |n\rangle \langle n|\alpha\rangle$$
$$= \langle 0|\alpha\rangle \sum_n \frac{\alpha^n}{\sqrt{n!}} |n\rangle . \tag{8.48}$$

Hence by fixing the arbitrary phase factor so that

$$\langle 0|\alpha\rangle = \exp\left[-\tfrac{1}{2}|\alpha|^2\right] , \tag{8.49}$$

we find that a coherent state of a single mode is given by the expansion

$$|\alpha\rangle = \exp\left[-\tfrac{1}{2}|\alpha|^2\right] \sum_n \frac{\alpha^n}{\sqrt{n!}} |n\rangle . \tag{8.50}$$

It is worth noting in particular that the state $|n=0\rangle$ is identical to the state $|\alpha=0\rangle$; the ground state of a mode is one of the coherent states.

Clearly a coherent state of a mode can be constructed for all values of the complex parameter α. Hence a coherent state of the field can be constructed, at least formally, for any sequence of complex parameters $\{\alpha_k\}$ or equivalently any eigenvalue function $\mathcal{E}(\mathbf{r},t)$. It follows then that there is a one-to-one correspondence between the positive frequency solutions to the Maxwell equations $\mathcal{E}(\mathbf{r},t)$ and the coherent states of the field $|\{\alpha_k\}\rangle$. There are just as many coherent states as there are classical (i.e. non-operator) solutions to the Maxwell equations.

8.6 The Coherent States

When a mode is in a coherent state the probability that it contains n quanta is

$$|\langle n|\alpha\rangle|^2 = \frac{|\alpha|^{2n}}{n!} \exp\left[-|\alpha|^2\right], \tag{8.51}$$

which is a Poisson distribution, in agreement with the theorem stated at the end of the last section. Indeed when each mode has a Poisson distribution it is easy to see that the total photon population of the field also has one. The mean number of quanta in the distribution Eq. (8.51) is

$$\langle n \rangle = |\alpha|^2. \tag{8.52}$$

The mode excitations which are strong enough to correspond to the limit of classical theory must therefore have $|\alpha|^2 \gg 1$. The classical fields to which these states correspond can be found by comparing the expansion Eq. (8.6) for the classical field $E^{(+)}$ with the expansion Eq. (8.42) for the eigenvalue field $\mathcal{E}(r,t)$. We see then the correspondence

$$C_k \leftrightarrow i\left(\tfrac{1}{2}\hbar\omega_k\right)^{1/2}\alpha_k \tag{8.53}$$

between the classical expansion coefficients C_k and the coherent state amplitudes α_k. Since $\hbar$ is in effect negligibly small for classical fields we see once again that the classical limit requires $|\alpha_k|^2 \gg 1$.

While the expansion (8.50) tells us most of what we need to know about the coherent states, there is a much more compact way of generating them which is often useful. Let us define a unitary operator $D(\beta)$ which carries out the translation

$$D^{-1}(\beta)\,a\,D(\beta) = a + \beta \tag{8.54}$$

$$D^{-1}(\beta)\,a^\dagger\,D(\beta) = a^\dagger + \beta^* \tag{8.55}$$

on the operators a and $a^\dagger$. It is easy to verify by using the commutation relations that a suitable operator is

$$D(\beta) = \exp[\beta a^\dagger - \beta^* a] \tag{8.56}$$

and its unitarity is evident from the relations

$$D(-\beta) = D^{-1}(\beta) = D^\dagger(\beta). \tag{8.57}$$

Now if we begin with the definition

$$a|\alpha\rangle = \alpha|\alpha\rangle$$

of a coherent state, multiply both sides of the equation by $D^{-1}(\alpha)$ and then use the relation Eq. (8.54), we see that

$$a D^{-1}(\alpha)|\alpha\rangle = 0. \tag{8.58}$$

Since the ground state of the oscillator is the unique solution of Eq. (8.58) we must have

$$D^{-1}(\alpha)|\alpha\rangle = |0\rangle,$$

or

$$|\alpha\rangle = D(\alpha)|0\rangle. \tag{8.59}$$

The coherent states, in other words, are just displaced versions of the ground state of the harmonic oscillator. When we introduce position and momentum operators for the oscillator we will see that the ground state wavepacket is displaced in general both in position and momentum space. It was because of this simple property that the states were first discussed by Schrödinger[9] in 1926. The coherent states have an extremely simple form of time dependence. If we subtract the additive constant from the Hamiltonian (8.17) (a change which only alters a phase factor in the state vector) we find that the time dependent form of a coherent state is

$$\begin{aligned}|\alpha,t\rangle &= \exp\left[-i\frac{Ht}{\hbar}\right]|\alpha\rangle \\ &= \exp\left[-ia^\dagger a\omega t\right]\exp\left[-\tfrac{1}{2}|\alpha|^2\right]\sum_n \frac{\alpha^n}{\sqrt{n!}}|n\rangle \\ &= |\alpha\exp[-i\omega t]\rangle.\end{aligned} \tag{8.60}$$

The state remains a coherent one at all times. Its complex amplitude, $\alpha(t) = \alpha\exp[-i\omega t]$, simply describes circles in the complex plane; the projections of this motion upon the real or imaginary axes form the familiar picture of simple harmonic motion.

We may digress for a moment by asking a somewhat broader question about time-dependent states. If the field Hamiltonian were different from that of a set of free harmonic oscillators we could not expect in general that a state which is initially coherent would remain so at later times. Are there any other Hamiltonians that do have the property of preserving the coherent character of states which are initially coherent? The answer is that there are a good many, including ones for which the oscillators are coupled. It is easy to show[10] that all Hamiltonians taking the form

$$H = \sum_{jk} a_j^\dagger f_{jk}(t) a_k + \sum_k \left\{g_k(t) a_k^\dagger + g_k^*(t) a_k\right\}, \tag{8.61}$$

where $f_{jk} = f_{kj}^*$ and the f_{jk} and g_k are arbitrary functions of time, have the desired property. These Hamiltonians in fact form the most general class (apart from the trivial possibility of adding functions of the time to H) which preserves the coherent character of state. Whenever the Hamiltonian takes the form of Eq. (8.61) the amplitudes $\{\alpha_k(t)\}$ of the time-dependent coherent states carry out the motion one would expect in a classical theory[10]. The wavepacket of the system in such states is always similar to that of the ground state of the free oscillators and simply moves along the classical trajectories without spreading.

It may help in forming a picture of the coherent states to construct their wave functions in position and momentum space. The position q and momentum p which we introduce for this purpose can be defined by

$$q = \left(\frac{\hbar}{2\omega}\right)^{1/2}(a^\dagger + a), \tag{8.62}$$

$$p = i\left(\frac{\hbar\omega}{2}\right)^{1/2}(a^\dagger - a). \tag{8.63}$$

The condition which defines a coherent state can then be written as

$$a|\alpha\rangle = \frac{1}{(2\hbar\omega)^{1/2}}(\omega q + ip)|\alpha\rangle = \alpha|\alpha\rangle. \tag{8.64}$$

In the representation in which q is diagonal this relation is the first order differential equation

$$\frac{1}{(2\hbar\omega)^{1/2}}\left(\omega q' + \hbar\frac{d}{dq'}\right)\langle q'|\alpha\rangle = \alpha\langle q'|\alpha\rangle. \tag{8.65}$$

It is convenient to write the wave function which solves the equation in the form

$$\langle q'|\alpha\rangle = \left(\frac{\omega}{\pi\hbar}\right)^{1/4}\exp\left[-\left(\left(\frac{\omega}{2\hbar}\right)^{1/2}q' - \alpha\right)^2 + \tfrac{1}{2}\alpha(\alpha - \alpha^*)\right]. \tag{8.66}$$

In so doing we have normalized the function so that

$$\int_{-\infty}^{\infty}|\langle q'|\alpha\rangle|^2\, dq' = 1 \tag{8.67}$$

and chosen its phase in a special way which we shall explain presently. (What we have done is to require that $\exp[\tfrac{1}{2}|\alpha|^2]\langle q'|\alpha\rangle$ be a function of α and not of α^*.)

By an analogous calculation in the momentum representation we can show that the momentum space wave function for a coherent state is

$$\langle p'|\alpha\rangle = \frac{1}{(\pi\hbar\omega)^{1/4}}\exp\left[-\left(\frac{p}{(2\hbar\omega)^{1/2}} + i\alpha\right)^2 - \tfrac{1}{2}\alpha(\alpha + \alpha^*)\right]. \tag{8.68}$$

(The phase of this wave function has been chosen so that $\exp[\tfrac{1}{2}|\alpha|^2]\langle p'|\alpha\rangle$ is independent of α^*).

If $\langle q\rangle$ and $\langle p\rangle$ are the mean values of the coordinate and momentum of a wavepacket then the variances of these quantities are defined as

$$(\Delta q)^2 = \langle q^2\rangle - \langle q\rangle^2, \tag{8.69}$$

$$(\Delta p)^2 = \langle p^2\rangle - \langle p\rangle^2. \tag{8.70}$$

The uncertainty principle requires that for all quantum states we have

$$(\Delta q)^2 (\Delta p)^2 \geq \tfrac{1}{4}\hbar^2 . \tag{8.71}$$

It is easy to see from the wave functions Eq. (8.66) and (8.68)) that for a coherent state we have

$$(\Delta p)^2 = \omega^2 (\Delta q)^2 = \tfrac{1}{2}\hbar\omega \tag{8.72}$$

and hence

$$(\Delta q)^2 (\Delta p)^2 = \tfrac{1}{4}\hbar^2 . \tag{8.73}$$

The coherent states, in other words, reduce the uncertainty product to its minimum value and in that sense are as nearly classical in character as the quantum theory allows any state to be.

There exist other wavepackets which satisfy the minimum uncertainty condition Eq. (8.73), in fact a whole family of them often referred to as Kennard[12] packets. The familiar derivation of the uncertainty principle[13] shows that the condition for the state $|\ \rangle$ to have the minimal value for $(\Delta q)^2 (\Delta p)^2$ is that it satisfy the eigenstate relation

$$(p - \langle p \rangle)|\ \rangle = i\mu(q - \langle q \rangle)|\ \rangle , \tag{8.74}$$

where μ is an arbitrary real positive constant. The ratio

$$\frac{(\Delta p)^2}{(\Delta q)^2} = \mu^2 \tag{8.75}$$

therefore may assume any positive value for these states. Only when $\mu = \omega$ is the state a coherent one in the sense that it is an eigenstate of the annihilation operator $a = (2\hbar\omega)^{-1/2}(\omega q + ip)$. It is clear from Eq. (8.74), on the other hand, that each member of the family of states is the eigenstate of an operator which we can define as

$$A(\mu) = (2\hbar\mu)^{-1/2}(\mu q + ip) \tag{8.76}$$

$$= \tfrac{1}{2}(\mu\omega)^{-1/2}\left\{(\mu + \omega)a + (\mu - \omega)a^\dagger\right\} . \tag{8.77}$$

The operator $A(\mu)$ can be regarded as a species of annihilation operator, but for an oscillator with frequency μ rather than ω. We see thus that the minimum uncertainty states, as a broad class, are quite similar to the coherent states; their wavepackets differ from those of the coherent states in general only by an obvious scale transformation, but their physical interpretation as harmonic oscillator states is rather different for $\mu \neq \omega$.

A somewhat different sense in which the coherent states are minimum uncertainty states can be constructed as a corollary of the uncertainty principle. The arithmetic mean-geometric mean inequality tells us that

$$\tfrac{1}{2}\left\{(\Delta p)^2 + \omega^2 (\Delta q)^2\right\} \geq \omega \left\{(\Delta q)^2 (\Delta p)^2\right\}^{1/2}$$

and hence

$$\tfrac{1}{2}\{(\Delta p)^2 + \omega^2 (\Delta q)^2\} \geq \tfrac{1}{2}\hbar\omega . \tag{8.78}$$

The lower bound for the uncertainty in this relation is only attained for the coherent states of the oscillator. That is clear, for example, if $\langle q \rangle = 0$ and $\langle p \rangle = 0$ since the lower bound condition is then the definition of the ground state of the oscillator. If $\langle q \rangle$ and $\langle p \rangle$ are different from zero we can always find a unitary operator $D(\alpha)$ which displaces the state without changing its variances $(\Delta q)^2$ and $(\Delta p)^2$ so that $\langle q \rangle$ and $\langle p \rangle$ are reduced to zero. It follows then that the only states which satisfy the minimum condition in Eq. (8.78) are displaced forms of the ground state, or coherent states as we see from Eq. (8.59).

8.7
Expansions in Terms of the Coherent States

The coherent states do not form an orthogonal set as the eigenstates of Hermitian operators do. We easily see from the expansion Eq. (8.50) for a coherent state of a single mode that the scalar product of two such state vectors is

$$\langle \alpha | \beta \rangle = \exp\left[\alpha^* \beta - \tfrac{1}{2}|\alpha|^2 - \tfrac{1}{2}|\beta|^2\right] .$$

The squared modulus of this expression,

$$|\langle \alpha | \beta \rangle|^2 = \exp\left[-|\alpha - \beta|^2\right] ,$$

shows us, in fact, that no two coherent states are completely orthogonal, even though their scalar product is extremely small for $|\alpha - \beta| \gg 1$.

The coherent states do, however, form a complete set. There exists a very simple way of resolving the unit operator into a sum of projection operators of the form $|\alpha\rangle\langle\alpha|$, and that is all we need to demonstrate completeness[14,4]. To show this we begin by writing the (real) element of area of the complex plane as

$$d^2\alpha = d(\operatorname{Re}\alpha)\, d(\operatorname{Im}\alpha) \tag{8.79}$$

and noting the integral identity

$$\int (\alpha^*)^n \alpha^m \exp\left[-|\alpha|^2\right] d^2\alpha = 2\pi \delta_{nm} \int_0^\infty |\alpha|^{2n+1} \exp\left[-|\alpha|^2\right] d|\alpha|$$
$$= \pi n!\, \delta_{nm} . \tag{8.80}$$

It is clear therefore, from the expansion Eq. (8.50), that if we integrate the dyadic $|\alpha\rangle\langle\alpha|$ over the complex plane we have

$$\frac{1}{\pi}|\alpha\rangle\langle\alpha|\,d^2\alpha = \frac{1}{\pi}\sum_{nm}(n!\,m!)^{-1/2}\int(\alpha^*)^n\alpha^m\exp\left[-|\alpha|^2\right]d^2\alpha\,|m\rangle\langle n|$$
$$= \sum_n |n\rangle\langle n|$$
$$= 1. \tag{8.81}$$

The completeness of the coherent states follows in the last step from the known completeness of the n-quantum states.

The completeness relation assures us that the set of coherent states is sufficiently large so that any state vector can be expanded in terms of them. The fact of completeness does not tell us, however, that the expansion can be carried out in only one way; the set of states may be larger than is necessary for completeness and some of its states may be expressible as linear combinations of others. The two-dimensional continuum of coherent states $|\alpha\rangle$ is in fact vastly overcomplete in this sense. Discrete sequences of the coherent states can be shown to be sufficient to form complete sets[15], so there must be a great deal of linear dependence present in the full two-dimensional set. We shall exhibit some of this linear dependence shortly.

An arbitrary state $|\ \rangle$ of a single mode can be expanded in terms of the n-quantum states by writing

$$|\ \rangle = \sum_n f_n |n\rangle$$
$$= \sum_n \frac{f_n(a^\dagger)^n}{(n!)^{1/2}}|0\rangle, \tag{8.82}$$

where the coefficients f_n obey the normalization condition $\sum |f_n|^2 = 1$. We can use the series which occurs in Eq. (8.82) to define a function of a complex variable z as follows:

$$f(z) = \sum_{n=0}^{\infty}\frac{f_n z^n}{(n!)^{1/2}}. \tag{8.83}$$

It is clear from the normalization condition on the f_n that this series converges for all values of z in the finite plane, and therefore represents an entire function. Bargmann[16] has discussed the representation of the quantum states of an oscillator which is furnished by these functions. He has shown that the functions form the elements of a Hilbert space and has developed rules for calculating with them. We can make use of the function f to write the arbitrary state in Eq. (8.82) as

$$|\ \rangle = f(a^\dagger)|0\rangle. \tag{8.84}$$

8.7 Expansions in Terms of the Coherent States

To expand the arbitrary state in terms of the coherent states we can multiply it from the left by the resolution of the unit operator given by Eq. (8.81). We then have

$$|\ \rangle = \frac{1}{\pi}\int |\beta\rangle \langle\beta|\ \rangle\, d^2\beta, \tag{8.85}$$

and the expansion coefficient in the integrand is given by

$$\langle\beta|\ \rangle = \langle\beta|f(a^\dagger)|0\rangle$$
$$= f(\beta^*)\langle\beta|0\rangle = f(\beta^*)\exp\left[-\tfrac{1}{2}|\beta|^2\right]. \tag{8.86}$$

The expansion coefficient, apart from the factor $\exp\left[-\tfrac{1}{2}|\beta|^2\right]$ which serves as a weight function, is the entire function which represents the state in the Bargmann space. (We have already made use of the condition represented by Eq. (8.86) in determining the wave functions $\langle q'|\alpha\rangle$ and $\langle p'|\alpha\rangle$ in the preceding section. The phase factors left undetermined by the differential equations and normalization conditions for these functions were chosen so that the complex conjugate form of Eq. (8.86) is satisfied.)

Since there is clearly a one-to-one correspondence between the functions f and the states of the mode, we have exhibited a construction for the coherent state expansion which is unique. The essential step in our construction is the use of the resolution of unity given by Eq. (8.81) and it is this which is responsible for the uniqueness of the resulting expansion. Because the coherent states form an overcomplete set other dyadic resolutions of unity exist, and their use could lead to different kinds of expansions of the same state. In practice, one universally applicable way of expanding the state vectors seems to be sufficient, at least for present purposes.

To give a simple illustration of the expansion Eq. (8.85) we may take the arbitrary state to be a coherent state. We then have the relation

$$|\alpha\rangle = \frac{1}{\pi}\int |\beta\rangle \langle\beta|\alpha\rangle\, d^2\beta$$
$$= \frac{1}{\pi}\int |\beta\rangle \exp\left[\beta^*\alpha - \tfrac{1}{2}|\alpha|^2 - \tfrac{1}{2}|\beta|^2\right] d^2\beta, \tag{8.87}$$

which shows explicitly that each coherent state may be expressed as a linear combination of all of the others.

As we have mentioned earlier, our knowledge of the state of a field usually requires that we describe it as a mixed state rather than a pure one. It is quite important therefore to be able to expand the density operator in terms of the coherent states. The most direct procedure for expanding the operator ρ is the immediate generalization of the one we have used for pure states. By multiplying ρ from both sides by the resolution of the unit operator in Eq. (8.81) we may write

$$\rho = \frac{1}{\pi^2}\int |\alpha\rangle \langle\alpha|\rho|\beta\rangle \langle\beta|\, d^2\alpha\, d^2\beta, \tag{8.88}$$

which is an expansion in terms of the dyadic products $|\alpha\rangle\langle\beta|$. To find the properties of the weight function in this expansion we write the density operator in terms of its expansion in the n-quantum states

$$\rho = \sum_{n,m} |n\rangle \rho_{nm} \langle m|$$

$$= \sum_{n,m} \rho_{nm} (n!\, m!)^{-1/2} (a^\dagger)^n |0\rangle \langle 0| a^m \,. \tag{8.89}$$

The form taken by this sum suggests that we define the complex-valued function

$$R(\alpha^*, \beta) = \sum_{n,m} \rho_{nm} (n!\, m!)^{-1/2} (\alpha^*)^n \beta^m \,, \tag{8.90}$$

which permits us to write the weight function in Eq. (8.88) as

$$\langle \alpha | \rho | \beta \rangle = R(\alpha^*, \beta) \langle \alpha | 0 \rangle \langle 0 | \beta \rangle$$

$$= R(\alpha^*, \beta) \exp\left[-\tfrac{1}{2}|\alpha|^2 - \tfrac{1}{2}|\beta|^2\right] \,. \tag{8.91}$$

Since the density operator is a bounded operator in the sense that

$$\mathrm{Tr}\{\rho^2\} = \sum_{n,m} |\rho_{nm}|^2 \leq 1 \,, \tag{8.92}$$

the series (8.90) for the function R converges for all finite values of α^* and β and R must be an entire function of both its arguments. Dyadic expansions of the type we are considering may be made for other types of operators than density operators. The weight function can be shown to retain the same analyticity property for a broad class of unbounded operators as well.

The advantage of expanding the density operator in terms of the coherent states becomes evident when we calculate the correlation functions $G^{(n)}$. The matrix element between two coherent states of any normally ordered product of field operators reduces to a product of the appropriate field eigenvalues, multiplied by the scalar product of the states. The non-commuting operators, in other words, become replaced by ordinary complex numbers and evaluating the average value of their product boils down to a problem of integration. There are evidently two complex amplitudes to be integrated over for each mode of the field.

The unit operator also possesses a simple dyadic expansion analogous to the one we have found for ρ. We may write it as

$$1 = \frac{1}{\pi^2} \int |\alpha\rangle \langle \alpha | \beta \rangle \langle \beta | \, d^2\alpha \, d^2\beta$$

$$= \frac{1}{\pi^2} \int |\alpha\rangle \exp\left[\alpha^* \beta - \tfrac{1}{2}|\alpha|^2 - \tfrac{1}{2}|\beta|^2\right] \langle \beta | \, d^2\alpha \, d^2\beta \,. \tag{8.93}$$

Because this is an expansion in terms of products of the form $|\alpha\rangle\langle\beta|$, it has a structure rather different from the resolution of unity which we have used earlier. We may make use of Eq. (8.87), however, to carry out the integration over the variable α and in this way arrive once more at the reduced or "diagonal" form of the expansion

$$1 = \frac{1}{\pi} \int |\beta\rangle \langle \beta | \, d^2\beta \,,$$

8.7 Expansions in Terms of the Coherent States

which we derived in Eq. (8.81). In effect the linear dependence of the coherent states supplies us with an alternative expansion for the unit operator and a much simpler one which happens already to be familiar to us.

The existence of a reduced or "diagonal" form for the coherent state expansion of the unit operator suggests that an analogous representation might exist for density operators. We may, accordingly, look for a representation of the density operator which takes the simple form

$$\rho = \int P(\alpha) \, |\alpha\rangle \langle\alpha| \, d^2\alpha \,. \tag{8.94}$$

The Hermitian character of the expansion is secured by requiring that the weight function $P(\alpha)$ be real-valued, and the normalization of the density operator, $\text{Tr}\,\rho = 1$, is secured by the condition

$$\int P(\alpha) \, d^2\alpha = 1 \,. \tag{8.95}$$

An example of a density operator which clearly possesses a representation of this type is the one for a pure coherent state $|\beta\rangle\langle\beta|$. For this case it suffices to let the function P be a two-dimensional delta function

$$P(\alpha) = \delta^{(2)}(\alpha - \beta) \,, \tag{8.96}$$

which we may define via

$$\delta^{(2)}(\alpha) = \delta(\text{Re}\,\alpha) \, \delta(\text{Im}\,\alpha) \,. \tag{8.97}$$

Other examples of the representation Eq. (8.94), which we have called the P representation, are easy to construct and we shall see some useful ones presently. The existence of these examples does not, however, answer the question of whether it it possible to construct a P representation for all density operators. We shall postpone discussing that question until we can consider it in a somewhat broader context in the next section. For the present let us simply assume that a P representation does exist and note some of its properties.

The role of the function $P(\alpha)$ as a weight function in a sum over the projection operators $|\alpha\rangle\langle\alpha|$ gives it some of the appearance of a probability density, as does the normalization condition Eq. (8.95). A literal interpretation of $P(\alpha)$ as a probability density however, would require that the projection operators $|\alpha\rangle\langle\alpha|$ form an orthogonal set, and we know that no such thing is true. There are many more or less related ways of seeing that $P(\alpha)$ is not in general a probability density. Perhaps the simplest is through the positive-definiteness property of the density operator, or the condition that its eigenvalues can never be negative. It is easy to see that functions $P(\alpha)$ which are nowhere negative lead to positive definite density operators, but the converse statement is, in fact, false. The positive definiteness conditions on ρ freely admit the possibility that the function $P(\alpha)$ takes on negative values (though not over very broad areas of the α plane). It is worth noting too that there is no physical sense in which we could truly measure a probability distribution for the complex eigenvalues of the operator a. The real and imaginary parts of a correspond to the position and momentum of an oscillator and since these operators do not commute they can not simultaneously be measured precisely.

In spite of all the reasons why $P(\alpha)$ is not a probability density there are a number of calculations which can be carried out very much as if it were one. Let us ask for the average value of a normally ordered operator such as $(a^\dagger)^n a^m$. Then we can use the P representation to write

$$\mathrm{Tr}\left\{\rho (a^\dagger)^n a^m\right\} = \int P(\alpha) (\alpha^*)^n \alpha^m \, d^2\alpha, \qquad (8.98)$$

and the latter expression looks a great deal like a classical average calculated with a probability density $P(\alpha)$.

A state of affairs not unlike the one we are describing has arisen at least once before in quantum mechanics. In 1932 Wigner[17] introduced a function which serves as a quantum mechanical analog of the classical phase space distribution in evaluating the averages of a particular class of ordered operators. He also showed that the function may take on negative values. The function $P(\alpha)$ is different from Wigner's function, though related to it, and evidently also deserves to be called a quasiprobability density, as the Wigner function is. We shall discuss the interrelationship of these functions and some others as well in the next section.

When the P representation is used for the density operator of a multimode field it takes the form

$$\rho = \int P(\{\alpha_k\}) |\{\alpha_k\}\rangle \langle\{\alpha_k\}| \prod_k d^2\alpha_k. \qquad (8.99)$$

If we now use this expression for the density operator in evaluating the correlation functions $G^{(n)}$ we find that they reduce to integrals of precisely the same form as we encountered in our discussion of classical correlation functions. If the state $|\{\alpha_k\}\rangle$ satisfies the eigenstate condition

$$E^{(+)}(x) |\{\alpha_k\}\rangle = \mathcal{E}(x\{\alpha_k\}) |\{\alpha_k\}\rangle, \qquad (8.100)$$

then we see that the function $G^{(1)}$, for example, is given by

$$G^{(1)}(x_1, x_2) = \int P(\{\alpha_k\}) \mathcal{E}^*(x_1\{\alpha_k\}) \mathcal{E}(x_2\{\alpha_k\}) \prod_k d^2\alpha_k, \qquad (8.101)$$

which is an integral having exactly the same structure as the classical one in Eq. (8.11). Part of this correspondence has already been indicated in Eq. (8.53); the classical amplitudes C_k correspond to $i(\frac{1}{2}\hbar\omega_k)^{1/2}\alpha_k$. The remainder of it lies in the correspondence of the classical probability density $\mathcal{P}(\{C_k\})$ with the quasiprobability density

$P(\{\alpha_k\})$. Indeed, in the classical limit of strongly excited fields we may expect the function $P(\{\alpha_k\})$ to become more and more accurately interpretable as a probability density and to approach the function $\mathcal{P}(\{C_k\})$ asymptotically.

What is remarkable about Eq. (8.101) and the corresponding expression for all of the functions $G^{(n)}$ is that anything like the classical prescription for averaging field quantities should hold, no matter how weak the fields may be, e.g. even when the fields contributed by the light sources are much weaker than the vacuum fluctuations that are always present. The fields dealt with in optical experiments are typically just that weak, but their intensities have traditionally been evaluated as averages of the products of classical fields. The existence of the P representation for such fields evidently gives that familiar procedure a precise and non-trivial justification. It works for intensity measurements based on the absorption of photons since those are the ones described by normally ordered operator products (and therefore ones to which vacuum fluctuations do not contribute).

8.8
Characteristic Functions and Quasiprobability Densities

We have emphasized the special role played by the weight function P in the averaging of normally ordered products. It is interesting to discuss two other forms of operator ordering and the corresponding weight functions which are associated with them. These are antinormal ordering, which is simply the reverse of normal ordering and symmetrical ordering, which we may define as the arithmetic mean of all the different orderings of a given product.

Before discussing questions of operator ordering, let us recall that one of the familiar ways of calculating classical probability densities is through the evaluation of their Fourier transforms or characteristic functions. The corresponding function, for quantum mechanical purposes, must be a kind of Fourier transform of the density operator ρ. We can define such a function as the average value of the unitary exponential operator D given by Eq. (8.56). The characteristic function then is

$$\chi(\lambda) = \text{Tr}\{\rho D(\lambda)\}$$
$$= \text{Tr}\left\{\rho \exp\left[\lambda a^\dagger - \lambda^* a\right]\right\}. \tag{8.102}$$

An upper bound for this function can be found by evaluating the trace in the representation which diagonalizes the density operator. It is clear then that the trace is the average value of the diagonal elements of $D(\lambda)$ and each of these has modulus less than or equal to one since $D(\lambda)$ is unitary. We thus have

$$|\chi(\lambda)| \leq 1, \tag{8.103}$$

for all values of the complex variable λ.

In close correspondence with $\chi(\lambda)$ we can define another species of characteristic function which is the average of the normally ordered form of the exponential function,

$$\chi_N(\lambda) = \text{Tr}\left\{\rho \exp\left[\lambda a^\dagger\right] \exp\left[-\lambda^* a\right]\right\}, \tag{8.104}$$

and still another in which the exponential function is antinormally ordered,

$$\chi_A(\lambda) = \text{Tr}\left\{\rho \exp\left[-\lambda^* a\right] \exp\left[\lambda a^\dagger\right]\right\}. \tag{8.105}$$

We shall refer to χ_N and χ_A as the normally ordered and antinormally ordered characteristic functions respectively.

The three characteristic functions we have defined may be regarded as generating functions for the averages of the appropriately ordered products. By expanding $\chi_N(\lambda)$ in powers of λ and λ^*, for example, we find

$$\frac{\partial^n}{\partial \lambda^n} \frac{\partial^m}{\partial (-\lambda^*)^m} \chi_N(\lambda) \bigg|_{\lambda=0} = \text{Tr}\left\{\rho (a^\dagger)^n a^m\right\}, \tag{8.106}$$

i.e. χ_N generates the averages of normally ordered products. Likewise by differentiating $\chi_A(\lambda)$ and $\chi(\lambda)$ we see that χ_A generates the averages of antinormally ordered products and χ generates the averages of products that are symmetrically ordered.

The identity

$$e^A e^B = e^{A+B+\frac{1}{2}[A,B]}, \tag{8.107}$$

which holds for all operators A and B which commute with their commutator $[A,B]$, can be used to express each characteristic function in terms of the others. The relationship we find is

$$\chi_N(\lambda) = \exp\left[\tfrac{1}{2}|\lambda|^2\right] \chi(\lambda) = \exp\left[|\lambda|^2\right] \chi_A(\lambda). \tag{8.108}$$

The inequality (8.103) then shows that $|\chi_A(\lambda)|$ always decreases at least as fast as $\exp[-\tfrac{1}{2}|\lambda|^2]$ for $|\lambda| \to \infty$, while $|\chi_N(\lambda)|$ on the other hand, may diverge as rapidly as $\exp[\tfrac{1}{2}|\lambda|^2]$ in that limit.

Let us turn away from the characteristic functions for a moment to examine the properties of some diagonal matrix elements of the density operator. The matrix elements $\langle \alpha | \rho | \alpha \rangle$, which are expectation values of ρ in the coherent states, have a number of interesting properties[5,18,19]. Like all diagonal elements of ρ they are real and constrained by the inequalities

$$1 \geq \langle \alpha | \rho | \alpha \rangle \geq 0. \tag{8.109}$$

They may be expressed in terms of the analytic function R defined in Eq. (8.90) as

$$\langle \alpha | \rho | \alpha \rangle = R(\alpha^*, \alpha) \exp\left[-|\alpha|^2\right]. \tag{8.110}$$

8.8 Characteristic Functions and Quasiprobability Densities

The most useful property of the function $\langle \alpha | \rho | \alpha \rangle$ becomes evident when we evaluate the average of an antinormally ordered product. Let us consider an average of the form

$$\text{Tr}\left\{\rho U(a) V(a^\dagger)\right\},$$

where U and V are arbitrary functions of the annihilation and creation operators, respectively. By using our resolution of unity and the cyclical property of the trace we may write the average as

$$\text{Tr}\left\{V(a^\dagger) \rho U(a)\right\} = \frac{1}{\pi} \int \text{Tr}\left\{|\alpha\rangle\langle\alpha| V(a^\dagger) \rho U(a)\right\} d^2\alpha$$
$$= \frac{1}{\pi} \int \left\langle \alpha \left| V(a^\dagger) \rho U(a) \right| \alpha \right\rangle d^2\alpha$$
$$= \frac{1}{\pi} \int \langle \alpha | \rho | \alpha \rangle U(\alpha) V(\alpha^*) d^2\alpha. \tag{8.111}$$

Thus we see that the function $\pi^{-1} \langle \alpha | \rho | \alpha \rangle$ plays the same role in averaging antinormally ordered products as $P(\alpha)$ plays in averaging normally ordered ones.

By setting $U = V = 1$ in Eq. (8.111) we see that the normalization condition on the density operator may be written as

$$1 = \text{Tr}\{\rho\}$$
$$= \frac{1}{\pi} \int \langle \alpha | \rho | \alpha \rangle d^2\alpha. \tag{8.112}$$

The function $\pi^{-1} \langle \alpha | \rho | \alpha \rangle$ evidently has many properties in common with a probability density, including the fact that it can never be negative. Still, the fact that it refers to a continuum of non-orthogonal states means that there is no well defined sense in which it can be measured physically as a probability density. We must regard it as still another form of quasiprobability function to add to our growing list.

We can now use the rule given by Eq. (8.111) as a means of evaluating the antinormally ordered characteristic function χ_A. What we find is

$$\chi_A(\lambda) = \frac{1}{\pi} \int \exp[\lambda \alpha^* - \lambda^* \alpha] \langle \alpha | \rho | \alpha \rangle d^2\alpha. \tag{8.113}$$

We note that the argument $\lambda \alpha^* - \lambda^* \alpha$ of the exponential function which occurs in the integrand is purely imaginary. In fact, if we use the real and imaginary parts of α and λ as coordinates in writing the integral we see that it is just a two-dimensional Fourier integral; $\chi_A(\lambda)$ is the Fourier transform of $\langle \alpha | \rho | \alpha \rangle$.

The inversion of the transform may be carried out by using the Fourier integral representation of the delta function,

$$\frac{1}{\pi^2} \int \exp[\lambda(\alpha^* - \beta^*) - \lambda^*(\alpha - \beta)] d^2\lambda = \delta^{(2)}(\alpha - \beta), \tag{8.114}$$

which is easily proved by introducing real and imaginary parts of the variables. The solution for $\langle \alpha | \rho | \alpha \rangle$ is simply

$$\langle \alpha | \rho | \alpha \rangle = \frac{1}{\pi} \int \exp[\alpha \lambda^* - \alpha^* \lambda] \, \chi_A(\lambda) \, d^2\lambda \,, \tag{8.115}$$

and we see that it is a two-dimensional Fourier transform of the antinormally ordered characteristic function.

Let us turn now to the characteristic function $\chi(\lambda)$ and make use of its Fourier transform to define the function

$$W(\alpha) = \frac{1}{\pi^2} \int \exp[\alpha \lambda^* - \alpha^* \lambda] \, \chi(\lambda) \, d^2\lambda \,. \tag{8.116}$$

This is in fact the quasiprobability function defined by Wigner[17,20]. As usually encountered it is expressed somewhat less compactly in terms of coordinate and momentum variables. By integrating by parts and using the identity (8.114) we see that the complex moments of the function W may be written as

$$\int (\alpha^*)^n \alpha^m W(\alpha) \, d^2\alpha = \int \delta^{(2)}(\lambda) \frac{\partial^n}{\partial \lambda^n} \frac{\partial^m}{\partial(-\lambda^*)^m} \chi(\lambda) \, d^2\lambda$$

$$= \frac{\partial^n}{\partial \lambda^n} \frac{\partial^m}{\partial(-\lambda^*)^m} \chi(\lambda) \bigg|_{\lambda=0} \,, \tag{8.117}$$

and by evaluating these derivatives as we did in the case of Eq. (8.106) we see that they are the averages of the symmetrically ordered products of n factors of $a^\dagger$ and m factors of a. The Wigner function is thus the quasiprobability density appropriate to symmetric ordering. In particular for $n = m = 0$ we see that it obeys the normalization condition

$$\int W(\alpha) \, d^2\alpha = \chi(0) = 1 \,. \tag{8.118}$$

After these brief surveys of the treatment of antinormal and symmetric ordering let us return once more to the discussion of normal ordering. If we assume that the density operator ρ possesses a P representation then we can write its normally ordered characteristic function as

$$\chi_N(\lambda) = \int P(\alpha) \langle \alpha | \exp[\lambda a^\dagger] \exp[-\lambda^* a] | \alpha \rangle \, d^2\alpha$$

$$= \int \exp[\lambda \alpha^* - \lambda^* \alpha] \, P(\alpha) \, d^2\alpha \,, \tag{8.119}$$

which is the two-dimensional Fourier transform of $P(\alpha)$. The inverse of this transform, when it possesses one, is given according to Eq. (8.114), by the Fourier integral

$$P(\alpha) = \frac{1}{\pi^2} \int \exp[\alpha \lambda^* - \alpha^* \lambda] \, \chi_N(\lambda) \, d^2\lambda \,, \tag{8.120}$$

which is an explicit construction of the function $P(\alpha)$. If the function χ_N is quadratically integrable, as it is in many examples, then we see that the weight function $P(\alpha)$ must be quadratically integrable as well. The cases in which χ_N is not quadratically integrable, however, take us out of the classic domain of Fourier integral theory. For the ground state of the oscillator, for example, it is easy to see that $\chi_N = 1$ for all λ. The integral formula Eq. (8.114) shows that in this case $P(\alpha)$ is the delta function $\delta^{(2)}(\alpha)$. The delta function and its derivatives are examples of a class of generalized functions known as tempered distributions; they are more or less familiar to physicists to the extent that they can be dealt with as ordinary functions. It can be shown that $P(\alpha)$ is a tempered distribution whenever the moduli of $\chi_N(\lambda)$ and of its derivatives are dominated for $\lambda \to \infty$ by real-valued polynomials of finite degree in λ and λ^*. It is altogether possible for $|\chi_N(\lambda)|$ to increase a good deal more rapidly, however. The relations (8.103) and (8.108) allow $|\chi_N(\lambda)|$ to increase as $\exp[\frac{1}{2}|\lambda|^2]$ for larger $|\lambda|$ and states for which this actually happens are not too hard to find[21]. For such states the integral Eq. (8.120) is so highly divergent that no simple summation technique can handle it; $P(\alpha)$ becomes, in effect, more singular than any tempered distribution.

It is not difficult to see, at least, in rather crude physical terms, why the P representation exhibits this singular behavior. The characteristic functions are Fourier transforms of the quasiprobability densities and so we may surmise that their remaining large for $|\lambda| \to \infty$ is connected with sharply peaked behavior of the quasiprobability densities over small intervals in the α plane. Let us therefore consider as an example one of the minimum uncertainty states we discussed earlier. It is sufficient to consider the eigenstate of the operator $A(\mu)$ defined by Eq. (8.76) which corresponds to the eigenvalue zero. The states with non-zero eigenvalues are simply displaced forms of the same state. If the state $|\ \rangle$ satisfies

$$A(\mu)|\ \rangle = 0, \tag{8.121}$$

then a very simple calculation shows that

$$\begin{aligned}\chi_N(\lambda) &= \langle\ |\exp\left[\lambda a^\dagger\right]\exp\left[-\lambda^* a\right]|\ \rangle \\ &= \exp\left\{\frac{\omega-\mu}{2}\left[\frac{(\operatorname{Re}\lambda)^2}{\omega}-\frac{(\operatorname{Im}\lambda)^2}{\mu}\right]\right\}.\end{aligned} \tag{8.122}$$

For $\omega > \mu$ this expression diverges as $|\operatorname{Re}\lambda| \to \infty$; for $\mu > \omega$ it diverges as $|\operatorname{Im}\lambda| \to \infty$. In neither case does it possess a Fourier transform in any of the familiar senses.

The state we are trying to represent differs from the coherent states in that it has the variances

$$(\Delta p)^2 = \tfrac{1}{2}\hbar\mu, \qquad (\Delta q)^2 = \frac{\hbar}{2\mu}. \tag{8.123}$$

According to whether μ is greater or less than ω either one of these variances or the other must be smaller than the values given by Eq. (8.72), which are the variances common to all of the coherent states. The physical reason for the breakdown of the P representation in these cases is now evident. What we are trying to do is to find something akin to a probabilistic mixture of states which has altogether $(\Delta p)^2 = \frac{1}{2}\hbar\mu$ when each of the states that we are mixing has $(\Delta p)^2 = \frac{1}{2}\hbar\omega$. For $\mu < \omega$ no such mixture exists; the divergence of χ_N shows up, as we expect, in its dependence on $\operatorname{Re}\lambda$, which is the variable conjugate to the momentum. Similarly we see that for $\mu > \omega$ we are trying to represent a state with a coordinate variance smaller than that of the coherent states.

While the example of the minimum uncertainty states is a rather special one there is nothing unphysical about it. The eigenstate defined by Eq. (8.121) for example is the ground state the mode oscillator would have if its frequency were changed from ω to μ. It could, in other words, represent the ground state of the same field mode in a dielectric medium rather than the vacuum for which our coherent states have been defined.

In view of the difficulty of extending the P representation to all states of the field two approaches seem open. One is to further broaden the class of mathematical structures we are willing to accept in place of a function $P(\alpha)$. Sudarshan[22,19,24], Klauder[21,23,24] and their coworkers have pointed out that a class of generalized distributions has been defined which is extensive enough to admit all Fourier transforms such as that of Eq. (8.120). The implicit drawback of this approach is that it sacrifices the last resemblances between $P(\alpha)$ and an ordinary function. Statistical averages taken over $P(\alpha)$ can only be defined rather abstractly as linear functionals, and no means is known of calculating them in general apart from carrying out limiting processes in each case.

The alternative approach is to acknowledge that the quasiprobability distribution $P(\alpha)$ is mainly of use as a computational device and as a way of seeing correspondences with the classical limit. When the function ceases to serve either of these ends and becomes too difficult to define we may as well say that the P representation ceases to exist. The general coherent state representation given by Eq. (8.88) of course, still offers a convenient way of averaging normally ordered operators for these cases.

Fortunately the P representation exists and is quite well behaved, at least for many of the fields which interest us. When the Fourier transform which defines $P(\alpha)$ does exist we can exhibit some very simple relations between $P(\alpha)$ and the other two quasiprobability densities we have discussed. Equation (8.108) shows that the three different characteristic functions are related to one another through Gaussian factors. By taking Fourier transforms of the characteristic functions and using the convolution theorem we find the three relations

$$\frac{1}{\pi}\langle\alpha|\rho|\alpha\rangle = \frac{2}{\pi}\int \exp\left[-2|\alpha-\beta|^2\right] W(\beta)\, d^2\beta, \tag{8.124}$$

$$W(\alpha) = \frac{2}{\pi}\int \exp\left[-2|\alpha-\beta|^2\right] P(\beta)\, d^2\beta \tag{8.125}$$

and
$$\frac{1}{\pi}\langle\alpha|\rho|\alpha\rangle = \frac{1}{\pi}\int \exp\left[-|\alpha-\beta|^2\right] P(\beta)\,d^2\beta\,. \tag{8.126}$$

The Wigner function W is a Gaussian average of the function P taken over a mean radius $|\alpha-\beta| \approx 2^{-1/2}$ and $\pi^{-1}\langle\alpha|\rho|\alpha\rangle$ is a similar average of W. It follows then, as we see in the third relation, that $\pi^{-1}\langle\alpha|\rho|\alpha\rangle$ is an average of P taken over a mean radius $|\alpha-\beta| \approx 1$. The averaging processes which take us from P to W to $\langle\alpha|\rho|\alpha\rangle$ tend to wash out any unruly behavior of P and transform it into the smooth and bounded behavior of $\langle\alpha|\rho|\alpha\rangle$. They also tend to wash away the areas of the α plane in which P and W are negative; the function $\langle\alpha|\rho|\alpha\rangle$ which results from the averaging is nowhere negative. The functions $W(\alpha)$ and $\langle\alpha|\rho|\alpha\rangle$ exist and remain well defined for all quantum states. The cases in which $P(\alpha)$ becomes undefinable are precisely the ones in which the convolution integrals Eq. (8.125) and (8.126) can not be inverted.

8.9
Some Examples

It may be of help in visualizing the quasiprobability functions to cite a few examples. The simplest mode excitation to discuss is a pure coherent state $|\beta\rangle$. For this state we clearly have

$$\begin{aligned}\chi_N(\lambda) &= \langle\beta|\exp[\lambda a^\dagger]\exp[-\lambda^* a]|\beta\rangle \\ &= \exp[\lambda\beta^* - \lambda^*\beta]\,,\end{aligned} \tag{8.127}$$

and the two other characteristic functions are then given by Eq. (8.108). The quasiprobability functions which are their Fourier transforms are found to be

$$P(\alpha) = \delta^{(2)}(\alpha-\beta)\,, \tag{8.128}$$

$$W(\alpha) = \frac{2}{\pi}\exp\left[-2|\alpha-\beta|^2\right]\,, \tag{8.129}$$

$$\frac{1}{\pi}\langle\alpha|\rho|\beta\rangle = \frac{1}{\pi}\exp\left[-|\alpha-\beta|^2\right]\,. \tag{8.130}$$

The latter two relations, of course, follow immediately from the first if we use Eqns. (8.125) and (8.126).

The state which is by far the most common of all states occurring in nature is best described as the state of chaos. For a single mode of the field, it is the state with the maximum value of the entropy,

$$S = -\operatorname{Tr}\{\rho\log\rho\}\,, \tag{8.131}$$

for a fixed mean occupation number

$$\langle n\rangle = \operatorname{Tr}\{\rho a^\dagger a\}\,. \tag{8.132}$$

The problem of maximizing the entropy subject to the condition of fixed $\langle n \rangle$ (and $\operatorname{Tr} p = 1$, of course) is a familiar one. Its solution can be written as

$$\rho = \exp\left[-\zeta_1 - \zeta_2 a^\dagger a\right], \tag{8.133}$$

where ζ_1 and ζ_2 are Lagrange multipliers to be determined from the normalization condition and Eq. (8.132). When these constants are evaluated we find

$$\rho = \frac{1}{1+\langle n\rangle} \exp\left[a^\dagger a \log\left(\frac{\langle n\rangle}{1+\langle n\rangle}\right)\right] \tag{8.134}$$

$$= \frac{1}{1+\langle n\rangle} \left(\frac{\langle n\rangle}{1+\langle n\rangle}\right)^{a^\dagger a} \tag{8.135}$$

$$= \frac{1}{1+\langle n\rangle} \sum_j \left(\frac{\langle n\rangle}{1+\langle n\rangle}\right)^j |j\rangle\langle j|, \tag{8.136}$$

where in the last relation we have expanded ρ in terms of the j-quantum states.

It is not at all difficult to construct the three differently ordered forms for the density operator ρ in this case, but instead of pausing to do so let us proceed directly to the evaluation of the quasiprobability functions. The function $R(\alpha^*, \beta)$ defined by Eq. (8.90) is clearly given by

$$R(\alpha^*, \beta) \frac{1}{1+\langle n\rangle} \exp\left[\frac{\langle n\rangle}{1+\langle n\rangle} \alpha^*\beta\right], \tag{8.137}$$

hence the function $\langle \alpha|\rho|\alpha\rangle$, as we see from Eq. (8.110), is given by

$$\langle \alpha|\rho|\alpha\rangle = \frac{1}{1+\langle n\rangle} \exp\left[-\frac{|\alpha|^2}{1+\langle n\rangle}\right]. \tag{8.138}$$

The function $\chi_A(\lambda)$ which is the Fourier transform of this function, is

$$\chi_A(\lambda) = \exp\left[-(1+\langle n\rangle)|\lambda|^2\right] \tag{8.139}$$

and that in turn determines $\chi(\lambda)$ and $\chi_N(\lambda)$ through Eq. (8.108). When the functions $W(\alpha)$ and $P(\alpha)$ are determined by Fourier inversion we then find

$$W(\alpha) = \frac{1}{\pi(\frac{1}{2}+\langle n\rangle)} \exp\left[-\frac{|\alpha|^2}{\frac{1}{2}+\langle n\rangle}\right] \tag{8.140}$$

and

$$P(\alpha) = \frac{1}{\pi\langle n\rangle} \exp\left[-\frac{|\alpha|^2}{\langle n\rangle}\right]. \tag{8.141}$$

The definition of chaotic states which we have given is a great deal more general than the notion of thermal equilibrium; there are all sorts of ways of generating chaos which are not associated with any temperature at all. Thermal equilibrium states are chaotic states, however, with

$$\langle n \rangle = \frac{1}{\exp\left[\dfrac{\hbar \omega}{\kappa T}\right] - 1} \tag{8.142}$$

and substituting this expression into Eq. (8.134) or (8.136) leads to familiar forms for the thermal density operator.

If the field has a multimode excitation the chaotic state with mean occupation numbers $\langle n_k \rangle$ is represented by the P function

$$P(\{\alpha_k\}) = \prod_k \frac{1}{\pi \langle n_k \rangle} \exp\left[-\sum_k \frac{|\alpha_k|^2}{\langle n_k \rangle}\right]. \tag{8.143}$$

Chaos in electromagnetic fields is synonymous with noise; Eq. (8.143) describes a field of pure noise with a spectrum specified by the $\{\langle n_k \rangle\}$. Any light source which is divisible in principle into many parts, each of which radiates independently of the others, will almost inevitably radiate a chaotic field. The range of applicability of Eq. (8.143) is therefore presumably quite broad.

The higher order correlation functions have a particularly simple structure for chaotic fields. It is not difficult to show[5] that they can all be expressed in terms of the first order correlation function via the relations

$$G^{(n)}(x_1 \ldots x_n, y_1 \ldots y_n) = \sum_P \prod_{j=1}^{n} G^{(1)}(x_j, y_{P_j}), \tag{8.144}$$

in which the sum is carried out over all permutations P_j of the integers $j = 1 \ldots n$. The particular value of $G^{(2)}$ which corresponds to a two-fold joint counting rate, for example, can be written as

$$G^{(2)}(x_1 x_2, x_2 x_1) = G^{(1)}(x_1, x_1) G^{(1)}(x_2, x_2) + |G^{(1)}(x_1, x_2)|^2. \tag{8.145}$$

If the field possessed second order coherence the second of the terms on the right would be absent, and two counters would record photons at the same rates they would measure independently. The second term is in fact equal to the first for $x_1 = x_2$ and decreases to zero in general as x_1 recedes from x_2. It represents the positive correlation in the delayed coincidence rate first detected by Hanbury Brown and Twiss[1]. Its origin as we see, lies entirely in the noisiness of the field, i.e. the fluctuations of its amplitudes.

A simple way of generating other examples of fields is through superposition. The fields generated by two sources may be superposed linearly as long as neither source can be affected by the field produced by the other. To describe the superposed field we must be able to construct its density operator from a knowledge of the density operators of the two component fields. The rule which does this can be shown[4] to correspond to convolution of the functions $P(\alpha)$. If the functions $P_1(\alpha)$ and $P_2(\alpha)$ describe single mode excitations produced by the first and second sources respectively, then the total excitation is described by the function

$$P(\alpha) = \int P_1(\alpha') P_2(\alpha - \alpha') \, d^2\alpha' . \tag{8.146}$$

Other ways of stating the rule, in terms of the characteristic functions and quasiprobability functions, are readily found.

An example which frequently occurs nowadays is the superposition of a coherent signal and noise. For single mode excitations a coherent signal corresponds to a pure coherent state $|\beta\rangle$, and this is represented by the delta function given in Eq. (8.128). The noise excitation is represented by the Gaussian function given in Eq. (8.141). The superposed fields are therefore represented by the convolution integral

$$P(\alpha) = \int \delta^{(2)}(\alpha' - \beta) \frac{1}{\pi \langle n \rangle} \exp\left[-\frac{|\alpha - \alpha'|^2}{\langle n \rangle}\right] d^2\alpha'$$

$$= \frac{1}{\pi \langle n \rangle} \exp\left[-\frac{|\alpha - \beta|^2}{\langle n \rangle}\right], \tag{8.147}$$

which is simply a displaced form of the Gaussian function.

Apart from rather idealized cases, such as those we have been discussing, there is no simple way in general of guessing what density operator characterizes a field. Our understanding of what makes a density operator is still so elementary that the only quantum mechanically reliable way we have of finding ρ is to construct a mathematical model for the source and its coupling to the field and to solve for the time dependent description of the whole system in statistical terms. Analyses of this type have been and are being made for a number of elementary models of laser oscillators, amplifiers and attenuators. The ways of representing density operators which we have been discussing here have turned out to be quite useful in this new area of statistical mechanics.

8.10
Photon Counting Distributions

Let us return now to the problem we stated at the outset. A photon counter is exposed to a given field from time zero to time t. What is the statistical distribution of the number of photons it records when this experiment is repeated many times? It is the tendency of the photon counts to occur at correlated times which makes an interesting question of this, and one which is not always easy to answer.

To find the distribution, let us begin by defining an operator $C(t)$ which represents the number of photons counted in time t. Such an operator clearly has integer eigenvalues $n = 0, 1, 2, \ldots$ It will depend on the variables which describe the counter system, as opposed to the field, but we do not need for present purposes any explicit construction of the operator. The introduction of an appropriately defined generating function makes it possible to give a fairly concise derivation of the unknown distribution. We let λ be an arbitrary parameter and define the generating function

$$Q(\lambda,t) = \mathrm{Tr}\left\{\rho(1-\lambda)^{C(t)}\right\}$$
$$= \left\langle (1-\lambda)^{C(t)} \right\rangle, \qquad (8.148)$$

where ρ is a density operator which describes the counter as part of the system.

As a complete set of states to use in describing the final state of the counter and field it is clearly convenient to take the eigenstates of $C(t)$. These are defined to satisfy

$$C(t)|n,l,t\rangle = n|n,l,t\rangle, \qquad (8.149)$$

in which l represents all of the quantum numbers other than the eigenvalue of $C(t)$. The trace which defines the generating function may then be expressed in terms of the set of numbers

$$p(n,t) = \sum_l \langle n,l,t|\rho|n,l,t\rangle, \qquad (8.150)$$

for $n = 0,1,2,\ldots$ These are simply the probabilities that the counter records n photons in time t. The generating function reduces therefore to the simple expression

$$Q(\lambda,t) = \sum_{n=0}^{\infty}(1-\lambda)^n p(n,t). \qquad (8.151)$$

If we are able to solve directly for the generating function we can use it to evaluate the probabilities since they are the coefficients

$$p(n,t) = \frac{(-1)^n}{n!}\frac{d^n}{d\lambda^n}Q(\lambda,t)\bigg|_{\lambda=1}, \qquad (8.152)$$

in its power series expansion about $\lambda = 1$. We can also find the factorial moments of the probability distribution easily since they are just the derivatives

$$\langle C(C-1)\ldots(C-n+1)\rangle = (-1)^n\frac{d^n}{d\lambda^n}Q(\lambda,t)\bigg|_{\lambda=0}, \qquad (8.153)$$

of the generating function evaluated at $\lambda = 0$.

To begin the process of finding the generating function, let us note that the sensitive element of a photon counter consists of many atoms ($N \gg 1$), any of which may undergo a photoabsorption process and give rise to a detected photon count. The operator C may be written therefore, as a sum

$$C = \sum_{j=1}^{N} c_j \qquad (8.154)$$

of operators c_j which represent the contributions of the individual atoms. The operators c_j are time-dependent in general, as is C, and have eigenvalues one and zero; one for final states to which j-th atom has contributed a photon count and zero for states to which it has contributed none.

Since the operators c_j refer to the variables of altogether different atoms they commute, and we can write

$$(1-\lambda)^C = (1-\lambda)^{\Sigma c_j}$$
$$= \prod_j (1-\lambda)^{c_j} .$$

The fact that the eigenvalues of the c_j are only zero and one means that

$$(1-\lambda)^{c_j} = 1 - \lambda c_j \tag{8.155}$$

and the generating function may be written therefore as

$$Q(\lambda,t) = \left\langle \prod_j (1-\lambda c_j) \right\rangle \tag{8.156}$$

$$= 1 - \lambda \sum_j \langle c_j \rangle + \lambda^2 \sum_{j \neq l} \langle c_j c_l \rangle - \ldots \tag{8.157}$$

Because the operators c_j are, in effect, projection operators on the states for which the j-th atom has contributed a count, an expectation value $\langle c_1 c_2 \ldots c_n \rangle$ is simply the probability that all of the atoms $1, 2 \ldots n$ have contributed counts in time t. That probability is one we have already calculated (in n-th order perturbation theory) and stated in Eq. (8.26). It is

$$\langle c_1 \ldots c_n \rangle = p^{(n)}(t \ldots t) \tag{8.158}$$

$$= s^n \int_0^t \ldots \int_0^t G^{(n)}(\boldsymbol{r}_1 t_1' \ldots \boldsymbol{r}_n t_n', \boldsymbol{r}_n t_n' \ldots \boldsymbol{r}_1 t_1') \prod_{j=1}^n dt_j' , \tag{8.159}$$

where $\boldsymbol{r}_1 \ldots \boldsymbol{r}_n$ are the positions of the n atoms.

We can simplify our calculation somewhat at this point by assuming that the geometry of our experiment is such that all the sensitive atoms of the detector experience the same field, e.g. that plane waves are incident normally on a plane photocathode. This assumption means simply that all atoms are equivalent in the summations in Eq. (8.157). (The assumption can easily be removed by replacing the sums over atoms by integrations over the detector volume.) If we assume furthermore that $N \gg n$, then we find

$$Q(\lambda,t) = \sum_{n=0} \frac{(-\lambda N s)^n}{n!} \int_0^t \ldots \int_0^t G^{(n)}(t_1' \ldots t_n', t_n' \ldots t_1') \prod dt_j' . \tag{8.160}$$

This expression for the generating function shows us immediately via Eq. (8.153) that the factorial moments of the photon counting distribution are

$$\langle C(C-1) \ldots (C-n+1) \rangle = (Ns)^n \int_0^t \ldots \int_0^t G^{(n)}(t_1' \ldots t_n', t_n' \ldots t_1') \prod dt_j' . \tag{8.161}$$

Since it is not too difficult, in practice, to measure the higher order moments of a distribution, we see that the integrals of the higher order correlation functions are more or less directly accessible to measurement.

We can now make use of our coherent state representation of the density operator to reduce the expression for the generating function to a closed form. Let us suppose, for example, that we can define a P representation for the density operator. Then the function $G^{(1)}(t'_1,t'_1)$ may be expressed as an integral over the amplitude parameters $\{\alpha_k\}$, as we have shown in Eq. (8.101), and analogous integral representations can be used for all of the higher order $G^{(n)}$. By substituting these representations in Eq. (8.160), we find that the sum over n can be carried out trivially, and leaves

$$Q(\lambda,t) = \int P(\{\alpha_k\}) \exp[-\lambda\Omega(\{\alpha_k\}),t]\, d^2\alpha_k, \qquad (8.162)$$

where Ω is a bilinear form in the amplitudes $\{\alpha_k\}$ given by

$$\omega(\{\alpha_k\},t) = Ns \int_0^t |\mathcal{E}(t',\{\alpha_k\})|^2 \, dt'. \qquad (8.163)$$

The simplest illustration of this result is the case of a pure coherent state. In that case the function P is a product of delta functions which constrains the amplitudes $\{\alpha_k\}$ to have a particular set of values, say $\{\beta_k\}$. Then the generating function reduces to

$$Q(\lambda,t) = \exp[-\lambda\Omega(\{\beta_k\},t)] \qquad (8.164)$$

and we see from Eq. (8.152) that the $p(n,t)$ form a Poisson distribution,

$$p(n,t) = \frac{\langle C \rangle^n}{n!} \exp[-\langle C \rangle], \qquad (8.165)$$

about the mean value $\langle C \rangle = \Omega(\{\beta_k\},t)$. The mean counting rate is just proportional to the instantaneous field intensity in this case

$$\frac{d}{dt}\langle C(t) \rangle = Ns|\mathcal{E}(t,\{\beta_k\})|^2,$$

and in coherent states that may vary arbitrarily with time.

To the extent that the function P resembles a probability distribution it is clear that the function $Q(\lambda)$ given by Eq. (8.162) will generate a kind of averaged Poisson distribution. We have already noted, however, that the function P may take on negative values and singularities unlike those of any probability distribution. It is not too surprising to find, therefore, that the distribution $p(n,t)$ need not actually have much resemblance to a Poisson distribution.

There is another way of constructing a closed form for the generating function which does not rely on the use of the P representation. The general form of the coherent state representation given by Eq. (8.88) can be used in much the same way and leads to an integral representation for $Q(\lambda t)$ analogous to Eq. (8.162)), but one in which two complex amplitudes must be integrated for each mode. When the function P is highly singular, as it is for example when the field has a fixed number of photons in a mode, this general representation is a good deal simpler to use; it leads to a binomial or Bernoulli distribution for the number of photons counted.

Another particularly useful example is one in which a single mode of the field is excited to a chaotic, or noisy state. The function $P(\{\alpha_k\})$ then is given by Eq. (8.143) with all of the $\{\langle n_k \rangle\}$ equal to zero save one. If we define an average counting rate w by writing

$$\langle C(t) \rangle = \langle \Omega(\{\alpha_k\},t) \rangle = wt, \tag{8.166}$$

then we find in this case that the Gaussian integral for the generating function has the value

$$Q(\lambda,t) = \frac{1}{1+\lambda wt}. \tag{8.167}$$

It follows that the photon counting probabilities have the geometrical distribution

$$p(n,t) = \frac{(wt)^n}{(1+wt)^{n+1}}. \tag{8.168}$$

A somewhat less elementary case is the superposition of a coherent signal and a noise excitation in a single mode. For this case we use the P function given by Eq. (8.147). The average counting rate w defined by Eq. (8.166) is seen to consist of two terms in this case, one contributed by the noise excitation and the other by the signal. We may write

$$w = w_S + w_\mathcal{N}, \tag{8.169}$$

where w_S and $w_\mathcal{N}$ are proportional to the intensities of their respective fields. Their ratio, in other words, is

$$\frac{w_S}{w_\mathcal{N}} = \frac{|\beta|^2}{\langle n \rangle}, \tag{8.170}$$

which is the signal-to-noise ratio. Now if we let

$$\begin{aligned} S &= w_S t, \\ \mathcal{N} &= w_\mathcal{N} t \end{aligned} \tag{8.171}$$

be the numbers of quanta which would be counted in the signal and noise fields measured individually, then we find on evaluating the elementary integral for the generating function that it can be written as

$$Q(\lambda,t) = \frac{1}{1+\lambda\mathcal{N}} \exp\left[-\frac{\lambda S}{1+\lambda\mathcal{N}}\right]. \tag{8.172}$$

This expression may be recognized as the generating function for the Laguerre polynomials L_m. By expanding it about $\lambda = 1$ we find

$$p(m,t) = \frac{\mathcal{N}^m}{(1+\mathcal{N})^{m+1}} L_m\left[-\frac{S}{\mathcal{N}(1+\mathcal{N})}\right] \exp\left[-\frac{S}{1+\mathcal{N}}\right]. \tag{8.173}$$

When the signal strength $|\beta|^2$ vanishes this distribution assumes the geometrical form given in Eq. (8.168). When, in the opposite extreme, the noise excitation $\langle n \rangle$ vanishes, $p(m,t)$ reduces to the Poisson distribution characteristic of a coherent state.

Measurements which have been made in the beam of a gas laser operating well above threshold yield photon counting distributions which can be compared directly with the distributions in Eq. (8.173). The measured distributions resemble Poisson distributions a good deal more than the geometrical ones of Eq. (8.168), but have appreciably greater variances than Poisson distributions. They are found to be in excellent agreement[25] with the distribution Eq. (8.173) when the laser beam is assumed to be the superposition of a strong coherent excitation and a relatively weak chaotic one. It is possible to increase the strength of the superposed noise by separating part of the beam with a beam splitter, passing it through a random scatterer such as a moving ground glass screen and then rejoining the transmitted beam with the original one. Photon counting distributions measured in such synthetically noisy beams are also in good agreement[26] with the distribution Eq. (8.173).

It is only by letting the number of excited modes become infinite that we can deal with fields of arbitrary spectral intensities. The formalism we have constructed deals with multimode fields of all sorts including the continuum limit, but to evaluate the expression Eq. (8.162) for the generating function requires, naturally enough, somewhat more mathematics for the multimode cases than for the single mode ones. When the function P behaves like a probability density, as it does for all chaotic fields, the problem is very closely related to ones which have already been discussed in detail in classical noise theory. For some of the simpler cases, it is possible to find a closed and exact expression for $Q(\lambda,t)$. This does not complete the solution of the problem however, since there remains the task of finding the probabilities $p(m,t)$. In practice it is only possible to carry out exactly the sequence of differentiations required by Eq. (8.152) if the expression for the generating function is fairly simple in its analytic form. The alternative is to use approximation methods and several of these have been devised.

It is interesting to note that comparisons of the generating function and several of its properties may be made directly with experiment. Let us consider the value of the generating function

$$Q(1,t) = p(0,t). \tag{8.174}$$

That is the probability that no photons are recorded in the time t, a function which is quite measurable for sufficiently short time intervals, i.e. before it too closely approaches zero. Another type of experiment closely related to this one is the measurement of the distribution of the time intervals which elapse before the occurrence of the first photon count. If we call this interval distribution $w(t)$ we clearly have

$$w(t) = -\frac{d}{dt}p(0,t) = -\frac{d}{dt}Q(1,t). \tag{8.175}$$

If the field we are investigating has a mean intensity which varies with time, then we must be careful always to begin the measurements of our time intervals at the same zero of time relative to the development of the field. If the field is statistically stationary on the other hand, the zero points for the time intervals may be chosen arbitrarily.

It is interesting to contrast this experiment with another which measures simply the distribution of time intervals between successive photon counts. To find this distribution, let us assume that the field is stationary. Then the probability that an initial photon count occurs in the interval Δt_0 about time t_0 and that no further photons are counted prior to time t is

$$\Delta t_0 \frac{d}{dt_0} p(0, t - t_0).$$

The probability that an initial count occurs in Δt_0 and the first subsequent one in the interval Δt following time t is

$$-\Delta t_0 \frac{d}{dt_0} \frac{d}{dt} p(0, t - t_0) \Delta t = \Delta t_0 \frac{d^2}{dt^2} p(0, t - t_0) \Delta t. \tag{8.176}$$

The distribution function we seek is the conditioned probability density $w(t_0|t)$ that, given a count at time t_0, the next occurs at time t. If the average counting rate in the stationary field is w, then the probability for the compound event of a count in Δt_0, none in the interval from t_0 to t, and then a count in the interval Δt is evidently

$$w \Delta t_0 \, w(t_0|t) \, \Delta t = \Delta t_0 \frac{d^2}{dt^2} p(0, t - t_0) \Delta t. \tag{8.177}$$

It follows that

$$w(t_0|t) = \frac{1}{w} \frac{d^2}{dt^2} Q(1, t - t_0). \tag{8.178}$$

The distributions $w(t)$ and $w(0|t)$ are the same for a stationary coherent field, but are quite different in general for other types of fields. For a single-mode chaotic excitation, for example, the generating function $Q(1,t)$ is given by Eq. (8.167) and we find

$$w(t) = \frac{w}{(1+wt)^2}, \tag{8.179}$$

$$w(0|t) = \frac{2w}{(1+wt)^3}. \tag{8.180}$$

The conditioned distribution is twice as large as the unconditioned one for $t = 0$ and decreases more rapidly with increasing time. The physical reason for this behavior is easy to see. When we require that a photon be counted at $t = 0$ we are giving additional weight to those cases in which the field strength has fluctuated to abnormally large values at that time. When the field strength is large the second photoabsorption process tends naturally to be hastened.

References

1. R. Hanbury Brown, R. Q. Twiss, *Nature* **177**, 27 (1956); *Proc. Roy. Soc. (London) A* **242**, 300 (1957); *Proc. Roy. Soc. (London) A* **243**, 291 (1957).
2. R. J. Glauber, *Phys. Rev. Lett.* **10**, 84 (1963).
3. R. J. Glauber, *Phys. Rev.* **130**, 2529 (1963); reprinted as Chapter 1 in this volume.
4. R. J. Glauber, *Phys. Rev.* **131**, 2766 (1963); reprinted as Sect. 2.9 in this volume.
5. R. J.Glauber, in *Quantum Optics and Electronics*, C. deWitt, Eds., Gordon & Breach, New York 1965, p. 63; reprinted as Chapter 2 in this volume.
6. R. J. Glauber, in *Physics of Quantum Electronics*, Conference Proceedings, P. L. Kelley et al., Eds., Mc-Graw Hill, New York 1966, p. 788.
7. L. Mandel, *Phys. Rev.* **152**, 438 (1966).
8. U. M. Titulaer, R. J. Glauber, *Phys. Rev. B* **140**, 676 (1965); reprinted as Chapter 3 in this volume.
9. E. Schrödinger, *Naturwiss.* **14**, 644 (1926).
10. R. J. Glauber, *Phys. Lett.* **21**, 650 (1966); reprinted as Chapter 5 in this volume.
11. U. M. Titulaer, R. J. Glauber, *Phys. Rev.* **145**, 1041 (1966); reprinted as Chapter 4 in this volume.
12. E. H. Kennard, *Z. Phys.* **44**, 326 (1927).
13. E. Merzbacher, in *Quantum Mechanics*, Wiley, New York 1961, p. 154.
14. J. R. Klauder, *Ann. Phys. New York* **11**, 123 (1960).
15. K. E. Cahill, *Phys. Rev. B* **138**, 1566 (1965).
16. V. Bargmann, *Commun. Pure Appl. Math.* **14**, 187 (1961); *Proc. Natl. Acad. Sci. USA* **48**, 199 (1962).
17. E. Wigner, *Phys. Rev.* **40**, 749 (1932).
18. Y. Kano, *J. Math. Phys.* **6**, 1913 (1965).
19. C. L. Mehta, E. C. G. Sudarshan, *Phys. Rev. B* **138**, 753 (1965).
20. J. E. Moyal, *Proc. Cambridge Phil. Soc.* **45**, 99 (1948).
21. J. R. Klauder, J. McKenna, D. G. Currie, *J. Math. Phys.* **6**, 734 (1965).
22. E. C. G. Sudarshan, *Phys. Rev. Lett.* **10**, 277 (1963).
23. J. R. Klauder, *Phys. Rev. Lett.* **15**, 534 (1966).
24. J. R. Klauder, E. C. G. Sudarshan, in *Fundamentals of Quantum Optics*, Benjamin, to be published.
25. C. Freed, H. A. Haus, *IEEE J. Quantum Elect.* **QE-2**, 190 (1966).
26. F. T. Arecchi, A. Berne, P. Burlamacchi, *Phys. Rev. Lett.* **16**, 32 (1966).

9
Ordered Expansions in Boson Amplitude Operators[1]

9.1
Introduction

Ambiguities in the ordering of operator products were among the earliest questions to occur in the development of quantum mechanics. In more recent years we have come to understand a close relationship between particular types of measurements and the operator orderings best suited to describing them. In many quantum-mechanical problems, such as those of quantum optics, we are now concerned with the ordering of general operator functions rather than simple operator products.

In the present paper we shall discuss the representation of arbitrary operators that refer to a system which we choose, for simplicity, to have only a single degree of freedom. We shall describe this system in terms of a pair of complex operators, a and $a^\dagger$, which we refer to as the annihilation and creation operators. These operators, which obey the commutation relation $[a, a^\dagger] = 1$, play a fundamental role in descriptions of systems of harmonic oscillators and quantized fields. Operators with the same algebraic properties may be defined for a broad class of different dynamical systems by forming complex linear combinations of pairs of observables q and p that are canonically conjugate, $[q, p] = i\hbar$.

Because of ordering problems which arise from the noncommutativity of the operators a and $a^\dagger$, the representation of operators is a considerably richer subject than the representation of c-number functions. We examine several aspects of this structure in the present paper with particular emphasis upon the types of ordering that are most useful in the description of physical experiments. In the paper which follows, we shall apply this analysis to the closely related problem of expressing quantum-mechanical ensemble averages in forms that offer, as much as possible, the simplicity of classical ensemble averages. The present discussion of operator ordering will provide the basis for our discussion there of the P representation, the Wigner distribution, and other ways of representing density operators.

It is conventional to distinguish the products $(a^\dagger)^n a^m$ and $a^m (a^\dagger)^n$ by calling the first normally ordered and the second antinormally ordered. In general an operator is said to be normally (antinormally) ordered if the operator a stands always to the right

---

[1] Reprinted with permission from K. E. Cahill, R. J. Glauber, *Phys. Rev.* **177**, 1857–1881 (1969). Copyright 2006 by the American Physical Society.

Quantum Theory of Optical Coherence. Selected Papers and Lectures. Roy J. Glauber.
Copyright © 2007 WILEY-VCH Verlag GmbH & Co. KGaA, Weinheim
ISBN: 978-3-527-40687-6

(left) of the operator $a^\dagger$. It is clear that any polynomial in the operators a and $a^\dagger$ may be cast into normally or antinormally ordered form by using the commutation relation a finite number of times.

A problem we discuss in detail is that of expanding an arbitrary operator as an ordered power series in the operators a and $a^\dagger$. We show that virtually every operator of interest possesses a convergent power-series expansion in the normally ordered products $(a^\dagger)^n a^m$. Power-series expansions in the antinormally ordered products $a^m (a^\dagger)^n$ are, however, of considerably less generality. We show that for many well-behaved operators, ones that are bounded and of finite trace, the required c-number coefficients are infinite.

These two varieties of ordering have been discussed recently in a number of references and some discussion has also been given of other varieties of ordering[1–6]. The emphasis of the work published to date has been principally upon the development of formal expressions for ordered operators rather than upon their explicit meaning. Since those meanings are not in all cases self-evident and we want in any case to understand their limitations, we have found it necessary to exercise some mathematical care in the development of our arguments.

In order to examine more closely the role of operator ordering in power-series expansions and to be able to interpolate between normal and antinormal ordering, we have been led to introduce a parametrized ordering convention. This convention associates with every complex number s a unique way of ordering all products of the operators a and $a^\dagger$. Normal ordering, antinormal ordering, and a type of ordering that is symmetric in the operators a and $a^\dagger$ correspond to the values $s = +1, -1$, and 0, respectively, of the order parameter s. We are not attempting by introducing this continuum of orderings to deal with all possible forms of ordering nor do we know of any physical applications for arbitrarily ordered operator functions. We shall continue to center our attention on the three principal forms of ordering. By embedding them in a continuum of orderings, we provide for them a natural context for viewing their differences and interrelationships.

Our parametrized ordering convention enables us to consider the existence of ordered power-series expansions for intermediate orderings. We show that for bounded operators the required c-number series coefficients are finite whenever the ordering is closer to normal ordering then to antinormal ordering, i.e., for $\operatorname{Re} s > 0$. We show as well that for such operators the power series converge, in a sense which we make precise in Sect. 9.4, when the ordering is closer to normal than to symmetric ordering, or, more precisely, for $\operatorname{Re} s > \frac{1}{2} + \frac{1}{2}(\operatorname{Im} s)^2$.

These ordered power-series expansions and the associated formulas for the c-number coefficients afford for every value of the order parameter s a relatively direct type of correspondence between operators and c-number functions. The rule characterizing each correspondence is to replace the s-ordered products $\{(a^\dagger)^n a^m\}_s$, in the power-series expansion of a given operator by the monomials $(\alpha^*)^n \alpha^m$. The properties of these correspondences, which are one to one, depend markedly upon the order parameter s. The classes of operators for which they are appropriate and the types of functions which they associate with different classes of operators are discussed in Sect. 9.7. The correspondence specified by $s = 0$ and associated with symmetric ordering may be identified with a correspondence introduced by Weyl[6].

Our main interest in these correspondences is the possibility of using the function associated with a given operator as a weight function in an integral representation of that operator. This possibility is realized in Sections 9.6 and 9.7, where for each correspondence we introduce a set of operators that forms a basis for such an integral representation. We discuss the properties of these representations and of their weight functions in Sections 9.6–9.8.

We begin with two introductory sections. Section 9.2 summarizes the useful properties of the coherent states and of the displacement operators $\exp(\alpha a^\dagger - \alpha^* a)$. In Sect. 9.3 we discuss the properties of an integral representation for arbitrary operators that is based upon the displacement operators. This expansion, which is due to Weyl[6], provides the basis for much of our subsequent analysis.

9.2
Coherent States and Displacement Operators

An important part of our discussion will be based upon the use of a particular set of quantum states. If the system being studied is the electromagnetic field, these are the states that describe completely coherent fields. The systematic use of such coherent states has been found particularly well suited to the solution of a number of quantum-electrodynamical problems and problems of other types involving harmonic-oscillator degrees of freedom. In this section we shall indicate how states of a corresponding type may be defined for a broad range of physical systems. This extension of the definition of coherent states furnishes the opportunity to review their properties briefly.

Let us consider for simplicity a dynamical system that is described by a single pair of Hermitian observables q and p which are canonically conjugate, $[q, p] = i\hbar$, and have eigenvalues which range continuously from minus infinity to plus infinity. The operators we shall use in defining the coherent states are complex linear combinations of q and p. Corresponding to any real parameter λ, different from zero, we may form the combinations

$$a = \frac{1}{(2\hbar)^{1/2}} \left(\lambda q + \frac{i}{\lambda} p \right), \tag{9.1}$$

$$a^\dagger = \frac{1}{(2\hbar)^{1/2}} \left(\lambda q - \frac{i}{\lambda} p \right). \tag{9.2}$$

These operators satisfy the familiar commutation relation

$$[a, a^\dagger] = 1 \tag{9.3}$$

and therefore possess the same algebraic properties as the operators associated with the complex amplitude of a harmonic oscillator or the photon annihilation and creation operators of quantum electrodynamics.

The operator a possesses, as we shall see, a two-dimensional continuum of eigenstates. Let us first note that it has a particularly simple eigenstate corresponding to the eigenvalue zero. If we denote this state by $|0\rangle$, then its defining equation

$$a|0\rangle = 0 \qquad (9.4)$$

may be written in the representation in which q is diagonal as

$$\left(\lambda^2 q' + \hbar \frac{d}{dq'}\right)\langle q'|0\rangle = 0. \qquad (9.5)$$

The wave function for this state is therefore a Gaussian function which takes the normalized form

$$\langle q'|0\rangle = \left(\frac{\lambda^2}{\pi\hbar}\right)^{1/4} \exp\left[-\frac{(\lambda q')^2}{2\hbar}\right]. \qquad (9.6)$$

This state, incidentally, is one for which the uncertainty product $\Delta q \Delta p$ assumes its minimum value of $\tfrac{1}{2}\hbar$ with the ratio $\Delta p/\Delta q = \lambda^2$.

It is clear from the algebraic properties of the operators a and $a^\dagger$ that we may construct, just as in the case of the harmonic oscillator, a sequence of states for the system which correspond to integer eigenvalues for the product $a^\dagger a$. These states, which we label as $|n\rangle$, satisfy the equation

$$a^\dagger a |n\rangle = n|n\rangle \qquad (9.7)$$

for $n = 0, 1, 2, \ldots$ They may be generated from the state $|0\rangle$ by the rule

$$|n\rangle = \frac{1}{(n!)^{1/2}}(a^\dagger)^n |0\rangle. \qquad (9.8)$$

The wave functions $\langle q'|n\rangle$ which represent these states are easily seen to take the same form as the familiar stationary-state wave functions for the harmonic oscillator. They are the Hermite functions which form a complete basis for the expansion of any quadratically integrable function. These states $|n\rangle$ therefore form a complete set.

It may be noted that no specification has been made of the Hamiltonian of the system that we are discussing. Since the system may have arbitrary dynamical behavior, the states $|n\rangle$ will not in general be energy eigenstates. Only when the Hamiltonian is a function of the operator

$$a^\dagger a = \frac{1}{2\hbar}\left(\frac{p^2}{\lambda^2} + \lambda^2 q^2 - \hbar\right) \qquad (9.9)$$

will the states $|n\rangle$ be stationary. The Hamiltonian of a harmonic oscillator of mass m and angular frequency ω is

$$H = \frac{1}{2m}(p^2 + m^2\omega^2 q^2). \tag{9.10}$$

The states $|n\rangle$ become the stationary states of this Hamiltonian when the arbitrary parameter λ is given the value $(m\omega)^{1/2}$. A normal mode of the electromagnetic field in a dielectric possesses a Hamiltonian of the form of Eq. (9.9) with $\lambda = \sqrt{\epsilon\omega}$, where ϵ is the dielectric constant. In this case the state $|n\rangle$ is an n-photon state.

Let us now define for each complex number α the exponential operator[7]

$$D(\alpha) = \exp(\alpha a^\dagger - \alpha^* a), \tag{9.11}$$

which is unitary and obeys the relation

$$D^\dagger(\alpha) = D^{-1}(\alpha) = D(-\alpha). \tag{9.12}$$

These operators, which were introduced by Weyl[6], are easily written in forms in which the operator a appears exclusively to the right or to the left of the operator $a^\dagger$. By means of the identity[8]

$$\exp A \exp B = \exp(A + B + \tfrac{1}{2}[A,B]), \tag{9.13}$$

which holds whenever the commutator $[A,B]$ commutes with both A and B, we find as the normally ordered form

$$D(\alpha) = e^{-|\alpha|^2/2} e^{\alpha a^\dagger} e^{-\alpha^* a} \tag{9.14}$$

and as the antinormally ordered form

$$D(\alpha) = e^{|\alpha|^2/2} e^{-\alpha^* a} e^{\alpha a^\dagger}. \tag{9.15}$$

By differentiating these two ordered forms of the operators $D(\alpha)$ with respect to $-\alpha^*$, we find[2]

$$-\frac{\partial}{\partial \alpha^*} D(\alpha) = D(\alpha)(a + \tfrac{1}{2}\alpha)$$
$$= (a - \tfrac{1}{2}\alpha)D(\alpha), \tag{9.16}$$

which implies the relation

$$D^{-1}(\alpha) a D(\alpha) = a + \alpha \tag{9.17}$$

and its adjoint

$$D^{-1}(\alpha) a^\dagger D(\alpha) = a^\dagger + \alpha^*. \tag{9.18}$$

---

[2] In this and in all subsequent differentiations with respect to conjugate complex variables we observe the convention $d\alpha/d\alpha^* = d\alpha^*/d\alpha = 0$.

Because of their property of displacing the operators a and $a^\dagger$ in a conjugate fashion, the operators $D(\alpha)$ have been called displacement operators.

It is evident from their definition that the displacement operators obey a simple multiplication law. By applying the identity Eq. (9.13) we find

$$D(\alpha)D(\beta) = D(\alpha+\beta)\exp\left[\tfrac{1}{2}(\alpha\beta^* - \alpha^*\beta)\right].\tag{9.19}$$

Thus, apart from unimodular phase factors, the displacement operators form an Abelian group.

For each complex number α the coherent state $|\alpha\rangle$ is defined by[7]

$$|\alpha\rangle = D(\alpha)|0\rangle.\tag{9.20}$$

From Eqns. (9.4) and (9.17) it is clear that the state $|\alpha\rangle$ is an eigenstate of the operator a with eigenvalue α,

$$a|\alpha\rangle = \alpha|\alpha\rangle.\tag{9.21}$$

The displacement transformation (9.20) leaves the variances of the coordinate and momentum variables unaltered so that they have for all coherent states the values

$$(\Delta p)^2 = \lambda^4(\Delta q)^2 = \tfrac{1}{2}\lambda^2\hbar,\tag{9.22}$$

which are characteristic of the ground state Eq. (9.6).

By using Eqns. (9.4), (9.8), (9.14), and (9.20), we may relate the coherent states to the states $|n\rangle$:

$$\begin{aligned}|\alpha\rangle &= D(\alpha)|0\rangle \\ &= e^{-|\alpha|^2/2}\,e^{\alpha a^\dagger}\,e^{-\alpha^* a}\,|0\rangle \\ &= e^{-|\alpha|^2/2}\,e^{\alpha a^\dagger}\,|0\rangle \\ &= e^{-|\alpha|^2/2}\sum_{n=0}^{\infty}\frac{1}{(n!)^{1/2}}\alpha^n|0\rangle.\end{aligned}\tag{9.23}$$

This expansion and the orthonormality of the basis states $|n\rangle$ allow us to write the scalar product $\langle\beta|\alpha\rangle$ in the form

$$\langle\beta|\alpha\rangle = \exp(-\tfrac{1}{2}|\alpha|^2 - \tfrac{1}{2}|\beta|^2 + \beta^*\alpha),\tag{9.24}$$

which shows that no two coherent states are orthogonal. The extent to which they overlap

$$|\langle\beta|\alpha\rangle|^2 = e^{-|\alpha-\beta|^2}\tag{9.25}$$

is, however, negligibly small when the states are macroscopically distinguishable, i.e., when $|\alpha-\beta|\gg 1$.

The coherent states provide a convenient representation for the unit or identity operator 1. Let us define as a real element of area in the complex α plane

$$\frac{1}{\pi}d^2\alpha = \frac{1}{\pi}d(\operatorname{Re}\alpha)\,d(\operatorname{Im}\alpha)\,, \tag{9.26}$$

which, by writing

$$d\alpha = \frac{1}{(2\hbar)^{1/2}}\left(\lambda dq' + \frac{i}{\lambda}dp'\right),$$

we may recognize as the familiar differential element of phase space,

$$\frac{1}{\pi}d^2\alpha = \frac{1}{h}dq'\,dp'\,.$$

Then on using the expansion Eq. (9.23) and integrating over the complex plane, we find

$$\int |\alpha\rangle\langle\alpha|\pi^{-1}d^2\alpha = \sum_{n=0}^{\infty}|n\rangle\langle n|$$
$$= 1\,, \tag{9.27}$$

which illustrates the completeness of the coherent states. This relation affords for the trace of an arbitrary operator F the simple expression

$$\operatorname{Tr}\{F\} = \operatorname{Tr}\left\{F\int|\alpha\rangle\langle\alpha|\pi^{-1}d^2\alpha\right\}$$
$$= \int\langle\alpha|F|\alpha\rangle\pi^{-1}d^2\alpha\,. \tag{9.28}$$

9.3 Completeness of Displacement Operators

The displacement operators $D(\alpha)$, which were defined in Sect. 9.2, possess a number of simple properties which will be particularly useful in the analysis which follows. Principal among these is the representation that they afford for a certain class of operators as weighted integrals with square-integrable weight functions. In this representation, as we shall see, they play a role very much analogous to that of the unimodular exponential functions in the Fourier integral representation of square-integrable functions. Although the completeness property has only recently been formulated rigorously[2,9], the displacement operators have often been discussed in the literature[6,7,10–15]. Because our analysis is cast in terms of the complex operators a and $a^\dagger$, as opposed to their real parts q and p, we shall need a notation for the Fourier transform that is more suited to complex numbers than to pairs of real numbers. If $g(\xi)$

is a function of the complex variable ξ, then we define its complex Fourier transform $f(\alpha)$ by the relation

$$f(\alpha) = \int e^{\alpha\xi^* - \alpha^*\xi} g(\xi) \pi^{-1} d^2\xi, \tag{9.29}$$

where the integration is over the whole ξ plane, and the element $d^2\xi$ is defined by Eq. (9.26). By letting $\alpha = x + iy$ and $\xi = u + iv$, we may write $f(\alpha)$ in the form

$$f(x+iy) = \int_{-\infty}^{\infty}\int_{-\infty}^{\infty} e^{2i(yu-xv)} g(u+iv) \pi^{-1} du\, dv, \tag{9.30}$$

from which it is evident that $f(\alpha)$ differs from the usual Fourier transform only by a scale change of its arguments $\text{Re}\,\alpha$ and $\text{Im}\,\alpha$.

By recalling the familiar formula for the Fourier integral representation of the product of two δ functions

$$\delta(x)\delta(y) = \int_{-\infty}^{\infty}\int_{-\infty}^{\infty} e^{i(xk+yk')} \frac{dk\, dk'}{(2\pi)^2},$$

we may express the two-dimensional δ function

$$\delta^{(2)}(\alpha) = \delta(\text{Re}\,\alpha)\,\delta(\text{Im}\,\alpha)$$

in the more convenient form

$$\delta^{(2)}(\alpha) = \int e^{\alpha\xi^* - \alpha^*\xi} \pi^{-2} d^2\xi. \tag{9.31}$$

In the present notation the processes of Fourier transformation and Fourier inversion are completely symmetrical. For, as is easily demonstrated with the aid of Eq. (9.31), if the functions $f(\alpha)$ and $g(\alpha)$ stand in the relationship

$$f(\alpha) = \int e^{\alpha\xi^* - \alpha^*\xi} g(\xi) \pi^{-1} d^2\xi, \tag{9.32}$$

then the inverse relationship is

$$g(\alpha) = \int e^{\alpha\xi^* - \alpha^*\xi} f(\xi) \pi^{-1} d^2\xi. \tag{9.33}$$

Thus, if $f(\alpha)$ is the complex Fourier transform of $g(\alpha)$, then $g(\alpha)$ is the complex Fourier transform of $f(\alpha)$.

The form which other familiar relations assume in this notation may be similarly derived through the use of the definition Eq. (9.29) and Eq. (9.31). If the functions $f_1(\alpha)$ and $f_2(\alpha)$ are the complex Fourier transforms of the functions $g_1(\xi)$ and $g_2(\xi)$, then we have

$$\int f_1^*(\alpha) f_2(\alpha) d^2\alpha = \int g_1^*(\xi) g_2(\xi) d^2\xi, \tag{9.34}$$

and

$$\int f_1(\alpha) f_2(\alpha) \, d^2\alpha = \int g_1(\xi) g_2(-\xi) \, d^2\xi \, . \tag{9.35}$$

Let us digress for a moment and remind ourselves of the class of functions to which the definition Eq. (9.29) and the relations (9.32)–(9.35) apply. This class, which is known as the Lebesgue class L_2, consists of all ordinary (i.e., measurable) functions $f(\alpha)$ for which the norm $||f||$ defined by

$$||f|| = \left(\int |f(\alpha)|^2 \pi^{-1} \, d^2\alpha \right)^{1/2} \tag{9.36}$$

is finite[16,17]. (For our purposes the term *measurable* may be largely ignored, and the terms L_2 and square-integrable used interchangeably.) Under the operation of Fourier transformation the class of L_2 functions is mapped onto itself in a one-to-one fashion. If in Eq. (9.32) the function $g(\xi)$ is in L_2, so is $f(\alpha)$, and the inverse relation (9.33) is valid. If in Eqns. (9.34) and (9.35) the functions $g_1(\xi)$ and $g_2(\xi)$ are in L_2, then so are the functions $f_1(\alpha)$ and $f_2(\alpha)$, and Eqns. (9.34) and (9.35) hold. For L_2 functions the Schwarz inequality takes the form

$$\left| \int f(\alpha) g(\alpha) \pi^{-1} \, d^2\alpha \right| \leq \int |f(\alpha) g(\alpha)| \pi^{-1} \, d^2\alpha \leq ||f|| \, ||g|| \, . \tag{9.37}$$

The notation that we have introduced may be used to derive a useful property of the traces of the displacement operators. By using the trace relation (9.28) we may write the trace of the operator (α) in the form

$$\mathrm{Tr}\{D(\alpha)\} = \int \langle \xi | D(\alpha) | \xi \rangle \pi^{-1} \, d^2\xi \, .$$

If we now write the displacement operator in its normally ordered form Eq. (9.14) and employ the eigenvalue property of the coherent states, we find

$$\mathrm{Tr}\{D(\alpha)\} = e^{-|\alpha|^2/2} \int \left\langle \xi \left| e^{\alpha a^\dagger} e^{-\alpha^* a} \right| \xi \right\rangle \pi^{-1} \, d^2\xi$$

$$= e^{-|\alpha|^2/2} \int e^{\alpha \xi^* - \alpha^* \xi} \pi^{-1} \, d^2\xi \, .$$

By referring to the δ function formula Eq. (9.31), we recognize this as

$$\mathrm{Tr}\{D(\alpha)\} = \pi e^{-|\alpha|^2/2} \delta^{(2)}(\alpha)$$

or, since the exponential is unity at $\alpha = 0$,

$$\mathrm{Tr}\{D(\alpha)\} = \pi \delta^{(2)}(\alpha) \, . \tag{9.38}$$

From this relation and the multiplication law Eq. (9.19), we find

$$\mathrm{Tr}\{D(\alpha) D^{-1}(\beta)\} = \pi \delta^{(2)}(\alpha - \beta) \, , \tag{9.39}$$

which may be viewed as a species of orthogonality rule for the displacement operators[14].

Before presenting a derivation of the completeness property of the displacement operators, we shall use their orthogonality to suggest the main features of their completeness. Let us suppose that the operator F and the function $f(\xi)$ stand in the relation

$$F = \int f(\xi) D^{-1}(\xi) \pi^{-1} d^2\xi . \tag{9.40}$$

Then we may write

$$\text{Tr}\{FD(\alpha)\} = \int f(\xi) \, \text{Tr}\{D^{-1}(\xi) D(\alpha)\} \, \pi^{-1} d^2\xi ,$$

which, by virtue of the orthogonality rule Eq. (9.38), implies that

$$f(\alpha) = \text{Tr}\{FD(\alpha)\} . \tag{9.41}$$

To see whether this solution is well behaved, we write

$$\text{Tr}\{F^\dagger F\} = \int f(\xi) \, \text{Tr}\{F^\dagger D^{-1}(\xi)\} \, \pi^{-1} d^2\xi$$
$$= \int |f(\xi)|^2 \pi^{-1} d^2\xi , \tag{9.42}$$

which implies that the function $f(\xi)$ is square-integrable when the trace $\text{Tr}[F^\dagger F]$ is finite. This trace is closely related to the Hilbert–Schmidt norm $||F||$ for the operator F which is defined by

$$||F|| = \left(\text{Tr}\{F^\dagger F\}\right)^{1/2} . \tag{9.43}$$

We shall say that an operator is *bounded* if this norm is finite. With this terminology, we may draw from Eqns. (9.40)–(9.42) the conclusion that if a bounded operator F possesses the expansion (9.40), then the weight function $f(\xi)$ is given by Eq. (9.41), is unique, is in L_2, and has the same norm as the operator F, i.e.,

$$||f|| = ||F|| . \tag{9.44}$$

We shall now derive the expansion Eq. (9.40) and several other identities. Our method will utilize the properties of the matrix elements of the displacement operators between coherent states. Let us denote by the symbol $I(\alpha, \beta, \gamma, \delta)$ the integral

$$I(\alpha, \beta, \gamma, \delta) = \int \langle \beta|D(\xi)|\alpha\rangle \langle \gamma|D(-\xi)|\delta\rangle \, \pi^{-1} d^2\xi . \tag{9.45}$$

We note that by writing the displacement operators in their normally ordered form Eq. (9.14) and using the eigenvalue property of the coherent states, we may write this integral in the form

$$I(\alpha, \beta, \gamma, \delta) = \langle \beta|\alpha\rangle \langle \gamma|\delta\rangle \int \exp\left[\xi(\beta - \gamma)^* - \xi^*(\alpha - \delta) - |\xi|^2\right] \pi^{-1} d^2\xi .$$

The Gaussian integral which occurs here occurs in a number of other places in the course of this paper; it is evaluated explicitly in Appendix A, Eq. (9.A2). By applying that result we find

$$I(\alpha,\beta,\gamma,\delta) = \langle\beta|\alpha\rangle \langle\gamma|\delta\rangle \exp[(\beta-\gamma)^*(\delta-\alpha)] \;.$$

If we simplify this expression by using Eq. (9.24) for the scalar product of two coherent states, we find

$$I(\alpha,\beta,\gamma,\delta) = \langle\gamma|\alpha\rangle \langle\beta|\delta\rangle$$

and so obtain the integral identity

$$\langle\gamma|\alpha\rangle \langle\beta|\delta\rangle = \int \langle\beta|D(\xi)|\alpha\rangle \langle\gamma|D(-\xi)|\delta\rangle \, \pi^{-1} d^2\xi \,, \tag{9.46}$$

which holds for all complex numbers α, β, γ, and δ.

Let us now recall that a matrix element identity of the form $\langle b|F|c\rangle = \langle b|G|c\rangle$, if it holds for a complete set of states $|b\rangle, |c\rangle, \ldots$, is equivalent to the corresponding operator equation $F = G$. Thus, since the coherent states form a complete set of states as shown by Eq. (9.27)), we see that the integral identity Eq. (9.46) implies the operator identities

$$|\alpha\rangle\langle\beta| = \int \langle\beta|D(\xi)|\alpha\rangle \, D(-\xi) \, \pi^{-1} d^2\xi \tag{9.47}$$

and

$$1\langle\gamma|\alpha\rangle = \int D(\xi) \, |\alpha\rangle\langle\gamma| \, D(-\xi) \, \pi^{-1} d^2\xi \,, \tag{9.48}$$

where 1 is the identity operator.

We now note that by using twice the resolution Eq. (9.27) of the identity operator we may express an arbitrary bounded operator F as an integral over the coherent-state dyadics $|\alpha\rangle\langle\beta|$ in the form

$$F = \int |\alpha\rangle\langle\beta| \langle\alpha|F|\beta\rangle \, \pi^{-2} d^2\alpha \, d^2\beta \,. \tag{9.49}$$

Thus if we substitute in this integral the expression Eq. (9.47) for the dyadic $|\alpha\rangle\langle\beta|$, we arrive at the expansion

$$F = \langle\alpha|F|\beta\rangle \langle\beta|D(\xi)|\alpha\rangle \, D(-\xi) \, \pi^{-3} d^2\xi \, d^2\alpha \, d^2\beta$$

and by using Eqns. (9.27) and (9.28) we find

$$F = \langle\alpha|FD(\xi)|\alpha\rangle \, D(-\xi) \, \pi^{-2} d^2\xi \, d^2\alpha$$
$$= \int \mathrm{Tr}\{FD(\xi)\} \, D(-\xi) \, \pi^{-1} d^2\xi \,, \tag{9.50}$$

which is the expansion Eq. (9.40). This argument concludes our derivation of the completeness property of the displacement operators.

Let us now observe that by multiplying both sides of expansion Eq. (9.50) by an arbitrary bounded operator G and forming the trace, we obtain the identity

$$\text{Tr}\{FG\} = \int \text{Tr}\{FD(\xi)\}\,\text{Tr}\{GD(-\xi)\}\,\pi^{-1}\,d^2\xi\,. \tag{9.51}$$

When the operators F and G are both bounded, the left-hand side of this equation is finite because of the Schwarz inequality

$$|\text{Tr}\{FG\}| \leq ||F||\,||G||\,. \tag{9.52}$$

By referring to the inequality (9.37) and Eqns. (9.41) and (9.44), we may impose the same upper bound upon the right-hand side of Eq. (9.51).

If we multiply both sides of the identity (9.48) by the coherent-state matrix element $\langle \alpha|F|\gamma\rangle$ of an arbitrary operator F and integrate over α and γ by using Eqns. (9.27) and (9.28), then we arrive at the identity

$$1\,\text{Tr}\{F\} = \int D(\xi)\,F\,D^{-1}(\xi)\,\pi^{-1}\,d^2\xi\,. \tag{9.53}$$

By multiplying both sides of this equation by an arbitrary operator G, we may obtain the relations

$$G\,\text{Tr}\{F\} = \int GD(\xi)\,F\,D^{-1}(\xi)\,\pi^{-1}\,d^2\xi \tag{9.54}$$

and

$$\text{Tr}\{G\}\,\text{Tr}\{F\} = \int \text{Tr}\{GD(\xi)\,F\,D^{-1}(\xi)\}\,\pi^{-1}\,d^2\xi\,. \tag{9.55}$$

Let us now observe that if some set of states $|a\rangle, |b\rangle, |c\rangle, \ldots$ forms a complete orthonormal set, then the matrix elements $\langle a|D(\xi)|b\rangle \ldots$ form a complete orthonormal set of functions. For if we let $F = |c\rangle\langle d|$ and $G = |b\rangle\langle a|$ then Eq. (9.55) becomes

$$\delta_{a,b}\,\delta_{c,d} = \int \langle a|D(\xi)|c\rangle\,\langle d|D^{-1}(\xi)|b\rangle\,\pi^{-1}\,d^2\xi$$

$$= \int \langle a|D(\xi)|c\rangle\,\langle b|D(\xi)|d\rangle^*\,\pi^{-1}\,d^2\xi\,, \tag{9.56}$$

which verifies the orthonormality property. The statement of completeness follows from Eq. (9.39), which, when two factors of the identity operator in the form

$$1 = \sum_a |a\rangle\langle a|$$

are inserted, becomes

$$\sum_{a,b} \langle a|D(\xi)|b\rangle\,\langle a|D(\xi')|b\rangle^* = \pi\delta^{(2)}(\xi-\xi')\,. \tag{9.57}$$

9.3 Completeness of Displacement Operators

As we show in Appendix B, when the orthonormal states are the basis states $|n\rangle$, the matrix elements $\langle m|D(\xi)|n\rangle$ assume the form

$$\langle m|D(\xi)|n\rangle = \left(\frac{n!}{m!}\right)^{1/2} \xi^{m-n} e^{-|\xi|^2/2} L_n^{(m-n)}(|\xi|^2), \qquad (9.58)$$

where $L_q^{(p)}(x)$ is an associated Laguerre polynomial[18].

Let us now return to the expansion Eq. (9.40) which, as we have seen, affords a one-to-one correspondence

$$F \leftrightarrow f(\xi) = \text{Tr}\{FD(\xi)\} \qquad (9.59)$$

between the class of all bounded operators F and the class L_2 of all square-integrable functions $f(\xi)$. In view of Eq. (9.44) this correspondence is norm-preserving in the sense that the L_2 norm $||f||$ of the weight function $f(\xi)$ is equal to the Hilbert–Schmidt norm $||F||$ of the operator F being expanded. If we write the expansion Eq. (9.40) in the form

$$F = \int \exp\left[a\xi^* - a^\dagger \xi\right] \text{Tr}\{FD(\xi)\} \, \pi^{-1} \, d^2\xi, \qquad (9.60)$$

then we may draw several parallels between it and the complex Fourier transform expansion Eq. (9.29) of an L_2 function $f(\alpha)$. We may observe that the operators a and $a^\dagger$ in Eq. (9.60) correspond to the variables α and α^* in Eq. (9.29) and that the unitary operator $\exp(a\xi^* - a^\dagger \xi)$ in Eq. (9.60) plays the role of the unimodular function $\exp(\alpha \xi^* - \alpha^* \xi)$ in Eq. (9.29). The weight function $\text{Tr}\{FD(\xi)\}$ corresponds to the function $g(\xi)$ and may therefore be thought of as a species of Fourier transform for the operator F. Just as the ordinary Fourier-transform expansion Eq. (9.29) induces a one-to-one correspondence $f(\alpha) \leftrightarrow g(\xi)$ from L_2 onto itself that is norm-preserving, $||f|| = ||g||$, so too the correspondence Eq. (9.59), between L_2 and the class of all bounded operators is one-to-one and norm-preserving.

So far we have been talking about the expansion of operators that are bounded. In the remainder of this section we shall say a few things about the expansion of other types of operators.

We have been using, and shall continue to use, the term *bounded* to denote operators for which the Hilbert–Schmidt norm $||F|| = [\text{Tr}\{F^\dagger F\}]^{1/2}$ is finite. Another operator norm which is frequently used and which we shall denote by $||F||_1$ is the least upper bound (l.u.b.), taken over all states $|\psi\rangle$ of unit norm, $\langle \psi|\psi\rangle = 1$, of the quantity $\langle \psi|F^\dagger F|\psi\rangle$, i.e.,

$$||F||_1 = \underset{\langle\psi|\psi\rangle=1}{\text{l.u.b.}} \langle \psi|F^\dagger F|\psi\rangle. \qquad (9.61)$$

It is, loosely speaking, the largest eigenvalue of the operator $F^\dagger F$. Operators for which the norm $||F||_1$ is finite are often called bounded, but we shall call them finite.

All bounded operators are finite, and we have $||F||_1 < ||F||$. The most important unbounded but finite operators are the unitary operators for which $||U||_1 = 1$.

The displacement operator expansion Eq. (9.40) is not in general suited to the expansion of unbounded operators, even those that are finite. When the operator F is unbounded, the appropriate weight function $\text{Tr}\{FD(\xi)\}$, which must according to Eq. (9.42) lack square-integrability, is often singular. If, for example, we take as the operator F the displacement operator $D(\alpha)$, which being unitary is finite, then the weight function $f(\xi)$ is given by Eq. (9.39) as $\pi \delta^{(2)}(\alpha + \xi)$. For this reason we shall not in this paper attempt a careful formulation of the representation of unbounded operators.

We may observe, however, that the coherent-state expansion Eq. (9.49) affords a representation for a broad class of unbounded operators. In this representation, as in the one afforded by the displacement operators, we find quadratic integrability finked with boundedness. For by using Eq. (9.49) to form the Hilbert–Schmidt norm of the operator F, we secure the result

$$||F||^2 = \text{Tr}\left\{F^\dagger F\right\} = \int |\langle \alpha|F|\beta\rangle|^2 \, \pi^{-2} \, d^2\alpha \, d^2\beta \, . \qquad (9.62)$$

The weight function $\langle \alpha|F|\beta\rangle$ of the representation Eq. (9.49) has no singularities in the finite α, β planes unless the operator F is particularly pathological[7]. Moreover, when F is a finite operator the modulus of $\langle \alpha|F|\beta\rangle$ is bounded by $||F||_1$,

$$|\langle \alpha|F|\beta\rangle| \leq ||F||_1 \, , \qquad (9.63)$$

as may be seen from the definition Eq. (9.61).

For an arbitrary bounded operator F the weight function $\text{Tr}\{FD(\xi)\}$ need not be bounded or continuous. There is, however, a smaller class of operators for which both of these conditions are met. This is the trace class, also called the class of nuclear operators. The derivations of the statements that we shall now make about trace-class operators may be found in Refs[2,19].

Every trace-class operator F may be factored uniquely into the product of unitary operator U and an operator B which is positive definite and of finite trace,

$$F = UB \, . \qquad (9.64)$$

In terms of this decomposition, the trace-class norm is defined as

$$||F||_2 = \text{Tr} B \, . \qquad (9.65)$$

Every density operator ρ is a member of the trace class with $||\rho||_2 = 1$.

Every trace-class operator is bounded, as is shown by the inequality[19]

$$||F||_1 \leq ||F|| \leq ||F||_2 \qquad (9.66)$$

for the three norms which we have mentioned. The product of a finite operator G and a trace-class operator F is a trace-class operator and we have the inequality[19]

$$|\operatorname{Tr} FG| \leq ||F||_2 ||G||_1 . \tag{9.67}$$

From this relation it follows, since the displacement operators are finite with $||D(\xi)||_1 = 1$, that for every trace-class operator F the weight function $\operatorname{Tr}\{FD(\xi)\}$ is uniformly bounded by the trace-class norm $||F||_2$

$$|\operatorname{Tr}\{FD(\xi)\}| \leq ||F||_2 . \tag{9.68}$$

It may also be shown[2] that for every trace-class operator F the weight function $\operatorname{Tr}\{FD(\xi)\}$ is a (uniformly) continuous function of ξ.

9.4
Ordered Power-Series Expansions

The completeness property of the displacement operators, which we discussed in Sect. 9.3, affords a convenient framework for examining various ways in which bounded operators may be represented. In this section and in Sect. 9.5, we discuss the problem of expanding a bounded operator as an ordered power series in the operators a and $a^\dagger$. The cases of normal ordering, of antinormal ordering, and of a type of ordering that is symmetric in the creation and annihilation operators occupy most of the present section. For these orderings we obtain closed integral expressions for the coefficients of the power-series expansions of an arbitrary bounded operator. It will become evident that there is a marked contrast between the properties of normally and antinormally ordered power-series expansions. We show that normally ordered power-series expansions exist and converge in a well-defined sense for a very broad class of operators. Antinormally ordered power series, however, do not afford a completely satisfactory representation for all bounded operators; we show that the appropriate coefficients are singular for large classes of bounded operators. We also consider briefly the expansion of bounded operators as ordered power series in the operators q and p. Because these operators are Hermitian, the series coefficients, as we shall see, tend not to develop singularities.

To begin with, let us consider the possibility of expanding an arbitrary bounded operator F as a normally ordered power series, i.e., in the form

$$F = \sum_{n,m=0}^{\infty} c_{n,m} (a^\dagger)^n a^m , \tag{9.69}$$

where the coefficients $c_{n,m}$ are complex numbers. Since the operator F is assumed to be bounded, it possesses the expansion Eq. (9.40),

$$F = \int \operatorname{Tr}\{FD(\xi)\} D^{-1}(\xi) \pi^{-1} d^2\xi , \tag{9.70}$$

where the weight function $\text{Tr}\{FD(\xi)\}$ is square-integrable. We may formally generate the expansion Eq. (9.69) by writing the operator $D^{-1}(\xi)$ in its normally ordered form Eq. (9.14),

$$F = \text{Tr}\{FD(\xi)\} \, e^{-|\xi|^2/2} e^{-\xi a^\dagger} e^{\xi^* a} \pi^{-1} d^2\xi, \tag{9.71}$$

and then expanding the exponentials in powers of a and $a^\dagger$ so that we have

$$F = \sum_{n,m=0}^{\infty} (a^\dagger)^n a^m \frac{1}{n!\,m!} \int \text{Tr}\{FD(\xi)\} \, e^{-|\xi|^2/2} (-\xi)^n (\xi^*)^m \pi^{-1} d^2\xi. \tag{9.72}$$

We may therefore identify the coefficients $c_{n,m}$ with the integrals

$$c_{n,m} = \frac{1}{n!\,m!} \int \text{Tr}\{FD(\xi)\} \, e^{-|\xi|^2/2} (-\xi)^n (\xi^*)^m \pi^{-1} d^2\xi. \tag{9.73}$$

By using the inequality (9.37) and the relation (9.44), we find that for every bounded operator F the coefficients $c_{n,m}$ are finite and are bounded by the quantities

$$|c_{n,m}| \leq \frac{1}{n!\,m!} [(n+m)!]^{1/2} ||F||, \tag{9.74}$$

where $||F||$ is the Hilbert–Schmidt norm Eq. (9.43).

The presence of the factor $e^{-|\xi|^2/2}$ in the integral Eq. (9.73) for the coefficients $c_{n,m}$ suggests that normally ordered expansions may be appropriate for a larger class of operators than the one being considered. We shall take up this matter, as well as the question of the convergence of the series Eq. (9.69), after we have considered the cases of antinormal and symmetric order.

The coefficients $d_{n,m}$ of the antinormally ordered expansion

$$F = \sum_{n,m=0}^{\infty} d_{n,m} \, a^m (a^\dagger)^n \tag{9.75}$$

of an arbitrary bounded operator F may be identified by an argument entirely similar to that of Eqns. (9.69)–(9.73). (4.5). If we write the operator $D^{-1}(\xi)$ in Eq. (9.69) in its antinormally ordered form Eq. (9.15), we have

$$F = \int \text{Tr}\{FD(\xi)\} \, e^{|\xi|^2/2} e^{\xi^* a} e^{-\xi a^\dagger} \pi^{-1} d^2\xi, \tag{9.76}$$

and, by expanding the exponentials in powers of a and $a^\dagger$, we find

$$d_{n,m} = \frac{1}{n!\,m!} \int \text{Tr}\{FD(\xi)\} \, e^{|\xi|^2/2} (-\xi)^n (\xi^*)^m \pi^{-1} d^2\xi. \tag{9.77}$$

This expression differs from the corresponding one Eq. (9.73) for the coefficients $c_{n,m}$ of the normally ordered series in that the exponential $e^{-|\xi|^2/2}$ has been replaced by $e^{+|\xi|^2/2}$. Since for the class of bounded operators F the weight functions $\text{Tr}\{FD(\xi)\}$

form exactly the class L_2 of square-integrable functions, it is evident that the coefficients $d_{n,m}$ can become singular for certain classes of bounded operators. Examples of such operators are discussed at the end of this section.

So far we have made use of the normally and anti normally ordered expansions

$$D(\alpha) = e^{-|\alpha|^2/2} \sum_{n,m=0}^{\infty} \frac{(\alpha a^\dagger)^n}{n!} \frac{(-\alpha^* a)^m}{m!} \tag{9.78}$$

and

$$D(\alpha) = e^{|\alpha|^2/2} \sum_{n,m=0}^{\infty} \frac{(-\alpha^* a)^m}{m!} \frac{(\alpha a^\dagger)^n}{n!}. \tag{9.79}$$

Let us now examine the expansion

$$D(\alpha) = \exp(\alpha a^\dagger - \alpha^* a)$$

$$= \sum_{n=0}^{\infty} \frac{1}{n!} (\alpha a^\dagger - \alpha^* a)^n, \tag{9.80}$$

in which the operators a and $a^\dagger$ are on an equal footing with respect to order. There are $(n+m)!/n!m!$ different ways of ordering the product of n factors of $a^\dagger$ and m factors of a. Let us denote by the symbol $\{(a^\dagger)^n a^m\}$ the average of these $(n+m)!/n!m!$ differently ordered operator products.

Two examples of this average product, which we shall refer to as the symmetrically ordered product[2,20,21], are

$$\{a^\dagger a\} = \tfrac{1}{2}(a^\dagger a + aa^\dagger)$$

and

$$\{a^\dagger a^2\} = \tfrac{1}{3}(a^\dagger a^2 + aa^\dagger a + a^2 a^\dagger).$$

In this notation we have

$$(\alpha a^\dagger - \alpha^* a)^n = \sum_{r=0}^{n} \frac{n!}{r!(n-r)!} \alpha^r (-\alpha^*)^{n-r} \{(a^\dagger)^r a^{n-r}\}, \tag{9.81}$$

so that we may write the expansion Eq. (9.80) for the displacement operator as

$$D(\alpha) = \sum_{n,m=0}^{\infty} \frac{\alpha^n (-\alpha^*)^m}{n! m!} \{(a^\dagger)^n a^m\}. \tag{9.82}$$

By differentiating these relations we find for the operator $\{(a^\dagger)^n a^m\}$ the expression

$$\{(a^\dagger)^n a^m\} = \frac{\partial^{n+m} D(\alpha)}{\partial \alpha^n \partial (-\alpha^*)^m}\bigg|_{\alpha=0}, \tag{9.83}$$

which may be simplified to the form

$$\{(a^\dagger)^n a^m\} = \frac{\partial^{n+m}}{\partial x^n \partial y^m} \left. \frac{(xa^\dagger + ya)^{n+m}}{(n+m)!} \right|_{x=y=0}, \quad (9.84)$$

where the variables x and y are set to equal zero after the differentiation.

We may now proceed as in the cases of normal and antinormal order to construct for an arbitrary bounded operator F its symmetrically ordered power series expansion,

$$F = \sum_{n,m=0}^{\infty} b_{n,m} \{(a^\dagger)^n a^m\}. \quad (9.85)$$

When the expansion Eq. (9.82) is inserted into the general representation Eq. (9.70), we find

$$F = \sum_{n,m=0}^{\infty} \{(a^\dagger)^n a^m\} \frac{1}{n!\,m!} \int \text{Tr}\{FD(\xi)\} (-\xi)^n (\xi^*)^m \pi^{-1} d^2\xi, \quad (9.86)$$

which implies that the coefficients $b_{n,m}$ are given by the integrals

$$b_{n,m} = \frac{1}{n!\,m!} \int \text{Tr}\{FD(\xi)\} (-\xi)^n (\xi^*)^m \pi^{-1} d^2\xi. \quad (9.87)$$

The class of bounded operators F for which these integrals converge consists therefore of those whose weight functions $\text{Tr}\{FD(\xi)\}$ possess finite moments. While this is not the entire class of bounded operators, it is a much broader class than the class of operators for which the coefficients $d_{n,m}$ of the antinormally ordered expansions are finite. Speaking loosely, we might say that the coefficients $b_{n,m}$ tend to be finite for bounded operators.

Let us now return to the case of normal ordering and take up the question of convergence. A sequence of operators A_n is said to converge weakly to the operator A if for every pair of normalized states $|f\rangle$ and $|g\rangle$ the sequence $\langle f|A - A_n|g\rangle$ converges to zero. This type of convergence is clearly inappropriate as a definition for the convergence of power-series expansions. The state $|g\rangle$,

$$|g\rangle = \frac{\sqrt{6}}{\pi} \sum_{n=0}^{\infty} \frac{1}{n} |n\rangle,$$

for example, is normalized; but the quantities $\langle g|(a^\dagger)^n a^m|g\rangle$ are all infinite, except when $n + m < 2$. Thus, the power-series expansions that we have discussed do not converge weakly unless they terminate.

We shall adopt for simplicity the following criterion[2] for the convergence of ordered power-series expansions. The power series

$$F = \sum_{n,m=0}^{\infty} f_{n,m} \{(a^\dagger)^n a^m\}_{\text{ord}}, \quad (9.88)$$

where the symbol $\{(a^\dagger)^n a^m\}_{\text{ord}}$ denotes an arbitrarily ordered product, will be said to converge if for every pair of coherent states $|\alpha\rangle$ and $|\beta\rangle$ we have

$$\langle\beta|F|\alpha\rangle = \lim_{N,M\to\infty} \sum_{n=0}^{N} \sum_{m=0}^{M} \left\langle\beta\left|f_{n,m}\{(a^\dagger)^n a^m\}_{\text{ord}}\right|\alpha\right\rangle. \quad (9.89)$$

In effect this is weak convergence over the set of coherent states. Since the coherent states form a complete set, an operator is defined uniquely by its power-series expansion if that expansion converges.

In terms of this criterion the question of the convergence of the normally ordered power series Eq. (9.69) is a simple one. Our task is to show that the sequence Eq. (9.89) converges to the matrix element $\langle\beta|F|\alpha\rangle$. The series in question is

$$\langle\beta|F|\alpha\rangle = \langle\beta|\alpha\rangle \sum_{n,m=0}^{\infty} c_{n,m} (\beta^*)^n \alpha^m. \quad (9.90)$$

We note that this series converges if and only if the function $\langle\beta|F|\alpha\rangle / \langle\beta|\alpha\rangle$ is an entire function of the variables β^* and α.

In order to examine the analyticity of this function, let us use Eq. (9.23) to write it in the form

$$\frac{\langle\beta|F|\alpha\rangle}{\langle\beta|\alpha\rangle} = e^{-\beta^*\alpha} \sum_{n,m=0}^{\infty} \frac{\langle n|F|m\rangle}{(n!\,m!)^{1/2}} (\beta^*)^n \alpha^m.$$

This series converges and defines an entire function of β^* and α for an extremely broad class of operators. This is the case, for example, when for some M, R_1, R_2, and $\epsilon > 0$ the inequalities

$$|\langle n|F|m\rangle| \leq M R_1^n R_2^m (n!\,m!)^{1/2-\epsilon} \quad (9.91)$$

are satisfied. Finite operators fulfill these conditions with $R_1 = R_2 = 1$, $\epsilon = \frac{1}{2}$, and $M = \|F\|_1$.

We have shown that an operator F possesses a convergent normally ordered power-series expansion when the series Eq. (9.90) converges and that the latter series converges when the operator F satisfies the condition Eq. (9.91). The very general condition Eq. (9.91) is therefore a sufficient one for the convergence of the normally ordered power series Eq. (9.69).

It is easy to adapt the methods of this section to the expansion of bounded operators as power series in the Hermitian operators q and p. To begin with, we must write the displacement operator $D(\alpha)$ in forms in which the operators q and p appear in definite orders. By using Eqns. (9.1), (9.2) and (9.13), and writing $\alpha = (2\hbar)^{-1}(\lambda q' + i\lambda^{-1} p')$, we find, after putting $\hbar = 1$,

$$D(q', p') \equiv D(\alpha) = e^{i(qp' - pq')}$$
$$= e^{iqp'} e^{-ipq'} e^{-iq'p'/2}$$
$$= e^{-ipq'} e^{iqp'} e^{iq'p'/2}. \quad (9.92)$$

If we now write Eq. (9.70) in the form

$$F = \int \text{Tr}\{FD(q',p')\} e^{-i(qp'-pq')} (2\pi)^{-1} dq' dp', \qquad (9.93)$$

then we may derive in analogy with Eqns. (9.72)–(9.77) and (9.86) three expansions for an arbitrary bounded operator F. These are what we may call the p-ordered power-series expansion

$$F = \sum_{n,m=0}^{\infty} q^n p^m \frac{1}{n!\,m!} \int \text{Tr}\{FD(q',p')\} (-ip')^n (iq')^m e^{-iq'p'} (2\pi)^{-1} dq' dp',$$

$$(9.94)$$

the q-ordered expansion

$$F = \sum_{n,m=0}^{\infty} p^m q^n \frac{1}{n!\,m!} \int \text{Tr}\{FD(q',p')\} (-ip')^n (iq')^m e^{iq'p'} (2\pi)^{-1} dq' dp', \qquad (9.95)$$

and the q–p-symmetric expansion

$$F = \sum_{n,m=0}^{\infty} \{q^n p^m\} \frac{1}{n!\,m!} \int \text{Tr}\{FD(q'p')\} (-ip')^n (iq')^m (2\pi)^{-1} dq' dp', \qquad (9.96)$$

where the q–p-symmetric product is defined by analogy with Eq. (9.83) as

$$\{q^n p^m\} = i^{m-n} \frac{\partial^{n+m} D(q',p')}{\partial (p')^n \partial (q')^m}\bigg|_{q'=p'=0}. \qquad (9.97)$$

Because the commutator $[q,p] = i$ is purely imaginary, the integrals defining the coefficients of these expansions differ from one another only by unimodular factors in their integrands. A bounded operator that possesses one of the expansions (9.94)–(9.96) is therefore very likely to possess the other two. This class of operators is approximately the same as the class for which the symmetrically ordered expansion Eq. (9.85) is appropriate.

Let us now illustrate some of the results of this section by considering some simple examples. Our first one will serve to answer a question which was not raised in the foregoing discussion, namely, whether antinormally ordered expansions are suitable for trace-class operators (defined in Sect. 9.3). The operator $|\alpha\rangle\langle\beta|$, which is the outer product of two coherent states, is both bounded and in the trace class. By using Eqns. (9.14) and (9.21), we find as its weight function the exponential

$$\text{Tr}\{D(\xi)|\alpha\rangle\langle\beta|\} = \langle\beta|\alpha\rangle \exp\left[-\tfrac{1}{2}|\xi|^2 + \xi\beta^* - \xi^*\alpha\right]. \qquad (9.98)$$

Reference to Eqns. (9.73), (9.77) and (9.87) reveals that the coefficients of the normally and symmetrically ordered expansions are finite, while those of the antinormally ordered expansion are singular.

Another operator which is in the trace class is the outer product $|n\rangle\langle m|$ of two states with fixed numbers of quanta. The appropriate weight function is

$$\mathrm{Tr}\{|n\rangle\langle m| D(\xi)\} = \langle m|D(\xi)|n\rangle ,$$

and from the explicit formula Eq. (9.58) for this function and Eqns. (9.73), (9.77) and (9.87) it is evident that the coefficients $c_{n,m}$ and $b_{n,m}$ are finite, while the $d_{n,m}$ are singular. We conclude from these two examples that trace-class operators do not necessarily possess anti-normally ordered power-series expansions.

As a more useful example let us consider the operator

$$F(\lambda) = \lambda^{a^\dagger a} = \exp\left[a^\dagger a \ln \lambda\right] , \tag{9.99}$$

which depends analytically upon the complex parameter λ since the operator $a^\dagger a$ has integer eigenvalues. This operator is finite for $|\lambda| \leq 1$ and is both bounded and in the trace class for $|\lambda| < 1$. The operator $F(\lambda)$ has many applications in thermal equilibrium statistics where we usually have $\lambda = \exp(-\hbar\omega/kT) < 1$, where k is the Boltzmann constant and T is the temperature.

We may find directly the normally ordered power series for $F(\lambda)$ by differentiating it with respect to λ:

$$\frac{dF(\lambda)}{d\lambda} = a^\dagger a \lambda^{a^\dagger a - 1} . \tag{9.100}$$

Then since for any function f we have

$$af(a^\dagger a) = f(a^\dagger a + 1)a , \tag{9.101}$$

we may write

$$\frac{dF(\lambda)}{d\lambda} = a^\dagger \lambda^{a^\dagger a} a$$

$$= a^\dagger F(\lambda) a . \tag{9.102}$$

The operator $F(\lambda)$ thus satisfies a differential equation whose solution is a normally ordered exponential function of $a^\dagger a$. The initial condition $F(1) = 1$ is satisfied by the solution

$$F(\lambda) =: e^{(\lambda-1)a^\dagger a} : , \tag{9.103}$$

where the symbol : : means that the exponential is in normally ordered form. The coefficients $c_{n,m}$ of this expansion are given by

$$c_{n,m} = \delta_{n,m}(n!)^{-1}(\lambda-1)^n . \tag{9.104}$$

They are entire functions of λ. Since the operator $F(\lambda)$ as is easily shown, satisfies the conditions Eq. (9.91), its normally ordered expansion Eq. (9.103) converges for all λ.

By using Eqns. (9.15), (9.28) and (9.103), we may write the weight function for $F(\lambda)$ in the form

$$\mathrm{Tr}\{F(\lambda)D(\xi)\} = e^{|\xi|^2/2}\,\mathrm{Tr}\left\{e^{\xi a^\dagger} F(\lambda) e^{-\xi^* a}\right\}$$
$$= e^{|\xi|^2/2} \int \exp\left[\xi\alpha^* - \xi^*\alpha + (\lambda-1)|\alpha|^2\right](\pi)^{-1}\mathrm{d}^2\alpha, \quad (9.105)$$

an integral whose value is given by the general expression Eq. (9.A2) as

$$\mathrm{Tr}\{F(\lambda)D(\xi)\} = \frac{1}{(1-\lambda)} \exp\left[-\frac{(1+\lambda)|\xi|^2}{2(1-\lambda)}\right]. \quad (9.106)$$

According to Eq. (9.87) the coefficients $b_{n,m}$ of the symmetrically ordered power series for $F(\lambda)$ are given by the integrals

$$b_{n,m}(\lambda) = \frac{\delta_{n,m}}{(n!)^2\,(1-\lambda)} \int \exp\left[-\frac{(1+\lambda)|\xi|^2}{2(1-\lambda)}\right](-|\xi|^2)^n \pi^{-1}\,\mathrm{d}^2\xi,$$

which converge whenever $F(\lambda)$ is bounded, i.e., for $|\lambda| < 1$, and yield

$$b_{n,m}(\lambda) = \frac{2\delta_{n,m}}{n!\,(\lambda+1)} \left(2\frac{\lambda-1}{\lambda+1}\right)^n. \quad (9.107)$$

These coefficients are analytic functions of λ and may be continued outside the region in which the integrals defining them converge. They all have a pole at $\lambda = -1$, where the operator $F(\lambda)$ is finite but not bounded. Using the symbol $\{\ \}$ once more to denote symmetric ordering, we may express the symmetrically ordered series for $F(\lambda)$ in the closed form

$$F(\lambda) = \frac{2}{\lambda+1} \exp\left[2a^\dagger a \frac{\lambda-1}{\lambda+1}\right]. \quad (9.108)$$

Turning now to the case of antinormal ordering, we find from Eq. (9.77) the result

$$d_{n,m}(\lambda) = \frac{\delta_{n,m}}{(n!)^2\,(1-\lambda)} \int \exp\left[-\frac{\lambda|\xi|^2}{1-\lambda}\right](-|\xi|^2)^n \pi^{-1}\,\mathrm{d}^2\xi$$
$$= \frac{\delta_{n,m}}{n!\,\lambda}\left(1-\frac{1}{\lambda}\right)^n. \quad (9.109)$$

These coefficients all have poles at $\lambda = 0$ for which value the operator $F(\lambda)$ is the projection operator $|0\rangle\langle 0|$, which is both bounded and in the trace class. If we denote antinormal ordering by the symbol $\{\ \}_A$, then we may express the antinormally ordered series in the closed form

$$F(\lambda) = \lambda^{-1}\left\{\exp\left[(1-\lambda^{-1})a^\dagger a\right]\right\}_A. \quad (9.110)$$

It may be shown[2] that the antinormally ordered series for $F(\lambda)$ converges only for $|1-\lambda^{-1}| < 1$ or equivalently for $\mathrm{Re}\,\lambda > \frac{1}{2}$.

9.5
s-Ordered Power-Series Expansions

Section 9.4 contains some elementary observations about ordered power-series expansions. It was shown that normally ordered power series converge for virtually all operators of interest but that the coefficients of antinormally ordered expansions are singular even for some trace-class operators. In order to shed more light on this matter we introduce in this section a parametrized ordering convention according to which normal, symmetric, and antinormal ordering are distinguished by three distinct values of a continuous order parameter. By means of this convention we are able to vary the type of ordering in a continuous way from antinormal order to normal order and to see when the coefficients of the expansions become finite and when the expansions themselves become convergent. We show that for all bounded operators the coefficients are finite when the ordering is closer to normal than to antinormal ordering and that the series converge when the ordering is closer to normal than to symmetric ordering.

In introducing this ordering convention we do not suggest that the new types of orderings it defines have direct physical significance. It is intended instead as a useful device for understanding the problems associated with the three useful orderings – normal, symmetric, and antinormal ordering. Our main use of the ordering convention will be in connection with a parametrized integral representation which we introduce in Sect. 9.6. It will be helpful there to observe the way the structure of this representation changes with variations of the order parameter.

We have seen in Eqns. (9.11)–(9.15) how to express the displacement operators in various ordered forms. Let us define the s-ordered displacement operator $D(\alpha, s)$ by the relation

$$D(\alpha, s) = D(\alpha) e^{s|\alpha|^2/2}, \tag{9.111}$$

where s is a complex number. For the three discrete values of $s = +1, 0, -1$, the operator $D(\alpha, s)$ can be written as an exponential which is, respectively, normally ordered

$$D(\alpha, 1) = e^{\alpha a^\dagger} e^{-\alpha^* a}, \tag{9.112}$$

symmetrically ordered

$$D(\alpha, 0) = e^{\alpha a^\dagger - \alpha^* a}, \tag{9.113}$$

and antinormally ordered

$$D(\alpha, -1) = e^{-\alpha^* a} e^{\alpha a^\dagger}. \tag{9.114}$$

Proceeding as in our description of symmetric ordering, Eqns. (9.80)–(9.84), we define the s-ordered product $\{(a^\dagger)^n a^m\}_s$ by means of the Taylor series

$$D(\alpha, s) = \sum_{n,m=0}^{\infty} \{(a^\dagger)^n a^m\}_s \frac{1}{n! \, m!} \alpha^n (-\alpha^*)^m, \tag{9.115}$$

or, equivalently, as the derivative

$$\{(a^\dagger)^n a^m\}_s = \frac{\partial^{n+m} D(\alpha,s)}{\partial \alpha^n \partial(-\alpha^*)^m}\bigg|_{\alpha=0} \tag{9.116}$$

evaluated at $\alpha = 0$.

By applying this differential relation to the ordered exponentials (9.112)–(9.114), we find that the orderings specified by $s = +1, 0, -1$ are, respectively, normal,

$$\{(a^\dagger)^n a^m\}_1 = (a^\dagger)^n a^m, \tag{9.117}$$

symmetric in the sense of Eqns. (9.80)–(9.84),

$$\{(a^\dagger)^n a^m\}_0 = \{(a^\dagger)^n a^m\}, \tag{9.118}$$

and antinormal,

$$\{(a^\dagger)^n a^m\}_{-1} = a^m (a^\dagger)^n. \tag{9.119}$$

The s-ordered products can be simply expressed in terms of normally ordered products. Thus, for example, by writing

$$D(\alpha,s) = e^{(s-1)|\alpha|^2/2} D(\alpha,1), \tag{9.120}$$

we find for the simplest nontrivial s-ordered product $\{a^\dagger a\}_s$,

$$\{a^\dagger a\}_s = \frac{\partial^2}{\partial\alpha\,\partial(-\alpha^*)} e^{(s-1)|\alpha|^2/2} e^{\alpha a^\dagger} e^{-\alpha^* a}\bigg|_{\alpha=0}$$

$$= \frac{\partial}{\partial\alpha}\left[e^{\alpha a^\dagger} a + \tfrac{1}{2}(1-s)\alpha e^{\alpha a^\dagger}\right]\bigg|_{\alpha=0}$$

$$= a^\dagger a + \tfrac{1}{2}(1-s).$$

Similarly, for the s-ordered product $\{a^\dagger a^2\}$, we find

$$\{a^\dagger a^2\}_s = a^\dagger a^2 + (1-s)a.$$

It is not difficult to express an arbitrary s-ordered product as a polynomial in the t-ordered products, where t is also arbitrary. By writing

$$D(\alpha,s) = e^{(s-t)|\alpha|^2/2} D(\alpha,t) \tag{9.121}$$

and differentiating, we find

$$\{(a^\dagger)^n a^m\}_s = \sum_{k=0}^{(n,m)} k! \binom{n}{k}\binom{m}{k}\left(\frac{t-s}{2}\right)^k \{(a^\dagger)^{n-k} a^{m-k}\}_t, \tag{9.122}$$

where the symbol (n,m) denotes the smaller of the integers n and m and where

$$\binom{n}{k} \equiv \frac{n!}{k!(n-k)!}$$

is a binomial coefficient. These relations may be put more succinctly in terms of the associated Laguerre polynomials[18]. As we show in Appendix C, we have for $n \geq m$,

$$\{(a^\dagger)^n a^m\}_s = m! \left(\frac{t-s}{2}\right)^m \left\{(a^\dagger)^{n-m} L_m^{(n-m)}\left(\frac{2a^\dagger a}{s-t}\right)\right\}_t \quad (9.123)$$

and for $m \geq n$

$$\{(a^\dagger)^n a^m\}_s = n! \left(\frac{t-s}{2}\right)^n \left\{(a^\dagger)^{m-n} L_n^{(m-n)}\left(\frac{2a^\dagger a}{s-t}\right)\right\}_t, \quad (9.124)$$

where the polynomials within the curly brackets are in t-ordered form.

We are now in a position to consider the expansion of a bounded operator F as the s-ordered power series

$$F = \sum_{n,m=0}^{\infty} f_{n,m}(s) \{(a^\dagger)^n a^m\}_s, \quad (9.125)$$

where the coefficients $f_{n,m}(s)$ are complex numbers. We may identify the coefficients $f_{n,m}(s)$ by substituting in the general expansion Eq. (9.50) the expression

$$D^{-1}(\xi) = e^{-s|\xi|^2/2} D(-\xi, s),$$

so that we have

$$F = \int \text{Tr}\{FD(\xi)\} e^{-s|\xi|^2/2} D(-\xi, s) \pi^{-1} d^2\xi, \quad (9.126)$$

which we may expand in powers of a and $a^\dagger$ according to Eq. (9.115):

$$F = \sum_{n,m=0}^{\infty} \{(a^\dagger)^n a^m\}_s \frac{1}{n! m!} \int \text{Tr}\{FD(\xi, -s)\} (-\xi)^n (\xi^*)^m \pi^{-1} d^2\xi. \quad (9.127)$$

In this way we obtain the coefficients $f_{n,m}(s)$ as the integrals

$$f_{n,m}(s) = \frac{1}{n! m!} \int \text{Tr}\{FD(\xi, -s)\} (-\xi)^n (\xi^*)^m \pi^{-1} d^2\xi. \quad (9.128)$$

The parametrized relations (9.126)–(9.128) compactly express the previously derived results for normal, symmetric and antinormal ordering. We recover Eqns. (9.73), (9.87), and (9.77) by setting $s = +1, 0$, and -1, respectively.

By applying the inequality (9.37) and the relation Eq. (9.44) to the integral Eq. (9.128), we find that for all bounded operators F the coefficients are finite when $\text{Re}\,s > 0$ and are bounded by the quantities[2]

$$|f_{n,m}(s)| \leq \frac{[(n+m)!]^{1/2}}{n! m!} \frac{||F||}{(\text{Re}\,s)^{(n+m+1)/2}}, \quad (9.129)$$

where $||F||$ is the Hilbert–Schmidt norm Eq. (9.43). This inequality generalizes the upper bound Eq. (9.74), which was obtained for normal ordering, $s = 1$, to those orderings specified by $\text{Re}\, s > 0$, i.e., to those which may be thought of as closer to normal than to antinormal order.

We recall from Sect. 9.4 that the coefficients $f_{n,m}(0)$, corresponding to symmetric order, are singular for certain bounded operators. We may therefore say that the coefficients $f_{n,m}(s)$ are finite for all bounded operators F if and only if $\text{Re}\, s > 0$.

Before turning to the question of the convergence of the series Eq. (9.125), we may note[2] that if the operator F is bounded then the coefficients $f_{n,m}(s)$ are all analytic functions of the order parameter s throughout the half-plane $\text{Re}\, s > 0$. This analyticity is intuitively clear from the structure of the integrals Eq. (9.128) and from the square-integrability of the weight functions $\text{Tr}\{FD(\xi)\}$ belonging to bounded operators.

We defined a type of convergence that is suited to ordered power-series expansions in Sect. 9.4, Eqns. (9.88) and (9.89). We shall now show that for bounded operators F the series Eq. (9.125) converges according to this definition when $\text{Re}\, s > \frac{1}{2} + \frac{1}{2}(\text{Im}\, s)^2$.

The series Eq. (9.127) is the limit of the sequence of operators F_N,

$$F_N = \sum_{n,m=0}^{N} f_{n,m}(s)\, \{(a^\dagger)^n a^m\}_s ,\qquad(9.130)$$

of which the coherent state matrix elements are

$$\begin{aligned}
\langle\beta|F_N|\alpha\rangle &= \sum_{n,m=0}^{N} f_{n,m}(s) \left\langle\beta\left|\{(a^\dagger)^n a^m\}_s\right|\alpha\right\rangle \\
&= \sum_{n,m=0}^{N} f_{n,m}(s)\, \frac{\partial^{n+m}}{\partial\zeta^n \partial(-\zeta^*)^m}\, \langle\beta|D(\zeta,s)|\alpha\rangle\bigg|_{\zeta=0} \\
&= \sum_{n,m=0}^{N} \int \text{Tr}\{FD(\xi,-s)\}\, \frac{(-\xi)^n}{n!}\, \frac{(\xi^*)^m}{m!} \\
&\qquad \times \left(\frac{\partial^{n+m}\langle\beta|D(\zeta,s)|\alpha\rangle}{\partial\zeta^n \partial(-\zeta^*)^m}\bigg|_{\zeta=0}\right) \pi^{-1}\, d^2\xi. \quad (9.131)
\end{aligned}$$

The sequence $\langle\beta|F_N|\alpha\rangle$ therefore, converges to $\langle\beta|F|\alpha\rangle$ when in this expression the limits of summation and integration may be interchanged. The relevant sequence of functions is

$$\phi_N(\xi) = \sum_{n,m=0}^{N} \frac{(-\xi)^n}{n!}\, \frac{(\xi^*)^m}{m!} \left(\frac{\partial^{n+m}\langle\beta|D(\zeta,s)|\alpha\rangle}{\partial\zeta^n \partial(-\zeta^*)^m}\bigg|_{\zeta=0}\right),$$

which converges to

$$\begin{aligned}
\phi(\xi) &= \langle\beta|D(-\xi,s)|\alpha\rangle \\
&= \langle\beta|\alpha\rangle \exp\left[\tfrac{1}{2}(s-1)|\xi|^2 - \beta^*\xi + \alpha\xi^*\right].
\end{aligned}$$

9.5 s-Ordered Power-Series Expansions

It is not difficult to show that for all N the functions $\phi_N(\xi)$ are dominated in accordance with the inequality

$$|\phi_N(\xi)| \leq M(\xi) \equiv |\langle \beta | \alpha \rangle | \exp\left[\tfrac{1}{2}|s-1||\xi|^2 + (|\beta|+|\alpha|)|\xi|\right]. \tag{9.132}$$

Thus, according to the Lebesgue dominated convergence theorem[16], the interchange of limits is permissible when the integral

$$\int M(\xi) |\operatorname{Tr}\{FD(\xi,-s)\}| \pi^{-1} d^2\xi \tag{9.133}$$

converges. Now for an arbitrary bounded operator F this integral converges when $\operatorname{Re} s > |s-1|$, or, equivalently, when s lies in the parabolic region

$$\operatorname{Re} s > \tfrac{1}{2} + \tfrac{1}{2}(\operatorname{Im} s)^2. \tag{9.134}$$

This condition is therefore sufficient for the series Eq. (9.125) to converge for every bounded operator F. In particular, all bounded operators possess convergent s-ordered power series for $s > \tfrac{1}{2}$, i.e., when the ordering is closer to normal than to symmetric.

The example considered in Sect. 9.4,

$$F(\lambda) = \lambda^{a^\dagger a},$$

provides a simple illustration of some of the results that we have derived. By using Eq. (9.106), we find that the integrals Eq. (9.128) for the coefficients $f_{n,m}(s)$ assume the form

$$\begin{aligned}
f_{n,m}(s) &= \frac{1}{n!\,m!(1-\lambda)} \int \exp\left[\left(\frac{1-s}{2} + \frac{1}{\lambda-1}\right)|\xi|^2\right] (-\xi)^n (\xi^*)^m \pi^{-1} d^2\xi \\
&= \frac{\delta_{n,m}}{n!(\lambda-1)} \left(\frac{1-s}{2} + \frac{1}{1+\lambda}\right)^{-(n+1)},
\end{aligned} \tag{9.135}$$

where we have used a series expansion of the formula Eq. (9.A2) to do the integral. The coefficients $f_{n,m}(s)$ are analytic functions of s except for poles located at

$$s = \frac{(\lambda+1)}{(\lambda-1)}. \tag{9.136}$$

This relation is a familiar one; it maps the interior of the unit circle $|\lambda| < 1$ onto the left half-plane $\operatorname{Re} s < 0$. Thus, the location of the singularities of the coefficients $f_{n,m}(s)$ ranges over the half-plane $\operatorname{Re} s < 0$ as the parameter λ ranges over the region in which the operator $F(\lambda)$ is bounded. On summing the s-ordered power series for $F(\lambda)$, we obtain the expression

$$F(\lambda) = \lambda^{a^\dagger a} = \frac{2}{1+s+\lambda-s\lambda} \left\{ \exp\left[\frac{2(\lambda-1)a^\dagger a}{1+s+\lambda-s\lambda}\right] \right\}. \tag{9.137}$$

By examining the convergence of the integral (9.133), we find, taking s real for simplicity, that the s-ordered power series for $F(\lambda)$ converges when

$$s > 1 + \frac{\operatorname{Re}\lambda - 1}{|\lambda - 1|^2}. \tag{9.138}$$

Thus, when $\lambda = 0$ and $F(\lambda) = |0\rangle\langle 0|$, for example, the series Eq. (9.137) converges for $s > 0$.

As a final illustration let us note that we may write the s-ordered displacement operator $D(\alpha, s)$ in the form

$$\begin{aligned} D(\alpha, s) &= e^{s|\alpha|^2/2} D(\alpha) \\ &= \left\{ \exp(\alpha a^\dagger - \alpha^* a) \right\}_s \\ &= \{D(\alpha)\}_s, \end{aligned} \tag{9.139}$$

which justifies our calling it the s-ordered displacement operator.

9.6
Integral Expansions for Operators

The representation of operators as integrals over the displacement operators is in many respects analogous to the representation of functions as Fourier integrals. The displacement-operator expansion

$$F = \int e^{a\xi^* - a^\dagger \xi} f(\xi)\, \pi^{-1}\, d^2\xi$$

differs from the Fourier expansion

$$g(\alpha) = \int e^{\alpha\xi^* - \alpha^*\xi} \varphi(\xi)\, \pi^{-1}\, d^2\xi$$

because of the noncommutativity that distinguishes the variables a and $a^\dagger$ from their counterparts α and α^*. Distinctions in operator ordering lend an interesting structure to the Fourier representation of operators. We consider some elementary aspects of this structure in the present section. For each value of the order parameter s, we define a set of operators $T(\alpha, s)$ that forms a basis for an integral representation for arbitrary operators. Each of these representations has the desirable property that the associated weight function bears a particularly direct relationship to the operator being expanded.

These new representations are interesting from a number of standpoints. In particular, for the cage of antinormal ordering the new expansion expresses operators as integrals over projection operators upon the coherent states. For the case in which the operator being expanded is a density operator, this expansion takes the form of the P representation[7,22,23], which has been widely discussed in connection with problems

9.6 Integral Expansions for Operators

in quantum optics. When the ordering is symmetric and the operator being expanded is a density operator, the weight function in the expansion is the one introduced by Wigner[24] as a quantum-mechanical analog of the classical phase-space density function. We shall make a detailed application of the present formalism to the case of the density operator in the paper which follows. Let us recall that according to Eq. (9.51) the trace of the product of any two bounded operators can be written as the integral

$$\mathrm{Tr}\{FG\} = \int \mathrm{Tr}\{FD(\xi)\}\,\mathrm{Tr}\{GD(-\xi)\}\,\pi^{-1}\,d^2\xi, \tag{9.140}$$

in which the functions $\mathrm{Tr}\{FD(\xi)\}$ and $\mathrm{Tr}\{GD(-\xi)\}$ are both square-integrable functions. By using the definition Eq. (9.111) of the operator $D(\xi,s)$,

$$D(\xi,s) = e^{s|\xi|^2/2} D(\xi),$$

we may trivially reexpress this trace in the form

$$\mathrm{Tr}\{FG\} = \int \mathrm{Tr}\{FD(\xi,-s)\}\,\mathrm{Tr}\{GD(-\xi,s)\}\,\pi^{-1}\,d^2\xi. \tag{9.141}$$

Now if both of the traces appearing in this integral are square-integrable functions, then they possess the complex Fourier transforms

$$f(\alpha,-s) = \int \exp(\alpha\xi^* - \alpha^*\xi)\,\mathrm{Tr}\{FD(\xi,-s)\}\,\pi^{-1}\,d^2\xi \tag{9.142}$$

and

$$g(\alpha,s) = \int \exp(\alpha\xi^* - \alpha^*\xi)\,\mathrm{Tr}\{GD(\xi,s)\}\,\pi^{-1}\,d^2\xi, \tag{9.143}$$

which are also square-integrable, and we may apply the identity Eq. (9.35) to secure the result

$$\mathrm{Tr}\{FG\} = \int f(\alpha,-s)\,g(\alpha,s)\,\pi^{-1}\,d^2\xi. \tag{9.144}$$

The condition that the traces $\mathrm{Tr}\{FD(\xi,-s)\}$ and $\mathrm{Tr}\{GD(\xi,s)\}$ both be square-integrable is clearly satisfied for all bounded operators F and G when $\mathrm{Re}\,s = 0$. The relation (9.144) is consequently for $\mathrm{Re}\,s = 0$ an identity holding for all bounded operators. It is likewise true that for every value of s one of the traces, $\mathrm{Tr}\{FD(\xi,-s)\}$ or $\mathrm{Tr}\{GD(-\xi,s)\}$, will be square-integrable given that both operators are bounded. Because these traces depend upon the parameter s through the factors $\exp(\pm\frac{1}{2}s|\xi|^2)$, their mathematical properties depend primarily upon the real part of s. We shall discuss the occurrence of singularities in Eqns. (9.142)–(9.144) at a later point; we may infer for the present that for every two bounded operators F and G there will, in general, be a strip $x_1 < \mathrm{Re}\,s < x_2$ about the imaginary axis of the s plane in which Eqns. (9.142)–(9.144) are valid. Let us define the operator $T(\alpha,s)$ as the complex Fourier transform of the s-ordered displacement operator $D(\xi,s)$,

$$T(\alpha,s) = \int e^{\alpha\xi^* - \alpha^*\xi}\,D(\xi,s)\,\pi^{-1}\,d^2\xi. \tag{9.145}$$

This notation allows us to express Eqns. (9.142)–(9.144) in the more compact form

$$f(\alpha, -s) = \mathrm{Tr}\{FT(\alpha, -s)\}, \qquad (9.146)$$
$$g(\alpha, s) = \mathrm{Tr}\{GT(\alpha, s)\}, \qquad (9.147)$$

and

$$\mathrm{Tr}\{FG\} = \int \mathrm{Tr}\{FT(\alpha, -s)\}\, \mathrm{Tr}\{GT(\alpha, s)\}\, \pi^{-1}\, d^2\alpha. \qquad (9.148)$$

When the operator G in Eq. (9.148) is taken to be the outer product of two states $|\psi\rangle$ and $|\varphi\rangle$, we have

$$\langle \varphi|F|\psi \rangle = \int \mathrm{Tr}\{FT(\alpha, -s)\}\, \langle \varphi|T(\alpha, s)|\psi\rangle\, \pi^{-1}\, d^2\alpha$$
$$= \int f(\alpha, -s)\, \langle \varphi|T(\alpha, s)|\psi\rangle\, \pi^{-1}\, d^2\alpha \qquad (9.149)$$

If for some value of s this relation holds for all normalizable states $|\varphi\rangle$ and $|\psi\rangle$, then we may say that the operator F possesses the representation

$$F = \int f(\alpha, -s)\, T(\alpha, s)\, \pi^{-1}\, d^2\alpha$$
$$= \int \mathrm{Tr}\{FT(\alpha, -s)\}\, T(\alpha, s)\, \pi^{-1}\, d^2\alpha \qquad (9.150)$$

for that value of s.

There are, of course, many senses in which an operator F can be said to possess this representation for a particular value of s. We might, for example, require only that Eq. (9.149) hold for some complete set of states. A stronger condition would be that Eq. (9.148) be true for all bounded operators G. Alternatively, by interpreting Eqns. (9.148)–(9.150) in terms of limiting processes or in terms of generalized functions[2], it is possible to set down conditions that are so broad as to include all bounded operators F and all values of s.

By using Eqns. (9.139) and (9.31) we may express the operator $T(\alpha, s)$ in the suggestive forms

$$T(\alpha, s) = \left\{ \int \exp\left[(\alpha - a)\xi^* - (\alpha^* - a^\dagger)\xi\right] \pi^{-1}\, d^2\xi \right\}_s \qquad (9.151)$$
$$= \pi \left\{ \delta^{(2)}(\alpha - a) \right\}_s. \qquad (9.152)$$

In terms of this notation we may write the representation Eq. (9.150) in the form

$$F = \int f(\alpha, -s)\, \left\{ \delta^{(2)}(\alpha - a) \right\}_s d^2\alpha. \qquad (9.153)$$

The classical analog of this representation is therefore the trivial identity

$$g(\alpha) = \int g(\alpha')\, \delta^{(2)}(\alpha - \alpha')\, d^2\alpha' \qquad (9.154)$$

9.6 Integral Expansions for Operators

in which the relationship between the function being expanded and its weight function is one of identity. The considerably more interesting structure of the representation Eq. (9.153) arises, of course, from the noncommutativity of the operators a and $a^\dagger$.

On the basis of the analogy between Eqns. (9.153) and (9.154), we may reasonably expect to find a close relationship between the weight function $f(\alpha,-s)$ and the operator F in Eq. (9.153). We shall describe this relationship and other properties of these representations in Sect. 9.7. The remainder of the present section is devoted to the properties of the operators $T(\alpha,s)$.

By using the definition Eq. (9.145) to form the Hermitian adjoint

$$T^\dagger(\alpha,s) = \int D(-\xi,s^*) \exp(\alpha^*\xi - \alpha\xi^*) \pi^{-1} d^2\xi$$

and replacing ξ by $-\xi$, we find

$$T^\dagger(\alpha,s) = T(\alpha,s^*) . \tag{9.155}$$

Thus for real values of the order parameter s, the operator $T(\alpha,s)$ is Hermitian

$$T(\alpha,s) = T^\dagger(\alpha,s) \quad \text{for } s \text{ real} . \tag{9.156}$$

From the multiplication law Eq. (9.19), we find

$$D(\alpha) D(\xi,s) D^{-1}(\alpha) = D(\xi,s) \exp(\alpha\xi^* - \alpha^*\xi) ,$$

which, when substituted into the definition Eq. (9.145), yields

$$T(\alpha,s) = \int D(\alpha) D(\xi,s) D^{-1}(\alpha) \pi^{-1} d^2\xi$$
$$= D(\alpha) T(0,s) D^{-1}(\alpha) . \tag{9.157}$$

Thus, the α dependence of the operator $T(\alpha,s)$ is governed by the unitary transformation induced by the operator $D(\alpha)$. On using Eq. (9.19) once again we find

$$T(\alpha,s) = D(\alpha - \beta) T(\beta,s) D^{-1}(\alpha - \beta) . \tag{9.158}$$

To find the s dependence of the operator $T(\alpha,s)$ we note that the operator $T(0,s)$ is defined by Eq. (9.145) as the integral

$$T(0,s) = \int D^{-1}(\xi) e^{s|\xi|^2/2} \pi^{-1} d^2\xi . \tag{9.159}$$

This is a displacement operator expansion of the form of Eq. (9.50) with weight function

$$\text{Tr}\{T(0,s) D(\xi)\} = e^{s|\xi|^2/2} . \tag{9.160}$$

We encountered a similar weight function in our discussion of the operator $F(\lambda) = \lambda^{a^\dagger a}$ and, if we compare Eqns. (9.159) and (9.160) with Eq. (9.106) and make the identification

$$s = \frac{\lambda+1}{\lambda-1} \quad \text{or} \quad \lambda = \frac{s+1}{s-1}$$

then, since the correspondence $F \leftrightarrow \text{Tr}\{FD(\xi)\}$ is one-to-one, we secure the result

$$T(0,s) = \frac{2}{1-s}\left(\frac{s+1}{s-1}\right)^{a^\dagger a}. \tag{9.161}$$

By using Eq. (9.157) and the displacement property Eqns. (9.17) and (9.18) of the unitary operators $D(\alpha)$, we obtain for the operator $T(\alpha, s)$ the following expressions:

$$T(\alpha,s) = \frac{2}{1-s} D(\alpha)\left(\frac{s+1}{s-1}\right)^{a^\dagger a} D^{-1}(\alpha) \tag{9.162}$$

$$= \frac{2}{1-s}\left(\frac{s+1}{s-1}\right)^{(a^\dagger-\alpha^*)(a-\alpha)} \tag{9.163}$$

$$= \frac{2}{1-s}\exp\left[(a^\dagger-\alpha^*)(a-\alpha)\ln\left(\frac{s+1}{s-1}\right)\right]. \tag{9.164}$$

The expansion of $T(\alpha,s)$ in terms of the eigenstates $|n\rangle$ of the operator $a^\dagger a$ is, from Eq. (9.162),

$$T(\alpha,s) = \frac{2}{1-s}\sum_{n=0}^{\infty} D(\alpha)|n\rangle\left(\frac{s+1}{s-1}\right)^n \langle n|D^\dagger(\alpha). \tag{9.165}$$

The states $D(\alpha)|n\rangle$ thus form a complete orthonormal set of eigenstates of the operator $T(\alpha,s)$,

$$T(\alpha,s)D(\alpha)|n\rangle = \frac{2}{1-s}\left(\frac{s+1}{s-1}\right)^n D(\alpha)|n\rangle, \tag{9.166}$$

with eigenvalues

$$e_n(s) = \frac{2}{1-s}\left(\frac{s+1}{s-1}\right)^n \tag{9.167}$$

which are independent of α.

We note in particular that for all values of α all the eigenvalues $e_n(s)$ of the operator $T(\alpha,s)$ are infinite at $s=1$. At $s=-1$, on the other hand, the series (9.165) terminates after the first term and the operator $T(\alpha,-1)$ is simply the projection operator on the coherent state $|\alpha\rangle = D(\alpha)|0\rangle$,

$$T(\alpha,-1) = |\alpha\rangle\langle\alpha|. \tag{9.168}$$

Let us now consider to what classes of operators the operator $T(\alpha,s)$ belongs for different values of s. The three norms that we defined in Sect. 9.3 are invariant under unitary transformations and do not, therefore, for the case of the operator $T(\alpha,s)$ depend upon the parameter α. By using Eqns. (9.41)–(9.43) and (9.160), we may express the Hilbert–Schmidt norm $||T(\alpha,s)||$ in terms of the integral

$$||T(\alpha,s)|| = \int |e^{s|\xi|^2/2}|^2 \pi^{-1} d^2\xi$$
$$= \int e^{\operatorname{Re} s|\xi|^2} \pi^{-1} d^2\xi ,$$

so that we have

$$||T(\alpha,s)|| = \frac{1}{(-\operatorname{Re} s)^{1/2}} \quad \text{for} \quad \operatorname{Re} s < 0 . \tag{9.169}$$

The operator $T(\alpha,s)$ is therefore a bounded operator only for $\operatorname{Re} s < 0$. When $\operatorname{Re} s \leq 0$ the operator $T(\alpha,s)$ is a finite operator, and by using Eq. (9.165) we find for its norm $||T(\alpha,s)||_1$, defined by Eq. (9.61), the value

$$||T(\alpha,s)||_1 = \left|\frac{2}{1-s}\right| \quad \text{for} \quad \operatorname{Re} s \leq 0 . \tag{9.170}$$

For $\operatorname{Re} s < 0$ the operator $T(\alpha,s)$ is not only bounded but also in the trace class. By using Eq. (9.165) we obtain for its trace class norm Eq. (9.65) the series

$$||T(\alpha,s)||_2 = \frac{2}{|1-s|} \sum_{n=0}^{\infty} \left|\frac{s+1}{s-1}\right|^n$$
$$= \frac{2}{|1-s|-|1+s|} \quad \text{for} \quad \operatorname{Re} s < 0 . \tag{9.171}$$

Thus, for $\operatorname{Re} s < 0$ the operator $T(\alpha,s)$ is a member of all three classes of operators, for $\operatorname{Re} s > 0$ it is in none of them, and on the line $\operatorname{Re} s = 0$ it is in only the largest class, the class of finite operators.

By referring to the relation (9.103) between the operator $F(\lambda)$ and its normally ordered form we find

$$T(\alpha,s) = \frac{2}{1-s} : \exp\left[\left(\frac{2}{s-1}\right)(a^\dagger - \alpha^*)(a-\alpha)\right] : \tag{9.172}$$

$$= \exp\left[-\frac{2|\alpha|^2}{1-s}\right] \exp\left[\frac{2\alpha a^\dagger}{1-s}\right] T(0,s) \exp\left[\frac{2\alpha^* a}{1-s}\right] , \tag{9.173}$$

where the colons denote normal ordering. From Eq. (9.172) there follow easily for the coherent-state matrix elements the relations

$$\langle \beta|T(\alpha,s)|\gamma\rangle = \frac{2\langle\beta|\gamma\rangle}{1-s} \exp\left[\left(\frac{2}{s-1}\right)(\beta^* - \alpha^*)(\gamma-\alpha)\right] \tag{9.174}$$

and

$$\langle\beta|T(\alpha,s)|\beta\rangle = \frac{2}{1-s}\exp\left[-\frac{2|\alpha-\beta|^2}{1-s}\right], \tag{9.175}$$

from which, by taking the limit as s approaches unity from smaller real values, we find

$$\langle\beta|T(\alpha,1)|\beta\rangle = \pi\delta^{(2)}(\alpha-\beta). \tag{9.176}$$

By using Eqns. (9.175) and (9.28), we may express the trace of the operator $T(\alpha,s)$ as the integral

$$\begin{aligned}\mathrm{Tr}\{T(\alpha,s)\} &= \int \langle\beta|T(\alpha,s)|\beta\rangle\,\pi^{-1}\,d^2\beta \\ &= \frac{2}{1-s}\exp\left[-\frac{2|\alpha-\beta|^2}{1-s}\right]\pi^{-1}\,d^2\beta \\ &= 1 \quad \text{for} \quad \mathrm{Re}\,s < 1. \end{aligned} \tag{9.177}$$

That the operator $T(\alpha,s)$ is of unit trace can also be seen by summing its eigenvalues $e_n(s)$ which are given by Eq. (9.167). The procedure given above is a rearrangement of the series $\sum e_n(s)$ which for $\mathrm{Re}\,s \geq 0$ lies outside its radius of convergence but can still be *summed* to unity.

Another trace which will be useful in what follows is $\mathrm{Tr}\{T(\alpha,s)T(\beta,t)\}$. By using the definition Eq. (9.145) and the orthogonality rule Eq. (9.39) we find that

$$\begin{aligned}\mathrm{Tr}\{T(\alpha,s)T(\beta,t)\} &= \int \mathrm{Tr}\{D(\xi,s)D(\zeta,t)\}\exp(\alpha\xi^* - \alpha^*\xi + \beta\zeta^* - \beta^*\zeta) \\ &\qquad\qquad\qquad\qquad\qquad\qquad\qquad\qquad\qquad \times \pi^{-2}\,d^2\xi\,d^2\zeta \\ &= \int \exp\left[(\alpha-\beta)\xi^* - (\alpha-\beta)^*\xi + \tfrac{1}{2}(s+t)|\xi|^2\right]\pi^{-1}\,d^2\xi \\ &= -\frac{2}{s+t}\exp\left[\frac{2|\alpha-\beta|^2}{s+t}\right] \quad \text{for} \quad \mathrm{Re}(s+t) < 0. \end{aligned} \tag{9.178}$$

where we have used the formula Eq. (9.A2) to do the integral, which converges only for $\mathrm{Re}(s+t) < 0$. By letting t approach $-s$ from below we find

$$\mathrm{Tr}\{T(\alpha,s)T(\beta,-s)\} = \pi\delta^{(2)}(\alpha-\beta), \tag{9.179}$$

which is the counterpart for the operators $T(\alpha,s)$ to the orthogonality rule Eq. (9.39) for the operators $D(\xi,s)$.

If we now apply the expansion Eq. (9.150) to the operator $T(\alpha,s)$, we find, using the trace relation Eq. (9.178),

$$\begin{aligned}T(\alpha,s) &= \int T(\beta,t)\,\mathrm{Tr}\{T(\alpha,s)T(\beta,-t)\}\,\pi^{-1}\,d^2\beta \\ &= \frac{2}{t-s}\int \exp\left[-\frac{2|\alpha-\beta|^2}{t-s}\right]T(\beta,t)\,\pi^{-1}\,d^2\beta \quad \text{for} \quad \mathrm{Re}\,t > \mathrm{Re}\,s. \end{aligned} \tag{9.180}$$

This Gaussian convolution is the complex Fourier transform of the product $\exp[\frac{1}{2}(s-t)|\xi|^2]D(\xi,t)$. The differential form of this integral relation can be found by differentiating both sides of Eq. (9.145); it is

$$\frac{\partial T(\alpha,s)}{\partial s} = -\frac{1}{2}\frac{\partial^2 T(\alpha,s)}{\partial\alpha\,\partial\alpha^*}. \tag{9.181}$$

By further differentiation of Eq. (9.145), we obtain the relations

$$\left.\frac{\partial^{n+m}T(\alpha,s)}{\partial\alpha^n\,\partial(\alpha^*)^m}\right|_{\alpha=0} = \int D(\xi,s)\,(\xi^*)^n(-\xi)^m\,\pi^{-1}\,d^2\xi. \tag{9.182}$$

If we use Eqns. (9.32) and (9.33) to invert Eq. (9.145), we find that

$$D(\xi,s) = \int T(\alpha,s)\,\exp(\xi\alpha^* - \xi^*\alpha)\,\pi^{-1}\,d^2\alpha \tag{9.183}$$

and, by expanding both sides in powers of ξ and ξ^* and using the definition (9.116), we may express the s-ordered products as the integrals

$$\left\{(a^\dagger)^n a^m\right\}_s = \int T(\alpha,s)\,(\alpha^*)^n\,\alpha^m\,\pi^{-1}\,d^2\alpha. \tag{9.184}$$

The fact that under integration the operator $T(\alpha,s)$ turns the monomial $(\alpha^*)^n\alpha^m$ into the s-ordered product $\{(a^\dagger)^n a^m\}_s$ illustrates again the sense in which it is an s-ordered operator analog of the δ function. As in the use of singular functions, some discrimination is called for in the application of this relation, particularly for $\operatorname{Re} s > 0$, where $T(\alpha,s)$ is not a finite operator and at $s=1$ where it is explicitly singular.

When $n = m = 0$ Eq. (9.184) becomes

$$1 = \int T(\alpha,s)\,\pi^{-1}\,d^2\alpha, \tag{9.185}$$

which for $s = -1$ is the completeness relation Eq. (9.27)

$$1 = \int |\alpha\rangle\langle\alpha|\,\pi^{-1}\,d^2\alpha$$

for the coherent states.

By applying the orthogonality rule for the T operators (9.179) to Eq. (9.184), we obtain the relation

$$\operatorname{Tr}\left[\left\{(a^\dagger)^n a^m\right\}_s T(\alpha,-s)\right] = \int \operatorname{Tr}[T(\beta,s)\,T(\alpha,-s)]\,(\beta^*)^n\,\beta^m\,\pi^{-1}\,d^2\beta$$

$$= \int (\beta^*)^n\,\beta^m\,\delta^{(2)}(\alpha-\beta)\,d^2\beta$$

$$= (\alpha^*)^n\,\alpha^m, \tag{9.186}$$

which is the inverse of the relation Eq. (9.184).

From the relation Eq. (9.157) with α infinitesimal we find that

$$dT(\alpha,s) = \left[a^\dagger\, d\alpha - a\, d\alpha^*, T(\alpha,s)\right] \tag{9.187}$$

or, equivalently,

$$\frac{\partial T(\alpha,s)}{\partial \alpha} = \left[a^\dagger, T(\alpha,s)\right] \tag{9.188}$$

and

$$\frac{\partial T(\alpha,s)}{\partial \alpha^*} = -\left[a, T(\alpha,s)\right], \tag{9.189}$$

which are the complex Fourier transforms of the commutation relations implicit in Eqns. (9.17) and (9.18).

From Eq. (9.175) by a process of differentiation we may obtain the matrix elements $\langle n|T(\alpha,s)|m\rangle$. As we show in Appendix D, the result of that calculation is

$$\langle n|T(\alpha,s)|m\rangle = \left(\frac{n!}{m!}\right)^{1/2} \left(\frac{2}{1-s}\right)^{m-n+1} \left(\frac{s+1}{s-1}\right)^n (\alpha^*)^{m-n}$$
$$\times \exp\left[-\frac{2|\alpha|^2}{1-s}\right] L_n^{(m-n)}\left(\frac{4|\alpha|^2}{1-s^2}\right), \tag{9.190}$$

where $L_n^{(m)}(x)$ is an associated Laguerre polynomial[18].

9.7
Correspondences Between Operators and Functions

In Sect. 9.6 we introduced the operators $T(\alpha,s)$ which form, for each value of the order parameter s, a basis for the expansion of operators as weighted integrals. In this section we show that each of these integral representations maintains a close relationship between the operator being expanded and its weight function. We show that the mathematical properties of the weight functions change substantially as the order parameter is varied from $s = -1$, antinormal order, to $s = 1$, normal order.

We have seen that, at least in the vicinity of the line $\operatorname{Re} s = 0$, every bounded operator F possesses the representation

$$F = \int f(\alpha, -s)\, T(\alpha,s)\, \pi^{-1}\, d^2\alpha, \tag{9.191}$$

where the weight function $f(\alpha, -s)$ is given by the trace

$$f(\alpha, -s) = \operatorname{Tr}\{F T(\alpha, -s)\}. \tag{9.192}$$

Let us note that when the expansion is in terms of the operators $T(\alpha,s)$ the parameter s appears in the weight function $f(\alpha, -s)$ with a minus sign.

9.7 Correspondences Between Operators and Functions

We now observe that the trace Eq. (9.192) is the unique weight function for the operator F in the expansion Eq. (9.191). For if the operator F and the function $g(\alpha)$ stand in the relationship

$$F = \int g(\alpha') \, T(\alpha', s) \, \pi^{-1} \, d^2\alpha' \,, \tag{9.193}$$

then by using the trace relation Eq. (9.179) we find that

$$\begin{aligned}
\mathrm{Tr}\{FT(\alpha,-s)\} &= \int g(\alpha') \, \mathrm{Tr}\{T(\alpha',s) \, T(\alpha,-s)\} \, \pi^{-1} \, d^2\alpha' \\
&= \int g(\alpha') \, \delta^{(2)}(\alpha - \alpha') \, d^2\alpha' \\
&= g(\alpha) \,.
\end{aligned} \tag{9.194}$$

We may regard Eqns. (9.191) and (9.192) as defining, for every value of s, a correspondence $F \leftrightarrow f(\alpha, -s)$ between operators and their weight functions. Since the weight function corresponding to a given operator is unique, these correspondences are one-to-one.

It is in terms of power-series expansions that the correspondences, which we have just introduced, take their simplest form. If we assume that the operator F possesses the s-ordered power-series expansion

$$F = \sum_{n,m=0}^{\infty} f_{n,m}(s) \left\{ (a^\dagger)^n a^m \right\}_s, \tag{9.195}$$

where the coefficients are given by Eq. (9.128), then by using Eq. (9.186) we may secure for the weight function $f(\alpha, -s)$ the power-series expansion

$$\begin{aligned}
f(\alpha, -s) &= \mathrm{Tr}\{FT(\alpha, -s)\} \\
&= \sum_{n,m=0}^{\infty} f_{n,m}(s) \, \mathrm{Tr}\left[\left\{ (a^\dagger)^n a^m \right\}_s T(\alpha, -s) \right] \\
&= \sum_{n,m=0}^{\infty} f_{n,m}(s) \, (\alpha^*)^n \alpha^m \,,
\end{aligned} \tag{9.196}$$

with the same coefficients $f_{n,m}(s)$. Conversely, if we assume for the function $f(\alpha, -s)$ this series expansion, then by using Eq. (9.184) we may obtain the operator F in the form

$$\begin{aligned}
F &= \int f(\alpha, -s) \, T(\alpha, s) \, \pi^{-1} \, d^2\alpha \\
&= \sum_{n,m=0}^{\infty} f_{n,m}(s) \int (\alpha^*)^n \alpha^m \, T(\alpha, s) \, \pi^{-1} \, d^2\alpha \\
&= \sum_{n,m=0}^{\infty} f_{n,m}(s) \left\{ (a^\dagger)^n a^m \right\}_s \,.
\end{aligned} \tag{9.197}$$

Thus, for every complex number s the operator–function correspondence

$$F \underset{s}{\longleftrightarrow} f(\alpha,-s) = \text{Tr}\{FT(\alpha,-s)\} \tag{9.198}$$

involves simply the interchange

$$\{(a^{\dagger})^n a^m\}_s \underset{s}{\longleftrightarrow} (\alpha^*)^n \alpha^m, \tag{9.199}$$

which is affected by the reciprocal relations Eq. (9.184) and Eq. (9.186).

We shall refer to this correspondence as the correspondence $C(s)$ or, when it is clear which value of the order parameter is meant, as the C-correspondence. Another way of expressing the correspondence $C(s)$ is to say that the weight function $f(\alpha,-s)$ is a generating function for the coefficients $f_{n,m}(s)$ of the s-ordered power-series expansion for F according to the rule

$$\left. \frac{\partial^{n+m} f(\alpha,-s)}{\partial \alpha^n \partial (\alpha^*)^m} \right|_{\alpha=0} = n!\, m!\, f_{n,m}(s). \tag{9.200}$$

The consistency of this prescription for the coefficients $f_{n,m}(s)$ with the earlier one, Eq. (9.128), follows from Eq. (9.182). Two simple examples of the correspondence $C(s)$ are

$$D(\xi,s) \underset{s}{\longleftrightarrow} e^{\xi \alpha^* - \xi^* \alpha} \tag{9.201}$$

and

$$T(\xi,s) \underset{s}{\longleftrightarrow} \pi \delta^{(2)}(\alpha - \xi). \tag{9.202}$$

There are two ways in which singularities can occur in the representation Eq. (9.191). For $\text{Re}\, s > 0$ the operator $T(\alpha,s)$ is not a finite operator and when we form a matrix element $\langle \varphi | F | \psi \rangle$ of the representation (9.191) singularities can arise in the function $\langle \varphi | T(\alpha,s) | \psi \rangle$ which appears in the integral Eq. (9.149). On the other hand, when $\text{Re}\, s < 0$ the weight function $f(\alpha,-s)$ can become singular because the operator $T(\alpha,-s)$ in the trace $f(\alpha,-s) = \text{Tr}\{FT(\alpha,-s)\}$ is not a finite operator. Such behavior on the part of the weight function $f(\alpha,-s)$ for $\text{Re}\, s < 0$ is of course related via Eq. (9.200) to singularities in the coefficients $f_{n,m}(s)$, the occurrence of which we discussed in Sections 9.4 and 9.5.

The trace relation

$$\text{Tr}\{FG\} = \int f(\alpha,-s)\, g(\alpha,s)\, \pi^{-1}\, d^2\alpha$$

$$= \int \text{Tr}\{FT(\alpha,-s)\}\, \text{Tr}\{GT(\alpha,s)\}\, \pi^{-1}\, d^2\alpha \tag{9.203}$$

illustrates the need for caution in departing far from the line $\text{Re}\, s = 0$ since both operators $T(\alpha,s)$ and $T(\alpha,-s)$ appear explicitly. We note that the functions $f(\alpha,-s)$ and

$g(\alpha,s)$ are associated with the operators F and G by the correspondences $C(s)$ and $C(-s)$, respectively. Only for $s = 0$, symmetric order, do the two correspondences coincide.

The correspondence $C(1)$ associated with normal ordering, $s = 1$, is particularly simple. As we have shown in Sect. 9.4, virtually every operator F possesses a convergent normally ordered power-series expansion

$$F = \sum_{n,m=0}^{\infty} f_{n,m}(1) (a^\dagger)^n a^m . \tag{9.204}$$

The corresponding function

$$f(\alpha,-1) = \text{Tr}\{FT(\alpha,-1)\} \tag{9.205}$$

is, according to Eq. (9.168), just the diagonal coherent-state matrix element

$$f(\alpha,-1) = \langle \alpha|F|\alpha \rangle . \tag{9.206}$$

Although the correspondence $C(1)$ is well defined for an extremely broad class of operators, at $s = 1$ the integral expansion Eq. (9.191),

$$F = \int \langle \alpha|F|\alpha \rangle T(\alpha,1) \pi^{-1} d^2\alpha , \tag{9.207}$$

is of decidedly less generality since it involves the operator $T(\alpha,1)$ all of whose eigenvalues are infinite, as shown by Eq. (9.167). In fact, it may be shown[2] that for no states $|\psi\rangle$ and $|\varphi\rangle$ is the matrix element $\langle \varphi|T(\alpha,1)|\psi\rangle$ a square integrable function of α.

For the case of antinormal ordering, $s = -1$, the correspondence $C(-1)$ associates with an operator F the function

$$f(\alpha,1) = \text{Tr}\{FT(\alpha,1)\} , \tag{9.208}$$

which is clearly not in general free of singularities. These singularities are intimately related to those which occur in the coefficients of the antinormally ordered expansion

$$F = \sum_{n,m=0}^{\infty} f_{n,m}(-1) a^m (a^\dagger)^n . \tag{9.209}$$

For only when the function $f(\alpha,1)$ is infinitely differentiable at $\alpha = 0$ are the coefficients $f_{n,m}(-1)$ finite, as is shown by Eq. (9.200). When, however, the function $f(\alpha,1)$ is well behaved, then according to Eq. (9.168) the expansion (9.191) assumes the simple form

$$F = \int f(\alpha,1) T(\alpha,-1) \pi^{-1} d^2\alpha$$
$$= \int f(\alpha,1) |\alpha\rangle \langle \alpha| \pi^{-1} d^2\alpha \tag{9.210}$$

For the case of the density operator ρ, this is the P representation[2,3,7,23]

$$\rho = \int P(\alpha) |\alpha\rangle \langle\alpha| \, d^2\alpha, \tag{9.211}$$

where the weight function $P(\alpha)$ is given by[2,3]

$$P(\alpha) = \pi^{-1} \operatorname{Tr}\{\rho T(\alpha,1)\}. \tag{9.212}$$

Applications of the present results to the representation of density operators and to P representation in particular are discussed in the following paper.

There are no serious problems in the correspondence $C(0)$ associated with symmetric order, $s = 0$. The function $f(\alpha,0)$ associated with the operator F,

$$f(\alpha,0) = \operatorname{Tr}\{FT(\alpha,0)\}, \tag{9.213}$$

is the weight function for the expansion

$$F = \int f(\alpha,0) T(\alpha,0) \pi^{-1} d^2\alpha. \tag{9.214}$$

The operator $T(\alpha,0)$ appearing in these relations is finite, though not bounded; and, as we observed in Sect. 9.6, the function $f(\alpha,0)$ is in L_2 (i.e., is square-integrable) when the operator F is bounded.

We recall, from Eqns. (9.40)–(9.44) that the correspondence $F \leftrightarrow \operatorname{Tr}\{FD(\xi)\}$ is one-to-one between L_2 and the class of all bounded operators and, from Eqns. (9.32)–(9.34), that the Fourier transform maps L_2 onto L_2 in a one to one fashion. Thus, since we defined the weight function $f(\alpha,0)$ as the complex Fourier transform Eq. (9.142) of the function $\operatorname{Tr}\{FD(\xi)\}$, it follows that the symmetric correspondence $C(0)$ maps L_2 onto the class of all bounded operators in a one-to-one fashion. The correspondence $C(0)$ is also, in view of Eqns. (9.34) and (9.44), norm-preserving in that the L_2 norm Eq. (9.36) of the weight function $f(\alpha,0)$ is equal to the Hilbert–Schmidt norm Eq. (9.43) of the associated operator F; i.e., we have

$$||f(\alpha,0)|| = ||F||. \tag{9.215}$$

In terms of power-series expansions, the symmetric correspondence $C(0)$ associates the operator

$$F = \sum_{n,m=0}^{\infty} f_{n,m}(0) \left\{(a^\dagger)^n a^m\right\}_0 \tag{9.216}$$

with the function

$$f(\alpha,0) = \sum_{n,m=0}^{\infty} f_{n,m}(0) (\alpha^*)^n \alpha^m, \tag{9.217}$$

where, as we have seen in Sections 9.4 and 9.5, the coefficients $f_{n,m}(0)$ are finite for most though not all bounded operators F.

The correspondence $C(s)$ can easily be written in terms of complex Fourier transforms. By using Eqns. (9.35) and (9.145), we see that the correspondence $C(s)$ associates the operator

$$F = \int f(\alpha, -s) T(\alpha, s) \pi^{-1} d^2\alpha$$
$$= \int g(\xi, -s) D(-\xi, s) \pi^{-1} d^2\xi \qquad (9.218)$$

with the function

$$f(\alpha, -s) = \text{Tr}\{FT(\alpha, -s)\}$$
$$= \int g(\xi, -s) \exp(\alpha\xi^* - \alpha^*\xi) \pi^{-1} d^2\xi . \qquad (9.219)$$

By setting $s = 0$ and using Eqns. (9.1) and (9.2) to write these relations in real rather than complex notation, we may express the symmetric correspondence $C(0)$ in the form in which it was introduced by Weyl[6]:

$$F = \int g(q', p') e^{-i(qp' - pq')} (2\pi)^{-1} dq' dp' , \qquad (9.220)$$
$$f(x, y) = \int g(q', p') e^{-i(xp' - yq')} (2\pi)^{-1} dq' dp' , \qquad (9.221)$$

where we have set $\hbar = 1$.

It is a straightforward matter to derive a number of the properties of the weight functions $f(\alpha, -s)$ by using the analysis of the operators $T(\alpha, s)$ which was presented in Sect. 9.6. To avoid unnecessary minus signs we shall discuss the function $f(\alpha, s)$ rather than the weight function $f(\alpha, -s)$ of the expansion Eq. (9.191).

By multiplying both sides of Eq. (9.180) by the operator F and forming the trace of the resulting relation, we find that for different values of the order parameter s the functions $f(\alpha, s)$ are related by the simple Gaussian convolution

$$f(\alpha, s) = \frac{2}{t-s} \int f(\alpha', t) \exp\left[-\frac{2|\alpha - \alpha'|^2}{t-s}\right] \pi^{-1} d^2\alpha' \quad \text{for} \quad \text{Re}\, s < \text{Re}\, t . \quad (9.222)$$

This relation makes it clear that if the function $f(\alpha, t)$ is well behaved, then so is the function $f(\alpha, s)$ for $\text{Re}\, s < \text{Re}\, t$.

By performing a similar operation upon Eq. (9.181) we find that the function $f(\alpha, s)$ satisfies the differential equation

$$\frac{\partial f(\alpha, s)}{\partial s} = -\frac{1}{2} \frac{\partial^2 f(\alpha, s)}{\partial \alpha \, \partial \alpha^*} , \qquad (9.223)$$

which is the differential form of Eq. (9.222). The relations (9.223) and (9.222) have the same form as the heat-diffusion equation and its solution. This analogy becomes more complete when the operator F is a density operator; we shall discuss it in that context in the paper which follows.

From Eq. (9.185) we find that the function $f(\alpha,s)$ is normalized in the sense that

$$\operatorname{Tr} F = \int f(\alpha,s)\, \pi^{-1}\, d^2\alpha, \qquad (9.224)$$

subject to the existence of the trace and the convergence of the integral. This expression for the trace of an operator is a generalization of the earlier one Eq. (9.28)) which may be recovered by putting $s = -1$.

Let us now use the properties of the operator $T(\alpha,s)$ which were discussed in Sect. 9.6 to characterize the behavior of the function $f(\alpha,s) = \operatorname{Tr}\{FT(\alpha,s)\}$ as a function of the order parameter s. We first focus our attention primarily upon the left half-plane $\operatorname{Re} s < 0$ where, as we shall show, the function $f(\alpha,s)$ is a bounded, square-integrable, and infinitely differentiable function of α for all bounded operators F. We shall then examine the changes in the properties of the function $f(\alpha,s)$ as the real part of s becomes positive.

We have noted earlier that the operator $T(\alpha,s)$ is bounded for $\operatorname{Re} s < 0$, as is shown by the estimate Eq. (9.169) for its Hilbert–Schmidt norm $\|T(\alpha,s)\|$. It follows, therefore, from the inequality (9.52) that if the operator F is bounded with norm $\|F\|$, then the modulus of the function $f(\alpha,s)$ is bounded for $\operatorname{Re} s < 0$ by the quantity

$$|f(\alpha,s)| = |\operatorname{Tr}\{FT(\alpha,s)\}| \le \|F\|\,\|T(\alpha,s)\| \le \frac{\|F\|}{(-\operatorname{Re} s)^{1/2}}, \qquad (9.225)$$

which is independent of α.

Let us now observe that by using the identity Eq. (9.34) and the definition Eq. (9.142) we may obtain the relation

$$\int |f(\alpha,s)|^2\, d^2\alpha = \int |\operatorname{Tr}\{FD(\xi,s)\}|^2\, d^2\xi$$

$$= \int e^{\operatorname{Re} s |\xi|^2}\, |\operatorname{Tr}\{FD(\xi)\}|^2\, d^2\xi, \qquad (9.226)$$

which for $\operatorname{Re} s < 0$ implies the inequality

$$\int |f(\alpha,s)|^2\, d^2\alpha \le \int |\operatorname{Tr}\{FD(\xi)\}|^2\, d^2\xi. \qquad (9.227)$$

When the operator F is bounded, the integral on the right-hand side of this equation converges, as is shown by Eq. (9.44). Thus for all bounded operators F, the function $f(\alpha,s)$ is a square-integrable function of α for $\operatorname{Re} s \le 0$. By comparing Eq. (9.227) with Eq. (9.44), we find for the norm $\|f(\alpha,s)\|$ of the function $f(\alpha,s)$, defined by Eq. (9.36), the inequality

$$\|f(\alpha,s)\| \le \|F\| \quad \text{for} \quad \operatorname{Re} s \le 0. \qquad (9.228)$$

We may infer from the relation Eq. (9.226) that the regions in the s plane in which the functions $f(\alpha,s)$ and $\text{Tr}\{FD(\xi,s)\}$ are square-integrable are identical and are bounded by a straight line on which the real part of s is constant. Let us denote this line by $\text{Re}\, s = x(F)$. Then for bounded operators F we have $x(F) \geq 0$ since the function $f(\alpha,s)$ is in L_2 at least for $\text{Re}\, s \leq 0$.

By differentiating both sides of the relation (9.142), we may express the derivatives of the function $f(\alpha,s)$ as the integrals

$$\frac{\partial^{l+n+m} f(\alpha,s)}{\partial s^l \, \partial \alpha^n \, \partial (\alpha^*)^m} = \int (\tfrac{1}{2}|\xi|^2)^l (\xi^*)^n (-\xi)^m \, \text{Tr}\{FD(\xi,s)\} \, e^{\alpha \xi^* - \alpha^* \xi} \pi^{-1} d^2\xi. \qquad (9.229)$$

The moduli of these derivatives are, accordingly, bounded by the integrals

$$\left| \frac{\partial^{l+n+m} f(\alpha,s)}{\partial s^l \, \partial \alpha^n \, \partial (\alpha^*)^m} \right| \leq (\tfrac{1}{2})^l \int |\xi|^{2l+n+m} e^{\text{Re}\, s |\xi|^2} |\text{Tr}\{FD(\xi,s)\}| \, \pi^{-1} d^2\xi, \qquad (9.230)$$

which are independent of α. According to the Schwarz inequality for functions Eq. (9.37), the convergence of these integrals for $\text{Re}\, s < 0$ and for all bounded operators F is ensured by the exponential factor $\exp(\text{Re}\, s |\xi|^2)$ and the square-integrability of the function $\text{Tr}\{FD(\xi)\}$. A similar argument[2] shows that the derivatives Eq. (9.229) exist and are bounded for $\text{Re}\, s < x(F)$.

We may conclude, therefore, that for $\text{Re}\, s < x(F)$, a region which includes the half-plane $\text{Re}\, s < 0$ if F is bounded, the function $f(\alpha,s)$ possesses derivatives of all orders with respect to s, α, and α^* and that these derivatives are bounded by quantities that are independent of α. In particular the function $f(\alpha,s)$ is an analytic function of s for $\text{Re}\, s < x(F)$ and its modulus is bounded by a quantity $M(s)$ which depends upon s but not upon α; i.e., we have

$$|f(\alpha,s)| \leq M(s) \quad \text{for} \quad \text{Re}\, s < x(F). \qquad (9.231)$$

It may be shown[2] that the Taylor series in s, α, and α^* formed with the derivatives Eq. (9.229) converges for all α and all s such that $\text{Re}\, s < x(F)$.

We recall that the existence of derivatives of all orders with respect to α and α^* is required if the function $f(\alpha,-s)$ is to be a generating function for the coefficients $f_{n,m}(s)$ according to the rule Eq. (9.200). We also note that the convergence of the power series for $f(\alpha,-s)$ guarantees that the function $f(\alpha,-s)$ is well defined by the correspondence $C(s)$ in the form of Eqns. (9.195)–(9.197). Finally, since for F bounded and $\text{Re}\, s < 0$ the derivatives of $f(\alpha,-s)$ with respect to α and α^* are analytic functions of s, our earlier observation that for F bounded and $\text{Re}\, s > 0$ the coefficients $f_{n,m}(s)$ are analytic is confirmed.

As the parameter s crosses the line $\text{Re}\, s = 0$, the class of operators F for which the function $f(\alpha,s) = \text{Tr}\{FT(\alpha,s)\}$ is bounded shrinks dramatically. We have seen in Eq. (9.225) that this function is bounded for all bounded operators when $\text{Re}\, s < 0$. Let us now note that the function $f(\alpha,s)$ is bounded even for all finite operators for $\operatorname{Re} s < 0$. According to Eq. (9.171)) the operator $T(\alpha,s)$ is in the trace class for $\operatorname{Re} s < 0$. It follows, therefore, from the inequality (9.67) that if the operator F is finite with norm $||F||_1$, as defined by Eq. (9.61), then the modulus of the function $f(\alpha,s)$ is bounded for $\operatorname{Re} s < 0$ by the quantity

$$|f(\alpha,s)| = |\operatorname{Tr}\{FT(\alpha,s)\}| \leq ||F||_1 ||T(\alpha,s)||_2 \leq \frac{2||F||_1}{|1-s|-|1+s|}, \qquad (9.232)$$

which is independent of α.

On the line $\operatorname{Re} s = 0$, however, the operator $T(\alpha,s)$ is finite but not bounded and the function $f(\alpha,s)$ is not necessarily bounded for finite operators or even for bounded operators. It is in general necessary for the operator F to be a trace-class operator if the function $f(\alpha,s)$ is to be bounded for $\operatorname{Re} s = 0$. In this case, by using the inequality (9.67) and the estimate Eq. (9.170) for the norm $||T(\alpha,s)||_1$, defined in Eq. (9.61)), we may obtain for the function $f(\alpha,s)$ the upper bound

$$|f(\alpha,s)| = |\operatorname{Tr}\{FT(\alpha,s)\}| \leq ||F||_2 ||T(\alpha,s)||_1 \leq \frac{2||F||_2}{|1-s|} \quad \text{for} \quad \operatorname{Re} s \leq 0, \quad (9.233)$$

where $||F||_2$ is the trace-class norm of the operator F, a norm which we defined in Eq. (9.65). For $\operatorname{Re} s > 0$ the function $f(\alpha,s)$ is not in general bounded even for trace-class operators.

We have seen that for all bounded operators F the function $f(\alpha,s)$ possesses derivatives of all orders with respect to the variables s, α, and α^* for $\operatorname{Re} s < 0$. For $\operatorname{Re} s = 0$, however, the function $f(\alpha,s)$ is not even a continuous function of α for all bounded operators. It is in general necessary for the operator F to be a trace-class operator for the function $f(\alpha,s)$ to be continuous on the line $\operatorname{Re} s = 0$, where it may be shown[2] to be uniformly continuous.

For $\operatorname{Re} s > 0$ the operator $T(\alpha,s)$ is not a finite operator and the three norms $||T(\alpha,s)||$, $||T(\alpha,s)||_1$, $||T(\alpha,s)||_2$, which are defined by Eqns. (9.43), (9.61), and (9.65), are all infinite. At $s = 1$ all of the eigenvalues $e_n(s)$ of the operator $T(\alpha,s)$ are infinite as may be seen from Eq. (9.167). For these reasons the function

$$f(\alpha,s) = \operatorname{Tr}\{FT(\alpha,s)\}$$

typically develops singularities of some type at one or more points in the half-plane $\operatorname{Re} s > 0$. Since the function $f(\alpha,s)$ is bounded, square-integrable, and infinitely differentiable for $\operatorname{Re} s < x(F)$, this singularity must occur at a value of $s = s_0$ for which $\operatorname{Re} s_0 > x(F)$. The useful properties that we have attributed to the function $f(\alpha,s)$ cease to hold if not at the line $\operatorname{Re} s = x(F)$ then certainly at the appearance of the first singularity. For larger values of $\operatorname{Re} s$ the function $f(\alpha,s)$ is simply too singular to be used either as a weight function for the operator F in the expansion Eq. (9.191) or as a generating function for the coefficients of its s-ordered power-series expansion. A further discussion of this problem in terms of distribution theory is given in Ref.[2], where it is shown that the function $f(\alpha,s)$ lies outside the space of tempered distributions[3] for $\mathrm{Re}\, s > \mathrm{Re}\, s_0$ where s_0 is the first singular point.

The locations of the line $\mathrm{Re}\, s = x(F)$ and of the singularities of the function $f(\alpha,s)$ vary considerably from one operator to another. As our example in Sect. 9.8 will show, the quantities $\mathrm{Re}\, s_0$ and $x(F)$ may be arbitrarily small even when the operator F is in the trace class.

9.8
Illustration of Operator–Function Correspondences

We shall now illustrate the results of Sections 9.6 and 9.7 on the representation of operators by considering a simple example in some detail. This example will show how the weight functions $f(\alpha,s) = \mathrm{Tr}\{FT(\alpha,s)\}$, which are extremely well behaved for $\mathrm{Re}\, s < 0$, can develop singularities for $\mathrm{Re}\, s > 0$ even when the operator F is both bounded and in the trace class. As we mentioned earlier, these singularities reflect the fact that $T(\alpha,s)$ is a finite operator only for $\mathrm{Re}\, s < 0$.

Let us consider again the example provided by the operator $F(\lambda) = \lambda^{a^\dagger a}$. Since the operator $F(\lambda)$ may be written, according to Eq. (9.161), as

$$F(\lambda) = \tfrac{1}{2}(1-t)\,T(0,t), \tag{9.234}$$

with

$$t = \frac{\lambda+1}{\lambda-1}, \tag{9.235}$$

we may write the function

$$f_\lambda(\alpha,s) = \mathrm{Tr}\{F(\lambda)\,T(\alpha,s)\} \tag{9.236}$$

in the form

$$f_\lambda(\alpha,s) = \tfrac{1}{2}(1-t)\,\mathrm{Tr}\{T(0,t)\,T(\alpha,s)\}. \tag{9.237}$$

The trace may be evaluated by means of Eq. (9.178), which yields

$$\begin{aligned} f_\lambda(\alpha,s) &= \frac{t-1}{t+s}\exp\!\left[\frac{2|\alpha|^2}{t+s}\right] \\ &= \frac{2}{1+\lambda-s+s\lambda}\exp\!\left[\frac{2(\lambda-1)|\alpha|^2}{1+\lambda-2+s\lambda}\right]. \end{aligned} \tag{9.238}$$

---

[3] Tempered distributions form a class of continuous linear functional which includes the δ function and its derivatives. A distribution is said to be tempered if it can be expressed as a derivative of finite order of a continuous function that is bounded by a polynomial. See, e.g., I. M. Gel'Fand, G. E. Shilov, *Generalized Functions*, translated by E. Saletan, Academic Press, New York 1964, Vol. I.

The function $f_\lambda(\alpha,s)$ is an analytic function of s except for an essential singularity at

$$s = s_0 = \frac{1+\lambda}{1-\lambda}. \tag{9.239}$$

As a function of α, it is for $\mathrm{Re}\, s < \mathrm{Re}\, s_0$ a Gaussian function with its maximum at $\alpha = 0$. The function $f_\lambda(\alpha,s)$ is accordingly a bounded, square-integrable, and infinitely differentiable function of α for $\mathrm{Re}\, s < \mathrm{Re}\, s_0$. Thus, for the operator $F(\lambda)$ the quantity $x[F(\lambda)]$, introduced in Sect. 9.7, is given by

$$x[F(\lambda)] = \mathrm{Re}\, s_0 = \mathrm{Re}\left(\frac{1+\lambda}{1-\lambda}\right)$$
$$= \frac{1-|\lambda|^2}{|1-\lambda|^2}. \tag{9.240}$$

Let us recall that the operator $F(\lambda)$ is both bounded and in the trace class for $|\lambda| < 1$, that it is finite but not bounded for $|\lambda| = 1$, and that it is neither finite nor bounded for $|\lambda| > 1$. From Eq. (9.240) we see that the parameter x is positive for $|\lambda| < 1$, zero for $|\lambda| = 1$, and negative for $|\lambda| > 1$. The region $\mathrm{Re}\, s < x$, in which the function $f_\lambda(\alpha,s)$ is well behaved, therefore includes the half-plane $\mathrm{Re}\, s < 0$ when and only when the operator $F(\lambda)$ is both bounded and in the trace class. Under the transformation Eq. (9.239), the location s_0 of the singularity in the function $f_\lambda(\alpha,s)$ assumes every value in the half-plane $\mathrm{Re}\, s > 0$ as the parameter λ ranges over the disk $|\lambda| < 1$. Thus there are values of λ for which the operator $F(\lambda)$ is in the trace class but for which the line $\mathrm{Re}\, s = x$, on which the singular point s_0 falls, lies arbitrarily close to the imaginary axis $\mathrm{Re}\, s = 0$.

According to Eq. (9.150) the operator $F(\lambda)$ may be expanded in the form

$$F(\lambda) = \lambda^{a^\dagger a}$$
$$= \int f_\lambda(\alpha,-s)\, T(\alpha,s)\, \pi^{-1} d^2\alpha, \tag{9.241}$$

where the weight function $f_\lambda(\alpha,-s)$, which is given by Eq. (9.238), is the one associated with the operator $F(\lambda)$ by the correspondence $C(s)$. By comparing Eq. (9.238)), after substituting $-s$ for s, with the s-ordered power-series expansion for the operator $F(\lambda)$, Eq. (9.137), we may verify that the association of the weight function $f_\lambda(\alpha,-s)$ with the operator $F(\lambda)$ is in accordance with the rules Eq. (9.195) and (9.196) of the correspondence $C(s)$.

We see from Eq. (9.238) that for $\mathrm{Re}\, s > \mathrm{Re}\, s_0$ the function $f_\lambda(\alpha,s)$ increases for large values of $|\alpha|$ as an exponential function of $|\alpha|^2$. For this reason the function $f_\lambda(\alpha,s)$ is a tempered distribution only for $\mathrm{Re}\, s < \mathrm{Re}\, s_0$. Now we have seen that as the parameter λ ranges over the disk $|\lambda| < 1$ in which the operator $F(\lambda)$ is both bounded and in the trace class, the singularity at $s_0 = (1+\lambda)(1-\lambda)^{-1}$ ranges over the half-plane $\mathrm{Re}\, s > 0$. From this counterexample we may conclude that for no value of $\mathrm{Re}\, s <$

0 do the operators $T(\alpha,s)$ afford a basis for the integral representation Eq. (9.150) of an arbitrary bounded operator F even if the whole class of tempered distributions is admitted as weight functions. Thus, the operators $T(\alpha,s)$ for $\operatorname{Re} s < 0$ contrast sharply with the displacement operators $D(\alpha)$ in terms of which every bounded operator may be expanded with a square-integrable weight function. In this sense the operators $T(\alpha,s)$ for $\operatorname{Re} s < 0$ must be regarded as undercomplete.

We may mention that it is not difficult to show[2] that the weight function $f(\alpha,-s)$ may, for all values of s and all bounded operators F, be interpreted as a member of the space[4] of ultradistributions Z'. In order to accommodate such weight functions, however, the structure of the representation Eq. (9.150) must be changed in a major way. When the operator F is in the trace class, the additional terms required to regularize the representation Eq. (9.150) may be found by generalizing a procedure formulated[25] for the case of the P representation of the density operator, Eqns. (9.211) and (9.212).

Appendix A

Our object here is to derive the useful integral identity[7]

$$z^{-1} f(z^{-1} y) = \int f(\alpha) \exp(\alpha^* y - z|\alpha|^2) \pi^{-1} d^2\alpha, \qquad \operatorname{Re} s > 0, \qquad (9.\mathrm{A}1)$$

which holds for all entire functions $f(\alpha)$ subject to appropriate conditions on the convergence of the integral. We evaluate first the integral K_n:

$$\begin{aligned}
K_n &= \int \alpha^n \exp(\alpha^* y - z|\alpha|^2) \pi^{-1} d^2\alpha \\
&= \frac{y^n}{n!} \int |\alpha|^{2n} e^{-z|\alpha|^2} \pi^{-1} d^2\alpha \\
&= z^{-1} (z^{-1} y)^n .
\end{aligned}$$

Then by writing

$$f(\alpha) = \sum_{n=0}^{\infty} c_n \alpha^n$$

we find that

$$\begin{aligned}
\int e^{\alpha^* y - z|\alpha|^2} f(\alpha) \pi^{-1} d^2\alpha &= \sum_{n=0}^{\infty} c_n K_n \\
&= z^{-1} f(z^{-1} y),
\end{aligned}$$

---

4) The space Z' of ultradistributions is a class of generalized functions which includes but is much larger than the space of tempered distributions. It is discussed in the work by Gel'Fand and Shilov (see footnote 3).

which is Eq. (9.A1), provided that the integral may be done term by term. In particular, if we let $f(\alpha) = \exp(\alpha x)$, then Eq. (9.A1) becomes

$$z^{-1} \exp(z^{-1}xy) = \int \exp(\alpha x + \alpha^* y - z|\alpha|^2) \pi^{-1} d^2\alpha, \tag{9.A2}$$

provided $\operatorname{Re} z > 0$.

Appendix B

We shall now express the matrix elements of the displacement operator $D(\alpha)$ in the n-quantum representation in terms of the associated Laguerre polynomials $L_n^{(m)}(x)$. We first note that if $|\alpha\rangle$ is a coherent state then, by using Eqns. (9.19) and (9.20), we may write

$$\begin{aligned} D(\xi)|\alpha\rangle &= D(\xi)D(\alpha)|0\rangle \\ &= D(\xi+\alpha)|0\rangle \exp\left[\tfrac{1}{2}(\xi\alpha^* - \xi^*\alpha)\right] \\ &= |\xi+\alpha\rangle \exp\left[\tfrac{1}{2}(\xi\alpha^* - \xi^*\alpha)\right]. \end{aligned}$$

Thus, by using Eq. (9.23), we find that

$$\begin{aligned} \langle m|D(\xi)|\alpha\rangle &= \frac{1}{(m!)^{1/2}}(\xi+\alpha)^m \exp\left[\tfrac{1}{2}(\xi\alpha^* - \xi^*\alpha) - \tfrac{1}{2}|\xi+\alpha|^2\right] \\ &= \frac{1}{(m!)^{1/2}}(\xi+\alpha)^m \exp\left[-\tfrac{1}{2}|\xi|^2 - \tfrac{1}{2}|\alpha|^2 - \xi^*\alpha\right]. \end{aligned} \tag{9.B1}$$

Another expression for this same matrix element also follows from Eq. (9.23):

$$\langle m|D(\xi)|\alpha\rangle = e^{-|\alpha|^2/2} \sum_{m=0}^{\infty} \frac{\alpha^n}{(n!)^{1/2}} \langle m|D(\xi)|n\rangle. \tag{9.B2}$$

If we now combine Eqns. (9.B1) and (9.B2) and put $y = \xi^{-1}\alpha$, we arrive at the relation

$$(1+y)^m e^{-y|\xi|^2} = e^{|\xi|^2/2} \sum_{n=0}^{\infty} \left(\frac{m!}{n!}\right)^{1/2} \xi^{n-m} \langle m|D(\xi)|n\rangle y^n. \tag{9.B3}$$

The left-hand side of this equation is a generating function for the associated Laguerre polynomials $L_n^{(m)}(x)$ according to the identity[18]

$$(1+y)^m e^{-xy} = \sum_{n=0}^{\infty} L_n^{(m-n)}(x) y^n, \tag{9.B4}$$

which holds for $|y| < 1$. Thus by comparing Eqns. (9.B3) and (9.B4) we obtain the expression

$$\langle m|D(\xi)|n\rangle = \left(\frac{n!}{m!}\right)^{1/2} \xi^{m-n} e^{-|\xi|^2/2} L_n^{(m-n)}(|\xi|^2). \tag{9.B5}$$

Appendix C

Our definition Eq. (9.116) of the s-ordered product was as the derivative

$$\left\{(a^\dagger)^n a^m\right\}_s = \left.\frac{\partial^{n+m} D(\alpha,s)}{\partial \alpha^n \partial(-\alpha^*)^m}\right|_{\alpha=0}. \tag{9.C1}$$

By using Eq. (9.111) we may write the operator $D(\alpha,s)$ as a t-ordered exponential

$$\begin{aligned} D(\alpha,s) &= D(\alpha,t)\, e^{(s-t)|\alpha|^2/2} \\ &= \left\{\exp\left[\alpha a^\dagger - \alpha^* a + \tfrac{1}{2}(s-t)|\alpha|^2\right]\right\}_t. \end{aligned} \tag{9.C2}$$

We note the following equivalent expansions:

$$e^{\lambda w + \mu z + \nu w z} = \sum_{m=0}^{\infty} \frac{z^m}{m!}(\mu + \nu w)^m e^{\lambda w} \tag{9.C3a}$$

$$= \sum_{n=0}^{\infty} \frac{w^n}{n!}(\lambda + \nu z)^n e^{\mu z}. \tag{9.C3b}$$

By using the identity Eq. (9.B4), we may write these expansion in the forms

$$e^{\lambda w + \mu z + \nu w z} = \sum_{n,m=0}^{\infty} \frac{z^m}{m!} w^n \mu^{m-n} \nu^n L_n^{(m-n)}(-\lambda\mu/\nu) \tag{9.C4a}$$

$$= \sum_{n,m=0}^{\infty} \frac{w^n}{n!} z^m \lambda^{n-m} \nu^m L_m^{(n-m)}(-\lambda\mu/\nu). \tag{9.C4b}$$

Let us make the identifications

$$w = \alpha, \quad \lambda = a^\dagger, \quad \nu = -\tfrac{1}{2}(s-t), \quad z = \alpha^*, \quad \mu = a,$$

in Eqns. (9.C4a) and (9.C4b) and specify t ordering of both sides of these equations. Then according to Eq. (9.C2) we have found the t-ordered form for the displacement operator $D(\alpha,s)$. By carrying out the differentiations indicated in Eq. (9.C1)) we find from the expansion Eq. (9.C4a) the result

$$\left\{(a^\dagger)^n a^m\right\}_s = n!\left(\frac{t-s}{2}\right)^n \left\{a^{m-n} L_n^{(m-n)}\left(\frac{2a^\dagger a}{s-t}\right)\right\}_t, \tag{9.C5}$$

which is useful for $m \geq n$, and from Eq. (9.C4b)

$$\left\{(a^\dagger)^n a^m\right\}_s = m!\left(\frac{t-s}{2}\right)^m \left\{(a^\dagger)^{n-m} L_m^{(n-m)}\left(\frac{2a^\dagger a}{s-t}\right)\right\}_t, \tag{9.C6}$$

which is useful for $n \geq m$. These two expressions correspond to the explicit expansion

$$\left\{(a^\dagger)^n a^m\right\}_s = \sum_{k=0}^{(n,m)} k!\binom{n}{k}\binom{m}{k}\left(\frac{t-s}{2}\right)^k \left\{(a^\dagger)^{n-k} a^{m-k}\right\}_t, \tag{9.C7}$$

which is noted in Eq. (9.122).

Appendix D

By expanding the coherent states $|\beta\rangle$ and $\langle\beta|$ of Eq. (9.175) in terms of the n-quantum states, we find

$$e^{|\beta|^2} \langle\beta|T(\alpha,s)|\beta\rangle = \sum_{n,m=0}^{\infty} \frac{1}{(n!\,m!)^{1/2}} (\beta^*)^n \beta^m \langle n|T(\alpha,s)|m\rangle$$

$$= \frac{2}{1-s} \exp\left[|\beta|^2 - \frac{2|\alpha-\beta|^2}{1-s}\right].$$

If we now make use of the expansion Eq. (9.C4a) with the identifications

$$w = \beta^*, \quad \lambda = \frac{2\alpha}{1-s}, \quad \nu = -\frac{1+s}{1-s}, \quad z = \beta, \quad \mu = \frac{2\alpha^*}{1-s},$$

then we find

$$e^{|\beta|^2} \langle\beta|T(\alpha,s)|\beta\rangle = \frac{2}{1-s} \exp\left[-\frac{2|\alpha|^2}{1-s}\right]$$

$$\times \sum_{n,m=0}^{\infty} \frac{(-\beta^*)^n \beta^m}{m!} \left(\frac{2\alpha^*}{1-s}\right)^{m-n} \left(\frac{1+s}{1-s}\right)^n L_n^{(m-n)}\left(\frac{4|\alpha|^2}{1-s^2}\right).$$

Hence we have

$$\langle n|T(\alpha,s)|m\rangle = \frac{n!}{m!} \left(\frac{2}{1-s}\right)^{m-n+1} \left(\frac{s+1}{s-1}\right)^n (\alpha^*)^{m-n}$$

$$\times \exp\left[-\frac{2|\alpha|^2}{1-s}\right] L_n^{(m-n)}\left(\frac{4|\alpha|^2}{1-s^2}\right),$$

which is Eq. (9.190).

References

1 W. H. Louisell, *Radiation and Noise in Quantum Electronics*, McGraw-Hill, New York 1965, pp. 104–119.

2 K. E. Cahill, thesis, Harvard University 1967 (University Microfilms, Ann Arbor), unpublished. Many of the results of the present paper are presented there.

3 M. Lax, W. H. Louisell, *J. Quantum Electron.* **QE3**, 47 (1967).

4 M. Lax, *Phys. Rev.* **172**, 350 (1968).

5 G. S. Agarwal, E. Wolf, *Phys. Lett.* **26A**, 485 (1968).

6 H. Weyl, *The Theory of Groups and Quantum Mechanics*, Dover Publications, New York 1950, pp. 272–276.

7 R. J. Glauber, *Phys. Rev.* **131**, 2766 (1963); reprinted as Sect. 2.9 in this volume.

8 A. Messiah, *Quantum Mechanics*, North-Holland, Amsterdam 1961), Vol. I, p. 442.

9 J. C. T. Pool, *J. Math. Phys.* **7**, 66 (1966).

10 J. E. Moyal, *Proc. Cambridge Phil. Soc.* **45**, 99 (1948).

11 M. S. Bartlett, J. E. Moyal, *Proc. Cambridge Phil. Soc.* **45**, 545 (1949).

12 U. Fano, *Rev. Mod. Phys.* **29**, 74 (1957).

13 J. Schwinger, *Proc. Natl. Acad. Sci. USA* **46**, 883 (1960).

14 A. E. Glassgold, D. Holliday, *Phys. Rev. A* **139**, 1717 (1965).

15 J. R. Klauder, J. McKenna, D. G. Currie, *J. Math. Phys.* **6**, 743 (1965).

16 W. Rudin, *Principles of Mathematical Analysis*, McGraw-Hill, New York 1953. In Chapter 10 the Lebesgue theory is discussed in elementary terms.

17 S. Bochner, K. Chandrasekharan, *Fourier Transforms*, Princeton University Press, Princeton 1949, Chapter 4.

18 W. Magnus, F. Oberhettinger, *Formulas and Theorems for the Functions of Mathematical Physics*, Chelsea Publishing, New York 1954, p. 85.

19 M. Gel'Fand, N. Ya. Vilenkin, *Generalized Functions*, translated by A. Feinstein, Academic Press, New York 1964, Vol. 4. Chap. 1, Sec. 2, is a detailed and readable account of the properties of trace-class operators and of Hilbert–Schmidt operators.

20 B. R. Mollow, *Phys. Rev.* **162**, 1256 (1967).

21 R. J. Glauber, in *Proceedings of the Second International Summer School on Fundamental Problems of Statistical Mechanics, Noordwijk aan Zee, The Netherlands 1967*, E. G. D. Cohen, Ed., North-Holland, Amsterdam 1968.

22 R. J. Glauber, *Phys. Rev. Lett.* **10**, 84 (1963).

23 E. C. G. Sudarshan, *Phys. Rev. Lett.* **10**, 277 (1963).

24 E. P. Wigner, *Phys. Rev.* **40**, 749 (1932).

25 K. E. Cahill, to be published.

10
Density Operators and Quasiprobability Distributions[1]

10.1
Introduction

The statistical description of a microscopic system may usually be formulated in terms of its density operator ρ. The familiar expression for the statistical average of measurements of a microscopic observable F is the trace of the product of the operators ρ and F,

$$\langle F \rangle = \text{Tr}\{\rho F\} . \tag{10.1}$$

In this paper we discuss ways of writing this statistical average for certain physically important classes of operators $\langle F \rangle$ as integrals similar to the phase-space integrals of classical probability theory. The preceding Chapter[1] on integral representations for operators and on correspondences between operators and c-number functions provides the framework for our analysis[2] of this problem.

For simplicity we consider only systems which have a single degree of freedom. Further, since we have in mind applications to quantum optics, we describe these systems in terms of the complex operators a and $a^\dagger$, which satisfy the commutation relation $[a, a^\dagger] = 1$, rather than in terms of their real and imaginary parts q and p, for which $[q, p] = i\hbar$. Our discussion is thus directly applicable to a single mode of the electromagnetic field.

A number of procedures for simplifying the evaluation of certain classes of expectation values have been put forward, the first of them by Wigner[3] and by Moyal[4]. A common feature of these methods is the transfer of statistical information from the density operator ρ to a weight function $w(\alpha)$ which refers to the density operator and whose complex argument a represents a point (q', p') in the phase space of the system. An expectation value $\langle F \rangle$ is then written as an integral of the product of the weight function $w(\alpha)$ and a function $f(\alpha)$ which refers to the operator F,

$$\text{Tr}\{\rho F\} = \int w(\alpha) f(\alpha) \, d^2\alpha . \tag{10.2}$$

---

[1] Reprinted with permission from K. E. Cahill, R. J. Glauber, *Phys. Rev.* **177**, 1882–1902 (1969). Copyright 2006 by the American Physical Society.

Quantum Theory of Optical Coherence. Selected Papers and Lectures. Roy J. Glauber.
Copyright © 2007 WILEY-VCH Verlag GmbH & Co. KGaA, Weinheim
ISBN: 978-3-527-40687-6

The integration is carried out over all possible states of the system, i.e., over the complex α plane, and the differential $d^2\alpha$ is a real element of area proportional to the phase-space element $dq'\,dp'$. The function $w(\alpha)$ is not, in general, interpretable as a probability distribution, but it plays so closely the role of one that we refer to it as a quasiprobability distribution.

The various procedures for expressing expectation values in the form of Eq. (10.2) differ principally in the way in which the functions $w(\alpha)$ and $f(\alpha)$ correspond to the operators ρ and F. Because quantum-mechanical operators do not, in general, commute, there are many ways of associating operators with functions. With the function $|\alpha|^2$, for example, we could associate either the normally ordered operator $a^\dagger a$ or the antinormally ordered operator $a\,a^\dagger$ or, as well, the symmetrized product $\frac{1}{2}\{a^\dagger a + a a^\dagger\}$. The most useful correspondences are based upon these three types of ordering.

In Chapter 9 we have analyzed various ways of defining correspondences between operators and functions. We summarize these results together with some related ones on series and integral expansions for operators in Sect. 10.2.

In Sect. 10.3 we discuss a procedure for simplifying expectation values that is based upon the P representation[5–16] for the density operator. In this representation the density operator assumes the form of a weighted integral over the projection operators upon the coherent states. We show that the weight function $P(\alpha)$ of the P representation is related to the density operator by a correspondence defined in terms of anti-normal ordering[2,14]. The P representation is particularly convenient for dealing with operators that are written in normal order; it affords for the expectation values of the normally ordered products $(a^\dagger)^n a^m$ the simple integral expressions

$$\mathrm{Tr}\left\{\rho(a^\dagger)^n a^m\right\} = \int P(\alpha)\,(\alpha^*)^n\alpha^m\,d^2\alpha\,. \tag{10.3}$$

This relation is a special case of Eq. (10.2) in which the function $w(\alpha)$ associated with the density operator is the weight function $P(\alpha)$ and in which the correspondence between the operator F and the function $f(\alpha)$ is based upon normal ordering.

When the correspondences that associate the functions $w(\alpha)$ and $f(\alpha)$ with the operators ρ and F are both defined in terms of symmetric ordering, the integral Eq. (10.2) may be identified with the procedure introduced by Wigner[3] and by Moyal[4]. We discuss this procedure in Sect. 10.4. We show that the Wigner distribution $W(\alpha)$ affords for the density operator an integral representation[2,16,17] that is considerably more regular than the P representation. In particular, we show that the Wigner distribution is continuous and bounded, being the expectation value of a Hermitian operator whose eigenvalues are ± 2. The weight function $P(\alpha)$, by contrast, may be exceedingly singular, being the expectation value of a Hermitian operator all of whose eigenvalues are infinite.

In Sect. 10.5 we discuss a procedure[8,9,18] for dealing with operators written in antinormal order. In this procedure the function corresponding to the weight function $w(\alpha)$ in Eq. (10.2) is the function $\langle\alpha|\rho|\alpha\rangle$, where $|\alpha\rangle$ is a coherent state.

In Sect. 10.6 we unify and relate the procedures discussed in Sections 10.3 and 10.5 by using the parametrized ordering convention introduced in Chapter 9. We construct a parametrized representation for the density operator in which the weight function $W(\alpha,s)$ may be identified with the function $\langle \alpha|\rho|\alpha\rangle$, with the Wigner distribution $W(\alpha)$ and with the weight function $P(\alpha)$ when the order parameter s assumes the values $s = -1, 0, +1$, respectively. By varying the order parameter s in a continuous way from $s = -1$ to $s = +1$, we observe an orderly progression of changes in the mathematical properties of this quasiprobability distribution. In this way we are able to see when and how singularities appear in the function $W(\alpha,s)$ and are thereby in a better position to understand, in particular, why the P representation does not exist for important classes of density operators.

We illustrate in Sect. 10.7 the results of the earlier sections by means of several examples which are worked out in detail. We use one of these examples to develop a physical characterization of those density operators for which the P representation is appropriate.

In Sect. 10.8 we show that the quasiprobability distribution $W(\alpha,s)$ satisfies a partial differential equation similar in form to the heat-diffusion equation. This fact leads to an instructive analogy between the function $W(\alpha,s)$ and a temperature distribution on an infinite plane. Finding the function $W(\alpha,s)$ in terms of the function $W(\alpha,t)$ for $\mathrm{Re}\,s < \mathrm{Re}\,t$ corresponds to Poisson's solution for the heat-diffusion equation. The inverse problem, of extending the function $W(\alpha,s)$ in the other direction, corresponds to extending a temperature distribution backwards in time, and this problem is solved in Sect. 10.9.

In Sect. 10.10 we return to Eq. (10.2) and discuss the extent to which any procedure for expressing expectation values in that form must resemble the general procedure discussed in Sect. 10.6.

10.2
Ordered Operator Expansions

The annihilation and creation operators a and $a^\dagger$ may be defined for any system that is described by a single pair of Hermitian observables q and p, which are canonically conjugate, $[q,p] = i\hbar$. If we form the complex linear combinations

$$a = \frac{1}{(2\hbar)^{1/2}}\left(\lambda q + \frac{i}{\lambda}p\right), \tag{10.4}$$

$$a^\dagger = \frac{1}{(2\hbar)^{1/2}}\left(\lambda q - \frac{i}{\lambda}p\right), \tag{10.5}$$

where λ is an arbitrary real parameter, then the operators a and $a^\dagger$ obey the familiar commutation relation

$$[a,a^\dagger] = 1. \tag{10.6}$$

A particularly useful complete set of states is the coherent states $|\alpha\rangle$ which are the eigenstates of the operator a, i.e., for each complex number α we have

$$a|\alpha\rangle = \alpha|\alpha\rangle .\tag{10.7}$$

The basic properties of the coherent states and of the unitary displacement operators

$$D(\alpha) = \exp(\alpha a^\dagger - \alpha^* a) ,\tag{10.8}$$

which generate them from the state $|0\rangle$ are described in Sections 9.2 and 9.3.

The space of eigenvalues or the phase space for our dynamical system is the infinite plane of eigenvalues (q',p') of the Hermitian observables q and p. An equivalent phase space is the complex plane of eigenvalues[2] $\alpha = (2\pi\hbar)^{-1}(\lambda q' + i\lambda^{-1}p')$ of the annihilation operator a. As in Chapter 9, we make use, in integrating over the complex plane, of the convenient differential element of area

$$\pi^{-1} d^2\alpha = \pi^{-1} d(\mathrm{Re}\,\alpha)\, d(\mathrm{Im}\,\alpha) = (2\pi\hbar)^{-1} dq'\, dp' ,\tag{10.9}$$

which we see is also the familiar (dimensionless) element of phase space.

In Chapter 9 we examined the problem of representing an arbitrary operator F both as a power series in the operators a and $a^\dagger$ and as an integral over various complete and quasicomplete sets of operators. An important device in our treatment was a parametrization of the usual ways of ordering operators according to which normal order, antinormal order, and a type of ordering that is symmetric in the operators a and $a^\dagger$ correspond to three discrete values of a continuous order parameter s.

We defined the s-ordered products $\{(a^\dagger)^n a^m\}_s$ of the operators $(a^\dagger)^n$ and a^m by means of a Taylor-series expansion of the operator

$$\begin{aligned}D(\alpha,s) &\equiv D(\alpha)\,e^{s|\alpha|^2/2} \\ &= \exp(\alpha a^\dagger - \alpha^* a + \tfrac{1}{2}s|\alpha|^2) \\ &= \sum_{n,m=0}^\infty \frac{\alpha^n(-\alpha^*)^m}{n!\,m!} \left\{(a^\dagger)^n a^m\right\}_s .\end{aligned}\tag{10.10}$$

Since the operators $D(\alpha,1)$ and $D(\alpha,-1)$ are the normally and antinormally ordered exponentials

$$D(\alpha,1) = \exp(\alpha a^\dagger)\exp(-\alpha^* a) ,\tag{10.11}$$
$$D(\alpha,-1) = \exp(-\alpha^* a)\exp(\alpha a^\dagger) ,\tag{10.12}$$

---

[2] The coherent state $|\alpha\rangle$ is not an eigenstate of either q or p. The quantities q' and p' in this expression may be interpreted as the expectation values of q and p in the state $|\alpha\rangle$.

the normally and antinormally ordered products are distinguished by the values $s = \pm 1$:

$$\{(a^\dagger)^n a^m\}_1 = (a^\dagger)^n a^m , \tag{10.13}$$

$$\{(a^\dagger)^n a^m\}_{-1} = a^m (a^\dagger)^n . \tag{10.14}$$

The operator $\{(a^\dagger)^n a^m\}_0$ corresponding to the value $s = 0$ is the average of all ways of ordering the product of n factors of $a^\dagger$ and m factors of a. We have therefore designated the type of ordering which emerges for $s = 0$ as symmetric ordering.

In Sections 9.4 and 9.5, we discussed the problem of expanding an operator F as a power series in the s-ordered products, i.e., in the form

$$F = \sum_{n,m=0}^{\infty} f_{n,m}(s) \{(a^\dagger)^n a^m\}_s , \tag{10.15}$$

where the coefficients $f_{n,m}(s)$ are complex numbers. We derived explicit formulas for the coefficients $f_{n,m}(s)$ and defined a type of convergence that is appropriate for the power series Eq. (10.15). For the case of normal ordering, $s = 1$, we found the expansion Eq. (10.15) converges according to this definition for an extremely broad class of operators F. For $\mathrm{Re}\, s > 0$ or, equivalently, for orderings closer to normal order than to antinormal order, we showed that the coefficients $f_{n,m}(s)$ are finite for a more restricted class of operators F. These operators, which we refer to as the class of bounded operators, are the ones for which the Hilbert–Schmidt norm

$$||F|| = \left[\mathrm{Tr}\{F^\dagger F\}\right]^{1/2} \tag{10.16}$$

is finite. For such operators we also showed that the power series Eq. (10.15) converges for $\mathrm{Re}\, s > \frac{1}{2} + \frac{1}{2}(\mathrm{Im}\, s)^2$, i.e., for orderings that are closer to normal order than to symmetric order.

For the case of antinormal ordering, $s = -1$, however, and in fact throughout the half-plane $\mathrm{Re}\, s \leq 0$, the coefficients $f_{n,m}(s)$ can and do develop singularities. These singularities, which raise questions as to the meaning of expansions such as Eq. (10.15), occur even for operators F that are both bounded and in the trace class. Broadly speaking, the more negative the real part of s the more restricted must be our interpretation of the expansion Eq. (10.15); this limitation is far more restrictive for antinormal order, $s = -1$, than for symmetric order, $s = 0$.

In Sections 9.6–9.8 we discussed a class of integral expansions for operators. These expansions are based upon the operators $T(\alpha, s)$ which we defined as the complex Fourier transforms[3] of the operators $D(\alpha, s)$,

$$T(\alpha, s) = \int D(\xi, s) \exp(\alpha \xi^* - \alpha^* \xi) \pi^{-1} d^2\xi . \tag{10.17}$$

---

[3] This type of Fourier transformation, which differs from the usual one only by a change of scale, is defined by Eq. (9.29).

The expansions for an arbitrary operator F take the form

$$F = \int f(\alpha, -s) \, T(\alpha, s) \, \pi^{-1} \, d^2\alpha \,, \tag{10.18}$$

in which the weight function $f(\alpha, -s)$ is given by the trace

$$f(\alpha, -s) = \mathrm{Tr}\{F T(\alpha, -s)\} \,. \tag{10.19}$$

An important property of these expansions is the close relationship which exists between the operator F being expanded and its weight function $f(\alpha, -s)$. The power-series expansion of the weight function $f(\alpha, -s)$ possesses the same coefficients $f_{n,m}(s)$ that occur in the s-ordered power-series expansion Eq. (10.15) of the operator F, i.e., we have

$$f(\alpha, -s) = \sum_{n,m=0}^{\infty} f_{n,m}(s) \, (\alpha^*)^n \, \alpha^m \,. \tag{10.20}$$

In other words, $f(\alpha, -s)$ is the same function of α and α^* as the s-ordered version of F is of a and $a^\dagger$.

We may regard this identity of the coefficients $f_{n,m}(s)$ in the series (10.15) and (10.20) as defining, for each value of the order parameter s, a one-to-one correspondence between operators and functions. We have referred to this correspondence between operators F and their weight functions $f(\alpha, -s)$ as the correspondence $C(s)$. The rule of the correspondence $C(s)$

$$F \underset{s}{\longleftrightarrow} f(\alpha, -s) = \mathrm{Tr}\{F T(\alpha, -s)\} \tag{10.21}$$

is that the s-ordered product $\{(a^\dagger)^n a^m\}$ is associated with the monomial $(\alpha^*)^n \alpha^m$

$$\{(a^\dagger)^n a^m\}_s \underset{s}{\longleftrightarrow} (\alpha^*)^n \alpha^m \,. \tag{10.22}$$

As in the case of the power series expansion (10.15), the properties of the integral representation Eq. (10.18) and of the weight function $f(\alpha, -s)$ depend critically upon the order parameter s. In general for bounded operators F the weight functions $f(\alpha, -s)$ are well-behaved functions of α for $\mathrm{Re}\, s \geq 0$ but for $\mathrm{Re}\, s < 0$ they can develop singularities. These singularities are closely related to those that occur in the coefficients $f_{n,m}(s)$ of the power-series expansion Eq. (10.15). For according to Eq. (10.20), the weight function $f(\alpha, -s)$ is a generating function for the coefficients $f_{n,m}(s)$. When this function is infinitely differentiable, then we may write the operator F in the form

$$F = \sum_{n,m=0}^{\infty} \frac{1}{n! \, m!} \{(a^\dagger)^n a^m\}_s \left(\frac{\partial^{n+m} f(\alpha, -s)}{\partial(\alpha^*)^n \partial\alpha^m} \right)\bigg|_{\alpha=0}, \tag{10.23}$$

which is another way of expressing the operator–function correspondence $C(s)$.

10.3
The *P* Representation

The utility of normal ordering has long been recognized. It is used, for example, in the Dyson–Wick expansion of the scattering matrix in quantum electrodynamics[19]. Because of the appropriateness of normally ordered products for the description of photon absorption processes, normal ordering has had new applications, recently, in the interpretation of photon counting and coherence experiments[6,8,20]. The correlation functions $G^{(n)}$ for the quantized field are defined, for example, as expectation values of normally ordered products of the positive-frequency part of the electric field operator $E_\mu^{(+)}(x)$ and the negative-frequency part $E_\mu^{(-)}(x)$. The *n*-th-order correlation function may be written as

$$G^{(n)}(x_1, x_2, \ldots, x_n, x_{n+1}, \ldots, x_{2n}) = \text{Tr}\left\{\rho E^{(-)}(x_1) \ldots E^{(-)}(x_n), E^{(+)}(x_{n+1}) \ldots E^{(+)}(x_{2n})\right\}, \quad (10.24)$$

where we have denoted $2n$ sets of space-time coordinates as $x_1 \ldots x_{2n}$ and have suppressed the vector polarization indices of the field operators.

We have mentioned that the *P* representation for the density operator,

$$\rho = \int P(\alpha) |\alpha\rangle \langle\alpha| \, d^2\alpha, \quad (10.25)$$

affords a convenient way of evaluating the ensemble averages of normally ordered operators. It has apart from this important use a variety of additional applications[17,21–24] because of the elementary character of the coherent states $|\alpha\rangle$. Among these applications is the particularly simple form which it provides for a quantum-mechanical analog[6] of the classical superposition principle for the electromagnetic field.

We may use the *P* representation to write the expectation value of an arbitrary operator *F* as the integral

$$\text{Tr}\{\rho F\} = \int P(\alpha) \, \text{Tr}\{|\alpha\rangle \langle\alpha| F\} \, d^2\alpha$$
$$= P(\alpha) \langle\alpha|F|\alpha\rangle \, d^2\alpha. \quad (10.26)$$

The association of the function

$$f(\alpha, -1) = \langle\alpha|F|\alpha\rangle \quad (10.27)$$

with the operator *F* defines a particularly simple correspondence between operators and functions.

Let us suppose that the operator *F* possesses a normally ordered power series of the form

$$F = \sum_{n,m=0}^{\infty} f_{n,m}(1) \, (a^\dagger)^n a^m. \quad (10.28)$$

As we showed in Chapter 9, such series exist and converge for an extremely broad class of operators F, including all that satisfy the condition Eq. (9.91). Now because of the eigenvalue property of the coherent states, the function $f(\alpha) = \langle \alpha | F | \alpha \rangle$ may be obtained from the operator F by replacing the operators a and $a^\dagger$ in the series Eq. (10.28) by their complex eigenvalues α and α^*, so that we have

$$f(\alpha, -1) = \sum_{n,m=0}^{\infty} f_{n,m}(1) (\alpha^*)^n \alpha^m . \tag{10.29}$$

It is this function which occurs in the integral Eq. (10.26) for the ensemble average of the operator F,

$$\text{Tr}\{\rho F\} = \int P(\alpha) F(\alpha, -1) \, d^2\alpha . \tag{10.30}$$

Since the series Eq. (10.28) and (10.29), as was shown in Chapter 9, converge for virtually all operators of interest, this expression is completely reliable provided that the P representation exists for the density operator ρ.

We observe that, since all density operators are Hermitian and of unit trace, the weight function $P(\alpha)$, when it exists, can be chosen to be real and normalized to unity,

$$1 = \text{Tr}\,\rho = \int P(\alpha) \, d^2\alpha . \tag{10.31}$$

Having defined the P representation and touched upon its usefulness, we shall now consider it from a variety of viewpoints each of which reveals a way of constructing the weight function $P(\alpha)$. All of these ways show a certain lack of generality which reflects the fact that the P representation does not exist for all density operators.

Let us note that by using the resolution of the identity operator in terms of the coherent states, Eq. (9.27), we may write the antinormally ordered products $a^m (a^\dagger)^n$ in the form

$$\begin{aligned} a^m (a^\dagger)^n &= a^m \int |\alpha\rangle \langle \alpha| \, \pi^{-1} d^2\alpha \, (a^\dagger)^n \\ &= \int (\alpha^*)^n \alpha^m |\alpha\rangle \langle \alpha| \, \pi^{-1} d^2\alpha . \end{aligned} \tag{10.32}$$

Let us now suppose that the density operator ρ possesses the antinormally ordered power series

$$\rho = \sum_{n,m=0}^{\infty} \rho_{n,m}(-1) a^m (a^\dagger)^n . \tag{10.33}$$

It then follows from Eq. (10.32) that this density operator also possesses the P representation Eq. (10.25) with the function

$$P(\alpha) = \pi^{-1} \sum_{n,m=0}^{\infty} \rho_{n,m}(-1) (\alpha^*)^n \alpha^m \tag{10.34}$$

as its weight function[2,14]. The function $P(\alpha)$ is thus a generating function for the coefficients of the anti-normally ordered power-series expansion Eq. (10.33) of the density operator ρ. Equivalently, we may say that the correspondence that associates a density operator ρ with its weight function $P(\alpha)$ is the one based upon antinormal ordering, $s = -1$.

This correspondence does not, however, constitute a general prescription for constructing the weight function $P(\alpha)$. For, as we have seen in Sections 9.4 and 9.5, the coefficients $f_{n,m}(-1)$ of antinormally ordered power-series expansions are sometimes singular, even for operators that are both bounded and in the trace class.

As another approach to finding the function $P(\alpha)$ let us note that the function $f(\alpha, -s)$ associated with an arbitrary operator F by the correspondence $C(s)$ based upon s-ordering is in general given by Eq. (10.21) as the trace

$$f(\alpha, -s) = \text{Tr}\{FT(\alpha, -s)\} . \tag{10.35}$$

Since all density operators are both bounded and in the trace class, there is no intrinsic difficulty in applying this relation to an arbitrary density operator, $F = \rho$. If, therefore, we set $s = -1$, corresponding to antinormal ordering, then we may write the weight function $P(\alpha)$ as the trace

$$P(\alpha) = \pi^{-1} \text{Tr}\{\rho T(\alpha, 1)\} . \tag{10.36}$$

This trace is simply the expectation value of the operator $T(\alpha, 1)$.

Let us recall that the operator $T(\alpha, s)$, defined by Eq. (10.17), is Hermitian for real values of s, as shown by Eq. (9.156) and is in the trace class for $\text{Re}\, s < 0$, as shown by Eq. (9.171). According to Eq. (9.167), the eigenvalues $e_n(s)$ of the operator $T(\alpha, s)$ are independent of α and are given by

$$e_n(s) = \frac{2}{1-s}\left(\frac{s+1}{s-1}\right)^n . \tag{10.37}$$

We see that they are all infinite for $j = 1$.

The fact that the function $P(\alpha)$ is the expectation value of a Hermitian operator $T(\alpha, 1)$ all of whose eigenvalues $e_n(1)$ are infinite is related to the occurrence of singularities in the coefficients $\rho_{n,m}(-1)$ of the anti-normally ordered power series Eq. (10.33) for the density operator ρ. Furthermore, it sheds some light on why the weight function $P(\alpha)$ is, for certain classes of density operators, exceedingly singular[10].

We shall provide a fuller discussion of these singularities at the end of this section and also in Sections 10.6 and 10.7 in which we make use of the continuous order parameter s. For the present let us merely note that Eq. (10.36) is not well defined as it stands. It may be made meaningful for a broad class of density operators ρ if we use a limiting process in which the order parameter s approaches one along the real axis from smaller values, i.e.,

$$P(\alpha) = \pi^{-1} \lim_{s \to 1-} \text{Tr}\{\rho T(\alpha, s)\} . \tag{10.38}$$

Let us recall that, according to Eq. (9.168), the operator $T(\alpha,s)$ is for $s=-1$ simply the projection operator upon the coherent state $|\alpha\rangle$,

$$T(\alpha,-1) = |\alpha\rangle\langle\alpha| \,. \tag{10.39}$$

We may thus write the P representation in the form

$$\rho = \int P(\alpha)\, T(\alpha,-1)\, d^2\alpha \,. \tag{10.40}$$

We note that this expansion conforms to Eqns. (10.18) and (10.19) when the weight function $P(\alpha)$ is given by Eq. (10.36). If we now multiply both sides of the expansion Eq. (10.40) by an arbitrary operator F and form the trace of the resulting expression, then we arrive at the relation

$$\mathrm{Tr}\{\rho F\} = \int \mathrm{Tr}\{\rho T(\alpha,1)\}\, \mathrm{Tr}\{FT(\alpha,-1)\}\, \pi^{-1}\, d^2\alpha \,, \tag{10.41}$$

where we have used Eq. (10.36) for $P(\alpha)$. In this integral for the quantity $\mathrm{Tr}\{\rho F\}$ the order parameter s occurs twice, as $s=+1$ and as $s=-1$, in accordance with the occurrence of both s and $-s$ in the general relation Eq. (10.18).

We shall now derive what is perhaps the simplest way of constructing the weight function $P(\alpha)$ when it is nonsingular. By substituting our definition Eq. (10.17) of the operator $T(\alpha,s)$ into Eq. (10.38), we may express the function $P(\alpha)$ in terms of the limiting process

$$P(\alpha) = \pi^{-2} \lim_{s\to 1_-} \int \exp[\alpha\xi^* - \alpha^*\xi]\, \mathrm{Tr}\{\rho D(\xi,s)\}\, d^2\xi \,. \tag{10.42}$$

Let us now define the normally ordered characteristic function $\chi_N(\xi)$ as the expectation value[24]

$$\chi_N(\xi) = \mathrm{Tr}\{\rho D(\xi,1)\}$$
$$= \mathrm{Tr}\left\{\rho \exp[\xi a^\dagger] \exp[-\xi^* a]\right\} \,. \tag{10.43}$$

For many density operators ρ the function $\chi_N(\xi)$ possesses a complex Fourier transform[4]; and, when that is the case, the limiting process Eq. (10.42) is unnecessary. For these density operators the weight function $P(\alpha)$ is the complex Fourier transform of the normally ordered characteristic function[24]

$$P(\alpha) = \pi^{-2} \int \exp(\alpha\xi^* - \alpha^*\xi)\, \chi_N(\xi)\, d^2\xi \tag{10.44}$$

and by using Eqns. (9.32) and (9.33) we find as the inverse relation

$$\chi_N(\xi) = \int \exp(\xi\alpha^* - \xi^*\alpha)\, P(\alpha)\, d^2\alpha \,. \tag{10.45}$$

---

[4] See footnote 3 on p. 387.

For other density operators, the limiting process Eq. (10.42) typically leads to a generalized function that is too singular to be used as a weight function for the P representation.[5] For the case of pure states, for example, it has been shown[25] that the function $P(\alpha)$ is always singular and that it is a tempered distribution only when the state may be expressed as a linear combination of a finite number of the states $(a^\dagger)^n |\alpha\rangle$ for a single, arbitrary value of α.

We have seen that the P representation provides for the ensemble averages of the normally ordered products $(a^\dagger)^n a^m$ the simple expression Eq. (10.3). Let us now observe that by using Eq. (9.124) for example we may use the P representation to write the ensemble averages of the symmetrically and antinormally ordered products in the forms

$$\mathrm{Tr}\left[\rho\left\{(a^\dagger)^n a^m\right\}_0\right] = \frac{n!}{2^n}\int \alpha^{m-n} L_n^{(m-n)}(-2|\alpha|^2) P(\alpha)\, d^2\alpha \tag{10.46}$$

and

$$\mathrm{Tr}\left[\rho a^m (a^\dagger)^n\right] = n!\int \alpha^{m-n} L_n^{(m-n)}(-|\alpha|^2) P(\alpha)\, d^2\alpha, \tag{10.47}$$

where $L_n^{(m)}(x)$ is an associated Laguerre polynomial[26].

10.4
Wigner Distribution

In this section we discuss a representation for the density operator that is particularly suited for the averaging of operators written in symmetric order. This representation exists for all density operators and affords a way of expressing the expectation value of every bounded operator as a convergent integral. The weight function of this representation is the quasiprobability distribution function introduced by Wigner[3]. Our discussion will begin with the displacement-operator expansion in which operators are expressed in a form that is implicitly symmetric in the operators a and $a^\dagger$.

We have seen in Sect. 9.3 that the displacement operators $D(\alpha)$ form a complete set of operators. They afford for every operator F that is bounded in the sense of Eq. (10.16) an expansion of the form

$$F = \int f(\xi) D^{-1}(\xi) \pi^{-1}\, d^2\xi, \tag{10.48}$$

in which the weight function

$$f(\xi) = \mathrm{Tr}\{FD(\xi)\} \tag{10.49}$$

---

[5] K. E. Cahill, to be published. It is shown there that it is possible to regularize the P representation so as to accommodate such generalized functions but only at the expense of adding to it three two-dimensional integrals over outer products of non-identical coherent states. This regularized P representation exists with four nonsingular weight functions for all density operators ρ. Only when the three supplementary weight functions can be set equal to zero, however, does the P representation exist.

is unique and square-integrable. Thus since every density operator is bounded, we may write an arbitrary density operator in the form

$$\rho = \int \chi(\xi) D^{-1}(\xi) \pi^{-1} d^2\xi, \tag{10.50}$$

where the weight function $\chi(\xi)$ is given by the trace

$$\chi(\xi) = \text{Tr}\{\rho D(\xi)\} \tag{10.51}$$

which is the expectation value of the displacement operator $D(\xi)$. This relation is the familiar definition of the characteristic function, as contrasted with the normally ordered form considered in Sect. 10.3. It plays a role in quantum-statistical theory analogous to that of the characteristic function of classical probability theory. According to Eq. (9.42) we have

$$\int |\chi(\xi)|^2 \pi^{-1} d^2\xi = \text{Tr}\rho^2 \leq 1, \tag{10.52}$$

which shows that the characteristic function $\chi(\xi)$ is always square-integrable. It is clear from the expansion Eq. (10.50), that the function $\chi(\xi)$ uniquely determines the density operator ρ.

Since the displacement operator $D(\xi)$ is unitary for all ξ and unity for $\xi = 0$, the characteristic function $\chi(\xi)$, which is its expectation value, must satisfy the conditions

$$|\chi(\xi)| \leq 1, \quad \chi(0) = 1. \tag{10.53}$$

The boundedness condition on $\chi(\xi)$ contrasts sharply with the growth condition that applies to the normally ordered characteristic function $\chi_N(\xi)$. If we compare the definitions (10.43) and (10.51) with Eq. (9.14), then we may deduce from the condition Eq. (10.53) that the function $\chi_N(\xi)$ must obey the growth condition

$$|\chi_N(\xi)| \leq e^{|\xi|^2/2}. \tag{10.54}$$

The possibility of such exponential growth at large values of $|\xi|$ is the reason why the function $\chi_N(\xi)$ does not naturally possess a Fourier transform, even one that is a tempered distribution.

In the case of the characteristic function $\chi(\xi)$, however, which is both bounded and square-integrable, we encounter no difficulty in defining its complex Fourier transform as

$$W(\alpha) = \int \exp(\alpha \xi^* - \alpha^* \xi) \chi(\xi) \pi^{-1} d^2\xi. \tag{10.55}$$

This function differs only in normalization from the distribution $W(q',p')$ introduced by Wigner[3] as a quantum-mechanical analog of the phase-space distribution of classical statistical mechanics. By using the Fourier inversion relations Eq. (9.32) and (9.33), we may express the function $\chi(\xi)$ in terms of the Wigner function as

$$\chi(\xi) = \int \exp(\xi \alpha^* - \xi^* \alpha) W(\alpha) \pi^{-1} d^2\alpha. \tag{10.56}$$

If we now substitute this expression for the characteristic function into the expansion Eq. (10.50) and use the definition Eq. (10.17) of the operator $T(\alpha,0)$, we arrive at the representation[2,16,17]

$$\rho = \int W(\alpha) T(\alpha,0) \pi^{-1} d^2\alpha . \tag{10.57}$$

The Wigner function $W(\alpha)$ is therefore a weight function for the expansion of the density operator in terms of the operators $T(\alpha,0)$ which, according to Eq. (9.162), are given by

$$T(\alpha,0) = 2D(\alpha)(-1)^{a^\dagger a} D^{-1}(\alpha) . \tag{10.58}$$

By writing Eq. (10.55) in the form

$$W(\alpha) = \mathrm{Tr}\left\{\rho \int \exp(\alpha\xi^* - \alpha^*\xi) D(\xi) \pi^{-1} d^2\xi\right\} ,$$

and using the definition Eq. (10.17), we find that the Wigner distribution is the expectation value of the operator $T(\alpha,0)$:

$$W(\alpha) = \mathrm{Tr}\{\rho T(\alpha,0)\} . \tag{10.59}$$

Since the operator $T(\alpha,0)$ is Hermitian, as shown by Eq. (10.58), the Wigner function $W(\alpha)$ is real-valued,

$$W(\alpha) = W(\alpha)^* . \tag{10.60}$$

We mentioned in Chapter 9 that both the displacement operators $D(\alpha)$ and the operators $T(\alpha,s)$ for $\mathrm{Re}\, s \leq 0$ have the property[2] that if $\mathcal{T}$ is any trace class operator then the function $\mathrm{Tr}\{\mathcal{T} D(\alpha)\}$ and $\mathrm{Tr}\{\mathcal{T} T(\alpha,s)\}$ are uniformly continuous functions of α. Thus, since every density operator is in the trace class, both the characteristic function $\chi(\xi)$ and the Wigner function $W(\alpha)$ are uniformly continuous functions of α. By using Eq. (10.52) and Eq. (9.34), we find

$$\int |W(\alpha)|^2 \pi^{-1} d^2\alpha = \mathrm{Tr}\,\rho^2 \leq 1 , \tag{10.61}$$

which shows that the Wigner distribution is a square-integrable function of α for all density operators ρ.

The Hermitian operators $T(\alpha,0)$ possess the same type of completeness as do the unitary operators $D(\alpha)$. For, according to Eqns. (10.18) and (10.19), we may expand any bounded operator F in the form

$$F = \int f(\alpha,0) T(\alpha,0) \pi^{-1} d^2\alpha , \tag{10.62}$$

where the weight function $f(\alpha,0)$, which is given by the trace

$$f(\alpha,0) = \mathrm{Tr}\{FT(\alpha,0)\} , \tag{10.63}$$

is unique and square-integrable. We see from Eq. (9.215) that this function satisfies the relation

$$\int |f(\alpha,0)|^2 \pi^{-1} d^2\alpha = \text{Tr}\left\{F^\dagger F\right\}. \tag{10.64}$$

This relation is identical in content to the relation Eq. (9.42), which characterizes the completeness property of the displacement operators. The expansion Eq. (10.57) for the density operator and the relations (10.59) and (10.61) are a special case of Eqns. (10.62)–(10.64).

Let us now note that by using either of the expansions Eqns. (10.57) and (10.62) we may write the expectation value of any bounded operator F in the form

$$\text{Tr}\{\rho F\} = \int W(\alpha) f(\alpha,0) \pi^{-1} d^2\alpha. \tag{10.65}$$

We may conclude from the Schwarz inequality for functions Eq. (9.37) that this integral converges for all bounded operators F and all density operators ρ since, for such operators, the functions $F(\alpha,0)$ and $W(\alpha)$ are both square-integrable. It will become evident in Sect. 10.6 that this property of the Wigner distribution is a consequence of the fact that the real part of the order parameter is zero.

In Sect. 9.7 we noted that the correspondence between operators and functions $C(0)$ based upon symmetric ordering $s=0$ is the same as the Weyl correspondence[27], Eqns. (9.220) and (9.221). By comparing Eqns. (10.59) and (10.63) with Eq. (10.21), we see that it is this correspondence which associates the functions $W(\alpha)$ and $f(\alpha,0)$ with the operators ρ and F. Thus if the operators ρ and F possess the symmetrically ordered power-series expansions

$$\rho = \sum_{n,m=0}^{\infty} \rho_{n,m}(0) \left\{(a^\dagger)^n a^m\right\}_0 \tag{10.66}$$

and

$$F = \sum_{n,m=0}^{\infty} f_{n,m}(0) \left\{(a^\dagger)^n a^m\right\}_0 \tag{10.67}$$

then the functions $W(\alpha)$ and $f(\alpha,0)$ are given by

$$W(\alpha) = \sum_{n,m=0}^{\infty} \rho_{n,m}(0) (\alpha^*)^n \alpha^m \tag{10.68}$$

and

$$f(\alpha,0) = \sum_{n,m=0}^{\infty} f_{n,m}(0) (\alpha^*)^n \alpha^m. \tag{10.69}$$

The functions $W(\alpha)$ and $f(\alpha,0)$ may be thought of as generating functions for the coefficients of the symmetrically ordered power-series expansions of the operators ρ

and F. It is a unique property of the Wigner distribution, among all the distribution functions considered in this paper, that in the integral Eq. (10.65) for the ensemble average $\text{Tr}\{\rho F\}$ both the function $W(\alpha)$ representing the density operator and the function $f(\alpha,0)$ representing the operator F are associated with their respective operators ρ and F by the same correspondence.

The integral expression Eq. (10.65) for the expectation value $\text{Tr}\{\rho F\}$ becomes particularly simple when the operator F is written in symmetric order. The expectation value of the symmetrically ordered product $\{(a^\dagger)^n a^m\}_0$, for example, is given by the integral

$$\text{Tr}\left[\rho\{(a^\dagger)^n a^m\}_0\right] = \int W(\alpha)(\alpha^*)^n \alpha^m \pi^{-1} d^2\alpha. \tag{10.70}$$

Three simple illustrations of this relation are

$$\tfrac{1}{2}\langle a^\dagger a + aa^\dagger \rangle = \int |\alpha|^2 W(\alpha) \pi^{-1} d^2\alpha, \tag{10.71}$$

$$\tfrac{1}{2}\langle a^\dagger a^2 + aa^\dagger a + a^2 a^\dagger \rangle = \int \alpha^* \alpha^2 W(\alpha) \pi^{-1} d^2\alpha, \tag{10.72}$$

and

$$\text{Tr}\{\rho\} = 1 = \int W(\alpha) \pi^{-1} d^2\alpha, \tag{10.73}$$

which shows that the Wigner function is normalized.

By using Eq. (9.124) we may express the ensemble averages of the normally and antinormally ordered products in terms of the Wigner distribution as the integrals

$$\text{Tr}\left\{\rho (a^\dagger)^n a^m\right\} = n! \, (-\tfrac{1}{2})^n \int \alpha^{m-n} L_n^{(m-n)}(2|\alpha|^2) W(\alpha) \pi^{-1} d^2\alpha \tag{10.74}$$

and

$$\text{Tr}\left\{\rho a^m (a^\dagger)^n\right\} = n! \, (\tfrac{1}{2})^n \int \alpha^{m-n} L_n^{(m-n)}(-2|\alpha|^2) W(\alpha) \pi^{-1} d^2\alpha. \tag{10.75}$$

A number of properties of the Wigner distribution follow directly from the explicit form of the operator $T(\alpha,0)$ given by Eq. (10.58). The eigenvalues $e_n(0)$ of this Hermitian operator are $2(-1)^n$. It follows then from Eq. (10.59) that the Wigner function, being the expectation value of $T(\alpha,0)$, is bounded according to the inequalities

$$-2 \leq W(\alpha) \leq 2 \tag{10.76}$$

for all density operators.

The operator $(-1)^{a^\dagger a} = \exp(i\pi a^\dagger a)$ is a reflection operator in the sense indicated by the relations

$$\exp(i\pi a^\dagger a) a \exp(-i\pi a^\dagger a) = -a \tag{10.77}$$

and
$$\exp(i\pi a^\dagger a)a^\dagger \exp(-i\pi a^\dagger a) = -a^\dagger, \qquad (10.78)$$

which imply that it also reflects the Hermitian operators q and p. By applying these relations to the displacement operator $D(\alpha)$ we find

$$\exp(i\pi a^\dagger a) D(\alpha) \exp(-i\pi a^\dagger a) = D^{-1}(\alpha), \qquad (10.79)$$

so that we have

$$T(\alpha,0) = 2D(2\alpha) \exp(i\pi a^\dagger a) \qquad (10.80)$$
$$= 2 \exp(i\pi a^\dagger a) D^{-1}(2\alpha). \qquad (10.81)$$

By multiplying together these two forms for $T(\alpha,0)$, we find

$$[T(\alpha,0)]^2 = 4. \qquad (10.82)$$

By using Eq. (10.80) we may write the Wigner function as

$$W(\alpha) = 2 \operatorname{Tr}\left\{\rho D(2\alpha) \exp(i\pi a^\dagger a)\right\}, \qquad (10.83)$$

which, for a state of precisely n quanta, i.e., $\rho = |n\rangle\langle n|$, is simply

$$W_n(\alpha) = 2(-1)^n \langle n|D(2\alpha)|n\rangle. \qquad (10.84)$$

By referring to the explicit form Eq. (9.58) for this matrix element, we find

$$W_n(\alpha) = 2(-1)^n L_n(4|\alpha|^2) \exp(-2|\alpha|^2), \qquad (10.85)$$

where $L_n(x)$ is a Laguerre polynomial[26]. Since half of the eigenvalues of the operator $T(\alpha,0)$ are negative, the Wigner function often assumes negative values as is illustrated by

$$W_n(0) = 2(-1)^n. \qquad (10.86)$$

Among the functions $P(\alpha)$, $W(\alpha)$, and $\langle\alpha|\rho|\alpha\rangle$, only the last, which we discuss in Sect. 10.5, is positive definite for all density operators.

By using Eqns. (10.83) and (10.58), we may write the Wigner distribution in the following additional forms:

$$W(\alpha) = 2 \sum_{n=0}^{\infty} (-1)^n \langle n|\rho D(2\alpha)|n\rangle \qquad (10.87)$$

$$= 2 \operatorname{Tr}\left\{D^{-1}(\alpha)\rho D(\alpha) \exp(i\pi a^\dagger a)\right\} \qquad (10.88)$$

$$= 2 \sum_{n=0}^{\infty} (-1)^n \langle n|D^{-1}(\alpha)\rho D(\alpha)|n\rangle \qquad (10.89)$$

The last two forms make clear that the dependence of the Wigner function $W(\alpha)$ upon the variable α is related to a unitary transformation of the density operator by the displacement operator $D(\alpha)$. By applying Eq. (9.157) to Eq. (10.36), we may similarly interpret the a dependence of the weight function $P(\alpha)$ of the P representation.

10.5
The Function $\langle \alpha|\rho|\alpha\rangle$

In Sections 10.3 and 10.4 we considered distribution functions that are useful for finding the expectation values of operators written in normal or symmetric order. We turn now to a function that may be used to express the ensemble averages of antinormally ordered operators as simple integrals.

Let us suppose that the operator F possesses the antinormally ordered power-series expansion

$$F = \sum_{n,m=0}^{\infty} f_{n,m}(-1)\, a^m \left(a^\dagger\right)^n . \tag{10.90}$$

Then by proceeding as in Eqns. (10.32)–(10.34), we may write the operator F in the form

$$F = \int f(\alpha, 1)\, |\alpha\rangle\langle\alpha|\, \pi^{-1}\, d^2\alpha , \tag{10.91}$$

where the function $f(\alpha, 1)$ is given by

$$f(\alpha, 1) = \sum_{n,m=0}^{\infty} f_{n,m}(-1)\, (\alpha^*)^n\, \alpha^m . \tag{10.92}$$

By referring to Eq. (10.21), we see that this function may be expressed in terms of the operator $T(\alpha, 1)$ as the trace

$$f(\alpha, 1) = \mathrm{Tr}\{F T(\alpha, 1)\} . \tag{10.93}$$

The appearance within this trace of the operator $T(\alpha, 1)$ all of whose eigenvalues are infinite reflects the fact that antinormally ordered power series do not exist for all operators F.

We may use the expansion Eq. (10.91) to write the expectation value of the operator F in the form

$$\mathrm{Tr}\{\rho F\} = \int f(\alpha, 1)\, \mathrm{Tr}\{\rho|\alpha\rangle\langle\alpha|\}\, \pi^{-1}\, d^2\alpha \tag{10.94}$$

$$= \int f(\alpha, 1)\, \langle\alpha|\rho|\alpha\rangle\, \pi^{-1}\, d^2\alpha . \tag{10.95}$$

Thus the function $\langle\alpha|\rho|\alpha\rangle$ has for antinormally ordered operators the useful property[8,9,18] exhibited in Eqns. (10.30) and (10.65) by the functions $P(\alpha)$ and $W(\alpha)$ for the cases of normally and symmetrically ordered operators.

The function $\langle\alpha|\rho|\alpha\rangle$ clearly is non-negative and bounded according to the relation

$$0 \leq \langle\alpha|\rho|\alpha\rangle \leq 1 . \tag{10.96}$$

Its normalization,

$$1 = \mathrm{Tr}\,\rho = \int \langle\alpha|\rho|\alpha\rangle\, \pi^{-1} d^2\alpha , \tag{10.97}$$

is evident from Eq. (9.28)). By recalling that the operator $T(\alpha,-1)$ is a projection operator upon the coherent state $|\alpha\rangle$, as shown by Eq. (10.39), we may write the function $\langle\alpha|\rho|\alpha\rangle$ in the form

$$\langle\alpha|\rho|\alpha\rangle = \mathrm{Tr}\{\rho T(\alpha,-1)\} . \tag{10.98}$$

This relation is the analog for antinormal order, $s = -1$, of Eqns. (10.36) and (10.59) for the functions $P(\alpha)$ and $W(\alpha)$.

By using Eqns. (10.93) and (10.98) we may write Eq. (10.95) in the form

$$\mathrm{Tr}\{\rho F\} = \int \mathrm{Tr}\{FT(\alpha,1)\}\, \mathrm{Tr}\{\rho T(\alpha,-1)\}\, \pi^{-1} d^2\alpha . \tag{10.99}$$

If we now contrast this relation with its counterpart Eq. (10.41) for the case of the P representation, we see that the roles of the operators $T(\alpha,1)$ and $T(\alpha,-1)$ have been interchanged. In the corresponding relation for the case of the Wigner distribution, Eq. (10.65), both $T(\alpha,-1)$ and the singular operator $T(\alpha,1)$ are replaced by $T(\alpha,0)$ whose eigenvalues are ± 2. For this reason only the Wigner function of the three quasiprobability distributions considered thus far may be used to write the ensemble averages of all bounded operators as convergent integrals.

We now note the very limited sense in which the function $\langle\alpha|\rho|\alpha\rangle$ is a weight function for the density operator. By using Eqns. (10.98), (10.18) and (10.19) we may formally express the density operator as the integral

$$\rho = \int \langle\alpha|\rho|\alpha\rangle\, T(\alpha,1)\, \pi^{-1} d^2\alpha . \tag{10.100}$$

The appearance in this expansion of the singular operator $T(\alpha,1)$ may be interpreted in terms of the properties of the function $\langle\alpha|\rho|\alpha\rangle$. As we have noted, this function is bounded, non-negative, and normalized. It is also an infinitely differentiable function, as may be seen by using Eq. (9.23) to expand the coherent states $|\alpha\rangle$ and $\langle\alpha|$ in powers of α and α^*. The function $\langle\alpha|\rho|\alpha\rangle$ thus has all the properties that one would expect of the smoothest classical probability distribution. The expansion Eq. (10.100) therefore suggests the interpretation that only so singular an operator as $T(\alpha,1)$ can be used to construct arbitrary density operators when weighted with so well behaved a weight function as $\langle\alpha|\rho|\alpha\rangle$. The opposite interpretation evidently applies to the P representation, in which the roles of the operators $T(\alpha,1)$ and $T(\alpha,-1) = |\alpha\rangle\langle\alpha|$ are interchanged. It appears, then, that the nonexistence of the P representation for certain density operators is a consequence of the extreme "smoothness" of the coherent state projection operator $|\alpha\rangle\langle\alpha|$.

In analogy with Eqns. (10.43) and (10.51), let us define the antinormally ordered characteristic function $\chi_A(\xi)$ as the expectation value

$$\chi_A(\xi) = \text{Tr}\{\rho D(\xi, -1)\}$$
$$= \text{Tr}\left\{\rho \exp(-\xi^* a) \exp(\xi a^\dagger)\right\}. \tag{10.101}$$

By using Eqns. (10.10)–(10.12) we see that the characteristic functions $\chi_N(\xi)$, $\chi(\xi)$, and $\chi_A(\xi)$ stand in the simple relationship

$$\chi_N(\xi) = e^{|\xi|^2/2} \chi(\xi) = e^{|\xi|^2} \chi_A(\xi). \tag{10.102}$$

Since, as we noted in Sect. 10.4, the function $\chi(\xi)$ is a continuous function of ξ, it follows that all three characteristic functions are continuous. We see from Eq. (10.53) that the modulus of the antinormally ordered characteristic function $\chi_A(\xi)$ is bounded by

$$|\chi_A(\xi)| \leq e^{-|\xi|^2/2}. \tag{10.103}$$

If we use Eq. (10.95) to evaluate the antinormally ordered expectation value Eq. (10.101), then we find

$$\chi_A(\xi) = \int \exp(\xi\alpha^* - \xi^*\alpha) \langle\alpha|\rho|\alpha\rangle \pi^{-1} d^2\alpha, \tag{10.104}$$

which shows that the function $\chi_A(\xi)$ is the complex Fourier transform of the function $\langle\alpha|\rho|\alpha\rangle$. By means of Eq. (9.32)) and Eq. (9.33) we obtain the inverse relation

$$\langle\alpha|\rho|\alpha\rangle = \int \exp(\alpha\xi^* - \alpha^*\xi) \chi_A(\xi) \pi^{-1} d^2\xi. \tag{10.105}$$

As a simple illustration of Eq. (10.95), we note that the expectation values of the antinormally ordered products can be written in the form

$$\text{Tr}\left\{\rho a^m (a^\dagger)^n\right\} = \int (\alpha^*)^n \alpha^m \langle\alpha|\rho|\alpha\rangle \pi^{-1} d^2\alpha. \tag{10.106}$$

The function $\langle\alpha|\rho|\alpha\rangle$ may also be used to evaluate the mean values of normally and symmetrically ordered products. By using Eq. (9.124) we may write them as the integrals

$$\text{Tr}\left[\rho (a^\dagger)^n a^m\right] = (-1)^n n! \int a^{m-n} L_n^{(m-n)}(|\alpha|^2) \langle\alpha|\rho|\alpha\rangle \pi^{-1} d^2\alpha \tag{10.107}$$

and

$$\text{Tr}\left[\rho \left\{(a^\dagger)^n a^m\right\}_0\right] = (-\tfrac{1}{2})^n n! \int a^{m-n} L_n^{(m-n)}(2|\alpha|^2) \langle\alpha|\rho|\alpha\rangle \pi^{-1} d^2\alpha. \tag{10.108}$$

In Sect. 9.4 we showed that normally ordered power series exist and converge for virtually all operators of interest, i.e., for those that satisfy the condition Eq. (9.91). In particular, every density operator ρ possesses the expansion

$$\rho = \sum_{n,m=0}^{\infty} p_{n,m}(1) (a^{\dagger})^n a^m, \tag{10.109}$$

and by using the eigenvalue property of the coherent states we find

$$\langle \alpha | \rho | \alpha \rangle = \sum_{n,m=0}^{\infty} p_{n,m}(1) (\alpha^*)^n \alpha^m. \tag{10.110}$$

Thus the function $\langle \alpha | \rho | \alpha \rangle$ is the one associated with the density operator ρ by the correspondence $C(1)$ based upon normal ordering. Equivalently, the function $\langle \alpha | \rho | \alpha \rangle$ is a generating function for the coefficients of the normally ordered power-series expansion of the density operator. The convergence of the series Eq. (10.109) and (10.110) is consistent with the fact, mentioned earlier, that the function $\langle \alpha | \rho | \alpha \rangle$ is infinitely differentiable. We note that Eqns. (10.109) and (10.110) correspond to Eqns. (10.33) and (10.34) for the weight function $P(\alpha)$ and to Eqns. (10.66) and (10.68) for the Wigner distribution $W(\alpha)$.

Let us now define the function $R(\alpha^*, \beta)$ by the relation[6]

$$R(\alpha^*, \beta) = e^{(|\alpha|^2 + |\beta|^2)/2} \langle \alpha | \rho | \beta \rangle. \tag{10.111}$$

The function $R(\alpha^*, \beta)$ is an entire function of the two complex variables α^* and β. By twice using Eq. (9.27), we may write the density operator in the form

$$\rho = \int |\alpha\rangle \langle \beta | \langle \alpha | \rho | \beta \rangle \pi^{-2} d^2\alpha \, d^2\beta. \tag{10.112}$$

The function $R(\alpha^*, \beta)$ therefore affords for every density operator ρ the two-variable representation

$$\rho = \int |\alpha\rangle \langle \beta | R(\alpha^*, \beta) e^{(|\alpha|^2 + |\beta|^2)/2} \pi^{-2} d^2\alpha \, d^2\beta, \tag{10.113}$$

which has been called the R representation. The relationship between the functions $R(\alpha^*, \beta)$ and $\langle \alpha | \rho | \alpha \rangle$ is given by

$$\langle \alpha | \rho | \alpha \rangle = e^{-|\alpha|^2} R(\alpha^*, \alpha). \tag{10.114}$$

10.6
Ensemble Averages and s Ordering

In the present section we draw together the results of the Sections 10.3–10.5 by making use of the parametrized ordering convention introduced in Chapter 9. We define a general representation for the density operator in which the weight function $W(\alpha,s)$ may be identified with the weight functions $\langle\alpha|\rho|\alpha\rangle$, $W(\alpha)$, and $P(\alpha)$ when the order parameter s assumes the values $s = -1, 0,$ and $+1$, respectively. In characterizing the behavior of the function $W(\alpha,s)$ we will see the unfolding of an orderly progression of mathematical properties as the parameter s is advanced from $s = -1$ corresponding to the function $\langle\alpha|\rho|\alpha\rangle$ to $s = +1$ corresponding to the function $P(\alpha)$.

In Chapter 9 we introduced for every bounded operator F the representation

$$F = \int f(\alpha, -s) T(\alpha, s) \pi^{-1} d^2\alpha , \qquad (10.115)$$

in which the weight function $f(\alpha, -s)$ is given by

$$f(\alpha, -s) = \text{Tr}\{F T(\alpha, -s)\} . \qquad (10.116)$$

The operator F and its weight function $f(\alpha, -s)$ are associated by the correspondence $C(s)$ which is based upon the power-series expansions

$$F = \sum_{n,m=0}^{\infty} f_{n,m}(s) \left\{(a^\dagger)^n a^m\right\}_s \qquad (10.117)$$

and

$$f(\alpha, -s) = \sum_{n,m=0}^{\infty} f_{n,m}(s) (\alpha^*)^n \alpha^m . \qquad (10.118)$$

By multiplying both sides of the expansion Eq. (10.115) by an arbitrary density operator ρ and forming the trace of the resulting relation we find

$$\text{Tr}\{\rho F\} = \int f(\alpha, -s) \, \text{Tr}\{\rho T(\alpha, s)\} \, \pi^{-1} d^2\alpha . \qquad (10.119)$$

Let us introduce the function $W(\alpha,s)$ as the expectation value of the operator $T(\alpha,s)$,

$$W(\alpha,s) = \text{Tr}\{\rho T(\alpha,s)\} . \qquad (10.120)$$

This function permits us to express the expectation value of the operator F as the integral

$$\text{Tr}\{\rho F\} = \int f(\alpha, -s) W(\alpha,s) \pi^{-1} : d^2\alpha , \qquad (10.121)$$

in which F is represented by the function $f(\alpha, -s)$ associated with it by the correspondence $C(s)$.

By referring to Eqns. (10.115) and (10.116) we find that the function $W(\alpha,s)$ is also the weight function for the density operator ρ in the expansion

$$\rho = \int W(\alpha,s) T(\alpha, -s) \pi^{-1} d^2\alpha . \qquad (10.122)$$

This relation, we note, affords an alternative way of deriving the integral expression Eq. (10.121) for the ensemble average $\text{Tr}\{\rho F\}$.

By comparing the definition Eq. (10.120) with Eqns. (10.36), (10.59), and (10.98) we may identify the function $W(\alpha,s)$ for $s = -1, 0,$ and $+1$ with the functions $\langle \alpha|\rho|\alpha\rangle$, $W(\alpha)$, and $\pi P(\alpha)$, respectively:

$$\langle \alpha|\rho|\alpha\rangle = W(\alpha,-1), \tag{10.123}$$

$$W(\alpha) = W(\alpha,0), \tag{10.124}$$

$$P(\alpha) = \pi^{-1} W(\alpha,1). \tag{10.125}$$

If in Eqns. (10.121) and (10.122) we let $s = -1, 0, +1$, we then obtain, respectively, Eqns. (10.95) and (10.100) for the case of the function $\langle \alpha|\rho|\alpha\rangle$, Eqns. (10.65) and (10.57) for the case of the Wigner distribution $W(\alpha)$, and Eqns. (10.26) and (10.25) for the case of the P representation.

By comparing the definition Eq. (10.120) with Eqns. (10.117) and (10.118), we see that the function $W(\alpha,s)$ and the density operator ρ are related by the correspondence $C(-s)$:

$$\rho = \sum_{n,m=0}^{\infty} \rho_{n,m}(-s) \left\{ (a^\dagger)^n a^m \right\}_{-s} \tag{10.126}$$

$$W(\alpha,s) = \sum_{n,m=0}^{\infty} \rho_{n,m}(-s) (\alpha^*)^n \alpha^m. \tag{10.127}$$

Thus in the integral Eq. (10.121) for the ensemble average $\text{Tr}\{\rho F\}$ the functions $f(\alpha,-s)$ and $W(\alpha,s)$ bear to the operator F and ρ the relationships $C(s)$ and $C(-s)$, respectively. If in Eqns. (10.126) and (10.127) we put $s = -1, 0, +1$, then we may recover the earlier relations (10.109), (10.110), (10.66), (10.68), (10.33), and (10.34), which obtain for normal, symmetric, and antinormal ordering.

A simple illustration of the weight function $W(\alpha,s)$ is provided by the density operator for a pure coherent state, $\rho = |\beta\rangle\langle\beta|$. According to the definition Eq. (10.120) the function $W(\alpha,s)$ is given by the matrix element

$$W(\alpha,s) = \langle\beta|T(\alpha,s)|\beta\rangle,$$

which is evaluated in Eq. (9.175). By using that result we find

$$W(\alpha,s) = \frac{2}{1-s} \exp\left[-\frac{2|\alpha-\beta|^2}{1-s}\right], \tag{10.128}$$

which shows that the function $W(\alpha,s)$ has an essential singularity at $s=1$ and is for $\text{Re}\,s < 1$ a Gaussian exponential with a maximum at $\alpha = \beta$. By referring to Eqns. (10.123)–(10.125) we see that the weight functions $\langle\alpha|\rho|\alpha\rangle$, $W(\alpha)$, and $P(\alpha)$

are given by

$$\langle \alpha | \rho | \alpha \rangle = e^{-|\alpha-\beta|^2}, \tag{10.129}$$

$$W(\alpha) = 2e^{-2|\alpha-\beta|^2}, \tag{10.130}$$

$$P(\alpha) = \delta^{(2)}(\alpha-\beta), \tag{10.131}$$

where the last expression is obtained by using the limiting process Eq. (10.38) for $P(\alpha)$. By using Eq. (9.155) we find

$$W(\alpha, s^*) = W(\alpha, s)^*, \tag{10.132}$$

which shows that the function $W(\alpha, s)$ is real for real values of s. By using either the general expression Eq. (10.121) or the relation Eq. (9.184), we may write the expectation values of the s-ordered products as the integrals

$$\mathrm{Tr}\left[\rho\left\{(a^\dagger)^n a^m\right\}_s\right] = \int (\alpha^*)^n \alpha^m W(\alpha, s)\, \pi^{-1}\, d^2\alpha. \tag{10.133}$$

This expression is a generalization to arbitrary s of Eqns. (10.106), (10.70), and (10.3) which apply for $s = -1, 0, +1$, respectively. By setting $n = m = 0$ in this relation, we find, as the normalization condition for $W(\alpha, s)$,

$$1 = \mathrm{Tr}\,\rho = \int W(\alpha, s)\, \pi^{-1}\, d^2\alpha. \tag{10.134}$$

The expectation value of the s-ordered displacement operator $D(\xi, s)$ may be obtained from Eqns. (9.201) and (10.121) as the integral

$$\mathrm{Tr}\{\rho D(\xi, s)\} = \int e^{\xi\alpha^* - \xi^*\alpha} W(\alpha, s)\, \pi^{-1}\, d^2\alpha. \tag{10.135}$$

Let us define what we may call the s-ordered characteristic function $\chi(\xi, s)$ by the relation

$$\chi(\xi, s) = \mathrm{Tr}\{\rho D(\xi, s)\} = e^{s|\xi|^2/2}\chi(\xi), \tag{10.136}$$

where $\chi(\xi)$ is the characteristic function introduced in Sect. 10.4. We see from Eq. (10.135) that it is the complex Fourier transform of the function $W(\alpha, s)$,

$$\chi(\xi, s) = \int e^{\xi\alpha^* - \xi^*\alpha} W(\alpha, s)\, \pi^{-1}\, d^2\alpha, \tag{10.137}$$

and by referring to Eqns. (9.32) and (9.33) we find as the inverse relation

$$W(\alpha, s) = \int e^{\alpha\xi^* - \alpha^*\xi} \chi(\xi, s)\, \pi^{-1}\, d^2\xi. \tag{10.138}$$

These Fourier transform relations generalize to arbitrary s the earlier relations (10.104) and (10.105), (10.55) and (10.56), and (10.44) and (10.45), which are realized when $s = -1, 0, +1$.

We noted in Sect. 10.4 that the function $\chi(\xi)$ is a uniformly continuous function that is bounded by unity. The function $\chi(\xi,s)$ is therefore a continuous function of ξ and, of course, an entire function of s. It is bounded by

$$|\chi(\xi,s)| \leq \exp(\tfrac{1}{2}\operatorname{Re} s|\xi|^2). \tag{10.139}$$

By using Eqns. (10.133) and (10.136), we easily find the relation

$$\operatorname{Tr}\left[\rho\left\{(a^\dagger)^n a^m\right\}_s\right] = \left.\frac{\partial^{n+m}\chi(\xi,s)}{\partial \xi^n \partial(-\xi^*)^m}\right|_{\xi=0}. \tag{10.140}$$

According to Eq. (9.167) the eigenvalues $e_n(s)$ of the operator $T(\alpha,s)$ are given by

$$e_n(s) = \frac{2}{1-s}\left(\frac{s+1}{s-1}\right)^n. \tag{10.141}$$

We note that they are all positive and less than unity for s real and < -1, with the limiting values $e_n(-1) = \delta_{n,0}$. Thus the function $W(\alpha,s)$, being the expectation value of $T(\alpha,s)$, satisfies the inequality

$$0 \leq W(\alpha,s) \leq \frac{2}{1-s} \tag{10.142}$$

for s real and ≤ -1. By examining $e_0(s)$ and $e_1(s)$ we see, furthermore, that the full set of eigenvalues is not positive unless s is real and ≤ -1. The function $W(\alpha,s)$ is therefore positive definite for all density operators only when s is real and ≤ -1.

Since every density operator is in the trace class, we know from the analysis of Sect. 9.7 that the function $W(\alpha,s)$ is a uniformly continuous function of α for $\operatorname{Re} s \leq 0$. By using Eq. (9.233) and the fact that $\operatorname{Tr}\rho = 1$, we find that the modulus of the function $W(\alpha,s)$ is bounded by

$$|W(\alpha,s)| \leq \frac{2}{|1-s|} \tag{10.143}$$

for $\operatorname{Re} s \leq 0$.

Since every density operator is a bounded operator, we may obtain from Eq. (9.228) the relation

$$1 \geq \operatorname{Tr}\rho^2 \geq \int |W(\alpha,s)|^2 \pi^{-1} d^2\alpha \tag{10.144}$$

for $\operatorname{Re} s \leq 0$, in which the first and second inequalities are equalities only when the density operator represents a pure state and when $\operatorname{Re} s = 0$, respectively. From the analysis of Sect. 9.7 we may assert further that the function $W(\alpha,s)$ is analytic in s and infinitely differentiable with respect to α and α^* throughout the half-plane $\operatorname{Re} s < 0$. The function $W(\alpha,s)$ may be represented by a power series[2] in the variables s, α, and α^* for $\operatorname{Re} s < 0$.

As the line $\operatorname{Re} s = 0$ is crossed, however, the function $W(\alpha,s)$ retains, in general, none of the properties which we have observed in the half-plane $\operatorname{Re} s \leq 0$. As we noted in Sect. 9.7, for each density operator ρ there is a non-negative number $x(\rho)$ such that the function $W(\alpha,s)$ remains square-integrable and infinitely differentiable for $\operatorname{Re} s < x(\rho)$, but not for larger values of $\operatorname{Re} s$. Our examples in Sect. 10.7 will show that the quantity $x(\rho)$ is arbitrarily small for certain classes of density operators. The function $W(\alpha,s)$ typically has a singularity of some sort on the line $\operatorname{Re} s = x(\rho)$ and beyond that singularity it lies outside the class of tempered distributions[2]. In Sect. 10.3 we mentioned the result that the weight function $P(\alpha)$ of the P representation is singular for all pure states. Thus for such density operators the quantity $x(\rho)$ lies in the interval

$$0 \leq x(\rho) \leq 1 . \tag{10.145}$$

The dependence of the function $W(\alpha,s)$ upon the parameter s is expressed by the convolution law Eq. (9.222), according to which we have

$$W(\alpha,s) = \frac{2}{t-s} \int \exp\left[-\frac{2|\alpha-\beta|^2}{t-s}\right] W(\beta,t) \pi^{-1} d^2\beta \tag{10.146}$$

for $\operatorname{Re} s < \operatorname{Re} t$. This Gaussian convolution tends to smooth out any unruly behavior of the function $W(\beta,t)$. It shows clearly why the quasiprobability distribution $W(\alpha,s)$ becomes progressively better behaved as $\operatorname{Re} s$ decreases. We note also from Eq. (10.146) that the function $W(\alpha,s)$ obeys a second-order partial differential equation analogous to the heat-diffusion equation. This property leads to an instructive analogy which we take up in Sect. 10.9.

The convolution law Eq. (10.146) provides us with a number of integral relations between the functions considered in Sections 10.3–10.5. If we assume that the P representation exists for the density operator ρ, then by letting $t = -1$, and $s = 0$ and -1 we may express the Wigner function $W(\alpha)$ and the function $\langle \alpha | \rho | \alpha \rangle$ as the integrals[8]

$$W(\alpha) = 2 \int e^{-2|\alpha-\beta|^2} P(\beta) d^2\beta \tag{10.147}$$

and

$$\langle \alpha | \rho | \alpha \rangle = \int e^{-|\alpha-\beta|^2} P(\beta) d^2\beta . \tag{10.148}$$

Similarly, since the Wigner function $W(\alpha)$ exists for all density operators ρ, we have in general the relation[2]

$$\langle \alpha | \rho | \alpha \rangle = 2 \int e^{-2|\alpha-\beta|^2} W(\alpha) \pi^{-1} d^2\beta . \tag{10.149}$$

Let us now suppose that we wish to find the ensemble average of the s-ordered product $\{(a^\dagger)^n a^m\}_s$, but that we know only the function $W(\alpha,t)$ and not the function $W(\alpha,s)$ or, alternatively, that the function $W(\alpha,s)$ is too singular for the relation Eq. (10.133) to apply. We may in such a case use the relations Eqns. (9.123) and (9.124) to express the s-ordered product in terms of t-ordered products and so obtain the desired expectation value as the integrals

$$\mathrm{Tr}\left[\rho\left\{(a^\dagger)^n a^m\right\}_s\right]$$

$$= m!\left(\frac{t-s}{2}\right)^m \int (\alpha^*)^{n-m} L_m^{(n-m)}\left(\frac{2|\alpha|^2}{s-t}\right) W(\alpha,t)\,\pi^{-1}\,d^2\alpha \qquad (10.150)$$

$$= n!\left(\frac{t-s}{2}\right)^n \int \alpha^{m-n} L_n^{(m-n)}\left(\frac{2|\alpha|^2}{s-t}\right) W(\alpha,t)\,\pi^{-1}\,d^2\alpha. \qquad (10.151)$$

10.7
Examples of the General Quasiprobability Function $W(\alpha,s)$

We now illustrate the results of the preceding sections by constructing the function $W(\alpha,s)$ for some simple density operators and by using it to evaluate the ensemble averages of a variety of observables. We have shown in Sect. 10.6 that the function $W(\alpha,s)$ is a well-behaved function of both α and s when the real part of s is negative. The examples of the present section will verify this behavior and will illustrate some of the singularities that occur for $\mathrm{Re}\,s > 0$.

By referring to Eqns. (9.156) and (10.141), and (9.177), we see that for t real and ≤ -1 each operator $T(\gamma,t)$ is Hermitian, positive definite, and of unit trace. The operators $T(\gamma,t)$, which we have introduced for other purposes, therefore form for $t \leq -1$ a two-parameter family of density operators.

This family of density operators in fact describes a broad class of physically important fields. For $t = -1$ the operator $T(\gamma,-1)$ describes a coherent field of amplitude γ,

$$T(\gamma,-1) = |\gamma\rangle\langle\gamma|, \qquad (10.152)$$

as is shown by Eq. (9.168). According to Eq. (9.165) it characterizes at the other extreme, when $\gamma = 0$ and $t \leq -1$, the general chaotic field[6]

$$T(0,t) = \frac{1}{\langle n\rangle + 1}\sum_{n=0}^{\infty}\left(\frac{\langle n\rangle}{\langle n\rangle + 1}\right)^n |n\rangle\langle n|, \qquad (10.153)$$

in which the mean number of quanta $\langle n\rangle$ is given by

$$\langle n\rangle = -\tfrac{1}{2}(t+1). \qquad (10.154)$$

10.7 Examples of the General Quasiprobability Function W(α,s)

The operator $T(\gamma,t)$, regarded as a density operator, describes more generally the field that results from the superposition of a coherent signal of amplitude γ,

$$T(\gamma,t) = D(\gamma)\, T(0,t)\, D^{-1}(\gamma), \tag{10.155}$$

upon a chaotic field with a mean member of quanta $\langle n \rangle = -\frac{1}{2}(t+1)$. By using Eq. (9.164)) we may write the density operator $T(\gamma,t)$ in the suggestive form

$$T(\gamma,t) = (1 - e^{-\beta})\exp[-\beta \mathcal{N}(\gamma)], \tag{10.156}$$

where

$$\beta(t) = -\ln\left(\frac{t+1}{t-1}\right) \tag{10.157}$$

and

$$\mathcal{N}(\gamma) = D(\gamma)\, a^\dagger a\, D^{-1}(\gamma)$$
$$= a^\dagger a - \gamma a^\dagger - \gamma^* a + |\gamma|^2. \tag{10.158}$$

For $t \leq -1$ the parameter β is real and positive with the limiting value of $\beta = \infty$ for $t = -1$. The operator Eq. (10.156) for $\gamma = 0$ is of the type which describes the thermal equilibrium state of an electromagnetic field mode. If this state is exposed to the radiation from a prescribed, c-number current distribution then the appropriate density operator[28] is given by Eq. (10.156), with $\gamma \neq 0$.

By using Eq. (9.178) we may obtain the function $W(\alpha,s)$ for the density operator $T(\gamma,t)$ as the trace

$$W(\alpha,s) = \mathrm{Tr}\{T(\gamma,t)\, T(\alpha,s)\}$$
$$= -\frac{2}{t+s}\exp\left[\frac{2|\alpha - \gamma|^2}{t+s}\right]$$
$$= \frac{2}{2\langle n \rangle + 1 - s}\exp\left[\frac{-2|\alpha - \gamma|^2}{2\langle n \rangle + 1 - s}\right]. \tag{10.159}$$

The weight function $P(\alpha)$ of the P representation, the Wigner distribution $W(\alpha)$, and the function $W(\alpha,-1)$ correspond to the values $s = 1, 0, -1$ and are given by

$$W(\alpha,1) = \pi P(\alpha) = \frac{1}{\langle n \rangle}\exp\left[-\frac{|\alpha - \gamma|^2}{\langle n \rangle}\right], \tag{10.160}$$

$$W(\alpha,0) = W(\alpha) = \frac{1}{\langle n \rangle + \frac{1}{2}}\exp\left[-\frac{|\alpha - \gamma|^2}{\langle n \rangle + \frac{1}{2}}\right], \tag{10.161}$$

$$W(\alpha,-1) = \langle \alpha|\rho|\alpha \rangle = \frac{1}{\langle n \rangle + 1}\exp\left[-\frac{|\alpha - \gamma|^2}{\langle n \rangle + 1}\right]. \tag{10.162}$$

For the density operator $T(\gamma,t)$ the function $W(\alpha,s)$ has an essential singularity at $s = -t = 1 + 2\langle n \rangle$. The parameter $x(\rho)$, introduced in Sect. 10.6, clearly assumes the value

$$x(\rho) = -t = 2\langle n \rangle + 1 \geq 1 . \tag{10.163}$$

The function $W(\alpha,s)$ is a tempered distribution on the line $\operatorname{Re} s = x(\rho)$ but not for larger values of $\operatorname{Re} s$, where it increases as an exponential function of $|\alpha - \gamma|^2$.

The family of density operators $T(\gamma,t)$ provides a physical characterization of the fields which may be described by means of the P representation. For if ρ is a density operator for which the function $W(\alpha,s)$ is well behaved for $\operatorname{Re} s \geq 1$, then by using Eqns. (10.122) and (10.156) we may write it in the form

$$\rho = (1 - e^{-\beta(-s)}) \int W(\alpha,s) \exp[-\beta(-s)\mathcal{N}(\alpha)]\, \pi^{-1} d^2\alpha , \tag{10.164}$$

where $s \geq 1$. Such a density operator is therefore a linear combination, or, if $W(\alpha,s) \geq 0$, a statistical average of the density operators $T(\alpha,-s)$ which describe the superposition of a coherent excitation upon a thermal equilibrium state. We see from Eq. (10.157) that the larger the value of s for which the function $W(\alpha,s)$ is regular, the higher are the temperature and entropy of the thermal or chaotic component of the mixtures Eq. (10.156) which may appear in the expansion Eq. (10.164).

For the density operator $T(\gamma,t)$ the ensemble average of any operator F is the function $f(\gamma,t)$ associated with F by the correspondence $C(-t)$,

$$\operatorname{Tr}\{T(\gamma,t) F\} = f(\gamma,t) , \tag{10.165}$$

as shown by Eq. (10.116). Thus in particular by referring to Eq. (9.124) we secure for the expectation value of the general s-ordered product $\{(a^\dagger)^n a^m\}$, the result

$$\operatorname{Tr}\left[T(\gamma,t) \left\{ (a^\dagger)^n a^m \right\}_s \right] = n! \left(\frac{t+s}{2}\right)^n \gamma^{m-n} L_n^{(m-n)}\left(\frac{2|\gamma|^2}{s+t}\right) . \tag{10.166}$$

In terms of the mean number of quanta $\langle n \rangle$ of the chaotic component of the field, this ensemble average may be written as

$$\operatorname{Tr}\left[T(\gamma,t) \left\{ (a^\dagger)^n a^m \right\}_s \right] = n! \langle n \rangle^n \left(1 + \frac{1-s}{2\langle n \rangle}\right)^n \gamma^{m-n}$$

$$\times L_n^{(m-n)}\left[-\frac{|\gamma|^2}{\langle n \rangle} \left(1 + \frac{1-s}{2\langle n \rangle}\right)^{-1} \right] . \tag{10.167}$$

When the coherent signal or prescribed current vanishes, i.e., γ is zero, we have

$$\operatorname{Tr}\left[T(0,t) \left\{ (a^\dagger)^n a^m \right\}_s \right] = \delta_{n,m} n! \langle n \rangle^n \left(1 + \frac{1-s}{2\langle n \rangle}\right)^n . \tag{10.168}$$

10.7 Examples of the General Quasiprobability Function $W(\alpha,s)$

For large numbers of quanta, i.e., $\langle n \rangle \gg |1-s|$, we have the approximation

$$\text{Tr}\left[T(0,t)\left\{(a^\dagger)^n a^m\right\}_s\right] \sim \delta_{n,m}\, n!\, \langle n \rangle^n \left[1 + \frac{n}{\langle n \rangle}\left(\frac{1-s}{2}\right)\right]. \tag{10.169}$$

For the case of normal ordering, $s = 1$, Eq. (10.168) becomes

$$\text{Tr}\left\{T(0,t)\,(a^\dagger)^n a^m\right\} = \delta_{n,m}\, n!\, \langle n \rangle^n . \tag{10.170}$$

We note that the general expectation value Eq. (10.167) factors into a part referring to the chaotic field alone and a part representing the effect of the coherent signal of amplitude γ:

$$\text{Tr}\left[T(\gamma,t)\left\{(a^\dagger)^n a^m\right\}_s\right] = \text{Tr}\left[T(0,t)\left\{(a^\dagger)^n a^m\right\}_s\right]\gamma^{m-n}$$

$$\times L_n^{(m-n)}\left[-\frac{|\gamma|^2}{\langle n \rangle}\left(1 + \frac{1-s}{2\langle n \rangle}\right)^{-1}\right]. \tag{10.171}$$

By using the definition[26] of the associated Laguerre polynomial we find for high signal-to-noise ratios, i.e., for $|\gamma|^2 \gg \langle n \rangle$, that the two leading terms of the Laguerre polynomial yield

$$\text{Tr}\left[T(\gamma,t)\left\{(a^\dagger)^n a^m\right\}_s\right] \sim (\gamma^*)^n \gamma^m \left[1 + \frac{nm\,\langle n \rangle}{|\gamma|^2}\left(1 + \frac{1-s}{2\langle n \rangle}\right)\right]. \tag{10.172}$$

Specializing now, first to the case of normal ordering, $s = 1$, and then to the case of a pure coherent field, $t = 1$, we find

$$\text{Tr}\left\{T(\gamma,t)\,(a^\dagger)^n a^m\right\} \sim (\gamma^*)^n \gamma^m \left[1 + \frac{nm\,\langle n \rangle}{|\gamma|^2}\right] \tag{10.173}$$

and

$$\text{Tr}\left\{T(\gamma,-1)\,(a^\dagger)^n a^m\right\} = (\gamma^*)^n \gamma^m , \tag{10.174}$$

respectively.

According to Eq. (9.190), the function $W(\alpha,s)$ for the pure state of precisely n quanta, $\rho = |n\rangle\langle n|$, is given by

$$W_n(\alpha,s) = \langle n|T(\alpha,s)|n\rangle$$

$$= \frac{2}{1-s}\left(\frac{s+1}{s-1}\right)^n \exp\left[-\frac{2|\alpha|^2}{1-s}\right] L_n\left(\frac{4|\alpha|^2}{1-s^2}\right). \tag{10.175}$$

The function $W_n(\alpha,s)$ is analytic in s except for an essential singularity at $s = 1$, the pole at $s = -1$ in the polynomial being canceled by the zero of the factor which multiplies it. By setting $s = 1 - \epsilon, 0, -1$, we find

$$W_n(\alpha, 1) = \pi P_n(\alpha)$$
$$= (-1)^n e^{|\alpha|^2} \frac{d^n}{d(|\alpha|^2)^n} \delta(|\alpha|^2), \quad (10.176)$$

$$W(\alpha, 0) = W_n(\alpha) = 2(-1)^n e^{-2|\alpha|^2} L_n(4|\alpha|^2), \quad (10.177)$$

$$W_n(\alpha, -1) = \langle \alpha | \rho_n | \alpha \rangle = \frac{1}{n!} |\alpha|^{2n} e^{-|\alpha|^2}, \quad (10.178)$$

where $\delta(x)$ is the one-dimensional δ function.

The two examples considered thus far have had the property that the weight function $P(\alpha) = \pi^{-1} W(\alpha, 1)$ of the P representation exists either as a well-behaved function or as a tempered distribution. This property cannot be true for the pure state that is the superposition[25] of two coherent states,

$$|\psi\rangle = N^{-1/2}(|\gamma\rangle + |\delta\rangle), \quad (10.179)$$

since it is not of the form noted in Sect. 10.3. The normalization factor N is given by

$$N = 2 + 2\exp(-\tfrac{1}{2}|\gamma - \delta|^2) \cos[\mathrm{Im}(\gamma^* \delta)] \quad (10.180)$$

and by using Eq. (9.174) we find easily

$$W(\alpha, s) = \langle \psi | T(\alpha, s) | \psi \rangle$$
$$= \frac{2N^{-1}}{1-s} \left[\exp\left(-\frac{2|\alpha - \delta|^2}{1-s}\right) + \exp\left(-\frac{2|\alpha - \gamma|^2}{1-s}\right) \right.$$
$$+ \langle \gamma | \delta \rangle \exp\left(-\frac{2}{1-s}(\gamma^* - \alpha^*)(\delta - \alpha)\right)$$
$$\left. + \langle \delta | \gamma \rangle \exp\left(-\frac{2}{1-s}(\gamma - \alpha)(\delta^* - \alpha^*)\right) \right]. \quad (10.181)$$

For this state $x(\rho) = 1$, but because of the second two exponential terms $W(\alpha, s)$ is not a tempered distribution on the line $\mathrm{Re}\, s = 1$.

For this density operator the functions $W(\alpha, 0)$ and $W(\alpha, -1)$ are given by

$$W(\alpha, 0) = W(\alpha) = \frac{2}{N} \left[e^{-2|\alpha - \delta|^2} + e^{-2|\alpha - \gamma|^2} + \langle \gamma | \delta \rangle e^{-2(\gamma^* - \alpha^*)(\delta - \alpha)} + \right.$$
$$\left. \langle \delta | \gamma \rangle e^{-2(\gamma - \alpha)(\delta^* - \alpha^*)} \right] \quad (10.182)$$

and by

$$W(\alpha, -1) = \langle \alpha | \rho | \alpha \rangle = \frac{1}{N} \left[e^{-2|\alpha - \delta|^2} + e^{-2|\alpha - \gamma|^2} + \langle \gamma | \delta \rangle e^{-(\gamma^* - \alpha^*)(\delta - \alpha)} + \right.$$
$$\left. \langle \delta | \gamma \rangle e^{-(\gamma - \alpha)(\delta^* - \alpha^*)} \right], \quad (10.183)$$

which are clearly well behaved.

10.7 Examples of the General Quasiprobability Function W(α,s)

Let us next consider the parametrized family of density operators

$$\rho_{\langle n \rangle} = \frac{2}{\langle n \rangle + 2} \sum_{n=0}^{\infty} \left(\frac{\langle n \rangle}{\langle n \rangle + 2} \right)^n |2n\rangle \langle 2n| , \qquad (10.184)$$

where $\langle n \rangle$ is the mean number of quanta and only even numbers of quanta are present. If we set

$$y = \frac{\langle n \rangle}{\langle n \rangle + 2}$$

and by

$$t = -\frac{1+\sqrt{y}}{1-\sqrt{y}} \leq -1 ,$$

then, by using Eq. (9.161), we may write

$$\rho_{\langle n \rangle} = \frac{1}{1-t} \left[T(0, t^{-1}) - t T(0, t) \right] ,$$

so that we have, on using Eq. (9.178),

$$W(\alpha, s) = \text{Tr}\{\rho_{\langle n \rangle} T(\alpha, s)\}$$

$$= \frac{2}{1-t} \left[\frac{t}{s+t} \exp\left(\frac{2|\alpha|^2}{s+t} \right) - \frac{1}{s+t^{-1}} \exp\left(\frac{2|\alpha|^2}{s+t^{-1}} \right) \right] . \qquad (10.185)$$

The function $W(\alpha, s)$ has essential singularities at $s = -t > 1$ and at $s = -t^{-1} < 1$. As the mean number of quanta $\langle n \rangle$ tends to infinity, the parameter

$$x(\rho) = -t^{-1}$$

approaches zero. The P representation does not exist for any member of this family of density operators. The Wigner distribution and the function $\langle \alpha | \rho | \alpha \rangle$, on the other hand, are given by

$$W(\alpha, 0) = W(\alpha) = \frac{2}{1-t} \left[\exp\left(\frac{2|\alpha|^2}{t} \right) - t \exp\left(2t|\alpha|^2 \right) \right] \qquad (10.186)$$

and by

$$W(\alpha, -1) = \langle \alpha | \rho | \alpha \rangle = \frac{2}{1-t} \left[\frac{t}{t-1} \exp\left(-\frac{2|\alpha|^2}{1-t} \right) + \frac{1}{1-t^{-1}} \exp\left(-\frac{2|\alpha|^2}{1-t^{-1}} \right) \right] , \qquad (10.187)$$

respectively.

In Eqns. (10.4) and (10.5), the operators a and $a^\dagger$ are defined as linear combinations of the operators q and p with coefficients involving an arbitrary real parameter λ.

Let us alter the scale parameter λ to the new value λ' and consider the new set of operators[29]

$$a' = \frac{1}{(2\hbar)^{1/2}}\left(\lambda' q + \frac{ip}{\lambda'}\right) \qquad (10.188)$$

and

$$a'^\dagger = \frac{1}{(2\hbar)^{1/2}}\left(\lambda' q - \frac{ip}{\lambda'}\right), \qquad (10.189)$$

which are appropriate, for example, to the description of a field amplitude in a medium with a different dielectric constant. These operators have the same algebraic properties as the operators a and $a^\dagger$. In particular, the operator a' possesses a complete set of eigenstates $|\beta\rangle'$:

$$a'|\beta\rangle' = \beta|\beta\rangle'. \qquad (10.190)$$

We now consider the ground state $|0\rangle'$ in the primed system. In terms of the operator

$$D'(\xi) \equiv \exp(\xi a'^\dagger - \xi^* a'),$$

the characteristic function $\chi'(\xi)$ for this density operator is given by

$$\chi'(\xi) \equiv {}'\langle 0|D'(\xi)|0\rangle'$$
$$= {}'\langle 0|\xi\rangle' = e^{-|\xi|^2/2}.$$

By putting $r = \lambda^{-1}\lambda'$ and $\xi = x + iy$, we see that

$$D(x+iy) = D'(rx + ir^{-1}y).$$

Thus we find for the unprimed characteristic function

$$\chi(\xi) \equiv {}'\langle 0|D(\xi)|0\rangle'$$

the value

$$\chi(x,y) = \exp\left[-\tfrac{1}{2}(r^2 x^2 + r^{-2} y^2)\right], \qquad (10.191)$$

where

$$\chi(x,y) \equiv \chi(x+iy).$$

If we now use Eq. (10.138), we find

$$W(\alpha,s) = \int \exp(\alpha\xi^* - \alpha^*\xi + \tfrac{1}{2}s|\xi|^2)\,\chi(\xi)\,\pi^{-1}\,d^2\xi,$$

10.7 Examples of the General Quasiprobability Function W(α,s)

or, in real notation with $W(u,v,s) \equiv W(u+\mathrm{i}v, s)$

$$W(u,v,s) = \int \exp\left[2\mathrm{i}(xv-yu) + \tfrac{1}{2}(s-r^2)x^2 + \tfrac{1}{2}(s-r^{-2})y^2\right]\pi^{-1}\,dx\,dy$$

$$= \frac{2}{[(r^2-s)(r^{-2}-s)]^{1/2}} \exp\left[-\frac{2u^2}{r^{-2}-s} - \frac{2v^2}{r^2-s}\right]. \tag{10.192}$$

The function $W(\alpha,s)$ has essential singularities at $s = r^2$ and at $s = r^{-2}$. By letting the ratio $r = \lambda^{-1}\lambda'$ depart sufficiently far from unity, we may move one of the singularities down to the line $\operatorname{Re} s = 0$ and make the parameter

$$x(\rho) = \min(r^2, r^{-2}) \tag{10.193}$$

arbitrarily small.

The P representation does not exist for any member of this family of density operators, except for the special cases $r = \pm 1$, for which we have

$$P(\alpha) = \delta^{(2)}(\alpha). \tag{10.194}$$

By setting $s = 0$ and $s = -1$ in Eq. (10.192), we find that the Wigner function $W(\alpha)$ and the function $\langle\alpha|\rho|\alpha\rangle$ are given by

$$W(u,v,0) = 2\exp(-2r^2 u^2 - 2r^{-2} v^2) \tag{10.195}$$

and by

$$W(u,v,-1) = \frac{2}{|r|+|r|^{-1}} \exp\left(\frac{-2r^2 u^2 - 2v^2}{r^2 + 1}\right), \tag{10.196}$$

respectively.

As a final illustration, let us consider the expectation values of the s-ordered products for the state $|0\rangle'$. They are given by the integrals

$${}'\langle 0|\{(a^\dagger)^n a^m\}_s|0\rangle' = (\mu\nu)^{1/2}\int (u-\mathrm{i}v)^n (u+\mathrm{i}v)^m e^{-\mu u^2 - \nu v^2}\pi^{-1}\,du\,dv, \tag{10.197}$$

where we have put

$$\mu^{-1} = \tfrac{1}{2}(r^{-2}-s), \qquad \nu^{-1} = \tfrac{1}{2}(r^2-s).$$

If we put $n = m$ we may find a generating function for these quantities by forming the series

$$\sum_{n=0}^{\infty} \frac{1}{n!} y^n {}'\langle 0|\{(a^\dagger)^n a^n\}_s|0\rangle' = (\mu\nu)^{1/2}\int \exp\left[-(\mu-y)u^2 - (\nu-y)v^2\right]\pi^{-1}\,du\,dv$$

$$= \left(1 - \frac{(\mu+\nu)y}{\mu^2} + \frac{y^2}{\mu\nu}\right)^{1/2}.$$

If we now compare this relation with the generating function[30] for the Legendre polynomials $P_n(z)$,

$$(1 - 2hz + h^2)^{-1/2} = \sum_{n=0}^{\infty} h^n P_n(z),$$

then we find, on making the identifications

$$h = (\mu\nu)^{-1/2} y, \qquad z = \tfrac{1}{2}\left[\left(\frac{\mu}{\nu}\right)^{1/2} + \left(\frac{\nu}{\mu}\right)^{1/2}\right],$$

the result

$$\begin{aligned}
'\langle 0|\{(a^\dagger)^n a^n\}_s|0\rangle' &= n!\,(\mu\nu)^{-n/2} P_n\left(\tfrac{1}{2}\left[\left(\frac{\mu}{\nu}\right)^{1/2} + \left(\frac{\nu}{\mu}\right)^{1/2}\right]\right) \\
&= \frac{n!}{2^n}[(r^2-s)(r^{-2}-s)]^{-n/2} \\
&\quad \times P_n\left(\tfrac{1}{2}\left[\left(\frac{r^2-s}{r^{-2}-s}\right)^{1/2} + \left(\frac{r^{-2}-s}{r^2-s}\right)^{1/2}\right]\right).
\end{aligned} \qquad (10.198)$$

In particular, for $s = 0$ we have

$$'\langle 0|\{(a^\dagger)^n a^n\}_0|0\rangle' = \frac{n!}{2^n} P_n(\tfrac{1}{2}(r^2 + r^{-2})), \qquad (10.199)$$

from which the mean number of quanta may be shown to be

$$\langle n \rangle = '\langle 0|a^\dagger a|0\rangle' = \tfrac{1}{4}(r - r^{-1})^2 = \frac{1}{4}\left(\frac{\lambda}{\lambda'} - \frac{\lambda'}{\lambda}\right)^2. \qquad (10.200)$$

10.8
Analogy with Heat Diffusion

It is possible to draw an instructive analogy between the function $W(\alpha, s)$ and the temperature distribution on a plane. We discuss this analogy in the present section and draw from it the conclusion that for every density operator the function $W(\alpha, s)$ must at some point in the half-plane $\mathrm{Re}\, s > 0$ either assume negative values or cease to be normalized.

By differentiating the Gaussian convolution Eq. (10.146) or, equivalently, by using Eq. (9.223), we find that the function $W(\alpha, s)$ obeys the second-order partial differential equation

$$\frac{\partial W(\alpha, s)}{\partial s} = -\frac{1}{2}\frac{\partial^2 W(\alpha, s)}{\partial(\alpha^*)\partial\alpha}. \qquad (10.201)$$

10.8 Analogy with Heat Diffusion

If we express this equation in terms of real coordinates by writing $\alpha = x + iy$ and $W(x,y,s) = W(x+iy,s)$, then it assumes the form

$$\frac{\partial W(x,y,s)}{\partial s} = -\frac{1}{8}\left(\frac{\partial^2}{\partial x^2} + \frac{\partial^2}{\partial y^2}\right) W(x,y,s) . \tag{10.202}$$

This is a species of heat-diffusion equation in which the function $W(x,y,s)$ plays the role of the temperature at the point (x,y) and the variable s corresponds in a *negative* sense to the time variable, $s = -t$. In this analogy the convolution integral Eq. (10.146) is Poisson's solution for the temperature in terms of the temperature distribution at an earlier time.

The differential equation Eq. (10.202) may be viewed as a conservation law expressing the assumed absence of sources or sinks of heat. It is a well-known feature of solutions to the heat-diffusion equation that, in general, they cannot be extrapolated into the past without the appearance of singularities at some point. These singularities, which represent the sources that originally supplied heat to the plane, correspond to the singularities exhibited by the function $W(\alpha, s)$ for $\operatorname{Re} s > 0$.

We have seen in Eqns. (10.134) and (10.142) that the function $W(\alpha, s)$ is normalized and non-negative when s is real and ≤ -1. It therefore corresponds to a temperature distribution that for times later than $t = -s = 1$ is non-negative and describes a unit amount of heat on the plane. With the passage of time the heat diffuses over the plane according to the convolution integral Eq. (10.146). Since there is only a unit amount of heat at time $t = -s = +1$ and since the plane is of infinite extent, the temperature at each point on the plane tends ultimately to zero after a sufficiently long time. This property of temperature distributions is true also of the function $W(\alpha, s)$ since, according to the inequality (10.143), the modulus of $W(\alpha, s)$ goes to zero as $-s = t$ tends to infinity.

It is evident that a temperature distribution that is normalized and non-negative at the time $t = -s = 1$ cannot have been normalized and non-negative at all prior times. For if this were the case, then by the argument given in the preceding paragraph the temperature would be zero at all points of the plane. We may prove the corresponding statement for the function $W(\alpha, s)$ by using the convolution law Eq. (10.146).

Let us assume that the function $W(\alpha, s)$ is non-negative and normalized at some real and positive value of $s = u > 0$. Then for $\operatorname{Re} s < u$, the modulus of the exponential which appears in the integral Eq. (10.146), where we replace t by u to avoid confusion with the time variable, is less than unity. We have, therefore, since $W(\beta, u)$ is non-negative by assumption, the inequality

$$|W(\alpha, s)| \leq \left|\frac{2}{u-s}\right| \int W(\beta, u) \pi^{-1} d^2\beta .$$

If the function $W(\beta, u)$ is also normalized in the sense of Eq. (10.134), then the integral over β is unity and we may write

$$|W(\alpha, s)| \leq \frac{2}{|u-s|} \tag{10.203}$$

for $\operatorname{Re} s < u$. Thus if the function $W(\alpha,u)$ were non-negative and normalized for all positive values of u, then, by letting u approach infinity in this inequality, we would find that $W(\alpha,s)$ is identically zero. We have shown, therefore, that the function $W(\alpha,s)$ must either assume negative values or cease to be normalized at some point $s = u > 0$.

If the modulus of the function $W(\alpha,s)$ exceeds the value $2|u-s|^{-1}$ specified by Eq. (10.203) for any value of $u > \operatorname{Re} s$, then we may conclude that the function $W(\alpha,s)$ either assumes negative values or is not normalized for $s = u$. The smallest value of u for which the inequality (10.203) is violated may therefore be used as an estimate (or, more precisely, an upper bound) for the smallest value of $\operatorname{Re} s$ at which the function $W(\alpha,s)$ takes on negative values or ceases to be normalized.

We may illustrate the use of this estimate by considering the density operator $T(\gamma,t)$ for which the function $W(\alpha,-1)$ is given by Eq. (10.162) as

$$W(\alpha,-1) = \frac{1}{\langle n \rangle + 1} \exp\left(-\frac{|\alpha-\gamma|^2}{\langle n \rangle + 1}\right),$$

where $\langle n \rangle = -\frac{1}{2}(t+1)$ is the mean number of quanta. This function attains a maximum value of $(\langle n \rangle + 1)^{-1}$ at $\alpha = \gamma$. Thus by setting $s = -1$ in Eq. (10.203), we find that the inequality

$$\frac{1}{\langle n \rangle + 1} \leq \frac{2}{u+1}$$

breaks down at

$$u = 1 + 2\langle n \rangle.$$

We estimate then that the function $W(\alpha,s)$ will misbehave in some way at this point or at a smaller value of $\operatorname{Re} s$. According to Eq. (10.159) the function $W(\alpha,s)$ has an essential singularity at $s = 1 + 2\langle n \rangle$ for which value it may be written as

$$W(\alpha, 1+2\langle n \rangle) = \pi \delta^{(2)}(\alpha-\gamma).$$

This singularity corresponds to a unit amount of heat concentrated at a point $\alpha = \gamma$ at the time $t = -1 - 2\langle n \rangle$.

10.9
Time-Reversed Heat Diffusion and W(α,s)

According to the analogy developed in Sect. 10.8, the extension of the function $W(\alpha,s)$ from one value of s to another with a larger real part corresponds to the continuation of an initial temperature distribution to prior times. We derive various procedures for doing this in the present section.

One method is to make use of the simple exponential dependence of the function $\chi(\xi,s)$ upon the variable s,

$$\chi(\xi,s) = e^{(s-t)|\xi|^2/2}\chi(\xi,t),$$

in conjunction with the complex Fourier transform relations Eq. (10.137) and (10.138) which connect the functions W and χ. In this way we express the function $W(\alpha,s)$ in terms of $W(\beta,t)$ as the double integral

$$W(\alpha,s) = \iint \exp\left[(\alpha-\beta)\xi^* - (\alpha-\beta)^*\xi + \tfrac{1}{2}(s-t)|\xi|^2\right] W(\beta,t)\, \pi^{-2}\, d^2\beta\, d^2\xi, \tag{10.204}$$

in which the integration over β must be done first unless $\operatorname{Re} s \leq \operatorname{Re} t$, in which case the double integral reduces to the convolution Eq. (10.146).

A second method permits us to avoid integrating twice at the expense of introducing another complex variable. Let us define the function $W(\alpha,\beta,s)$ as the integral

$$W(\alpha,\beta,s) = \int e^{\alpha\xi^* - \beta\xi}\chi(\xi,s)\, \pi^{-1}\, d^2\xi. \tag{10.205}$$

This function may be shown[2] to be analytic in all three variables for $\operatorname{Re} s < x(\rho)$. The function $W(\alpha,s)$ is $W(\alpha,-\alpha^*,s)$. Let us consider the integral

$$\frac{2}{s-t}\int W(\alpha+\gamma,\beta+\gamma^*,t)\exp\left[-\frac{2}{s-t}|\gamma|^2\right]\pi^{-1}\,d^2\gamma$$

By writing W in terms of χ, we express it as the double integral

$$\frac{2}{s-t}\int \chi(\xi,t)\exp\left[(\alpha+\gamma)\xi^* + (\beta+\gamma^*)\xi - \frac{2}{s-t}|\gamma|^2\right]\pi^{-2}\,d^2\gamma\,d^2\xi.$$

If $\operatorname{Re} s > \operatorname{Re} t$, we may use the formula Eq. (9.A2) to integrate over the variable γ, after which it reduces to

$$\int \chi(\xi,s)\,e^{\alpha\xi^* + \beta\xi}\,\pi^{-1}\,d^2\xi,$$

which we recognize as the expression Eq. (10.205) for $W(\alpha,\beta,s)$. Thus we have the integral formula

$$W(\alpha,\beta,s) = \frac{2}{s-t}\int W(\alpha+\gamma,\beta+\gamma^*,t)\exp\left(\frac{-2|\gamma|^2}{s-t}\right)\pi^{-1}\,d^2\gamma \tag{10.206}$$

for $\operatorname{Re} s > \operatorname{Re} t$, which if we set $\beta = -\alpha^*$ becomes

$$W(\alpha,s) = \frac{2}{s-t}\int W(\alpha+\gamma,\gamma^* - \alpha^*,t)\exp\left(\frac{-2|\gamma|^2}{s-t}\right)\pi^{-1}\,d^2\gamma. \tag{10.207}$$

In a similar fashion, by expressing $W(\gamma,\gamma^*,t)$ in terms of χ, we may verify the relation

$$W(\alpha,\beta,s) = \frac{2}{s-t}\exp\left(\frac{2\alpha\beta}{t-s}\right)\int W(\gamma,\gamma^*,t)\exp\left(\frac{2(\alpha\gamma^*+\beta\gamma-|\gamma|^2)}{s-t}\right)\pi^{-1}d^2\gamma \tag{10.208}$$

for $\mathrm{Re}\,s > \mathrm{Re}\,t$. At $\beta = -\alpha^*$ this relation reduces to

$$W(\alpha,s) = \frac{2}{s-t}\exp\left(\frac{2|\alpha|^2}{s-t}\right)\int W(\gamma,\gamma^*,t)\exp\left(\frac{2(\alpha\gamma^*-\alpha^*\gamma-|\gamma|^2)}{s-t}\right)\pi^{-1}d^2\gamma . \tag{10.209}$$

We may illustrate these relations if we observe that according to Eq. (10.111) the weight function $R(\beta^*,\alpha)$ of the R representation and the function $W(\alpha,\beta,s)$ stand in the relationship

$$R(\beta^*,\alpha) = e^{\beta^*\alpha} W(\alpha,-\beta^*,-1) . \tag{10.210}$$

Thus by putting $s = +1$ and $t = -1$ in Eqns. (10.207) and (10.209), we may express the weight function of the P representation as the complex Fourier transforms

$$P(\alpha) = \pi^{-2} e^{-|\alpha|^2}\int e^{\alpha\gamma^*-\alpha^*\gamma} R(\alpha^*-\gamma^*,\alpha+\gamma) d^2\gamma \tag{10.211}$$

and[2,31]

$$P(\alpha) = \pi^{-2} e^{|\alpha|^2}\int e^{\alpha\gamma^*-\alpha^*\gamma} R(-\gamma^*,\gamma) d^2\gamma . \tag{10.212}$$

10.10
Properties Common to all Quasiprobability Distributions

In Sect. 10.6 we introduced a parametrized distribution function $W(\alpha,s)$ in terms of which ensemble averages may be expressed as weighted integrals. It is instructive to consider what features of this scheme are necessary ones that must be shared by any scheme for expressing quantum-mechanical ensemble averages as weighted integrals. The present section is a discussion of this point.

Classically, the ensemble average of a quantity $f(\alpha)$ may be written as an integral over a phase-space probability distribution $p(\alpha)$ in the form

$$\langle f \rangle = \int f(\alpha) p(\alpha) \pi^{-1} d^2\alpha . \tag{10.213}$$

Our method for carrying over some of the simplicity of this expression into the quantum theory is represented by the relation Eq. (10.121)

$$\mathrm{Tr}\{\rho F\} = \int f(\alpha,-s) W(\alpha,s) \pi^{-1} d^2\alpha , \tag{10.214}$$

in which the functions $f(\alpha,-s)$ and $W(\alpha,s)$ are related to the operators F and ρ by the correspondences $C(s)$ and $C(-s)$, respectively.

Let us now assume that we have an alternative scheme for expressing the quantities $\text{Tr}(\rho F)$ in forms resembling the classical expression Eq. (10.213), i.e., that for every density operator ρ and every operator F we have

$$\text{Tr}\{\rho F\} = \int f(\alpha) w(\alpha) \pi^{-1} d^2\alpha, \qquad (10.215)$$

where $f(\alpha)$ refers to F and $w(\alpha)$ to ρ. The correspondence between F and $f(\alpha)$ implicit in the scheme Eq. (10.215) must be linear since the trace $\text{Tr}\{\rho F\}$ is linear.

Let us denote by $f_{n,m}(\alpha)$ the function associated with the operator $|n\rangle\langle m|$ and by $X(\alpha)$ the operator defined by $f_{n,m}(\alpha) = \langle n|X(\alpha)|m\rangle$. Since the correspondence $F \leftrightarrow f(\alpha)$ is linear, it must associate with every operator F

$$F = \sum_{n,m=0}^{\infty} |n\rangle\langle n|F|m\rangle\langle m|$$

the function

$$f(\alpha) = \sum_{n,m=0}^{\infty} \langle n|F|m\rangle f_{n,m}(\alpha)$$
$$= \text{Tr}\{FX(\alpha)\}. \qquad (10.216)$$

Thus the correspondence implicit in the alternative scheme Eq. (10.215) assumes the general form of the correspondence $C(s)$:

$$F \underset{s}{\longleftrightarrow} f(\alpha, -s) = \text{Tr}\{FT(\alpha, -s)\}.$$

If we now substitute the expression Eq. (10.216) for the function $f(\alpha)$ into Eq. (10.215), then we find

$$\text{Tr}\{\rho F\} = \int \text{Tr}\{FX(\alpha)\} w(\alpha) \pi^{-1} d^2\alpha. \qquad (10.217)$$

Thus letting $F = |n\rangle\langle m|$, we have for all states $|n\rangle$ and $|m\rangle$ the result

$$\langle m|\rho|n\rangle = \int \langle m|X(\alpha)|n\rangle w(\alpha) \pi^{-1} d^2\alpha, \qquad (10.218)$$

which implies that

$$\rho = \int X(\alpha) w(\alpha) \pi^{-1} d^2\alpha. \qquad (10.219)$$

This expansion corresponds to Eq. (10.122).

By again invoking the linearity of the trace in Eq. (10.215) or by referring directly to Eq. (10.219), we see that the correspondence $\rho \leftrightarrow w(\alpha)$ must also be linear and must therefore assume the form

$$w(\alpha) = \text{Tr}\{\rho Y(\alpha)\} \qquad (10.220)$$

for some set of operators $Y(\alpha)$. This relation, which corresponds to Eq. (10.120), in turn implies, by the argument we have given, that Eq. (10.215) is equivalent to the representation

$$F = \int Y(\alpha) f(\alpha) \pi^{-1} d^2\alpha, \qquad (10.221)$$

which corresponds to Eq. (10.115).

We now observe that from Eqns. (10.219) and (10.220) we have

$$w(\alpha) = \int w(\beta) \operatorname{Tr}\{X(\beta) Y(\alpha)\} \pi^{-1} d^2\beta \qquad (10.222)$$

and from Eqns. (10.216) and (10.221),

$$f(\alpha) = \int f(\beta) \operatorname{Tr}\{X(\alpha) Y(\beta)\} \pi^{-1} d^2\beta. \qquad (10.223)$$

Since these relations are assumed to hold for all ρ and F, we must have

$$\operatorname{Tr}\{X(\beta) Y(\alpha)\} = \pi \delta^{(2)}(\alpha - \beta), \qquad (10.224)$$

which is the analog of the orthogonality rule

$$\operatorname{Tr}\{T(\alpha, s) T(\beta, -s)\} = \pi \delta^{(2)}(\alpha - \beta).$$

So far we have exploited only the linearity of the scheme Eq. (10.215). If we now assume that it offers some of the simplicity of the classical procedure Eq. (10.213) then for some species of ordered product, let us write it as $[(a^\dagger)^n a^m]_q$, where q denotes the type of ordering, we must have[6]

$$\operatorname{Tr}\left\{\rho \left[(a^\dagger)^n a^m\right]_q\right\} = \int (\alpha^*)^n \alpha^m w(\alpha) \pi^{-1} d^2\alpha. \qquad (10.225)$$

It is the type of ordering involved in this relation that distinguishes the arbitrary scheme Eq. (10.215) from the procedure of Eq. (10.121). There is a great deal of latitude here; for, although we have considered a complex plane of orderings in the procedure of Sect. 10.6, there exists a considerable variety of other types of ordering. The three ways of ordering the operators p and q considered in Sect. 9.4, for example, are easily generalized to a complex plane of q, p orderings.

If, however, the relation Eq. (10.225) should apply to the s-ordered products for some value of s, then, by forming the ensemble average of the series Eq. (10.10) for $D(\xi, s)$, we would obtain the relation

$$\chi(\xi, s) = \int e^{\xi \alpha^* - \alpha^* \xi} w(\alpha) \pi^{-1} d^2\alpha, \qquad (10.226)$$

which upon Fourier inversion would imply $w(\alpha) = W(\alpha, s)$.

[6] The characteristic function $\chi(\alpha)$ and the displacement operators $D(\alpha)$, discussed in Sect. 9.3, form the basis for a scheme of the form of Eq. (10.215). The roles of the function $w(\alpha)$ and of the operators $X(\alpha)$ and $F(\alpha)$ are played by $\chi(\alpha)$, $D(-\alpha)$, and $D(\alpha)$, respectively. Equations (10.215), (10.216), (10.219)–(10.221), and (10.224) correspond to Eqns. (9.51), (10.49)–(10.51), (10.48), and (9.39), respectively. This scheme does not, however, conform to Eq. (10.225) as is shown, for example, by Eq. (10.140).

References

1. K. E. Cahill, R. J. Glauber, *Phys. Rev.* **176**, 1857 (1968); reprinted as Chapter 9 in this volume.
2. K. E. Cahill, thesis, Harvard University 1967 (University Microfilms, Arm Arbor), unpublished. Many of the results of the present paper are presented there.
3. E. P. Wigner, *Phys. Rev.* **40**, 749 (1932).
4. J. E. Moyal, *Proc. Cambridge Phil. Soc.* **45**, 99 (1948); **45**, 545 (1949).
5. R. J. Glauber, *Phys. Rev. Lett.* **10**, 84 (1963).
6. R. J. Glauber, *Phys. Rev.* **131**, 2766 (1963); reprinted as Sect. 2.9 in this volume.
7. E. C. G. Sudarshan, *Phys. Rev. Lett.* **10**, 277 (1963).
8. R. J. Glauber, in *Quantum Optics and Electronics*, C. de Witt et al., Eds., Gordon and Breach, New York 1965, p. 63; reprinted as Chapter 2 in this volume.
9. C. L. Mehta, E. C. G. Sudarshan, *Phys. Rev. B* **138**, 274 (1965).
10. K. E. Cahill, *Phys. Rev. B* **138**, 1566 (1965).
11. J. R. Klauder, *Phys. Rev. Lett.* **15**, 534 (1966).
12. R. Bonifacio, L. M. Narducci, E. Montaldi, *Phys. Rev. Lett.* **16**, 1125 (1966).
13. R. Bonifacio, L. M. Narducci, E. Montaldi, *Nuovo Cimento* **47**, 890 (1967).
14. M. Lax, W. H. Louisell, *J. Quantum Electron.* **QE3**, 47 (1967).
15. M. M. Miller, E. A. Mishkin, *Phys. Rev.* **164**, 1610 (1967).
16. U. Fano, *Rev. Mod. Phys.* **29**, 74 (1957).
17. B. R. Mollow, *Phys. Rev.* **162**, 1256 (1967).
18. Y. Kano, *J. Math. Phys.* **6**, 1913 (1965).
19. G. C. Wick, *Phys. Rev.* **80**, 268 (1950).
20. R. J. Glauber, *Phys. Rev.* **130**, 2529 (1963); reprinted as Chapter 1 in this volume.
21. R. J. Glauber, *Phys. Lett.* **21**, 650 (1966); reprinted as Chapter 5 in this volume.
22. B. R. Mollow, R. J. Glauber, *Phys. Rev.* **160**, 1076 (1967); **160**, 1097 (1967); reprinted as Chapters 6 and 7, respectively, in this volume.
23. Y. R. Shen, *Phys. Rev.* **155**, 921 (1967).
24. R. J. Glauber, in *Proceedings of the Physics of Quantum Electronics Conference, San Juan, Puerto Rico 1965*, P. L. Kelly el al., Eds., McGraw-Hill, New York 1966, p. 788.
25. K. E. Cahill, to be published.
26. W. Magnus, F. Oberhettinger, *Formulas and Theorems for the Functions of Mathematical Physics*, Chelsea Publishing, New York 1954, p. 85.
27. H. Weyl, *The Theory of Groups and Quantum Mechanics*, Dover Publications, New York 1950, pp. 272–276.
28. R. J. Glauber, *Phys. Rev.* **84**, 395 (1951).
29. J. R. Klauder, J. McKenna, D. G. Currie, *J. Math. Phys.* **6**, 734 (1965).
30. Ref.[26], p. 51.
31. C. L. Mehta, *Phys. Rev. Lett.* **18**, 752 (1967).

11
Coherence and Quantum Detection[1]

11.1
Introduction

The portions of the electromagnetic spectrum explored to date cover a vast range of frequencies, and the phenomena that we observe throughout the high-frequency end of the spectrum (X-rays and beyond) are dominated by the corpuscular behavior of light. We have become thoroughly accustomed to considering the interactions of short-wave-length quanta with matter in quantum-mechanical terms. Still it is not yet appropriate to apply the term "quantum optics" to studies in the very-high-frequency range since the experiments we can perform in it lack so many of the features we associate with optics. Such sources as we have for high-frequency radiation tend to be quite uncontrollable both in the spatial and temporal behavior of their outputs; they are furthermore so feeble in terms of the number of quanta they emit that the statistical properties of the fields they radiate are trivial. No practical X-ray source, for example, need ever be regarded as emitting simultaneously two identical quanta.

In the low-frequency end of the electromagnetic spectrum the picture is altogether different. In the radio-frequency range, for example, we have available extremely intense sources over which we can exercise a remarkable degree of control. But because the quantum energies, $\hbar\omega$, are small the corpuscular character of the radiation remains unobservable, in effect, and classical electromagnetic theory suffices for the description of all experiments.

The central range of the electromagnetic spectrum, which includes visible or optical frequencies shows some of the features of both the high- and low-frequency ranges. The light sources which have been used in optical experiments have, in effect, been weak ones until recently. Although quantum phenomena are not difficult to observe in this frequency range, they have usually not been the subject of investigation in optical experiments. Most of optical theory has accordingly been developed in classical terms.

It is only within roughly the last ten years that we have become very deeply interested in experiments based on the detection of individual quanta at optical frequencies.

---

[1] Reprinted from *Quantum Optics*, R. J. Glauber, Ed., Academic Press, New York 1969, pp. 15–56. Copyright by Società Italiana di Fisica.

Quantum Theory of Optical Coherence. Selected Papers and Lectures. Roy J. Glauber.
Copyright © 2007 WILEY-VCH Verlag GmbH & Co. KGaA, Weinheim
ISBN: 978-3-527-40687-6

The experiment which began this trend was one in which Hanbury Brown and Twiss showed that the photons in a normal light beam of narrow spectral bandwidth have a distinct tendency to arrive at a detector in correlated pairs. This experiment and others like it have been repeated a great many times by now. The counting techniques on which they are based are simple extensions into the domain of optical frequencies of the techniques which have been developed for quantum detection in the much higher frequency domain of elementary-particle physics.

Another sort of change in optics, in many ways a more far-reaching if less fundamental one, has been brought about in the same period through the development of coherent light sources. These are sources which are vastly more intense than ordinary sources in their output of nearly similar photons. Their quantum outputs are furthermore controllable in many of the same senses that radio-frequency fields are. We now have in hand the means of reducing by orders of magnitude the spectral band width and spatial divergence of light beams and even of greatly reducing their intrinsic noisiness. These possibilities have opened altogether new experimental areas in optics. The methods used in several of these areas and their results to date form the common theme of the lectures of our school.

Some of the new phenomena which have been observed arise simply because we have for the first time in any region of the electro-magnetic spectrum the possibility both of generating fields coherently and of detecting them via their individual quanta. We shall devote this set of five lectures to discussing the theoretical background of such experiments. Broadly speaking we shall discuss the ways in which optical fields may be characterized in quantum-mechanical terms and the kinds of quantities which are measured in photon detection experiments.

In the first three lectures (Sections 11.2–11.7) we shall give a general survey of the quantum theory of optical coherence. Most of the material of these introductory lectures has already been published in one place or another; we present them mainly as a review of things which we hope are already more or less familiar. In the fourth lecture, (Sections 11.8–11.12), we discuss the statistical properties of an oscillating mode of the electromagnetic field whose amplitude decays with time. We shall present there a new approach to the problem which is, in effect, a new way of dealing with the quantum mechanics of the damped harmonic oscillator. In the last lecture, (Sections 11.13 and 11.14), we shall show how the solution of the damped-oscillator problem can be used in a seemingly very different context to provide a new approach to the problem of calculating photodetector responses.

11.2
The Statistical Properties of the Electromagnetic Field

In all of the older optical experiments it was quite sufficient, crudely speaking, to describe an incident light beam simply in terms of its spectral data. We have become accustomed, for example, to thinking of a beam of plane waves with specified polarization as being described fully by its spectral intensity function $I(\omega)$. If such descriptions have been adequate in the past it is because optical experiments have been confined simply to the measurement of average field intensities. They were measurements which could have been made on arbitrarily weak fields simply by increasing the integration time of the detectors. In quantum-mechanical terms, measurements of this type are sensitive only to the single-photon properties of the field.

In actual optical fields we usually have many photons present at a time. By using the equipment now available to investigate the statistical distributions of these photons we can discover properties of the field and of its sources which were altogether hidden from view in classic optics. Fields which have identical spectral properties may have altogether different statistical properties as revealed by photon counting experiments.

In order to avoid the purely formal difficulties associated with infinitely extended electromagnetic fields, let us confine the field to a finite volume, say cubical in shape. If we take the volume to be large enough, the precise nature of the boundary conditions will not matter, and we could consider the volume to have perfectly reflecting walls or perfectly transparent walls leading on to an infinity of identical boxes. In the first case the boundary conditions lead to normal modes which are standing waves; in the second, the normal modes are traveling waves.

The electromagnetic field inside this box might be in some quantum-mechanical state $|\psi\rangle$. We rarely have precise information about the nature of this state, however, and we must usually be content with fairly rudimentary information about the statistical properties of the field. In such cases the best predictions we can make for the results of experiments are averages over our uncertainties about the state of the field. Thus the expectation value of an operator O will be

$$\langle O \rangle = \{\langle \psi | O | \psi \rangle\}_{av} , \qquad (11.1)$$

where the average is to be taken over the states $|\psi\rangle$ which occur in repeated preparations of the same field. We can formally rearrange this expression by introducing the trace

$$\langle O \rangle = \{\text{Tr}[|\psi\rangle\langle\psi| O]\}_{av} = \text{Tr}[\{|\psi\rangle\langle\psi|\}_{av} O] . \qquad (11.2)$$

The operator $\{|\psi\rangle\langle\psi|\}_{av}$ is the density operator for the system

$$\rho = \{|\psi\rangle\langle\psi|\}_{av} ;$$

it embodies all the information we have about the state of the field. Measurements of quantities related to the field, when averaged, are to be compared with expressions of the form

$$\langle O \rangle = \text{Tr}\{\rho O\} . \qquad (11.3)$$

We must now ask what kinds of operators O correspond to the measurements we actually make upon the field.

11.3
The Ideal Photon Detector

One of the important quantities for describing any field is its intensity. The photon flux of a field should be measured ideally by a detector whose response does not depend upon the frequency of the radiation, and whose size is much smaller than any spatial variation of the field. Such a detector could then be said to measure the intensity of the field at a single point[1]. It is not impossible to imagine good approximations to such a detector: an atom which detects photons by being excited to a continuum state will have a fairly flat response over a considerable frequency range and is much smaller in size than the wave length of optical-frequency radiation.

The interaction of the electromagnetic field with an atom is given, in the electric-dipole approximation by the interaction energy

$$H_\mathrm{I} = -\boldsymbol{E}(\boldsymbol{r}t,) \cdot \boldsymbol{d} , \tag{11.4}$$

where $\boldsymbol{E}$ is the electric field at the position, $\boldsymbol{r}$, of the nucleus and $\boldsymbol{d}$ is the dipole moment of the atom. The detection of a photon is accomplished through an absorption process which brings the field from its initial state $|i\rangle$ to a final state $|f\rangle$. At the same time the atom is raised from its ground state, $|g\rangle$, to an excited state $|e\rangle$. The probability amplitude for this process is easily computed in first-order perturbation theory. It is simply proportional to the matrix element of H_I,

$$\langle f|\boldsymbol{E}(\boldsymbol{r},t)|i\rangle \cdot \langle e|\boldsymbol{d}|g\rangle$$

between these states.

The atomic dipole matrix element oscillates in time as $\exp[(i/\hbar)(E_e - E_g)t]$, where $E_e - E_g$ is the positive energy difference between the states $|e\rangle$ and $|g\rangle$. Since energy is conserved in the transition the only components of the electric field which can actually contribute to the transition amplitude are those with the complementary time-dependence $\exp[-i\omega t]$ for $\hbar\omega \approx E_e - E_g$. It is convenient, therefore, to separate the electric field into its positive- and negative-frequency parts

$$\boldsymbol{E}(\boldsymbol{r},t) = \boldsymbol{E}^{(+)}(\boldsymbol{r},t) + \boldsymbol{E}^{(-)}(\boldsymbol{r},t) . \tag{11.5}$$

The positive-frequency part $\boldsymbol{E}^{(+)}$ is defined to contain all terms which oscillate as $\exp[-i\omega t]$ for $\omega > 0$, and the negative frequency part is its Hermitian conjugate

$$\boldsymbol{E}^{(-)}(\boldsymbol{r},t) = \left\{\boldsymbol{E}^{(+)}(\boldsymbol{r},t)\right\}^\dagger . \tag{11.6}$$

Since only the positive frequency part of the field contributes significantly to the photoabsorption process, the matrix element which interests us can be written as

$$\boldsymbol{M} \cdot \left\langle f \left|\boldsymbol{E}^{(+)}(\boldsymbol{r},t)\right| i \right\rangle , \tag{11.7}$$

where we have used M as an abbreviation for the atomic matrix element. (Since the vector nature of the electromagnetic field will not play any significant role in our analysis we will feel free from this point on to suppress the use of vector and tensor notation in most of our formulae. It is a trivial matter to restore it if that is necessary, in order to deal with fields having impure polarizations.)

The transition probabilities for the particular photoabsorption process which leaves the field in the final state $|f\rangle$ is proportional, crudely speaking, to the squared modulus of the matrix element Eq. (11.7). Since the final state of the field remains unobserved, what we actually measure corresponds to a sum of transition probabilities taken over all possible final states $|f\rangle$, that is to the sum

$$|M|^2 \sum_f \left|\langle f|E^{(+)}(r,t)|i\rangle\right|^2 .$$

Since this sum may be freely extended over a complete set of final states it reduces to

$$|M|^2 \langle i|E^{(-)}(r,t)\sum_f|f\rangle\langle f|E^{(+)}(r,t)|i\rangle = |M|^2 \langle i|E^{(-)}(r,t)E^{(+)}(r,t)|i\rangle . \tag{11.8}$$

As we have mentioned, the initial state $|i\rangle$ of the field is rarely known with certainty. We must in general average the absorption probability over the possible initial states of the field; the output of our idealized photodetector will therefore be proportional to[1]

$$I(r,t) = \mathrm{Tr}\left\{\rho E^{(-)}(r,t) E^{(+)}(r,t)\right\} . \tag{11.9}$$

11.4
Correlation Functions and Coherence

If we consider experiments which measure the intensity of the field produced by superposing the fields at two or more space-time points, (as we do for example in Young's double-slit experiment) we find that our photodetector will in general measure values of the field correlation function

$$G^{(1)}(x_1,x_2) = \mathrm{Tr}\left\{\rho E^{(-)}(x_1) E^{(+)}(x_2)\right\} . \tag{11.10}$$

We have introduced here the abbreviation of letting x_j stand for the coordinates of the space-time point r_j, t_j.

A more detailed analysis than that we have given here[2] shows that the probability of counting a single photon between the times 0 and t will take the form

$$p^{(1)}(t) = \int_0^t dt' \int_0^t dt'' S(t'-t'') G^{(1)}(r,t',r,t'') . \tag{11.11}$$

The function $S(t'-t'')$ is proportional to the square of the matrix element M in the expression Eq. (11.7). Its time dependence is determined by the frequency dependence of the counter sensitivity. For the ideal broad-band detector we have been considering, the sensitivity $S(t'-t'')$ will reduce in effect to the delta function $s\delta(t'-t'')$, where s is a constant which describes the sensitivity of the device. We then have

$$p^{(1)}(t) = s \int_0^t dt' G^{(1)}(\boldsymbol{r},t',\boldsymbol{r},t') \tag{11.12}$$

and the counting rate is just

$$w^{(1)}(t) = \frac{dp^{(1)}(t)}{dt} = sG^{(1)}(\boldsymbol{r},t,\boldsymbol{r},t),$$

which is equivalent to Eq. (11.9).

We can consider more complicated counting experiments, in which we determine the correlations in the response of several counters placed in the radiation field. Let us consider n ideal broad-band counters, each equipped with a shutter which couples the k-th counter to the radiation field at time t_0, and removes the coupling at time t_k. The probability of having each counter record precisely one photon during the time it is coupled to the field can be shown in n-th order perturbation theory to be[2]

$$p^{(n)}(t_1,\ldots,t_n) = s^n \int_{t_0}^{t_1} dt'_1 \ldots \int_{t_0}^{t_n} dt'_n$$
$$\times \text{Tr}\left\{\rho E^{(-)}(\boldsymbol{r}_1,t'_1)\ldots E^{(-)}(\boldsymbol{r}_n,t'_n) E^{(+)}(\boldsymbol{r}_n,t'_n)\ldots E^{(+)}(\boldsymbol{r}_1,t'_1)\right\},$$

in which $\boldsymbol{r}_1\ldots \boldsymbol{r}_n$ represent the positions of the counters. If we introduce the n-th order correlation function[1]

$$G^{(n)}(x_1,\ldots,x_n,y_1,\ldots,y_n) = \text{Tr}\left\{\rho E^{(-)}(x_1)\ldots E^{(-)}(x_n) E^{(+)}(y_1)\ldots E^{(+)}(y_n)\right\}, \tag{11.13}$$

we can write the n-fold joint counting rate, $w^{(n)}(t_1\ldots t_n)$, as

$$w^{(n)}(t_1\ldots t_n) = \frac{\partial^n p^{(n)}(t_1\ldots t_n)}{\partial t_1 \ldots \partial t_n} = s^n G^{(n)}(\boldsymbol{r}_1 t_1 \ldots \boldsymbol{r}_n t_n, \boldsymbol{r}_n t_n \ldots \boldsymbol{r}_1 t_1). \tag{11.14}$$

In a typical field the joint counting rate $w^{(n)}(t_1\ldots t_n)$ will be altogether different from the product $\prod_{j=1}^n w^{(1)}(t_j)$ of the counting rates for the n counters each recording individually in the absence of the others. Since we ordinarily have

$$G^{(n)}(x_1,\ldots,x_n,x_n,\ldots,x_1) \neq \prod_{j=1}^n G^{(1)}(x_j,x_j),$$

there will be a certain tendency for the counting events which occur in the different detectors to occur at correlated times. It is only for special classes of fields that the joint counting rates will factorize, in general, and we shall speak of these fields as possessing certain new types of coherence properties. Before discussing these further let us recall the meaning of the most familiar type of optical coherence.

The patterns of fringes observed in optical interference experiments, may be described as having a certain degree of contrast or visibility. When this contrast is a maximum, the fields which are superposed are said to be coherent[3]. It is not difficult to show that the condition of maximum visibility for the fringes is equivalent to the requirement that

$$\left|G^{(1)}(x_1,x_2)\right|^2 = G^{(1)}(x_1,x_1)\,G^{(1)}(x_2,x_2)\,, \tag{11.15}$$

which is the largest value that the modulus of the cross-correlation function can actually have, according to the Schwarz inequality.

It is obvious that this condition for full coherence between the fields at points x_1 and x_2 is satisfied when $G^{(1)}$ factorizes into the form

$$G^{(1)}(x_1,x_1) = \mathcal{E}^*(x_1)\,\mathcal{E}(x_2) \tag{11.16}$$

for some suitable function $\mathcal{E}(x)$ which contains only positive-frequency components. Thus, factorization of the correlation function is a sufficient condition for coherence. It is not too difficult to show[4] that factorization of the correlation function is also a necessary condition for ordinary optical coherence. Thus Eq. (11.16) is completely equivalent to Eq. (11.15).

Since it is the first-order correlation function which is involved in these statements, we shall call the equivalent conditions Eq. (11.15) and (11.16) the conditions for first-order coherence of the field. Although first-order coherence imposes very stringent conditions on the density operator for the fields[5], electromagnetic fields which satisfy these conditions quite accurately in large space-time volumes are no longer difficult to generate in practice.

The conditions Eq. (11.15) and (11.16) have an obvious generalization for the higher-order correlation functions. We shall say that a field exhibits n-th order coherence if and only if

$$G^{(m)}(x_1\ldots x_m, y_1\ldots y_m) = \mathcal{E}^*(x_1)\ldots\mathcal{E}^*(x_m)\,\mathcal{E}(y_1)\ldots\mathcal{E}(y_m)\,, \tag{11.17}$$

for all points $x_1\ldots x_m$, $y_1\ldots y_m$, and for all $m \leq n$.

It can be shown[4] that fields which exhibit first-order coherence must have correlation functions of the form

$$G^{(n)}(x_1\ldots x_n, y_1\ldots y_n) = g_n\,\mathcal{E}^*(x_1)\ldots\mathcal{E}^*(x_n)\,\mathcal{E}(y_1)\ldots\mathcal{E}(y_n)\,, \tag{11.18}$$

where the g_n are a set of nonnegative constants and $g_1 = 1$, of course. The conditions for higher-order coherence are thus simply equivalent to

$$g_n = 1 \quad \text{for} \quad n \leq m\,. \tag{11.19}$$

11.5
Other Correlation Functions

The correlation functions $G^{(n)}$ are by no means the only ones that we can define for the radiation field or even the only ones that can in principle be measured. We can, for instance, define a more general set of functions

$$G^{(n,m)}(x_1\ldots x_n, y_1\ldots y_m) = \text{Tr}\left\{\rho E^{(-)}(x_1)\ldots E^{(-)}(x_n) E^{(+)}(y_1)\ldots E^{(+)}(y_m)\right\}.$$
(11.20)

However, these functions will usually vanish for $n \neq m$ unless rather special conditions prevail. Let us examine these conditions.

Usually, our knowledge of the fields remains stationary in time, so that the density operator (which is always fixed in the Heisenberg picture) actually commutes with the field Hamiltonian H. For such a stationary density operator, any expectation value of the form $\text{Tr}\{\rho A(t_1) B(t_2)\ldots Q(t_n)\}$ must satisfy the relation

$$\text{Tr}\{\rho A(t_1+\tau) B(t_2+\tau)\ldots Q(t_n+\tau)\}$$
$$= \text{Tr}\{\rho \exp[-iH\tau/\hbar] A(t_1) B(t_2)\ldots Q(t_n) \exp[iH\tau/\hbar]\}$$
(11.21)
$$= \text{Tr}\{\rho A(t_1) B(t_2)\ldots Q(t_n)\},$$

i.e., it must remain unchanged under time displacements.

Let us suppose that the field has Fourier components at the frequencies $\omega_1, \omega_2\ldots$ etc. Then we can deduce from Eq. (11.21) that $G^{(m,n)}$ can take on nonvanishing values only if among the frequencies present it is possible to form two sets, an n-fold set $\{\omega_i\}$ and an m-fold set $\{\omega_j\}$ such that

$$\sum_{i=1}^{n} \omega_i - \sum_{j=1}^{m} \omega_j = 0,$$
(11.22)

and further if the field amplitudes at these frequencies are suitably correlated in their statistical behavior (e.g. the density operator ρ must not factorize into independent density operators for these modes or else $G^{(m,n)}$ will vanish).

It is clear from the frequency condition Eq. (11.22) for example that $G^{(1,0)}$ will always vanish in stationary fields which have no static ($\omega = 0$) components. For larger values of n and m the conditions that must be satisfied to have $G^{(n,m)}$ different from zero for $n \neq m$ in stationary fields are only easy to achieve, an a rule, in nonlinear optical experiments. For $n = 2$, $m = 1$ for example we must have three modes excited at frequencies ω_1, ω_2 and ω_3 such that

$$\omega_1 + \omega_2 = \omega_3,$$

but that is precisely the condition satisfied by the three coupled modes of a parametric amplifier. For the case $\omega_3 = 2\omega_1 = 2\omega_2$ it is the condition satisfied in second harmonic generation of light. To have $G^{(2,1)} \neq 0$ we must furthermore require in these examples that there be phase correlations between the mode excitations at the different frequencies. The necessary phase correlations are in fact supplied dynamically in the parametric amplifier or in harmonic generation by the nonlinear mechanism which couples the modes.

Since $G^{(2,1)}$ can be different from zero in a stationary field, it is interesting to ask how the function can be measured. Let us imagine a harmonic generation experiment in which a light beam of frequency ω_1 is incident upon a nonlinear medium. We assume that two superposed beams leave the medium, the fundamental beam with propagation vector $\mathbf{k}_1$ and frequency ω_1, and the second harmonic beam with frequency $2\omega_1$ and propagation vector $\mathbf{k}_2$, which is fairly close in value to $2\mathbf{k}_1$. Let us now introduce into these superposed beams a photon counter which is insensitive to photons of frequency ω_1 or lower. We can do this by using for the photocathode of the detector a substance with a work function somewhat larger than $\hbar\omega_1$. Our counter will therefore freely detect the photons in the harmonic beam at frequency $2\omega_1$.

While the photons of the intense beam at frequency ω_1 cannot affect the counter individually they can nonetheless make it register by acting in pairs to liberate photoelectrons. Since the fundamental and harmonic beams have fixed phase relations, the two different varieties of photoabsorption processes, absorption of single second-harmonic photons and pairs of fundamental ones will in fact interfere with one another. The interference of the amplitudes for single- and double-photon absorption is described by the function $G^{(2,1)}$. If the plane of the photocathode is oriented perpendicular to the vector $\mathbf{k}_2 - 2\mathbf{k}_1$ and moved along this direction, the interference term in the photoemission probability will oscillate as the photocathode position changes. The amplitude of this interference oscillation will be proportional to the real part of $G^{(2,1)}$.

For coherent fields the condition that our knowledge be stationary leads to the condition most familiarly associated with first-order coherence: that the fields be monochromatic. If we are to have both

$$G^{(1)}(t_1,t_2) = G^{(1)}(t_1+\tau,t_2+\tau)$$

and

$$G^{(1)}(t_1,t_2) = \mathcal{E}^*(t_1)\,\mathcal{E}(t_2)\,,$$

then we obtain a functional equation for $\mathcal{E}(t)$ whose only solution is

$$\mathcal{E}(t) = \mathcal{E}(0)\,e^{-i\omega t}$$

for some frequency $\omega \geq 0$. Conversely, any field that contains only a single excited mode will exhibit first-order coherence. Single-mode excitations, however, need not have higher-order coherence.

11.6
The Coherent States

For our further discussion of the electromagnetic field, it will be convenient to introduce a set of pure states which have the property that the fields they describe are ones for which all the correlation functions factorize. We have called such fields fully coherent, and will therefore call these special states coherent states of the field[6].

Since the measured field operators we will be dealing with, such as the products of the negative- and positive-frequency parts of the electromagnetic fields that occur in $G^{(n,m)}$ are normally ordered, states which are eigenstates of the positive-frequency part of the field will be particularly useful. In fact, if we can find a state $|\ \rangle$ for which

$$E^{(+)}(x)|\ \rangle = \mathcal{E}(x)|\ \rangle,$$

it is easy to see that all the correlation functions, which are simply expectation values in this state, will indeed factorize.

Since the positive-frequency electric field contains photon annihilation operators, we can interpret the eigenvalue condition as saying that removing a photon from the field leaves the field state completely unchanged. This remarkable property of the coherent states shows immediately that they cannot be described in terms of any fixed number of photons.

To find explicit forms for the coherent states it is convenient to expand the field in terms of the normal modes appropriate to the quantization volume. Let us consider a set of vector mode functions $\boldsymbol{u}_k(\boldsymbol{r})$ for the field which are orthonormal,

$$\int \boldsymbol{u}_k^*(\boldsymbol{r}) \cdot \boldsymbol{u}_{k'}(\boldsymbol{r}) \, d\boldsymbol{r} = \delta_{kk'},$$

and which satisfy the Helmholtz equation,

$$\nabla^2 \boldsymbol{u}_k(\boldsymbol{r}) + \frac{1}{c^2}\omega_k^2 \boldsymbol{u}_k(\boldsymbol{r}) = 0,$$

and a transversality condition $\nabla \cdot \boldsymbol{u}_k(\boldsymbol{r}) = 0$. We can then write the operator and c-number fields as expansions of the form

$$\boldsymbol{E}^{(+)}(\boldsymbol{r},t) = i\sum_k \sqrt{\frac{\hbar\omega_k}{2}} a_k \boldsymbol{u}_k(\boldsymbol{r}) \exp[-i\omega_k t], \tag{11.23}$$

$$\boldsymbol{\mathcal{E}}(\boldsymbol{r},t) = i\sum_k \sqrt{\frac{\hbar\omega_k}{2}} \alpha_k \boldsymbol{u}_k(\boldsymbol{r}) \exp[-i\omega_k t]. \tag{11.24}$$

(We have used Heaviside–Lorentz units, i.e. rationalized Gaussian units, in which, for instance, Gauss' law is $\nabla \cdot \boldsymbol{E} = \rho$.)

The operators a_k and their adjoints satisfy the commutation relations

$$[a_k, a_{k'}^\dagger] = \delta_{kk'}, \tag{11.25a}$$

$$[a_k, a_{k'}] = 0 = [a_k^\dagger, a_{k'}^\dagger]. \tag{11.25b}$$

11.6 The Coherent States

It is clear that if the coherent states are eigenstates of the field $E^{(+)}$ corresponding to the eigenvalue $\mathcal{E}$, then they will also be eigenstates of all the operators a_k, corresponding to eigenvalues α_k, and we will therefore denote them by $|\{\alpha_k\}\rangle$. We can, in fact, express $|\{\alpha_k\}\rangle$ as a direct product of states referring to the individual modes:

$$|\{\alpha_k\}\rangle = \prod_k |\alpha_k\rangle_k , \tag{11.26}$$

with

$$a_k |\alpha_k\rangle_k = \alpha_k |\alpha_k\rangle_k , \tag{11.27}$$

so that, of course,

$$a_{k'} |\{\alpha_k\}\rangle = \alpha_{k'} |\{\alpha_k\}\rangle \tag{11.28}$$

holds for all k'. We shall confine ourselves here to cataloging briefly some of the more important properties of the states $|\{\alpha_k\}\rangle$. A more complete discussion will be found in Ref.[6,7]. Let us consider for simplicity only single-mode states, and let us forget about the subscripts k. Then we have states $|\alpha\rangle$ with the property $a|\alpha\rangle = \alpha|\alpha\rangle$; the dual states $\langle\alpha|$ are eigenvectors of $a^\dagger$, with $\langle\alpha|a^\dagger = \alpha^*\langle\alpha|$. Because of the non-Hermitian nature of the operators a it is impossible to construct eigenbras $\langle\beta|$ of a or eigenkets of $a^\dagger$. The states $|\alpha\rangle$ can, of course, be expressed in terms of the n-quantum states, which we define as

$$|n\rangle = \frac{1}{\sqrt{n!}} (a^\dagger)^n |0\rangle , \tag{11.29}$$

where $|0\rangle$ is the ground state of the mode. By using the commutation relations and the definition Eq. (11.27) it is easy to show that

$$|\alpha\rangle = \exp\left[-\tfrac{1}{2}|\alpha|^2\right] \sum_{n=0}^\infty \frac{\alpha^n}{\sqrt{n!}} |n\rangle . \tag{11.30}$$

The probability of having n photons in a coherent state $|\alpha\rangle$ is just a Poisson distribution:

$$p(n) = \frac{|\alpha|^{2n}}{n!} \exp\left[-|\alpha|^2\right] , \tag{11.31}$$

whose mean is $\langle n \rangle = |\alpha|^2$.

By expressing the states $|n\rangle$ of Eq. (11.30) in terms of the vacuum state, we can see that the coherent states are just displaced forms of the vacuum state. If we introduce the unitary operator

$$D(\alpha) = \exp\left[\alpha a^\dagger - \alpha^* a\right] , \tag{11.32}$$

which simply displaces the operators a and $a^\dagger$ by the complex numbers α and α^* respectively, we find that we can write the coherent state $|\alpha\rangle$ as

$$|\alpha\rangle = D(\alpha)|0\rangle . \tag{11.33}$$

This characterization of the states is, in fact, implicitly the one used by Schrödinger when he first mentioned them in discussing the classical behavior of the quantized harmonic oscillator. The coherent states have localized wavepackets both in position and momentum space, and the variances of these conjugate variables obey the minimum uncertainty condition,

$$(\Delta p)^2 (\Delta q)^2 = \tfrac{1}{4}\hbar^2 .$$

(This minimum condition defines a more general family of states which are described by Kennard wavepackets. The coherent states are the members of this class for which $(\Delta p)^2 = \omega^2 (\Delta q)^2$.) The classical limit is, of course, approached only when $|\alpha| \gg 1$.

The states $|\alpha\rangle$ are not orthogonal, since

$$\langle \alpha | \beta \rangle = \exp\left[\alpha^*\beta - \tfrac{1}{2}|\alpha|^2 - \tfrac{1}{2}|\beta|^2\right] . \tag{11.34}$$

However, as one can see from the fact that

$$|\langle \alpha | \beta \rangle|^2 = \exp\left[-|\alpha - \beta|^2\right] , \tag{11.35}$$

the overlap of any two states falls off rapidly with the separation of their amplitudes in the complex plane. If $|\alpha\rangle$ and $|\beta\rangle$ are two different classical states, for example, ($|\alpha| \gg 1$, $|\beta| \gg 1$) they will be almost completely orthogonal in general unless α and β happen nearly to coincide.

The coherent states are, however, complete. We can express their completeness by the relationship

$$\int |\alpha\rangle\langle\alpha|\, \pi^{-1}\, d^2\alpha = \mathbf{1} , \tag{11.36}$$

where $\mathbf{1}$ is the unit operator and $d^2\alpha = d(\operatorname{Re}\alpha)\, d(\operatorname{Im}\alpha)$ is a real element of area in the complex plane. This completeness property of the coherent states can be shown to follow immediately from Eq. (11.30) and the completeness of the states $|n\rangle$. Since a coherent state $|\alpha\rangle$ exists for every complex number α, while the states $|n\rangle$ form only a denumerably infinite set, we may well suspect that the set of coherent states is a larger one than we need for completeness. In fact, Cahill has shown[8] that any set of distinct complex eigenvalues α having a limit point in the finite α plane defines a complete set of coherent states. In practice, however, there seems to be no advantage at all to discarding any members of the overcomplete set.

11.7
Expansions in Terms of Coherent States

It is possible, in view of the completeness of the coherent states, to express any state or any operator in terms of them. Thus, for example, for any state $|\psi\rangle$ we may use the completeness relation Eq. (11.36) to construct the expansion

$$|\psi\rangle = \int \frac{d^2\alpha}{\pi} \langle\alpha|\psi\rangle |\alpha\rangle . \tag{11.37}$$

It is not difficult to show that $\langle\alpha|\psi\rangle$ can be written in the form

$$\psi(\alpha^*) \exp[-\tfrac{1}{2}|\alpha|^2] ,$$

where $\psi(\alpha^*)$ is an entire function of α^*. (The exponential factor has been separated from $\psi(\alpha^*)$ since $|\alpha|^2 = \alpha^*\alpha$ is not an analytic function of α^*.) The correspondence between the analytic functions of α^* and the states of the harmonic oscillator has been studied by Bargmann[9]. In a similar fashion we can express any operator T as

$$\begin{aligned}T &= \int \frac{d^2\alpha}{\pi} \frac{d^2\beta}{\pi} |\alpha\rangle \langle\alpha|T|\beta\rangle \langle\beta| \\ &= \int \frac{d^2\alpha}{\pi} \frac{d^2\beta}{\pi} T(\alpha^*,\beta) \exp\left[-\tfrac{1}{2}|\alpha|^2 - \tfrac{1}{2}|\beta|^2\right] |\alpha\rangle \langle\beta| ,\end{aligned} \tag{11.38}$$

where again we have separated the matrix element into the exponential factors and an analytic function of both α^* and β.

We will be particularly interested of course in the density operator for the radiation field. For a single field mode we can express it in the form of Eq. (11.38) as follows:

$$\rho = \int \frac{d^2\alpha}{\pi} \frac{d^2\beta}{\pi} R(\alpha^*,\beta) \exp\left[-\tfrac{1}{2}|\alpha|^2 - \tfrac{1}{2}|\beta|^2\right] |\alpha\rangle \langle\beta| . \tag{11.39}$$

The analytic function R must satisfy some fairly stringent conditions, since ρ is positive definite and has unit trace.

The representation of the density operator in terms of the coherent states makes the computation of expectation values of normally ordered operators particularly convenient. In all such expectation values the field operators may simply be replaced by their eigenvalues. For example, the correlation function $G^{(n)}$ is given by the expression

$$G^{(n)}(x_1 \ldots x_n, y_1 \ldots y_n) =$$
$$\int \mathcal{E}^*(x_1,\{\alpha_k\}) \ldots \mathcal{E}^*(x_n,\{\alpha_k\}) \mathcal{E}(y_1,\{\beta_k\}) \ldots \mathcal{E}(y_n,\{\beta_k\}) R(\{\alpha_k^*\},\{\beta_k\})$$
$$\times \exp\left[-\sum_k |\alpha_k|^2 - \sum_k |\beta_k|^2\right] \prod_k \left(\frac{d^2\alpha_k}{\pi}\right) \prod_k \left(\frac{d^2\beta_k}{\pi}\right) , \tag{11.40}$$

in which we have generalized the notation to include the many-mode case, and have written

$$\mathcal{E}(x, \{\alpha_k\}) = i \sum_k \sqrt{\frac{\hbar \omega}{2}} \alpha_k u_k(\boldsymbol{r}) \exp[i\omega_k t] \tag{11.41}$$

for the eigenvalues of the positive-frequency field.

The representation of Eq. (11.39) is perfectly general, and can be used to specify any state of the field. For many fields it is possible to find a considerably simpler representation for the density operator as an expansion in terms of the coherent-state projection operators $|\alpha\rangle\langle\alpha|$ rather than the more general dyadics $|\alpha\rangle\langle\beta|$. We may, in other words, seek a representation in the general form[6,10],

$$\rho = \int d^2\alpha\, P(\alpha) |\alpha\rangle\langle\alpha|, \tag{11.42}$$

where $P(\alpha)$ is a real-valued weight function. We have called this the P representation for the density operator. The question of how broad the range of applicability of this representation is has been the subject of considerable discussion in the literature[11,12]. The most direct approach to the question is to try to solve for the function $P(\alpha)$.

To find the weight function $P(\alpha)$ it is convenient to introduce the function

$$\chi_N(\lambda) = \mathrm{Tr}\left\{\rho \exp[\lambda a^\dagger] \exp[-\lambda^* a]\right\}, \tag{11.43}$$

which we shall call the normally-ordered characteristic function[12]. By using the P representation to evaluate $\chi_N(\lambda)$ we find

$$\chi_N(\lambda) = \int d^2\alpha\, P(\alpha) \exp[\lambda \alpha^* - \lambda^* \alpha], \tag{11.44}$$

and by writing out the integrand of this expression in Cartesian coordinates we see that it expresses χ_N simply as a two-dimensional Fourier transform of the function P. If the Fourier transform can be inverted the solution for $P(\alpha)$ is given simply by

$$P(\alpha) = \int \frac{d^2\lambda}{\pi} \chi_N(\lambda) \exp[\lambda^* \alpha - \lambda \alpha^*]. \tag{11.45}$$

Since $\chi_N(\lambda)$ can be computed directly from our knowledge of the density operator, Eq. (11.45) can be used to determine $P(\alpha)$ whenever the function $\chi_N(\lambda)$ possesses a two-dimensional Fourier transform. Now χ_N is in general an extremely well-behaved function throughout the finite λ plane. It is bounded in fact by $|\chi_N(\lambda)| \leq \exp[\frac{1}{2}|\lambda|^2]$ but as this inequality indicates it can be a function which diverges quite rapidly for $\lambda \to \infty$. The Fourier integral Eq. (11.45) loses meaning when $|\chi_N(\lambda)|$ increases in this way; it diverges at all points in the α plane. Unfortunately there is nothing at all unphysical about states for which $|\chi_N|$ has this rapid increase. It is easy to show, for example, that for minimum uncertainty states other than the coherent states $|\chi_N(\lambda)|$ will in general diverge as $\exp[c(\operatorname{Re}\lambda)^2]$ or $\exp[c(\operatorname{Im}\lambda)^2]$ where $c > 0$. For such states there does not appear to be any way of defining a reasonably behaved weight function $P(\alpha)$. Since the difficulty is connected with the behavior of χ_N for large λ it tends to arise primarily for states which are characterized by small variances for the quantities which can be constructed as linear combinations of a and $a^\dagger$, e.g. position and momentum.

The attempts which have been made, chiefly by Sudarshan and Klauder, to provide a basis for using the P representation universally have resorted to more formal procedures. Physicists are already familiar with the class of mathematical objects known as tempered distributions (e.g. the delta function and its derivatives), and these are the basis of a well known way in which the Fourier integral can be generalized. Unfortunately the generalization which is needed in order to give universal meaning to Eq. (11.45) would require that we go a great deal further and work with distributions vastly more singular than tempered distributions. While it appears that this route is open, at least in a formal sense, the rules for doing calculations with such singular structures remain unclear. The only suggestion which has been made to date requires an infinite limiting process to evaluate each matrix element of ρ. This difficulty is of course completely absent when other means are used to represent the density operator. The general representation Eq. (11.39), in particular, offers most of the same convenience that the P representation does in evaluating the averages of normally ordered products of field operators and encounters no singularity problems of any sort.

11.8
A Few General Observations

Our ultimate aim is to be able to predict the results of a variety of measurements which can be made on the electromagnetic field. By using the correlation functions we have defined earlier as the basis for doing this, we are in effect splitting all of our prediction problems into two parts. The first part concerns the way in which the field is radiated. The device, whatever it may be, which generates the field brings it to a state which can only be described in statistical terms in general, by means of a density operator ρ. Our first question is: how do we find ρ?

If we are somehow given knowledge of the operator q, then we can turn to the second part of the prediction problem which is: how do we describe a specific set of measurements upon the field? For the simple sorts of measurements which we can actually carry out, the second part of the problem is typically a good deal simpler than the first. For the case of photon counting experiments, for example, a fairly accurate and comprehensive theory of what is measured can be constructed by beginning with the perturbation-theoretical expressions for many-atom transition probabilities which we discussed in Sect. 11.4.

Of course the problem of predicting the response of an actual photon counter is rather different from that of dealing with only a few photosensitive atoms at a time. For a macroscopic detector, one must find the probabilities that arbitrary numbers of atoms within the great collection which make up its sensitive element undergo detected transitions and that the remaining atoms do not do so. The solution to this problem[2] is fairly familiar by now. The results it leads to are discussed at length by Arecchi, Haus and Pike in their lectures at this school. So rather than repeat the familiar approach to the photon-counting problem we shall try later in these lectures to present a somewhat novel one.

Before going further with the subject of measurement, however, let us return to the problem which preceded it. How do we know what density operator ρ characterizes the field produced by an actual source? For a large class of sources whose output is statistically chaotic (including black-body sources, for example), the problem has a simple solution, but these are hardly the most interesting sources. For the kinds of sources which do behave in interesting ways the problem of finding the density operator can be an imposing one and quite a few of the lectures of our school are therefore devoted to it.

Since we have no way of guessing in general what density operators ought to look like, about all we can do to find them is to construct as simple a theoretical model as we can for a given type of source and then to solve explicitly for the time-dependent statistical properties of the field it radiates. For a certain class of devices which function as amplifiers, attenuators and frequency converters, the equations of motion for the field amplitudes may be cast in linear form, and as a result, the density operators for the fields radiated can be solved for in simple closed forms. When the fundamental equations of motion become nonlinear however, as they necessarily do in the case of the laser, the problem of finding the density operator assumes greater proportions. A number of the lectures of our school have been devoted to this problem; it is discussed from various standpoints by Scully, Haken, Weidlich, Louisell and Gordon.

To give a concrete illustration of what is involved in solving for a density operator, we shall discuss the quantum-mechanical problem of field attenuation. The fact that the field is damped means in effect that each of its modes is a damped harmonic oscillator. The model of the linearly damped harmonic oscillator has been a fundamental one for ages in classical mechanics. We have only recently begun to see how to treat the damped oscillator quantum-mechanically, however, and so the problem possesses a good deal of interest in its own right. We shall find, furthermore, that our discussion of the problem is quite useful in developing a refined theory of photon counting measurements, which we shall do toward the end of these lectures.

11.9
The Damped Harmonic Oscillator

Let us consider a single mode of the electromagnetic field, say a mode in a cavity with reflecting walls. We know that the amplitude of the mode may be regarded as the coordinate of a harmonic oscillator. If the walls of the cavity are perfectly reflecting, the oscillator is uncoupled to any other dynamical system. Once its oscillation is started, it will continue forever.

In practice, of course, the walls of a cavity are never perfectly reflecting. What happens is that the amplitude of the field mode tends to decrease fairly rapidly while at the same time the walls of the cavity become slightly heated. If alternatively the field mode were completely unexcited, i.e. at temperature absolute zero, then the walls, if they were at a finite temperature, would tend to warm up the field. These are clear-cut examples of processes which are irreversible in the thermodynamic sense; developing a macroscopic theory of damping means at the very least coming to grips with the problem of irreversibility.

To sum up the problem of irreversibility in a few words, we may say that the equations of motion which describe events occurring on a microscopic scale are intrinsically reversible, in sharp contrast to the equations which describe, for example, such large scale phenomena as the damping noted earlier. The broad outlines of the solution to this puzzle have been known since the times of Gibbs and Ehrenfest. It is only in a rather idealized limit, the limit in which the number of degrees of freedom of a system becomes infinite, that its behavior becomes truly irreversible. Systems with a large but finite number of degrees of freedom tend to exhibit extremely close approximations to irreversible behavior over enormous lengths of time, and it is this behavior which we are quite content to identify as irreversible in practice.

Unfortunately, there are not many eases in which this interpretation of irreversibility can be illustrated in mathematically concrete terms. To do so requires that we be able to solve for the motion of a system with infinitely many degrees of freedom and that is an imposing task for any but the simplest systems. It is a perfectly possible one for many kinds of systems of coupled harmonic oscillators, however, and the realization that such simple systems can illustrate irreversibility has led recently to a great many studies of soluble models of this type[13,14].

The discussion of damping which I shall present has been worked out in collaboration with Arecchi. It, too, is a coupled oscillator model, but one which is particularly simple and appropriate for the discussion of damping problems. It is a quantum-mechanical model which can be analyzed in very close correspondence with classical theory by making use of the coherent states.

We assume that the harmonic oscillator which interests us most, for example the one which represents the cavity mode mentioned earlier, is coupled linearly to each member of a large collection of independent oscillators which constitute both the damping mechanism and a "heat bath". We shall refer to the oscillator which interests us as the A oscillator and denote its annihilation and creation operators by $a(t)$ and $a^\dagger(t)$. The remaining oscillators which make up the heat bath we shall refer to as the B oscillators and denote their annihilation and creation operators by $b_k(t)$ and $b_k^\dagger(t)$, $k = 1, 2 \ldots N$. We take the frequency of the A oscillator to be ω and the frequencies of the B oscillators to be ω_k. Then following Gordon, Walker and Louisell[14], we take the Hamiltonian for the coupled oscillators to be

$$H = \hbar\omega\, a^\dagger a + \sum_k \hbar\omega_k b_k^\dagger b_k + \hbar \sum_k [\lambda_k a^\dagger b_k + \lambda_k^* b_k^\dagger a]\,, \tag{11.46}$$

where the λ_k are a set of coupling constants which may in general be complex. Two points are worth noting about this Hamiltonian. The fact that we have coupled the A oscillator only to other oscillators may seem to be an unrealistic element in the model. The walls of our resonant cavity, for example, need not be made of harmonic oscillators. But as we shall note in the last lecture, large systems which are not by any means made up of harmonic oscillators quite frequently possess modes of collective excitation whose amplitudes behave dynamically like oscillator amplitudes. The Hamiltonian Eq. (11.46) can represent a broad class of damping mechanisms involving collective excitations.

The second point worth noting about the Hamiltonian is that its coupling contains no terms of the form $a\, b_k$ or $b_k^\dagger a^\dagger$. Various physical coupling mechanisms which one can imagine do include such antiresonant terms, but their dynamical effect is typically rather small because of their rapid oscillation. They are, in fact, usually neglected by making the so-called rotating-wave approximation.

The Heisenberg equations of motion

$$\dot{a} = \frac{1}{i\hbar}[a,H]\,, \qquad \dot{b}_k = \frac{1}{i\hbar}[b_k,H] \tag{11.47}$$

for our model take the form

$$\dot{a} = -i\omega a - i\sum_k \lambda_k b_k\,, \tag{11.48}$$

$$\dot{b}_k = -i\omega_k b_k - i\lambda_k^* a\,. \tag{11.49}$$

Since all of the operators in these equations commute with one another at any given time, the equations can be integrated as if they were equations for classical quantities rather than operators. Eventually we will have to deal fairly explicitly with the solution to these equations, but for the moment let us just note that it must take the general form

$$a(t) = a(0)\, u(t) + \sum_k b_k(0)\, v_k(t)\,, \tag{11.50}$$

$$b_k(t) = \sum_{k'} b_{k'}(0)\, x_{k'k}(t) + a(0)\, y_k(t)\,, \tag{11.51}$$

where the functions $u(t)$, $v_k(t)$, $x_{k'k}(t)$ and $y_k(t)$ satisfy the initial conditions

$$u(0) = 1\,, \qquad v_k(0) = 0\,, \tag{11.52}$$

$$x_{k'k}(0) = \delta_{k'k}\,, \qquad y_k(0) = 0\,. \tag{11.53}$$

11.9 The Damped Harmonic Oscillator

The simplicity of the Hamiltonian Eq. (11.46) becomes clearest when we express the states of the system in terms of coherent states. This Hamiltonian is one of a class[15] which preserves the coherence of initially coherent states. Thus if we begin the motion of the system with the A oscillator and all of the B oscillators in coherent states, they will remain in coherent states at all later times. To verify this property, let us assume that the state of the system in the Heisenberg picture is the direct product of coherent states

$$|\alpha\rangle \prod_k |\beta_k\rangle_k \equiv |\alpha, \{\beta_k\}\rangle , \tag{11.54}$$

which is defined to have the properties

$$a(0)|\alpha, \{\beta_k\}\rangle = \alpha |\alpha, \{\beta_k\}\rangle , \tag{11.55}$$

$$b_k(0)|\alpha, \{\beta_k\}\rangle = \beta_k |\alpha, \{\beta_k\}\rangle . \tag{11.56}$$

Then it is clear from Eq. (11.50) that this state is also an eigenstate of $a(t)$ and the $b_k(t)$ for all times t. Let us define the time-dependent functions

$$\alpha(t) = \alpha u(t) + \sum_k \beta_k v_k(t) , \tag{11.57}$$

$$\beta_k(t) = \sum_{k'} \beta_{k'} x_{k'k}(t) + \alpha y_k(t) , \tag{11.58}$$

that obey the c-number differential equations

$$\dot{\alpha}(t) = -i\omega \alpha(t) - i\sum_k \lambda_k \beta_k(t) , \tag{11.59}$$

$$\dot{\beta}_k(t) = -i\omega_k \beta_k(t) - i\lambda_k^* \alpha(t) , \tag{11.60}$$

which are closely analogous to the operator equations (11.48) and (11.49). Then the Heisenberg state for the system obeys the eigenstate conditions

$$a(t)|\alpha, \{\beta_k\}\rangle = \alpha(t)|\alpha, \{\beta_k\}\rangle , \tag{11.61}$$

$$b_k(t)|\alpha, \{\beta_k\}\rangle = \beta_k(t)|\alpha, \{\beta_k\}\rangle \tag{11.62}$$

at all times t.

To relate these results to the Schrödinger picture, let us note that there exists in general a unitary operator $U(t)$ which relates the time-dependent operators of the Heisenberg picture to their initial values, e.g.

$$a(t) = U^{-1}(t) a(0) U(t) . \tag{11.63}$$

The initial values of the operators, $a(0)$ and $b_k(0)$, can be taken to be their forms in the Schrödinger picture. The Eqns. (11.61) and (11.62) are therefore equivalent to

$$a(0) U(t) |\alpha, \{\beta_k\}\rangle = \alpha(t) U(t) |\alpha, \{\beta_k\}\rangle , \tag{11.64}$$

$$b_k(0) U(t) |\alpha, \{\beta_k\}\rangle = \beta_k(t) U(t) |\alpha, \{\beta_k\}\rangle . \tag{11.65}$$

The state $U(t)|\alpha, \{\beta_k\}\rangle$ is simply the time-dependent Schrödinger state for the system and we see that when it begins as a coherent state ($U(0) = 1$), it remains a coherent state at all later times[15]. We are able, in other words, to find the Schrödinger state of our system (apart from an unimportant phase factor) without having to solve the Schrödinger equation at all. This is true because the motion of the system is, in effect, as nearly classical as it can be. The system is described by a minimum-uncertainty wavepacket which is simply carried along the classical trajectories given by Eqns. (11.57) and (11.58) and no spreading of the packet takes place.

Because the Schrödinger state for the system is a coherent state, we can label it by means of its time-dependent eigenvalues by writing

$$|\alpha(t), \{\beta_k(t)\}\rangle \equiv U(t)|\alpha, \{\beta_k\}\rangle . \tag{11.66}$$

It is clear from Eqns. (11.52), (11.53), (11.57) and (11.58) that $\alpha(0) = \alpha$ and $\beta_k(0) = \beta_k$.

11.10
The Density Operator for the Damped Oscillator

In order to make an optimally precise specification of the state of the A oscillator, let us assume that it is initially in the pure coherent state $|\alpha\rangle$. There would be no point, however, in attempting so precise a specification of the states of the oscillators in the heat bath. All we know about the elements of a heat bath in general is that they are in highly random, or mixed states. We can discuss a broad class of examples by assuming that the B oscillators are initially all in chaotic states. In a chaotic state the partial density operator ρ_k for the k-th heat-bath oscillator is given by the Gaussian P representation[6]

$$\rho_k = \frac{1}{\pi \langle n_k \rangle} \int \exp\left[\frac{-|\beta_k|^2}{\langle n_k \rangle}\right] |\beta_k\rangle_k {}_k\langle\beta_k| \, d^2\beta_k , \tag{11.67}$$

in which $\langle n_k \rangle$ is the mean occupation number for the mode. If the heat bath is initially at a specified temperature T, then the mean occupation numbers are given by the Planck formula

$$\langle n_k \rangle = \left\{\exp\left[\frac{\hbar \omega_k}{\kappa T}\right] - 1\right\}^{-1} , \tag{11.68}$$

in which κ is the Boltzmann constant.

If we let $\rho(t)$ be the time-dependent density operator for the system in the Schrödinger picture, then, for the initial state we have specified, the initial value of $\rho(t)$ is the direct product

$$\rho(0) = |\alpha\rangle\langle\alpha| \prod_k \rho_k = \int |\alpha, \{\beta_k\}\rangle \langle\alpha, \{\beta_k\}| \prod_k \exp\left[\frac{-|\beta_k|^2}{\langle n_k \rangle}\right] \frac{d^2\beta_k}{\pi \langle n_k \rangle} . \tag{11.69}$$

Now since we know that the state which evolves from each pure state $|\alpha\{\beta_k\}\rangle$ of the system is the coherent state given by Eq. (11.66), we can solve the equation of motion for the density operator, the quantum-mechanical Liouville equation, simply by inspection. The time-dependent density operator which evolves from $\rho(0)$ can only be given by

$$\rho(t) = \int |\alpha(t), \{\beta_k(t)\}\rangle \langle \alpha(t), \{\beta_k(t)\}| \prod_k \exp\left[\frac{-|\beta_k|^2}{\langle n_k \rangle}\right] \frac{d^2\beta_k}{\pi \langle n_k \rangle} . \quad (11.70)$$

Since the functions $\alpha(t)$ and $\beta_k(t)$ depend linearly on the initial parameters $\{\beta_k\}$, we see that the density operator does not factorize in general for $t \neq 0$; the coupling of all the oscillators removes the statistical independence which their states are taken to possess initially.

Since we are interested mainly in the behavior of the A oscillator and not the heat bath, it is convenient to define a reduced density operator $\rho_A(t)$ for that oscillator by taking the trace of $\rho(t)$ over the variables which describe the heat bath:

$$\rho_A(t) \equiv \text{Tr}_B \, \rho(t) . \quad (11.71)$$

The reduced density operator which corresponds to Eq. (11.70) is just

$$\rho_A(t) = \int |\alpha(t)\rangle \langle \alpha(t)| \prod_k \exp\left[\frac{-|\beta_k|^2}{\langle n_k \rangle}\right] \frac{d^2\beta_k}{\pi \langle n_k \rangle} , \quad (11.72)$$

in which the coherent states $|\alpha(t)\rangle$ refer only to the A oscillator. By noting the explicit form of $\alpha(t)$ given by Eq. (11.57), we can write this operator as

$$\rho_A(t) = \int |\alpha u(t) + \sum \beta_k v_k(t)\rangle \langle \alpha u(t) + \sum \beta_k v_k(t)| \prod_k \exp\left[\frac{-|\beta_k|^2}{\langle n_k \rangle}\right] \frac{d^2\beta_k}{\pi \langle n_k \rangle} . \quad (11.73)$$

It is convenient at this point to introduce the two-dimensional delta function

$$\delta^{(2)}(\zeta) = \delta(\text{Re}\,\zeta)\,\delta(\text{Im}\,\zeta) , \quad (11.74)$$

and use it to write Eq. (11.74) in the form

$$\rho_A(t) = \int |\alpha u(t) + \zeta\rangle \langle \alpha u(t) + \zeta| \, F(\zeta, t)\, d^2\zeta , \quad (11.75)$$

where $F(\zeta, t)$ is given by

$$F(\zeta, t) = \int \delta^{(2)}\left(\zeta - \sum_k \beta_k v_k(t)\right) \prod_k \exp\left[\frac{-|\beta_k|^2}{\langle n_k \rangle}\right] \frac{d^2\beta_k}{\pi \langle n_k \rangle} . \quad (11.76)$$

This Gaussian integral is easily shown to have the value

$$F(\zeta, t) = \frac{1}{\pi \sum_k \langle n_k \rangle |v_k(t)|^2} \exp\left[-\frac{|\zeta|^2}{\sum_k \langle n_k \rangle |v_k(t)|^2}\right] . \quad (11.77)$$

By introducing the variable $\gamma = \zeta + \alpha u(t)$, it is clear from Eq. (11.75) that the reduced density operator can be written as the P representation

$$\rho_A(t) = \int P(\alpha,0|\gamma,t)\,|\gamma\rangle\langle\gamma|\,d^2\gamma, \tag{11.78}$$

in which the weight function is given according to Eq. (11.77) by

$$P(\alpha,0|\gamma,t) = \frac{1}{\pi\sum_k \langle n_k\rangle |v_k(t)|^2}\exp\left[-\frac{|\gamma - \alpha u(t)|^2}{\sum_k \langle n_k\rangle |v_k(t)|^2}\right]. \tag{11.79}$$

We have used for the quasiprobability density P in this case the same sort of notation as is used for ordinary conditioned probability densities; $P(\alpha,0|\gamma,t)$ is the weight associated with the coherent-state amplitude γ at time t when the A oscillator is known to have begun with the amplitude α at $t=0$. Since the values of the functions $v_k(t)$ all vanish for $t \to 0$, we see that P is a delta function at $t=0$. At all later times it is Gaussian in form with its central point following the trajectory $\gamma = \alpha u(t)$.

Let us assume for a moment that the coupling constants λ_k all vanish so that the A oscillator is uncoupled to any others. Then the functions $v_k(t)$ vanish for all t and $u(t) = \exp[-i\omega_a t]$. The function P given by Eq. (11.79) therefore remains a delta function at all times and just moves uniformly around a closed circle in the γ plane. One of the obvious effects of coupling the A oscillator to the chaotically excited oscillators of the heat bath is to broaden its P distribution, or in other words to make its quantum state considerably less certain. To discuss our solution for the density operator in greater detail, we shall in the next Section undertake the explicit evaluation of the functions $u(t)$ and $v_k(t)$ which we have used to formulate the solutions to the equations of motion.

We have assumed to this point that the initial state of the A oscillator is the pure coherent state $|\alpha\rangle$. A much more general class of initial states for the A oscillator is the set of mixtures which can be described by P representations Eq. (11.42) with arbitrary weight functions $P(\alpha,0)$. For such initial states we see, by linearly superposing solutions of the form of Eq. (11.78), that the reduced density operator for the A oscillator has the time-dependent P representation

$$\rho_A(t) = \int P(\gamma,t)\,|\gamma\rangle\langle\gamma|\,d^2\gamma, \tag{11.80}$$

in which

$$P(\gamma,t) = \int P(\alpha,0)\,P(\alpha,0|\gamma,t)\,d^2\alpha. \tag{11.81}$$

If the functions P were classical probability densities we would recognize this relation as the definition of a continuous Markoff process, one which is stationary in time since the properties of the heat bath are time-independent. The fact that the functions P are quasiprobability functions rather than ordinary probability functions means that we are making a natural extension of the notion of a Markoff process into the quantum domain.

11.11
Irreversibility and Damping

It is worth emphasizing that our explicit solution for the density operator has involved no approximations whatever up to this point. That result has, been based on the assumption that we know the solution to the equations of motion Eqns. (11.48) and (11.49), or equally well the solution to the c-number differential equations (11.59) and (11.60). If the number of coupled modes in our problem were not very large it would not be too difficult to solve Eqns. (11.59) and (11.60) numerically, and in that way to retain arbitrarily high accuracy. Since we are explicitly interested in letting the number of heat bath modes become infinite, however, we shall be content with an approximate but highly accurate solution.

Let us first note that the differential equations (11.59) and (11.60)) are quite familiar in form if not in notation to anyone who has studied atomic radiation theory. They take, in fact, the same form as the approximate equations of radiation damping theory[16] which relate the amplitude that an atom remains in an excited state to the amplitudes for the presence of a photon in the various modes of the radiation field with the atom in its ground state. In Eqns. (11.59) and (11.60) $\alpha(t)$ plays (in spite of its rather different physical meaning) the role of the atomic excitation amplitude, the $\beta_k(t)$ play the roles of the single-photon-state amplitudes and the λ_k play the roles of transition matrix elements. If we assume that the frequency spectrum ω_k of the heat-bath oscillators is essentially a continuous one in the neighborhood of the frequency ω, then the mathematical problem of solving the differential equations becomes precisely the one treated by Weisskopf and Wigner in their classic paper[17] on radiation damping.

By using the Laplace transform to construct the exact solution of the Eqns. (11.59) and (11.60) and then making the Weisskopf–Wigner approximation we find that the solutions for the functions $u(t)$ and $v_k(t)$ may be written in the form[14]

$$u(t) = \exp\left[-[\kappa + i(\omega + \delta\omega)]t\right], \tag{11.82}$$

$$v_k(t) = \frac{-i\lambda_k}{\kappa + i(\omega + \delta\omega - \omega_k)} \left\{\exp[-i\omega_k t] - \exp\left[-[\kappa + i(\omega + \delta\omega)]t\right]\right\}, \tag{11.83}$$

where the constants $\delta\omega$ and κ are defined by the relation

$$\delta\omega - i\kappa = \lim_{\epsilon \to +0} \sum_k \frac{|\lambda_k|^2}{\omega - \omega_k + i\epsilon}. \tag{11.84}$$

It is clear from Eqns. (11.82) and (11.83) that κ plays the role of a damping constant while $\delta\omega$ is a frequency shift. The Weisskopf–Wigner approximation requires that these quantities both be small compared to the frequency ω, and compared to the frequency intervals over which the coupling constants λ_k and the parameters describing the states of the heat-bath oscillators change appreciably.

The damped nature of the solutions Eq. (11.82) and (11.83) shows clearly their irreversible character. The element of irreversibility appears to have entered our equations through the Weisskopf–Wigner approximation, but if we look more closely we may see that it is really implied by one of the assumptions on which the approximation rests. The approximation assumes implicitly that the heat-bath oscillators are so numerous that their frequency spectrum is a continuous one. In any practical example, of course, only a finite number of heat bath oscillators is present and their frequency spectrum, however dense, is discrete rather than continuous. The Weisskopf–Wigner approximation therefore deals only with the asymptotic limit of the behavior of the system as the number of heat-bath oscillators becomes infinite. But as we have already noted it is precisely in that limit, and only in that limit, that we can expect to see truly irreversible behavior. The irreversible solutions Eqns. (11.82) and (11.83) are idealized ones in other words, but they must in fact be approached with great accuracy over long periods of time by the solutions for systems with large but finite numbers of degrees of freedom.

Let us assume that we can represent the frequency spectrum $\{\omega_k\}$ of the heat-bath oscillators by means of the spectral density function $g(\omega_k)$. Then by taking the imaginary part of Eq. (11.84) we see that the damping constant may be written as

$$\kappa = \pi \sum_k |\lambda_k|^2 \delta(\omega - \omega_k) = \pi \int |\lambda_k|^2 g(\omega_k) \delta(\omega - \omega_k) d\omega_k = \pi |\lambda_\omega|^2 g(\omega) , \quad (11.85)$$

where λ_ω is the value of the coupling constant for $\omega_k = \omega$. The functions $v_k(t)$ describe the contributions to the excitation of the A oscillator mode by the individual B oscillators. We see from Eq. (11.83) that the functions $|v_k(t)|^2$ are proportional to $[(\omega + \delta\omega - \omega_k)^2 + \kappa^2]^{-1}$. They are in other words sharply peaked at $\omega_k = \omega + \delta\omega$, and fall to small values outside a frequency band of width $\Delta\omega_k \sim 2\kappa$ about this peak.

We have stated in Eq. (11.79) a general expression for the weight function $P(\alpha, 0|\gamma, t)$ of the P representation of the density operator, and we are almost in a position to evaluate it explicitly. To do so we must evaluate the sum $\sum_k \langle n_k \rangle |v_k(t)|^2$ and we can do that by taking advantage of the sharply peaked frequency dependence of $|v_k|^2$. If we assume that the mean occupation numbers $\langle n_k \rangle$ are approximately constant over a frequency band of width $\Delta\omega_k \sim 2\kappa$ about $\omega_k = \omega + \delta\omega$ then we may write

$$\sum_k \langle n_k \rangle |v_k(t)|^2 = \langle n_{\omega'} \rangle \sum_k |v_k(t)|^2 , \quad (11.86)$$

where $\langle n_{\omega'} \rangle$ is the value of the mean occupation number at frequency ω' and we have written $\omega' \equiv \omega + \delta\omega$. To evaluate $\sum_k |v_k(t)|^2$ we may use the solution Eq. (11.83) directly, but it is somewhat simpler to make use of the identity

$$|u(t)|^2 + \sum_k |v_k(t)|^2 = 1 , \quad (11.87)$$

which is easily seen to be an exact consequence of the equations of motion Eqns. (11.59) and (11.60), the definition Eq. (11.57) and the initial conditions

Eq. (11.52). By substituting this relation in Eq. (11.86) and using the solution Eq. (11.82) for $u(t)$ we find

$$\sum_k \langle n_k \rangle |v_k(t)|^2 = \langle n_{\omega'} \rangle \{1 - |u(t)|^2\} = \langle n_{\omega'} \rangle (1 - \exp[-2\kappa t]). \tag{11.88}$$

By using the various expressions we have evaluated we can now see that the weight function P for the damped oscillator given by Eq. (11.79) is

$$P(\alpha, 0|\gamma, t) = \frac{1}{\pi \langle n_{\omega'} \rangle (1 - \exp[-2\kappa t])}$$
$$\times \exp\left[-\frac{|\gamma - \alpha \exp[-(\kappa + i\omega')t]|^2}{\langle n_{\omega'} \rangle (1 - \exp[-2\kappa t])}\right]. \tag{11.89}$$

At any moment of time t the central point of this Gaussian function lies at

$$\gamma = \alpha \exp[-(\kappa + i\omega')t], \tag{11.90}$$

which represents the mean value of γ. As t increases the central point describes a spiral path with a steadily shrinking radius about the origin of the complex γ plane. The shrinking of the radius is of course the effect of damping.

At the initial instant $t = 0$, the variance of the Gaussian distribution Eq. (11.89) vanishes and so, as we have already noted,

$$P(\alpha, 0|\gamma, 0) = \delta^{(2)}(\gamma - \alpha). \tag{11.91}$$

At all later times however the variance takes on nonvanishing values, increasing steadily to the limit $\langle n_{\omega'} \rangle$ for times much longer than $(2\kappa)^{-1}$. This steadily increasing uncertainty of the state of the A oscillator is due to the cumulative effect of the fluctuating forces exerted on it by the heat-bath oscillators. For times $t \gg (2\kappa)^{-1}$ the A oscillator loses all recollection of its initial state, and its weight function takes on the stationary form

$$P(\alpha, 0|\gamma, \infty) = \frac{1}{\pi \langle n_{\omega'} \rangle} \exp\left[-\frac{|\gamma|^2}{\langle n_{\omega'} \rangle}\right], \tag{11.92}$$

which is identical to that of the heat-bath oscillators at the effective frequency $\omega' = \omega + \delta\omega$. The final state of the A oscillator, in other words, is just the equilibrium state described by the equipartition theorem.

11.12
The Fokker–Planck and Bloch Equations

The complex amplitude γ for the A oscillator has, in addition to its tendency to rotate with angular velocity ω', a tendency to move in a more or less random way toward the origin. The random motion of γ may be thought of as a kind of two-dimensional diffusion process in the complex plane. Indeed the function $P(\alpha,0|\gamma,t)$ has, according to Eqns. (11.81) and (11.91), at least two of the properties of a Green's function for a diffusion process. The question is: Is there any simple differential equation which governs the diffusion process?

As a matter of fact the solution Eq. (11.89) for $P(\alpha,0|\gamma,t)$ does obey a simple second-order differential equation, one which can be derived most easily in a coordinate system which rotates uniformly with angular velocity ω'. If we let

$$\gamma' = \gamma \exp[i\omega't] \tag{11.93}$$

then it is easy to verify that $P(\alpha,0|\gamma,t)$ obeys the equation

$$\frac{\partial}{\partial(2\kappa t)}P = \langle n_{\omega'}\rangle \frac{\partial^2}{\partial \gamma' \partial(\gamma'^*)}P + \frac{1}{2}\left(\gamma'\frac{\partial}{\partial \gamma'} + \gamma'^*\frac{\partial}{\partial \gamma'^*}\right)P + P. \tag{11.94}$$

This equation may be recognized somewhat more easily if we use vector notation to represent points in the γ plane, i.e. if we let the point $\gamma' = \gamma_1' + i\gamma_2'$ be represented by the vector $\boldsymbol{\gamma}' = (\gamma_1',\gamma_2')$. It then assumes the form

$$\frac{\partial}{\partial t}P = \tfrac{1}{2}\kappa\langle n_{\omega'}\rangle \nabla_{\gamma'}^2 P + \kappa \nabla_{\gamma'}\cdot(\boldsymbol{\gamma}'P), \tag{11.95}$$

which is a variant of the heat-diffusion equation known as the Fokker–Planck equation. The solution for P expressed by Eq. (11.89), when referred to the rotating coordinate system, is indeed its Green's function.

If we let $\langle n_{\omega'}\rangle = 0$, which corresponds to having the heat bath at temperature $T=0$, we see that the term in Eq. (11.95) which is linear in $\nabla_{\gamma'}$ describes the damped aspect of the motion of $\boldsymbol{\gamma}'$. The solution P given by Eq. (11.89) in that case remains a delta function at all times; the A oscillator remains in a pure coherent state with exponentially decreasing amplitude. For $\langle n_{\omega'}\rangle \neq 0$, on the other hand, the term quadratic in $\nabla_{\gamma'}$ clearly describes the diffusive aspect of the motion of $\boldsymbol{\gamma}'$ which is due to the random excitation of the heat bath.

By showing that the quasiprobability density P obeys the Fokker–Planck equation we have made contact with a considerable body of classical literature on the damped oscillator[18]. Quantum-mechanical studies of the statistics of the damped oscillator, on the other hand, have not until very recently begun to use the quasiprobability formulation. Instead they have proceeded, for the most part, via the direct construction of differential equations for the density operator itself.

We know, of course, of one differential equation of first order in $\partial/\partial t$, the Liouville equation, which the density operator must satisfy exactly. The density operator which satisfies it, however, is the one which describes the entire coupled system; the Liouville equation is, in effect, simply another way of stating the Heisenberg equations of motion Eq. (11.47). What we want instead is a closed equation of some sort for the reduced density operator for the one part of the entire system which interests us, the A oscillator. An analogous problem was considered by Bloch a number of years ago in connection with magnetic resonance experiments on nuclei embedded in dense matter. Bloch considered each nuclear spin to be interacting with a species of heat bath and used perturbation theory to construct a differential equation for the reduced density operator for the spin orientation[19]. A closely analogous procedure can be used to construct a differential equation for the reduced density operator for the damped oscillator. Bather than take this approach, however, we shall use the analysis we've already carried out to reach a somewhat deeper understanding of the equation.

What we shall do is to find a fairly explicit form for the reduced density operator as a function of the operators a and $a^\dagger$. We can do that on the basis of information already at hand by using a simple theorem which has been noted by Louisell and Lax[20]. Let us suppose that the reduced density operator ρ_A possesses a well-defined power-series expansion in the set of antinormally ordered products $a^m (a^\dagger)^n$ for all $m, n = 0, 1 \ldots$; i.e. we can write

$$\rho_A = \sum_{n,m} d_{nm} a^m (a^\dagger)^n \tag{11.96}$$

$$\equiv \rho_{\text{ant}}(a, a^\dagger), \tag{11.97}$$

where ρ_{ant} stands for a particular function of a and $a^\dagger$ which possesses the power series expansion given on the line above.

By using the resolution of the unit operator given in Eq. (11.36) we may now write

$$a^m (a^\dagger)^n = \frac{a^m}{\pi} \int |\alpha\rangle \langle\alpha| d^2\alpha (a^\dagger)^n = \frac{1}{\pi} \int \alpha^m (\alpha^*)^n |\alpha\rangle \langle\alpha| d^2\alpha. \tag{11.98}$$

If the operator ρ_A, therefore, is expanded as in Eq. (11.96) we must have

$$\rho_A = \frac{1}{\pi} \int \rho_{\text{ant}}(\alpha, \alpha^*) |\alpha\rangle \langle\alpha| d^2\alpha. \tag{11.99}$$

We see, in other words that the function $\pi^{-1}\rho_{\text{ant}}(\alpha, \alpha^*)$ is the weight function of the P representation[20];

$$P(\alpha) = \frac{1}{\pi} \rho_{\text{ant}}(\alpha, \alpha^*). \tag{11.100}$$

If we know the antinormally ordered form of ρ_A we can immediately construct its P representation, and conversely by knowing the P representation, as we do for the damped oscillator, we can construct an antinormally ordered form for ρ_A. Since the function P given by Eq. (11.89) is Gaussian for $t > 0$, the function $\pi^{-1}\rho_{\text{ant}}(\alpha, \alpha^*)$ clearly possesses a convergent power-series expansion. (It may be worth noting that the use of the antinormally ordered expression for ρ_A is not superior in generality to the approach we have used in Sect. 11.7 for the definition of the P representation. Cases in which the function P is highly singular are ones in which ρ_A does not possess a well-defined expansion in antinormal order.)

To make use of the function $P(\alpha,0|\gamma,t)$ given by Eq. (11.89), let us rewrite it as

$$P(\alpha,0|\gamma,t) = \frac{1}{\pi\bar{n}(t)} \exp\left[-\frac{1}{\bar{n}(t)}|\gamma-\bar{\alpha}(t)|^2\right], \qquad (11.101)$$

where

$$\bar{\alpha}(t) = \alpha \exp[-(\kappa+i\omega')t], \qquad (11.102)$$

$$\bar{n}(t) = \langle n_{\omega'}\rangle (1-\exp[-2\kappa t]). \qquad (11.103)$$

The reduced density operator ρ_A is then given according to Eqns. (11.100) and (11.97) by the expression

$$\rho_A(t) = \frac{1}{\pi\bar{n}(t)}\left\{\exp\left[-\frac{1}{\bar{n}(t)}[a-\bar{\alpha}(t)][a^\dagger-\bar{\alpha}^*(t)]\right]\right\}_{\text{ant}}, \qquad (11.104)$$

in which the brackets $\{\ \}_{\text{ant}}$ indicate that the function they enclose is to be defined by writing the terms of its power series in antinormal order.

It is not difficult to evaluate the antinormally ordered form of ρ_A in terms of more familiar operator expressions, but that is not at all necessary for the purpose at hand. What we want do to is to find the differential equation which $\rho_A(t)$ satisfies and having it in a form in which its operator arguments are explicitly ordered is a great convenience since that tells us quite directly how to evaluate its derivatives. We see, for example, that derivatives of $\rho_A(t)$ with respect to the parameters $\bar{\alpha}^*$, $\bar{\alpha}$ and $\bar{n}$ must be written as

$$\frac{\partial}{\partial\bar{\alpha}^*}\rho_A = -\frac{1}{\bar{n}(t)}\{a-\bar{\alpha}(t)\}\rho_A \qquad (11.105)$$

$$\frac{\partial}{\partial\bar{\alpha}}\rho_A = -\frac{1}{\bar{n}(t)}\rho_A\{a^\dagger-\bar{\alpha}^*(t)\} \qquad (11.106)$$

and

$$\frac{\partial}{\partial\bar{n}}\rho_A = -\frac{1}{\bar{n}(t)}\rho_A + \frac{1}{[\bar{n}(t)]^2}[a-\bar{\alpha}(t)]\rho_A[a^\dagger-\bar{\alpha}^*(t)]. \qquad (11.107)$$

By carrying out the differentiations in this way it is easy to show that $\rho_A(t)$ as given by Eq. (11.104) obeys the relation

$$\frac{d}{d(2\kappa t)}\rho_A(t) = \tfrac{1}{2}\langle n_{\omega'}\rangle\left\{[a^\dagger,\rho_A a]+[a^\dagger\rho_A,a]\right\} + $$
$$\tfrac{1}{2}(1+\langle n_{\omega'}\rangle)\left\{[a\rho_A,a^\dagger]+[a,\rho_A a^\dagger]\right\}, \qquad (11.108)$$

which is precisely the Bloch equation for the damped oscillator[21]. Bonifacio and Haake, in a recent paper[22], have taken a somewhat more direct route to demonstrate the equivalence of this equation to the Fokker–Planck equation for the function P.

What we have done instead is to show that both equations are equivalent to using the Weisskopf–Wigner approximation and are therefore a good deal more accurate than their perturbation-theoretical origins suggest.

The Bloch equation (11.108) may not be too familiar in structure, but its diagonal matrix elements, at least, could hardly be more familiar. If we take the expectation value of both sides of Eq. (11.108) in the m-quantum state of the A oscillator we find the relation

$$\frac{d}{d(2\kappa t)} \langle m|\rho_A|m\rangle = (1+\langle n_{\omega'}\rangle)(1+m)\langle m+1|\rho_A|m+1\rangle$$
$$+ \langle n_{\omega'}\rangle m \langle m-1|\rho_A|m-1\rangle - \{\langle n_{\omega'}\rangle(1+m) + (1+\langle n_{\omega'}\rangle)m\}\langle m|\rho_A|m\rangle , \tag{11.109}$$

which is variously called the Pauli equation, the rate equation or the master equation for the populations $\langle m|\rho_A|m\rangle$. It is precisely the equation one would construct by using the well-known transition probabilities for the oscillator system (calculated to lowest order in perturbation theory) to describe the way in which the occupation number probabilities change with time. It is clear from the arguments we have given that the rate equation (11.109), though it describes only the diagonal part of the density operator ρ_A, has the same domain of accuracy as the Weisskopf–Wigner approximation.

11.13
Theory of Photodetection.
The Photon Counter Viewed as a Harmonic Oscillator

Let us turn back now the problem of predicting the response of a photon counter, which we mentioned last in Sect. 11.8. That problem may seem quite unrelated to what we have done in Sections 11.9–11.12, but in fact we shall describe a new way of approaching it which leans heavily on our discussion of the damped harmonic oscillator.

To define the photon counting problem in fairly concrete terms, let us assume that a photon counter equipped with a shutter of some sort is placed in a light beam. The shutter is opened at time zero and closed again at time t. We ask then: What is the probability distribution for the random numbers of photons recorded over the interval in which the shutter is open?

We assume that each photon absorbed in the sensitive element of the counter excites some atom in it and that it is the atomic excitations which are ultimately detected by the device. For such a counter it is not difficult in general to construct a Hermitian operator $C(t)$ having integer eigenvalues which represent the possible numbers of photons recorded in time t. To find the probability that any particular photon number is realized it is convenient to define a function Q whose dependence upon a parameter

λ generates these probabilities. We define this generating function as

$$Q(\lambda,t) = \left\langle (1-\lambda)^{C(t)} \right\rangle = \mathrm{Tr}\left\{\rho(1-\lambda)^{C(t)}\right\}, \qquad (11.110)$$

where ρ is the density operator for the system of atoms in the counter. If we know the probabilities $p(n,t)$ that n photons are counted in the interval t then we clearly have

$$Q(\lambda,t) = \sum_{n=0}^{\infty}(1-\lambda)^n p(n,t). \qquad (11.111)$$

If, on the other hand, we can find the generating function Q by some independent means, the probabilities $p(n,t)$ are given by its derivatives at $\lambda = 1$:

$$p(n,t) = \frac{1}{n!}\left(-\frac{d}{d\lambda}\right)^n Q(\lambda,t)\bigg|_{\lambda=1}. \qquad (11.112)$$

The factorial moments of the distribution $p(n,t)$ may be found by differentiating the same generating function at $\lambda = 0$ since

$$\langle C(C-1)\ldots(C-n+1)\rangle = \left(-\frac{d}{d\lambda}\right)^n Q(\lambda,t)\bigg|_{\lambda=0}. \qquad (11.113)$$

Conversely, if we happen to know the factorial moments, we can find the function Q through its power-series expansion about $\lambda = 0$. By continuing the function to $\lambda = 1$ we can then derive the probabilities $p(n,t)$.

That is precisely the method we used to evaluate the distribution $p(n,t)$ in the Les Houches summer school notes[2] of 1964. The key point to the procedure was, of course, the evaluation of the factorial moments. What we showed is that the m-th factorial moment is proportional to a sum, taken over all combinations of m atoms in the detector, of the probabilities that all m atoms in each combination absorb quanta from the light beam in the time interval from 0 to t. We have already shown in Sect. 11.4 how such probabilities are found in m-th order perturbation theory and how they may be written as integrals of the m-th order correlation function for the field. When the factorial moments were evaluated in this way and we then solved for the distribution $p(n,t)$, what we found was that it takes quite generally the form of a compound Poisson distribution, that it to say a weighted average of Poisson distributions. The way in which the set of Poisson distributions is weighted was determined by the P representation of the density operator for the incident light beam.

The mathematical approach we have referred to is fairly simple in practice, and the results it furnishes have been verified experimentally in a number of ways. (Some of the verifications are discussed in the lectures of Arecchi, Haus and Pike.) Still the method is not entirely free from objection. Its dependence on the use of lowest-order perturbation theory to evaluate the factorial moments inevitably raises the question of just how accurate the results can be. Let us therefore try to find another approach to the problem, one which avoids the use of perturbation theory altogether.

11.13 Theory of Photodetection. The Photon Counter Viewed as a Harmonic Oscillator

To simplify that task a bit, let us assume that the atoms which make up the photodetector have only two states, a ground state $|g\rangle$ and an excited state $|e\rangle$. A photon is detected, we shall assume, whenever a photoabsorption process raises an atom from its ground state to its excited state. It is convenient to define a pair of raising and lowering operators $\tau_\pm$ with the properties

$$\tau_+|g\rangle = |e\rangle, \qquad \tau_+|e\rangle = 0,$$
$$\tau_-|g\rangle = 0, \qquad \tau_-|e\rangle = |g\rangle, \tag{11.114}$$

$$\tau_+\tau_- + \tau_-\tau_+ = 1. \tag{11.115}$$

Furthermore, by analogy with the operators used in nuclear physics to represent isobaric spin $\frac{1}{2}$, we can define the operator

$$\tau_3 = [\tau_+, \tau_-], \tag{11.116}$$

which has the properties

$$\tau_3|e\rangle = |e\rangle,$$
$$\tau_3|g\rangle = -|g\rangle. \tag{11.117}$$

Let us suppose now that the sensitive element of the photodetector contains N atoms labeled $j = 1\ldots N$. If we define a set of operators $\tau_{\pm j}$, τ_{3j} for each atom we can write the Hamiltonian for the system of atoms uncoupled to the field as

$$H_0 = \tfrac{1}{2}N\hbar\omega + \tfrac{1}{2}\hbar\omega\sum_j \tau_{3j} \tag{11.118}$$

$$= \hbar\omega\sum_j \tau_{+j}\tau_{-j}, \tag{11.119}$$

where $\hbar\omega$ is the energy difference between the excited and ground states of an atom.

We shall assume, in order to simplify the spatial properties of our model, that the atoms are all located at equivalent positions in the field. (That condition is satisfied for example when a beam of plane waves is normally incident on a plane photocathode). Let us assume furthermore that the incident field has some fixed polarization so that the detector can be regarded as interacting with only one component of the field. If we then write the matrix element for an electric-dipole transition between the two atomic states as M, we can write the interaction Hamiltonian which describes the coupling of the atoms to the field as

$$H' = -eM\left\{\left(\sum_j \tau_{+j}\right)E^{(+)}(\mathbf{r},t) + \left(\sum_j \tau_{-j}\right)E^{(-)}(\mathbf{r},t)\right\}, \tag{11.120}$$

where $E^{(\pm)}$ are the appropriate vector components of the positive- and negative-frequency parts of the field. The first term in the interaction describes the excitation of an atom together with the annihilation of a photon, that is to say a photoabsorption process. The second describes the reverse process, of photoemission. (Note that in constructing the Hamiltonian Eq. (11.119) we have made the so-called "rotating wave" approximation, i.e. we have omitted from the interaction terms containing such products as $\tau_{-j} E^{(+)}$ and $\tau_{+j} E^{(-)}$. These are terms which tend to oscillate quite rapidly with time and to have negligible effect on the real transitions which interest us.)

The interaction Hamiltonian Eq. (11.120) depends on the atomic variables only through the sum $\sum \tau_{+j}$ and its adjoint $\sum \tau_{-j}$. These two operators, we shall show, play a fundamental role in the theory. Their commutation relation, according to Eqns. (11.115) and (11.116) is

$$\left[\sum_j \tau_{-j}, \sum_j \tau_{+j}\right] = -\sum_j \tau_{3j} \tag{11.121}$$

$$= N - 2\sum_j \tau_{+j} \tau_{-j}. \tag{11.122}$$

When any product of the operators $\sum \tau_{-j}$ and $\sum \tau_{+j}$ is applied to the initial-state vector for the atomic system (in which all the atoms are in the ground state), it generates a state in which the excitations are symmetrically distributed among the atoms. There exist, of course, a great many unsymmetrical ways in which the atoms can be excited, but the interaction Eq. (11.120) will never lead to such states. To an excellent approximation, as we shall see shortly, the unsymmetrically excited states of the system can simply be ignored.

Now the symmetrically excited states of a set of identical two-level atoms may not appear to have much physical resemblance to the excited states of a harmonic oscillator, but in fact there can be a considerable mathematical resemblance between them. The all-important point is that the collection of atoms we are considering has a set of equally spaced energy levels just as does a harmonic oscillator. The level spectrum of the atoms is of course bounded from above by $N\hbar\omega$ while that of the oscillator of frequency ω is unbounded, but for values of N which are as large as those we encounter in practice, the difference is not very important. If our model photon counter is anything like the ones which are used in the laboratory, a very small fraction of its atoms ever become excited. The atomic system never becomes excited to any degree even remotely approaching the top of its spectrum and so there is no noticeable error involved in approximating the spectrum as one which rises indefinitely.

What we are suggesting is that the system of atoms can be treated collectively with very little error as a species of harmonic oscillator. If that is so, then it should be possible to consider $\sum \tau_{+j}$ as a creation operator for oscillator excitations and $\sum \tau_{-j}$ as an annihilation operator. Their commutator, however, as we can see from Eqns. (11.121) and (11.122) is not a c-number as it should be for creation and annihilation operators. It is at this point that we can take advantage of the fact that relatively few of the atoms ever become excited. The eigenvalues of the operator $\sum \tau_{+j} \tau_{-j}$ represent the number of excited atoms present in a state. As long as this number remains negligible compared to N we can drop the term $2\sum \tau_{+j} \tau_{-j}$ from the right-hand side of the commutator Eq. (11.122). The operators $N^{-1/2} \sum \tau_{-j}$ and $N^{-1/2} \sum \tau_{+j}$ then obey the

11.13 Theory of Photodetection. The Photon Counter Viewed as a Harmonic Oscillator

approximate commutation relation

$$\left[N^{-1/2} \sum_j \tau_{-j},\ N^{-1/2} \sum_j \tau_{+j} \right] \approx 1, \tag{11.123}$$

which is indeed the commutation relation for annihilation and creation operators.

To complete the formulation of our problem in terms of annihilation and creation operators we must show that the Hamiltonian H_0 of the uncoupled atoms can be expressed in terms of them. In order to do that, it is convenient to introduce creation and annihilation operators which correspond to the unsymmetrical excitations of the atomic system as well as the symmetrical ones. Let us define the set of operators $a_1 \ldots a_N$ as linear combinations of $\tau_{-1} \ldots \tau_{-N}$, by writing

$$a_i = \sum_j U_{ij} \tau_{-j}, \tag{11.124}$$

where the U_{ij} are elements of an $N \times N$ unitary matrix. Then the adjoint operators are given by

$$a_i^\dagger = \sum_j U_{ij}^* \tau_{+j}. \tag{11.125}$$

We shall choose the matrix U so that $U_{1j} = N^{-1/2}$ for $j = 1 \ldots N$. The operators a_1 and $a_1^\dagger$ are chosen in this way to be

$$a_1 = N^{-1/2} \sum_j \tau_{-j}, \tag{11.126}$$

$$a_1^\dagger = N^{-1/2} \sum_j \tau_{+j}, \tag{11.127}$$

which are the operators occurring in the commutation relation Eq. (11.123) and the interaction Eq. (11.120). The matrix elements U_{ij} for $i \neq 1$ can remain unspecified. The commutation relation for the operators a_i and $a_i^\dagger$ is given by

$$[a_i, a_j^\dagger] = \sum_{m,n} U_{im} U_{jn}^* [\tau_{-m}, \tau_{+n}] = -\sum_{m,n} U_{im} U_{nj}^* \delta_{mn} \tau_{3m} = -\sum_m U_{im} U_{mj}^* \tau_{3m}. \tag{11.128}$$

Now, if all of the atoms or nearly all of them are in the ground state we have to a good approximation $\tau_3 = -1$ for each of them. The commutation relation Eq. (11.128) then reduces to the approximate form

$$[a_i, a_j^\dagger] \approx \sum_m U_{im} U_{mj}^* \approx \delta_{ij}, \tag{11.129}$$

which is the commutation relation for a set of linearly independent annihilation and creation operators. The relation Eq. (11.123) which we discussed earlier corresponds to the case $i = j = 1$.

Because of the assumed unitarity of the matrix U, the operators a_i and $a_i^\dagger$ obey the identity

$$\sum_i a_i^\dagger a_i = \sum_j \tau_{+j} \tau_{-j} . \tag{11.130}$$

The Hamiltonian Eq. (11.119) for the uncoupled atoms may therefore be written as

$$H_0 = \hbar\omega \sum_{i=1}^{N} a_i^\dagger a_i , \tag{11.131}$$

which may be regarded, because of the approximate commutation relation Eq. (11.129), as the Hamiltonian for a set of independent harmonic oscillators. We now see from Eq. (11.120) that only one of these oscillators, the one we have labeled with the index 1 is actually coupled to the field. The oscillators labeled by $i = 2 \ldots N$ play only a passive role in our problem; the sum $\sum_{i=2}^{N} a_i^\dagger a_i$ is a constant of the motion of the system which vanishes in the initial state and therefore in all subsequent states as well. To the extent that our approximate commutation relation holds, in other words, we can simply omit the terms $a_i^\dagger a_i$ for $i = 2 \ldots N$ from H_0. The effective Hamiltonian for the atoms and the field can be written then, according to Eqns. (11.120) and (11.126) and (11.127), as

$$H = \hbar\omega a_1^\dagger a_1 - eMN^{1/2} \left\{ a_1^\dagger E^{(+)}(\boldsymbol{r},t) + a_1 E^{(-)}(\boldsymbol{r},t) \right\} + H_\mathrm{f} , \tag{11.132}$$

where H_f is the Hamiltonian for the free field.

Since we have already used the symbols a and $a^\dagger$ to describe the atomic excitations, let us write the quantum annihilation and creation operators for the k-th mode of the electromagnetic field as b_k and $b_k^\dagger$. Then, if we expand the electromagnetic field in terms of the mode functions $\boldsymbol{u}_k$ by writing

$$E^{(+)}(\boldsymbol{r},t) = i \sum_k \left(\frac{\hbar\omega_k}{2} \right)^{1/2} b_k(t) \boldsymbol{u}_k(\boldsymbol{r}) , \tag{11.133}$$

and introduce the parameters λ_k defined by

$$\lambda_k = -ieM \left(\frac{\omega_k N}{2\hbar} \right) \boldsymbol{u}_k(\boldsymbol{r}) , \tag{11.134}$$

we can write the effective Hamiltonian Eq. (11.132) as

$$H = \hbar\omega a_1^\dagger a_1 + \hbar \sum \left\{ \lambda_k a_1^\dagger b_k + \lambda_k^* b_k^\dagger a \right\} + \sum_k \hbar\omega_k b_k^\dagger b_k . \tag{11.135}$$

This Hamiltonian is precisely the one we considered earlier in Eq. (11.46)) as the starting point for our discussion of the damped harmonic oscillator. The photodetector plays the role of the A oscillator while the electromagnetic field is cast in the role of the heat bath.

11.14
The Density Operator for the Photon Counter

Since the sensitive element of our photodetector has approximately the mathematical properties of a harmonic oscillator we can use the coherent states to describe it. We can in fact make use of most of the calculations we carried out for the coupled oscillator system in Sections 11.9–11.11. The only significant change we have to make is to remember that the excitation of the electromagnetic field is not necessarily thermal in character. Its initial density operator need not at all correspond to the chaotic state we assumed for the heat bath.

When the exposure of the photodetector to the light beam begins at $t = 0$, its atoms are all in the ground state. Its initial state is thus the oscillator ground state $|0\rangle$ which is determined by the condition $a_1 |0\rangle = 0$. The initial atomic state, may in other words, be regarded as the coherent state $|\alpha\rangle$ with $\alpha = 0$. The solution to the equation of motion Eq. (11.57) for the amplitude $\alpha(t)$ at later times is therefore

$$\alpha(t) = \sum_k \beta_k v_k(t) . \tag{11.136}$$

Let us assume that the initial density operator for the electromagnetic field is given by the P representation

$$\rho_\beta = \int P(\{\beta_k\}) |\{\beta_k\}\rangle \langle\{\beta_k\}| \prod_k d^2\beta_k . \tag{11.137}$$

In effect the function $P(\{\beta_k\})$ simply replaces a product of Gaussian functions in our earlier calculations. We can now construct the reduced density operator for the atomic system at time t just as we constructed it for the A oscillator in Eq. (11.73). It is given by

$$\rho_A(t) = \int |\alpha(t)\rangle \langle\alpha(t)| P(\{\beta_k\}) \prod_k d^2\beta_k , \tag{11.138}$$

in which the amplitude function $\alpha(t)$ is given by Eq. (11.136).

If we write the n-th quantum state of the A oscillator as $|n\rangle$, then the probability that n atoms are excited at time t is given by

$$\langle n|\rho_A(t)|n\rangle = \int |\langle n|\alpha(t)\rangle|^2 P(\{\beta_k\}) \prod_k d^2\beta_k$$

$$= \int \frac{|\alpha(t)|^{2n}}{n!} \exp[-|\alpha(t)|^2] P(\{\beta_k\}) \prod_k d^2\beta_k . \tag{11.139}$$

When we are dealing with a fully coherent field, e.g. when the function $P(\{\beta_k\})$ is simply a product of delta functions, these probabilities clearly form a Poisson distribution. For more general sorts of fields they are described by a superposition of Poisson distributions which is in fact quite analogous to the one found for the number of photocounts by applying perturbation theory[2].

To make the comparison somewhat more compact let us construct the generating function $Q(\lambda,t)$ which corresponds to the distribution Eq. (11.139). It is given according to Eq. (11.111) by

$$Q(\lambda,t) = \sum_n (1-\lambda)^n \langle n|\rho_A(t)|n\rangle = \int P(\{\beta_k\}) \exp[-\lambda\Omega(t,\{\beta_k\})] \prod_k d^2\beta_k ,$$

(11.140)

where Ω is given by

$$\Omega(t,\{\beta_k\}) = |\alpha(t)|^2 = \left|\sum_k \beta_k v_k(t)\right|^2 ,$$

(11.141)

which is a positive definite quadratic form in the parameters β_k. We already know the functions $v_k(t)$. They are given in the Weisskopf–Wigner approximation by Eq. (11.83).

The physical significance of the function Q can be made more evident if we introduce the eigenvalue field

$$\mathcal{E}(\mathbf{r},t,\{\beta_k\}) = i\sum_k \left(\frac{\hbar\omega_k}{2}\right)^{1/2} \beta_k u_k(\mathbf{r}) \exp[-i\omega_k t] ,$$

(11.142)

since by using Eqns. (11.83) and (11.124) we can then write

$$\alpha(t) = \sum_k \beta_k v_k(t) = \frac{ieMN^{1/2}}{\hbar} \int_0^t \exp[-(\kappa+i\omega')(t-t')] \,\mathcal{E}(\mathbf{r},t',\{\beta_k\}) \, dt' ,$$

(11.143)

in which $\omega' = \omega + \delta\omega$ and the frequency shift $\delta\omega$ and the damping constant κ are defined by Eq. (11.84). The function Ω is therefore given by

$$\Omega(t,\{\beta_k\}) = N\left(\frac{eM}{\hbar}\right)^2 \left|\int_0^t \exp[-(\kappa+i\omega')(t-t')] \,\mathcal{E}(\mathbf{r},t',\{\beta_k\}) \, dt'\right|^2 .$$

(11.144)

If the field were the fully coherent one characterized by a fixed set of amplitudes $\{\beta_k\}$, then Ω would be the mean number of atoms in the detector excited at time t. For fields which are not fully coherent Ω must be averaged over the distribution $P(\{\beta_k\})$ in order to find the mean number of excited atoms.

Two properties of the integral Eq. (11.143) which occurs in the expressions for Ω are worth noting. First, it is clear from the occurrence of the factor $\exp[-i\omega'(t-t')]$ in the integrand that the counter tends to probe the components of the field near the frequency $\omega' = \omega + \delta\omega$ rather than ω. That is because the interaction of each atom

11.14 The Density Operator for the Photon Counter

with the radiation field alters its resonant frequency by $\delta\omega$, the radiative line shift. Second, the presence in the integrand of the damping factor $\exp[-\kappa(t-t')]$ means that our photon counter tends only to be sensitive to the values of the field $\mathcal{E}$ which occur within a time interval $\Delta t \sim \kappa^{-1}$ immediately prior to the time t. That is because the time $(2\kappa)^{-1}$ is the radiative lifetime of the atoms in their excited states. For times $t \gg \kappa^{-1}$ the counter retains little memory of the field values occurring near time zero since most of the atoms which were excited then have radiatively decayed back to their ground states. (The re-emitted photons are, in effect, just resonantly scattered out of the incident beam.)

When the generating function $Q(\lambda,t)$ for a detector consisting of two-level atoms is calculated, on the other hand, by means of the perturbation-theoretical approach we mentioned in Sect. 11.13, we find[2] a result of precisely the same form as Eq. (11.140), but with Ω given by

$$\Omega(t,\{\beta_k\}) = N\left(\frac{eM}{\hbar}\right)^2 \left|\exp[-i\omega(t-t')]\mathcal{E}(\mathbf{r},t',\{\beta_k\})\,dt'\right|^2 \tag{11.145}$$

rather than by Eq. (11.144). Since the perturbation calculation assumes that once an atom is excited it remains excited it should, strictly speaking, only be compared with our more exact calculation for the case of weak damping, $\kappa \ll t^{-1}$. In that case the two results only differ by the inclusion of the radiative line shift in Eq. (11.144). The effect of the line shift is typically quite tiny. Most applications of the formula Eq. (11.145) would include it in any case simply by taking ω to be the observed atomic frequency.

The inconsistency of the perturbation-theoretical approach is illustrated in rather extreme form by the fact that for a steady light beam the mean values of Ω as given by Eq. (11.19) increase indefinitely with time. The number of photons which such a simple model of a counter can record, however, is clearly bounded since it contains a finite number of atoms and no atom is allowed in the perturbation approach to be excited twice. It is also true that the perturbation-theory approach does not account for the way in which the detection process attenuates the field. In the latter connection an approach to the photon counting problem similar to the one we have presented here has been developed independently by Mollow[23].

In all of the photon counting experiments which have been done to date the probability that any given detector atom undergoes a photoabsorption process remains quite small. Furthermore the probability that any photon in the field is eventually detected is also small. The inconsistencies of the perturbation-theoretical approach have therefore not had any quantitative importance thus far. It is interesting none the less to see that they can be removed to an excellent approximation just by taking damping processes and line shifts into account.

References

1. R. J. Glauber, *Phys. Rev.* **130**, 2529 (1963); reprinted as Chapter 1 in this volume.
2. R. J. Glauber, in *Quantum Optics and Electronics, Les Houches 1964*, C. de Witt, A. Blandin, C. Cohen-Tannoudji, Eds., Gordon & Breach, New York 1965, p. 63; reprinted as Chapter 2 in this volume.
3. M. Born, E. Wolf, *Principles of Optics*, London 1959, Chap. X.
4. U. M. Titulaer, R. J. Glauber, *Phys. Rev. B* **140**, 676 (1965); reprinted as Chapter 3 in this volume.
5. U. M. Titulaer, R. J. Glauber, *Phys. Rev.* **145**, 1041 (1966); reprinted as Chapter 4 in this volume.
6. R. J. Glauber, *Phys. Rev.* **131**, 2766 (1963); reprinted as Sect. 2.9 in this volume.
7. R. J. Glauber, in *Fundamental Problems in Statistical Mechanics II*, E. G. D. Cohen, Ed., North-Holland, Amsterdam 1968, p. 140; reprinted as Chapter 8 in this volume.
8. K. E. Cahill, *Phys. Rev. B* **138**, 1566 (1965).
9. V. Bargmann, *Commun. Pure Appl. Math.* **14**, 187 (1961); *Proc. Natl. Acad. Sci. USA* **48**, 199 (1962).
10. R. J. Glauber, *Phys. Rev. Lett.* **10**, 84 (1963).
11. E. C. G. Sudarshan, *Phys. Rev. Lett.* **10**, 277 (1963); D. Holliday, M. Sage, *Phys. Rev. B* **138**, 485 (1965); J. Klauder, J. McKenna, D. Currie, *J. Math. Phys.* **6**, 734 (1965); C. L. Mehta, E. C. G. Sudarshan, *Phys. Rev. B* **138**, 753 (1965); J. Klauder, *Phys. Rev. Lett.* **15**, 534 (1966); R. Bonifacio, L. Narducci, E. Montaldi, *Nuovo Cimento* **47**(S), 890 (1967).
12. R. J. Glauber, in *Physics of Quantum Electronics*, P. L.Kelley, B. Lax, P. E. Tannenwald, Eds., McGraw-Hill, New York 1966, p. 788.
13. P. Mazur, E. Montroll, *J. Math. Phys.* **1**, 70 (1960); Y. Kogure, *J. Phys. Soc. Jpn.* **16**, 14 (1961); **17**, 36 (1962); G. W. Ford, M. Kac, P. Mazur, *J. Math. Phys.* **6**, 504 (1965); P. Ullersma, *Physica* **32**, 27, 56 (1966); A. López, *Z. Phys.* **192**, 63 (1966).
14. P. Gordon, L. R. Walker, W. H. Louisell, *Phys. Rev.* **130**, 806 (1963).
15. R. J. Glauber, *Phys. Lett.* **21**, 650 (1966); reprinted as Chapter 5 in this volume.
16. W. Heitler, *The Quantum Theory of Radiation*, 3rd Ed., Oxford 1954, p. 181.
17. V. Weisskopf, E. Wigner, *Z. Phys.* **63**, 54 (1930).
18. S. Chandrasekhar, *Rev. Mod. Phys.* **15**, 1 (1943); M. C. Wang, G. E. Uhlenbeck, *Rev. Mod. Phys.* **17**, 323 (1945). Both papers are reprinted in *Selected Papers on Noise and Stochastic Processes*, N. Wax, Ed., New York 1954.
19. K. Wangsness, F. Bloch, *Phys. Rev.* **89**, 728 (1953).
20. M. Lax, W. H. Louisell, *J. Quantum Electron.* **QE3**, 47 (1967).
21. W. Weidlich, F. Haake, *Z. Phys.* **185**, 30 (1965).
22. R. Bonifacio, F. Haake, *Z. Phys.* **200**, 526 (1967).
23. B. R. Mollow, private communication.

12
Quantum Theory of Coherence[1]

12.1
Introduction

Quantum electrodynamics is a fairly mature field these days; it is about forty-two years old. The curious thing is that although it has become an immensely accomplished and versatile study as it has developed from optics, there has been very little connection maintained between quantum electrodynamics and optics. As you know, the two studies overlap completely in principle, but until a few years ago the connection between them remained only rudimentary. We shall consider first the question of why that situation existed for so long. The much closer liaison which we have now between these two fields developed at approximately the same time as the laser appeared on the scene, and one of its by-products has been a particularly useful way of describing laser beams. The primary impetus to develop the theory, however, lay in a number of interesting questions about the behavior of photons which dated from earlier years: we shall see below what those questions were.

Perhaps the easiest way to obtain an overall picture of quantum electrodynamics is to think of the electromagnetic spectrum. Just how far we say the spectrum extends toward its low-frequency end is more or less a matter of taste. For example, it could be maintained that we have seen fields oscillating at something like 10 Hz or thereabouts; on the other hand, at the violet end of the spectrum, the quanta become more and more energetic as we go through the regions of X-rays, and γ-rays, etc. The γ-ray spectrum is rather extensive and it is not unlikely that we have seen the effects of γ-ray quanta whose frequencies are of the order of 10^{27} Hz. So we have something like 26 powers of 10 covering the spectrum with the visible band of the spectrum more or less in the middle as we observe it. The vast range of phenomena that radiations of these different frequencies produce, of course, recapitulates the entire history of physics.

Probably the greatest of the accomplishments of present day physics lies in the fact that one single theory explains everything that occurs over the vast range of the electromagnetic spectrum. Exactly the same things are happening, from a mathematical standpoint, in the radio frequency range and in the highest frequency range of the

---

[1] Reprinted, with permission from the author, from *Quantum Optics*, S. M. Kay, A. Maitland (Eds.), p. 53–125; Academic Press, London 1970. This publication had been prepared with a minimum of editing, directly from tape recordings of R. J. Glauber's lectures.

Quantum Theory of Optical Coherence. Selected Papers and Lectures. Roy J. Glauber.
Copyright © 2007 WILEY-VCH Verlag GmbH & Co. KGaA, Weinheim
ISBN: 978-3-527-40687-6

γ-rays produced only in cosmic radiation. It is all one theory; it has to be one theory because of the character of the Lorentz transformation, because we know that quanta of one frequency can be turned into quanta of any other frequency simply by observing them from a moving system. While that does not sound like a very practical thing for people to do (turning radio waves into γ-rays, let us say), it is a very practical thing for electrons; electrons are easily accelerated to the appropriate velocities. In fact one way we have these days of creating a monochromatic γ-ray beam with energies of hundreds of millions of volts is to let ordinary visible quanta collide with electron beams in a linear accelerator. That has been done at Stanford and it does provide a usable γ-ray beam. The electrons thus "see" ordinary light quanta as energetic γ-rays and in colliding with them transfer just a little momentum to them, but that little momentum looks like an awful lot in the laboratory frame, and, indeed, those visible quanta come out at very substantial energies. That ought to be evidence enough that the spectrum is all one unit, and it must be one theory which embraces the whole.

The different physical qualities which characterize the different ends of the spectrum are completely at variance with one another. From a quantum mechanical standpoint, in the low frequency end of the spectrum we have fantastic quantum densities. If we asked, "How many quanta are there, in, for example, a cubic wavelength of radio broadcast radiation?" (of course, the wavelength gets a little large, for that kind), the answer tends to be an astronomical number. It is not unusual to see 10^{30}–10^{40} quanta per cubic wavelength. We have, of course, very efficient sources of such radiation, ones which produce vast numbers of essentially identical quanta. So at the lower frequency end of the spectrum we have lots of feeble-energy quanta. At the other end of the spectrum we have extremely energetic quanta; individual ones can do simply catastrophic things to atomic or nuclear systems. We have no trouble at all detecting such quanta by almost macroscopic means, but we have very very few quanta there at all. Now, in a sense we have still fewer quanta if we compare the number of quanta with how many we could have. The idea of mode density in the field is considered elsewhere in this chapter. It is slightly artificial for optics to consider the field within a finite volume, but it is very convenient to do so. It is convenient because it turns the continuum of oscillation of the field into a discrete set of modes which can be counted. It is a little artificial because we scarcely know what a perfect reflector is for, say, γ-rays at 100 MeV. But let us imagine such a thing just for the sake of defining the modes of the field mathematically. If we do that, the density of the modes per unit frequency interval rises as the square of the frequency and so the modes become enormously dense at very high frequencies.

But as the frequency increases the number of quanta produced by all reasonable sources goes down dramatically. Since nearly all the modes are thus empty there is a sense in which all high frequency quantum beams are exceedingly weak. We are familiar, on the other hand, with quantum beams of enormous intensity at the low frequency end of the spectrum where we are dealing with quantum energies which are exceedingly small compared even to thermal excitation energies. We have to go up to frequencies corresponding to wavelengths of 0.1 mm, or thereabouts, before we get to quantum energies which are comparable even with rather low thermal energies. So, quantum phenomena are masked by thermal noise and are not visible in the low frequency end of the spectrum which thus remains the natural domain of classical theory. We should have a rather hard time indeed persuading a radio frequency engineer that there is any need to deal in quantum mechanical terms with the output of a broadcast station. On the other hand, we could not be more familiar with the use of quantum theory in dealing with the high frequency region, essentially infra-red frequencies and higher, and that has always been the domain of quantum mechanics.

There are two other qualities of these two regions which are worth mentioning. One is the degree of control we have of the radiation. Our knowledge of classical radiation and how to control it in macroscopic terms is about 100 years old or so and we have achieved a great degree of versatility in being able to control radiation of relatively low frequencies. We can make the fields in oscillators do just about anything that the Maxwell equations will allow them to do, as a function both of the space and time variables. But, what happens, crudely speaking, as we proceed to higher frequencies is that we lose more and more control over what the field does. At low frequencies we have temporal control; we are able to produce, for example, quite monochromatic oscillations with steady amplitudes. As the frequency rises we sacrifice that possibility rather quickly. We still have monochromatic radiations available to us in the visible region, but only because of the accident of nature that atoms happen to have very well-defined frequencies of oscillation. As we go on into the X-ray region there is more and more radiation that just comes from the continuum of atomic excitations. Beyond the energies of a few million volts, we have even gone past the discrete levels of the nuclei and we then deal only with continuum radiation. We have virtually no monochromatic sources in the extremely high frequency region, except, perhaps, the strange one I mentioned which involves the Compton scattering of monochromatic visible photons at a particular angle. That is just about the only monochromatic source that has been proposed in the multi-million volt region. In other words, we lose control of the time dependence of the fields as we increase the frequency. We also lose, somewhat more slowly, but none-the-less positively, control of the spatial behavior of the fields. Optics in the visible region is a field in which we are able to maintain very accurate control of the spatial behavior of the field, for example, in the creation of interference patterns and the like. But it is quite difficult at the same time to maintain very accurate control over the time dependence of optical fields. Even though optical fields may be fairly monochromatic their amplitudes tend to fluctuate uncontrollably. We have thus to ask: "How does the field behave? What does the indefiniteness of the frequency of an optical field mean? What kind of time variations of the field are responsible for the line-broadening?" We shall see something very specific about that presently. What we shall find is that all radiation which does not come from specifically stabilized sources, i.e. all natural radiation is, in a fundamental sense, noise. It may have any sort of spectrum but it is what the electrical engineer would simply call colored noise.

The whole science of optics, in a sense, has just consisted of the constructive use of noise.

Quantum phenomena are easily observed in the visible region. They are observed in the sense that a photon with an energy of a few volts is perfectly detectable in a photomultiplier. Not very much work, historically, was done by detecting individual quanta in the visible region; the technique was thought of mainly as a way of detecting rather weak light beams. It was thought of, in other words, only as a means of intensity measurement and there are other and simpler ways of measuring low intensities, for example by integrating over a long period of time using photographic detection. There was, in all, remarkably little interest in quantum detection. That has changed a good deal within the last few years with the entry of a certain amount of technology into the picture – technology and a few very clever ideas.

Historically, the first development based on quantum detection was in an experiment by Hanbury Brown and Twiss in 1956[5]. (To be perfectly precise, it was not quantum counting in the original experiment, it was the examination of fluctuations in the photo-current, but in all the subsequent experiments it has literally been quantum counting.) In counting quanta we can detect something more than just the intensity of the field; we can detect correlations within the field, and out of that fact has grown a whole idea of statistical studies of photons in the field. The techniques of measurement are ones which are carried down from the high frequency end of the spectrum. They are the techniques of high energy physics, a gift, if you like, from the people who have preoccupied themselves with that end of the spectrum. And there is one more great development, which is the development of the laser itself.

The laser radiates in approximately the same way as the classical radiator to which we have been accustomed for a long time at the low frequency end of the spectrum. That is done by exploiting atomic transitions and macroscopic techniques as well. What results, in effect, is a transfer of the ideas and the techniques of the low frequency end of the spectrum upwards in frequency by a factor of perhaps 10^5 or so. The result is that now there is a region of the spectrum extending over the visible frequencies and into the near ultra violet in which we can both generate quanta coherently and detect them quantum mechanically, i.e. with devices which detect the individual photons. Now, lasers are extremely intense sources of essentially identical photons. Such sources were previously available only at radio frequencies but, at optical frequencies, the photons were so sparse that we never had to think of two identical photons as actually being in the field at the same time. That, crudely speaking, is still the case at X-ray frequencies and beyond. There are so many modes available to distribute the photons into, that we simply never have, to any noticeable degree, two photons occupying a single mode of the field at the same time, and so the statistical properties of the fields at X-ray frequencies and beyond are in a deep sense rather trivial – nothing very interesting goes on.

The laser produces almost arbitrarily intense photon beams in the visible region of the spectrum and experiments with these beams can lead to all sorts of quantum populations of the modes of the electromagnetic field. We can use quantum detection to investigate the statistical properties of these populations. That is, in essence, the change that has taken place. It has been necessary, therefore, for theory to respond in a way which permits us to deal fairly easily with many-photon problems. The development of the theory, until a few years ago, really went no further than the use of classical theory in the low frequency domain and at high frequencies quantum mechanics in its simplest form, dealing with one or two photons at one time. There are many reasons why the classical limit of quantum electrodynamics was never very fully developed, even though we had every reason to be confident that quantum electrodynamics has the correct classical limit. The most fundamental of these reasons is the smallness of the number $e^2/(4\varepsilon_0 hc)$. Its smallness means that we can proceed with great accuracy in doing calculations with individual charged particles by a perturbation expansion. In point of fact, it has been very difficult in the past in quantum electrodynamics (q.e.d.) to do anything other than this sort of approximation, but the perturbation expansion becomes vastly more difficult as you proceed from one order to the next. The tendency, therefore, has been to restrict our attention in q.e.d. to problems involving very few quanta at a time, and that is frequently a good approximation. So, if we are dealing with individual electrons, as we often are in this region of the spectrum, it is quite sufficient to say that we have an electron with only one or two photons in the field. If we are terribly ambitious and work for half a lifetime we might do a problem with three or four photons in the field, but scarcely more. There was therefore no great enthusiasm, nor was there a great need for dealing quantum mechanically with problems in which we had vast numbers of photons in the field at the same time. These days, however, we have very little alternative to doing just that in the visible region for the reasons explained above, and that simply means that we must develop other approximations. The approximations cannot be precisely the approximations of familiar perturbation theory. We have no choice but to deal with very considerable numbers of photons at once.

The sort of optics which existed until recently (an immensely successful study) has been largely constructed in classical terms. The description of interference experiments, diffraction experiments, and the like, has usually been constructed in classical terms that were observed to work, and the advent of the quantum theory really lent a certain support to those rather *ad hoc* procedures of calculation in the following sense. It is very easy to calculate diffraction patterns in the one-photon approximation. We say the field has one photon in it and ask for the probability that one photon is seen in a particular position on the screen. Then we have to interpret that in statistical terms; we have to imagine that the one-photon experiment has been done time and again and ask where, statistically, that photon turns out to be. The average density of the photon, the statistical density, then agrees with the classical diffraction pattern and there can be no great surprise about that. Calculating the expectation value of the field intensity in single-photon states can be reduced in fact to much the same mathematical procedure as the classical intensity calculation. All of the calculations that we could do in single photon terms were quickly found thus to agree with the results of the classical calculations and to lend a good deal of support to the notion that the classical procedures were correct.

The kind of experiment we have described gives us a good way of characterizing the older optics, the optics prior to the developments of the past few years. In all of these traditional optical experiments we imagine a limiting process in which the field intensity is allowed to go to zero, but the detection time is allowed to become infinite – in other words, the experiment of Taylor[2] described by Sillitto in his introductory remarks in this volume[1]. All of the traditional optical experiments (perhaps that is a good way of defining tradition) have the quality that we could go to zero intensity and just increase the detection efficiency and we would detect basically the same thing, the same diffraction pattern. Those are one-photon experiments. They are the type of experiment we can describe by saying there is only one photon in the field at a time. But that description does not hold, for example, for the Hanbury Brown–Twiss type of experiment which detects correlations between pairs of photons.

The Hanbury Brown–Twiss experiment is really the first one in the higher frequency range which stays completely outside that tradition. Since that experiment, there have been a great many which step outside the tradition and which cannot be dealt with at all in terms of one-photon optics. To begin with, we can understand in such terms essentially nothing of the functioning of the laser. The laser is intrinsically a nonlinear device; it only works when the field intensities become so high that the photons know a great deal about each other's presence. That could not be less typical of one-photon optics. More generally, the availability of laser beams has made it possible to develop a new area of optics, nonlinear optics. In nonlinear optics, we do dramatic things such as sewing pairs of photons together to make new photons of twice the frequency. Needless to say, we learn nothing of phenomena like that from one-photon optics, and, as a matter of fact, a great deal in the phenomenon of harmonic generation depends upon the tendency of photon packets to overlap, or not to overlap, in any given beam. What we are beginning to suggest is that the description of the electromagnetic field which was used in the traditional optics is really inadequate. While we can describe the one-photon experiments by means of a very small group of parameters, in general, or a very restricted class of correlation functions, to characterize an electromagnetic field more fully, and, in particular, to do it in quantum mechanical terms, we need to know a great deal more.

12.2
Classical Theory

What we shall do now is to develop something of the newer and fuller description of the field that we need in order to deal with this much broader class of phenomena. Let us do this by first making contact with the classical noise theory and let me introduce the notation for the mode functions. Imagine that we are dealing with the field inside an enclosed volume. The specification of the boundary conditions of that volume is really not essential to us but we need some sort of boundary condition in order to complete the specification of the problem. The boundary conditions could be those of a perfectly conducting surface; that, however, condemns us to talking about standing waves and makes it difficult to deal with progressive waves if we are really dealing with a problem in free space in which the very idea of boundaries is unnatural. A simpler thing to do is to use periodic boundary conditions. (All we need agree upon here is the existence of the mode functions; the boundary conditions are really immaterial, that is why they are left unspecified.)

We consider, then, the field inside a closed volume and, for an appropriate set of frequencies ω_k, we introduce a set of vector mode functions $\boldsymbol{u}_k(\boldsymbol{r})$ with mode index k which satisfy the wave equation

$$\left(\nabla^2 + \frac{\omega_k^2}{c^2}\right)\boldsymbol{u}_k(\boldsymbol{r}) = 0, \tag{12.1}$$

the transversality condition

$$\nabla \cdot \boldsymbol{u}_k(\boldsymbol{r}) = 0, \tag{12.2}$$

and the boundary conditions on the electric fields. The set of mode functions are orthogonal if the modes are of different frequency, or they can be made orthogonal if they are part of a degenerate subspace. The orthonormality condition is

$$\int \boldsymbol{u}_k^*(\boldsymbol{r}) \cdot \boldsymbol{u}_l(\boldsymbol{r})\, d\boldsymbol{r} = \delta_{kl}. \tag{12.3}$$

The electric field may be broken into two terms; we shall call them the positive and negative frequency parts and define them by convention thus

$$\boldsymbol{E}(\boldsymbol{r},t) = \boldsymbol{E}^{(+)}(\boldsymbol{r},t) + \boldsymbol{E}^{(-)}(\boldsymbol{r},t), \tag{12.4}$$

where $\boldsymbol{E}^{(+)}$ is the sum of all terms which vary as $\exp(-i\omega_k t)$ for all $\omega_k > 0$. All the positive frequency terms are in $\boldsymbol{E}^{(+)}$ and all the complex conjugate terms are in $\boldsymbol{E}^{(-)}$. Therefore, we have

$$\boldsymbol{E}^{(-)} = (\boldsymbol{E}^{(+)})^*. \tag{12.5}$$

These are classical fields, we are not dealing with operators yet.

Now we expand a component of $\boldsymbol{E}^{(+)}$ in terms of the set of mode functions $\boldsymbol{u}_k$ and obtain

$$\boldsymbol{E}^{(+)}(\boldsymbol{r},t) = \sum_k C_k \boldsymbol{u}_k(\boldsymbol{r}) \exp(-i\omega_k t). \tag{12.6}$$

If there is no source connected to the field, these Fourier coefficients C_k are constants of motion. We shall deal with a field remote from its source; let us say the field is radiated from a distant nebula; we can regard it, to an excellent approximation, as being source-free. The field is then described completely by this set of coefficients (complex numbers) C_k. The question is, "What are they?" If the field is perfectly well defined, if there is no randomness in it whatever, then we know, in principle, precisely what those numbers C_k are. To a communications engineer, a noiseless field would be one for which we know all the complex numbers, C. In practice, there is no such thing as a completely noiseless field, we do not know the numbers C_k, and the most accurate specification of a classical field, in statistical terms, can only be through a specification of the probability distribution of these complex numbers. Let us call the probability distribution P and whenever we are dealing with a sequence of numbers we shall gather them using face brackets. We then have the notation

$$P(C_1, C_2, \ldots) \equiv P(\{C_k\}) . \tag{12.7}$$

The random variables describing the signal, then, are the complex numbers C, and the space over which the signal is defined is the space of the complete set of complex numbers. An element of area in the subspace corresponding to each complex number can be defined in this way. We define $d^2 C_k$ as the product,

$$d^2 C_k = d(\operatorname{Re} C_k) d(\operatorname{Im} C_k) . \tag{12.8}$$

The probability distribution has the following normalization

$$\int P(\{C_k\}) \prod_k d^2 C_k = 1 \tag{12.9}$$

when we integrate over the entire space of the complex Fourier coefficients.

So we are beginning to develop a classical noise theory. What happens when we measure the intensity of the field? The intensity of the field is given by the squared modulus of the positive frequency part (say) of the field. That is the same as multiplying the negative frequency part by the positive frequency part,

$$\left\{ |E^{(+)}(r,t)|^2 \right\}_{av} = \left\{ E^{(-)}(r,t) E^{(+)}(r,t) \right\}_{av} . \tag{12.10}$$

This quantity, however, is a random one, and, of course, if we make any measurement of a random field, we get a random result. To compare with theory, we have to take an ensemble average. This is the meaning of the average we have written. We take the average over the random coefficients C_k and obtain an intensity. When we measure the intensity of a field we are measuring the ensemble average.

More generally, we can use not just one detector to sample the field, but use two different detectors and superpose their output. In that case we measure an interference pattern which contains the interference of the amplitudes which are brought in by those two detectors, classically. The amplitudes superposed in finding the total intensity can correspond to fields at different spatial positions or delay times. It is thus useful to define the function

$$G_{cl}^{(1)}(rt, r't') = \left\{ E^{(-)}(rt) E^{(+)}(r't') \right\}_{av} . \tag{12.11}$$

This is a classical correlation function for the field values at two different space-time points. It becomes the field intensity if r and r' are the same and the times t and t' are the same. The field $E^{(+)}$ is linear in the random amplitude parameters $\{C_k\}$ and $E^{(-)}$ is linear in $\{C_k^*\}$. The ensemble average may also be expressed by

$$G_{\text{cl}}^{(1)}(rt, r't') = \int \mathcal{P}(\{C_k\}) E^{(-)}(rt) E^{(+)}(r't') \prod_k \mathrm{d}^2 C_k \,, \tag{12.12}$$

where the integration is over the entire space of random coefficients. The study of these random functions of time ($E^{(+)}$, $E^{(-)}$) in which the Fourier coefficients have specified probability distributions is really the subject of stochastic processes, or noise theory. Any such random continuous function of the time is called a stochastic process (a continuous stochastic process in this case). There is a vast body of theory about these processes and there is really no need to go into further details of that theory here. We have simply noted a few formulae for eventual comparison with the quantum theory.

12.3
Quantum Theory

In the quantum theory we deal, not with numbers, but with operators. The field is represented by an operator, but that operator can be split into a positive and negative frequency part just as the classical c-number fields, (c-number is a term invented by Dirac for an ordinary number to distinguish ordinary numbers from operators which he called q-numbers.)

$$\boldsymbol{E}(\boldsymbol{r},t) = \boldsymbol{E}^{(+)}(\boldsymbol{r},t) + \boldsymbol{E}^{(-)}(\boldsymbol{r},t) \,. \tag{12.13}$$

We are no longer dealing with c-number fields, we are dealing with Dirac's q-numbers, his operators. The negative frequency part of the field is the Hermitian conjugate of the positive frequency part. This is not simply the complex conjugate, but the transposed complex conjugate, i.e. the operator analog of complex conjugation. Thus, we have

$$\boldsymbol{E}^{(-)} = \left(\boldsymbol{E}^{(+)}\right)^\dagger \tag{12.14}$$

i.e. $\boldsymbol{E}^{(\pm)}$ are mutually adjoint.

We can make the same mode decomposition of the positive frequency part of the field as we did classically. The operator is expanded in terms of the set of mode functions as discussed above, and the role of the classical Fourier coefficients $\{C_k\}$ is taken by a sequence of quantum mechanical amplitude operators. By convention, we write these amplitude operators as $\{a_k\}$ and normalize them so that we have $\boldsymbol{E}^{(+)}$ in the form given by

$$\boldsymbol{E}^{(+)}(\boldsymbol{r},t) = \mathrm{i} \sum_k (\tfrac{1}{2}\hbar\omega)^{1/2} a_k \boldsymbol{u}_k(\boldsymbol{r}) \exp(-\mathrm{i}\omega_k t) \,, \tag{12.15}$$

where $\boldsymbol{u}_k(\boldsymbol{r})$ is a vector mode function. The operators a_k are the annihilation operators, one for each mode of the electromagnetic field. For plane waves, a mode of the electromagnetic field is specified by a propagation vector and a polarization for the field which is embedded in the mode function. For example, for plane wave modes

$$\boldsymbol{u}_k(\boldsymbol{r}) = \hat{\boldsymbol{e}}(\boldsymbol{k}) \exp(i\boldsymbol{k}\cdot\boldsymbol{r})$$

where $\hat{\boldsymbol{e}}(\boldsymbol{k})$ is the polarization vector. The Hermitian conjugates of the annihilation operators a_k are the creation operators $a_k^\dagger$. The normalization factors are inserted to secure the familiar canonical commutation relations between the creation and annihilation operators for the different modes.

$$\begin{aligned}\left[a_k, a_{k'}^\dagger\right] &= \delta_{kk'}\,, \\ \left[a_k, a_{k'}\right] &= \left[a_k^\dagger, a_{k'}^\dagger\right] = 0\,.\end{aligned} \quad (12.16)$$

In other words, the one change which we make in going from the classical to the quantum theory is that the complex coefficients (ordinary numbers) become operators and that these operators obey the familiar commutation relations i.e. have the same algebraic properties as the complex amplitude operators which are used to describe the quantum mechanical form of the simple harmonic oscillator. The commutation relations (12.16) thus define the amplitude operators for an infinite set of oscillators, one for each of the field modes. For harmonic oscillators, it is well known that the products $a_k^\dagger a_k$ have as eigenvalues, the integers $n_k = 0, 1, 2, \ldots$ The integer n_k is just the number of photons in the k-th mode.

It is possible to go through a great deal of theory just constructing equations for operators, but, sooner or later, we have to say something about quantum states. Physical quantities are always scalar products of quantum state vectors. We are concerned with states of the radiation field. The ground state of the field is defined as the state $|\text{vac}\rangle$. Here we have used the notation of writing vacuum explicitly in the vacuum state because that is the state in which there are no quanta in any mode. The vacuum state can also be written as

$$|\text{vac}\rangle = |\{0\}\rangle\,. \quad (12.17)$$

Here we have the set of quantum numbers zero in all modes of the field. That is to say, if we apply an annihilation operator to the vacuum state, we get absolutely nothing because there are no lower energies available; so applying the annihilation operator annihilates the state thus,

$$a_k |\text{vac}\rangle = 0\,. \quad (12.18)$$

We can generate the n-quantum states, $|\{n_k\}\rangle$, by applying the creation operators $a_k^\dagger$ to the vacuum state, thus

$$|\{n_k\}\rangle = \prod_k \frac{(a_k^\dagger)^{n_k}}{\sqrt{(n_k)!}} |\text{vac}\rangle\,, \quad (12.19)$$

where we see that we construct such a state by creating n quanta in the k-th mode, then normalizing, and then doing this for all modes.

We now consider the description of general states of the electromagnetic field. The $|\{n_k\}\rangle$ form a complete orthonormal set (we shall see below some sets which are not normalized and are certainly not orthogonal) spanning the familiar Fock space of quantum field theory. This is a good set of basis vectors with which to describe a general state of the electromagnetic field and, indeed, it was the only one which was ever used in quantum electrodynamical calculations until just a few years ago. Their use was natural enough; in dealing with perturbation theory and never with more than two photons in the field, we might as well confine ourselves to discussion of these particular states, which are very simple in their orthonormality properties, and very simple in algebraic terms. But the most general state of the field, obviously, is the most general superposition of such n-quantum states, and so we have

$$| \rangle = \sum_{\{n_k\}} f_{\{n_k\}} |\{n_k\}\rangle . \tag{12.20}$$

The situation is thus rather different in the quantum mechanical specification of states from the situation in classical theory. The most precise specification in classical theory was achieved by specifying one complex number for every mode of the field. Here this notation indicates a set of coefficients with as many indices as there are modes. If we had one mode only of the field, we should still have an infinite set of coefficients in that specification of a state. For one mode, the state may be expanded as a sum of n-quantum states with arbitrary coefficients summed from $n = 0$ to ∞, thus

$$| \rangle = \sum_{n=0}^{\infty} f_n |n\rangle , \tag{12.21}$$

where the only restriction on the f_n is the normalization condition,

$$\sum_n |f_n|^2 = 1 . \tag{12.22}$$

Thus, we see that, in quantum theory, there is an infinite set of complex numbers which specifies the state of a single mode. This is in contrast to classical theory where each mode may be described by a single complex number. This shows that there is vastly more freedom in the quantum theory to invent states of the world than there is in the classical theory. We cannot think of quantum theory and classical theory in one-to-one terms at all. In quantum theory there exist whole spaces which have no classical analogs, whatever.

The correspondence between classical and quantum theory can only be drawn in an asymptotic sense, and usually only in the familiar sense of the correspondence principle of Bohr, which is to say that as we go to very large quantum numbers (i.e. high intensities), we obtain classical results. Now that is a slightly strange statement, familiar as it is, in view of some of the things we have said earlier, because the close correspondence that was observed long ago between classical theory and quantum electrodynamics in the area of optics is really concerned with one-quantum phenomena. In traditional optics, the light beams have been very weak indeed. Zero point vibrations are very strong compared with the strengths of ordinary light beams. (Zero point vibrations may be observed in such devices as the parametric amplifier.) So we have been dealing with a situation in which the fields are actually quite weak. The quantum numbers are very small, and yet that is a situation in which classical theory is applicable. This is a remarkable situation from a quantum mechanical stand point. We should not expect classical theory to give universally correct answers for situations in which we are dealing with such weak fields. The only way in which quantum theory corresponds to classical theory is, ordinarily, in the limit of large quantum numbers and very strong fields.

12.4
Intensity and Coincidence Measurements

Let us consider an ideal device, an ideal photon counter. An ideal photon counter is meant to be approximately the simplest thing we can imagine. We also simplify the calculations by showing their skeletal characteristics, by leaving out many of the detailed steps and just trying to convey something of the spirit of the calculation. A photodetector is a device in which there is, typically, a transition when a photon is absorbed. That may seem trivial but it is a very deep and a very far reaching characterization of virtually all the photon detection devices that we use. All the practical devices we have, work by absorbing photons from the field and that has some very important consequences for what they detect. In particular, since it is not possible to absorb energy from the vacuum fluctuations, it means, for example, that all of these devices remain completely insensitive to vacuum fluctuations, and, as we mentioned above, that is rather fortunate. As an ideal detector, we shall use a single atom. That is a very good detector, though not a very efficient one, and we might have to work rather hard with the single atom, but the good thing about it is that it is very small compared with the wavelength of visible radiation, and so the chances are very good that we can use the electric dipole approximation to describe a transition. In this approximation, the interaction between the atom and the field is given by

$$H_I = -e\sum_{\gamma} \boldsymbol{q}_{\gamma} \cdot \boldsymbol{E}(\boldsymbol{r},t) \tag{12.23}$$

where $\boldsymbol{q}_{\gamma}$ is the spatial coordinate of the γ-th electron relative to the nucleus located at $\boldsymbol{r}$. Because we are using the electric dipole approximation, it is sufficient to write the electric field operator just at the nuclear position rather than at the position of all the individual electrons. When the atom absorbs a photon, it makes a transition from a certain initial state to a certain final state.

12.4 Intensity and Coincidence Measurements

The transition amplitude is given in first order perturbation theory as

$$A_{fi} = M \langle f | E^{(+)}(\mathbf{r},t) | i \rangle . \tag{12.24}$$

There are several little details we have left out, for example, we have paid no particular attention to the conservation of energy in writing down this transition amplitude and we have lumped all of the factors which enter the transition amplitude into the factor M which is the atomic matrix element for going between the initial and final states, whatever they may be. The initial and final state vectors of the electromagnetic field are $|i\rangle$ and $|f\rangle$, respectively. The operator $E^{(+)}$ is just the positive frequency part of the electric field i.e. $E = E^{(+)} + E^{(-)}$. Here we have just retained the positive frequency part; that is already an approximation we are making. There are two terms in the transition amplitude one of which can easily be resonant and the other one of which never is. The atom, let us say, begins in its ground state, that is in the situation where only absorption takes place. In our detector we begin with very low temperature detectors (all atoms in their ground state). Then all those atoms can do is absorb and therefore only the annihilation operator enters the calculation. That is true because of other properties of the correct expression which we shall come to presently. The conservation of energy is not part of this expression because we have not written down its time integral structure at all. The conservation comes about because we are dealing with functions that oscillate rapidly as a function of time, and as soon as we integrate over a considerable number of cycles of oscillation of the internal motions of the atom and the field, we find we get zero in the integrals unless there is a precise phase matching between the amplitudes of the field and the atomic amplitudes. We shall put in the conservation of energy presently but, for the moment, we do not consider it. It is that which leads us to select only the absorption part (only the annihilation part of the electric field operator) in doing this calculation. If we were dealing with a detection interaction which only lasted one or two atomic cycles, (that would be a ridiculously short time for a detector, 10^{-15} s or so) then, indeed, we would have to take account of both the positive and negative frequency parts of the field. Usually, we deal with detection processes which are a factor of 10^5 or 10^7 times slower, and so the statement that energy is conserved in the transitions involved has at least an accuracy of one part in 10^5 or 10^7, and it is correspondingly that accurate to discard the negative frequency part of the field.

The transition probability (if you are willing to believe that the transition amplitude is of the form Eq. (12.24), and in its mathematical skeleton it certainly is of that form) is given by

$$T_{fi} = \left\{ |M|^2 \sum_f \left| \langle f | E^{(+)}(\mathbf{r},t) | i \rangle \right|^2 \right\}_{\text{av}} \tag{12.25}$$

where the sum is taken over all final states because when we detect the photon we make no measurement of the final state of the field. All possible final states of the field can be summed over, but not all conceivable final states of the field are reached in this transition; no matter how many photons there are in the initial state, only one is taken away by the absorption transition. Since we are not observing the final state, we may as well sum over all of the possible final states. All we have said is that most of them do not contribute to the sum but we may as well throw them all in to carry out the sum (mathematically that is simplest by far). If we do that, we obtain[2]

$$T_{fi} = \left\{ |M|^2 \sum_f \langle i | E^{(-)}(\mathbf{r},t) | f \rangle \langle f | E^{(+)}(\mathbf{r},t) | i \rangle \right\}_{av} \qquad (12.26)$$

Again the summation is taken over final states. However, the completeness relation for the set of all possible final states is that the sum over final states is just the unit operator, thus we have

$$\sum_f |f\rangle\langle f| = 1 . \qquad (12.27)$$

From Eqns. (12.26) and (12.27) we get

$$T_{fi} = \left\{ |M|^2 \langle i | E^{(-)}(\mathbf{r},t) E^{(+)}(\mathbf{r},t) | i \rangle \right\}_{av} . \qquad (12.28)$$

Another way of writing this expression is to think of the outer product (the dyadic product) of the two state vectors as a kind of operator in Hilbert space. We then obtain

$$T_{fi} = |M|^2 \mathrm{Tr}\left\{ \{|i\rangle\langle i|\}_{av} E^{(-)}(\mathbf{r},t) E^{(+)}(\mathbf{r},t) \right\} . \qquad (12.29)$$

Here we have written the two state vectors $|i\rangle$ and $\langle i|$ side by side and we are now dealing with something which has a square array of components if we take matrix elements. The structure of Eq. (12.29) (precisely the expression (12.28)) is such that we are evaluating its trace.

We have assumed here that we know the initial state of the electromagnetic field. However, in practice, we rarely have this information. The devices that radiate are always systems with many particles and the most we ever know about them is their statistical behavior. Thus we are always uncertain of the initial state of the field and the transition probability which we measure is something that is defined only in a statistical sense. We are making a measurement with one atom and we get a particular number for the length of time it takes that atom to make a particular transition in a field; that number will vary from one repetition of the experiment to another. We have to keep repeating the experiment; in other words, we evaluate an ensemble average. That means that we must average Eq. (12.29) over the uncertain initial states of the field. The advantage of the somewhat cumbersome notation of Eq. (12.29) is that we

---

[2] What we have done, is simply to write out the absolute value of the matrix element squared using the familiar theorem that the complex conjugate of a matrix element is the matrix element with the states written in the reverse order of that in which the hermitian conjugate of the operators were taken.

12.4 Intensity and Coincidence Measurements

have isolated the two quantities over which we are averaging. We are averaging the dyadic product.

Let us define the density operator, ρ, by the dyadic product averaged over many preparations of the field. We take whatever system it is which produces the field and let it produce that field time and time again, thus producing a certain ensemble of initial states of the field. We then average the dyadic product over those indefinite initial states of the field and thus we determine the density operator

$$\rho = \{|i\rangle\langle i|\}_{\text{av}}\,.\tag{12.30}$$

The field intensity that we measure is $I(r,t)$ and is given by

$$I(r,t) = \text{Tr}\{\rho E^{(-)}(r,t)E^{(+)}(r,t)\}\,.\tag{12.31}$$

The intensity is the trace of the density operator times the product of the two field operators, the negative frequency part times the positive frequency part. This explicit structure is very important. Notice that the negative frequency part contains all the creation operators, $a_k^\dagger$ when we are quantizing the field and that the positive frequency part contains all the annihilation operators a_k. It follows then that in the vacuum state we get

$$\rho = |\text{vac}\rangle\langle\text{vac}|\,,$$

$$\text{Tr}\left(\rho E^{(-)}E^{(+)}\right) = \left\langle \text{vac}\left|E^{(-)}E^{(+)}\right|\text{vac}\right\rangle = 0\,.\tag{12.32}$$

Thus, in the vacuum state the intensity is zero; this is a consequence of the normal ordering of the product of field operators (i.e. the fact that annihilation operators always stand to the right of creation operators). The vacuum state is virtually the only case in which we ever really have a pure state.

The fact that the intensity in vacuum vanishes is very important in avoiding the kind of confusion which arises when the zero-point energy (the zero-point oscillation) is mixed in with observations. If we write expressions other than the normally ordered one, we inevitably run into problems with the zero-point oscillations.

Let us now examine the structure of the kind of statistical average which measures the intensity of the field. Firstly, we shall abbreviate the coordinates by using x instead of a position and time variable. We can define a first order correlation function of the field by

$$G^{(1)}(x_1,x_2) = \text{Tr}\left\{\rho E^{(-)}(x_1)E^{(+)}(x_2)\right\}\,.\tag{12.33}$$

Here it is defined for two arbitrary space-time points,

$$x_1(r_1,t_1)\,,\qquad x_2(r_2,t_2)\,.$$

Had we done the perturbation calculation more fully, including the effects of energy conservation (i.e., putting the time integrals and everything else in) we should have found for the transition probability (the probability that by time t, a single atom undergoes a photoabsorption process) an expression which looks like this more generally

$$p^{(1)}(t) = \int_0^t dt' \int_0^t dt'' S(t'-t'') G^{(1)}(\mathbf{r}t', \mathbf{r}t''). \tag{12.34}$$

This is still an electric dipole approximation, where S is a sensitivity factor which contains the atomic matrix elements. We should find that the process is not local in the time variable. There are two different times in this expression (t' and t''). The individual transition amplitude is a time integral and when we square the transition amplitude, the probability obtained is a double time integral. Typically, our device has differing sensitivities for various frequencies of light quanta. All the different sensitivities of the device are easily accommodated in this function S, which can take a variety of forms, and which we shall not discuss here in any detail. (A more complete discussion of photon counting probabilities and sensitivity problems is given in the Les Houches notes, 1964.) For the simplest case of a detector which does not discriminate between frequencies, S is a delta function.

For this broadband counter, we have

$$S(t'-t'') \approx s\delta(t'-t''). \tag{12.35}$$

This cannot be a precise statement because the function s has a one-sided frequency spectrum, but it can be an excellent approximation relative to the relaxation times of the actual field, (i.e. relative to the characteristic times of the function $G^{(1)}$). For this case, with a broadband detector[3] the integral is a little simpler. The sensitivity is described by a constant and the double integral becomes a single integral.

Then we can think of the absorption process as being local in time, that is, it involves only one time variable.

$$p^{(1)}(t) = s \int_0^t dt' G^{(1)}(\mathbf{r}t', \mathbf{r}t'). \tag{12.36}$$

The rate at which the photoabsorption takes place is given by

$$w^{(1)}(t) = \frac{dp^{(1)}(t)}{dt} = sG^{(1)}(\mathbf{r}t, \mathbf{r}t)$$

$$= sI(\mathbf{r}t). \tag{12.37}$$

The electric dipole approximation has neglected the size of the atom and for this reason we have evaluated the field at the position of the nucleus. If we took the size of the atom into account, then we really could not use the electric dipole approximation, and we should probably then have written these correlation functions in terms of the

---

[3] A broadband detector is one whose sensitivity as a function of frequency is flatter than the frequency distribution of the radiation to be detected. Typically this radiation is in rather a narrow band.

vector potential rather than the electric field operator, but their structure would be identical. We should find, much more generally, that these transition probabilities involve volume integrals over the correlation functions. If we were taking the size of the atom into account then we would have the function $G^{(1)}(\mathbf{r},\mathbf{r}')$ and we would have to integrate over the variables $\mathbf{r}$ and $\mathbf{r}'$ as well. There would be two more integrals added to the expression and, of course, then there would be the atomic wave functions in the expression. All of these are complications which we really do not need in order to understand the essential points of the theory.

Suppose we have n atoms forming n different broadband detectors with a shutter in front of each atom. We shall open all the shutters at time 0 and let each of them remain open until time t_j where $j = 1, 2, \ldots, n$. The probability, that all n atoms have undergone the photoabsorption transition and the transition has been detected for each of them is given by,

$$p^{(n)}(t_1 \ldots t_n) = s^n \int_0^{t_1} dt'_1 \int_0^{t_2} \ldots \int_0^{t_n} dt'_n$$
$$\times \text{Tr}\left\{\rho E^{(-)}(\mathbf{r}_1, t'_1) \ldots E^{(-)}(\mathbf{r}_n, t'_n) E^{(+)}(\mathbf{r}_n, t'_n) \ldots E^{(+)}(\mathbf{r}_1, t'_1)\right\}. \quad (12.38)$$

This is the sort of structure which can be obtained, formally at least, only in n-th order perturbation theory. But, n-th order perturbation theory is really only an n-fold iteration of first order perturbation theory. There is not really very much added. In other words, to absorb n photons, we require n annihilation operators. We square the matrix element, therefore we have n creation operators (the negative frequency components of the field). We still have the structure we had before in terms of the unknown initial state of the field which has to be averaged over, so we include the density operator. Finally, we have n integrals over n time variables. Each photoabsorption process has been made local in each time variable by the assumption that each atom is a broadband detector. We may define an n-th order correlation function by

$$G^{(n)}(x_1 \ldots x_{2n}) = \text{Tr}\left\{\rho E^{(-)}(x_1) \ldots E^{(-)}(x_n) E^{(+)}(x_{n+1}) \ldots E^{(+)}(x_{2n})\right\} \quad (12.39)$$

Again we note that it is a product of $2n$ field operators written explicitly in normal order. The annihilation operators (n of them) standing to the right, the creation operators (n of them) standing to the left. Now, having defined these functions, there are a great many identities we could prove about them, but we won't need any of them for our present purposes. However, one obvious thing to point out is this. We can define a kind of joint counting rate when we have more than one photon counter in the field by differentiating Eq. (12.39) (the probability that by the succession of times $t_1, \ldots, t_n$ all of the atoms have undergone photoabsorption processes which have been detected) with respect to all of the times $t_1, \ldots, t_n$. If we do that, we turn the n-fold integral into its integrand and obtain

$$w^{(n)}(t_1\ldots t_n) = \frac{\partial^n p^{(n)}(t_1\ldots t_n)}{\partial t_1\ldots \partial t_n}$$

$$= s^n G^{(n)}(r_1 t_1 \ldots r_n t_n, r_n t_n \ldots r_1 t_1). \tag{12.40}$$

The expression is therefore the n-fold correlation function, but notice that the arguments $r_1 t_1 \ldots r_n t_n$ are repeated. This n-th order correlation function is indeed arrived at in perturbation theoretical terms by an n-th order iteration of first order perturbation theory. But, on the other hand, the result is different from simply the product of first-order perturbation theory taken with itself n times. To put it a little differently, the n-th order correlation function, even with its arguments repeated in this way, does not generally factorize. Physically, this non-factorization is caused by interference effects which take place in the absorption of photons by the different counters. We cannot generally tell which photon is absorbed by which counter and all of the interference terms are implicitly summed in the expression we have written involving the trace with the $2n$ factors of the field. The interference terms would all be omitted if we simply took the product of n first-order correlation functions to represent the right-hand side of Eq. (12.40). The first order correlation function with its arguments repeated is a counting rate for a single counter, and what we are saying is that the n-fold joint counting rate does not factorize into the product of the counting rates that we would measure with each of these counters present in the absence of all the others. In other words, there is generally a tendency towards statistical correlation (correlation, of course, including the idea of anti-correlation as well). There is generally a lack of statistical independence in the counting rates of the n counters in the field.

We can imagine using other sorts of detectors, one possibility suggested by Mandel[3] among others, is to use the process of stimulated emission as a basis for detection. We could put into the field a collection of atoms which are not in their ground state but in the excited state. In this case, the sort of argument we have given would still hold but the first thing we would have to do is to emit photons rather than absorb them and so the operators would occur in anti-normal order rather than normal order, thus

$$p^{(n)}(t_1\ldots t_n) = s^n \int_0^{t_1} dt'_1 \ldots \int_0^{t_n} \mathrm{Tr}\left\{\rho E^{(+)}(r_1 t_1)\ldots E^{(-)}(r_1 t_1)\right\}. \tag{12.41}$$

The operators occur in anti-normal order for a detector based on stimulated emission. There are certain practical problems in constructing a detector based on stimulated emission. It is perfectly true that the atoms in an excited state will go to the ground state a little faster when we shine fight upon them, and that in principle is detectable, but most of what we would detect would be just the noise of spontaneous emission. The effect of shining the light on these spontaneously decaying atoms would be rather small so such a detector is not very practical. The point of this argument is simply to emphasize that different operator orderings are associated with different sorts of experiments; it is the experiment which determines the ordering which is convenient. In particular, the stimulated emission device would use anti-normal order. Similarly, we could conceive a device based on scattering. The photons, in being scattered from free charged particles, would make them recoil. Detecting the recoiling particles would be another form of photon detection, a very inefficient one because the cross-sections are small, but one that is used at high energies in the Compton effect detectors. This detector measures a sort of combination which is different again, one which is not at all as simple either as normal ordering or anti-normal ordering. At low frequencies we can probably say that what it detects corresponds more closely to the symmetrical ordering of the annihilation and creation operators. The experiment determines the order. The ordering really does mean something, at least in terms of the convenience of describing particular experiments. Of course, we can always change the orders of operators by using commutation relations but we may increase the complication of the expressions considered by doing that. Typically, for the experiments which have a null result in vacuum, we ought to use normal order; that is the convenient one. For most of what we say, normal ordering will play a particularly important role.

We have not exhausted the possibilities of defining correlation functions. There are many expressions we can define. For example, we can define a new set of correlation functions based on anti-normal order, or any kind of funny order we please. We can, furthermore, define more correlation functions which are based on normal order, and there has been a certain amount of discussion of those, and so let us just write the definition. Here is such a definition,

$$G^{(n,m)}(x_1 \ldots x_n Y_1 \ldots y_m) = \mathrm{Tr}\left\{\rho E^{(-)}(x_1) \ldots E^{(-)}(x_n) E^{(+)}(y_1) \ldots E^{(+)}(y_m)\right\}. \quad (12.42)$$

This function differs in general from the normally ordered correlation functions we have written earlier by including the possibility of having different numbers of annihilation and creation operators. For $n = m$ we have

$$G^{(n,m)} \equiv G^{(n)}. \quad (12.43)$$

Expressions of this sort may be defined, but what we find under most of the conditions we study, is that these additional expressions for $n \neq m$ (i.e. the extra ones gained by the definition Eq. (12.42)) typically vanish. Let us see why this is so.

Most optical experiments are performed in essentially a steady state sort of light beam. A steady stochastic process is called a stationary stochastic process. This does not mean that nothing happens as a function of time, what it means is that all of the statistical statements that we make are independent of the origin of time. That is, our statistical knowledge of whatever time dependent behavior may be going on is invariant under time displacement. That is what stationarity means for us. Now we can be talking about fields which are oscillating very rapidly; there is no question about the presence of the oscillation, but what we know about the oscillations is independent of the time of day. For example, for a stationary field, the correlation function which generally depends on two times, t_1 and t_2, is invariant under time displacement. This means if we add a time τ to both arguments, we get the same function for all times, τ,

$$G^{(1)}(t_1,t_2) = G^{(1)}(t_1+\tau,t_2+\tau)$$
$$= G^{(1)}(t_1-t_2) . \qquad (12.44)$$

That means, in practice, that the correlation function depends only on the difference of the two times. Now, of course, all statistical statements about a stationary field have to be time-independent and that means that there must be a corresponding property for the higher order correlation functions as well. So here, for the general function $G^{(n,m)}$, we may write the same statement

$$G^{(n,m)}(t_1+\tau,\ldots,t_{n+m}+\tau) = G^{(n,m)}(t_1\ldots t_{n+m}) . \qquad (12.45)$$

Now what does that statement mean from a microscopic standpoint? What does it mean about the density operator for the field? Considering any expectation value in the Heisenberg picture, the trace of the density operator times some product of operators (that, of course, is the structure of all our correlation functions) we get,

$$\mathrm{Tr}\{\rho A(t_1+\tau)B(t_2+\tau)\ldots Q(t_n+\tau)\} =$$
$$\mathrm{Tr}\{\rho \exp(-iH\tau/\hbar) A(t_1) B(t_1)\ldots Q(t_n) \exp(iH\tau/\hbar)\} . \qquad (12.46)$$

Here, the operators are displaced in time. In the Heisenberg picture, we use an exponential based on the Hamiltonian operator to displace operators in time; thus we obtain the right-hand side of Eq. (12.46). So displacement by time, τ, means performing a familiar translation displacement under time, a translation on all of the operators, and when we perform this same unitary transformation on all of the operators, of course all the exponentials cancel except the first one, and the last one. Now, the trace has cyclic symmetry, so the last exponential can be moved around freely to the first position, and the right-hand side of Eq. (12.46) becomes

$$\mathrm{Tr}\{\exp(iH\tau/\hbar)\rho \exp(-iH\tau/\hbar) A(t_1)\ldots Q(t_n)\} . \qquad (12.47)$$

All of the remaining operators are evaluated at their original times $t_1,\ldots,t_n$. We can write Eq. (12.47) in the form

$$\mathrm{Tr}\{\rho A(t_1)\ldots Q(t_n)\} , \qquad (12.48)$$

provided the transformed version of the density operator is equal to the density operator itself. In other words, provided the Hamiltonian commutes with the density operator,

$$[H, \rho] = 0 .$$

Commutation of the density operator with the Hamiltonian is the necessary and sufficient condition, finally, for stationarity.

Now supposing we have a stationary field, let us see what we can say about the correlation functions, and, in particular, these correlation functions with different numbers of creation and annihilation operators. Suppose the field is one in which the set of modes which we label by "k" is occupied. The first of the more general correlation functions may be constructed as one in which we have no creation operators, and one annihilation operator. It then just has one time argument. This function is, of course, the trace of ρ times $E^{(+)}$ at the time t. It is just the expectation value of the field. The form it takes is necessarily this,

$$G^{(0,1)}(t) = \text{Tr}\left\{\rho E^{(+)}(t)\right\} = \sum_k F_k \exp(i\omega_k t) \ . \tag{12.49}$$

If we have certain modes of the field which are occupied, we imagine the mode decomposition of the operator $E^{(+)}$ and calculate the expectation value and, indeed, we have a sum of periodic terms which can only be the exponentials associated with the time dependences of the occupied modes. Now, we are assuming that the field is stationary, and if the field is stationary, we simply cannot have oscillations like that. Everyone of those coefficients F_k must vanish. If the field is to be stationary, the only possibility is that the correlation function $G^{(0,1)}(t)$ vanishes. That is not surprising. For a stationary field, the expectation value of the field amplitude itself must vanish, so the first of these more general correlation functions vanishes.

Now, how about the other correlation functions? It is always possible that we have equal numbers of creation and annihilation operators to pair-off the oscillating time-dependencies and in that way to get a time-independent function. However, if we have unequal numbers of creation and annihilation operators, we are dealing with a condition which is very much more difficult to satisfy. The more general situation for a field with $\{k\}$ modes occupied is,

$$G^{(n,m)}(t_1 \ldots t_{n+m}) = \sum_{k_1} \cdots \sum_{k_{n+m}} F_{k_1 \ldots k_{n+m}} \exp\left[i\left\{\sum_{j=1}^{n} \omega_j t_j - \sum_{l=n+1}^{n+m} \omega_l t_l\right\}\right] \ . \tag{12.50}$$

In general, the (n,m)-th order correlation function is a summation for which in all the terms of the summand, there occur all possible choices of the exponential time dependencies which characterize the modes: there appears in the positive frequency sense, m such factors, and in the negative frequency sense there appear n such factors. If there is any unbalanced oscillation in these two exponents, and if we are asserting that the field is stationary, viz. that this function is independent of time displacements, then the only way that that can be secured is to say that the corresponding coefficient vanishes. A necessary condition, then, to have a non-vanishing function $G^{(n,m)}$ is that it be possible to put together sets of frequencies of n frequencies and m frequencies, (and we can allow repetitions), so that

$$\sum_{j=1}^{n} \omega_j = \sum_{j=n+1}^{n+m} \omega_l \,. \tag{12.51}$$

We take the set of occupied modes, write down their frequencies, and choose n of them with possible repetitions, and choose m of them allowing for possible repetitions. However, if it is not possible to do that so that the one sum is precisely equal to the other, then there is going to be unbalanced oscillation present in the corresponding term for the correlation function, and that coefficient must vanish if the field is to be stationary. That means that the simplest way of securing a non-vanishing correlation function for a stationary field is just to have $m = n$ and to make identical choices of the sets of frequencies. Then the one time dependence will cancel the other and it is very easy to have a non-vanishing function. It is not so easy if we choose $m \neq n$. To satisfy a condition of this sort a very special set of frequencies must be occupied in the field. For example, if we have three frequencies present, then we see that if we call them ω_1, ω_2 and ω_3 we must have some such relationship as this,

$$\omega_1 + \omega_2 = \omega_3 \,, \tag{12.52}$$

among the three frequencies, if we are to get a non-vanishing result for what we have defined as $G^{(1,2)}$. In fact, a condition of this sort can hold, for example, in the parametric amplifier, or in the coherent Raman effect where the modes excited have this sort of property. But this is only a necessary condition in order to have a non-vanishing correlation function for these peculiar orders. It is only a necessary condition because, in fact, if these modes at different frequency oscillated independently of one another, we should have to look further at the structure of the coefficients, and then we still should not find any which were different from zero under the stationarity condition. There has, additionally, to be a statistical correlation between the density operators for these particular modes. If, on the other hand it were true that the density operator for the entire field factored into a product of density operators (stationary density operators for the individual modes) once again, we could show that the functions vanish. The actual physical situation in devices like the parametric amplifier or a medium in which the coherent Raman effect is amplified, or a substance in which second harmonic production is occurring, is one in which there do exist such statistical correlations between the modes of different frequencies. And so, it is true that these peculiar correlation functions take on non-vanishing values.

In the case of harmonic production which is the simplest, we send in red photons, and pairs of red photons join together to form blue photons. In the outgoing field we have two components present, one of which is at precisely twice the frequency of the other. Furthermore, there are phase relations between the two fields. There is statistical dependence between these two fields, and, in fact, in the region illuminated by such a crystal producing the second harmonic, we should find that the function $G^{(1,2)}$ is different from zero. But it takes all of the complication of nonlinear optics to bring that situation about. If we just produced two fields from independent lasers,

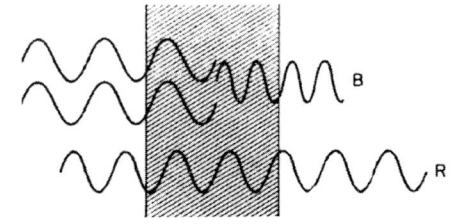

Figure 1 Second harmonic production in a medium; some red light is converted to blue light; some red light is transmitted.

say, which happened to have frequencies such that the one had precisely twice the frequency of the other, the function $G^{(1,2)}$ would still vanish, because there would then be no phase relations between the two fields. The dynamical process of harmonic generation is, here, precisely that which does generate the phase relations to give a non-vanishing function.

It is interesting to consider how we would detect the function $G^{(1,2)}$. We would have to perform a curious sort of interference experiment in which an amplitude based on $(E^{(+)})^2$ was somehow or other able to interfere with an amplitude based on a single factor of the field $E^{(-)}$. It is not impossible to do that and we may consider the following experiment as a suggestion as to how it may be done. Figure 1 shows a device which is producing harmonics. The problem is to make the two beams R, B interfere. There is a variety of ways. We can engage in harmonic production once more on the red beam. Of course, we shall then get some of the blue light through and a little bit more. The red light will go through and produce some blue light. This is a species of interference experiment because we can now let the blue beams interfere with one another. However, if we do that, we are measuring $G^{(2)}$ which is something quartic in the field and it is not what is wanted. This $G^{(2)}$ is the average of $(E^{(-)})^2(E^{(+)})^2$. What we have to do is some kind of experiment which literally measures the high frequency field against the low frequency field. It has to be an interference experiment of some sort. A possibility is the following. Suppose we use an atom as a photon detector and examine the photoelectrons emitted. Suppose also that atom has a threshold for photon detection which lies higher in frequency than the red frequency used. A single red photon will not give a photo-electron, but two red photons will. Now, two photons *can* be observed simultaneously and here we have a good strong beam of red photons, so there will be some photoelectric effect and that photoelectric effect will occur only in second order; it will be proportional to $(E^{(+)})^2$. Alternatively, the blue photons, which have, each of them, twice the frequency, can make the photoelectric transition occur by the absorption of a single quantum. These two processes are going to interfere with one another, and the amplitude of the interference term for the two processes will be proportional to the statistical average of $E_B^{(-)} E_R^{(+)} E_R^{(+)}$, i.e.

$$\text{Tr}\left(\rho E_B^{(-)} E_R^{(+)} E_R^{(+)}\right).$$

12 Quantum Theory of Coherence

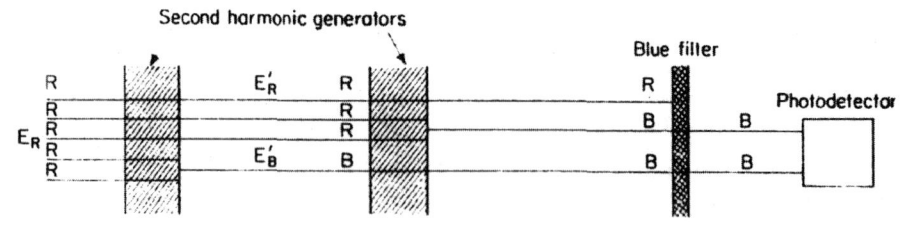

Figure 2

As we move the photodetector which is based on the interference of two photon absorption versus single photon absorption, we shall discover interference terms just because of a small difference of propagation vectors which typically occurs. The propagation vector of the blue photons will not be precisely equal to twice the propagation vector of the red photons ($k_B \neq 2k_R$ in general) because of dispersion. Therefore, as we move the detector based on the interference of these two different sorts of absorption processes, we shall find fringes in the interference of the two absorption processes. So, in principle, this is a possible measurement.

Another possibility is to have two converters as shown in Fig. 2. Many red photons enter the system, some are converted to blue ones, and some remain red. We have a second converter located as shown and some of the remaining red ones are turned blue. Finally, we have a blue filter, which removes the red. An ordinary photodetector is used. The total field between the two converters is the sum of the blue and red components. The final intensity measurement is $G^{(1)}$ in the region beyond the converters, and is given by

$$G^{(1)} = \text{Tr}\left\{\rho E''^{(-)} E''^{(+)}\right\}. \tag{12.53}$$

This expression is quadratic in the field E''. The field E'' in the region beyond the second converter contains the more-or-less unchanged field E'_B and a part which is proportional to the square of the red component, thus we have

$$E''^{(+)} = E'_B{}^{(+)} + \text{constant} \times (E'_R{}^{(+)})^2. \tag{12.54}$$

If we substitute the sum of these two terms for the doubly primed fields in the intensity we measure, we see that the function $G^{(1)}$ contains $E'_B{}^{(-)}$, $E'_R{}^{(+)}$, i.e. $G^{(1)}$ contains

$$\text{Tr}\left\{\rho E'_B{}^{(-)} E'_R{}^{(+)} E'_R{}^{(+)}\right\}.$$

In other words, $G^{(1)}$ contains the function $G^{(1,2)}$ when we are using fields between the two converters. There are other ways of constructing the intensity measured finally. For example, we can go back to the original fields and write the intensity in terms of those, but then it becomes a quartic expression. So it really all depends on which fields we are considering, whether we are measuring one order of the correlation function,

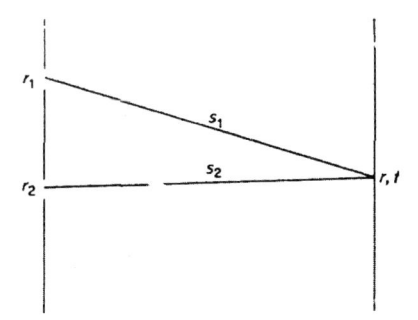

Figure 3 Young's experiment.

or another. Viewed in these terms, the measurement made is one of the (2,2) correlation function, or the ordinary 2nd order correlation function when dealing with fields before they have entered the system. For fields before entering the converters, $G^{(1)}$ contains the second order correlation function,

$$\text{Tr}\left\{\rho E^{(-)}E^{(-)}E^{(+)}E^{(+)}\right\} = G^{(2,2)} = G^{(2)}. \tag{12.55}$$

So in this sense, it is a matter of definition, just which of these correlation functions we are considering. But if we are dealing with the fields in between the two converters, then we can view the combination of the last converter, the filter, and the photodetector as being equivalent to the single atom photodetector with the high threshold which we considered above.

We have indicated that the higher order correlation functions with the differing numbers of creation and annihilation operators can, indeed, be different from zero and be measurable. However, this is not done to give too heavy an emphasis to these curious sorts of correlation functions; in fact under all ordinary experimental circumstances in stationary fields (whenever we have a steady beam, we effectively have a stationary field) for $m \neq n$, they vanish. Therefore, it is quite sufficient to retain the simpler set of correlation functions and deal just with those i.e. the ones for which the number of annihilation operators is equal to the number of creation operators. In practice those are almost always the functions which we measure.

12.5
Coherence

We have not said a word about coherence yet, so let us go back to the most elementary considerations and look at the first sort of experiment in which we would propose to use the term coherence. This goes way back to optics. Figure 3 is a crude representation of Young's experiment. The screen has two pinholes which are very tiny because we do not want to calculate the diffraction pattern of each hole. For all we care, the wave which spreads behind each of these holes is perfectly isotropic. We are looking at the intensity on a distant screen at a point r, t. The field at r is, to a good approximation by Huygens' principle, a superposition of the fields at these two pinholes.

The point considered lies at two different distances from the two pinholes and so we are effectively sampling the fields at these two pinholes at two different times. The distances are s_1 and s_2 and the times are $[t-(s_1/c)]$ and $[t-(s_2/c)]$. The field is given by

$$R^{(+)}(r,t) = \lambda \left\{ E^{(+)}\left(r_1 t - \frac{s_1}{c}\right) + E^{(+)}\left(r_2 t - \frac{s_2}{c}\right) \right\} . \tag{12.56}$$

All of diffraction theory is bound up in the parameter λ which tells us the transmission of the hole. (We are assuming that this is independent of the angle at which these propagation vectors lie on the other side of the hole). We are taking the positive frequency component of the field. Abbreviating the appropriately retarded space-time points by X_1 and X_2, we get

$$E^{(+)}(r,t) \equiv \lambda \{E^{(+)}(X_1) + E^{(+)}(X_2)\} . \tag{12.57}$$

The intensity is found from the correlation function by evaluating the fields $E^{(+)}$ and $E^{(-)}$ at the same space-time point which is the space-time point on the screen, thus

$$I(r,t) = \text{Tr}\left\{\rho E^{(-)}(rt)E^{(+)}(rt)\right\} . \tag{12.58}$$

But each of these fields is a superposition of the field operators (i.e. a sum) corresponding to the appropriately retarded times at the pinholes. Substituting the two sums, we get

$$I(r,t) = |\lambda|^2 \text{Tr}\left\{\rho \left(E^{(-)}(X_1) + E^{(-)}(X_2)\right)\left(E^{(+)}(X_1) + E^{(+)}(X_2)\right)\right\} . \tag{12.59}$$

This is a sum of correlation functions. Apart from the constant λ^2 which contains all the physical optics, it is the sum of four correlation functions,

$$I(r,t) = |\lambda|^2 \left\{ G^{(1)}(X_1,X_1) + G^{(1)}(X_2,X_2) + G^{(1)}(X_1,X_2) + G^{(1)}(X_2,X_1) \right\} . \tag{12.60}$$

The function $G^{(1)}(X_1,X_1)$ is the intensity obtained if the second hole is closed. The function $G^{(1)}(X_2,X_2)$ is the intensity if the first hole is closed. Then we have two terms which exist only because both holes are open. (We cannot tell which hole any given photon went through). These two terms are complex conjugates of one another so they are just twice the real part of $G^{(1)}$ with the arguments in either order,

$$G^{(1)}(X_1,X_2) + G^{(1)}(X_2,X_1) = 2 \operatorname{Re} G^{(1)}(X_1,X_2) . \tag{12.61}$$

If we write the function as its absolute value times an exponential of unit modulus (i.e. $e^{i\phi}$), then the real part is just $|G^{(1)}(X_1,X2)|$ times the cosine of the phase factor ϕ,

$$I(r,t) = |\lambda|^2 \left\{ G^{(1)}(X_1,X_1) + G^{(1)}(X_2,X_2) + 2|G^{(1)}(X_1,X_2)|\cos(X_2,X_1) \right\} , \tag{12.62}$$

where we have taken $G^{(1)} = |G^{(1)}|e^{i\phi}$. Let us now consider what happens as we move the point of observation up and down the screen. The positions of the holes at r_1 and r_2 (the spatial parts) of the space-time position X_1 and X_2, remain fixed. As the two distances s_1 and s_2 vary the times at which we sample the fields vary. The absolute value remains positive and represents the envelope of the sort of wiggle which is associated with the cosine. The wiggle represents the fringes; the intensity goes up and down. The rising and falling of the intensity (i.e. the degree of contrast) is governed by $|G^{(1)}(X_1, X_2)|$. The function $G^{(1)}(X_1, X_2)$ is a correlation function. If it vanishes, it means that the fields at the two pinholes are uncorrected. If there is no correlation, we say that the light at the two points is incoherent at the one point relative to the other (i.e. relatively incoherent). On the other hand, increase in coherence results in an increase in fringe contrast, i.e. the amplitude of the wiggles becomes large. Since $|\cos\phi|$ is never greater than unity, the fringe contrast is maximized by making $|G^{(1)}(X_1, X_2)|$ as large as possible, and that signifies maximum coherence. There is an upper bound because the total intensity of the expression can never become negative. It is very important that we have the intensity terms $G^{(1)}(X_1, X_2)$ and $G^{(1)}(X_2, X_1)$ because they are intrinsically positive whereas the interference term oscillates between positive and negative values. The upper bound to $|G^{(1)}(X_1, X_2)|$ is given by the Schwarz inequality,

$$G^{(1)}(X_1, X_1) G^{(1)}(X_2, X_2) \geq |G^{(1)}(X_1, X_2)|^2. \tag{12.63}$$

This inequality can be proved easily from the positive definite character of the density operator; the fact that the density operator can never have negative eigenvalues. The squared modulus of the cross-correlation function has to be less than, or equal to the product of the two intensities. We secure maximum fringe contrast by using the "equals" sign.

$$G^{(1)}(X_1, X_1) G^{(1)}(X_2, X_2) = |G^{(1)}(X_1, X_2)|^2. \tag{12.64}$$

That is a perfectly good characterization of optical coherence as it has always been used in all of the older contexts of optics. Let us agree that that is what we mean by optical coherence.

The fields are coherent at two space-time points (X_1, X_2) if the equality holds and, furthermore, we have optical coherence of the entire field if this equality holds for all space-time points X_1 and X_2. This is a slightly peculiar looking condition, and it is not immediately obvious how it is satisfied. Let us cite one *sufficient condition* for satisfying the optical coherence condition. We shall call the optical coherence condition Eq. (12.64) the first order coherence condition because it deals only with the first order correlation function. We have already seen that there is a hierarchy of correlation functions and we shall talk about other sorts of optical coherence in connection with those below. Suppose that the first order correlation function factorizes in the form,

$$G^{(1)}(X_1, X_2) = \mathcal{E}^*(X_1) \mathcal{E}(X_2). \tag{12.65}$$

If we write it in any factored form and then use the fact that the function with its arguments reversed is the complex conjugate of the function, then we see immediately that the two factors must be complex conjugates of one another. Furthermore, the field $\mathcal{E}$ which represents one of the factors must be a positive frequency field, that is $\mathcal{E}$ contains $\exp(-i\omega_k t)$. If this factorization condition holds, then by just noting that $G(X_1, X_1)$ is the squared modulus of this field and $G(X_2, X_2)$ is likewise, we see that the condition for maximum fringe contrast holds identically. Obviously, this is a sufficient condition for optical, or first order, coherence. What is a good deal less obvious, but nonetheless true, is that if the maximum fringe contrast condition holds everywhere in space-time, then *it* implies the factorization condition. It does not take any deep mathematics to prove that, it is just the Schwarz inequality but it does take about ten minutes to prove. It is proved in the Les Houches notes, and it is also proved by Titulaer and Glauber[4]. This definition of optical coherence goes a little bit beyond the definition that had earlier been used in optics because of the usual limitation of experimenting with stationary light beams, in fact, the usual definition of the correlation function in optical terms was made in terms of a time average rather than an ensemble average. If we are dealing with a time average, we have only one time left in the expression, so it is not easy to see then how to extend the definition to deal with non-stationary fields. The new definition is a definition which extends immediately to non-stationary fields. It is also a quantum mechanical one rather than a classical one.

One of the nice properties of the statement of factorization, once we know that the statement of factorization is equivalent to the statement of maximum fringe contrast, is that we can see clearly what the relationship is between monochromaticity and optical coherence. As we know, all of the efforts over many years to increase optical coherence were efforts to provide more and more monochromatic optical fields. The curious thing about the definition we have given here is that it says absolutely nothing about monochromaticity. In fact these fields $\mathcal{E}$ into which the correlation function factorizes can have absolutely arbitrary spectra. So an optically coherent field according to our definition need not be monochromatic at all. However, when we restrict ourselves to dealing with stationary fields, then indeed, the correlation function must be invariant under time displacements, i.e. it must depend only on the difference of two times

$$G^{(1)}(t_1, t_2) = G^{(1)}(t_1 - t_2) = \mathcal{E}^*(t_1)\mathcal{E}(t_2) . \tag{12.66}$$

If we require then that it factorizes as well, that is, if we impose the first-order coherence condition, then the only solution of this functional equation (the only way of writing a product which depends only on the difference in the two times) is to say that the field $\mathcal{E}$ depends exponentially on the time, and then since it must be a positive frequency field, it can only take the form

$$\mathcal{E}(t) = \mathcal{E}(0)\exp(-i\omega t) . \tag{12.67}$$

Are there any fields that obey our optical coherence condition precisely? One trivial sort of example is just a classical field which is perfectly well determined in its behavior, one whose Fourier coefficients are absolutely fixed. Then the procedure of averaging over unknown Fourier coefficients referred to above becomes a trivial one and it is clear that the correlation function factorizes in this case. Of course, we are dealing there with the classical correlation function and it is that that factorizes. We have said nothing about the quantum mechanical one and we leave open for the moment the question of whether there are quantum mechanical fields which do this; as we know, there is a vast variety of them. The reason we mention this particular classical example is to show the first of the many associations we shall see between the definiteness of the field, the fixed character of the (otherwise) random Fourier coefficients and coherence. Coherence tends to limit the noise in the field. The more coherence, we shall see, the less noise is allowable. Optical coherence is a rather weak limitation on noise in the field. We can have quite noisy fields which are still consistent with the optical coherence condition and only by adding further conditions can we get rid of much more noise.

We can go on; having defined first order coherence, and being equipped with a great many correlation functions, we can define more exotic sorts of coherence. An obvious thing to do is to look at the higher order correlation functions and require that they factorize in essentially the same way. Let us assume that the first m correlation functions $1,\ldots,m$ all factorize in essentially this way,

$$G^{(m)}(X_1,\ldots,X_{2m}) = \mathcal{E}^*(X_1)\ldots\mathcal{E}^*(X_m)\mathcal{E}(X_{m+1})\ldots\mathcal{E}(X_{2m}) , \qquad (12.68)$$

for $m \leq n$, where the function $\mathcal{E}$ is the same in all cases and is not an operator, it is an ordinary function which is a positive frequency solution of the Maxwell equations that obeys all the boundary conditions which the field must obey. A field which obeys this condition for the first n correlation functions will be called n-th order coherent. There is room for some variation in the choice of the particular definition of n-th order coherence; some other suggestions have been made. Cases in which we have, in practice, fields with precisely 17th order coherence (say), and no higher coherence are not known; it is intended to suggest by this definition that we are adding more and more conditions and, indeed, when we let $n \to \infty$, we shall then be talking about a fully coherent field; that is the other significant extreme. If n-th order coherence holds, then Eq. (12.68) implies that all of the correlation functions which represent joint counting rates for m-fold coincidence experiments and $m \leq n$ factorize in this simple way:

$$G^{(m)}(X_1,\ldots,X_m,X_m,\ldots,X_1) = \prod_{j=1}^{m} G^{(1)}(X_j,X_j) . \qquad (12.69)$$

Notice that the arguments are repeated here, X_1 occurs twice, and finally X_m occurs twice. In other words, fields with higher order coherence are the fields for which the registry of counts by the individual photon counters *is* statistically independent. We put n different counters in the field and they will each record photons in a way which

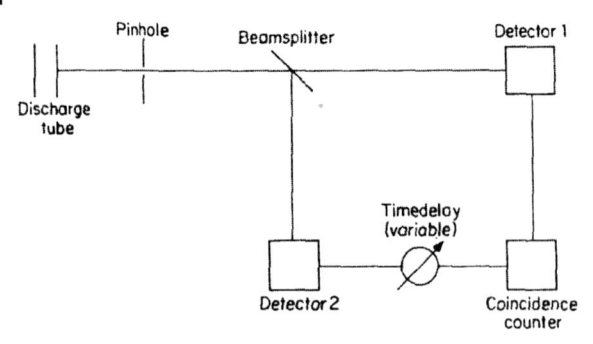

Figure 4 Hanbury Brown–Twiss experiment.

is statistically independent of all of the others with no special tendency towards coincidences, or correlations, or anti-correlations, except as is governed by the overall intensities of the fields at those particular points. (Of course, if the field strength essentially goes to zero all over the place, none of the counters will register.) Here we are considering correlation, not rates, and there will be no tendency towards correlation, given higher order coherence.

Now let us switch away for a moment and say why it is interesting to define such a curious idea. In the Hanbury Brown–Twiss experiment[5] (Fig. 4) light from a discharge tube (a narrow bandwidth spectral source) is passed through a pinhole and emerges as a very coherent beam in the first order sense. The beam is divided by a half-silvered mirror and recorded by two photodetectors. In the original experiment these were photomultiplier tubes which gave a more-or-less continuous output which was random. The two outputs were then multiplied together to see if there was any correlation between them and such a correlation was found. In all later versions of this experiment, the detectors are photon counters and coincidence detection is used. In the original experiment the detectors were moved relative to their zero points and in all subsequent versions of the experiment, the counters are kept more or less fixed and variable time delay is used in the coincidence gating. In this experiment the coincidence rate was plotted against time delay (the two counters have finite resolving times and so we expect some smearing-out of coincidences). We should expect that if the entry of the photons into the two counters were statistically independent events we should see a uniform background in the coincidence rate versus the time delay (Fig. 5). In the original experiment a very tiny deviation from the uniform coincidence rate versus time delay was observed. In fact, however, the effect is not a tiny one. If it is possible to get the band width of the incident radiation to be narrow enough, in other words, if the experiment is done in ideal circumstances, we find a coincidence rate corresponding to zero time delay which is twice the accidental background rate. So there is in the light entering the system, a distinct tendency for photons to occur in pairs, to enter the system in pairs, or at least for these two photon counters to register photons simultaneously. That is the Hanbury Brown–Twiss effect (HBT effect).

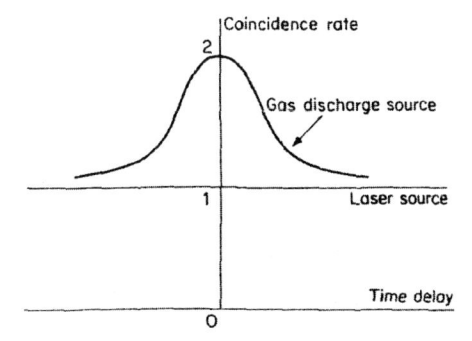

Figure 5 Results of Hanbury Brown–Twiss experiment.

Why do we see this effect? Is it, for example, some trick in the mirror? No, the only function of the mirror is to allow us to do something which would otherwise be very inconvenient, which is to put two counters at the same place in space. The mirror creates the same field in front of counter 1 as is in front of counter 2. We could perform the experiment with a single counter if we had a fast enough resolving time. We should investigate the distribution of intervals between photon counts and we should discover that there is an anomalously large number of short intervals between photon counts. There is again a distinct tendency for photons to clump as they enter a given counter. Well, here we have two photon counters in the field and they show results which are not statistically independent. This is a situation in which $G^{(2)}$ is definitely not equal to $G^{(1)}G^{(1)}$. There is no second order coherence. Then we might think that there is no sense in defining such a notion. But, in fact, this statement is not true for all fields and one of the ideas which led to all the work we have been describing is the surmise, very quickly born out by experiment, that in fact there is no HBT effect when a stabilized laser beam is used as the source. When we plot the coincidence rate, we find no bump at all (Fig. 5), regardless of the time delay. No HBT effect occurs in a stabilized laser beam. Well, that is the sense of defining it. There is now a good deal of verification that the factorization holds, certainly up to some quite high order (six) of the correlation functions. So up to sixth order there is this factorization in the stabilized output of a high quality gas laser; that is the rationale in. defining these correlation functions and discussing higher order coherence.

Let us now consider a little further the consequences of defining higher order coherence. We can prove some mathematical theorems and if we return to the papers of Titulaer and Glauber[4] one of the interesting things is this. There are more consequences of the first-order coherence notion (i.e. maximum fringe contrast, or the factorization condition, which are equivalent) than first appears. One consequence is that a species of factorization holds in all of the higher order correlation functions. They all factorize into approximately the form we should like but with a coefficient g_n which is generally different from unity.

$$G^{(n)}(X_1,\ldots,X_{2n}) = g_n \mathcal{E}^*(X_1)\ldots\mathcal{E}^*(X_n)\mathcal{E}(X_{n+1})\ldots\mathcal{E}(X_{2n}). \qquad (12.70)$$

In higher order coherence the coefficients g_n (which can generally be positive numbers) are all equal to 1,

$$g_n = 1, \quad \text{all } n.$$

So we can specify the coherence properties of a beam by considering first the factorization of the first order correlation function and then, if it factorizes, by discussing the discrete sequence of numbers g_n (they are quite measurable numbers). A problem in all definitions of coherence is that they can only be fulfilled in approximate terms. If we speak of factorization as a coherence condition, in actual fact, we should never fulfil those conditions everywhere in the universe. But a laser beam can do it over vast intervals of the space variable and intervals of the time variable, perhaps as long as 0.1 s.

There are a number of other consequences of the assumption of first order coherence. Let us consider the coherence conditions in still one more way. One of the ways of securing the first order coherent beam is to occupy just one mode. If only one mode of the field is occupied then we can always write the correlation function in factorized form, so one mode occupation does it. That will always allow us to write $G^{(1)}$ as a factorized form. We can broaden that example greatly if we define modes more generally than we are accustomed to, e.g. define a set of mode functions $v_l(r,t)$ in terms of our other mode functions $u_k(r)$, as

$$v_k(r,t) = \sum_k \gamma_{lk} u_k(r) \exp(-i\omega_k t). \tag{12.71}$$

The new set of mode functions are non-monochromatic mode functions. It is easy to define, as linear combinations of the familiar creation and annihilation operations, new creation and annihilation operators for these new non-monochromatic modes. These non-monochromatic modes are perhaps also describable as wavepackets (whether we should call them modes is open to question). The most general first order coherent field is one in which only one of these generalized modes, or only one of these wavepackets has any photons in it at all, other varieties are empty. That is a fairly simple characterization of first order coherence.

All coherence, or higher and higher order coherence represents more and more constraints upon the distribution of quantum numbers for the occupation of that one mode, and, finally, full coherence, implies that there is a Poisson distribution for the number of photons. These are the definitions of coherence. Again, are there any fields which will fill all of them? There is the trivial example of a classical field with no uncertainty in it. That is a classical field in which the Fourier coefficients are absolutely fixed. For such a field, all of the classical correlation functions factorize in just the way we are describing. So now it is obvious that any noiseless field, any perfect signal, is fully coherent. What our hierarchy of definitions of coherence has done is to approach in the limit, the sort of field which any communication engineer would say is coherent, viz. one which involves no noise, except, finally, the trivial sort of noise, which represents a phase arbitrariness in the field $\mathcal{E}$. Any overall phase in the expression just cancels out. So much for the subject of coherence.

12.6
Coherent States

We shall now consider quantum mechanical fields which have the property of full coherence. First we shall consider pure states since these are likely to be the simplest. That means that the density operator can be written as a trivial dyadic product of this state with itself,

$$\rho = |\ \rangle\langle\ |. \tag{12.72}$$

Then our correlation function $G^{(n)}$ takes the form of an expectation value in that state of a normally ordered product of operators with creation operators on the left, annihilation operators on the right,

$$G^{(n)} = \left\langle \left| E^{(-)} \ldots E^{(+)} \right| \right\rangle. \tag{12.73}$$

What kind of pure state leads to factorization of this expression? If it were possible to diagonalize the operators $E^{(+)}$ and $E^{(-)}$ simultaneously, then such an eigenstate of them would fill the bill very nicely, but we cannot do it because the operators $E^{(+)}$ and $E^{(-)}$ do not commute. Suppose we continue the restriction of dealing with just those experiments which can be described in terms of normally ordered products. As long as the products are normally ordered, we do not really need simultaneous eigenstates of $E^{(+)}$ and $E^{(-)}$ because all we ever find is $E^{(+)}$ at the right and $E^{(-)}$ at the left so it is quite sufficient to find an eigenstate in this sense. The $E^{(+)}(r,t)$ applied to our state gives an ordinary function, $\mathcal{E}(r,t)$ multiplied by that state,

$$E^{(+)}(r,t)|\ \rangle = \mathcal{E}(r,t)|\ \rangle. \tag{12.74}$$

Of course, if that is true, then there will be a dual relationship which is this: the dual state with the Hermitian adjoint operator applied will give the complex conjugate eigenvalue $\mathcal{E}^*(r,t)$ field multiplied by the dual state,

$$\langle\ |E^{(-)}(r,t) = \mathcal{E}^*(r,t)\langle\ |. \tag{12.75}$$

With these two relationships, we have secured, for all normally ordered products, an expectation value which factorizes, and therefore $G^{(m)}$ will be given by

$$G^{(m)} = \mathcal{E}^*(r_1,t_1)\ldots\mathcal{E}(r_{2m},t_{2m})\langle\ |\ \rangle. \tag{12.76}$$

The two relations (12.74) and (12.75) together provide a sufficient condition for factorization. We have not in this way determined the most general set of pure states which does this. That is done in the papers of Titulaer and Glauber[4]. The additional states that are found are not very interesting. They differ just in having some extra phase factors, which the states we shall discuss do not contain. The additional ones are ruled out by considering the more general set of correlation functions $G^{(n,m)}$. If we require that these factorize as well, the only pure state that will do it is the state that satisfies the eigenvalue condition. The eigenstates of $E^{(+)}(r,t)$ are the coherent states. It is obvious that the states of Eq. (12.74) are not states with a fixed number of quanta in them because $E^{(+)}$ is an annihilation operator and the statement here is that we are applying an annihilation operator taking away one photon and, even so, we have the same state left. That is not possible if we have a fixed number of photons, so the number of photons must be indefinite. Now to see what the eigenstate condition Eq. (12.74) means in greater detail let us expand the positive frequency field operator in terms of normal modes as we did in Eq. (12.15)

$$E^{(+)} = i \sum \left(\frac{\hbar \omega_k}{2}\right)^{1/2} a_k u_k(r) \exp(-i\omega_k t) . \tag{12.77}$$

A corresponding expansion must be available for the eigenvalue function

$$\mathcal{E} = i \sum \left(\frac{\hbar \omega_k}{2}\right)^{1/2} \alpha_k u_k(r) \exp(-i\omega_k t) . \tag{12.78}$$

Since the mode functions $u_k(r)$ form an orthogonal set the eigenstate condition Eq. (12.74) implies that

$$a_k | \ \rangle = \alpha_k | \ \rangle \tag{12.79}$$

for all values of the mode index k. Since it is possible to project out of the dependence on the spatial coordinate, r, all the various coefficients associated with the different mode functions, then it must obviously be true that a_k applied to this unknown state gives us the number α_k applied to the unknown state. In other words, these states are the right eigenstates of all of the annihilation operators and now we know the properties of those states. Equation (12.79) alone is sufficient to determine them for us. For one mode, it follows from this equation, that the state can be expanded in the following way in terms of the n-quantum states,

$$|\alpha\rangle = \exp\left(-\tfrac{1}{2}|\alpha|^2\right) \sum \frac{\alpha^n}{\sqrt{n!}} |n\rangle \tag{12.80}$$

where the normalization factor is $\exp(-i|\alpha|^2)$. Then we label this one mode state with the complex eigenvalue itself.

Now, obviously there is such a state for any one mode for any complex number α that we choose. For the full set of modes we can write

$$|\{\alpha_k\}\rangle = \prod_k |\alpha_k\rangle_k . \tag{12.81}$$

We secure all possible classical solutions to the Maxwell equations by substituting in Eq. (12.80) all possible complex coefficients α_k and, therefore, we see that there is a one-to-one correspondence between eigenstates of Eq. (12.74) (i.e. eigenstates of the positive frequency part of the field) and classical solutions to the Maxwell equations. In this very limited sense there is a one-to-one correspondence between classical electromagnetic theory and quantum theory. The coherent states (which have now to be characterized by the entire set $\{\alpha_k\}$) are in one-to-one correspondence with the classical positive frequency solutions to the Maxwell equations. Of course, the way we construct a state of the field for all possible modes is to take the outer product, the direct product (which is what is meant by the product symbol $\prod_k$) taken over all modes of the coherent states for each mode. Those are the coherent states.

The question now is, "Can we do something more with coherent states than simply say that this is a set of states which obeys our coherence conditions?" If that were all we could do with them, they would belong in a museum somewhere. The point is that they are very useful in doing calculations. First of all, it is useful to realize that any state of the field can be expressed in terms of the coherent states. Because there is such a vast number of coherent states, a two-dimensional continuum of them, and because they happen not to form an orthogonal set, there is a variety of ways in which we can expand any state, in terms of the coherent states. However, we are not going to expand states in terms of coherent states, we are not going to expand arbitrary state vectors, we are going to be concerned with operators (though the idea is basically the same). Suppose we want to expand any state in terms of the coherent states. We multiply by the unit operator which is $1/\pi$ times the dyadic product integrated over the entire complex plane, thus

$$|?\rangle = \frac{1}{\pi} \int |\alpha\rangle \langle \alpha | \, d^2\alpha | ?\rangle , \qquad (12.82)$$

where $d^2\alpha$ means the differential of the real part of α times the differential of the imaginary part of α.[4] Here we have explicitly an expansion of our unknown state in terms of coherent states. We can do that in general and we can expand operators in terms of the coherent states as well. The virtue of doing that, in particular for the density operator, is this. Once we have the density operator expressed in terms of the coherent states, then since everything we want to evaluate is in normally ordered form (at least for the moment) all of those operators applied to the coherent states on either side of them are turned into numbers and the operator aspect of the problem disappears. We then deal just with complex ordinary numbers. To expand the density operator, ρ, we multiply it by 1 on each side. Having used it twice, we have to integrate over two complex variables,

$$\rho = \frac{1}{\pi^2} \int |\alpha\rangle \langle \alpha | \rho | \beta \rangle \langle \beta | \, d^2\alpha \, d^2\beta . \qquad (12.83)$$

---

[4] $|?\rangle = \frac{1}{\pi} \langle \alpha | \langle \alpha | ?\rangle \, d^2\alpha$,
$\langle \alpha | ?\rangle = \exp(-\frac{1}{2}|\alpha|^2) f(\alpha^*)$, where $f(\alpha^*)$ is an entire function of α^*.

Again, if we know the operator ρ, we know the function which stands in the middle of this expression. More generally for many modes, we have to expand in terms of a whole set of α coefficients and a whole set of β coefficients, as follows:

$$\rho = \frac{1}{\pi^2} \iint |\{\alpha_k\}\rangle \langle\{\alpha_k\}|\rho|\{\beta_k\}\rangle \langle\{\beta_k\}| \prod_k d^2\alpha_k \, d^2\beta_k \,. \tag{12.84}$$

We do the same thing for every mode in the field. There are quite a few modes and quite a few integrals have to be done.

The point of this is that the expectation value of the field operators in a state like this (the matrix element between these states) reduces to a product of c-numbers.

$$\left\langle \{\beta_k\} \left| E^{(-)} \ldots E^{(+)} \right| \{\alpha_k\} \right\rangle = \prod \mathcal{E}^*(\{\beta_k\}) \mathcal{E}(\{\alpha_k\}) \langle\{\beta_k\}|\{\alpha_k\}\rangle \,. \tag{12.85}$$

All of the annihilation operators are applying to the right on a coherent state, so all of these annihilation operators become numbers, in other words this becomes just the product of eigenvalues $\mathcal{E}$. All of the creation operators are applying to the left on a coherent state, these turn into a product of complex conjugate eigenvalues. Then we have the integrations left which are implicit in this expression. The process of evaluating an expectation value becomes a process of integration over two complex variables for each mode. This is the most direct and the most obvious way of using the coherent states but there are other ways. The advantage of this particular procedure is that the sorts of functions that we deal with are always terribly well behaved. It is always true that the scalar product of the state α with anything can be expressed as $\exp(-\frac{1}{2}|\alpha|^2)$ times some function of the complex variable α^*, but it has to be an entire function. (An entire function is about the best behaved sort of function we can have – it has no singularities anywhere in the finite plane.) So those are non-singular functions, and, equivalently it is true in this representation that the weight function for the density operator, when multiplied by $\exp(\frac{1}{2}|\alpha|^2) \exp(\frac{1}{2}|\beta|^2)$ is likewise an entire function of α^* and β, again a very well behaved sort of function. We have that for the cost of doing two integrals for each mode.

There are other ways of proceeding. Let us first see something of the ambiguity in the use of the coherent states as an expansion basis. We can take the operator 1 and expand it in two ways in the coherent states, one way is this,[5]

$$1 = \frac{1}{\pi^2} \int |\alpha\rangle \langle\alpha|1|\beta\rangle \langle\beta| \, d^2\alpha \, d^2\beta$$

$$= \frac{1}{\pi^2} \int |\alpha\rangle \langle\beta| \exp(\alpha^*\beta - \tfrac{1}{2}|\alpha|^2 - \tfrac{1}{2}|\beta|^2) \, d^2\alpha \, d^2\beta \tag{12.86}$$

This is a representation of unity as a double integral in terms of the coherent state. But we know another representation, which is this,

$$1 = \frac{1}{\pi} \int |\alpha\rangle \langle\alpha| \, d^2\alpha \,. \tag{12.87}$$

[5] $\langle\alpha|1|\beta\rangle = \langle\alpha|\beta\rangle$,

$\langle\alpha|\beta\rangle = \exp(\alpha^*\beta - \tfrac{1}{2}|\alpha|^2 - \tfrac{1}{2}|\beta|^2)$.

If we do the β integration which is easy enough to do in this case, the first form of the representation degenerates into the second one. We can derive the second one from the first one. There are these two different ways of expanding this sample operator, unity, in terms of the dyadic product with different coherent state arguments and in terms of those with the same arguments, which is to say projection operators on the coherent states. These two varieties of expansion can be available for operators of all sorts and we shall next discuss the availability of the second kind of expansion for the density operator. We expand the density operator in the following way in terms of projection operators on the coherent states,

$$\rho = \int P(\alpha) |\alpha\rangle \langle\alpha| \, d^2\alpha , \qquad (12.88)$$

where $P(\alpha)$ is a weight function which must be a real valued function of α in order to keep ρ Hermitian (as it must be). Is such a representation available? Well, thereby hangs a tale. Of course, if it does exist, then, since the trace of ρ is 1 and the trace of the dyadic product of the α state with itself is just unity, the function $P(\alpha)$ must be normalized when integrated over the entire complex plane,

$$\mathrm{Tr}\,\rho = 1 = \int P(\alpha)\, d^2\alpha . \qquad (12.89)$$

If this representation exists, $P(\alpha)$ has some of the properties of a probability distribution. In fact, it is not one, and we shall see why below. But, is there such a representation as this and can we use it; and can we manage to survive what happens in the literature if we do?

12.7
The *P* Representation

The point of expanding in terms of coherent states is that it simplifies the evaluation of statistical averages (expectation values). The expansion in terms of coherent states can be carried out in two ways and we illustrated both of those above. We recall that for the unit operator there are two species of expansion. The one is an expansion in terms of pairs of different coherent states and the other is an expansion in terms of pairs of identical coherent states. Now this second form of expansion is the simpler one since it involves only one complex variable for each harmonic oscillator in the field (only one complex amplitude for each of them). Let us talk therefore about *that* particular representation, which, wanting a better name, we call the "*P* representation",

$$\rho = \int P(\alpha) |\alpha\rangle \langle\alpha| \, d^2\alpha . \qquad (12.90)$$

The function *P* should be real valued in order to keep ρ a Hermitian operator. There is a normalization condition on the density operator which is that the sum of its eigenvalues (which, as you know, represent probabilities) must be unity. The trace of ρ, therefore, is 1; that means in this representation that the integral of the function P taken over the entire complex plane has to be unity,

$$\int P(\alpha)\,d^2\alpha = 1 \,. \tag{12.91}$$

We have left completely open the question as to whether it is possible, in general, to construct such a representation. For some states obviously we can do it. For example, for a pure coherent state, the one based on the amplitude β,

$$\rho = |\beta\rangle\langle\beta| \,. \tag{12.92}$$

The way in which we do it is simply to choose for the function P, a δ function,

$$P(\alpha) = \delta^{(2)}(\alpha - \beta) \,, \tag{12.93}$$

where $\delta^{(2)}(\gamma)$ is a two dimensional δ function. A 2-dimensional δ function is the product of two one-dimensional δ functions,

$$\delta^{(2)}(\gamma) \equiv \delta(\mathrm{Re}\,\gamma)\,\delta(\mathrm{Im}\,\delta) \,, \tag{12.94}$$

and the arguments of those functions are real, i.e. one is the real part of the argument, the other is the imaginary part of the argument; both of those are real numbers. The great advantage of this representation, where we can use it, is the following. If we have a normally ordered product of operators and want its statistical average, then we substitute for p the integral expression given in Eq. (12.90) for the density operator and the trace is then the average expectation value in a coherent state of the normally ordered product,

$$\mathrm{Tr}\left(\rho\,(a^\dagger)^n\,a^m\right) = \int P(\alpha)\,\langle \alpha | (a^\dagger)^n a^m | \alpha \rangle\,d^2\alpha \,. \tag{12.95}$$

But now the operator a^m operates on a coherent state to the right and $(a^\dagger)^n$ on a coherent state to the left; each of those operators then turns into its eigenvalue and what we are averaging is just a pair of complex numbers,

$$\mathrm{Tr}\left(\rho\,(a^\dagger)^n\,a^m\right) = \int P(\alpha)\,(\alpha^*)^n \alpha^m\,d^2\alpha \,. \tag{12.96}$$

The averaging of operators turns into the averaging of numbers, but not only that, the averaging procedure looks a great deal like statistical averaging of the usual sort in probability theory.

There are a number of things to observe about this representation when we can use it. The temptation is very great to think of the function P as a probability density. All of the properties that we have indicated here are properties of a probability density. For example, when we have a perfectly definite state represented by a δ function which is a non-negative function, or can be chosen as such, the averaging procedure

12.7 The P Representation

we use is exactly what we would use in probabilistic terms. The only trouble is that this function P is *not* a probability density. It is *not* for many reasons which are all more or less related. The states that we label with the parameter α (the coherent states) are not an orthogonal set. So, for example, the projections in the directions of these coherent states in our entire space of states are not orthogonal projections. A harmonic oscillator can be in two different coherent states at once. We are not, therefore, summing mutually exclusive possibilities in summing these projection operators, so at least in that sense, P is not a probability density. It is not measurable experimentally as a probability density either, because α is an eigenvalue of a non-Hermitian operator, a, which has real and imaginary parts, or Hermitian and skew-Hermitian parts which do not commute and which are not, therefore, simultaneously measurable. They correspond to the position and momentum respectively of the oscillator and we cannot carry out a measurement which tells us anything definite about both position and momentum simultaneously. So there are no measurements we could make to determine a probability distribution (density) which would correspond to the function P. We have to think of the function P simply as a weight function in an expansion of this sort. Asymptotically, in the classical limit, it becomes a probability density. We shall see below a little more about what the classical limit is. In general, quantum mechanically, we have to face the possibility that the function P can take on negative values. (It is easy to find examples for which it does.) If P does happen to be a non-negative function it is easy to show that that is consistent with the positive definite character which the operator ρ must have, namely ρ as an operator can never have any negative eigenvalues. Indeed, positive functions P lead to *that* condition, but the converse is not true. Positive, definite and perfectly well behaved density operators do lead to functions P which freely take on negative values and there is in general, no avoiding that.

The function P is analogous to the probability density; asymptotically, in the classical limit, it is the probability density and it is probably best to give it the name, a name which was first used for another function, the name of quasiprobability density. The first such function which was known was that introduced by Wigner[6]. Wigner's interest was in finding the quantum mechanical analog of the phase space density function in classical statistical mechanics. In classical statistical mechanics we are accustomed to talking about systems in terms of a density function in the space whose coordinates are position and momentum variables simultaneously. Now there is obviously something peculiar about doing that in quantum mechanics. It is only in a very limited sense that we can discuss such a function. It is obvious that no such thing can have the interpretation of a probability density, in general. Well, Wigner found such a function. It plays, as we shall see, for certain orderings of operators, a role analogous to that of the probability density. The particular operator orderings involved are ones which are symmetrical in the p and q variables. That happens to be the same statement as the statement that the orderings are symmetrical in the a and $a^\dagger$ variables. The Wigner function has a particular range of uses. It is most useful, naturally, when we are talking about symmetrically ordered variables but they are not what comes out as a rule in describing the simplest experiments in quantum optics. Well, our function P is not the Wigner function, it is another quasiprobability density, but, as we shall see below, a fairly close relative of the Wigner function.

Now, let us see what happens when we deal with a slightly more complicated situation in which many modes are excited. Then, of course, we have to deal with a many variable function $P(\{\alpha_k\})$. We deal with a function in which one amplitude is specified for every mode of the field. If we just define a symbol now which is the field eigenvalue corresponding to a particular coherent state, we see, if the coherent state is the one labeled with all of the amplitudes α_k, we can write the eigenvalue of the positive frequency part of the fields as $\mathcal{E}$ and it will be an expression which is linear in the amplitudes α_k,

$$E^{(+)}(r,t)|\{\alpha_k\}\rangle = \mathcal{E}(r,t\{\alpha_k\})|\{\alpha_k\}\rangle . \tag{12.97}$$

The expression is a linear combination, an infinite sum, a sum over all the modes of terms which are linear in the amplitudes α_k. Now suppose we are told that such a P representation exists for the density operator of the entire field, how do we calculate the correlation functions? Let us illustrate this for the first order correlation function. The story, of course, is the same for all of them. The first order correlation function is the statistical average of the normally ordered product of the field operators, and because it is a normally ordered product, we can just replace the field operators by their eigenvalues. The first order correlation function expressed as an integral taken over all of the complex mode amplitudes α_k is given by

$$G^{(1)}(X_1,X_2) = \int P(\{\alpha_k\})\,\mathcal{E}^*(X_1\{\alpha_k\})\,\mathcal{E}(X_2\{\alpha_k\}) \prod d^2\alpha_k . \tag{12.98}$$

This is an exact quantum mechanical expression, and yet we see here something which looks a great deal like the classical procedure which we encountered in Eq. (12.12) for evaluating the classical first order correlation function. The correspondence is as follows:

$$\mathcal{E}(X\{\alpha_k\}) \longleftrightarrow E_{cl}^{(+)}(X\{C_k\}) , \tag{12.99}$$

$$P(\{\alpha_k\}) \longleftrightarrow \mathcal{P}(\{C_k\}) , \tag{12.100}$$

$$i(\hbar\omega_k/2)^{1/2}\alpha_k \longleftrightarrow C_k . \tag{12.101}$$

The field eigenvalue $\mathcal{E}$ corresponds simply to the positive frequency part of the classical field. The weight function in our representation of the density operator, $P(\alpha)$, corresponds to the probability density for the random Fourier coefficients C_k. Finally, the C's themselves correspond to particular given constants multiplying the amplitudes α_k. When we make these identifications, there is a precise correspondence between the formulae for the classical correlation function and the quantum mechanical one. Presumably, we use the classical procedure when the numbers are meant to describe strong fields. But, of course, if we are dealing with really strong fields, the number C_k is not an infinitesimal amplitude, it is substantial in typical classical units. Then since Planck's constant is so terribly small, that is going to mean that $|\alpha|$ is going to have to be very large, and so the classical limit, strictly speaking is the limit in which the dimensionless excitation moduli, $|\alpha|$, become very much larger than unity.

An illustration of the transition from the quantum to the classical limit may be seen in the analysis of the parametric amplifier by Mollow and Glauber[7]. The dynamical behavior of an amplifier is discussed. Initially any field is put into the amplifier – any field at all. Now, in general, for arbitrary quantum fields, the function P can take on negative values of course, but worse still, it can be quite singular; we see one example of it in the case of the coherent state itself in which P is a δ function. But, as we shall see in a little more detail below, the function P can be even more singular than a δ function, and in fact it can be so singular that there is a serious question whether it exists at all. We put such fields (for example for arbitrarily singular functions P) into our model amplifier and let the amplifier amplify and develop the function P as a function of time. The density operator of the field in the amplifier of course keeps changing as the amplifier amplifies it and what we discover is that, physically, the amplifier turns this, perhaps poorly behaved function P at time zero, into an exceedingly smooth and well-behaved function at large time. If P took on negative values at time zero, it turns out there is a time beyond which it can only take on positive values. The function P depends, in other words, on the initial field. The field in the amplifier is described by a function P which eventually becomes very smoothly behaved over distances in the complex plane of the order of one unit and takes on all of the properties of a probability density in the classical limit in which it really becomes one.

We have so far provided a precise, non-trivial justification for the procedure which has always been used in optical calculations, where it has been used correctly. The grand tradition in doing optical calculations in determining the intensities of interference patterns and the like, has been to assume that a classical field is incident on the system, to calculate the field amplitude where it is being observed and then to average that field amplitude over some sort of distribution. Typically, we do not deal with the absolute intensity anyway; we normalize it relative to the input intensity. In that case, we are carrying out exactly this averaging procedure and we have shown that for the class of experiments in which we are detecting what amounts to the averages of normally ordered operators, that is a precisely correct procedure to carry out. The general characterization of such experiments is that they are the ones which give null results in the vacuum. It is not really necessary to call that the semi-classical method, or quasi-classical, or anything of the sort, it is precisely quantum mechanical provided we know that the representation of the field we are dealing with exists in this particular form in terms of the coherent states (P representation). Now, of course, if we are doing a different experiment, which is based upon a different ordering of the operators, then the story is different. There is no statement about the strength of the field required in order to use this particular representation, it holds for arbitrarily weak fields. In that sense, it is a completely quantum mechanical result – we simply should not describe it as classical in any sense at all.

Now we have to face the question "Can we find such a representation?" and, "How, in particular, do we discover what the function P is for any given field?" There is a number of such questions which we can gather together and discuss as a unit. First let us note that there is a whole class, a whole family of functions which we ought to identify as quasiprobability densities. One of them is the function P, above, another one is the Wigner function which is now some 38–39 years old[6], and still another is a function which we can form very easily by taking the diagonal matrix element of the density operator in a pure coherent state; that, obviously, is something which is non-negative, it is greater than or equal to zero, because the density operator is a positive operator,

$$P(\alpha), \qquad W(\alpha), \qquad \langle\alpha|\rho|\alpha\rangle \geq 0 . \qquad (12.102)$$

The functions (12.102) are quasiprobability densities because again there is no way in which we could measure them having measured the real and imaginary parts of the coordinate a and observing the frequencies with which they occur (notwithstanding the fact, in other words, that $\langle\alpha|\rho|\alpha\rangle$ is a positive function, it is still a quasiprobability density). These are three particular examples of a more general family of quasiprobability densities, $W(\alpha,s)$, distinguished by the parameter s. This general family of quasiprobability densities is described in detail by Cahill and Glauber[10]. There are many mathematical theorems we can prove about them and we shall only discuss one or two of them.

One theorem which is very convenient to know about, if we want to order creation and annihilation operators, is the theorem for multiplying together two exponential functions of operators which have a rather simple commutation relation. We assume that their commutator commutes with each of them. Obviously this is true for creation and annihilation operators because the commutator is unity or zero. The product of two such exponentials is not simply $\exp(A+B)$ because A and B do not commute but all the correction we need is to add in the exponent the quantity $\frac{1}{2}[A,B]$, which can be treated as a c-number since it commutes with both A and B,

$$\exp A \, \exp B = \exp(A+B+\tfrac{1}{2}[A,B]) . \qquad (12.103)$$

This holds for all operators A and B such that

$$[A[A,B]] = [B[A,B]] = 0 . \qquad (12.104)$$

The simplest way to prove this theorem is just to put a parameter, say λ, in the exponent, differentiate with respect to the parameter and then observe that the differential equation obtained may be immediately integrated to give the result. The proof is in Messiah's book on quantum mechanics[8].

---

[6] Editor's note: Keep in mind that this article has been published in 1971.

Let us now define a unitary displacement operator $D(\lambda)$. There is in the exponent a skew-Hermitian operator. This is an operator analog of, for example, $\exp[i(\text{real number})]$. It is a unitary operator in the sense that $D^{-1} = D(-\lambda) = D^\dagger(\lambda)$. Now, when we consider ordinary probability densities, it is useful, frequently to define the Fourier transform of the probability density. There are a great many calculations done in probability theory which are done more easily that way; it is the so-called characteristic function. However, in considering a density operator in the present case we have to find a somewhat different analog for the construction of the characteristic function. We shall still call what we construct a "characteristic function" but we shall take it to be formed in this particular way,

$$X(\lambda) = \text{Tr}\{\rho D(\lambda)\} = \text{Tr}\left\{\rho \exp(\lambda a^\dagger - \lambda^* a)\right\}. \tag{12.105}$$

where the Fourier transform of the probability density is really the expectation value of the Fourier exponential function. Here we simply take that expectation value by taking the trace of the density operator times the unitary operator D.

It is implicit in this exponential function that operators a and $a^\dagger$ are treated in similar ways and if we were to expand this exponential function of the operators we would find that the operators are ordered in all possible ways in each term of the expansion. This, in other words, is a symmetrically ordered expression when we expand it.

Now, there are other characteristic functions that we can define in which the operators would not be ordered exactly symmetrically. For example, we can define another version of the characteristic function in which we separate the exponential functions of the two operators (that is of course simply a different function). Here the operator content of the expectation value is written in normally ordered form:

$$X_N(\lambda) = \text{Tr}\left\{\rho \exp\left(\lambda a^\dagger\right) \exp\left(-\lambda^* a\right)\right\}. \tag{12.106}$$

Of course, that has to be related to the characteristic function defined by Eq. (12.105) and, using the multiplication theorem for exponentials, we obtain the latter with the correction factor $\exp(\frac{1}{2}|\lambda|^2)$,

$$X_N(\lambda) = \text{Tr}\left\{\rho \exp\left(\lambda a^\dagger - \lambda^* a\right)\right\} \exp\left(\tfrac{1}{2}|\lambda|^2\right). \tag{12.107}$$

We could equally well define a characteristic function in which the operators are in antinormal form and the correction factor is $\exp(-\frac{1}{2}|\lambda|^2)$,

$$X_A(\lambda) = \text{Tr}\left\{\rho \exp\left(-\lambda^* a\right) \exp\left(\lambda a^\dagger\right)\right\} \tag{12.108}$$

$$= \text{Tr}\left\{\rho \exp\left(\lambda a^\dagger - \lambda^* a\right)\right\} \exp\left(-\tfrac{1}{2}|\lambda|^2\right). \tag{12.109}$$

We are considering here two examples of something which is obviously more general.

We can define a characteristic function which depends not only on the arbitrary parameter λ but on a new variable s; we shall define it with a correction factor $\exp(\frac{1}{2}s|\lambda|^2)$ thus,

$$X(\lambda, s) = X(\lambda) \exp\left(\tfrac{1}{2}s|\lambda|^2\right) . \tag{12.110}$$

Then, having done that, we can identify the three characteristic functions which correspond to different values of the parameter s. When $s = 1$, we have the normally ordered one, when $s = 1$ we have the anti-normally ordered one, when $s = 0$ we just have the one defined originally,

$$\begin{aligned}X(\lambda, 1) &\equiv X_N(\lambda) , \\ X(\lambda, 0) &\equiv X(\lambda) , \\ X(\lambda, -1) &\equiv X_A(\lambda) .\end{aligned} \tag{12.111}$$

We can use the exponential function to define s-ordered products of operators,

$$\exp\left(\lambda a^\dagger - \lambda^* a + \tfrac{1}{2}s|\lambda|^2\right) = \sum \frac{\lambda^n(-\lambda^*)^m}{n!\,m!} \{(a^\dagger)^n a^m\}_s . \tag{12.112}$$

This is very convenient for the purposes of dealing with all orderings at once. Suppose we write the exponential operator and add a correction factor. When $s = 1$ we know that this exponential is identically equal to the normally ordered exponential. When $s = -1$ we know that this exponential is identically equal to the anti-normally ordered exponential function. When $s = 0$ it is obviously the symmetrical one. Those three cases, in other words, are included and then, of course, by letting the parameter s take on other values, we include many other "orderings". In a formal sense they are orderings of operators but I do not know how we would write them if, for example $s = \sqrt{1/\pi}$; it is simply a formal notion, but it does allow us to deal with the above three orderings together, and to see something which we would not otherwise see of the relationship among the functions (the auxiliary functions) for each variety of operator ordering. The right-hand side of Eq. (12.112) is obtained by expanding the exponential function. We can use the terms of the power series expansion to define s-ordered operators and then for $s = 1$ we are dealing with normally ordered combinations (normally ordered products), for $s = -1$ we are dealing with anti-normally ordered ones, and for $s = 0$, this is a definition of symmetrically ordered ones (and what we mean by symmetrically ordered there, is really taking the arithmetic mean of all the possible ways of arranging these particular factors in each term of the expansion).

Now we can see the following things. If we want to evaluate the statistical average of such an s-ordered product, we can do that by going back to the appropriate characteristic function, differentiating with respect to the parameter λ, or λ^* an appropriate number of times, and then setting $\lambda = 0$,

$$\frac{\partial^n}{\partial \lambda^n} \frac{\partial^m}{\partial(-\lambda^*)^m} X(\lambda, s)\bigg|_{\lambda=0} = \text{Tr}\left[\rho\{(a^\dagger)^n a^m\}_s\right] . \tag{12.113}$$

This *s*-ordered characteristic function we have defined is really a generating function for the statistical averages of the arbitrarily ordered products of operators. If we know the characteristic function, then we find the averages just by evaluating all of its derivatives at the origin, at $\lambda = 0$.

Now let us look at the case which interests us most, the case of normal order, the normally ordered characteristic function, for which $s = 1$. Let us assume that P representation exists and, in that case, we evaluate that characteristic function by just taking the statistical average of the normally ordered exponential function, and that is a trivial thing to do. Since the exponential is normally ordered, each of the operators in it simply becomes an ordinary number multiplying a coherent state; the scalar product of the two coherent states is unity and they go away and we are left then with this explicit evaluation of the normally ordered characteristic function,

$$X(\lambda, 1) = X_N(\lambda) = \int P(\alpha) \left\langle \alpha \left| \exp(\lambda a^\dagger) \exp(-\lambda^* a) \right| \alpha \right\rangle d^2\alpha$$
$$= \int \exp(\lambda \alpha^* - \lambda^* \alpha) P(\alpha) \, d^2\alpha . \qquad (12.114)$$

The exponential function here has an imaginary argument; it is a function of the two complex variables λ and α and from an algebraic standpoint, it is simplest at this stage to use Cartesian coordinates. When we do that we see that we have a Fourier transform in two variables of the function P. The Fourier transform is, presumably, invertible (that is a bit of a question here, we will have to come back to it). Let us assume that the Fourier transform *is* invertible. A convenient way in which to write the Fourier integral here uses the Dirac–Fourier integral representation of the δ function written out in two dimensions and in complex notation. This particular form of integral representation is,

$$\frac{1}{\pi^2} \int \exp\left[\lambda(\alpha^* - \beta^*) - \lambda^*(\alpha - \beta)\right] d^2\lambda = \delta^{(2)}(\alpha - \beta) . \qquad (12.115)$$

By multiplying Eq. (12.114) by an exponential function of λ of the appropriate form and integrating, we, in effect, project out the function P,

$$P(\alpha) = \frac{1}{\pi^2} \int X_N(\lambda) \exp(\alpha \lambda^* - \alpha^* \lambda) \, d^2\lambda . \qquad (12.116)$$

The function P is thus the corresponding Fourier integral of the normally ordered characteristic function, providing we can invert the Fourier transform.

The problem in inverting the Fourier transform is that for some states (and these are perfectly reasonable ones,) the normally ordered characteristic function, though it is a continuous and very well-behaved function, becomes extremely large as the modulus of λ goes to infinity. The only general bound that we can put on the normally ordered characteristic function comes from this consideration: The ordinary characteristic function, the one which is the expectation value of the unitary operator, has to be bounded by unity, simply because it is the average of a unitary operator,

$$|X(\lambda)| \leq 1 . \qquad (12.117)$$

However, since the normally ordered characteristic function is equal to the ordinary one times a Gaussian factor, then we see that the normally ordered characteristic function is only bounded by a function which blows up as λ goes to infinity,

$$|X_N(\lambda)| \leq \exp\left(\tfrac{1}{2}|\lambda|^2\right) . \tag{12.118}$$

Now, we can indeed find quantum states, which at least come close to meeting this bound as $\lambda \to \infty$ and then we are in a terrible position; if we insist upon inverting the Fourier transform Eq. (12.114), we have to calculate the Fourier integral of a function which blows up as $\exp(+x^2)$, as it were. We can find examples, and many of them, for which that happens in at least one of the two variables of integration. In other words, there are cases in which it does not seem possible to define the function P in any familiar sense in Fourier transform theory. Certainly not in the sense of square integrable functions, and in fact not more generally in the much broader context of tempered distributions which is discussed these days in connection with Fourier transform theory.

Now, we can take either of two attitudes towards this problem which seem to have evolved in the literature. One attitude is that when we are dealing with quantum states for which the function becomes so enormously singular, there is really not much convenience left in defining the P representation or attempting to use it, and there is certainly very little sense in saying that we have any remaining resemblance between the quantum mechanical and classical pictures. A different attitude is that we can define, perhaps not a P representation in the sense that we have been describing, but we can, for example, do calculations by using limiting procedures by representing the density operator as a limit of an approximating sequence of P representations. That has been the attitude of Klauder and Sudarshan[9]. Their coworkers have gone on to identify integrals of this sort, not with distributions or tempered distributions in the ordinary sense, but with things which are now being called ultra-distributions. They are still a vastly more general notion of a generalized function. This is rather deep mathematical water because once we have defined integrals of this sort, it is not at all clear what we do with them. Thus far, nothing more than the definition seems to exist.

There does not seem to be a clear or simple answer to the question of whether the P representation can be constructed in general. It depends, at least in part, on definition. Now, for the other quasiprobability densities, the situation tends to be rather simpler and we can deal with them all at once in the following way.

We can define a function $W(\alpha,s)$, where s is the ordering parameter, and take that in each case to be the Fourier transform of the appropriately ordered characteristic function,

$$W(\alpha,s) \equiv \int X(\lambda,s) \exp(\alpha\lambda^* - \alpha^*\lambda) \frac{d^2\lambda}{\pi} . \tag{12.119}$$

Then the formula (12.114) is precisely this for the case $s = 1$ except that we have put a factor of π into the formula for W,

$$W(\alpha, 1) = \pi P(\alpha).$$

We can invert the transform Eq. (12.119), in general, (or let us assume that we can): then the ordered characteristic function is written as a Fourier integral

$$X(\lambda, s) = \int \exp(\lambda \alpha^* - \lambda^* \alpha) W(\alpha, s) \frac{d^2 \alpha}{\pi}. \tag{12.120}$$

Suppose we differentiate the inverted transform (12.120) in the neighborhood of the origin. We have already shown that is the statistical average of the s-ordered product of operators (12.113). Now it is obvious that if we differentiate the exponential in Eq. (12.120) and then set $\lambda = 0$, we are finding the moments of the function W,

$$\frac{\partial^n}{\partial \lambda^n} \frac{\partial^m}{\partial (-\lambda^*)^m} X(\lambda, s) \bigg|_{\lambda=0} = \text{Tr}\left[\rho\{(a^\dagger)^n a^m\}_s\right] = \int (\alpha^*)^n \alpha^m W(\alpha, s) \frac{d^2 \alpha}{\pi}. \tag{12.121}$$

Now for all of these orderings, for the three familiar orderings and for all of those that we define formally for all complex values of s, the following is true. The statistical average of the ordered product of operators is given by the ordinary c-number moments, the classical moments, so-to-speak, of the function W. In particular, for $n = m = 0$ we have just the normalization relation,

$$\int W(\alpha, s) \frac{d^2 \alpha}{\pi} = 1, \tag{12.122}$$

because, of course, the characteristic function at argument zero is just the trace of the density operator, that is, just unity. So Eq. (12.122) is a normalization condition for the entire family of quasiprobability distributions. For $s = 0$, we have precisely the function W which was defined by Wigner in terms of real variables, not in terms of complex amplitudes, many years ago. For $s = -1$ we have anti-normal order ($X(\lambda, -1)$ is the anti-normal characteristic function), that is, just the average of the pair of exponential functions which are written in anti-normal order

$$X(\lambda, -1) \equiv X_A(\lambda) = \text{Tr}\left\{\rho \exp(-\lambda^* a) \exp(\lambda a^\dagger)\right\}$$

$$= \text{Tr}\left\{\rho \exp(-\lambda^* a) \frac{1}{\pi} \int |\alpha\rangle \langle \alpha| \, d^2 \alpha \exp(\lambda \alpha^\dagger)\right\}. \tag{12.123}$$

Here we have written that same statistical average on the fine beneath but we have written the operator 1 in between the two exponential functions as an expansion in terms of the coherent states. Now we evaluate the trace. To evaluate the trace, of course, we push the integral sign to one side and the differential to the other side with $(1/\pi)$. To evaluate the trace we just take the expectation value of the expression inside the brackets due to the coherent state. But $\exp(\lambda a^\dagger)$ is applied from the right to the left on a coherent state, the exponential function $\exp(-\lambda^* a)$ is applied to a coherent state, so those turn into ordinary numbers and what remains finally is just an integral, in fact a Fourier integral of the diagonal matrix element of the density operator ρ,

$$X(\lambda, -1) = \int \exp(\lambda \alpha^* - \lambda^* \alpha) \langle \alpha | \rho | \alpha \rangle \frac{d^2\alpha}{\pi} . \tag{12.124}$$

So, in other words, the anti-normally ordered characteristic function is just the Fourier transform of the diagonal matrix element of ρ.

Now we defined the function W for the general case of arbitrary s as the Fourier transform of X, and X is consequently the Fourier transform of W. Well, we can conclude from this, that $W(\alpha, -1)$ is just the diagonal matrix element of the density operator in a coherent state,

$$W(\alpha, -1) = \langle \alpha | \rho | \alpha \rangle . \tag{12.125}$$

There are many properties of the Wigner function that we easily prove this way but we are not too interested in symmetrical ordering so let us pass over them. We have given three identifications of values of the function W. The first is that $W(\alpha, +1)$ is πP, the second is that $W(\alpha, 0)$ is just the Wigner function, and finally, $W(\alpha, -1)$ is simply the diagonal element of the density operator in a coherent state.

An example of the sort of relationship between these functions may be proved easily in the following way. The ordered characteristic function we have defined is just the ordinary one times $\exp(\frac{1}{2} s |\lambda|^2)$. So the relationship between two of these ordered characteristic functions is obviously given by

$$X(\lambda, s) = X(\lambda) \exp\left[\tfrac{1}{2} s |\lambda|^2\right] ,$$
$$X(\lambda, s) = X(\lambda, t) \exp\left[\tfrac{1}{2} (s - t) |\lambda|^2)\right] . \tag{12.126}$$

The quasiprobability functions W are the Fourier transforms of the characteristic functions. We take the Fourier transform of both sides of Eq. (12.126). On the right we must have a convolution integral, and the kernel of that convolution integral is just the Fourier transform of this particular Gaussian function. That means that we have the following general relation among these quasiprobability densities,

$$W(\alpha, s) = \frac{2}{t - s} \int \exp\left[-\frac{2|\alpha - \beta|^2}{t - s}\right] W(\beta, t) \frac{d^2\beta}{\pi} ,$$

for

$$\mathrm{Re}\, t > \mathrm{Re}\, s . \tag{12.127}$$

12.7 The P Representation

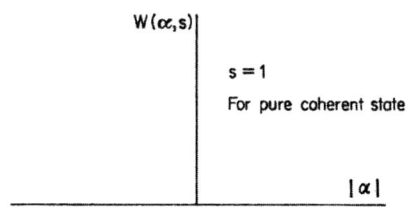

Figure 6

W(α,s), s = 1, For pure coherent state, |α|

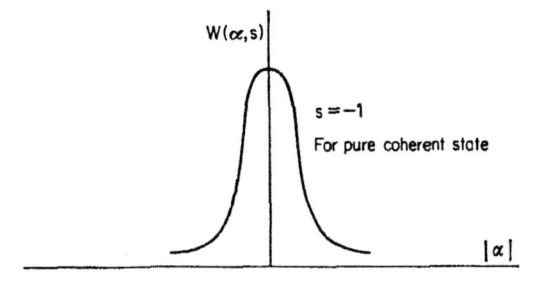

Figure 7

W(α,s), s = −1, For pure coherent state, |α|

The $W(\alpha,s)$ are all related to one another, they are all Gaussian convolution integrals of one another. For example, if $t = 1$ and $s = -1$, the diagonal matrix element of the density operator is just this particular integral,

$$\langle \alpha | \rho | \alpha \rangle = \int \exp\left(-|\alpha - \beta|^2\right) P(\beta) \, d^2\beta \, . \tag{12.128}$$

This is a trivial relationship, it is one which we see immediately by just taking the expectation value of the P representation.

There is a large family of such relationships. What happens, in fact is this. As we go along the succession of orderings from $s = 1$ to $s = -1$, we find smoother and smoother functions. A state, let us say a pure coherent state, for which we have a delta function corresponding to normal ordering, corresponds to a Gaussian function for $s = -1$, see Figs. 6 and 7. There is a tendency always to smooth the functions as s decreases. For a more general state, the function P might be something which wiggles a great deal. Figure 8 is an example with a good deal of wiggle in it, including negative values. As we go towards the other end of our spectrum of quasiprobability functions, and go eventually toward those which deal with anti-normally ordered operators we tend to find smoother versions of the functions, for example, we discover that the Gaussian function which carries out the smoothing process wipes out the negative values entirely and we have a function which is never negative as shown in Fig. 9.

There are many games we can play with these functions; we can answer a great many questions about how explicitly to order operators. For example, we might imagine that, given any particular operator, then it can easily be cast into anti-normally ordered form, or normally ordered form. If we are dealing with a polynomial in the creation and annihilation operators, it is obvious that that can be done by just using

12 Quantum Theory of Coherence

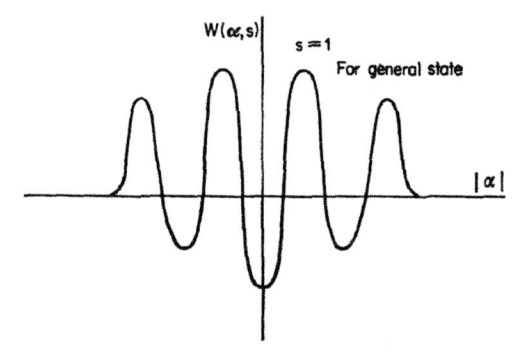

Figure 8

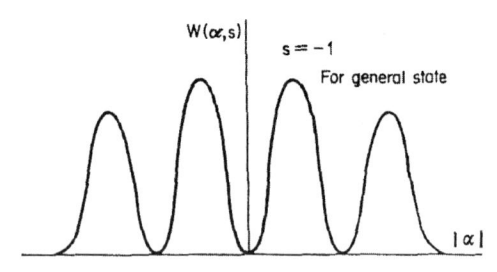

$W(\alpha, s)$, $s = 1$, For general state, $|\alpha|$

Figure 9

$W(\alpha, s)$, $s = -1$, For general state, $|\alpha|$

the commutation relations. But it is much less obvious whether those two operations can be carried out when we are dealing with infinite series in the creation and annihilation operators. In this case, the peculiar result that emerges is this: A very broad class of operators can, indeed, be expanded in terms of normally ordered products, but it is only a very much smaller class of operators, which can be expanded in terms of anti-normally ordered products of creation and annihilation operators. That is simply a mathematical result. It is one which again stands in the way of the general construction of the P representation. One of the advantages of this generalized treatment of orderings is that we have the parameter s with which we can attempt to define the P representation in singular examples. It turns out that that procedure actually fails, that there is no means by which we can define the P representation in general, for all possible states. At least, there is no method within this formalism which permits us to do this.

These questions of operator ordering, expansions, and the like are not physics. They are, at best, a matter of mathematical convenience, and at worst, a kind of preoccupation which consumes us as we discover more and more interesting theorems. Let us get back to physics.

Unhappily, the most typical state of the world is not a pure coherent state. We have lived for a long time without having to pay any attention to coherent states, physically. The most typical state of the world, the one which seems to prevail in the overwhelming proportion of cases, is the state of chaos. To define chaos we define the entropy of the field. The entropy of any chaotic system has this familiar definition which goes back to von Neumann,

$$S = -\text{Tr}(\rho \log \rho)_{\max} . \tag{12.129}$$

There are some constraints in maximizing the entropy. Obviously, the trace of the density operators has to remain 1. Suppose we fix the mean number of quanta present in the field, we then have

$$\text{Tr}\,\rho a^\dagger a = \langle n \rangle . \tag{12.130}$$

The problem then is to maximize Eq. (12.129) with the constraint of Eq. (12.130). In saying that the typical state of the world is chaos, all we are saying is what any statistical mechanician would say (they are all cynics). We have to quote some good reason to have the world in a state other than one of chaos. There are quite frequently good reasons. But if someone gives us a light source and asks for our best guess, of its state, saying that it has such and such an intensity, then the very best guess, in the absence of information is the one which maximizes entropy. That is the statement of information theory. It is very easy to carry out the mathematical operation using Lagrange multipliers to maximize the entropy and the density operator that we get for a single mode is this:

$$\rho = \frac{1}{1+\langle n \rangle} \left[\frac{\langle n \rangle}{1+\langle n \rangle} \right]^{a^\dagger a} . \tag{12.131}$$

That may not look like a familiar density operator, but in fact we are quite familiar with it for the case of thermal excitation, which, by the way, is only one particular example of chaos. (It is really making a mistake to say that all chaos is thermal, the radiation which emerges from an electric discharge tube at an energy of a few electron volts is certainly chaotic in character, but it is by no means thermal.) For the thermal case

$$\langle n \rangle = \frac{1}{\exp[(\hbar\omega/kT)-1]} .$$

Then Eq. (12.131), in the thermal case becomes the Planck density operator.

It is easy to evaluate the functions (12.102). To obtain the diagonal matrix element of the density operator, we just sandwich ρ between two coherent states,

$$\langle \alpha | \rho | \alpha \rangle = \frac{1}{1+\langle n \rangle} \exp(-|\alpha|^2) \sum_j \frac{|\alpha|^{2j}}{j!} \left(\frac{\langle n \rangle}{1+\langle n \rangle} \right)^j . \tag{12.132}$$

We can easily evaluate the infinite sum and get

$$\langle \alpha | \rho | \alpha \rangle = \frac{1}{1+\langle n \rangle} \exp\left[-\frac{|\alpha|^2}{1+\langle n \rangle} \right] = W(\alpha,-1) . \tag{12.133}$$

Taking the Fourier transform, we obtain the characteristic function for

$$X(\lambda, -1) = \exp\left[-(1 + \langle n \rangle)|\lambda|^2\right] . \tag{12.134}$$

If we then multiply by the exponential function which gives us the general form of the characteristic function for all values of the ordering parameter, we have the result:

$$X(\lambda, s) = \exp\left[-\left(\frac{1-s}{2} + \langle n \rangle\right)|\lambda|^2\right] . \tag{12.135}$$

Now, if we go back and take the Fourier transform once more we have the result finally, for the whole family of quasiprobability distributions,

$$W(\alpha, s) = \frac{2}{2\langle n \rangle + 1 - s} \exp\left(\frac{-2|\alpha|^2}{2\langle n \rangle + 1 - s}\right) . \tag{12.136}$$

In particular, for $s = 1$, we have the simple Gaussian function,

$$P(\alpha) = \frac{1}{\pi} W(\alpha, 1) = \frac{1}{\pi \langle n \rangle} \exp\left(-\frac{|\alpha|^2}{\langle n \rangle}\right) . \tag{12.137}$$

Nothing could be more obvious or simple for the case of chaos. One of the best ways, perhaps, of defining chaos, is to say it is what we must have when we have many independently contributing elements. If we take any light source, which is divisible in principle, into many small parts, and assume that they all contribute independently, then it is inevitable from the central limit theorem of probability theory (there is a simple generalization in the quantum mechanical case), that we have a Gaussian distribution of amplitudes. That is precisely what we mean by chaos. We shall see below what results from this chaos as far as measurements on the field are concerned.

12.8
Chaotic States

All of the older and more traditional measurements in optics are ones that we can describe in terms of the first order correlation function $G^{(1)}$. In a sense, the older optics by measuring intensities only, really looks no further into the field than into the contents of the function $G^{(1)}$. For the case of a stationary field, as we shall see, the first order correlation function is determined ordinarily just by the spectral data of the field, i.e. the intensities of excitation of the various modes of the field. When we say spectral data we mean polarization data as well as the usual sort of spectral data. We can imagine different sorts of fields, which have identical spectra. We might take them from different sources, such as a laser or a black-body; we might have intensity problems, but we can, presumably, filter black body radiation to achieve any spectral radiation distribution that we like, including the distribution which comes from a laser. Then we might ask the question, "Are these two fields distinguishable?"

12.8 Chaotic States

If their spectra are identical, the functions $G^{(1)}$ will be identical, and then no matter which of the traditional optical experiments that we perform, we shall never be able to distinguish between these two light beams.

On the other hand, the distinction becomes very clear when we begin to look at the effects that are involved in the higher order correlation functions. The first of these was the HBT effect. We have to measure the second order correlation function or some higher order correlation function in order to see the difference. For the particular case of chaotic fields, it must still be true that these higher order correlation functions are determined simply by the spectral data of the field. We know that for a multi-mode, chaotic state, the field (P representation) can be described in terms of a multivariate Gaussian function

$$P(\{\alpha_k\}) = \prod_k \frac{1}{\pi \langle n_k \rangle} \exp\left(-\frac{|\alpha_k|^2}{\langle n_k \rangle}\right). \tag{12.138}$$

The only information that we need in order to construct this function is just the mean number of photons in each mode. Those are exactly the same data that are contained in the first order correlation function. So, for the case of a stationary Gaussian field, i.e. a chaotic state, the entire density operator contains no more information than is contained in just the first order correlation function. If we measure the first order correlation function for such fields, we measure everything. Now that is not to say that the higher order correlation functions have no interesting structure, in fact they have some structure, but it must be possible for the case of chaotic fields to construct the higher order correlation functions from a knowledge simply of the first order correlation function. There is a theorem which permits us to do that. It is a generalization of a theorem which holds in the theory of stochastic processes quite generally. If we use this Gaussian distribution Eq. (12.138) to construct the correlation function of n th order, we find (remember it has altogether $2n$ different arguments, in general) that the higher order correlation function can always be expressed as the sum of products of the first order correlation functions and for the n-th order correlation function we have to form an n-fold product; we have to sum n-fold products over all possible permutations of the coordinates,

$$G^{(n)}(x_1 \ldots x_n y_1 \ldots y_n) = \sum_P \prod_{j=1}^n G^{(1)}(x_j y_{P_j}). \tag{12.139}$$

In other words, if the first coordinate in $G^{(1)}$ takes on all the values from $1 \to n$, the coordinates associated in the same functions with those coordinates constitute the $n!$ permutations of the set of numbers $1, \ldots, n$.

As an example of such a function, consider the two-fold joint counting rates, the second order correlation function, in other words, with its arguments repeated; x_1 occurs twice, x_2 occurs twice. That is a sum of two-fold products of correlation functions. The identity permutation corresponds to the first term, and then the simple interchange of the x_1 and x_2 coordinates determines the second term. The two factors are complex conjugates of one another so we simply write the squared absolute value, for the second term:

$$G^{(2)}(x_1 x_2 x_2 x_1) = G^{(1)}(x_1 x_1) G^{(1)}(x_2 x_2) + |G^{(1)}(x_1, x_2)|^2 . \tag{12.140}$$

If the field were a coherent one, only the first term would be present, the correlation function would factorize and all we would have is

$$G^{(2)}(x_1 x_2 x_2 x_1) = G^{(1)}(x_1 x_1) G^{(1)}(x_2 x_2) .$$

This means that there would be no correlation at all in the counting rates observed by two counters placed in the field; there would be no tendency towards photon coincidences. The additional term $|G^{(1)}(x_1, x_2)|^2$ is due to the fact that the field to be averaged has a Gaussian distribution of field amplitudes. The term $|G^{(1)}(x_1, x_2)|^2$ is the one responsible for the HBT effect.

Let us now see how we calculate first order correlation functions. In most of the stationary fields we work with, we can always choose a suitable set of modes such that the mean value of the products of the creation operator for the mode k and the annihilation operator for the mode k' vanishes unless the modes k and k' are the same and then, of course, it must be just the mean number of quanta in the mode,

$$\mathrm{Tr}\left\{\rho a_k^\dagger a_{k'}\right\} = \langle n_k \rangle \delta_{kk'} . \tag{12.141}$$

Now in the stationary field, the expectation value when these mode indices are different must vanish if the modes are at different frequencies. If the modes are of the same frequency, it is possible to have non-vanishing values for different modes, but then a suitable choice of the modes k and k' (in other words the use of the correct principal axes in the Hilbert space of modes) will lead to a reduction of the expectation value to the form Eq. (12.141) in any case. Now if this is true, then we find the first order correlation function by just substituting for the expectation values in the formula for the expansion of $G^{(1)}$ in terms of the mode amplitudes, and we obtain

$$G^{(1)}(rt, r't') = \tfrac{1}{2} \sum_k \hbar \omega_k \langle n_k \rangle u_k^*(r) u_k(r') \exp[i\omega_k(t-t')] . \tag{12.142}$$

We see how for the stationary field simply the set of spectral data (the set of mean occupation numbers) determines the first order correlation function.

Consider a specific example, say a beam of plane waves. As long as we have the symmetry which describes plane waves, i.e. a beam moving uniformly in the $+x$ direction for which the mode function is $u_k(x) \propto \exp(ikx)$, there need be only one argument in the first order correlation function. We shall call that argument τ,

$$G^{(1)} = G^{(1)}(\tau), \qquad \tau = t_1 - t_2 - \frac{1}{c}(x_1 - x_2) . \tag{12.143}$$

The function $G^{(1)}$ is independent of y and z if x is the direction of propagation of the waves. Let us make a spectral assumption about the plane waves; a simple one to make

12.8 Chaotic States

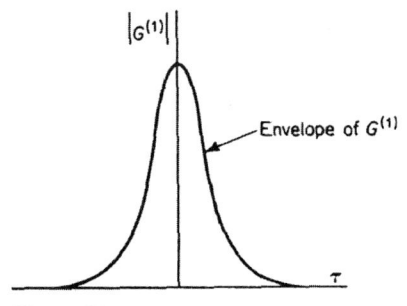

Figure 10

is that the occupation numbers have a Gaussian behavior, i.e. we have a Gaussian line profile such as may be given by Doppler broadening,

$$\langle n_\omega \rangle = \text{constant} \times \exp\left[-\frac{(\omega - \omega_0)^2}{2\gamma^2}\right] . \tag{12.144}$$

We could work with any line profile but since we have to take the Fourier integral of the spectral density, the choice of a Gaussian makes this easy. The integral is the summation over modes. We assume that the modes are closely spaced and therefore we can integrate over the propagation vectors or, equally well, frequencies (since we have specified the direction of propagation). We find the first order correlation function in terms of the interval variable τ is given by

$$G^{(1)}(\tau) = G^{(1)}(0) \exp\left[i\omega_0 \tau - \tfrac{1}{2}\gamma^2 \tau^2\right] . \tag{12.145}$$

We see that $G^{(1)}$ oscillates with the central frequency of the line and that it decreases for values of $\tau \neq 0$. The envelope of $G^{(1)}$ has a Gaussian sort of behavior (Fig. 10).

In a photon coincidence experiment we measure $G^{(2)}$. To calculate $G^{(2)}$ we note that $G^{(1)}(x_1,x_1)$ is the function for which τ vanishes so we have $G^{(1)}(0)\,G^{(1)}(0)$ as the first term of Eq. (12.140),

$$G^{(2)}(x_1 x_2, x_2 x_1) = G^{(1)}(x_1,x_1)\,G^{(1)}(x_2,x_2) + |G^{(1)}(x_1,x_2)|^2$$

$$= G^{(1)}(0)\,G^{(1)}(0) + |G^{(1)}(0)|^2 \exp(-\gamma^2 \tau^2) . \tag{12.146}$$

The cross correlation function gives the variable contribution. The square of the cross correlation function does not have the oscillating part, it is simply the squared envelope, as it were, of $G^{(1)}$, so we get

$$G^{(2)}(x_1 x_2, x_2 x_1) = \left\{G^{(1)}(0)\right\}^2 \{1 + \exp(-\gamma^2 \tau^2)\} . \tag{12.147}$$

By plotting $G^{(2)}$ (coincidence rate) against τ (Fig. 11), we see that $G^{(2)} \to 1$ as $\tau \to \pm\infty$ but is 2 at $\tau = 0$. This shows that the HBT effect is not a small effect. In the

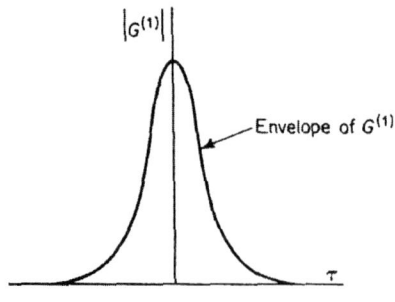

Figure 11

original experiment it was small because the counters had long resolution times. The HBT effect does not last over a long time interval. The time interval τ is of order of $1/\gamma$ where γ is of the order of the frequency band width, so it helps to have a light beam with very narrow frequency spread to make the effect last for a somewhat longer time. The shape of the curve here is Gaussian because that is the transform of the spectrum we used. Had we used a natural line-shape, then we would have had an exponential behavior of the HBT correlation on each side of $\tau = 0$ as shown in Fig. 11. Notice that we are describing the effect without putting any mirrors in the calculation at all. We are just considering detection of photons in a one dimensional beam with two counters which we assume have infinitely small resolving time. There is always a tendency towards coincidences and this is, in effect, a statement that the photons tend to clump in the beam. We can still ask why this happens. The argument we have given is mathematically complete but it may be that there is not enough in what we have considered to convey the full physical image. In other words, the theorem which says that the higher order correlation functions can be expanded in terms of the lower ones is the essential point of the derivation. Everything we ever need to know about the character of Gaussian stochastic processes really is included in that statement but it is not, perhaps, completely clear what that statement is in physical terms. So let us use a slightly different method. We need not use this fancy expansion to evaluate the higher order correlation function, we can go back to the Gaussian distribution itself and imagine doing the integral. Let us see how that works.

Consider the correlation functions for $\tau = 0$. The HBT effect is there already; we do not see how the effect relaxes over time, but we do see, for example, that the coincidence rate is twice the square of the ordinary counting rate, Eq. (12.149). For time interval zero, $G^{(1)}(0)$ is just the second moment of $P(\alpha)$,

$$G^{(1)}(0) = \int P(\alpha)|\mathcal{E}|^2 d^2\alpha = \text{constant} \int \exp\left(-\frac{|\alpha|^2}{\langle n \rangle}\right)|\alpha|^2 d^2\alpha . \quad (12.148)$$

Here for simplicity we have assumed that just one mode of the field is excited. Nothing in the argument really depends on the number of modes that is excited except the relaxation properties. The fact that there are many different modes excited which all

12.8 Chaotic States

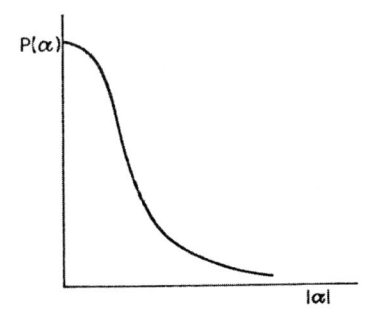

Figure 12

have slightly different frequencies obviously determines the way in which the HBT correlation disappears as a function of time, but here we are looking at the effect just at one moment of time, at time interval zero. In that case $G^{(1)}$ is just the second moment of the Gaussian distribution.

The function $G^{(2)}$ is just the fourth moment of the Gaussian distribution,

$$G^{(2)} = \int P(\alpha) |\mathcal{E}|^4 \, d^2\alpha = 2 \left(\int P(\alpha) |\mathcal{E}|^2 \, d^2\alpha \right)^2 = 2 \left\{ G^{(1)}(0) \right\}^2. \quad (12.149)$$

It is a property of Gaussian functions that the fourth moment is twice the square of the second moment. Does that have any physics in it? Well, it is a little hard to call it physics – it is so simple. The Gaussian function has a long tail, even though it decreases quickly (Fig. 12). The fact that the Gaussian has this tail means that the fourth moment is larger than the square of the second moment. If the Gaussian function were to cut off sharply, it would not be possible for the fourth moment to be so very much larger than the square of the second moment. So here we have it. It is because the Gaussian function has a tail that we find that these higher order correlation functions evaluated for time zero are generally larger by a factor of $n!$ than the n-th powers of the first order correlation functions. For a coherent field, on the other hand, with random phase, we have essentially a δ function structure. If we had a precisely determined phase, for a coherent state, then, indeed, we would be considering a δ function at one point in the complex plane. We essentially never have such information about laser beams; we never know, for example, when precisely they are turned on. So the actual representation of the function P for the most perfect of laser beams would be one in which there is a kind of circular symmetry in the complex α plane and we have a probability distribution which is zero, except on a cylindrical shell, a shell obtained by rotating the diagram Fig. 13 about its vertical axis. That is what P would look like for the most perfect of coherent states with random phase. Now we see that that kind of distribution has no variance at all in radial terms. For that kind of distribution it is simply true that the fourth power of $\mathcal{E}$ when averaged is precisely equal to the average of $\mathcal{E}$ squared, and squared again; that is why there is no HBT effect for the purely coherent field.

$P(\alpha)$

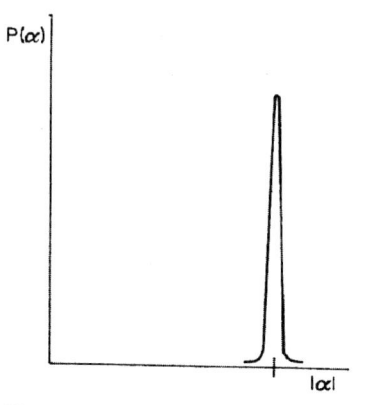

$|\alpha|$

Figure 13

Let us put it in still more elementary terms. In a chaotic field, the amplitude is fluctuating all the time. Sometimes the amplitude is large, more often, the amplitude is small. If we put two photon counters in the field, they will both of them tend to register when the amplitude happens to be large, and they will neither of them tend to register when the amplitude happens to be small. That automatically means that there will be a correlation in the behavior of the two counters. The two counters do tend to register together and they tend to keep quiet together. No such correlation is present in the coherent field because the amplitude tends to remain constant as a function of time. Both counters are seeing the same amplitude at all times. Not just the same as one another, but an amplitude constant in time. For that reason, their behavior is statistically independent of one another. Now that leaves us with a number of other questions, of course. It was surmised at an early stage that a laser beam was represented to quite good accuracy by one of these coherent fields. Why should this be so? It is a long story to analyze in detail. One reason is that, to a fair approximation, a laser is a macroscopic distribution of atomic polarizations equivalent to a macroscopic current distribution which does not recoil perceptibly when it emits photons. A laser is, to a good approximation, the kind of thing Kibble described by speaking of the radiation emitted by a specified current distribution. The situation is one in which the current is specified and the field is quantized, but that is not enough really to characterize the situation physically.

We could still imagine random behavior in this classical current which would lead to much variance in the amplitude radiated by the field and there is something intrinsic to the laser which tends to suppress that amplitude fluctuation. The laser is a nonlinear device which has a kind of feed-back built into it. The field is radiated by a population inversion which has to be maintained in order for the device to keep amplifying its own field. If the field ever gets stronger than a certain steady value, it tends to deplete the inversion, the amplification tends to decrease, and the device tends to stabilize its own amplitude. The theoretical analyses of the laser do not really show that a precisely coherent state is radiated by the device, but they do tend to show that, at sufficiently high power levels, quite a fair approximation to such a coherent state is what is radiated. The complications of dealing with the laser problem itself are quite imposing.

We see that to explain the HBT effect quantum mechanically we go back, in essence, to its classical explanation. The important point is that we have found the justification for using that essentially classical explanation in the quantum context.

Those are the best ways of explaining the HBT effect in elementary form. However, every time I present explanations of this sort, I find students come up later and ask, "What about the photons in all of this? What you have done really is to persuade us that it is not worth talking about photons at all. What you ought to do is go back to the same old classical discussions that were used long ago and simply give them their correct quantum mechanical interpretation." Now, all of us were taught somewhere in the early chapters of the quantum mechanics books that photons are little puffs of probability amplitude; that they have well defined wavepackets that move with the velocity of light and tell us whatever we are capable of knowing about the localization of the photon. In our discussions of photon counting experiments, of the HBT effect in particular, we do not see any of those wavepackets at all. My own feeling is that we are much happier for not having the wavepackets present. Conceptually, the wavepackets are certainly a very good way of looking at the localization of quantum mechanical particles but, in practice, we never know the various parameters that we would have to know in order to determine the wavepackets for individual particles. If a cyclotron produces a beam of charged particles, there is nobody in the cyclotron laboratory who can tell us what the wavepackets are of those emerging particles. What they *can* tell us is the degree of collimation of the beam, the energies of the particles in the beam and general data of that sort. The simplest description of that beam is a description analogous to the kind we have been using here for photons; it is really a density operator description which does not attempt to construct wavepackets for the individual particles because, to do it, we have to deal with information that we simply do not have. On the other hand, at least from a conceptual standpoint, it is an interesting thing to ponder the problem of just how we would go about constructing a description of a beam of particles if we did know everything about the wavepackets of the individual particles. That is what we shall do in the next section for the particular case of a chaotic field.

12.9
Wavepacket Structure of Chaotic Field[11]

The formula for the chaotic density operator is just the Gaussian mixture of coherent states,

$$\rho = \int \exp\left[-\sum_l \frac{|\alpha_l|^2}{\langle n_l \rangle}\right] |\{\alpha_k\}\rangle \langle\{\alpha_k\}| \prod \frac{d^2\alpha_k}{\pi \langle n_k \rangle}. \qquad (12.150)$$

Evaluating the Gaussian integrals we obtain,

$$\rho = \sum_{\{m_k\}} \prod_k \frac{\langle n_k \rangle^{m_k}}{(1 + \langle n_k \rangle)^{1+m_k}} |\{m_k\}\rangle \langle\{m_k\}| . \tag{12.151}$$

The density operator is the sum of state vector dyadics corresponding to all possible numbers of quanta in the individual modes. Because it is a stationary operator and has a separable or factorizable P representation it is, in fact, diagonal in terms of the occupation number representation. The coefficients of the various quantum states are just powers of elementary parameters. The parameters are just the ratios $\{\langle n_k \rangle / (1 + \langle n_k \rangle)\}$ which for the case of thermal radiation, when raised to the power m_k, become

$$\left(\frac{\langle n_k \rangle}{1 + \langle n_k \rangle}\right)^{m_k} = \exp\left(-\frac{m_k \hbar \omega_k}{kT}\right) . \tag{12.152}$$

Suppose we are given a collection of as many identical photon packets as we want. How do we put those photon wavepackets together to make the field which is described by the chaotic density operator? When we discover how to do that, we shall learn something about the microscopic structure of chaotic fields. We shall find out, within chaotic fields, something about where the photons are relative to one another and the distribution of their number. A one-photon packet is described by

$$\sum_k f(k) a_k^\dagger |\text{vac}\rangle . \tag{12.153}$$

This is the basic element in the calculation. We are going to create a single photon but, we may create a photon in any mode so we sum over all modes, k. It takes a certain set of amplitudes, one amplitude $f(k)$ for each mode in order to specify what the packet is. We can make the packet look like anything we please by appropriate choice of this set of numbers $f(k)$. Imagine that the packet is localized somewhere in space. In fact, it might be localized in a different way. Nothing we are saying really will bear on the question of just how it is localized but imagine that it is a wavepacket that has some particular spatial localization. Let us even assume that that packet is somewhere near the origin $r = 0$, just for simplicity. This state has to obey a normalization condition, which we obtain by just squaring the state, thus

$$\left\langle \text{vac} \left| \sum_k f^*(k) a_k \sum_{k'} f(k') a_{k'}^\dagger \right| \text{vac} \right\rangle = \sum_k |f(k)|^2 = 1 . \tag{12.154}$$

The square is the sum of the squared moduli of the coefficients $f(k)$. So those coefficients f must obey the normalization Eq. (12.154). Now let us imagine displacing this wavepacket from the origin to the point r. We do that by applying a displacement operator, which in quantum mechanics takes the particular exponential form shown:

$$\exp(-i\mathbf{P} \cdot \mathbf{r}/\hbar) \sum f(k) a_k^\dagger |\text{vac}\rangle = \sum f(k) \exp(-i\mathbf{k} \cdot \mathbf{r}) a_k^\dagger |\text{vac}\rangle$$

$$\equiv A^\dagger(\mathbf{r}) |\text{vac}\rangle . \tag{12.155}$$

12.9 Wavepacket Structure of Chaotic Field

We know what the momentum of any one of these photon states is. Let us assume now that it is a state with a particular propagation vector. Let us assume that the modes k, are plane-wave modes. So to apply the displacement operator means just including in our summation over modes, the particular phase factor associated with that propagation vector. The summation of creation operators which creates a photon packet in the neighborhood of $\mathbf{r}$ can be written as

$$A^{\dagger}(\mathbf{r}) \equiv \sum_k f(k) \exp(-i\mathbf{k}\cdot\mathbf{r}) a_k^{\dagger} . \tag{12.156}$$

These are the basic elements in the "do it yourself kit". We are given as many as we want of these packet creation operators referring to all possible points in space. The problem is to put these operators together so as to create a chaotic field. The method is not immediately obvious. If we just take a collection of the operators and apply them completely randomly (or in what we think is a completely random manner) to the vacuum state, we shall not, in fact, be creating chaos, we shall be creating order.

The state with n packets identical except that they are shifted relative to one another in position so that they are at the positions $\mathbf{r}_1,\ldots,\mathbf{r}_n$ is given by

$$\prod_{j=1}^{n} A^{\dagger}(\mathbf{r}_j) |\text{vac}\rangle . \tag{12.157}$$

Here we have applied the appropriate succession of creation operators to the vacuum state. It does not make any difference what order we apply those operators in, since the creation operators all commute with one another. In fact, that shows us that we are obeying the necessary symmetry rules; this state is symmetrical under the interchange of the coordinates of all the wavepackets, so we are describing bosons correctly.

We define the normalization function for the state Eq. (12.157) as

$$W^{(n)}(\{\mathbf{r}_j\}) \equiv \left\langle \text{vac} \left| \prod_j A(\mathbf{r}_j) \prod_l A^{\dagger}(\mathbf{r}_l) \right| \text{vac} \right\rangle . \tag{12.158}$$

Let us describe an essential problem that we encounter in defining the state given by Eq. (12.157). The state is not normalized and the normalization of the state depends on just where the photons are relative to one another. If we are going to normalize the state, we have to calculate the norm of the vector we have written (its square length, in other words). The squared length is written out on the right of Eq. (12.158). It is the vacuum expectation value of a product of packet annihilation operators and packet creation operators. To evaluate the function $W^{(n)}(\{\mathbf{r}_j\})$ it is a good idea to define another function which we can construct in terms of a single creation operator for a packet at one point and a single annihilation operator for a packet at a different point,

$$v(\mathbf{r}-\mathbf{r}') \equiv \left\langle \text{vac} \left| A(\mathbf{r}) A^{\dagger}(\mathbf{r}') \right| \text{vac} \right\rangle . \tag{12.159}$$

The operator product can be written equally as a commutator because if we consider the two operators in their commuting positions, we then have an annihilation operator on the right which applies to the vacuum state, and it is zero. Since the commutator is a *c*-number the function v reduces simply to the commutator itself,

$$v(\boldsymbol{r}-\boldsymbol{r}') = [A(\boldsymbol{r}), A^\dagger(\boldsymbol{r}')] \,. \tag{12.160}$$

By using the expression (12.156) and its adjoint we then find

$$v(\boldsymbol{r}-\boldsymbol{r}') = \sum_k |f(k)|^2 \exp[i\boldsymbol{k}\cdot(\boldsymbol{r}-\boldsymbol{r}')] \,. \tag{12.161}$$

This is the explicit expansion of the function. If we know the packet amplitude $f(k)$ then we know the function v.

The function v, however, does not tell us anything about the phases of the packet amplitudes. It is easy to prove the following properties of function v,

$$v(0) = 1 \,,$$
$$v(-\boldsymbol{r}) = v^*(\boldsymbol{r}) \,,$$
$$|v(\boldsymbol{r})| \leq 1 \,. \tag{12.162}$$

Firstly, the normalization condition; if $\boldsymbol{r} = \boldsymbol{r}'$, we have just the sum of the squared moduli of $f(k)$, that is, unity. If we invert the argument of the function we get the complex conjugate, obviously. Furthermore, this function is bounded by unity, which is the value it takes on for zero argument, and if we calculate this function, we find, typically, for any smooth behavior of the amplitudes $f(k)$, the function goes to zero as the points $\boldsymbol{r}$ and $\boldsymbol{r}'$ recede from one another. The packet, in other words, would have only a finite size. Any photon packet would have some characteristic packet dimension and v would vanish for $|\boldsymbol{r}-\boldsymbol{r}'|$ larger than that particular dimension.

The function $W^{(n)}$ which is the normalization function, is easily seen to be the sum of products given by

$$W^{(n)}(\{\boldsymbol{r}_j\}) = \sum_{\text{perm}} \prod_{j=1}^{n} v(\boldsymbol{r}_j - \boldsymbol{r}_{P_j}) \,.$$

We can use the same technique as we used to evaluate high order correlation functions for the Gaussian case to prove this theorem for the normalization functions. What that means then for the first few cases is worth illustrating. For a single photon packet we have

$$W^{(1)} = 1 \,. \tag{12.163}$$

If we have two packets in the field, $W^{(2)}$ contains the product $v(0)v(0)$ that is, just unity, and then we have the additional term with the arguments permuted which just adds $|v(\boldsymbol{r}_1 - \boldsymbol{r}_2)|^2$; thus

$$W^{(2)} = 1 + |v(\boldsymbol{r}_1 - \boldsymbol{r}_2)|^2 \,. \tag{12.164}$$

If we have a state with three packets in it we have a somewhat more complicated structure,

$$W^{(3)} = 1 + |v(\boldsymbol{r}_1 - \boldsymbol{r}_2)|^2 + |v(\boldsymbol{r}_2 - \boldsymbol{r}_3)|^2 + |v(\boldsymbol{r}_3 - \boldsymbol{r}_1)|^2 \\ + 2\operatorname{Re} v(\boldsymbol{r}_1 - \boldsymbol{r}_2) v(\boldsymbol{r}_2 - \boldsymbol{r}_3) v(\boldsymbol{r}_3 - \boldsymbol{r}_1) . \quad (12.165)$$

These two cases $W^{(2)}$ and $W^{(3)}$ illustrate the structure of the normalization function $W^{(n)}$ in its simplest form. If the packets are far apart, for example, $\boldsymbol{r}_1$ is very different from $\boldsymbol{r}_2$, then the overlap amplitude v vanishes and if all of the packets are far apart in the n-fold case and tend not to overlap, then all of these normalization functions are going to reduce, essentially, to unity $W^{(n)} \approx 1$. The normalization functions, however, tend to differ from unity when the packets are made to overlap. When $|\boldsymbol{r}_1 - \boldsymbol{r}_2|$ is small enough so that the packets do overlap, for example, the function v rises to its maximum value of one, and, more generally, we find that the maximum value of $W^{(n)}$ is $n!$. So that

$$n! \geq W^{(n)} \geq 0 . \quad (12.166)$$

If we imagine distributing wavepackets at random positions in the field, we shall find the normalization function is essentially always one, or nearly always one if we have a very dilute photon gas. But as the packets begin to overlap, the normalization moves away from unity and every time two packets precisely overlap there is a bump in the normalization function. We have to take that into account of course if we want to deal with a correctly normalized state. One of the hazards of dealing with wavepackets is that calculations based on them have not always been done with correctly normalized states.

A correctly normalized state for wavepackets is given by

$$|n\{\boldsymbol{r}_j\}\rangle \equiv \frac{1}{\sqrt{W^{(n)}(\{\boldsymbol{r}_j\})}} \prod_j A^\dagger(\boldsymbol{r}_j) |\text{vac}\rangle . \quad (12.167)$$

We have taken the state generated by the n packet creation operators and have put in the reciprocal of the square root of our normalization function.

In our process of constructing the chaotic field we naturally want to put these photon packets in all possible different positions in the field, that is, we naturally think of a chaotic field having properties we assume to be uniform as a function of spatial variables. We wish to consider a mixture of states in which these photon packets are in all possible positions. A pure state with one particular set of positions for the photon packets is given by Eq. (12.167). We mix these states together by multiplying by some kind of a weight function of the photon packet coordinates, then integrate over all photon packet coordinates. We know that the total number of photons in a chaotic field is quite indefinite because, in general, we do not know how many photons there are in any one mode so we have only a statistical notion of how many there are in the entire field. We therefore have to use a succession of these weight functions, one for every value of n and sum over all possible values of the number of photons in the field. The density operator we construct is given by

$$\rho = \sum_n \int F_n(\boldsymbol{r}_1 \ldots \boldsymbol{r}_n) |n\{\boldsymbol{r}_j\}\rangle \langle n\{\boldsymbol{r}_j\}| \frac{\prod d\boldsymbol{r}_j}{V^n} \,. \tag{12.168}$$

We now wish to know the kind of weight functions we should choose in order to obtain a chaotic field. In other words, we wish to generate real chaos, to scrupulously avoid order, to avoid having the entropy of the field at anything less than its maximum value. The crucial point is the location of the photons to maximize entropy. At this point it should be noted that we can prove that the correct weight function for a chaotic field is

$$F_n(\boldsymbol{r}_1 \ldots \boldsymbol{r}_n) = c_n W^{(n)}(\boldsymbol{r}_1 \ldots \boldsymbol{r}_n) \,, \tag{12.169}$$

where c_n is a constant. The proof is a bit long and is therefore omitted, but basically one way of proving it is as follows. We take the matrix elements of ρ between coherent states, and go back to the original chaotic Gaussian density operator Eq. (12.150) and take the matrix element of that between coherent states and then compare the two expressions obtained. We want them to assume the same form and the only way they can ever have the same form is by taking these weight functions F_n to be proportional (the constant of proportionality is c_n which is independent of packet positions) to the normalization function $W^{(n)}$. Now that is simply a mathematical result and the only way of justifying it is by going back and making the explicit comparison and showing algebraically that that is the only way we can do it. But we have there an interesting statement about the microscopic structure of a chaotic field because this function F_n is a weight function telling about the tendency of photon packets to overlap, (whether they really tend to overlap or not). The statement is here that they must have a distinct tendency to overlap. The function $W^{(n)}$ is something which is unity when no photon packets overlap. However, every time a pair of photon packets overlaps it jumps away from the value unity to a larger value. In other words, this weighting tends to favor the overlap of wavepackets. Many things are said about the HBT effect. A number of authors state that it is purely and simply a consequence of Bose–Einstein statistics. We can see here that that statement is an extremely incomplete one, and if that is the only statement which is made, it is incorrect. It is incorrect because we can imagine photon wavepackets having any distribution of positions we please. Boson statistics are extremely permissive, we can imagine constructing states with any kind of anti-correlation we like between photon packets. How we would generate those states and what the practical device is that would do it, is another question. It might be very difficult to think of a device which would create anti-correlation between photons.

The essential point is this. For Boson statistics and a requirement of maximum entropy, then the only way of meeting the requirement is to give the photon packets a distinct tendency to overlap and use the weight function $W^{(n)}$ in this construction which gives us an additional unit of weight every time two packets overlap. If we use this weight function in doing the calculation, then the density operator obtained is

$$\rho = \sum_n c_n \int A^\dagger(r_1) \ldots A^\dagger(r_n) |\text{vac}\rangle \langle \text{vac}| A(r_n) \ldots A(r_1) \frac{\prod dr_j}{V^n} . \tag{12.170}$$

The function $W^{(n)}$ put in as a weight, cancels the reciprocal of $W^{(n)}$ which is inserted in order to normalize, and so the density operator has a very much simpler structure. It has precisely the sort of structure that we would have guessed if we had guessed wrongly and forgotten all about the normalization problem to begin with. That is what has happened in some of the papers that have been written on the subject which purport to derive the HBT effect. It is one of the reasons why they say it results only from Boson statistics. What has happened is that, without any attempt at justification, they have used a structure like Eq. (12.170) which is not normalized (it has a variable normalization). In fact, the HBT effect has been put into the very first statement of the derivation, and, naturally enough, it is taken out at the end.

We still have to determine the coefficients c_n, and it is not clear at this stage what the relationship is between our packet amplitudes $f(k)$ and the mean occupation numbers n_k. It is easy enough to derive these relationships. We simply carry out the spatial integrations on the form of the density operator given by Eq. (12.170) and we obtain the following expression in terms of the n-quantum states,

$$\rho = \sum_{n=0}^\infty n! c_n \sum_{\sum m_k = n} \prod_k |f(k)|^{2m_k} |\{m_k\}\rangle \langle \{m_k\}| . \tag{12.171}$$

This is a double summation, a summation over the total number of quanta in the field and a summation over the set of occupation numbers of the individual modes. In the first summation the total number of quanta in the field has to be restricted to the value n which is then summed over its full range from zero to infinity. We now compare this expression with the one that goes back to the chaotic density operator Eq. (12.150). In Eq. (12.171) we see powers of the squared modulus of the packet amplitude. In the original one we had powers of the parameter $\langle n_k \rangle / (1 + \langle n_k \rangle)$ where $\langle n_k \rangle$ is the mean occupation number. Obviously these two have to be proportional to one another. So the identification of the one density operator with the other requires that

$$\frac{\langle n_k \rangle}{1 + \langle n_k \rangle} = Z |f(k)|^2 \tag{12.172}$$

where Z is the constant of proportionality. Then to complete the identification, the comparison of Eq. (12.171)) with Eq. (12.150), we have to say that the unknown coefficients c_n which occur in the weight functions are given by

$$c_n = \frac{Z^n}{n!} \prod_k \frac{1}{1 + \langle n_k \rangle}$$

$$= \frac{Z^n}{n!} \prod_k (1 - Z |f(k)|^2) . \tag{12.173}$$

12 Quantum Theory of Coherence

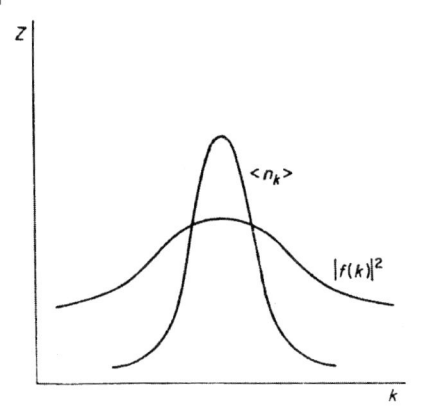

Figure 14

The parameter Z is then given by the normalization condition

$$\sum_k |f(k)|^2 = 1 \quad \Rightarrow \quad Z = \sum_k \frac{\langle n_k \rangle}{1 + \langle n_k \rangle}.$$

This means that the mean occupation number and the squared packet amplitude have slightly different shapes as functions of frequency or wave number as shown in Fig. 14. When we put the first photon packet in the field we may think that the spectrum of the field eventually is going to look like $|f(k)|^2$, but it will not because of the effect of wavepacket overlap. There will be the relationship we have written between $|f(k)|^2$ and the mean occupation number, which is a nonlinear relationship.

Finally, when we put all of this together, we have the following two expressions for the density operator. One in terms of the wavepacket amplitude squared and the n-quantum states and the other in which we see explicitly the sum of the averages of the pure states, which correspond to the presence of packets at all possible positions,

$$\rho = \prod_k (1 - Z|f(k)|^2) \sum_n \frac{Z^n}{n!} \sum_{\sum m_k = n} \prod_k |f(k)|^{2m_k} |\{m_k\}\rangle \langle\{m_k\}|,$$

$$\rho = \prod_k (1 - Z|f(k)|^2) \sum_n \frac{Z^n}{n! V^n} \int \prod_j A^\dagger(r_j) |\text{vac}\rangle \langle\text{vac}| \prod_j A(r_j) \prod_j dr_j. \quad (12.174)$$

Now can the latter expression be used for anything?

Its principal use has been given above and lies in the statement that, in constructing a chaotic field from wavepackets, it is necessary to put the packets together with a weighting $W^{(n)}$ which tends to favor the overlap of packets. We can use this form of the density operator for various sorts of calculations, for example, it is a relatively convenient form for finding the distribution of the number of photons in the field. But if we try doing photon counting calculations with this representation of the density operator, we find that the complications are immense. For example, to describe the absorption of a single photon from the field we have the problem that the photon may be absorbed from any of these packets and just to describe the single, most elementary process of absorption of one photon, we would have to take into account interference terms in all of these packets because we generally do not know from which packet the photon was absorbed. So the use of photon packets does lead to immense algebraic complications in actual calculation and it is something to be avoided. On the other hand, we *can* carry through the comparison of the density operator formalism that we have used with the photon packet scheme and in that way see something about the microscopic structure of the field.

One of the advantages of this approach is that the treatment of Fermions is really no different in principle from the treatment of Bosons. That is one of the real gains in using this particular approach, because we cannot use the coherent state approach at all in discussing Fermion beams. There is simply no such thing as a coherent state for Fermions. For any Fermion mode we can only have the occupation number zero or one, and there is no such thing as an eigenstate of the annihilation operator. An eigenstate cannot be constructed with those two possibilities. Try doing it algebraically and you will see what I mean. There are no coherent states for Fermions but it is easy to construct maximum entropy states for Fermions. We can do that just by inspection. Since we are specifying the mean number of Fermions present, we are saying, in other words, that $\text{Tr}(\rho a^\dagger a)$ for one mode is equal to a certain fixed expectation value,

$$\text{Tr}(\rho a^\dagger a) = \langle n_k \rangle, \qquad n_k = \begin{cases} 0 \\ 1 \end{cases} \qquad (12.175)$$

That means, as part of the maximization process for entropy, that the density operator has to be diagonal, so the density operator must be a linear combination of the projection on the zero quantum state for one mode and the one quantum state. It must have the specified expectation value, and it must have trace unity. We have only two unknown coefficients and that determines the two coefficients, so in fact, the density operator has to be this:

$$\rho = (1 - \langle n \rangle)|0\rangle\langle 0| + \langle n \rangle |1\rangle\langle 1|, \quad \text{where} \quad \text{Tr}\rho = 1. \qquad (12.176)$$

This may be generalized to fields containing many modes by the same arguments as for Bosons and we find the following relation between the occupation numbers for Fermions and the packet amplitudes,

$$\frac{\langle n_k \rangle}{1 - \langle n_k \rangle} = Z|f(k)|^2 \quad \Rightarrow \quad Z = \sum_k \frac{\langle n_k \rangle}{1 - \langle n_k \rangle}, \qquad (12.177)$$

where the coefficient Z is determined again from the normalization condition. The density operator obtained, finally, has almost the same structure as that for Bosons,

$$\rho = \prod_k \frac{1}{1 + Z|f(k)|^2} \sum_n \frac{Z^n}{n! V^n} \int \prod_j A^\dagger(r_j)|\text{vac}\rangle\langle\text{vac}|\prod_j A(r_j) \prod_j dr_j. \qquad (12.178)$$

As far as the appearance of the summation over states in which we have used the packet amplitudes is concerned, that is precisely the same. The normalization factor simply looks a little different. The point is that the operators $A(r)$ anti-commute, and so the function v is the anti-commutator of the two operators, the one for creation and the other for annihilation of a packet,

$$\{A(r),A^\dagger(r')\}_+ = v(r-r') . \tag{12.179}$$

The normalization function $W^{(n)}$ is an antisymmetrical sum over permutations of products of the functions v,

$$W^{(n)} = \sum_P (-1)^P \prod v(r_j - r_{P_j}) . \tag{12.180}$$

That means that the packets will tend to avoid one another. This is a normalization which will go to zero whenever two packets overlap. In other words, in attempting to put together the chaotic density operator for Fermions with our "do it yourself kit" we lose every time we set down two particle packets overlapping each other. We are only allowed to put states together with packets not overlapping and, indeed, there is an anti-HBT effect for Fermions, and that *does* come purely and simply from the Fermi statistics. That would be true whether we were considering a chaotic state for the Fermions or any other state. What we have shown is that a chaotic state for Fermions does have a particularly simple representation which is completely analogous to the one which we derived for Bosons.

References

1 See *Quantum Optics*, S. M. Kay, A. Maitland (Eds.), Academic Press, London 1970.

2 G. I. Taylor, *Proc. Camb. Phil. Soc. Math. Phys. Sci.* **15**, 114 (1909).

3 L. Mandel, *Phys. Rev.* **152**, 438 (1966).

4 U. M. Titulaer, R. J. Glauber, *Phys. Rev. B* **140**, 676 (1965), reprinted as Chapter 3 in this volume; *Phys. Rev. B* **145**, 1041 (1966), reprinted as Chapter 4 in this volume; *Physics of Quantum Electronics*, P. L. Kelley, B. Lax, P. E. Tannenwald, Eds., McGraw-Hill, New York 1966, pp. 812–821.

5 R. Hanbury Brown, R. Q. Twiss, *Proc. Roy. Soc. (London) A* **242**, 300 (1957); *Proc. Roy. Soc. (London) A* **243**, 291 (1957).

6 E. P. Wigner, *Phys. Rev.* **40**, 749 (1932).

7 B. R. Mollow, R. J. Glauber, *Phys. Rev.* **160**, 1076, 1097 (1967); reprinted as Chapters 6 and 7 in this volume.

8 A. Messiah, *Quantum Mechanics*, Vol. I, p. 442, North-Holland, Amsterdam 1961.

9 J. R. Klauder, E. C. G. Sudarshan, *Fundamentals of Quantum Optics*, Benjamin, 1968.

10 K. E. Cahill, R. J. Glauber, *Phys. Rev.* **177**, 1857, 1882 (1969).

11 The material of Sect. 12.9 was presented in a paper "Microstructure of Light Beams" delivered at the Seventh Session of the International Commission for Optics, Paris, May 2, 1966; R. J. Glauber, *Optica Acta* **13**, 375 (1966).

13
The Initiation of Superfluorescence[1]

13.1
Introduction

In the last few years an experimental picture of infrared superfluorescent radiation pulses has emerged which clearly calls for a new theoretical understanding[1–4]. All existing theories of the radiative behavior of distributions of completely excited atoms offer, at best, only qualitative agreement with the measured data[5].

The mathematical description of even the simplest superfluorescent system encounters three key difficulties.

(i) As in any spontaneous emission process quantum effects play an essential role. The first few photons emitted by initially fully inverted atoms arise as quantum noise. Even if the ensuing amplification eventually follows nearly classical laws, the radiated pulse consists of amplified quantum fluctuations. We propose, for this reason, to treat quantum effects in detail.

(ii) Since the linear dimensions of the radiating volume are typically much larger than the wavelength λ of the emitted light, the interaction between the atoms and the electromagnetic field will involve propagation effects. These effects have generally been omitted from earlier quantum mechanical discussions of superfluorescence. We find that it is quantitatively essential to retain them.

(iii) The interaction between the atoms and the field is intrinsically nonlinear, but that is a complication which only becomes important once the field intensity is quite high. In the initial phase of the radiation process, before the atomic inversion is depleted appreciably, the process is essentially linear in its behavior. Since this phase may last nearly the entire length of time before the appearance of the pulse peak, its treatment seems to us an especially simple and important part of the whole problem.

By restricting our attention, for the present, to the early part of the radiation process, we shall show that the corresponding linear problem can be solved in a closed form

---

[1] Reprinted from R. Glauber, F. Haake, *Phys. Lett. A* **68**, 29–32 (1978), copyright 2006, with permission from Elsevier.

Quantum Theory of Optical Coherence. Selected Papers and Lectures. Roy J. Glauber.
Copyright © 2007 WILEY-VCH Verlag GmbH & Co. KGaA, Weinheim
ISBN: 978-3-527-40687-6

which includes fully the effects of propagation and quantum fluctuations. This essentially rigorous analysis enables us to predict fluorescent pulses. In particular, we can infer that the maximum pulse intensity will occur after a delay time t_d which depends on the total number of atoms N as

$$t_d \propto \frac{1}{V}(\ln N)^2 . \tag{13.1}$$

This result differs from predictions of theories which neglect propagation effects[6,7]. Although the logarithmic factor in Eq. (13.1) is only a weak modification to the factor N^{-1} it is conceivable that experiments presently undertaken will discriminate between Eq. (13.1) and the rival predictions mentioned.

Finally, and perhaps most importantly, the present linear analysis of the onset of superfluorescence provides the decisive clue to the approximate solution of the full nonlinear problem which will be published elsewhere.

13.2
Basic Equations for a Simple Model

Let us consider N identical two-level atoms and describe them in terms of the raising operators s_μ^+ with $\mu = 1, 2, \ldots, N$, the lowering operators s_μ^-, and the inversion operators s_μ^z. Each s_μ^z has eigenvalues $\pm \frac{1}{2}$ and the $s_\mu^{\pm, z}$ obey angular momentum commutation relations. Since a continuum description will be convenient for us we introduce the inversion density

$$J_z(\mathbf{x}) = \sum_{\mu=1}^{N} s_\mu^z \delta^{(3)}(\mathbf{x} - \mathbf{x}_\mu) , \tag{13.2}$$

and the two fields

$$J_\pm(\mathbf{x}) = \sum_{\mu=1}^{N} s_\mu^\pm \delta^{(3)}(\mathbf{x} - \mathbf{x}_\mu) .$$

The Hamiltonian of the noninteracting atoms can then be written as

$$H_A = \int d^3x \, \hbar \omega J_z(\mathbf{x}) ; \tag{13.3}$$

here ω is the transition frequency for the two levels of any atom; the space integral extends over the active volume.

If only one polarization is assumed for the electric field, then in the electric dipole approximation the interaction Hamiltonian reads

$$H_{AF} = i \hbar g \int d^3x \, \{E^{(+)}(\mathbf{x}) + E^{(-)}(\mathbf{x})\} \{J_-(\mathbf{x}) - J_+(\mathbf{x})\} . \tag{13.4}$$

13.2 Basic Equations for a Simple Model

Note that we have split the electric field $E(x)$ into its positive and negative frequency parts. The coupling constant g is determined by the component of the atomic dipole matrix element parallel to the electric field; $i\hbar g\{J_-(x) - J_+(x)\}$ is the operator for the electric polarization density.

The remaining term in the Hamiltonian is the well-known free-field Hamiltonian.

Let the active volume be a rod extending between $x=0$ and $x=l$ in the x direction and let its transverse dimensions be fixed by the requirement that the Fresnel number be unity. The pulse will then be radiated in the positive or negative x direction and can conveniently be described in terms of slowly varying dimensionless fields which we define by

$$J_\pm(x,t) = \tfrac{1}{2} n e^{\mp i(kx-\omega t)} R_\pm(x/l, t/\tau),$$

$$J_z(x,t) = \tfrac{1}{2} n Z(x/l, t/\tau), \qquad (13.5)$$

$$E^{(\pm)}(x,t) = \frac{1}{2g\tau} e^{\pm i(kx-\omega t)} E_R^{(\pm)}(x/l, t/\tau).$$

Here n is the number density of atoms in the active volume, k is the wave number obeying $k = \omega/c$, and τ is a time constant characteristic for the process described by the above Hamiltonian,

$$\frac{1}{\tau} = 2\pi g^2 \hbar^2 k n l. \qquad (13.6)$$

The time τ is sometimes called the superfluorescence time; it is related to the cooperation length l_c by

$$l_c^2 = lc\tau. \qquad (13.7)$$

The Heisenberg equations of motion for the fields $R_\pm(x,t)$, $Z(x,t)$, and $E_R^{(\pm)}(x,t)$ read, if both second order derivatives of these slowly varying quantities and rapidly oscillating terms involving $e^{\pm i2\omega t}$ are neglected,

$$\frac{\partial R_\pm}{\partial t} = Z E_R^{(\mp)},$$

$$\frac{\partial Z}{\partial t} = -\tfrac{1}{2}\left[E_R^{(+)} R_+ + E_R^{(-)} R_-\right], \qquad (13.8)$$

$$\left(\frac{\partial}{\partial x} + \alpha \frac{\partial}{\partial t}\right) E_R^{(\pm)} = R_\mp,$$

with

$$\alpha = \frac{l}{c\tau} = \left(\frac{l}{l_c}\right)^2. \qquad (13.9)$$

Since all coordinates and fields have been referred to their "natural" scales by Eqns. (13.7) the above equations of motion (13.8) involve only one parameter, α. Note that here and subsequently x and t designate dimensionless coordinates.

The equations of motion (13.8) have to be complemented with the boundary condition that no external signal impinge at $x = 0$ and with initial conditions for the fields $E_R^{(\pm)}(x,t)$, $R_\pm(x,t)$, $Z(x,t)$ at $t = 0$.

Once such solutions of the Heisenberg equations are known we may calculate the radiated intensity as the expectation value of the product $E_R^{(-)}(x = 1, t) \times E_R^{(+)}(x = 1, t)$ with respect to the initial state of the system. Since we are interested in a fluorescence phenomenon the appropriate initial state is the vacuum of the electromagnetic field and the excited state for each atom, $|\text{vac}, \{\uparrow\}\rangle$. We have, up to a normalization factor,

$$I(t) = \left\langle \text{vac}, \{\uparrow\} \left| E_R^{(-)}(x=1,t) E_R^{(+)}(x=1,t) \right| \text{vac}, \{\uparrow\} \right\rangle . \tag{13.10}$$

13.3
Onset of Superfluorescence

In the initial state chosen the inversion $Z(x,0)$ is sharp, $Z(x,0) = 1$. During the initial stages of the radiation process we can neglect the depletion of the upper atomic state and put

$$Z(x,t) = 1 , \tag{13.11}$$

disregarding the second of Eqns. (13.8).

The resulting linear initial and boundary value problem has the solution

$$E_R^{(\pm)}(x,t) = \int_0^x dx' R_\mp(x-x',0) \Theta(t - \alpha x') I_0 2\sqrt{x'(t - \alpha x')} \\ + \int_0^x dx' E_R^{(\pm)}(x-x',0) G(x',t) . \tag{13.12}$$

Here I_0 is the modified Bessel function of order zero; the function $G(x,t)$ – which measures the contribution of the initial electric field to the electric field at time t – will not be needed since due to our initial condition it does not contribute to the radiated intensity nor to any other normally ordered correlation function of the electromagnetic field. $\Theta(r)$ is unit step function.

It is a consequence of the linearity of the early stages of the radiation process that the result (13.12) holds true both quantum mechanically and classically. In this sense the quantum mechanical operators $E_R^{(\pm)}(x,t)$ and $R_\pm(x,t)$ may be said to follow their respectively corresponding classical trajectories.

The quantum nature of the process considered becomes manifest, of course, in the prescription for evaluating expectation values of products of field operators. In a classical approximation for such expectation values we would have vanishing initial values to both the polarization and the electric field and thus be left without any subsequent pulse – unless we resorted to the *ad hoc* invention of fictitious nonzero initial values. The correct quantum mechanical prescription (13.10) relates the radiated intensity to the atomic initial-state average

$$\langle\{\uparrow\}|R_+(x,0)R_-(x',0)|\{\uparrow\}\rangle = \frac{4}{N}\delta(x-x'). \tag{13.13}$$

This latter equality follows from the definitions (13.2) and (13.5) upon coarse-graining the atomic fields so as to make them macroscopically continuous.

Our final result for the radiated intensity now follows from Eqns. (13.10), (13.12), and (13.13) and reads

$$I(t) = \frac{4}{N}\int_0^1 dx\,\Theta(t-\alpha x)\left[I_0 2\sqrt{x(t-\alpha x)}\right]^2. \tag{13.14}$$

This remarkably simple expression shows the slow early build-up of the radiated pulse. It holds as long as $I(t)$ is small compared to unity.

By extrapolating the above formula beyond its limit of validity we may expect the maximum intensity of the pulse to occur at a time at which the above $I(t)$ approaches values near unity. Since the Bessel function I_0 increases exponentially with large values of its argument we arrive at the prediction (13.1).

Let us consider the multi-point generalization of the initial expectation value given in Eq. (13.13),

$$\langle R_+(x_1)R_+(x_2)\ldots R_+(x_n)R_-(x_1')R_-(x_2')\ldots R_-(x_n')\rangle$$
$$= \sum_P \langle R_+(x_1)R_-(x_1')\rangle\langle R_+(x_2)R_-(x_2')\rangle\ldots\langle R_+(x_n)R_-(x_n')\rangle \tag{13.15}$$
$$= \left(\frac{4}{N}\right)^n \sum_P \delta(x_1-x_1')\delta(x_2-x_2')\ldots\delta(x_n-x_n'),$$

where the sum runs over all $n!$ permutations of the primed coordinates. Equation (13.15) is easily shown to hold if N very large. We conclude that the initial polarization field has gaussian statistics. An obvious consequence is that the early electric field will have gaussian statistics, too.

We can interpret our results by thinking of the operator fields as being represented by c-number fields which follow classical dynamics, starting out from certain initial values. For the fields corresponding to the operators $E^{(\pm)}$ the initial value is zero everywhere; the field corresponding to the atomic inversion Z has the initial value unity within the active volume; however, the fields associated with the polarization operators $R_\pm$ do not have sharp initial values but rather a gaussian probability distribution with width $4/N$ according to Eqns. (13.13) and (13.15). Quantum mechanical expectation values then take the form of averages of classical trajectories over all initial polarizations with a gaussian weight. The quantum fluctuations inherent in the radiated pulse thus appear as due not so much to the dynamics of the system as to the uncertainty the initial polarization must have in an eigenstate of the inversion since $R_\pm$ and Z do not commute.

The above interpretation of our present results will not carry over rigorously to the later nonlinear stages of the radiation process. We can, however, expect it to hold there to an excellent degree of approximation both on intuitive grounds and on the basis of earlier work on the one-mode model of superfluorescent systems[8].

References

1 N. Skribanowitz, I. P. Herman, J. C. MacGillivray, M. S. Feld, *Phys. Rev. Lett.* **30**, 309 (1973).

2 M. Gross, C. Fabre, P. Pillet, S. Haroche, *Phys. Rev. Lett.* **36**, 1035 (1976).

3 A. Flusberg, T. Mossberg, S. R. Hartmann, *Phys. Lett. A* **58**, 373 (1976).

4 H. M. Gibbs, Q. H. F. Vrehen, *Phys. Rev. Lett.* **39**, 547 (1977).

5 *Cooperative Effects in Matter and Radiation*, Eds. C. M. Bowden, D. W. Howgate, H. R. Robl; Plenum Press, New York 1977. See especially the articles by R. Bonifacio, L. A. Lugiato and A. M. Ricca, p. 193, by J. C. MacGillivray and M. S. Feld, p. 1, and by P. K. Bullough and R. Saunders, p. 209.

6 N. E. Rehler, J. H. Eberly, *Phys. Rev. A* **3**, 1735 (1971).

7 R. Bonifacio, P. Schwendimann, F. Haake, *Phys. Rev. A* **4**, 302, 854 (1971).

8 F. Haake, R. J. Glauber, *Phys. Rev. A* **5**, 1457 (1972).

14
Amplifiers, Attenuators and Schrödingers Cat[1]

14.1
Introduction: Two Paradoxes

When we describe measurements made in the quantum domain, we usually imagine them to be registered by large-scale classical instruments. A deep problem then besets us. Where in the account of those measurements is the transition to be made from the quantum mechanical to the classical description? Furthermore, how can any transition between two descriptions so different in principle be carried out in a mathematically consistent way?

Some work on the theory of quantum mechanical amplifying devices has persuaded me to believe that at least part of that question can be answered in largely physical terms. Arbitrary states of the electromagnetic field, for example, can be used as inputs for the amplification process, and the amplified output fields can be so intense that they qualify as fully classical-that is to say, accurately measurable without any significant disturbance of the field. Such amplification processes are, in fact, actually used in many experiments to render quantum phenomena observable, but even where they are not used in practice, it is interesting to imagine the effect their use would have. They suggest, in other words, an interesting class of *gedanken* experiments, and I shall try presently to illustrate a few of them.

There was a time, well over a century ago, when clever schemes to construct perpetual motion machines enjoyed considerable attention. None of course succeeded, but the effort spent on them was not all wasted; they did help teach us two important principles of thermodynamics. We are not so deeply concerned with that venerable subject any more, but we do seem still to be learning about two more recent areas of interest: relativity theory and quantum mechanics. Therefore, it should not be too surprising that the same infernal ingenuity that once went into devising perpetual motion machines is now suggesting means for communicating faster than light, and for confounding the principle of complementarity. Some of these are interesting schemes; they too might just be capable of teaching us something.

---

[1] Reprinted with permission from *New Techniques and Ideas in Quantum Measurement Theory*, Vol. 480, Annals of the New York Academy of Sciences, Blackwell Publishing, New York 1986, p. 336–372.

Quantum Theory of Optical Coherence. Selected Papers and Lectures. Roy J. Glauber.
Copyright © 2007 WILEY-VCH Verlag GmbH & Co. KGaA, Weinheim
ISBN: 978-3-527-40687-6

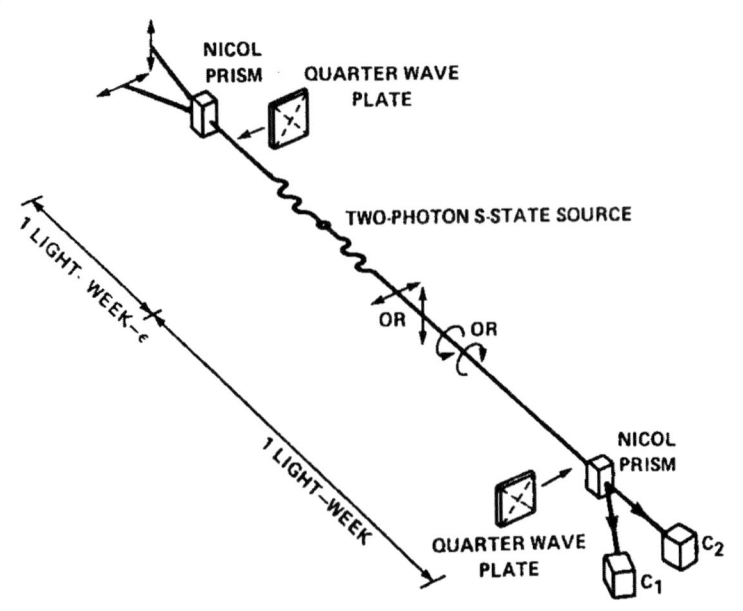

Figure 1 Projected setup for using pairs of photons with correlated polarizations for superluminal communication by means of the EPR paradox.

One interesting proposal amounts to using the Einstein–Podolsky–Rosen (EPR) paradox[1] as a means of communication. To be able to do that would mean, in fact, being able to communicate faster than light, and that is not a possibility to be altogether ignored. Such a thought must indeed have occurred to many of us as soon as we began reading of the EPR paradox and of the species of nonlocality it imposes on quantum mechanics.

Let us recall, therefore, the most elementary kind of proposal for superluminal communication that we might try to base on the EPR paradox. We begin with a light source that gives out photons in pairs. An ordinary atom that does that will give them off in random directions, but if we do not mind using the gamma rays of annihilation radiation, those photons can, in fact, be given off back-to-back-and with zero total angular momentum. That, of course, is an essential point. The two photons must have closely correlated angular momenta so that any statement made about the angular momentum of one of them, or any observation, conveys a unique implication about the other.

Now, what must we do to communicate by this strange medium? We may begin by setting up our own equipment at a certain distance from the source. We do not want it too far because we do not want to lose sight of the source, but let us just say it is one light-week away or perhaps a bit less. Our equipment is represented in the upper left of Fig. 1. The friend with whom we want to communicate may be at a comparable distance from the source; let us say down in the right foreground of Fig. 1. Now all we have to do is to make polarization measurements upon the photons that come our way.

It is a bit wasteful in practice, though, if we do that with polarizing filters; they just absorb half the photons and there is no point in letting that happen. It is a much better idea to use a Nicol prism or, better still, a Wollaston prism, which will send vertically plane polarized photons in one direction, and the others, with horizontal polarization, in another direction, so that none at all are lost. Then, of course, we can add to the equipment a quarter-wave plate, if we like, with its two axes at 45 degree angles with respect to the polarization directions that are separated by the prism. Putting the quarter-wave plate in front of the prism makes of it, in effect, a circular polarization detector for the two possible circular polarizations.

How do we send the message? We have no control whatsoever over the polarization of the photons being produced by our source. However, what we can do, of course, is to exercise our choice either to leave things as they stand in Fig. 1, and observe one or the other of two plane polarizations, or alternatively, to slide the quarter-wave plate into the beam we are detecting, and then we will inevitably observe one or the other of two circulation polarizations.

The correlation that exists between the polarizations of the photons going off in opposite directions then assures us that we, even at this great distance, have a certain control over the photons that are appearing virtually instantaneously at our friend's position. We are able to determine, simply by sliding the quarter-wave plate in or out, whether the individual photons our friend receives are going to be plane polarized or circularly polarized. If they are plane polarized, we cannot control which of the two plane polarizations they have. However, we can send him circularly polarized photons whenever we please and they are presumably distinguishable, in principle, from plane polarized ones. We can let photons polarized in those ways represent the dots and dashes, respectively, of Morse code, or the binary digits, and repeat our messages as often as our friend needs in order to overcome any statistical problems he may have.

The drawback of this communication scheme, of course, is that it does not work at all. Its failing lies in the very essence of what is necessary to send a message. Our friend has no idea, to begin with, whether any given photon he receives is polarized circularly or linearly. His ability to detect the message depends on his ability to determine the polarizations of those individual photons. Unfortunately, there is only a limited amount of measurement that he can make on any one photon before it is absorbed, or he otherwise loses track of its initial state. He can, for example, pass it through a plane-polarization filter. However, the fact that it has passed through the filter does not tell him whether its initial polarization was plane or one of the two circular varieties; that remains really quite undetermined. Let us say, for example, that a photon has passed through our own prism arrangement and has been revealed to have plane polarization. Our distant friend then receives a plane polarized photon, but he still cannot say with better than 50/50 likelihood that the arriving photon is either plane polarized or circularly polarized. Thus, there is no information transmitted at all, and the scheme collapses. That much has probably occurred to most of us as students of the EPR paradox.

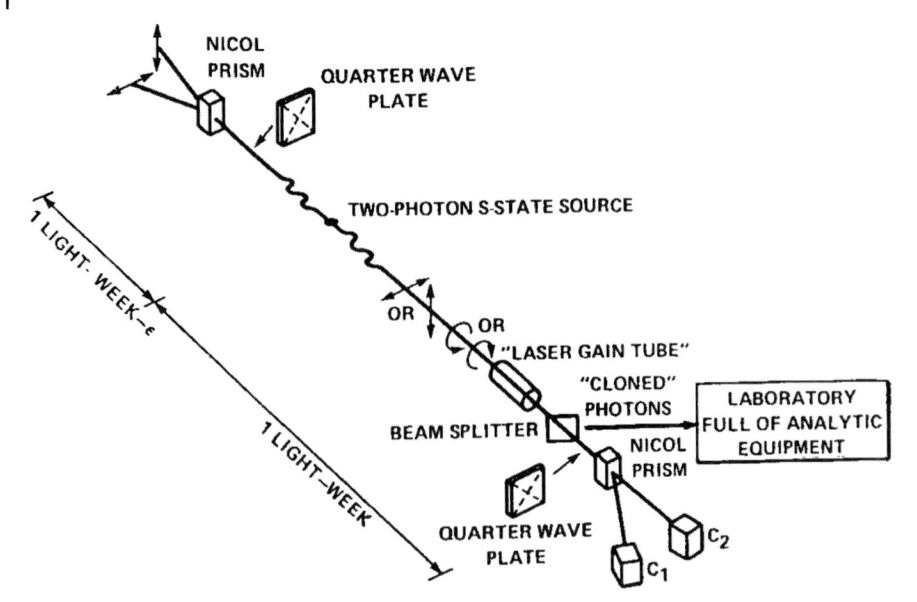

Figure 2 Improved version of the setup for superluminal communication via the EPR paradox. The addition of the "laser gain tube" is intended to facilitate polarization measurements.

A new and interesting suggestion has been added to this picture, however, by N. Herbert[2], who proposes to improve our friend's polarization measurements greatly by adding in effect, an amplifier to his system. He calls it a "laser gain tube", which you will recognize as a characteristically innocent and passive sounding name, and he uses another well-chosen term for its action; it "clones" the photons. Once the "gain tube", which is shown in place in Fig. 2, has "cloned" for our friend lots of photons identical to the original one, he should no longer find it difficult to determine their state of polarization. What he can do, for example, is to install one or more beam-splitters in order to send most of those cloned photons into another laboratory where he has all sorts of equipment available and where he can make measurements on as many of the identical photons as he likes; in that way, he can get an excellent idea of their polarization. He thereby determines precisely what the polarization was (whether plane or circular) of the photon that first entered his laser gain tube.

If we have found, at last, the means for him to do that (as the argument indicates), then we can indeed communicate outside the light cone and create miracles of all sorts. Well, does the scheme work? Let me postpone telling you the answer until we have developed the means of describing a bit better the role played by the quantum amplifier. Then we will be able to say what the insertion of an amplifier really does in that scheme, along with what it does in several other sorts of experiments as well.

In the meantime, I should like to tell you another story. This one is much older and much better known. It deals with one of the fundamental dilemmas of the quantum theory of measurement. Schrödinger[3], responding to questions about the completeness of quantum mechanics in 1935, indicated quite some dissatisfaction with the field he had pioneered. He was troubled by the fact that one never sees in our everyday world anything that looks like a superposition of grossly different quantum mechanical states. The microscopic world, on the other hand, is full of them.

You can easily imagine experimental arrangements in which events in the microscopic world would seem to imply the creation of such superpositions in our everyday world. Let us suppose that the quantum state, $|a\rangle$, of something in the microworld implies, by some kind of rigorous dynamics, that the state of something else that we can actually observe in the laboratory goes to a particular state, $|A\rangle$, and suppose too that the alternative microstate, $|b\rangle$, whatever it may be, leads to a large-scale state that we can label $|B\rangle$. Now, let us imagine a third microstate, $|c\rangle$, which is physically quite distinguishable from both $|a\rangle$ and $|b\rangle$, but can be expressed as a superposition of those two states with nonvanishing coefficients, α and β,

$$|c\rangle = \alpha|a\rangle + \beta|b\rangle \ .$$

That state should obviously lead to a similar linear combination,

$$|C\rangle = \alpha|A\rangle + \beta|B\rangle \ ,$$

as a final state of whatever it is we observe. That looks like no more than the postulate of the linearity of quantum mechanics, and it is hardly anything you could disagree with.

However, Schrödinger did indeed have some trouble with that idea, and he described a gedanken experiment that has become the definitive illustration of his dilemma. He imagined a diabolic arrangement in which a cat is confined to a box containing a lethal device that may be either triggered or left passive according to whether a radioactive nucleus decays or fails to decay. If the radioactive decay takes place, which he assumed would happen half the time, a hammer would strike a vial of cyanide, and the cat would be dispatched. It is not immediately clear, of course, how a radioactive nuclear decay process trips a hammer. Schrödinger evidently did not consider that a problem, but we may. The hammer will not budge unless the quantum signal is amplified greatly, and that amplification is what we want to discuss.

We can simplify Schrödinger's example a bit by thinking, not of radioactive decay processes, but of photons that can be polarized either horizontally or vertically. We can let the state of horizontal polarization for a single photon, $|h\rangle$, be the state $|a\rangle$, and let the state of vertical polarization, $|v\rangle$, be the state $|b\rangle$,

$$|a\rangle = |h\rangle \qquad (14.1)$$
$$|b\rangle = |v\rangle \qquad (14.2)$$

In addition for our two states in the macroscopic world, let us take as state $|A\rangle$, the cat, alive and healthy,

$$|A\rangle = |😺\rangle \ , \qquad (14.3)$$

and as the state $|B\rangle$, the cat, not so alive and not so healthy,

$$|B\rangle = \left|\text{🙀}\right\rangle . \tag{14.4}$$

Now, let us take our two coefficients, alpha and beta, both to be $1/\sqrt{2}$. The state

$$|c\rangle = \frac{1}{\sqrt{2}}(|h\rangle + |v\rangle) , \tag{14.5}$$

the superposition of two polarizations, is a state that is quite meaningful to us in the microworld. Such a superposition of the horizontal and the vertical polarizations is just another linear polarization inclined at 45 degrees to the two axes.

That gives us a simple way of carrying out Schrödinger's experiment. We can send a single photon polarized at 45 degrees (relative to the axes of a Wollaston prism) through the prism and let it go off in either of the two emergent polarized beams. In one of those two beams, we can place a photodetector that, when it registers, does something awful to the cat, and we can leave the other beam free, so that if the photon goes there, nothing happens to the cat. According to Schrödinger's analysis then, and according to our earlier argument, the cat is left finally in a linear superposition of two states,

$$|C\rangle = \frac{1}{\sqrt{2}}\left\{\left|\text{😺}\right\rangle + \left|\text{🙀}\right\rangle\right\} . \tag{14.6}$$

It has, to be precise, an amplitude of 0.7071 for being alive and an amplitude of 0.7071 for being dead. That is a very strange state of affairs because we do not recognize the existence of any such superposition in the everyday world. While we encounter superposition states of all sorts in the microscopic world, we certainly do not encounter states like $|C\rangle$ in the large-scale world. We shall have to examine the argument for their generation much more carefully.

The cat paradox may not seem too closely related to superluminal communication, but let us recall its requirement that the polarization of a single photon induce the tripping of a hammer. Clearly, it can only do that if some process of amplification intervenes. Thus, we are back once more to talking about amplifiers, and once more I would like to postpone further consideration of the problem until we have managed to discuss how they work.

14.2
A Quantum-Mechanical Attenuator: The Damped Oscillator

It may be a good idea, before discussing amplification, to say a few words about dissipation or attenuation, which is a much more natural state of affairs. Amplification, we will then show, can be described by making only a few small (but highly significant) changes in the basic mechanism. The model of the damped harmonic oscillator[4] I want to recall to you gives about the simplest description you can have of the attenuation process.

The damped oscillator is something that can easily be constructed from an assembly of simple harmonic oscillators in the following way: one oscillator, the central oscillator, which is described by the amplitude operators, a and $a^\dagger$, is coupled to a whole heat bath full of oscillators, described by the amplitude operators, b_k and $b_k^\dagger$, for $k = 1,\ldots,N$, where N is quite large, or perhaps even infinite. The Hamiltonian for the system need only couple the central oscillator to the heat bath oscillators through the familiar "rotating wave" terms. We write

$$H = \hbar\omega a^\dagger a + \sum_k \hbar\omega_k b_k^\dagger b_k + \hbar \sum_k (\lambda_k a^\dagger b_k + \lambda_k^* b_k^\dagger a) , \tag{14.7}$$

in which the frequencies, ω_k, cover a range in the neighborhood of ω fairly densely and where the λ_k are a set of coupling parameters. This Hamiltonian remains invariant when we make the phase transformation,

$$a \to a e^{i\theta} ,$$
$$b \to b e^{i\theta} ,$$

on all of the operators it contains. That invariance means that there is a corresponding conservation law. In the present case, the coupling obviously conserves the total number of quanta in the system. That property helps to make of it a highly soluble model.

The equations of motion, of course, are linear:

$$\dot{a} = -i\omega a - i \sum_k \lambda_k b_k ,$$
$$\dot{b}_k = -i\omega_k b_k - i\lambda_k^* a . \tag{14.8}$$

Furthermore, for this simple coupling, the Schrödinger states have the wonderful property that an initially coherent state remains coherent at all later times[5]. Therefore, in fact, the equations of motion could every bit as well have been written as equations for a set of c-number amplitudes $\alpha(t)$ and $\beta(t)$ for the coherent states of all the oscillators. The Schrödinger state of the oscillators, in other words, apart from a time-dependent phase factor of no interest, is just given by

$$|t\rangle = |\alpha(t)\rangle \prod_k |\beta_k(t)\rangle , \tag{14.9}$$

which is a product of coherent states in which the amplitudes, $\alpha(t)$ and $\beta_k(t)$, obey the equations of motion (14.8). There is a sense, then, in which this fully quantum mechanical system behaves as if it were classical. The fact that it is quantum mechanical even becomes a bit of a joke. The evolution of the system just carries its zero-point uncertainty cloud around the classical trajectory without altering it in any way. That is all that ever happens in such a system.

If the entire system is initially in a pure coherent state, its density operator in the Schrödinger picture at time t will be just

$$\rho(t) = |\alpha(t), \{\beta_k\}(t)\rangle \langle \alpha(t), \{\beta_k\}(t)| \, , \tag{14.10}$$

which is still a pure coherent state for all of the individual oscillators. Note that the phase factor left undetermined in the state vector has now canceled out.

It will be useful now to write the solutions to the equations of motion for the time-dependent amplitudes in an abbreviated form. The functions $\alpha(t)$ and $\beta_k(t)$ can be expressed always as linear combinations of their initial values, which we can write as α and β_k, respectively. The coefficients will be time-dependent functions, which we can write as $u(t)$ and $v_k(t)$, so that we have, for example

$$\alpha(t) = \alpha u(t) + \sum_k \beta_k v_k(t) \, , \tag{14.11}$$

and the initial conditions,

$$u(0) = 1, \qquad v_k(0) = 0 \, . \tag{14.12}$$

We shall presently determine the functions, $u(t)$ and $v_k(t)$, explicitly, but there is a good deal we can say before doing that.

The only one of the oscillators that really interests us is the central one. We can find the reduced density operator for that one by taking the trace over all of the variables for the heat bath oscillators. What we have left then is just

$$\rho_A = \mathop{\mathrm{Tr}}_A \rho(t) = ||\alpha(t)\rangle \langle \alpha(t)|| \, , \tag{14.13}$$

which still represents a pure state for the central oscillator.

Of course, if the heat bath begins in a mixed state rather than a pure coherent state, then we must mix together coherent states of this sort that correspond to different values of the initial amplitudes, β_k. The amplitudes, $\alpha(t)$, according to Eq. (14.11), depend linearly on the β_k, and so the appropriate reduced density operator should take the form,

$$\rho_A(t) = \int |\alpha(t)\rangle \langle \alpha(t)| \, P(\{\beta_k\}) \prod_k d^2 \beta_k \, . \tag{14.14}$$

When the heat bath is not at zero temperature, the amplitudes, β_k, have a characteristically chaotic distribution.

It is most convenient to express the reduced density operator for the central oscillator in terms of what I have called the P representation[6]; that is to say, a mixture of pure coherent states, $|\gamma\rangle\langle\gamma|$, with a weight function $P(\gamma)$, that is, strictly speaking, a quasiprobability distribution.

If we assume that the central oscillator is indeed initially in a coherent state with amplitude, α, then its reduced density operator can be written in the general form,

$$\rho_A(t) = \int P(\gamma, t | \alpha, 0) \, |\gamma\rangle \langle\gamma| \, d^2\gamma . \tag{14.15}$$

The weight function, $P(\gamma, t | \alpha, 0)$, in this expression can be thought of as a conditioned quasiprobability density for the occurrence of a coherent state amplitude γ at time t, given the amplitude α at time zero. If we then take the heat bath oscillators to be initially in chaotic states (i.e., Gaussian mixture states), with mean occupation numbers, $\langle n_k \rangle$, finding the function P is simply a matter of carrying out some Gaussian integrations. The result, we find, is always Gaussian in form:

$$P(\gamma, t | \alpha, 0) = \frac{1}{\pi D(t)} \exp\left\{ -\frac{|\gamma - \alpha u(t)|^2}{D(t)} \right\} . \tag{14.16}$$

The dispersion of this Gaussian function is given by

$$D(t) = \sum_k \langle n_k \rangle |v_k(t)|^2 , \tag{14.17}$$

and its mean value is given by $\alpha u(t)$. To evaluate the dispersion and the mean value of the Gaussian function we must, of course, solve the equations of motion to find the functions, $u(t)$ and $v_k(t)$.

According to the initial condition Eq. (14.12), the functions $v_k(t)$ must all vanish at $t = 0$, so the dispersion, $D(t)$, is zero initially and the function P begins life as a delta function. That of course, is no surprise because the central oscillator starts out in a pure coherent state, $|\alpha\rangle$. What is more interesting is that if we take the heat bath to be at zero temperature, that is, take all of the $\langle n_k \rangle$ to vanish, the dispersion, $D(t)$, remains zero at all times and the function P always remains a two-dimensional delta function,

$$P(\gamma, t | \alpha, 0) = \delta^{(2)}[\gamma - \alpha u(t)] . \tag{14.18}$$

In this case, in other words, the central oscillator always remains in a pure coherent state. It will typically lose its initial excitation to the heat bath oscillators or, to express that somewhat differently, we should expect the function $u(t)$ to decrease in modulus as the time t increases. That, at least, is the behavior we should anticipate in a dissipative system.

It is worth stressing two points, therefore, at this stage of the calculation. Firstly, the results we have reached at this point are exact; we have made no approximations. Secondly, because the result given by Eq. (14.18) is exact, it still reflects the intrinsic reversibility of the equations of motion. It is entirely possible for the heat bath oscillators to reexcite the central oscillator, and such Poincaré recurrences, though they may be enormously delayed, are in fact inevitable. The function, $u(t)$, then, when solved for exactly, will not in general decrease monotonically in modulus forever. However, the Poincaré recurrence times go rapidly to infinity as the number, N, of heat bath oscillators increases. In practice, N need not be very large before it becomes an excellent approximation (over lengths of time exceeding the age of the universe) to ignore the Poincaré recurrences altogether and use approximations in which the modulus of $u(t)$ does decrease steadily. Those are the approximate ways of approaching the equations of motion that also introduce the notion of irreversibility.

The equations of motion (14.8) have essentially the same structure as a set of coupled equations derived after a number of approximations by Weisskopf and Wigner[7] in order to describe the process of radiation damping. If the coupling constants, λ_k, are not too large in modulus and there are enough heat bath oscillators with frequencies, ω_k, near ω, then the function $u(t)$ can be approximated quite well by a complex exponential function,

$$u(t) = e^{-[\kappa + i(\omega + \delta)]t} = e^{-(\kappa + i\omega')t} . \tag{14.19}$$

This function has a certain frequency shift, $\delta\omega$, in it, and it has a certain damping constant, κ, as well. Those constants are given by the relation

$$\delta\omega - i\kappa = \lim_{t \to 0} \sum_k \frac{|\lambda_k|^2}{\omega - \omega_k + i\epsilon} , \tag{14.20}$$

which is characteristic of second-order perturbation theory, although the overall approximation retains terms of all orders in the coupling strengths.

With the heat bath at temperature zero then, the conditional quasiprobability density is given, according to Eqns. (14.18) and (14.19), by

$$P(\gamma, t | \alpha, 0) = \delta^{(2)}\left(\gamma - \alpha e^{-(\kappa + i\omega')t}\right) . \tag{14.21}$$

Because the state of the central oscillator, in this case, remains at all times a pure coherent state, it always has minimal uncertainty. This model for the dissipation process, in other words, is completely noise-free

The randomness we refer to as noise only enters the dissipation process when the heat bath possesses initial excitations that exert random driving forces on the central oscillator. For those cases, which correspond to $\langle n_k \rangle \neq 0$ in our model, the noise is described by the dispersion, $D(t)$. We can evaluate that expression approximately by noting that the functions $v_k(t)$ have a resonant character; they are only large in modulus for modes with frequencies, ω_k, within an interval of width κ, about the frequency $\omega' = \omega + \delta\omega$. If the mean occupation numbers, $\langle n_k \rangle$, take on the constant value $\langle n_{\omega'} \rangle$ within this narrow band, we can write Eq. (14.17) as

$$D(t) = \langle n_{\omega'} \rangle \sum_k |v_k(t)|^2 . \tag{14.22}$$

14.2 A Quantum-Mechanical Attenuator: The Damped Oscillator

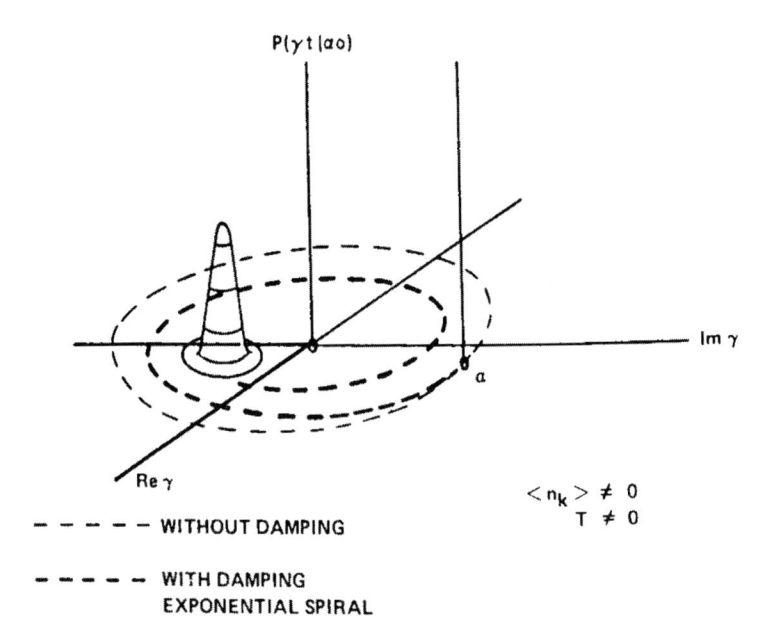

Figure 3 Conditioned quasiprobability density for the complex amplitude of the damped harmonic oscillator. The oscillator has begun at $t=0$ in a pure coherent state of amplitude α. The mean value of the amplitude moves on an exponential spiral, decreasing steadily in modulus, while its dispersion increases.

To evaluate the sum in this expression, we can note that the functions, u and v_k, obey an identity equivalent to the conservation law,

$$|u(t)|^2 + \sum_k |v_k(t)|^2 = 1, \tag{14.23}$$

which follows from the equations of motion Eq. (14.8) and the initial conditions Eq. (14.12). The dispersion can then be written as

$$D(t) = \langle n_{\omega'} \rangle \left(1 - |u(t)|^2\right), \tag{14.24}$$

which in the Weisskopf–Wigner approximation reduces to

$$D(t) = \langle n_{\omega'} \rangle \left(1 - e^{-2\kappa t}\right). \tag{14.25}$$

The dispersion, $D(t)$, then increases from the initial value zero, while the center of the Gaussian distribution given by Eq. (14.16) circles about on the exponential spiral given by Eq. (14.19). A three-dimensional portrait of the quasiprobability density as a function of the complex amplitude γ is shown in Fig. 3. For times t much greater than the damping time, κ^{-1}, the central oscillator comes to equilibrium with the heat bath oscillators. The dispersion then reaches the limiting value $\langle n_{\omega'} \rangle$ and the Gaussian distribution settles down into the stationary form shown at the center of Fig. 4.

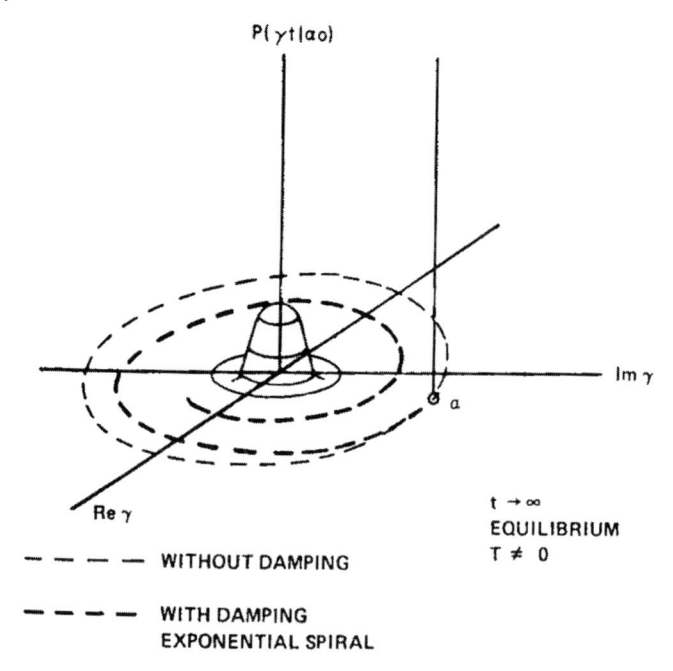

Figure 4 Limit for $\kappa t \to \infty$ of the quasiprobability distribution for the complex amplitude of the damped oscillator. The oscillator has come to equilibrium with its "heat bath".

14.3
A Quantum Mechanical Amplifier

How can we build a device that amplifies signals at the quantum level? In fact, we can do that too with harmonic oscillators, but at least one of them must be rather special[8] in nature. The energy levels and the potential of the special oscillator are illustrated in Fig. 5. As you can see, there is something a bit unusual about that harmonic oscillator. Its Hamiltonian is just the negative of the familiar one; both the potential and kinetic energies go down instead of up. That simple change of sign does not change the algebraic properties of the amplitude operators a and $a^\dagger$, however. They obey the same commutation relation they did before. In fact, only one thing changes: reversing the sign of the Hamiltonian implies reversing the sign of the frequency ω in the equation of motion. Thus, $a(t)$ varies as $a(0)\,e^{i\omega t}$ instead of $a(0)\,e^{-i\omega t}$.

Our inverted oscillator, strictly speaking, no longer has a ground state; it has a state of maximum energy that we must inevitably associate with quantum number $n = 0$. That leads to a certain dilemma of terminology. We need a name for the highest lying state and, lacking a better one, shall call it the ground state, as we usually do the $n = 0$ state. However, we must remember that we are describing a system in which everything that happens goes on underground.

14.3 A Quantum Mechanical Amplifier

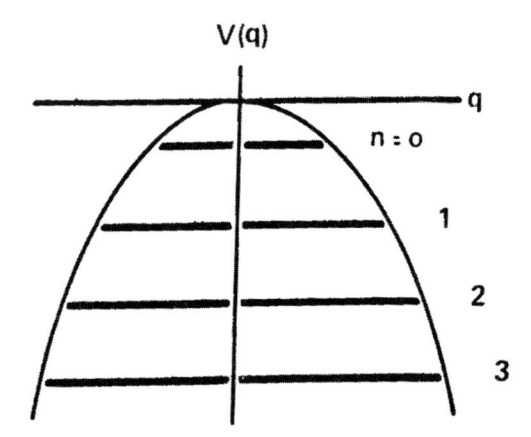

[Figure: V(q) potential diagram with energy levels labeled n=0, 1, 2, 3]

Figure 5 The potential and energy levels of the amplifying oscillator. (The kinetic energy is inverted in sign as well as the potential.)

The inverted oscillator still has a discrete succession of eigenstates that can be generated by applying powers of $a^\dagger$ to the "ground state"

$$|n\rangle = \frac{1}{\sqrt{n!}}(a^\dagger)^n |0\rangle . \tag{14.26}$$

One must remember, however, that the energies of these states have the negative sign in them, $E_n = -(n+\frac{1}{2})\hbar\omega$, so the creation operator, $a^\dagger$ really creates de-excitations, rather than positive energy quanta. Furthermore, the annihilation operator, a, actually raises the energy of the oscillator by decreasing the quantum number, n, by one unit.

Once we have absorbed those trifling changes, we are prepared to let the inverted oscillator take over the role of the central oscillator in the scheme we used earlier to discuss dissipation. The heat bath can remain precisely the same as we had it before, but when we couple the inverted oscillator to it, we must interchange the roles of a and $a^\dagger$ in the coupling terms in order to retain the "rotating wave" form. The coupling terms that tend to conserve energy, in other words, take the forms, ab_k and $b_k^\dagger a^\dagger$ rather than $ab_k^\dagger$ and $b_k a^\dagger$. When we make this interchange, the Hamiltonian for the coupled system becomes

$$H = -\hbar\omega a^\dagger a + \sum_k \hbar\omega_k b_k^\dagger b_k + \hbar \sum_k (\lambda_k a b_k + \lambda_k^* a^\dagger b_k^\dagger) . \tag{14.27}$$

This Hamiltonian has a certain invariance in it, which is a bit different from the one we saw earlier. If we alter the phase of a, then the b_k's must all undergo the complex conjugate transformation,

$$a \to a e^{i\theta}, \quad b_k \to b_k e^{-i\theta} . \tag{14.28}$$

That gives us a different sort of conservation law. The quantity conserved is the difference of the number of quanta in the central oscillator and the number in the heat bath,

$$a^\dagger a - \sum_k b_k^\dagger b_k = \text{constant} . \qquad (14.29)$$

The equations of motion, of course, are still linear,

$$a = \mathrm{i}\omega a - \mathrm{i}\sum_k \lambda_k^* b_k^\dagger$$
$$b_k = -\mathrm{i}\omega_k b_k - \mathrm{i}\lambda_k^* a^\dagger , \qquad (14.30)$$

but they mix the adjoint operators in with the nonadjoints and that changes the character of the solutions.

Before undertaking the explicit solution of those equations, however, it is useful once more to go as far as we can with the generic forms the solutions must take. Let us thus write the solution to the equations of motion for $a(t)$ in the form,

$$a(t) = a(0)U(t) + \sum_k b_k^\dagger(0) V_k(t) , \qquad (14.31)$$

and define thereby a set of functions, $U(t)$ and $V_k(t)$, that obey the initial conditions,

$$U(0) = 1 , \qquad V_k(0) = 0 . \qquad (14.32)$$

We shall eventually have to solve for those functions, but, as before, we can gain a number of insights before having to do that.

We can no longer use the trick of saying that a coherent state remains always a coherent state. In fact, for this model, a coherent state does not remain a coherent state; its intrinsic uncertainty cloud grows explosively. That is unavoidable as we will see in quantum mechanical linear amplifiers. Thus, if we are to evaluate the reduced density operator for the inverted oscillator, we must fall back on some more general analytic technique than we have used earlier.

It is sufficient for this purpose to introduce the characteristic function for the unknown density operator, in fact, it is sufficient to use the specific form in which the a and $a^\dagger$ operators are normally ordered,

$$\chi_N(\mu,t) = \text{Tr}\left\{\rho(0) e^{\mu a^\dagger(t)} e^{-\mu a(t)}\right\} \qquad (14.33)$$
$$= \text{Tr}\left\{\rho(t) e^{\mu a^\dagger(0)} e^{-\mu a(0)}\right\} . \qquad (14.34)$$

The trace that defines this characteristic function is written on the upper line in the Heisenberg picture and on the lower line in the Schrödinger picture. Let us again take the initial state of the system to be one in which the central (inverted) oscillator is in the coherent state $|\alpha\rangle$, while the heat bath oscillators are in chaotic states with occupation numbers $\langle n_k \rangle$:

$$\rho(0) = |\alpha\rangle\langle\alpha| \prod_k \int e^{-|\beta_k|^2/\langle n_k \rangle} |\beta_k\rangle\langle\beta_k| \frac{\mathrm{d}^2\beta_k}{\pi \langle n_k \rangle} . \qquad (14.35)$$

14.3 A Quantum Mechanical Amplifier

We can use this form for the density operator then, and we can use the expression for $a(t)$, given by Eq. (14.31), to evaluate the Heisenberg form for the characteristic function.

If we now write the reduced density operator in the Schrödinger picture in precisely the same form as we used in Eq. (14.15), that is to say, the P-representation form, we find that the characteristic function, $\chi_N(\mu,t)$, is just a two-dimensional Fourier transform[19] of the unknown weight function, $P(\gamma,t|\alpha,0)$. That transform is easily inverted and we then find a Gaussian form for the conditioned quasiprobability density,

$$P(\gamma,t|\alpha,0) = \frac{1}{\pi \mathcal{N}(t)} \exp\left[-\frac{|\gamma - \alpha U(t)|^2}{\mathcal{N}(t)}\right], \tag{14.36}$$

in which the dispersion, $\mathcal{N}(t)$, is given by

$$\mathcal{N}(t) = |U(t)|^2 - 1 + \sum_k \langle n_k \rangle |V_k(t)|^2 . \tag{14.37}$$

The functions, $U(t)$ and $V_k(t)$, obey an identity in this case too – one that is equivalent to the conservation law of Eq. (14.29) and states

$$|U(t)|^2 - \sum_k |V_k(t)|^2 = 1 . \tag{14.38}$$

We can use this relation to write the dispersion as

$$\mathcal{N}(t) = \sum_k (1 + \langle n_k \rangle) |V_k(t)|^2 . \tag{14.39}$$

This dispersion must vanish initially when we assume the central oscillator begins in a pure coherent state. However, at later times, the dispersion takes on positive values in general, and it does that even when the heat bath has no initial excitation, that is, when all $\langle n_k \rangle$ are zero.

The dispersion, $\mathcal{N}(t)$, is a measure of the noise present in the excitation of the central oscillator. There is no avoiding the occurrence of such noise. It results or example, from the spontaneous emission of quanta from the inverted oscillator into the heat bath, even when the latter is at zero temperature, and also when the initial amplitude vanishes as well.

The functions $U(t)$ and $V_k(t)$ can be approximated by the same sort of approximation we used earlier. This time, though, it should perhaps be called the anti-Weisskopf–Wigner approximation because the function $U(t)$ now blows up exponentially instead of decreasing. We find we can write $U(t)$ as

$$U(t) = e^{[\kappa + i(\omega - \delta\omega)]t} = e^{[\kappa + i\omega'']t}, \tag{14.40}$$

where the amplification constant, κ, and the line shift $\delta\omega$ are given once again in terms of the coupling strengths and the frequencies ω_k by Eq. (14.20).

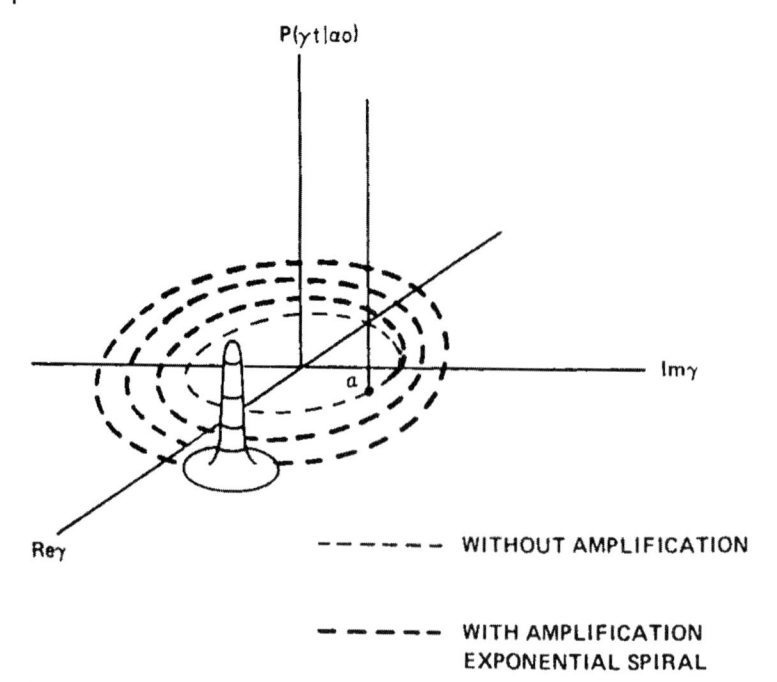

Figure 6 Conditioned quasiprobability density for the complex amplitude of the amplifying oscillator. The amplifier has begun at $t = 0$ in a pure coherent state of amplitude α. The mean value of the amplitude moves on a spiral, increasing exponentially in modulus, while its dispersion also increases exponentially.

The functions $V_k(t)$ have a resonant character for the case of amplification as well as dissipation. They are largest in modulus for frequencies ω_k close to ω. We can use that fact once more to approximate the dispersion function given by Eq. (14.38) as

$$\mathcal{N}(t) = (1 + \langle n_{\omega''} \rangle) \sum_k |V_k(t)|^2$$
$$= (1 + \langle n_{\omega''} \rangle)(|U(t)|^2 - 1) = (1 + \langle n_{\omega''} \rangle)(e^{2\kappa t} - 1) . \qquad (14.41)$$

The initial amplitude α may be regarded, in a sense, as an input signal for our amplifier. Then it is clear from the general form of the quasiprobability function Eq. (14.36) that the mean value of the distribution circles about the complex γ plane in an exponentially increasing spiral. That is to say, the amplified signal, $\alpha U(t)$, blows up exponentially in modulus. At the same time, furthermore, the dispersion $\mathcal{N}(t)$ or the noise present blows up just as dramatically. A somewhat subdued picture of the way the function, $P(\gamma, t|\alpha, 0)$, changes with time is given in Fig. 6. The dispersion tends to increase more rapidly than the figure indicates, but flattened-out Gaussian functions are very difficult to portray in three dimensions

Now that we know what happens when the amplifier begins in a pure coherent state, $|\alpha\rangle$, it is easy to find out what happens for a great variety of other initial states. To see what the amplifier does when it begins in the $n = 0$ state, for example, we just substitute $\alpha = 0$ in Eq. (14.36) to find the quasiprobability distribution,

$$P_0(\gamma,t) = \frac{1}{\pi \mathcal{N}(t)} \exp\left[-\frac{|\gamma|^2}{\mathcal{N}(t)}\right] . \tag{14.42}$$

This simple Gaussian function represents the purest sort of noise. A large part of it comes from the amplification of zero-point fluctuations or, equivalently, from amplified spontaneous emission. The remainder comes (when the $\langle n_k \rangle$ are not equal to zero) from random forcing of the central oscillator by the heat bath.

It is not much more work to find the P distribution that corresponds to any initial n-quantum state. To do that, we simply observe that the density operator for the pure coherent state $|\alpha\rangle$ can be written as

$$|\alpha\rangle \langle \alpha| = e^{-|\alpha|^2} \sum_{n,m} \frac{\alpha^n (\alpha^*)^m}{\sqrt{n! \, m!}} |n\rangle \langle m| . \tag{14.43}$$

It is, in other words, a species of generating function for the operators $|n\rangle\langle m|$. Included among these are the alternative initial density operators,

$$\rho_A(0) = |n\rangle \langle n| . \tag{14.44}$$

We can find the quasiprobability distributions that evolve from the entire set of these initial states just by expanding the function $P(\gamma,t|\alpha,0)$, given by Eq. (14.36) in a power series analogous to Eq. (14.43), and then evaluating the appropriate coefficients. The function $P(\gamma,t|\alpha,0)$ given by Eq. (14.36) is, in fact, a species of generating function[10] for the Laguerre polynomials, L_n, and so we easily find that an n-quantum initial state leads to

$$P_n(\gamma,t) = \frac{(-1)^n}{\pi \mathcal{N}^{n+1}(t)} L_n\left[\left(1 + \frac{1}{\mathcal{N}(t)}\right) |\gamma|^2\right] e^{-|\gamma|^2/\mathcal{N}(t)} . \tag{14.45}$$

Some inkling of the appearance of these functions at a time, t, of the order of κ^{-1} is given in Fig. 7. We may note that the function P_1, for example, takes on negative values for $|\gamma| < 1$, and then the higher order functions P_n do as well for values of $|\gamma|$ that are of the order of unity. The mean value of $|\gamma|$, on the other hand, increases exponentially as $e^{\kappa t}$, and for values of $|\gamma|$ large compared to unity, all of the functions P_n assume positive values and vary quite smoothly. While the quasiprobability densities, P_n, do not behave like ordinary probability densities within the quantum domain (for which $|\gamma|$ is of the order of unity), they do indeed behave like probability densities in the classical domain, $|\gamma| \gg 1$, to which the amplification process inevitably brings the excitation. When the amplification has taken place over a period several times κ^{-1} in duration, almost the entire normalization integral of P_n comes from the classical

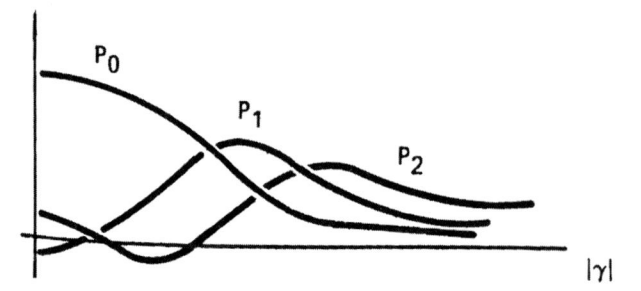

Figure 7 General appearance of the quasiprobability densities for the oscillator amplitude γ that evolves from amplification of initial n-quantum states.

domain of $|\gamma| \gg 1$, and it becomes quite correct asymptotically to interpret P_n as a classical probability distribution for the oscillator amplitude. When the heat bath is initially at zero temperature, for example, the functions $P_n(\gamma,t)$ become asymptotically

$$P_n(\gamma,t) = \frac{1}{\pi n!} \frac{|\gamma|^{2n}}{(|U(t)|^2 - 1)^{n+1}} e^{-|\gamma|^2/|U(t)|^2 - 1}. \tag{14.46}$$

Sometimes, we may be interested in finding the behavior of the amplifier for initial states that are superpositions of different n-quantum states. In fact, the examples we shall discuss involve some uncertainty in whether the amplifier begins in an $n = 0$ or an $n = 1$ state, so we need to know what the effect is of adding nondiagonal terms like $|0\rangle\langle 1|$ and $|1\rangle\langle 0|$ to the initial density operator. The density operator Eq. (14.43) for a pure coherent state is evidently a generating function for these operators too, and so their contribution to the time-dependent density operator can also be found from the appropriate terms of the same expansion of the function P.

Sometimes, we are also interested in knowing the matrix elements of the density operator in the n-quantum state basis. We can find those as well by using the generating function device. For an initially pure coherent state, $|\alpha\rangle$, for example, we thus find the matrix element for $\rho_A(t)$:

$$\langle n|\rho_A(t)|m\rangle = \sqrt{\frac{n!}{m!}} \frac{\mathcal{N}^m}{(1+\mathcal{N})^{n+1}} \left(\frac{\alpha U(t)}{1+\mathcal{N}}\right)^{m-n} L_n^{m-n}\left(-\frac{|\alpha|^2 |U|^2}{\mathcal{N}(1+\mathcal{N})}\right)$$

$$\times \exp\left[-\frac{|\alpha|^2 |U|^2}{1+\mathcal{N}}\right]. \tag{14.47}$$

It is interesting to examine the ratio of the off-diagonal terms of this matrix to the diagonal ones. For $m > n$ and $\kappa t \gg 1$, we find

$$\frac{\langle n|\rho_A(t)|m\rangle}{\langle n|\rho_A(t)|nm\rangle} \to \frac{1}{(m-n)!} \sqrt{\frac{n!}{m!}} \left(\frac{\alpha}{U^*(t)}\right)^{m-n}, \tag{14.48}$$

but this ratio goes to zero as $\exp[-(m-n)\kappa t]$. The off-diagonal matrix elements, therefore, tend to become quite small in comparison to the diagonal ones. That is not quite the same thing as saying that the matrix assumes a precisely diagonal or stationary form because there are so very many nonvanishing off-diagonal matrix elements present. It is a kind of asymptotic near-diagonality that the density matrix possesses, and it takes on that property eventually no matter what the initial state of the central oscillator may have been.

This asymptotic descent of the density matrix into quasi-diagonality is even more dramatic in the coordinate-space or momentum-space representations. The coordinate-space representation of the general expression Eq. (14.15) for the density operator is

$$\langle q'|\rho_A(t)|q''\rangle = \int P(\gamma,t|\alpha,0)\langle q'|\gamma\rangle\langle\gamma|q''\rangle\,d^2\gamma\,. \tag{14.49}$$

The coherent state wave functions, $\langle q'|\gamma\rangle$ and $\langle\gamma|q''\rangle$, are Gaussian in form[6], and they vanish quite rapidly except for values of q' and q'' that lie close to $(2\hbar/\omega)^{1/2}(\mathrm{Re}\,\gamma)$. The coordinate values for which the product $\langle q'|\gamma\rangle\langle\gamma|q''\rangle$ is significantly different from zero only allow $|q'-q''|$ to take on values comparable to the range of zero-point fluctuations, $(2\hbar/\omega)^{1/2}$. When this product of wave functions is multiplied by a smooth P function and integrated over γ, as in Eq. (14.49), the constraint becomes even tighter. With the function, P, given by the pure noise distribution of Eq. (14.42), for example, we find that the matrix element Eq. (14.49) approaches zero for $|q'-q''|$ larger than

$$\left(\frac{2\hbar(2\mathcal{N}+1)}{\omega\mathcal{N}(\mathcal{N}+1)}\right)^{1/2},$$

a distance that shrinks exponentially to zero.

This interesting property of asymptotic or macro-diagonality that the density matrix possesses will have an important consequence when we begin looking at experiments. It will imply that we cannot use our amplifier to generate any of those weird large-scale superposition states that quantum mechanics seems in principle to admit, but that we have never in our lives seen. Those are the states that Schrödinger maintained were intrinsically absurd consequences of quantum mechanics. They are not so much absurd as unrealizable.

How do we define the gain and noise figures for our amplifier? We can define the amplitude gain in an obvious way. It is just the exponential $e^{\kappa t}$. The power gain then is just the square of that:

$$G = |U(t)|^2 = e^{2\kappa t}\,. \tag{14.50}$$

The noise figure requires a bit more subtlety. A conventional way of defining the noise figure of an amplifier is to imagine a noiseless amplifier with the same gain and ask how much noise you would have had to put in initially in order for the noiseless amplification process to produce the same final noise that the real amplifier produces That figure then offers a fair measure of how strong your input signal ought to be if it is to be detectable above the final noise.

The amount of Gaussian noise you must add to an initially coherent state $|\alpha\rangle$ in order for this hypothetically noiseless amplification process to lead to the P distribution of Eq. (14.36) is

$$N_{\text{noise}} = \frac{\mathcal{N}(t)}{|U(t)|^2} = (1 + \langle n_{\omega''} \rangle)[1 - (1/G)] . \tag{14.51}$$

In a recent paper on amplifiers, Caves[11] has established a lower bound for this noise figure by invoking general principles rather than any specific model. His result for the noise figure was $N_{\text{noise}} > 1 - (1/G)$, which is entirely consistent with Eq. (14.51).

There is an interesting game that you can play with this sort of amplifier and with the attenuator that you can construct by turning the inverted oscillator back into a normal one. Let us say we put a microscopic signal of some sort into our amplifier and amplify it considerably. Then we stop the process and reinvert the central oscillator so that the signal is attenuated back down to its original strength. The question we now ask is, will we get the oscillator back to the quantum state in which it began? The attenuation process, we have noted, is completely noiseless as long as the heat bath is at temperature zero. Let us assume, therefore, that the heat bath always starts at temperature zero, even for the amplification process as well.

There would be some strange consequences if, in fact, it were possible to get back to the original state. The amplification process we are discussing is irreversible and this may be one of the better illustrations of that fact. The analysis is quite elementary, so we need not go through it in detail. Let us just note that by changing the sign of the frequency parameter, we can indeed turn the amplifier into an attenuator with a decay period equal to the prior amplification period. We can then decrease the oscillator excitation noiselessly, but we can never get its state back to the form in which it began. What happens is that the cycle of amplification and attenuation superposes on that state a certain amount of Gaussian noise. If the initial state of the amplifier is described by the coherent-state amplitude distribution $P(\alpha)$, and the amplifying and attenuating halves of the cycle each last for a length of time t, then the final state reached will have the reduced density operator,

$$\rho_A = \frac{1}{\pi M(t)} \int e^{-|\gamma - \alpha|^2 / M(t)} P(\alpha) \, |\gamma\rangle \langle\gamma| \, d^2\alpha \, d^2\gamma .$$

This operator describes a superposition of an average number of quanta $M(t)$ of Gaussian noise upon the original state. If the heat bath modes have an average occupation number $\langle n_\omega \rangle$ at the amplifier frequency ω, the average number of noise quanta may be shown to be

$$M(t) = (1 + 2\langle n_\omega \rangle)(1 - e^{-2\kappa t}) . \tag{14.52}$$

For a heat bath at zero temperature then, the number of noise quanta added is just $1 - e^{-2\kappa t}$, and for $\kappa t \gg 1$, each of our cycles adds, on the average, just a single quantum of noise.

Suppose it were true that we could carry out the cycle of amplification and attenuation and still return the system to precisely the state from which it began. We could put the central oscillator into some interesting quantum state and amplify it until all its variables assumed classical strength. We could then measure them all, including the complementary ones, without disturbing any of them significantly, and finally, we could attenuate the signal noiselessly to reestablish exactly the initial state. We would then find ourselves in possession of all sorts of information about the state that contradicts the uncertainty principle. It is the quantum mechanical nature of the amplification process ultimately that makes it both noisy and irreversible, and thus prevents the occurrence of any such miracle.

Our model for the amplifier may seem a bit unrealistic. It depends, after all, quite explicitly on the use of an inverted oscillator, and oscillators of that kind are not available "off-the-shelf". In fact, if we were in possession of just one such oscillator, it could solve the world's energy problems, so that does not sound too likely. However, there is no need, in fact, for us to find an inverted oscillator in the literal sense. All we need is something that behaves like one over the limited range of variables in which it is actually used. Many systems are available that do behave in essentially that way.

Let us consider, for example, a system (say, an atom) with a large total angular momentum, $j \gg 1$. If the atom has a magnetic moment in the direction of the angular momentum, $\mathbf{J}$, and we place it in a uniform magnetic field, it will have $2j+1$ equally spaced Zeeman levels. As long as the atom is in any of the states with J_z not too far from j, that is, near the state of maximum energy, its behavior will quite closely approximate that of an inverted oscillator. The operator, $(J_x + J_y)/\sqrt{2j}$, that raises the magnetic quantum number J_z by one unit then plays the role of the annihilation operator, a. Its commutator with its Hermitian adjoint has eigenvalues appropriately close to unity for the states with J_z near j. We can then identify the angular momentum states $|J_z = j - n\rangle$ with the states $|n\rangle$ of the inverted oscillator (see Fig. 8).

When this atom is in any state of small n and is coupled to the radiation field in the usual way, it will begin to emit quanta and descend to states of larger n in just the way that we have described for the inverted oscillator. Of course, the acceleration of the radiation rate will not continue indefinitely. When the quantum number, n, becomes comparable to j, it will no longer increase as rapidly. The rate of radiation reaches maximum for $n = j$ and then begins to decrease. All of this is only to say that a magnetic moment associated with a large value of j is not really a linear amplifier. It is a nonlinear one. However, there is a regime for large values of J_z in which it duplicates quite accurately the behavior of a linear amplifier.

The example of an atom with a huge angular momentum, j, may seem a bit far-fetched, but there is a context in which we often deal with such systems these days. A single two-level atom is equivalent algebraically to a system of spin $\frac{1}{2}$ in a mag-

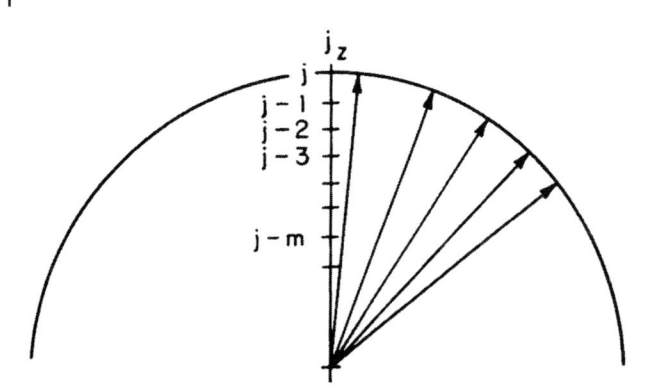

Figure 8 The succession of magnetic substates for large j and m near j in value. These are the states for which radiative transitions are amplified exponentially.

netic field. A system of N such atoms with identical couplings to the radiation field is therefore precisely equivalent to the angular momentum model we have just discussed with $j = N/2$. When all or nearly all of the N atoms are initially in their excited states, the emission process that ensues is superfluorescent in character. The initial phase of superfluorescence is one of linear amplification[12]. It continues long enough in practice to generate fields of essentially classical strength that are easily observed with large-scale laboratory equipment. The inverted oscillator model, in fact, is just an idealization of the way in which the superfluorescent radiation process begins.

Of course, the quanta that are being amplified in the inverted central oscillator are not the usual sorts of excitations of positive energy. They are, strictly speaking, deexcitations or, in effect, quanta of negative energy. We can easily alter that feature of the model without significantly changing its mathematical analysis by turning all of the oscillators upside down once more. We let the central oscillator, in other words, be a normal one with quanta of positive energy, and let the heat bath oscillators all be inverted. Any quanta initially present in the central oscillator will then be amplified to large positive energies. It is easy to verify that the only significant changes in our earlier results are inversions of the signs of the frequency parameters, ω and ω_k.

14.4
Specification of Photon Polarization States

One of the subjects we mentioned initially is a scheme for superluminal communication involving photon polarizations. That proposal gave a special role to states of plane and circular polarization, but any two distinguishable pairs of orthogonal polarization states would do equally well. It will be of some help, therefore, to have in hand a general procedure for dealing with photon polarization states of all sorts.

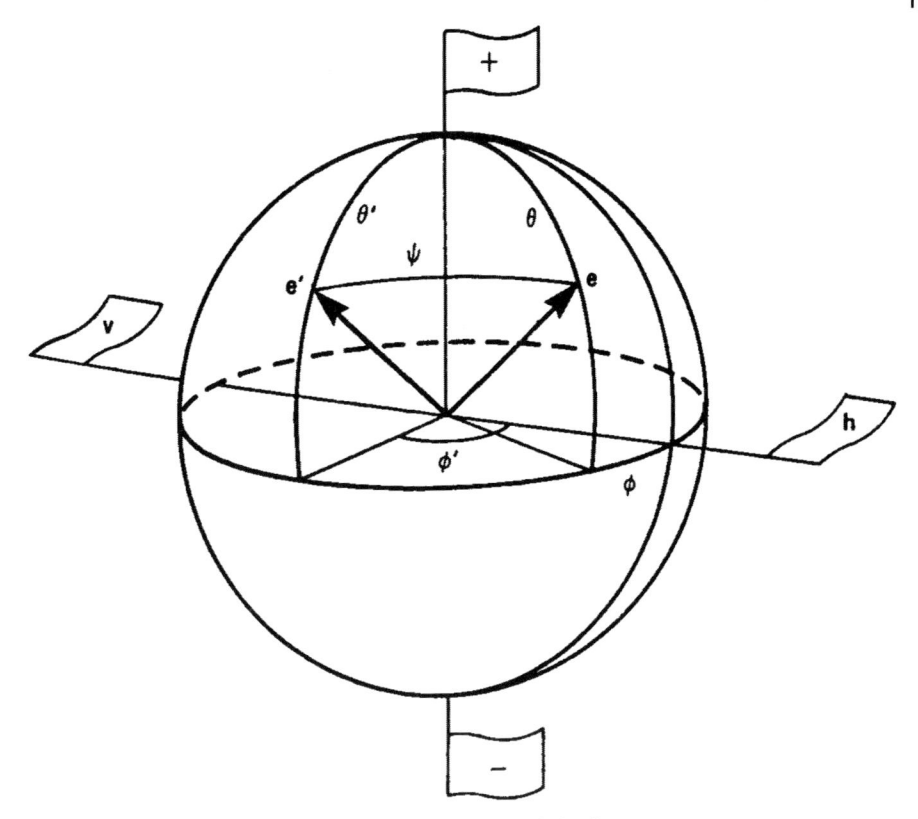

Figure 9 The Poincaré sphere for specifying photon polarizations. The north and south poles represent the two circular polarizations. All equatorial points represent linear polarizations. Antipodal points represent orthogonal elliptical polarizations, more generally.

For a light wave that propagates in a fixed direction, the polarization vectors are defined in a two-dimensional space transverse to that direction. Complex vectors of unit length in that two-dimensional space define a two-parameter family of elliptical polarization states; any two orthogonal vectors of that sort are eligible to be a pair of basis states. Transformations from one pair of basis states to another must then be unitary matrix transformations in two dimensions. They are, thus, a two-parameter subgroup of the group SU(2), which is the special unitary group in two dimensions.

There is a well-known correspondence between the transformation of SU(2) and real rotations in three dimensions that gives us a simple way of picturing the various polarization states. They can be put in one-to-one correspondence with the points on a unit sphere, so the transformations of SU(2) simply rotate them into one another. This way of dealing with all of the polarizations at once was invented by Poincaré[13] in 1892, well before the discovery of spinors by Cartan.

Figure 9 then is a picture of the Poincaré sphere. We have taken the north and south poles of the sphere to represent the two orthogonal states of circular polarization, and we have accordingly labeled them for the state $|+\rangle$, of helicity one, and for the state $|-\rangle$, of helicity minus one. Of course, you can equally well label them L and R for left and right circular polarizations (if you can remember which is which). An arbitrary polarization state, which we can call $|\theta, \varphi\rangle$, can then be written as

$$|\theta, \varphi\rangle = e^{i\varphi/2} \cos\frac{\theta}{2} |+\rangle + e^{-i\varphi/2} \sin\frac{\theta}{2} |-\rangle , \tag{14.53}$$

where the angles θ and φ are limited by

$$0 \leq \theta \leq \pi, \quad -\pi \leq \varphi < \pi. \tag{14.54}$$

Each point on the surface of the sphere thus defines a unique state of elliptical polarization. Antipodal points on the sphere always represent orthogonal polarization states.

Let us take $|h\rangle$ and $|v\rangle$ to represent the usual horizontal and vertical states of plane polarization; thus, the circular polarization states can be written as

$$|\pm\rangle = \frac{1}{\sqrt{2}} (|h\rangle \pm i |v\rangle) . \tag{14.55}$$

Then we find that in the $|\theta, \varphi\rangle$ scheme, these states are

$$|h\rangle = \left|\frac{\pi}{2}, 0\right\rangle , |v\rangle = \left|\frac{\pi}{2}, -\pi\right\rangle . \tag{14.56}$$

We have also indicated their locations on the Poincaré sphere in Fig. 9. They lie on the equator, which consists exclusively of plane polarizations.

One of the conveniences of the Poincaré sphere is that it gives a simple way of dealing with the scalar products of polarization vectors. The probability that a photon with polarization $|\theta, \varphi\rangle$, for example, is transmitted by a filter that selects polarizations $|\theta', \varphi'\rangle$ is the squared scalar product $|\langle \theta', \varphi' | \theta, \varphi \rangle|^2$. According to Eq. (14.55), the scalar product is

$$\langle \theta', \varphi' | \theta, \varphi \rangle = e^{(i/2)(\varphi - \varphi')} \cos\frac{\theta'}{2} \cos\frac{\theta}{2} + e^{-(i/2)(\varphi - \varphi')} \sin\frac{\theta'}{2} \sin\frac{\theta}{2} . \tag{14.57}$$

The squared modulus of this expression depends only on the angle between the radii to the points (θ, φ) and (θ', φ') on the sphere. If we call that angle ψ, as in Fig. 9, and recall the spherical law of cosines,

$$\cos\psi = \cos\theta \cos\theta' + \sin\theta \sin\theta' \cos(\varphi - \varphi') , \tag{14.58}$$

then we find that the transmission probability of our photon is just

$$|\langle \theta', \varphi' | \theta, \varphi \rangle|^2 = \tfrac{1}{2}(1 + \cos\psi) = \cos^2 \tfrac{1}{2}\psi . \tag{14.59}$$

14.5
Measuring Photon Polarizations

If we are to communicate by means of photon polarizations, we must be prepared to answer questions like this: A photon goes through a filter that transmits with 100 % efficiency the polarization $|\theta', \varphi'\rangle$; what is the probability that such a transmitted photon really had the initial polarization $|\theta', \phi'\rangle$? That is a question that can only be posed probabilistically because you cannot go back and verify what the state was. The possibility of communication, on the other hand, depends critically on the determination of those probabilities. How do we find them? We will presently show that Bayes' theorem gives a convenient way of determining such *a posteriori* probabilities, but we do not need all of that generality quite yet.

Let us assume we have some arbitrary beam of photons. We place in the beam a filter that transmits only the polarization $|\theta', \varphi'\rangle$. When a photon is transmitted, we can obviously say that the probability that its initial polarization was $|\theta, \varphi\rangle$ is proportional to the probability that a photon of polarization $|\theta, \varphi\rangle$ is transmitted by a filter that transmits polarizations $|\theta', \varphi'\rangle$. That probability, in other words, is proportional to the squared scalar product in Eq. (14.59), and has the angular dependence of $\cos^2(\psi/2)$. The $\cos^2(\psi/2)$ distribution is spread out quite smoothly all over the sphere, but it does convey some information. When you make a measurement on one photon, what you find is a new way of weighting whatever information you had initially about where its polarization vector might have been pointing. The new weighting multiplies whatever distribution you knew of before by the factor $\cos^2(\psi/2)$; if you knew nothing at all beforehand, if the *a priori* distribution were uniform, for example, then the final distribution over the surface of the sphere would just be a constant times $\cos^2(\psi/2)$.

If you were to attempt to perform the Einstein–Podolsky–Rosen experiment we described initially, and to use it to communicate via measurements made on individual photon polarizations, you would quickly find that you cannot distinguish between circular polarizations and linear polarizations. In fact, you cannot distinguish between any one pair of antipodal points on the Poincaré sphere and any other. The angles ψ for any pair of antipodal points are naturally supplementary, so the sum of the two probabilities is proportional to

$$\cos^2(\psi/2) + \cos^2\left(\frac{\pi - \psi}{2}\right) = \cos^2(\psi/2) + \sin^2(\psi/2) = 1 ,$$

and that is constant and quite independent of which pair of orthogonal polarization states was chosen. Thus, the task is hopeless. One polarization measurement on a single photon cannot ever provide the information desired.

We have the alternative suggestion, however, that we use a "laser gain tube" to multiply the number of photons to be processed. What such a laser gain tube does, presumably, is just to amplify in similar ways photons that are in any of the possible polarization states. Those polarization states can all be considered to be superpositions of one pair of basis states. Then everything the laser gain tube does can be represented

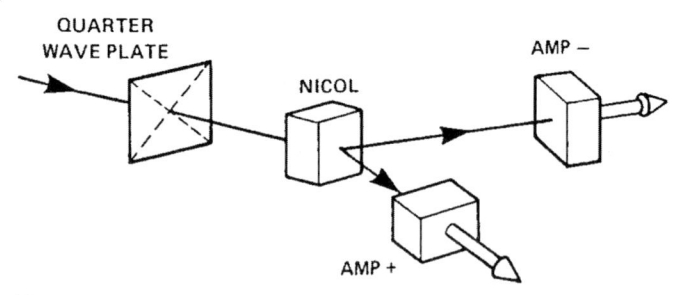

14 Amplifiers, Attenuators and Schrödingers Cat

Figure 10 A setup illustrating the use of a pair of amplifiers as a photon polarization detector.

by the action of two identical amplifiers, each of which is fed by the appropriate polarization component. Because we want to amplify normal positive energy photons, we should use as amplifiers normal positive energy oscillators that are coupled, as noted earlier, to heat baths consisting of inverted oscillators.

An appropriate sort of amplifier–detector arrangement is shown in Fig. 10. The quarter-wave plate converts circular polarizations into plane polarizations and the Nicol or Wollaston prism separates the latter into two beams, each of which is sent into its own amplifier and emerges in considerably strengthened form. In effect, then, one of those amplifiers amplifies the right-handed circularly polarized component of the incident beam and the other amplifies the left-handed component. The two output beams are so strong that we may describe them classically, and we can measure them to our heart's content without disturbing them in any way.

Let us agree, however arbitrarily, to carry out our calculations by using circularly polarized basis states. Let $a_+^\dagger$ be a creation operator for the state of helicity one and let $a_-^\dagger$ be a creation operator for the state of helicity minus one. Then a state with n_+ photons of the first variety and n_- of the second can be generated from the vacuum state, $|n_+ = 0, n_- = 0\rangle \equiv |0,0\rangle$, by applying the appropriate creation operators,

$$|n_+, n_-\rangle = \frac{(a_+^\dagger)^{n_+}}{\sqrt{n_+!}} \frac{(a_-^\dagger)^{n_-}}{\sqrt{n_-!}} |0,0\rangle \ . \tag{14.60}$$

We want still to be able to deal with photons in arbitrary polarization states. For a photon in a polarization state, $|\theta, \varphi\rangle$, we can define the creation and annihilation operators as

$$a_{\theta\varphi}^\dagger = e^{i\varphi/2} \cos\frac{\theta}{2} a_+^\dagger + e^{-i\varphi/2} \sin\frac{\theta}{2} a_-^\dagger \ , \tag{14.61}$$

$$a_{\theta\varphi} = e^{-i\varphi/2} \cos\frac{\theta}{2} a_+ + e^{i\varphi/2} \sin\frac{\theta}{2} a_- \ , \tag{14.62}$$

so that a single-photon state, $|\theta, \varphi\rangle$, can be written as

$$|\theta, \varphi\rangle = a_{\theta\varphi}^\dagger |0,0\rangle \ . \tag{14.63}$$

The initial density operator that represents this state is just this vector multiplied by its dual,

$$\rho(0) = a^{\dagger}_{\theta\varphi} |0,0\rangle \langle 0,0| a_{\theta\varphi} . \tag{14.64}$$

The amplifiers we are discussing are, of course, transient amplifiers. It would be nicer in some ways to imagine this experiment as carried out with continuously operating linear amplifiers. However, it is somewhat more complicated to construct models of CW amplifiers, and it is not really very essential for conceptual purposes. The time-dependent scheme we are discussing, in fact, suits the practical needs of a *gedanken* experiment quite well. One can adjust the various initial times and amplification times so that one secures an appropriately strong output signal from a single photon arriving at time zero. That does, however, require a somewhat delicate adjustment because these amplifiers are devices that provide an output signal even if there is no photon present at time zero. One must be careful about the coordination of such equipment, but there is nothing difficult about that, at least, in principle.

14.6
Use of the Compound Amplifier

We can deal with our pair of amplifiers very much as we dealt with a single one before. If the photon incident upon our detector is in the state $|\theta, \varphi\rangle$, the initial density operator for the two amplifiers is given by Eq. (14.64). The final reduced density operator for the two amplifiers can then be written as a two-mode P representation that has the general form,

$$\rho(t) = \int P(\gamma_+, \gamma_-, t|\theta, \varphi) |\gamma_+\rangle \langle \gamma_+| |\gamma_-\rangle \langle \gamma_-| \, d^2\gamma_+ \, d^2\gamma_- . \tag{14.65}$$

Evaluating the function P precisely is now, for the most part, a repetition of the calculation described earlier for the single-mode case. The only new points to be observed are that the function depends on two variables, $\gamma_\pm$ and that the initial density operator given by Eq. (14.64) contains off-diagonal terms in the quantum numbers $n_\pm$. The effect of such terms as we have noted can easily be found by the same generating function devices that we used for the diagonal ones.

The function P that we find in this way is

$$P(\gamma_+, \gamma_-, t|\theta, \varphi) = \frac{1}{\pi^2 \mathcal{N}^2} \left\{ 1 - \frac{|U|^2}{\mathcal{N}} + \frac{|U|^2}{\mathcal{N}^2} \left| \gamma_+ e^{-i\varphi/2} \cos\frac{\theta}{2} + \gamma_- e^{i\varphi/2} \sin\frac{\theta}{2} \right|^2 \right\}$$

$$\times \exp\left[-\frac{|\gamma_+|^2 + |\gamma_-|^2}{\mathcal{N}} \right], \tag{14.66}$$

where the functions U and $\mathcal{N}$ for the amplifier are those defined earlier and approximated by Eqns. (14.40) and (14.41), respectively. We can express this result a bit more simply by imagining the two amplified output fields to be superposed and then defining a polarization vector for the superposed fields. An appropriate unit polarization vector is

$$|\hat{e}_{\gamma_+\gamma_-}\rangle = \frac{1}{\sqrt{|\gamma_+|^2+|\gamma_-|^2}}\left\{\gamma_+|+\rangle+\gamma_-|-\rangle\right\}. \tag{14.67}$$

We can define the intensity associated with any polarization state, $|\theta',\varphi'\rangle$, of the superposed output fields as

$$I(\theta',\varphi') = \text{Tr}\left\{\rho(t)\, a^\dagger_{\theta'\varphi'}\, a_{\theta'\varphi'}\right\} \tag{14.68}$$

$$= \int P(\gamma_+,\gamma_-,t|\theta,\varphi)\left|\gamma_+\, e^{-i\varphi'/2}\cos\frac{\theta'}{2}+\gamma_-\, e^{i\varphi'/2}\sin\frac{\theta'}{2}\right|^2 d^2\gamma_+\, d^2\gamma_-$$

$$= \int |\langle\theta',\varphi'|\hat{e}_{\gamma_+\gamma_-}\rangle|^2 (|\gamma_+|^2+|\gamma_-|^2)\, P(\gamma_+,\gamma_-,t|\theta,\varphi)\, d^2\gamma_+\, d^2\gamma_-. \tag{14.69}$$

We can use the vector $|\hat{e}_{\gamma_+\gamma_-}\rangle$ furthermore to write the expression for the function P a bit more compactly as

$$P(\gamma_+,\gamma_-,t|\theta,\varphi) = \frac{1}{\pi^2\mathcal{N}^2}\left\{\frac{|U|^2}{\mathcal{N}^2}|\langle\hat{e}_{\gamma_+\gamma_-}|\theta\varphi\rangle|^2(|\gamma_+|^2+|\gamma_-|^2)-\frac{1}{\mathcal{N}}\right\}$$

$$\times \exp\left[-\frac{|\gamma_+|^2+|\gamma_-|^2}{\mathcal{N}}\right]. \tag{14.70}$$

The Gaussian integral for the polarization dependence of the amplified intensity leads to the simple zero-temperature result

$$I(\theta',\varphi') = |U(t)|^2\left\{1+|\langle\theta',\varphi'|\theta,\varphi\rangle|^2\right\}-1 \tag{14.71}$$

$$= e^{2\kappa t}\left\{1+|\langle\theta',\varphi'|\theta,\varphi\rangle|^2\right\}-1. \tag{14.72}$$

The compound amplifier, we see, does indeed "remember" the initial polarization state of the photon in the classical output it generates. While the output field follows the distribution of Eq. (14.70) and is therefore highly random, it is indeed polarized, on the average, in the direction, $|\theta,\varphi\rangle$. The fact that we can measure the polarization of the classical field as precisely as we like may seem to favor the EPR communication scheme, so we had better say a bit more about polarizations.

To define the (θ,φ) polarization of the amplified beam, we must compare the intensity, $I(\theta,\varphi)$, with the intensity for the orthogonal polarization state. Let $\bar{\theta} = \pi - \theta$ and $\bar{\varphi} = \varphi \pm \pi$ be the polar coordinates of the antipodal point on the Poincaré sphere.

Then, the polarization in the (θ, φ) direction is

$$p(\theta,\varphi) = \frac{I(\theta,\varphi) - I(\bar{\theta},\bar{\varphi})}{I(\theta,\varphi) + I(\bar{\theta},\bar{\varphi})}, \tag{14.73}$$

$$= \frac{|U(t)|^2}{3|U(t)|^2 - 2}, \tag{14.74}$$

$$= \frac{1}{3 - 2e^{-2\kappa t}}. \tag{14.75}$$

The last of these expressions shows that though the compound amplifier does remember the initial photon polarization, it does not have the clearest of memories. The strong field that emerges for $\kappa t \gg 1$ only has polarization $1/3$. That happens because of the amplified noise output that both amplifiers generate even with no initial photon present.

The amplified intensity given by Eq. (14.71) has a fully polarized component, $|U(t)|^2 |\langle \theta', \varphi' | \theta, \varphi \rangle|^2$, which we may consider as the amplified signal due to the incident photon. However, it also contains an unpolarized component, $|U(t)|^2 - 1$, which represents the noise output contributed equally by the two amplifiers. The amplification of the noise keeps pace with the signal, and in the long run, there is usually even somewhat more noise present than signal.

The compound amplifier that is part of our detector scheme duplicates precisely the action of a "laser gain tube" on the incident photons. However, it is not at all clear that such action deserves to be called "cloning". Both the compound amplifier and the laser gain tube are bound to generate more or less random numbers of photons in two orthogonal polarization states. There is no alternative, clearly, to suffering the presence of two varieties of photons. The description of polarization states requires two basis states, and spontaneous emission alone assures us that both varieties of photons will be present in general.

In a note inspired by the EPR communication scheme, Wootters and Zurek[14] have shown that it is not possible, by using a single amplifier, to clone photons of arbitrary polarization. Their analysis takes the definition of cloning quite literally, requiring all photons to be identical, and places the further restriction that the initially pure one-photon state has to always remain pure. It is not related, therefore, to the action of any real amplifier, let alone the pair of them that is necessary for the measurement of arbitrary polarizations. That a pair of analyzing systems is sufficient for the unbiased analysis of polarizations has been pointed out by Mandel[15], who discussed a detector consisting of two atoms.

14.7
Superluminal Communication?

Let me now return to our superluminal communication problem. Having introduced the quantum optical means one might use to make the measurements, I would like to

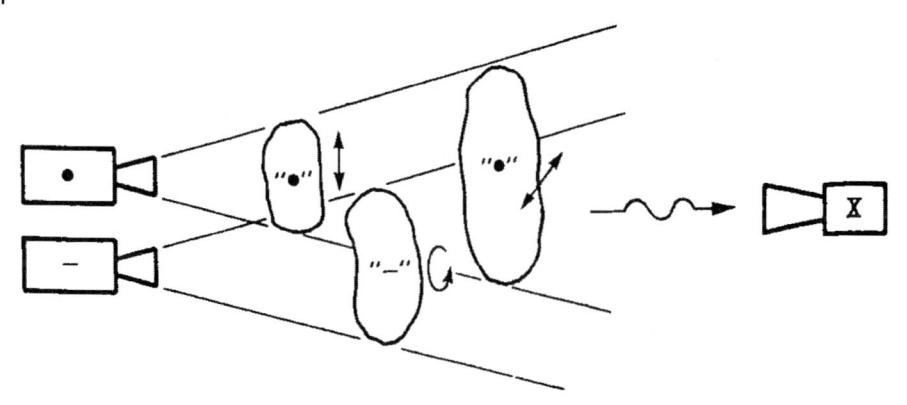

Figure 11 Schematic picture of the use of polarized photon wavepackets for communication purposes. One signal generator produces linearly polarized packets that are represented by "dots". The other produces circularly polarized packets that are represent by "dashes".

persuade you that we can, in fact, dismiss the "laser gain tube" scheme very quickly. I will nonetheless go on to analyze it a bit further as it is interesting to see in somewhat more detail how and why it does not work; there are also some more practical uses for the analysis as well.

Let us try to describe some devices that boil the communication problem down to its barest essentials. Let us say we have two devices, which could even be different states of the same device producing two varieties of signal. These are wave packets being sent off to a distant receiver. You can see them in Fig. 11. One variety I will call a dot, and the other one, a dash (Morse code). Our distant observer listens with whatever equipment he has and measures some property, X, of those wave packets. He hopes, on the basis of a measurement of X, to determine which variety of wave packet it is that he has received; that is, whether it is a dot or a dash. He must choose X to be a quantity that makes the clearest distinction between the two. Yet, he must inevitably face the problem, given the observation $X = x$, what is the probability that the signal is a dot? And given the observation $X = x$, what is the probability that it is a dash? Now, here, I do not know of any alternative to the use of Bayes' theorem or more sophisticated decision theoretical notions.

What are the dashes and dots in our communication model? The dashes might be represented by circularly polarized photon packets and the dots by linearly polarized packets. We want, if possible, to distinguish clearly between them. However, we must carefully note one thing more. In the class of linearly polarized packets that represent dots, we have no control over which of two polarizations is sent. Some of those dots are going to be vertically polarized, and an equal number, on the average, are going to be horizontally polarized. In the class of dashes, some of the packets are going to be right-handed circularly polarized, and some are going to be left-handed circularly polarized.

14.7 Superluminal Communication?

To phrase the decision problem more formally, our signaling devices have to be described statistically. Each of our signal generators puts the field into a state described by a particular density operator. The detection problem is to make measurements that offer a clear distinction between those two density operators. By measuring X, or whatever it is, in each wave packet, the detector device asks, in effect: do you represent the density operator for a dot, or the density operator for a dash?

We can phrase the mathematical part of the problem as follows: Let the probability that a given packet represents a dash or a dot be $p(-)$ and $p(\cdot)$, respectively. Let the probability that the detector observes the value x, if the input is a dash, be $P(x|-)$. Then, the joint probability that a packet represents a dash and the detector observes x is

$$P(x,-) = P(x|-)\,p(-)\,. \tag{14.76}$$

Bayes' theorem asserts that this joint probability can be also be written as the product,

$$P(x,-) = P(-|x)\,p(x)\,, \tag{14.77}$$

in which $p(x)$ is the probability that the detector registers x no matter which sort of packet arrives, and $P(-|x)$ is the probability, given the measurement of x, that the incident packet represents a dash. The probability $p(x)$ is evidently given by

$$p(x) = P(x|-)\,p(-) + P(x|\cdot)\,p(\cdot)\,, \tag{14.78}$$

while the probability we seek, namely the probability that the incident packet represents a dash, is evidently

$$P(-|x) = \frac{P(x|-)\,p(-)}{P(x|-)\,p(-) + P(x|\cdot)\,p(\cdot)}\,. \tag{14.79}$$

An analogous expression of course holds for $P(\cdot|x)$.

However, now that we have created all this machinery, we have to face the fundamental difficulty that if we look at the ensemble of wave packets that represent dots, we will find them unpolarized. What is the density matrix that represents an unpolarized beam? It is one-half of the unit matrix; the density operator is just one-half of the unit operator. Now, let us consider the ensemble that represents dashes. We have said they are circularly polarized, but in fact they are circularly unpolarized; they have no net polarization either. What is the density matrix that the dash generator turns out? That is once again one-half of the unit matrix. Therefore, we are asking our distant friend, whatever detection device he may be using, to make a distinction between things that are absolutely identical. The two density operators he must recognize individually are identical twins. This is a classic example of a distinction without a difference. The probabilities that result must satisfy the identity,

$$P(x|-) = P(x|\cdot)\,, \tag{14.80}$$

and thus,
$$P(-|x) = p(-) . \tag{14.81}$$

The *a posteriori* probability, in other words, is equal to the raw *a priori* probability. Measuring x, whatever it may be, does not add to our knowledge of the message at all.

That difficulty is bound to frustrate our distant friend. It cannot be too encouraging either to the theorists who insist on examining the detection problem microscopically rather than surveying it from this more global viewpoint. Our argument assures us that whatever useful possibilities we seem to see present from a microscopic standpoint are going to cancel out before the calculation is finished, and that it is just not possible to use the polarization scheme as a means of communication. Having said that, I want nevertheless to look a bit further at the scheme because it is an interesting one; most particularly, it is interesting as an illustration of what amplifiers do. We will find that our calculations do indeed have practical applications, but only in other experimental contexts.

Let us go back to the pair of amplifiers we are using to amplify each photon that is received and see what additional information we can elicit from them. What we do is let the amplifiers amplify for several gain periods, so that the field intensity is increased substantially. We are then talking about classical outputs. We need have no embarrassment about measuring those field amplitudes because they can be just as strong as we like; we can measure them without disturbing anything.

In this strong-field limit, we have

$$\frac{1}{\mathcal{N}(t)} = \frac{1}{e^{2\kappa t} - 1} \to 0$$

and

$$\frac{|U(t)|^2}{\mathcal{N}(t)} = \frac{e^{2\kappa t}}{e^{2\kappa t} - 1} \to 1 .$$

When the output fields are that strong, the P distribution for their amplitudes becomes, in effect, a classical probability density; Eq. (14.70) then reduces to the form

$$P(\gamma_+, \gamma_-, t | \theta, \varphi) = \frac{1}{\pi^2 \mathcal{N}^2} |\langle \hat{e}_{\gamma_+ \gamma_-} | \theta \varphi \rangle|^2 (|\gamma_+|^2 + |\gamma_-|^2) \exp\left[-\frac{|\gamma_+|^2 + |\gamma_-|^2}{\mathcal{N}}\right] . \tag{14.82}$$

We can use this expression now to determine what classical measurements made on the amplified fields can tell us about the original polarization state of any incident photon.

In any given detection process, we insert a photon in the compound system and let the two amplifiers amplify their initial fields. Then, we observe a pair of classical output field amplitudes, γ_+ and γ_-, that are governed by the probability distribution of Eq. (14.82). Once we have measured any such pair of amplitudes, we can ask the question: what is the probability that the initial photon had a polarization $|\theta, \varphi\rangle$? The answer is, in fact, staring at us in Eq. (14.82). It is the very same weighting that we would have arrived at by making a measurement on a single photon with a single polarizer. We are now no longer talking simply about the superposition of different polarization states. The amplification process, combined with the corruption of the information by noise, has brought us back, however, to exactly the same position we were in before. The compound amplifier, as far as polarization is concerned, has contrived to tell us nothing at all new because of the noise it has added. We can only make the same inferences about initial polarization states that we made without it. In fact, from that standpoint, the compound amplifier is no better than a single polarizer.

14.8
Interference Experiments and Schrödinger's Cat

Since we have some experience now at using amplifiers in pairs, there are all kinds of interesting games we can play. One is to perform Young's classic double-pinhole interference experiment with an amplifier placed behind each of the pinholes, as in Fig. 12.

Conceivably, we could use such a scheme to determine which pinhole any given photon has really passed through. Let us agree that we are going to look only at cases in which a single photon packet falls symmetrically on the two pinholes. That means that we are starting our amplifiers out in an initially pure state that is a superposition of the two states in which a photon enters one, while the other remains in its empty or "ground" state. We can use the plus and minus signs now just to label the two

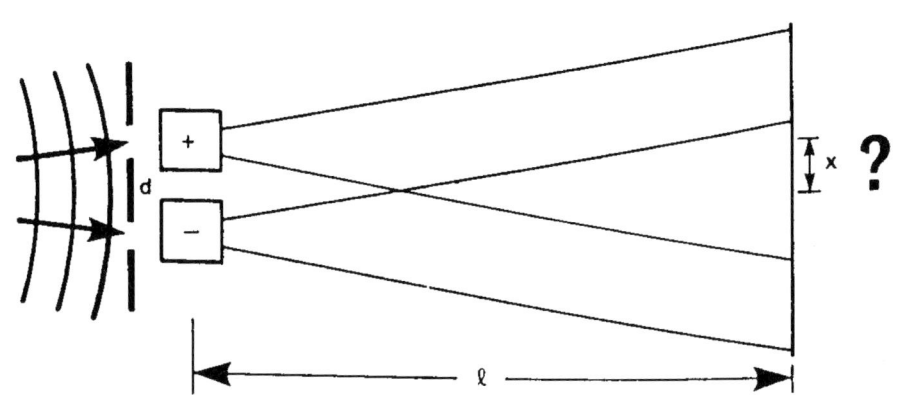

Figure 12 Young's double-pinhole interference experiment carried out with a single incident photon and an amplifier behind each pinhole. What sort of pattern appears where the emergent beams are superposed on the distant screen?

amplifiers. Those signs then no longer refer to polarizations. With that one change, we can make use of the same calculations we carried out earlier.

Now, what shall we expect to see when the amplified fields are superposed by projecting them onto the distant screen? Let me, for the sake of argument, list three bad guesses. Bad guess number one: A classical physicist who just believes in probability theory and no more will say that the photon is bound to go through one hole or the other. It will have a probability of 1/2 for going through the upper hole, and if it does that, the upper amplifier will produce a strong output field, while the lower one will produce nothing. Of course, it may happen the other way around, but we will never see interference fringes. That is what this person would say, however, benightedly.

Bad guess number two might be made by somebody who has studied Young's experiment as it is usually described, and takes the classical view of it. He will note that the photon wave enters those two holes coherently and is strengthened symmetrically by passage through both amplifiers. The fields projected by the two amplifiers, he will say, should simply show Young's interference fringes on the screen. In saying that, he too is of course overlooking something quite important.

This may not be an exhaustive list of bad guesses; naiveté can take many forms. However, a naively sophisticated view we should not overlook would be this: The photon begins in a well-defined initial quantum state. You may remember the Schrödinger cat example; the situation is not unlike that. Here, we have a well-defined initial superposition state. The photon may alternatively, but coherently, enter either of the two amplifiers. If their action has the elegant simplicity Schrödinger assumed his diabolical cat-killing machine to have, we should then find, as their output, a quantum mechanical superposition of two amplified fields.

What is the meaning of such a quantum mechanical superposition of two quantum states containing highly amplified fields? The two output states involved are orthogonal and quite dissimilar. They can easily be distinguished from one another classically. A quantum mechanical superposition of those states regards the two amplified outputs as alternatives. That feature is quite characteristic of superpositions of states in quantum mechanical Hilbert space. When you superpose two orthogonal states, $|A\rangle$ and $|B\rangle$, in Hilbert space, you are saying in a certain sense, "either $|A\rangle$ or $|B\rangle$, but not both". Therefore, you may just see in this third, slightly more sophisticated view of the amplification problem, the ghost of Schrödinger's cat.

Here, we indeed have a device in which we can trace what happens all the way from the quantum to the classical domain. It permits us, at last, to answer the question of whether we will see the quantum mechanical superposition of two macroscopic states as envisioned by Schrödinger. That pure state, if it were to occur, would be characterized again by a strong output from either one amplifier or the other, but not both. The product of the two classical output field strengths would always be zero, and thus there would be no interference fringes.

What is the correct way to treat the interference problem? When the photon impinges on the two pinholes, we can think of the two amplifiers, labeled $+$ and $-$, as beginning in the pure state,

$$|\psi\rangle = \frac{1}{\sqrt{2}}\left\{|1\rangle_+|0\rangle_- + |0\rangle_+|1\rangle_-\right\}, \tag{14.83}$$

in which $|0\rangle_\pm$ and $|1\rangle_\pm$ are the $n_\pm = 0$ and $n_\pm = 1$ states, respectively. This state corresponds precisely to the plane-polarization state, $|h\rangle = |\pi/2, 0\rangle$, for a single photon in the analysis we have just described, and so the results of that calculation may be taken over directly. If $\gamma_\pm$ are the complex amplitudes of the classical fields generated by the two amplifiers, we can write the probability distribution for those amplitudes, according to Eq. (14.82), as

$$P(\gamma_+, \gamma_-, t|\psi) = P(\gamma_+, \gamma_-, t|\pi/2, 0)$$
$$= \frac{1}{2\pi \mathcal{N}^3} |\gamma_+ + \gamma_-|^2 \exp\left[-\frac{|\gamma_+|^2 + |\gamma_-|^2}{\mathcal{N}}\right]. \tag{14.84}$$

This result shows us immediately that the two amplifiers will have positively corrected outputs; they will, to some degree, tend to radiate coherently, and thus to create fringe patterns on the screen.

How do we calculate the intensity of the light on the screen? We can assume that the fields emitted by the two amplifiers are nearly parallel plane waves, and then try to recall the geometric approximations that we used so often as students. When the pinholes are a distance d apart, the screen is at a distance ℓ, and we observe at a point x on the screen, as opposed to the central point (see Fig. 12), we find that the waves arriving from the two amplifiers undergo shifts of phase by $\pm \pi x d / \lambda \ell$.

When a single photon enters the system and the output fields of the amplifiers have amplitudes γ_+ and γ_-, the intensity on the screen will be proportional to

$$I(x, \gamma_+, \gamma_-) = \left|\gamma_+ \exp\left(\frac{i\pi dx}{\lambda \ell}\right) + \gamma_- \exp\left(\frac{-i\pi dx}{\lambda \ell}\right)\right|^2. \tag{14.85}$$

We should emphasize that this is an intensity distribution for an entire interference pattern, and not a probability distribution for the appearance of a single photon on the screen. A single photon arriving at the front end of our system produces an intense, classical interference pattern on the distant screen. (In fact, we would even find an interference pattern there, in general, when no photon arrives.)

To the extent that the amplitudes γ_+ and γ_- are more or less random variables, the interference pattern of Eq. (14.85) will also have some random features. It will always contain parallel intensity fringes, unless one or the other of $\gamma_\pm$ happens to vanish, but the fringes will shift in position from one repetition of the experiment to another. The fringes, furthermore, will usually not have the strong contrast typical of Young's experiment. The intensity has no zeroes unless $|\gamma_+| = |\gamma_-|$.

When we repeat the experiment many times, the random fringe system will have the average intensity,

$$I(x) = \int \left| \gamma_+ \exp\left(\frac{i\pi dx}{\lambda \ell}\right) + \gamma_- \exp\left(\frac{-i\pi dx}{\lambda \ell}\right) \right|^2 P(\gamma_+, \gamma_-, t|\psi)\, d^2\gamma_+ \, d^2\gamma_-$$

$$= 2\left(1 + \cos^2 \frac{\pi dx}{\lambda \ell}\right). \tag{14.86}$$

Taking the average, in other words, leaves us with Young's fringe pattern set against a constant background intensity. That constant background, of course, is the intensity due to spontaneous emission noise. It is just the average intensity we would find for the random fringes generated when no photon enters the system. The average visibility of the fringes when a photon does enter is

$$\mathcal{V}(x) = \frac{I_{max} - I_{min}}{I_{max} + I_{min}} = \frac{1}{3}. \tag{14.87}$$

This result corresponds, loosely speaking, to a mixture of all three of the wrong guesses we listed earlier, and probably some others as well. In any case, it does not correspond at all closely to the guess Schrödinger would have made by means of the same reasoning he used for his cat. The final state with an amplified wave coming from the upper amplifier and none from the lower one could correspond, according to Schrödinger's picture, to finding the cat alive, and the opposite configuration to finding the cat dead. The actual state of the amplified fields, however, is far from being Schrödinger's superposition of those two states. It is not any pure state at all; in the coherent state representation, it is a Gaussian mixture with an enormous variance. In the n-quantum state representation, it is a quasi-diagonal mixture with a vast dispersion too. The error in this projection of Schrödinger's argument is its omission of the effects of noise, and the noise in unavoidable; it is there for intrinsically quantum mechanical reasons.

It is occasionally said that quantum mechanics deals only with averages taken over ensembles of experiments, but here we have an explicit counterexample. The outputs of our amplifiers and the appearance of the fringes on the screen vary greatly from one repetition of the experiment to another, but they are all observable individually. In addition, we have no trouble measuring them without disturbing them in any way. In any one repetition, for example, the amplifiers may give us the two amplitudes, γ_+ and γ_-. The visibility of the set of fringes that results is then

$$\mathcal{V}(\gamma_+, \gamma_-) = \frac{2|\gamma_+||\gamma_-|}{|\gamma_+|^2 + |\gamma_-|^2}. \tag{14.88}$$

This expression only takes on the value one, which expresses strong fringe contrast, when the two amplitudes $\gamma_\pm$ are equal in modulus.

The question that always fascinates us about Young's experiment is: Which pinhole did the photon really go through? We can, in fact, say something about that in the present version of the experiment. If it should turn out in one repetition of the experiment that $|\gamma_+|$ is far larger than $|\gamma_-|$, we would have a certain indication that the photon went through the upper pinhole and amplifier rather than the lower ones. Of course, we can only make such a statement on a probabilistic basis. The way to make it is to use Bayes' theorem again to define the probability $P(+|\gamma_+,\gamma_-)$ that the photon passed through the upper pinhole, given that the two field amplitudes are $\gamma_\pm$. The result, you can easily see, is

$$P(+|\gamma_+,\gamma_-) = \frac{|\gamma_+|^2}{|\gamma_+|^2 + |\gamma_-|^2}. \tag{14.89}$$

Let us abbreviate the two probabilities $P(\pm|\gamma_+,\gamma_-)$ as $p_\pm$ so that we have

$$p_\pm = \frac{|\gamma_\pm|^2}{|\gamma_+|^2 + |\gamma_-|^2}. \tag{14.90}$$

The product of these probabilities, according to Eq. (14.89), is proportional to the square of the fringe visibility:

$$p_+ p_- = \tfrac{1}{4} \mathcal{V}^2. \tag{14.91}$$

An alternative way of phrasing the same relation is to write

$$(p_+ - p_-)^2 = \left\{ \frac{|\gamma_+|^2 - |\gamma_-|^2}{|\gamma_+|^2 + |\gamma_-|^2} \right\}^2 = 1 - \mathcal{V}^2. \tag{14.92}$$

Both of these relations show that when we have any degree of certainty about which way the photon went (i.e., $p_\pm = 1$), the fringe contrast, $\mathcal{V}$, goes to zero. It is only when we have no inkling of which way the photon went (i.e., $p_+ = p_- = \tfrac{1}{2}$) that we can see fringes with strong contrast, $\mathcal{V} = 1$. Complementarity, in short, has won again. The incident photon can behave either as a particle or a wave, but it never exhibits both extremes of behavior at once. What one usually sees is neither the one behavior nor the other, and Eqns. (14.91) and (14.92) describe a whole continuum of possible compromises.

Young's experiment could be a bit difficult, in practice, to carry out with two amplifiers. The pinholes, after all, must be quite close together and the space available for the amplifiers is rather cramped. There are other interference experiments, however, in which the geometry is less constraining. Plenty of room would be available for the amplifiers in the two arms of an interferometer, for example, and one might entertain the hope of using their outputs to determine which arm a photon actually entered. A suggestion of just such an experiment, using a single amplifier in one arm of a Mach–Zehnder interferometer, has been made recently by Gozzini[16]. That arrangement, with a second amplifier added in the other arm, is shown in Fig. 13. Placing the photodetectors, C_1, C_2, and C_3, in the positions shown and letting the mirror M_+ be slightly transparent make it possible to carry out several interesting experiments.

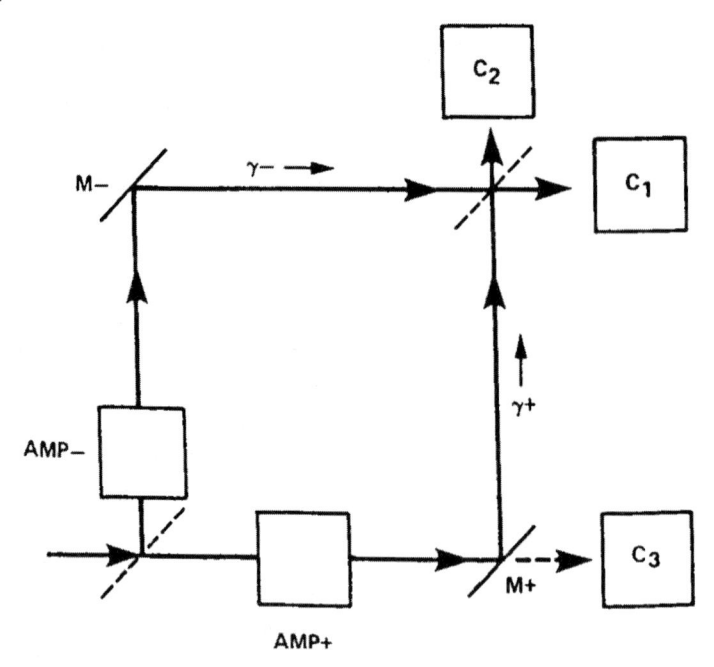

14 Amplifiers, Attenuators and Schrödingers Cat

Figure 13 Proposed setup for the inclusion of an amplifier in each arm of a Mach–Zehnder interferometer. When the mirror M_+ is made partially transparent, detection of photons by counter C_3 and measurement of C_1–C_3 and C_2–C_3 coincidences furnish information on the path chosen by the incident photon.

If the geometry is such that the two interferometer paths are precisely symmetric, and if the amplifiers were absent, then the detector C_1 would register photons, while C_2 would detect none. The two waves reaching the latter would interfere destructively.

When the amplifiers are present, we may assume that their outputs are described in precisely the same way as in Young's experiment. In other words, the two output fields have random amplitudes, γ_+ and γ_-, that are governed by the probability distribution in Eq. (14.84).

The total field amplitude at the counter C_1, we may assume, is proportional to the symmetric sum $(1/\sqrt{2})(\gamma_1 + \gamma_2)$, while that at the counter C_2 is the antisymmetric sum $(1/\sqrt{2})(\gamma_1 - \gamma_2)$. Then, the average counting rates at the two counters will be

$$I_{C_1} = \left\{ \tfrac{1}{2}|\gamma_+ + \gamma_-|^2 \right\}_{\text{av.}} = 2\mathcal{N}(t), \tag{14.93}$$

$$I_{C_2} = \left\{ \tfrac{1}{2}|\gamma_+ - \gamma_-|^2 \right\}_{\text{av.}} = \mathcal{N}(t). \tag{14.94}$$

The effects of spontaneous emission noise are again evident in these results; they provide a nonvanishing intensity for C_2 and contribute two-thirds of the total intensity recorded by the two detectors.

Making the mirror M_+ slightly transparent and placing the detector C_3 behind it offers an interesting, if somewhat sneaky, way of trying to determine which of the two interferometer paths the incident photon actually took. If the amplifiers were noiseless the detection of any light by C_3 would mean the initial photon penetrated the first beam-splitter and set out on the route we have labeled +. The amplitude γ_- would then be zero, and the counters C_1 and C_2 would receive equal amplitudes $(1/\sqrt{2})\gamma_+$. The coincidence rate of C_1 and C_3, in other words, would be equal to he coincidence rate of C_2 and C_3. Deviations from that prediction, it has been suggested, might then mean that there was still some interference involving a pilot wave or other mysterious goings-on in the half of the interferometer that the photon avoided.

We do expect the prediction of equal coincidence rates to prove false, but that is because the amplifiers are anything but noiseless. A signal recorded by C_3 does not tell us uniquely which path the original photon took. The actual C_1–C_3 coincidence rate is proportional to

$$\left\{ \tfrac{1}{2}|\gamma_+ + \gamma_-|^2 |\gamma_+|^2 \right\}_{\text{av.}} = 4\mathcal{N}^2 , \tag{14.95}$$

while the C_2–C_3 rate is proportional to

$$\left\{ \tfrac{1}{2}|\gamma_+ - \gamma_-|^2 |\gamma_+|^2 \right\}_{\text{av.}} = 2\mathcal{N}^2 . \tag{14.96}$$

The two rates are not equal; the C_1–C_3 rate is twice the C_2–C_3 rate. That should not be too difficult a result to verify experimentally.

References

1 A. Einstein, B. Podolsky, N. Rosen, *Phys. Rev.* **47**, 777 (1935).

2 N. Herbert, *Found. Phys.* **12**, 1171 (1982).

3 E. Schrödinger, *Naturwiss.* **23**, 807, 823, 844 (1935).

4 R. J. Glauber, in *Quantum Optics. Proceedings of the Enrico Fermi International School of Physics*, R. J. Glauber, Ed., Academic Press, New York 1969, Course 42, p. 32.

5 R. J. Glauber, *Phys. Lett.* **21**, 650 (1966); reprinted as Chapter 5 in this volume.

6 R. J. Glauber, *Phys. Rev.* **131**, 2766 (1963); reprinted as Sect. 2.9 in this volume.

7 V. Weisskopf, E. Wigner, *Z. Phys.* **63**, 54 (1930).

8 I have presented descriptions of this model for an amplifier at several conferences: *University of Texas Workshop on Irreversible Processes in Quantum Mechanics and Quantum Optics*, San Antonio, March 14–18, 1982; R. J. Glauber, in *Group Theoretical Methods in Physics: Proceedings of the International Seminar*, Zvenigorod, November, 24–26, 1982, Vol. II, Nauka,Moscow 1983, p. 165; R. J. Glauber, in *Group Theoretical Methods in Physics*, Vol. I, Harwood Academic Publishers, 1986, p. 137; R. J. Glauber, in *Proceedings of the VIth International School on Coherent Optics*, Ustron, Poland, September 19–26, 1985,in press. The inverted oscillator has also been used in connection with a laser model and with a different coupling by F. Schwabl, W. Thirrino, *Ergeb. Exakten Naturwiss.* **36**, 219 (1964).

9 K. E. Cahill, R. J. Glauber, *Phys. Rev.* **177**, 1882 (1969); reprinted as Chapter 10 in this volume. See in particular, Eqns. (10.44) and (10.45) on p. 392.

10 See, for example, B. R. Mollow, R. J. Glauber, *Phys. Rev.* **160**, 1076 (1967); reprinted as Chapter 6 in this volume. See in particular the appendix on p. 260.

11 C. Caves, *Phys. Rev. D* **23**, 1693 (1981).

12 R. J. Glauber, F. Haake. *Phys. Lett. A* **68**, 29 (1978); reprinted as Chapter 13 in this volume.

13 M. Born, E. Wolf, *Principles of Optics*, 6th ed., Pergamon Press, New York p. 31.

14 W. K. Wootters, W. H. Zurek, *Nature* **299**, 802 (1982).

15 L. Mandel, *Nature* **304**, 188 (1983).

16 A. Gozzini, in *The Wave-Particle Dualism*, S. Diner et al., Eds., Reidel, Dordrecht 1984, p. 129.

15
The Quantum Mechanics of Trapped Wavepackets[1]

15.1
Introduction

The technique of trapping ions in oscillating electromagnetic fields has many foreseeable applications in high-precision spectroscopy. It offers the possibility of our being able to exercise detailed control over the state of motion of the radiating ions, and thereby minimizing the uncertainties associated with random motion. It may even offer the possibility of cooling the ions very nearly to a state of rest and of eliminating the Doppler and recoil effects that ordinarily broaden spectral resonances. The achievement of these goals will clearly require a quantum-mechanical analysis of the motions of the trapped ions and of the way in which they interact with the radiation field.

Since no configuration of static fields is capable of permanently trapping charged particles in free space, use is inevitably made of time-varying fields. Dealing with the motion of charged particles in time-dependent fields, however, presents a number of novel quantum-mechanical problems. The absence of stationary states of the motion, for example, leaves us without many of the analytical methods we are accustomed to using in treating motion in static fields. Developing systematic ways of dealing with the wave functions of trapped ions, therefore, has a certain methodological interest in addition to the discussion of trapping itself.

The Paul trap[1], which is based on an oscillating electric-quadrupole field, offers a particularly simple example to analyze. The equations of motion of a particle in such a field are linear in their structure, notwithstanding the presence of a time-dependent coefficient. They can be solved by construction of an elementary constant of the motion[2]. The wave functions that are found as a result, though nonstationary in character, have nonetheless a one-to-one correspondence with the stationary states of an ordinary, static-field, harmonic oscillator. They possess a unique ordering according to a variable we shall call the quasi-energy, and that ordering is the same as the energy ordering of the static-field oscillator states.

---

[1] Reprinted, with permission from the author, from *Laser Manipulation of Atoms and Ions: Proceedings of the International Enrico Fermi School, Course 118, Varenna, Italy, July 1–19, 1992*, E. Arimondo, W. D. Phillips, F. Strumia, Eds., North Holland, Amsterdam, 1992, p. 643–660.

Quantum Theory of Optical Coherence. Selected Papers and Lectures. Roy J. Glauber.
Copyright © 2007 WILEY-VCH Verlag GmbH & Co. KGaA, Weinheim
ISBN: 978-3-527-40687-6

When the ions are taken to interact with the quantized radiation field as well as the classical trapping field, they emit electromagnetic quanta spontaneously while making transitions from one excitation state to another. Each type of transition, as we shall see, allows the ion to emit a quantum at any of an infinite sequence of frequencies. Furthermore, spontaneous emission can be accompanied by an increase of excitation instead of the more familiar decrease. Energy conservation cannot exclude such occurrences since the emission process takes place in a time-varying field that can feed the ion energy in arbitrary amounts. The spontaneous-emission processes that are usually identified with radiation damping can, therefore, play an opposite role for trapped ions. We shall show that they can lead instead to the steady growth of the ion excitation. Although this instability may be exceedingly slow to take effect at radio frequencies, we show that it is inevitable in the Paul trap.

15.2
Equations of Motion and Their Solutions

The time-dependent quadrupole potential in the Paul trap is at all times quadratic in each of the three Cartesian coordinates of the charged particle. The problem of treating the motion in three dimensions is, therefore, separable into three one-dimensional problems of similar structure. Each of these is, in effect, the problem of a one-dimensional harmonic oscillator with a time-varying spring constant. We can write its Hamiltonian as

$$H = \frac{p^2}{2m} + \frac{m}{2}W(t)q^2, \tag{15.1}$$

where p and q are the momentum and coordinate variables, respectively, and the time dependence of the function $W(t)$ can be left completely unspecified for the present. In the Paul trap, of course, $W(t)$ has a periodic dependence on time, which plays an important role in the trapping process. We can solve the problem posed by the Hamiltonian Eq. (15.1), however, in much more general terms and so there is no need to begin by specializing the function $W(t)$. The analysis prior to the explicit discussion of trapping in Sect. 15.4, in other words, applies equally well to problems like the passage of a particle through one or more quadrupole fields in which there is no trapping whatever.

The equations of motion that follow from the Hamiltonian Eq. (15.1) are

$$\dot{q} = \frac{p}{m} \tag{15.2}$$

and

$$\dot{p} = -mW(t)q. \tag{15.3}$$

They can be combined into the second-order equation

$$\ddot{q} + W(t)q = 0, \tag{15.4}$$

which must be solved subject to initial conditions on $q(0)$ and $\dot{q}(0)$ or $p(0) = m\dot{q}(0)$.

A familiar way of dealing with both the coordinates q and $\dot{q}$ at once is to introduce a complex coordinate with its real part proportional to $q(t)$ and its imaginary part proportional to $\dot{q}(t)$. For the case of an ordinary or static harmonic oscillator, for example, with $W(t) = \text{const.} = \omega^2$, we can form the complex linear combination

$$Z(t) = [\omega q(t) + i\dot{q}(t)]\exp[i\omega t], \tag{15.5}$$

which is easily verified to be constant in time. It follows then that

$$Z(t) = Z(0) \tag{15.6}$$

and

$$\omega q(t) + i\dot{q}(t) = Z(0)\exp[-i\omega t]. \tag{15.7}$$

We shall now develop an analogous procedure for solving Eq. (15.4) in the fully time-dependent case, that is for arbitrary $W(t)$. Let us consider, to this end, complex solutions $u(t)$ to the differential equation

$$\ddot{u} + W(t)u = 0. \tag{15.8}$$

If u_1 and u_2 are any two solutions of this equation, they must obey the Wronskian identity

$$u_1(t)\dot{u}_2(t) - \dot{u}_1(t)u_2(t) = \text{constant}, \tag{15.9}$$

and that relation, we shall see, is a convenient source of constants of the motion. It will be useful to define a standard solution $u(t)$ (analogous in several ways to the exponential function $\exp[i\omega t]$). We take this solution to satisfy the initial conditions

$$u(0) = 1 \tag{15.10}$$

and

$$\dot{u}(0) = i\omega, \tag{15.11}$$

where ω is to be regarded as a completely arbitrary real parameter, at least for the present. It need not be related in any way to the function $W(t)$. By using these initial conditions, for example, we see that, as long as $W(t)$ is real, the functions $u(t)$ and $u^*(t)$ must obey the Wronskian identity

$$u^*(t)\dot{u}(t) - u(t)\dot{u}^*(t) = 2i\omega, \tag{15.12}$$

which will be useful at several later points.

Since the unknown coordinate $q(t)$ and the function $u(t)$ both satisfy the same differential equation, they too must obey a Wronskian identity. It is convenient, therefore, to define the complex linear combination

$$Z(t) = i\{u(t)\dot{q}(t) - \dot{u}(t)q(t)\}, \qquad (15.13)$$

which must consequently also be constant in time, i.e.

$$Z(t) = Z(0) = \omega q(0) + i\dot{q}(0). \qquad (15.14)$$

To find the explicit solution for $q(t)$ we note that

$$Z(t)u^*(t) + Z^\dagger(t)u(t) = -i(u^*\dot{u} - u\dot{u}^*)q(t), \qquad (15.15)$$

and make use of the identities Eqns. (15.12) and (15.14), with the result

$$q(t) = \frac{1}{2\omega}\{Z(0)u^*(t) + Z^\dagger(0)u(t)\}. \qquad (15.16)$$

To find $\dot{q}(t)$ we either differentiate this expression or construct the expression

$$Z(t)\dot{u}^*(t) + Z^\dagger(t)\dot{u}(t) = -(u^*\dot{u} - u\dot{u}^*)\dot{q}(t), \qquad (15.17)$$

from which it likewise follows that

$$p(t) = m\dot{q}(t) = \frac{m}{2\omega}\{Z(0)\dot{u}^*(t) + Z^\dagger(0)\dot{u}(t)\}. \qquad (15.18)$$

We have thus expressed the solutions to the equations of motion in terms of a standard solution $u(t)$ to Eq. (15.8), which we assume to be known. We shall now show that this knowledge permits us to find the wave functions as well.

Since the equations we have been discussing are all linear, we have been able to solve them without having to specify whether the variables $q(t)$ and $p(t) = m\dot{q}(t)$ are classical functions or quantum-mechanical operators. It is, of course, the quantum-mechanical case that interests us most, so we can now take them to be operators in the Heisenberg picture, and that gives us another way of representing their time dependence. There must exist, we know, a unitary time evolution operator $U(t)$ which transforms the initial values of these operators into their values at time t according to the relations

$$q(t) = U^{-1}(t)q(0)U(t), \quad \text{etc.} \qquad (15.19)$$

The values of those operators at $t = 0$ may be identified with the fixed values they assume in the Schrödinger picture.

It is then the same operator $U(t)$ that transforms the fixed state vectors $|\ \rangle_H$ of the Heisenberg picture into the time-dependent ones $|t\rangle_S$ of the Schrödinger picture via the relation

$$|t\rangle_S = U(t)|\ \rangle_H. \qquad (15.20)$$

The time-dependent wavepackets that we want to describe are dealt with most easily and naturally in the Schrödinger picture, so we will need somehow to carry the solutions we have found in the Heisenberg picture over into the Schrödinger picture. We shall see that there is a simple way of accomplishing that without any need to solve explicitly for the operator $U(t)$.

15.3
The Wave Functions

In the Schrödinger picture we typically specify the state vector $|t\rangle_S$ at time $t = 0$ (taking it to be the same as the fixed Heisenberg state $|\ \rangle_H$) and let the Hamiltonian generate its time dependence via the Schrödinger equation

$$i\hbar \frac{\partial}{\partial t} |t\rangle_S = H(t) |t\rangle_S . \tag{15.21}$$

How can we specify the initial state vector? We can, of course, write it as an expansion in terms of any complete orthogonal set of basis states. A particularly convenient choice for these basis states is suggested by Eq. (15.14), which shows the constancy of $Z(t)$. Let us define the constant operator $C(t)$ by simply changing the normalization of $Z(t)$ to one more appropriate to the quantum theory,

$$C(t) = \sqrt{\frac{m}{2\hbar\omega}} \sqrt{Z(t)} , \tag{15.22}$$

$$C(t) = C(0) = \frac{i}{\sqrt{2m\hbar\omega}} \{u(t)p(t) - m\dot{u}(t)q(t)\} . \tag{15.23}$$

Then $C(t)$ assumes the constant value

$$C(t) = C(0) = \frac{i}{\sqrt{2m\hbar\omega}} (m\omega q(0) + ip(0)) , \tag{15.24}$$

which we recognize as the annihilation operator for excitations of an ordinary harmonic oscillator of mass m and frequency ω,

$$C(t) = C(0) = a . \tag{15.25}$$

It follows that

$$[C(t), C^\dagger(t)] = [a, a^\dagger] = 1 , \tag{15.26}$$

a relation that we may deduce directly, either from Eq. (15.24) or from the Wronskian relation Eq. (15.12).

The solutions to the Heisenberg equations of motion are given by Eqns. (15.16) and (15.18). We can express them in terms of the creation and annihilation operators simply by applying the scale change Eq. (15.22) to the operators $Z(0)$ and $Z^\dagger(0)$. The resulting expressions are

$$q(t) = \sqrt{\frac{\hbar}{2m\omega}} \left\{ a u^*(t) + a^\dagger u(t) \right\}, \tag{15.27}$$

$$p(t) = \sqrt{\frac{m\hbar}{2\omega}} \left\{ a \dot{u}^*(t) + a^\dagger \dot{u}(t) \right\}. \tag{15.28}$$

The identities (15.25) and (15.26) make it clear that there is a special convenience to choosing as basis states the stationary states of an ordinary oscillator of mass m and frequency ω. The ground state, $|n=0\rangle_\omega$, of such an oscillator, for example, obeys the condition

$$a |n=0\rangle_\omega = C(t) |n=0\rangle_\omega = 0, \tag{15.29}$$

and we can easily use this identity to find the time-dependent state that evolves from the initial state $|n=0\rangle_\omega$. To do that we note that the Heisenberg operator $C(t)$ is related to its Schrödinger picture counterpart $C_S(t)$ by the relation

$$C(t) = U^{-1}(t) C_S(t) U(t). \tag{15.30}$$

(The time dependence of the Schrödinger operator $C_S(t)$ is due entirely to the explicit time dependence of the functions $u(t)$ and $\dot{u}(t)$ included in the definition Eq. (15.23).) If we substitute the expression (15.30) for $C(t)$ into Eq. (15.29) and multiply by $U(t)$, we see that

$$C_S(T) U(t) |n=0\rangle_\omega = 0, \tag{15.31}$$

but $U(t) |n=0\rangle_\omega$ is the Schrödinger state of the time-dependent oscillator that evolves from the ground state of the static oscillator. If we call that state $|n=0,t\rangle$, then it obeys the relation

$$C_S(t) |n=0,t\rangle = 0, \tag{15.32}$$

or

$$\left(u(t) p - m \dot{u}(t) q \right) |n=0,t\rangle = 0, \tag{15.33}$$

at all times.

It is a simple matter now to find the wave function for the state $|n=0,t\rangle$. If we express the eigenstate condition Eq. (15.33) in coordinate space, it becomes the first-order differential equation

$$\left\{ u(t) \frac{\hbar}{i} \frac{\partial}{\partial q'} - m \dot{u}(t) q' \right\} \langle q' | n=0,t \rangle = 0. \tag{15.34}$$

The solution to this equation, in normalized form, is

$$\langle q' | n = 0, t \rangle = \left(\frac{m\omega}{\pi\hbar}\right)^{1/4} \frac{1}{\{u(t)\}^{1/2}} \exp\left[\frac{im\,\dot{u}(t)}{2\hbar\,u(t)} q'^2\right]. \quad (15.35)$$

This wave function differs from the ground-state wave function for a static-field oscillator by a complex time-dependent scale change of the coordinate. It reduces to the latter wave function for the constant field specified by $W(t) = \omega^2$, since for that case $u(t) = \exp[i\omega t]$.

The wave function that evolves from any other stationary state of the static-field oscillator, say the n-th excitation state, can be found by generating that state from the ground state via the relation

$$|n\rangle_H = \frac{(a^\dagger)^n}{\sqrt{n!}} |0\rangle_H. \quad (15.36)$$

It follows then that

$$U(t) |n\rangle_H = |n,t\rangle_S = \frac{[C_S^\dagger(t)]^n}{\sqrt{n!}} U(t) |0\rangle_H, \quad (15.37)$$

$$U(t) |n\rangle_H = \frac{[C_S^\dagger(t)]^n}{\sqrt{n!}} |n=0,t\rangle_S, \quad (15.38)$$

and we may find the corresponding spatial wave function by expressing this relation in coordinate space. The general expression for the wave function we find in this way is

$$\langle q' | nt \rangle = \frac{1}{\sqrt{n!}} \left(\frac{m\omega}{\pi\hbar}\right)^{1/4} \frac{1}{\{u(t)\}^{1/2}} \left\{\frac{u^*(t)}{2u(t)}\right\}^{n/2} H_n\left(\left\{\frac{m\omega}{\hbar |u(t)|^2}\right\}^{1/2} q'\right)$$

$$\times \exp\left[\frac{im\,\dot{u}(t)}{2\hbar\,u(t)} q'^2\right]. \quad (15.39)$$

More specialized wave functions of this form for the case in which the function $W(t)$ is periodic have been derived by Combescure[3] and Brown[4].

The forms that the wave functions Eqns. (15.35) and (15.39) take in momentum space can easily be derived with methods analogous to those we have outlined by using a derivative with respect to the momentum variable to represent the coordinate operator. We have presented more detailed derivations of these wave functions, both in coordinate and momentum space, and have discussed a number of their properties including uncertainty relations in papers presented at recent meetings[5].

It is interesting now to broaden the set of initial-state wave functions that we consider. Having found the wave functions that evolve from all the stationary states of a static-field oscillator, we can equally well ask for the wave functions that evolve from the coherent states of the oscillator. We have presented the form of these wave functions in the papers noted earlier, and will only note here that they share some of the remarkable properties of the more familiar coherent states of a static-field oscillator. The wave functions are simply displaced forms of the Gaussian wave function Eq. (15.35) that evolves from the ground state. The displacement of the Gaussian consists of a time-dependent translation in coordinate and momentum space which just represents the classical motion of the center of the initial-state wavepacket. The Gaussian wavepackets thus follow along a classical trajectory, hewing to all its loops and epicycles in the position–momentum plane. Furthermore, when the trapping field is periodic, we have shown[5] that the Gaussian packets remain confined in size. They generally pulsate periodically in width, but have no tendency to spread out over the classical orbits.

These generalized coherent states share another property of the more familiar ones. When a time-dependent uniform external field is applied in addition to the trapping field, such states undergo a very simple evolution. An initially coherent state, in particular, is always carried over into another coherent state, and the two differ by a translation that corresponds to the classical motion of the center of the packet induced by the external field.

15.4
Periodic Fields and Trapping

Since we have had no need to specialize the function $W(t)$ appearing in Eq. (15.1) up to this point, we have implicitly been discussing a much broader class of problems than those involving trapping. The applications we have in mind at present, however, are mainly to problems in which the function $W(t)$ oscillates periodically, say with period T,

$$W(t+T) = W(t), \qquad (15.40)$$

and with a functional form that allows our charged particle to become trapped. But to be more specific about trapping, we must first define it.

For a classical particle, the trapping condition is that the coordinate $q(t)$ remain bounded, $|q(t)| < \infty$, at all times, including $t \to \infty$. The wave functions we have discussed for the quantum-mechanical case have all been based on the construction of a certain standard solution $u(t)$ of the differential equation (15.8). This solution is chosen to obey the initial conditions Eqns. (15.10) and (15.11), the latter of which contains the parameter w, which is arbitrary; we are still free to give it any positive value. No matter what value we give it, however, the coordinate operator $q(t)$ is given at all times in terms of the initial values of q and $\dot{q}$ or p by Eq. (15.16). It is clear from that equation that the matrix elements of $q(t)$ will become unbounded in general unless the function $u(t)$ remains bounded. The trapping condition then is $|u(t)| < \infty$ at all times t, including $t \to \infty$, and that condition must hold for all values of the arbitrary parameter w.

15.4 Periodic Fields and Trapping

When the function $W(t)$ is periodic, the differential equation (15.8) is an example of Hill's equation, which has received considerable study[6]. It possesses in general two linearly independent solutions, $u_1(t)$ and $u_2(t)$, that can be written in the Floquet form

$$u_1(t) = \exp[i\mu t]\,\varphi_1(t)\,, \tag{15.41}$$

$$u_2(t) = \exp[-i\mu t]\,\varphi_2(t)\,, \tag{15.42}$$

where the functions of φ_j are periodic with the same period as the trapping field,

$$\varphi_j(t+T) = \varphi_j(t)\,, \qquad j = 1, 2, \tag{15.43}$$

and the parameter μ, the characteristic exponent, is determined by the properties of the function $W(t)$.

The solutions $u(t)$ that satisfy the initial conditions Eqns. (15.10) and (15.11) as well as the differential equation (15.8) are linear combinations in general of the two functions $u_1(t)$ and $u_2(t)$. The trapping condition that $|u(t)|$ remain bounded then is the condition that the characteristic exponent μ be real-valued. To reach a more explicit expression for $u(t)$, let us assume, for simplicity, that the function $W(t)$ is even-valued:

$$W(-t) = W(t)\,. \tag{15.44}$$

Then the solution $u_2(t)$ can be taken simply to be $u_1(-t)$ and we can write any solution $u(t)$ to Eq. (15.8) in the form

$$u(t) = Cu_1(t) + Du_1(-t)\,, \tag{15.45}$$

where C and D are two constants to be determined by the initial conditions. If we normalize the function φ_1 by letting $\varphi_1(0) = 1$, then the initial conditions become

$$u(0) = 1 = C + D\,, \tag{15.46}$$

$$\dot{u}(0) = i\omega = (C - D)\dot{u}_1(0)\,, \tag{15.47}$$

from which we find

$$C = \frac{1}{2}\left(1 + \frac{i\omega}{\dot{u}_1(0)}\right), \tag{15.48}$$

$$D = \frac{1}{2}\left(1 - \frac{i\omega}{\dot{u}_1(0)}\right). \tag{15.49}$$

As long as μ is real-valued, the solution for $u(t)$ given by Eq. (15.45) will remain bounded, and can thus describe trapped motion.

While the parameter ω has remained arbitrary to this point, it is clear that there is a particular value we can choose for it that greatly simplifies our description of the motion. If we choose the value

$$\omega = -i\dot{u}_1(0), \qquad (15.50)$$

then the constant D given by Eq. (15.49) vanishes, and the function $u(t)$ reduces to a single Floquet solution. The set of basis functions that corresponds to this particular oscillator frequency, in other words, offers a uniquely simple description of the motion in the trap.

Given the fact that the function $u(t)$ can be expressed in terms of just one of the two Floquet solutions, the factorization of that solution evident in Eq. (15.41) leads to a clear separation of two types of motion that the charged particle may be said to be carrying out simultaneously. The exponential function $\exp[i\mu t]$ describes the slow part of the motion, a simple oscillation at frequency $\mu/2\pi$. The function $\varphi_1(t)$, on the other hand, adds to it the rapid oscillation at the frequency of the trapping field and its harmonics that is often referred to as the "micromotion". When the special choice given by Eq. (15.50) is made for the frequency ω of our reference oscillator, the time-dependent wave functions given by Eqns. (15.35) and (15.39) simplify considerably in form. In particular the functions they contain, $|u(t)|^2$ and

$$\frac{\dot{u}(t)}{u(t)} = \frac{d}{dt}\log u(t) = i\mu + \frac{\dot{\varphi}_1(t)}{\varphi_1(t)}, \qquad (15.51)$$

both become periodic in time. The wave function that evolves from the n-th excitation state of the reference oscillator can thus be written as

$$\langle q'|n,t\rangle = \exp\left[-i\left(n+\tfrac{1}{2}\right)\mu t\right]\chi_n(t), \qquad (15.52)$$

where the function χ_n is given by

$$\chi_n(t) = \frac{1}{\sqrt{2^n n!}}\left(\frac{m\omega}{\pi\hbar}\right)^{1/4}\frac{\exp[-in\arg\varphi_1(t)]}{[\varphi_1(t)]^{1/2}} \times H_n\left(\left\{\frac{m\omega}{\hbar|\varphi_1(t)|^2}\right\}^{1/2}q'\right)$$
$$\times \exp\left[-\frac{m\mu}{2\hbar}\left(1-\frac{i}{\mu}\frac{\dot{\varphi}_1(t)}{\varphi_1(t)}\right)q'^2\right]. \qquad (15.53)$$

The special choice of ω thus produces a factorization of the wave function analogous to the separation of the classical motions implicit in the Floquet solution. The function $\chi_n(t)$ is periodic in time with period T. The classical micromotion appears in the wave function as a periodic pulsation or breathing, a kind of squeezing and stretching of the stationary-state functions. The generalized coherent-state wave functions we have mentioned earlier simply carry the pulsating ground-state density around the classical trajectory of its center.

The oscillation of the trapping field and the consequent time dependence of the Hamiltonian Eq. (15.1) imply that the energy of our moving charged particle does not remain constant. The particle has no stationary states in the trapping field, and its energy variable is of remarkably little use in discussing its quantum mechanics. Those facts notwithstanding, there is still a certain analogy we can draw between the trapping states and stationary states.

Energy eigenstates are called stationary states because, for time-independent interactions, arbitrary shifts of the time variable only change them by a phase factor. They are in this sense irreducible representations of the continuous group of time translations. When a periodic trapping field is present, the Hamiltonian Eq. (15.1) is only invariant under the discrete set of time translations through multiples of the period T. The wave functions that are irreducible representations of this discrete Abelian group can likewise only change by phase factors under these transformations. It follows then that such wave functions $\psi(t)$ can be factorized into the form

$$\psi(t) = \exp[i\nu t]\,\chi(t)\,, \tag{15.54}$$

where ν is a characteristic exponent of some sort and $\chi(t)$ is periodic in time with period T. Wave functions of this form represent the closest analogues one can find to stationary states. The quantity $\hbar\nu$ has been called the quasi-energy[7]. Once the frequency ω of our static-field reference oscillator is shown to satisfy Eq. (15.40), the resulting wave functions (15.52) fall precisely into the factorized form of those special analogue states, which we may speak of as quasi-stationary. Since they are generated by unitary transformations on the stationary states of the reference oscillator, they form a complete, orthonormal set.

15.5
Interaction With the Radiation Field

Whatever may be the quantum state in which a charged particle is trapped, it never ceases to interact with the radiation field. It may, therefore, change its state by emitting light quanta, and that is an interaction that exerts a familiar clamping action on most forms of microscopic motion. In the case of trapped particles, as we shall see, the effect of radiative interactions may be quite different from that.

Let us assume that the amplitude of the motion of the trapped particle is small enough in relation to the wavelength of the radiated field that we may describe the interaction in the electric-dipole approximation. Then we can write the interaction Hamiltonian as

$$H_{\text{int}}(t) = -eq(t)\,\hat{\boldsymbol{i}}\cdot\boldsymbol{E}(t)\,, \tag{15.55}$$

in which e is the magnitude of the electric charge, $\hat{\boldsymbol{i}}$ is a unit vector along the axis of its motion, and $\boldsymbol{E}(t)$ is the electric-field operator evaluated at the center of the motion.

The first-order transition amplitude of the system from an initial state $|i\rangle$ to a final state $|f\rangle$ in the Heisenberg picture is then

$$M = \frac{1}{i\hbar}\left\langle t \left| \int_{-\infty}^{\infty} H_{\text{int}}(t)\,dt \right| i \right\rangle . \tag{15.56}$$

The initial state of the radiation field, we assume, is the vacuum state, while its final state is one with a single photon present with propagation vector $\mathbf{k}$ and polarization vector $\hat{\mathbf{e}}^{(\lambda)}$ ($\lambda = 1, 2$). The appropriate matrix element for a single photon emission is then

$$\langle 1_{k,\lambda} | \hat{\mathbf{i}} \cdot \mathbf{E}(t) | 0 \rangle = i\sqrt{\frac{\hbar \omega_k}{2}}\, \hat{\mathbf{i}} \cdot \hat{\mathbf{e}}^{(\lambda)*} \exp[i\omega_k t] , \tag{15.57}$$

with $\omega_k = ck$, the angular frequency of the photon. If now the particle is initially in the n-th excitation state in the Heisenberg picture, that is, the reference oscillator is in the state $|n\rangle$, and it is finally in the m-th state, we can write the transition amplitude as

$$M = -e\sqrt{\frac{\omega_k}{2\hbar}}\, \hat{\mathbf{i}} \cdot \hat{\mathbf{e}}^{(\lambda)*} \int_{-\infty}^{\infty} \langle m | q(t) | n \rangle \exp[i\omega_k t]\, dt . \tag{15.58}$$

The expression Eq. (15.27) for the time-dependent coordinate operator $q(t)$ depends only linearly on the creation and annihilation operators a and $a^\dagger$. It is clear, therefore, that in this lowest-order approximation transitions can only take place to the states with $m = n \pm 1$. For the case $m = n - 1$ we will have

$$M_{n-1} = -\frac{e}{2}\sqrt{\frac{\omega_k}{m\omega}}\, \hat{\mathbf{i}} \cdot \hat{\mathbf{e}}^{(\lambda)*} \sqrt{n} \int_{-\infty}^{\infty} u^*(t) \exp[i\omega_k t]\, dt , \tag{15.59}$$

while for $m = n + 1$ we will have

$$M_{n+1} = -\frac{e}{2}\sqrt{\frac{\omega_k}{m\omega}}\, \hat{\mathbf{i}} \cdot \hat{\mathbf{e}}^{(\lambda)*} \sqrt{n+1} \int_{-\infty}^{\infty} u(t) \exp[i\omega_k t]\, dt . \tag{15.60}$$

The matrix elements are thus given by Fourier components of the functions $u^*(t)$ and $u(t)$, respectively.

With the appropriate choice of the parameter w, given by Eq. (15.50), the function $u(t)$ takes the Floquet form of Eq. (15.41). If we let Ω be the angular frequency

$$\Omega = \frac{2\pi}{T} \tag{15.61}$$

associated with the periodic function $\varphi_1(t)$, then we may expand the latter function in a Fourier series of the form

$$\varphi_1(t) = \sum_{j=-\infty}^{\infty} C_j \exp[ij\Omega t] . \tag{15.62}$$

15.5 Interaction With the Radiation Field

The integral that occurs in the matrix element Eq. (15.59) for transition to the $n-1$ state is then

$$\int_{-\infty}^{\infty} u^*(t) \exp[i\omega_k t]\,dt = 2\pi \sum_{j=-\infty}^{\infty} \delta(\omega_k - \mu - j\Omega) C_j^* \,. \tag{15.63}$$

The photon emitted in the transition may evidently have any frequency

$$\omega_k = \mu + j\Omega \,, \tag{15.64}$$

for integral values of j. Since the frequencies ω_k, μ and Ω are all intrinsically positive, and since the trapping frequency Ω generally exceeds μ, the only frequencies actually present will be those for $j \geq 0$, for which the coefficients C_j are different from zero.

The matrix element Eq. (15.60) for the transition to the $n+1$ state, on the other hand, is proportional to the integral

$$\int_{-\infty}^{\infty} u(t) \exp[i\omega_k t]\,dt = 2\pi \sum_{j=-\infty}^{\infty} \delta(\omega_k + \mu + j\Omega) C_j \,. \tag{15.65}$$

The photons emitted can thus have the frequencies

$$\omega_k = -\mu - j\Omega \,, \tag{15.66}$$

but only the values for $j \leq -1$ can be realized since the ω_k can never be negative.

The initial and final states of the transitions we are discussing are not, as we have stated earlier, energy eigenstates. They can be uniquely ordered, however, according to their quasi-energies $(n + \tfrac{1}{2})\hbar\mu$. It may appear somewhat surprising, therefore, that the trapped particle can rise in its quasi-energy while spontaneously emitting a photon of positive energy. There is no contradiction involved in that because neither the energy nor the quasi-energy is conserved in general in transitions that take place in an explicitly time-dependent field.

It is an easy matter now to square the transition amplitudes and sum over all possible photon states. When we do that, we find that the total transition probability from the n to the $n-1$ state can be written as

$$W_{n \to n-1} = nA \,, \tag{15.67}$$

with

$$A = \frac{e^2}{6\pi m\omega c^3} \sum_{j=0}^{\infty} (\mu + j\Omega)^3 |C_j|^2 \,. \tag{15.68}$$

The j-th term in this series evidently represents the partial rate for emission of a photon of frequency $\mu + j\Omega$.

For the upward transitions, on the other hand, the total transition rate is given by

$$W_{n \to n+1} = (n+1)B, \tag{15.69}$$

with

$$B = \frac{e}{6\pi m\omega c^3} \sum_{j=1}^{\infty} (j\Omega - \mu)^3 |C_{-j}|^2 . \tag{15.70}$$

When we couple an ordinary static-field harmonic oscillator to the radiation field, the spontaneous emissions of quanta that take place will inevitably damp its motion and bring it down to its ground state. For the case of our trapped particle, however, the emission of quanta may lead to a rise of its excitation as well as to a fall. Which of these processes predominates depends on the relative magnitudes of the constants A and B. Fortunately we can derive some sum rules that tell us what to expect.

15.6
Sum Rules

We shall find it useful to establish several properties of the Fourier coefficients C_j defined by Eq. (15.62). Since the function $W(t)$ that occurs in the Hamiltonian Eq. (15.1) is both an even function of time and real-valued, the function $u_1(t)$ must obey the identity

$$u_1^*(t) = u_1(-t) \tag{15.71}$$

and the function $\varphi_1(t)$ likewise the identity

$$\varphi_1^*(t) = \varphi_1(-t) . \tag{15.72}$$

It follows that the Fourier coefficients C_j are all real:

$$C_j^* = C_j . \tag{15.73}$$

Since the function $\varphi_1(t)$ has the normalization $\varphi_1(0) = 1$, the coefficients C_j obey the relation

$$\sum_j C_j = 1 . \tag{15.74}$$

The condition (15.50), which defines the frequency of the reference oscillator, on the other hand, corresponds to the relation

$$\omega = \mu - i\dot{\varphi}_1(0) = \mu + \sum_j j\Omega C_j , \tag{15.75}$$

$$\omega = \sum_j (\mu + j\Omega) C_j . \tag{15.76}$$

Another relation between the same parameters can be obtained by substituting the Fourier expansion for $u_1(t)$ into the Wronskian identity Eq. (15.12). The result is

$$\omega = \sum_j (\mu + j\Omega) C_j^2 , \tag{15.77}$$

a statement which contrasts interestingly with Eq. (15.76).

Another potentially useful relation is obtained by noting that the function $|u_1(t)|^2$ is periodic in time and averaging it over a period T. We then find

$$\frac{1}{T} \int_{-\infty}^{\infty} |u_1(t)|^2 \, dt = \sum_j C_j^2 . \tag{15.78}$$

When we likewise average the quantity $|\dot{u}_1(t)|^2$ over one period, we find

$$\frac{1}{T} \int_{-\infty}^{\infty} |\dot{u}_1(t)|^2 \, dt = \sum_j (\mu + j\Omega)^2 C_j^2 . \tag{15.79}$$

We can obtain still another relation by making use of the identity

$$\ddot{u}^*(t)\dot{u}(t) - \dot{u}^*(t)\ddot{u}(t) = -W(t)\{u^*(t)\dot{u}(t) - \dot{u}^*(t)u(t)\} = -2i\omega W(t) , \tag{15.80}$$

which follows from the Wronskian relation Eq. (15.12). When we average both sides of this identity over one period of length T, we find

$$\sum_{j=-\infty}^{\infty} (\mu + j\Omega)^3 C_j^2 = \omega \frac{1}{T} \int_0^T W(t) \, dt , \tag{15.81}$$

$$\sum_{j=-\infty}^{\infty} (\mu + j\Omega)^3 C_j^2 = \omega \overline{W} , \tag{15.82}$$

in which $\overline{W}$, as indicated, is the time average of $W(t)$. This relation provides us with a particularly useful sum rule, one that relates the two transition rates A and B defined by Eqns. (15.67) and (15.70). The difference of the rates A and B is evidently given by

$$A - B = \frac{e^2}{6\pi m\omega c^3} \sum_{j=-\infty}^{\infty} (\mu + j\Omega)^3 C_j^2 , \tag{15.83}$$

and, according to Eq. (15.82), that amounts to

$$A - B = \frac{e^2}{6\pi mc^3} \overline{W} . \tag{15.84}$$

Thus the sign of the difference of A and B is governed by the sign of the time-averaged potential. The transitions toward increased excitations will be faster than those toward lesser excitation if the time average of the potential is negative.

15.7
Radiative Equilibrium and Instability

The quantum number of a trapped particle, as we have seen, may either increase or decrease as a result of a spontaneous-emission process. To determine the probability that it has any particular quantum number, we must find the diagonal elements of its density matrix. These obey a simple rate equation. If we let P_n be the probability that the particle is in the n-th excitation state, then the transition rates Eqns. (15.67) and (15.69) imply that

$$\frac{d}{dt}P_n = (n+1)AP_{n+1} + nBP_{n-1} - [nA + (n+1)B]P_n \,. \tag{15.85}$$

In the most familiar cases of radiation damping, the coefficient B is small or equal to zero and the distribution P_n eventually comes to statistical equilibrium. If there is to be any equilibrium distribution, we can find it easily by solving the recursion relation that results from setting $\dot{P}_n = 0$ for all n. We find then that equilibrium can only be reached if the coefficient A exceeds B. The equilibrium distribution in that case is

$$P_n = \left(1 - \frac{B}{A}\right)\left(\frac{B}{A}\right)^n \,. \tag{15.86}$$

It is quite independent of the initial distribution and has the same form as a thermal distribution.

It is not difficult to construct a general solution to the set of rate equations (15.85), given an arbitrary initial distribution of the P_n, but we have no need for all of that detail here. It is sufficient simply to note the way in which the mean value of n

$$\langle n \rangle_t = \sum_{n=0}^{\infty} nP_n(t) \tag{15.87}$$

varies with time. We find that this mean occupation number is given in terms of its initial value $\langle n \rangle_0$ by the relation

$$\langle n \rangle_t = \langle n \rangle_0 \exp[(B-A)t] + \frac{B}{B-A}\left[\exp[(B-A)t] - 1\right] \,. \tag{15.88}$$

Thus, when B exceeds A, that is, whenever the time-averaged potential $\overline{W}$ is negative according to Eq. (15.84), the excitation will undergo exponential increase. In fact, no equilibrium can be reached even for $B = A$ since in that case

$$\langle n \rangle_t = \langle n \rangle_0 + At \,, \tag{15.89}$$

and the excitation still increases without bound.

Having $\overline{W}$ positive means that in the absence of radiative interactions the time-averaged potential alone is sufficient to trap the particle. That does tend to confine it more tightly than having $\overline{W} = 0$, or negative, but it is by no means a necessary condition for such a trapping. In fact, the true three-dimensional Paul trap depends critically on the fact that trapping can still occur for $\overline{W}(t) \leq 0$.

Our analysis of a one-dimensional problem up to this point is justified by the fact that the problem of motion in the three-dimensional trap is completely separable. With a choice of the x, y and z axes that exploits the symmetry of the field-generating electrodes, we can think of the coordinate q we have been discussing as, say, the x coordinate of the charged particle. Analyses similar to those we have presented would then be necessary for dealing with the y and z coordinates, and the overall wave functions would be threefold products of the one-dimensional functions.

Let us assume that the electric potential energy of the charged particle, when it is at a position $\boldsymbol{r}$, relative to the center of symmetry of the field is $V(\boldsymbol{r},t)$. Then the function $W(t)$ in the Hamiltonian Eq. (15.1) for motion along the x-axis is given by

$$W_x(t) = \frac{1}{m}\frac{\partial^2}{\partial x^2}V(\boldsymbol{r},t)\bigg|_{\boldsymbol{r}=0}, \qquad (15.90)$$

as long as the displacements are not too large. Now it is not difficult to provide a field configuration that makes the time average of this function positive, and the same thing can be said for the analogous functions for the motions along the y and z axes. The sum of those three functions, however, is given by

$$W_x(t) + W_y(t) + W_z(t) = \frac{1}{m}\nabla^2 V(\boldsymbol{r},t)\bigg|_{\boldsymbol{r}=0}, \qquad (15.91)$$

and, since the electric potential in free space must obey the Laplace equation, we must have at all times

$$W_x(t) + W_y(t) + W_z(t) = 0. \qquad (15.92)$$

Hence $\overline{W}$ is bound to be zero or negative for motion along at least one of the axes. This condition is quite consistent with stable trapping when radiative interactions are neglected, but once they are taken into account, as we can see, their effect is inevitably destabilizing.

Having raised the spectre of instability, we must hasten to point out that the processes that cause it can be exceedingly slow. For the ion traps that have been constructed to date, the photons we have been discussing are typically in the radiofrequency range and their spontaneous-emission rate is so low that the time scale of the instability may amount to weeks or even years. That allows more than enough time for experiments on particles that are, to all practical intents, quite stably trapped. Experiments on the cooling of ions, by the inelastic scattering of optical photons, for example, can be carried out on quite short time scales[8].

References

1 For a review, see W. Paul, *Rev. Mod. Phys.* **62**, 531 (1990).

2 V. V. Dodonov, V. I. Man'ko, in *Invariants and the Evolution of Nonstationary Quantum Systems*, M. A. Markov, Ed., Nova Scientific Publishers, Commack 1989, p. 103.

3 M. Combescure: *Ann. Inst. Henri Poincaré* **44**, 293 (1986).

4 L. S. Brown, *Phys. Rev. Lett.* **66**, 527 (1991).

5 R. J. Glauber, in *Proceedings of the International Conference on Quantum Optics, Hyderabad, India, January 5–10*, in press; in *Quantum Measurements in Optics*, Proceedings of the NATO Advanced Research Workshop, Cortina d'Ampezzo, Italy, January 21–25, 1991, P. Tombesi, D. Walls, Eds., Plenum Press, New York 1992, p. 3.

6 E. T. Whittaker, G. N. Watson, *Modern Analysis*, Macmillan, New York 1943, p. 406 *et seq.*.

7 Ya. B. Zel'dovich: *Z. Eksp. Teor. Fiz.* **51**, 1492 (1966); *Sov. Phys. JETP* **24**, 1006 (1967).

8 F. Diedrich, J. C. Bergquist, W. M. Itano, D. J. Wineland, *Phys. Rev. Lett.* **62**, 403 (1989).

16
Density Operators for Fermions[1]

16.1
Introduction

The Pauli exclusion principle plays an essential role in describing the behavior of the particles, both simple and complex, that we now call fermions. It is known to play a key role in determining the structure of the most fundamental elements of matter. These are systems like atoms, in which the phase-space density of fermions, electrons in this case, is quite high. But when fermionic atoms move freely in space or even when they are trapped electromagnetically, their phase-space density is usually so low that the effects of the exclusion principle remain completely hidden. A number of recent developments, however, point to the possibility of achieving much higher densities of fermionic atoms both in electromagnetic traps and in free space.

The various methods of optical cooling that have been developed for atomic beams work as well for fermions as they do for bosons and produce beams with temperatures of the order of 100 µK. Cooling fermions evaporatively to still lower temperatures poses a problem that requires a less direct solution. Evaporative cooling becomes inefficient for fermions since the exclusion principle tends to suppress collisions of identical atoms. It may be implemented nonetheless by sympathetic means[1], e.g., by cooling bosonic atoms at the same time, so that energy exchange still takes place freely. It thus seems possible that the realization of degenerate Fermi gases may become an important byproduct of Bose–Einstein condensation.

The detection methods that will be used in measurements on beams of cold fermionic atoms will be essentially the same as those now used on bosonic atoms cooled by optical or evaporative means. The measurements on bosons can be most conveniently described, in fact, by mathematical methods that were introduced in the context of quantum optics[2,3].

Much of the work in quantum optics, we may recall, is couched in the language of coherent states, which are eigenstates of the photon annihilation operators. They contain an intrinsically indefinite number of quanta but can nonetheless be used as a basis for describing all states of the electromagnetic field. While pure coherent states are not

---

[1] Reprinted with permission from K. E. Cahill, R. J. Glauber, *Phys. Rev. A* **59**, 1538–1555 (1999). Copyright 2006 by the American Physical Society.

Quantum Theory of Optical Coherence. Selected Papers and Lectures. Roy J. Glauber.
Copyright © 2007 WILEY-VCH Verlag GmbH & Co. KGaA, Weinheim
ISBN: 978-3-527-40687-6

physically attainable in bosonic systems with fixed numbers of particles, it likewise remains useful to describe boson fields in terms of suitably weighted superpositions and mixtures of coherent states. The weight functions associated with these combinations may be regarded as quasiprobability densities in the spaces of coherent-state amplitudes. The function P in the coherent-state representation of the density operator[2,3] plays this role; other quasiprobability densities including the Wigner function[3,4] and the Q function[3,4] play similarly convenient roles in representing the density operator.

In the case of fermion fields, the vacuum state is the only physically realizable eigenstate of the annihilation operators. It is possible, however, to define such eigenstates in a formal way and to put them to many of the same analytical uses as are made of the bosonic coherent states. Since fermion field variables anticommute, their eigenvalues must, as noted by Schwinger[5], be anticommuting numbers. Such numbers are Grassmann variables. They can be handled by means of the simple rules of Grassmann algebra[6], which we include here so that the calculations may be self-contained.

Within this context we formulate ways of expressing and evaluating a broad range of the correlation functions that are measured in experiments involving the counting of fermions. Central to this task is the expression of the quantum-mechanical density operator in terms of Grassmann variables. We develop a number of ways of doing that in general terms and present a detailed discussion of the density operators for chaotically excited fields. Included among the latter is a particularly useful Gaussian representation of the grand-canonical density operator for fermion fields. Having evaluated the statistically averaged correlation functions, we apply them to fermion-counting experiments and illustrate their use in determining the counting distributions.

We find throughout this work that notwithstanding great mathematical differences, many close parallels can be established between the expressions evaluated for fermion fields and the more familiar ones for boson fields. In particular, for example, we can construct a family of quasiprobability densities, as functions of the Grassmann variables, with properties parallel to those of the entire family of quasiprobability densities for bosons, including the P, Q, and Wigner functions. We can then evaluate the mean values of ordered products of fermion creation and annihilation operators by performing integrations over the Grassmann variables while using the appropriate quasiprobability density as a weight function. In both cases, we trade an inhomogeneous commutation relation and an ordering rule for a homogeneous commutation relation and a quasiprobability density. For boson fields the integrations are taken over commuting variables, which may be treated as if they were classical variables. For fermions, on the other hand, the integrations are over anti-commuting variables, which have no classical analogs. The weight functions for these integrations are nevertheless in one-to-one correspondence with the quasiprobability densities for bosons, so it seems appropriate to give them similar names. We have followed that convention for several other parallels as well.

16.2 Notation

Let us consider a system of fermions which may be described by the creation $a_n^\dagger$ and annihilation a_m operators which satisfy the familiar but ever mysterious relations

$$\{a_n, a_m^\dagger\} = \delta_{nm}, \tag{16.1}$$
$$\{a_n, a_m\} = 0, \tag{16.2}$$
$$\{a_n^\dagger, a_m^\dagger\} = 0, \tag{16.3}$$
$$a_n |0\rangle = 0, \tag{16.4}$$

in which $|0\rangle$ is the vacuum state.

We shall use lower-case Greek letters to denote Grassmann variables. These anticommuting numbers γ_n and their complex conjugates γ^* satisfy the convenient relations

$$\{\gamma_n, \gamma_m\} = 0, \tag{16.5}$$
$$\{\gamma_n^*, \gamma_m\} = 0, \tag{16.6}$$
$$\{\gamma_n^*, \gamma_m^*\} = 0. \tag{16.7}$$

We shall also assume that Grassmann variables anticommute with fermionic operators

$$\{\gamma_n, a_m\} = 0 \tag{16.8}$$

and commute with bosonic operators. In our notation a Grassmann number β_n and its complex conjugate β_n^* are independent variables. We make the arbitrary choice that Hermitian conjugation reverses the order of all fermionic quantities, both the operators and the anticommuting numbers. Thus, for instance, we have

$$(a_1 \beta_2 a_3^\dagger \gamma_4^*)^\dagger = \gamma_4 a_3 \beta_2^* a_1^\dagger. \tag{16.9}$$

Because the square of every Grassmann monomial vanishes, no nonzero Grassmann monomial can be an ordinary real, imaginary, or complex number.

16.3 Coherent States for Fermions

16.3.1 Displacement Operators

For any set $\gamma = \{\gamma_i\}$ of Grassmann variables, let us define the unitary displacement operator $D(\gamma)$ as the exponential

$$D(\gamma) = \exp\left(\sum_i (a_i^\dagger \gamma_i - \gamma_i^* a_i)\right). \tag{16.10}$$

16 Density Operators for Fermions

One of the useful properties of Grassmann numbers is that when, as in the preceding definition, they multiply fermionic annihilation or creation operators, their anticommutativity cancels that of the operators. Thus the operators $a_i^\dagger \gamma_i$ and $\gamma_j^* a_j$ simply commute for $i \neq j$. So we may rewrite the displacement operator as the product

$$D(\gamma) = \prod_i \exp\left(a_i^\dagger \gamma_i - \gamma_i^* a_i\right) \tag{16.11}$$

$$= \prod_i \left[1 + a_i^\dagger \gamma_i - \gamma_i^* a_i + (a_i^\dagger a_i - \tfrac{1}{2})\gamma_i^* \gamma_i\right]. \tag{16.12}$$

By the same token, the annihilation operator a_n commutes with all the operators $a_i^\dagger \gamma_i$ and $\gamma_i^* a_i$ when $n \neq i$, and so we may compute the displaced annihilation operator by ignoring all modes but the n-th:

$$D^\dagger(\gamma) a_n D(\gamma) = \prod_i \exp(\gamma_i^* a_i - a_i^\dagger \gamma_i) a_n \prod_j \exp(a_j^\dagger \gamma_j - \gamma_j^* a_j)$$

$$= \exp(\gamma_n^* a_n - a_n^\dagger \gamma_n) a_n \exp(a_n^\dagger \gamma_n - \gamma_n^* a_n)$$

$$= (1 - a_n^\dagger \gamma_n + \tfrac{1}{2}\gamma_n^* a_n a_n^\dagger \gamma_n) a_n (1 + a_n^\dagger \gamma_n - \tfrac{1}{2}a_n^\dagger \gamma_n \gamma_n^* a_n)$$

$$= (1 - a_n^\dagger \gamma_n - \tfrac{1}{2}\gamma_n^* \gamma_n) a_n (1 + a_n^\dagger \gamma_n + \tfrac{1}{2}\gamma_n^* \gamma_n)$$

$$= a_n - a_n^\dagger \gamma_n a_n + a_n a_n^\dagger \gamma_n = a_n + \gamma_n. \tag{16.13}$$

Similarly

$$D^\dagger(\gamma) a_n{}^\dagger D(\gamma) = a_n^\dagger + \gamma_n^*. \tag{16.14}$$

We may use the Baker–Hausdorff identity

$$e^{A+B} = e^A e^B e^{-[A,\,B]/2}, \tag{16.15}$$

which holds whenever the commutator $[A, B]$ commutes with both A and B, to write the displacement operator $D(\alpha)$ in forms that are normally ordered,

$$\exp\left(\sum_i a_i^\dagger \gamma_i\right) \exp\left(-\sum_i \gamma_i^* a_i\right) = \exp\left(\sum_i (a_i^\dagger \gamma_i - \gamma_i^* a_i)\right) e^{\gamma^* \cdot \gamma/2}, \tag{16.16}$$

$$D_N(\gamma) = D(\gamma) \exp\left(\tfrac{1}{2}\sum_i \gamma_i^* \gamma_i\right),$$

and antinormally ordered,

$$\exp\left(-\sum_i \gamma_i^* a_i\right) \exp\left(\sum_i a_i^\dagger \gamma_i\right) = \exp\left(\sum_i (a_i^\dagger \gamma_i - \gamma_i^* a_i)\right) e^{-\gamma^* \cdot \gamma/2}, \tag{16.17}$$

$$D_A(\gamma) = D(\gamma) \exp\left(-\tfrac{1}{2}\sum_i \gamma_i^* \gamma_i\right),$$

in which we have employed the concise notation

$$\gamma^* \cdot \gamma \equiv \sum_i \gamma_i^* \gamma_i , \qquad (16.18)$$

an abbreviation which we shall use occasionally but not exclusively. The identity Eq. (16.15) also allows one to show that the displacement operators form a ray representation of the additive group of Grassmann numbers,

$$D(\alpha)D(\beta) = D(\alpha+\beta) \exp\left(\tfrac{1}{2} \sum_i (\beta_i^* \alpha_i - \alpha_i^* \beta_i) \right) . \qquad (16.19)$$

16.3.2
Coherent States

For any set $\gamma = \{\gamma_i\}$ of Grassmann numbers, we define the *normalized* coherent state $|\gamma\rangle$ as the displaced vacuum state

$$|\gamma\rangle = D(\gamma)|0\rangle . \qquad (16.20)$$

By using the displacement relation Eq. (16.13), we may show that the coherent state is an eigenstate of every annihilation operator a_n:

$$\begin{aligned} a_n |\gamma\rangle &= a_n D(\gamma)|0\rangle = D(\gamma) D^\dagger(\gamma) a_n D(\gamma)|0\rangle \\ &= D(\gamma)(a_n + \gamma_n)|0\rangle = D(\gamma)\gamma_n |0\rangle = \gamma_n D(\gamma)|0\rangle \\ &= \gamma_n |\gamma\rangle . \end{aligned} \qquad (16.21)$$

By using the product formula (16.12) for the displacement operator, we may write the coherent state in the form

$$\begin{aligned} |\gamma\rangle = D(\gamma)|0\rangle &= \prod_i \left[1 + a_i^\dagger \gamma_i - \gamma_i^* a_i + (a_i^\dagger a_i - \tfrac{1}{2})\gamma_i^* \gamma_i \right] |0\rangle \\ &= \prod_i (1 + a_i^\dagger \gamma_i - \tfrac{1}{2}\gamma_i^* \gamma_i)|0\rangle \\ &= \exp\left(\sum_i (a_i^\dagger \gamma_i - \tfrac{1}{2}\gamma_i^* \gamma_i) \right) |0\rangle . \end{aligned} \qquad (16.22)$$

It may be worth emphasizing that in this formula the creation operator $a_i^\dagger$ stands to the *left* of the Grassmann number γ_i. Apart from these ordering considerations, this formula takes a form closely analogous to the one that defines bosonic coherent states.

The adjoint of the coherent state $|\gamma\rangle$ is

$$\langle\gamma| = \langle 0|D^\dagger(\gamma) = \langle 0|\exp\left(\sum_i (\gamma_i^* a_i - \tfrac{1}{2}\gamma_i^* \gamma_i) \right) , \qquad (16.23)$$

and it obeys the relation

$$\langle\gamma|a_n^\dagger = \langle\gamma|\gamma_n^* . \tag{16.24}$$

The inner product of two coherent states is

$$\langle\gamma|\beta\rangle = \exp\left(\sum_i [\gamma_i^*\beta_i - \tfrac{1}{2}(\gamma_i^*\gamma_i + \beta_i^*\beta_i)]\right), \tag{16.25}$$

so that

$$\langle\beta|\gamma\rangle\langle\gamma|\beta\rangle = \exp\left[-\sum_i(\beta_i^* - \gamma_i^*)(\beta_i - \gamma_i)\right]$$
$$= \prod_i[1 - (\beta_i^* - \gamma_i^*)(\beta_i - \gamma_i)] . \tag{16.26}$$

In contrast to the case of bosons, we may for fermions define for any set $\boldsymbol{\alpha} = \{\alpha_i\}$ of Grassmann numbers the normalized eigenstate $|\boldsymbol{\alpha}\rangle'$ of the fermion creation operators $a_i^\dagger$ as the displaced state

$$|\boldsymbol{\alpha}\rangle' = D(\boldsymbol{\alpha})|\mathbf{1}\rangle , \tag{16.27}$$

where $|\mathbf{1}\rangle$ is the state in which every mode is filled:

$$|\mathbf{1}\rangle = \prod_n a_n^\dagger|0\rangle . \tag{16.28}$$

By using the displacement relation Eq. (16.14), we may show that the state $|\boldsymbol{\alpha}\rangle'$ is an eigenstate of every creation operator $a_n^\dagger$:

$$a_n^\dagger|\boldsymbol{\alpha}\rangle' = a_n^\dagger D(\boldsymbol{\alpha})|\mathbf{1}\rangle = D(\boldsymbol{\alpha})D^\dagger(\boldsymbol{\alpha})a_n^\dagger D(\boldsymbol{\alpha})|\mathbf{1}\rangle$$
$$= D(\boldsymbol{\alpha})(a_n^\dagger + \alpha_n^*)|\mathbf{1}\rangle = D(\boldsymbol{\alpha})\alpha_n^*|\mathbf{1}\rangle = \alpha_n^* D(\boldsymbol{\alpha})|\mathbf{1}\rangle$$
$$= \alpha_n^*|\boldsymbol{\alpha}\rangle' . \tag{16.29}$$

The adjoint relation is

$$'\langle\boldsymbol{\alpha}|a_n = {'\langle\boldsymbol{\alpha}|}\alpha_n . \tag{16.30}$$

An explicit formula for the eigenstate $|\boldsymbol{\alpha}\rangle'$ follows from its definition Eq. (16.27):

$$|\boldsymbol{\alpha}\rangle' = \prod_i(1 - \alpha_i^*\alpha_i + \tfrac{1}{2}\alpha_i^*\alpha_i)|\mathbf{1}\rangle . \tag{16.31}$$

16.3.3
Intrinsic Descriptions of Fermionic States

The occupation-number description of states of fermions has well-known ambiguities. For $n \neq m$, for example, the state $|1_n, 1_m\rangle$ may be interpreted as $a_n^\dagger a_m^\dagger|0\rangle$ or as $a_m^\dagger a_n^\dagger|0\rangle = -a_n^\dagger a_m^\dagger|0\rangle$.

The creation operators themselves provide an unambiguous description of fermionic states,

$$|\psi\rangle = \sum_{\{n\}} c(n_1, n_2, \ldots) a_{n_1}^\dagger a_{n_2}^\dagger \cdots a_{n_m}^\dagger |0\rangle , \qquad (16.32)$$

which transfers to the coherent-state representation

$$\langle \alpha | \psi \rangle = \exp\left(-\tfrac{1}{2} \sum_n \alpha_n^* \alpha_n\right) \sum_{\{n\}} c(n_1, n_2, \ldots) \alpha_{n_1}^* \alpha_{n_2}^* \cdots \alpha_{n_m}^* , \qquad (16.33)$$

without any ambiguity or extra minus signs. Because coherent states are defined in terms of bilinear forms in anticommuting variables, there is no need to adopt a standard ordering of the modes.

16.4 Grassmann Calculus

16.4.1 Differentiation

Since the square of any Grassmann variable vanishes, the most general function $f(\xi)$ of a single anticommuting variable ξ is linear in ξ,

$$f(\xi) = u + \xi t . \qquad (16.34)$$

We define the left derivative of the function $f(\xi)$ with respect to the Grassmann variable ξ as

$$\frac{df(\xi)}{d\xi} = t . \qquad (16.35)$$

Note that if the variable t is anticommuting, then we may also write the function $f(\xi)$ in the form

$$f(\xi) = u - t\xi . \qquad (16.36)$$

Now to form the left derivative, we first move ξ past t, picking up a minus sign and obtaining the form Eq. (16.34) and the result Eq. (16.35). In this case, the right derivative is $-t$. In the present work, we shall use left derivatives exclusively and shall refer to them simply as derivatives.

16.4.2 Even and Odd Functions

It is useful to distinguish between functions that commute with Grassmann variables and ones that do not. We shall say that a function $f(\alpha)$ that commutes with Grassmann variables is *even* and that a function $f(\alpha)$ that anticommutes with Grassmann variables is *odd*.

Suppose that the only anticommuting quantities in a function $f(\alpha)$ are those of its argument α. Then if the function $f(\alpha)$ is even, it has even parity,

$$f(-\alpha) = f(\alpha), \qquad (16.37)$$

and if it is odd, it has odd parity,

$$f(-\alpha) = -f(\alpha), \qquad (16.38)$$

We shall often note the evenness or oddness of the functions we introduce.

16.4.3
Product Rule

To compute the derivative of the product of two functions $f(\alpha)$ and $g(\alpha)$ with respect to a particular variable α_i, one may explicitly move the α_i in $g(\alpha)$ through $f(\alpha)$ or one may move the operator representing differentiation through the function $f(\alpha)$. In either case if the function $f(\alpha)$ is odd, then one picks up a minus sign. The product rule is thus

$$\frac{\partial}{\partial \alpha_i}[f(\alpha)g(\alpha)] = \frac{\partial f(\alpha)}{\partial \alpha_i}g(\alpha) + \sigma(f)f(\alpha)\frac{\partial g(\alpha)}{\partial \alpha_i}, \qquad (16.39)$$

where the sign $\sigma(f)$ of $f(\alpha)$ is -1 if $f(\alpha)$ is an odd function and $+1$ if $f(\alpha)$ is even.

16.4.4
Integration

We define a sort of integration over the complex Grassmann variables by the following rules:

$$\int d\alpha_n = \int d\alpha_n^* = 0, \qquad (16.40)$$

$$\int d\alpha_n \, \alpha_m = \delta_{nm}, \qquad (16.41)$$

$$\int d\alpha_n^* \, \alpha_m^* = \delta_{nm}. \qquad (16.42)$$

This integration due to Berezin[6] is exactly equivalent to left differentiation.

We shall typically be concerned with pairs of anticommuting variables α_i and α_i^*, and for such pairs we shall adhere to the notation

$$\int d^2\alpha_n = \int d\alpha_n^* \, d\alpha_n, \qquad (16.43)$$

in which the differential of the conjugated variable comes first. Note that

$$d\alpha_n \, d\alpha_n^* = -d\alpha_n^* \, d\alpha_n. \qquad (16.44)$$

We have been using boldface type to denote sets of Grassmann variables; we shall extend that use to write multiple integrals over such sets in the succinct form

$$\int d^2\boldsymbol{\alpha} = \int \prod_i d^2\alpha_i \,. \tag{16.45}$$

We shall also occasionally employ the concise notation

$$\boldsymbol{\alpha}^* \cdot \boldsymbol{\beta} \equiv \sum_n \alpha_n^* \beta_n \tag{16.46}$$

for sums of simple products over all the modes of the system.

The simple integral formula

$$\int d^2\alpha_n \, e^{-\alpha_p^* \alpha_q} = \int d\alpha_n^* \, d\alpha_n (1 - \alpha_p^* \alpha_q)$$
$$= -\int d\alpha_n^* \, d\alpha_n \, \alpha_p^* \, \alpha_q = \int d\alpha_n^* \alpha_p^* \, d\alpha_n \alpha_q$$
$$= \int d\alpha_n^* \alpha_p^* \delta_{nq} = \delta_{np} \delta_{nq} \tag{16.47}$$

provides a useful example of Grassmann integration. We also note the general rule

$$\int d^2\alpha \, f(\lambda\alpha) = |\lambda|^2 \int d^2\beta \, f(\beta) \,, \tag{16.48}$$

in which λ is an arbitrary complex number and in which $f(\lambda\alpha)$ is an abbreviation for a function which necessarily depends on both $\lambda\alpha$ and $\lambda^*\alpha^*$. This rule owes its strange appearance to the definition of integration as differentiation.

Some further examples are the integral of the exponential function,

$$\int d^2\alpha \, \exp(\beta^*\alpha + \alpha^*\gamma + \alpha\alpha^*) = \exp(\beta^*\gamma) \,, \tag{16.49}$$

and the Fourier transform of a Gaussian,

$$\int d^2\xi \, \exp(\alpha\xi^* - \xi\alpha^* + \lambda\xi\xi^*) = \lambda \exp\left(\frac{\alpha\alpha^*}{\lambda}\right) , \tag{16.50}$$

where λ is an arbitrary complex number. The latter integral can be written in a somewhat more-general form which is no longer a Fourier transform:

$$\int d^2\xi \, \exp(\alpha\xi^* - \xi\beta^* + \lambda\xi\xi^*) = \lambda \exp\left(\frac{\alpha\beta^*}{\lambda}\right) . \tag{16.51}$$

16.4.5
Integration by Parts

Let us first observe that the integral of a derivative vanishes,

$$\int d^2\alpha \, \frac{\partial f(\alpha)}{\partial \alpha_i} = 0 \,, \tag{16.52}$$

because the derivative with respect to the variable α_i lacks the variable α_i. In particular, the integral of the derivative of the product of two functions also vanishes, and so by using the product rule Eq. (16.39), we have

$$\int d^2\alpha \frac{\partial}{\partial \alpha_i}[f(\alpha)g(\alpha)] = \int d^2\alpha \left[\left(\frac{\partial f(\alpha)}{\partial \alpha_i}\right)g(\alpha) + \sigma(f)f(\alpha)\left(\frac{\partial g(\alpha)}{\partial \alpha_i}\right)\right] = 0,$$
(16.53)

which is the formula for integration by parts,

$$\int d^2\alpha \left(\frac{\partial f(\alpha)}{\partial \alpha_i}\right)g(\alpha) = -\sigma(f)\int d^2\alpha\, f(\alpha)\frac{\partial g(\alpha)}{\partial \alpha_i},$$
(16.54)

where the sign $\sigma(f)$ is $+1$ if the function $f(\alpha)$ is even and -1 if it is odd.

16.4.6
Completeness of the Coherent States

We may use our Grassmann calculus to show that the coherent states are complete. Let us consider the state

$$|f\rangle = (c + da^\dagger)|0\rangle,$$
(16.55)

which for arbitrary complex numbers c and d is an arbitrary single-mode state. Then its inner product $\langle \gamma | f \rangle$ with the coherent state $|\gamma\rangle$ is the correct weight function for the coherent-state expansion since

$$\int d^2\gamma\, \langle\gamma|f\rangle |\gamma\rangle = \int d^2\gamma\, (c + d\gamma^*)(1 + \gamma\gamma^* - \gamma a^\dagger)|0\rangle$$
$$= (c + da^\dagger)|0\rangle = |f\rangle.$$
(16.56)

The reader may generalize this example to the multimode case. The coherent states in fact are overcomplete.

16.4.7
Completeness of the Displacement Operators

For a single mode, the identity operator I and the traceless operators a, $a^\dagger$ and $\frac{1}{2} - a^\dagger a$ form a complete set of operators. Since by using the expression Eq. (16.12) and our Grassmann calculus, we may write each of these operators as an integral over the displacement operators

$$I = \int d^2\gamma\, \gamma\gamma^* D(\gamma),$$
(16.57)

$$a = \int d^2\gamma\, (-\gamma) D(\gamma),$$
(16.58)

$$a^\dagger = \int d^2\gamma\, \gamma^* D(\gamma),\tag{16.59}$$

$$\tfrac{1}{2} - a^\dagger a = \int d^2\gamma\, D(\gamma),\tag{16.60}$$

it follows that the displacement operators form a complete set of operators for that mode. It is easy to generalize this proof to the multimode case. The displacement operators are overcomplete.

16.5 Operators

Some operators can be written as sums of products of even numbers of creation and annihilation operators; we shall call such operators *even*. Operators that can be written as sums of products of odd numbers of creation and annihilation operators we shall speak of as *odd*. Although most operators are neither even nor odd, the operators of physical interest are either even or odd. The number operator $a^\dagger a$, for example, is even, while the creation and annihilation operators, $a^\dagger$ and a, are odd.

The operators of quantum mechanics and of quantum field theory do not themselves involve Grassmann variables. Thus even operators commute with Grassmann variables, while odd ones anticommute.

16.5.1 The Identity Operator

If we compare the integral

$$\int d^2\alpha\, |\alpha\rangle\langle\alpha|\beta\rangle = \int d^2\alpha\, \exp\left(\sum_i (a_i^\dagger \alpha_i + \alpha_i^* \beta_i + \alpha_i \alpha_i^* + \tfrac{1}{2}\beta_i\beta_i^*)\right)|0\rangle,\tag{16.61}$$

with the integral formula Eq. (16.49) and identify β^* and γ in that formula with $a^\dagger$ and β in this integral, then we have

$$\int d^2\alpha\, |\alpha\rangle\langle\alpha|\beta\rangle = \exp\left(\sum_i (a_i^\dagger \beta_i + \tfrac{1}{2}\beta_i\beta_i^*)\right)|0\rangle = |\beta\rangle.\tag{16.62}$$

Since the coherent states form a complete set of states, as shown by the expansion Eq. (16.56), it follows that the identity operator is given by the integral

$$I = \int d^2\alpha\, |\alpha\rangle\langle\alpha|.\tag{16.63}$$

The corresponding expression for the identity operator in terms of the eigenstates $|\alpha\rangle'$ of the creation operators is

$$I = \int \prod_i (-d^2\alpha_i)\, |\alpha\rangle'\,'\langle\alpha|.\tag{16.64}$$

16.5.2
The Trace

The trace of an arbitrary operator B is the sum of the diagonal matrix elements of B in the n-quantum states,

$$\operatorname{Tr} B = \sum_n \langle n|B|n\rangle , \qquad (16.65)$$

which shows that the trace of an operator that is odd vanishes. By inserting the preceding formula Eq. (16.63) for the identity operator, we have

$$\operatorname{Tr} B = \sum_n \int d^2\alpha \, \langle n|\alpha\rangle \langle \alpha|B|n\rangle . \qquad (16.66)$$

If we move the coherent-state matrix element $\langle n|\alpha\rangle$ to the right of the matrix element of the even operator B, then we see from the formula (16.22) that minus signs arise that can be absorbed into the argument of either of the two coherent states,

$$\operatorname{Tr} B = \sum_n \int d^2\alpha \, \langle \alpha|B|n\rangle \langle n|-\alpha\rangle$$

$$= \sum_n \int d^2\alpha \, \langle -\alpha|B|n\rangle \langle n|\alpha\rangle , \qquad (16.67)$$

in which the sum $\sum_n |n\rangle\langle n| = I$ is the identity operator. The resulting multimode trace formula is

$$\operatorname{Tr} B = \int d^2\alpha \, \langle \alpha|B|-\alpha\rangle = \int d^2\alpha \, \langle -\alpha|B|\alpha\rangle , \qquad (16.68)$$

which holds also for odd operators, both sides vanishing. An important example is the trace of the dyadic operator $|\beta\rangle\langle\gamma|$,

$$\operatorname{Tr} |\beta\rangle\langle\gamma| = \int d^2\alpha \, \langle \alpha|\beta\rangle \langle \gamma|-\alpha\rangle = \int d^2\alpha \, \langle \gamma|-\alpha\rangle \langle \alpha|\beta\rangle$$

$$= \int d^2\alpha \, \langle -\gamma|\alpha\rangle \langle \alpha|\beta\rangle = \langle -\gamma|\beta\rangle = \langle \gamma|-\beta\rangle , \qquad (16.69)$$

in which we have used the completeness relation Eq. (16.63). Since the coherent states are complete, we may replace in this formula either the ket $|\beta\rangle$ or the bra $\langle\gamma|$ with its image $F|\beta\rangle$ or $\langle\gamma|F$ under the action of the arbitrary operator F and obtain the trace formula

$$\operatorname{Tr}\left(F|\beta\rangle\langle\gamma|\right) = \operatorname{Tr}\left(|\beta\rangle\langle\gamma|F\right) = \langle -\gamma|F|\beta\rangle = \langle \gamma|F|-\beta\rangle . \qquad (16.70)$$

16.5.3
Physical States and Operators

A state $|\psi\rangle$ is *physical* if it changes at most by a phase when subjected to a rotation of angle 2π about any axis,

$$U(\hat{n}, 2\pi)|\psi\rangle = e^{i\theta}|\psi\rangle . \qquad (16.71)$$

Since fermions carry half-odd-integer spin, a state of one fermion or of any odd number of fermions changes by the phase factor -1. States that contain no fermions or only even numbers of fermions are invariant under such 2π rotations.

Thus physical states are linear combinations of states with odd numbers of fermions or linear combinations of states with even numbers of fermions. But a state that is a linear combination of a state that contains an odd number of fermions and another that contains an even number of fermions is excluded. For instance, the state

$$\frac{1}{\sqrt{2}}(|0\rangle + |1\rangle) \qquad (16.72)$$

is unphysical because under a 2π rotation it changes into a different state:

$$U(\hat{n}, 2\pi)\frac{1}{\sqrt{2}}(|0\rangle + |1\rangle) = \frac{1}{\sqrt{2}}(|0\rangle - |1\rangle)$$

$$\neq e^{i\theta}\frac{1}{\sqrt{2}}(|0\rangle + |1\rangle). \qquad (16.73)$$

We define an operator as *physical* if it maps physical states onto physical states. Physical operators are either even or odd.

In all physical contexts that have been explored experimentally, the number of fermions (or more generally the number of fermions minus the number of antifermions) is strictly conserved. That conservation law leads to certain further restrictions on the permissible states of the field. If we let $N = \sum_k a_k^\dagger a_k$ be the fermion number, the law requires that any state arising from an eigenstate of N must remain an eigenstate of N. This law can be derived from an assumed $U(1)$ invariance of all the interactions under the transformation $U(\theta) = \exp(i\theta N)$, which changes a and $a^\dagger$ to

$$e^{-i\theta N} a\, e^{i\theta N} = e^{i\theta} a \qquad (16.74)$$

and

$$e^{-i\theta N} a^\dagger e^{i\theta N} = e^{-i\theta} a^\dagger. \qquad (16.75)$$

Fermion conserving interactions involving the a_k and $a_k^\dagger$ are ones in which the phase factors $e^{\pm i\theta}$ all cancel. If a system begins in a state with a fixed number of fermions, the conservation law restricts the set of accessible states considerably more than the 2π superselection rule mentioned earlier. Transitions cannot be made, for example, between states with different even fermion numbers or between states with different odd fermion numbers.

16.5.4
Physical Density Operators

A physical density operator can be written as a sum of dyadics of physical states with positive coefficients that add up to unity. It follows that a physical density operator ρ

is a positive Hermitian operator of unit trace: for any state $|\psi\rangle$

$$\langle\psi|\rho|\psi\rangle \geq 0, \tag{16.76}$$

$$\rho^\dagger = \rho, \tag{16.77}$$

$$\mathrm{Tr}\,\rho = 1. \tag{16.78}$$

Physical density operators are invariant under a 2π rotation. Thus the one-mode operator

$$\rho = \tfrac{1}{2}(|0\rangle\langle 0| + |1\rangle\langle 1|), \tag{16.79}$$

for example, is a physical density operator, but the dyadic

$$\rho = \tfrac{1}{2}(|0\rangle + |1\rangle)(\langle 0| + \langle 1|) \tag{16.80}$$

is not. In this work we shall consider only density operators that are physical in this sense.

The dynamical problems we solve do not always begin with a fixed number of fermions. More generally they begin with a mixture of states with different fermion numbers, that is, with density operators of the form

$$\rho = \sum_{N'} p_{N'} |N'\rangle\langle N'|, \tag{16.81}$$

where the $p_{N'}$ are real and non-negative. Such density operators are invariant under the transformation $U(\theta) = \exp(i\theta N)$, and the fermion conservation law assures us that they will always remain so,

$$e^{-i\theta N} \rho e^{i\theta N} = \rho. \tag{16.82}$$

The coherent states do undergo a simple change under this transformation,

$$U(\theta)|\alpha\rangle = e^{i\theta N}|\alpha\rangle = |e^{i\theta}\alpha\rangle, \tag{16.83}$$

$$\langle\alpha|U^\dagger(\theta) = \langle\alpha|e^{-i\theta N} = \langle e^{i\theta}\alpha|, \tag{16.84}$$

which leaves their scalar product invariant,

$$\langle e^{i\theta}\alpha|e^{i\theta}\alpha\rangle = \langle\alpha|\alpha\rangle. \tag{16.85}$$

16.6
δ Functions and Fourier Transforms

We can define a function

$$\delta(\xi - \zeta) \equiv \int d^2\alpha \, \exp\left(\sum_n [\alpha_n(\xi_n^* - \zeta_n^*) - (\xi_n - \zeta_n)\alpha_n^*]\right) \tag{16.86}$$

$$= \prod_n (\xi_n - \zeta_n)(\xi_n^* - \zeta_n^*), \tag{16.87}$$

which plays the role of a Dirac δ function in that if $f(\boldsymbol{\xi})$ is any function of the set $\boldsymbol{\xi}$ of Grassmann variables $\{\xi_1, \xi_2, \ldots\}$, then

$$\int d^2\xi\, \delta(\boldsymbol{\xi} - \boldsymbol{\zeta})\, f(\boldsymbol{\xi}) = f(\boldsymbol{\zeta}). \tag{16.88}$$

The δ function is doubly even: it commutes with Grassmann numbers and $\delta(\xi - \zeta) = \delta(\zeta - \xi)$.

We have been using the term Fourier transform to denote an integral of the form

$$\widetilde{f}(\alpha) = \int d^2\xi\, e^{\alpha\xi^* - \xi\alpha^*} f(\xi). \tag{16.89}$$

The δ-function identity Eq. (16.86) implies that the inverse Fourier transform is given by the similar formula

$$f(\xi) = \int d^2\alpha\, e^{\xi\alpha^* - \alpha\xi^*} \widetilde{f}(\alpha). \tag{16.90}$$

The identity Eq. (16.86) also leads to two forms of Parseval's relation:

$$\int d^2\alpha\, \widetilde{f}(\alpha) [\widetilde{g}(\alpha)]^* = \int d^2\xi\, f(\xi)\, g^*(\xi) \tag{16.91}$$

and

$$\int d^2\alpha\, \widetilde{f}(\alpha)\, \widetilde{g}(-\alpha) = \int d^2\xi\, f(\xi)\, g(\xi), \tag{16.92}$$

which apply also to operator-valued functions provided that complex conjugation is replaced by Hermitian conjugation.

We may use the formula Eq. (16.86) for the δ function to derive a fermionic analog of the convolution theorem:

$$\int d^2\xi\, e^{\alpha\xi^* - \xi\alpha^*} f(\xi)\, g(\xi) = \int d^2\beta\, d^2\xi\, e^{(\alpha-\beta)\xi^* - \xi(\alpha^* - \beta^*)} f(\xi) \int d^2\eta\, e^{\beta\eta^* - \eta\beta^*} g(\eta)$$

$$= \int d^2\beta\, \widetilde{f}(\alpha - \beta)\, \widetilde{g}(\beta), \tag{16.93}$$

which expresses the Fourier transform of the product of the two functions $f(\xi)$ and $g(\xi)$ as the convolution of their Fourier transforms $\widetilde{f}(\alpha - \beta)$ and $\widetilde{g}(\beta)$.

By using the normally ordered form Eq. (16.16) of the displacement operator, the eigenvalue property of the coherent states, and the preceding formula Eq. (16.86) for the δ function we find

$$\int d^2\gamma\, \langle \gamma | D(\alpha) | \gamma \rangle = \int d^2\gamma\, \langle \gamma | e^{a^\dagger \alpha} e^{-\alpha^* a} | \gamma \rangle e^{\alpha\alpha^*/2}$$

$$= \int d^2\gamma\, e^{\gamma^*\alpha - \alpha^*\gamma + \alpha\alpha^*/2}$$

$$= \delta(\alpha)\, e^{\alpha\alpha^*/2} = \delta(\alpha). \tag{16.94}$$

The addition rule Eq. (16.19) for successive displacements now implies that for the multimode case

$$\int d^2\gamma \, \langle \gamma | D(\alpha) D(-\beta) | \gamma \rangle = \delta(\alpha - \beta) \, . \tag{16.95}$$

16.7
Operator Expansions

The preceding δ-function identity Eq. (16.95) and the completeness Eqns. (16.57)–(16.60) of the displacement operators give us a means of expanding an arbitrary operator F in the form

$$F = \int d^2\xi \, f(\xi) D(-\xi) \, . \tag{16.96}$$

We may solve for the weight function $f(\xi)$ by multiplying on the right by the displacement operator $D(\alpha)$ and then taking the diagonal coherent-state matrix element in the state $|\beta\rangle$ and integrating over β:

$$\int d^2\beta \, \langle \beta | FD(\alpha) | \beta \rangle = \int d^2\beta \int d^2\xi \, f(\xi) \, \langle \beta | D(-\xi) D(\alpha) | \beta \rangle$$

$$= \int d^2\xi \, f(\xi) \, \delta(\alpha - \beta) = f(\alpha) \, . \tag{16.97}$$

The full expansion is thus

$$F = \int d^2\xi \int d^2\beta \, \langle \beta | FD(\xi) | \beta \rangle D(-\xi) \, . \tag{16.98}$$

Such expansions will prove useful in the sections that follow.

The formula Eq. (16.86) for the delta function $\delta(\xi - \zeta)$ may be interpreted as a trace identity. From the eigenvalue property of the coherent states, it follows that

$$\delta(\xi - \zeta) = \int d^2\alpha \, e^{\alpha \xi^* - \xi \alpha^*} e^{\zeta \alpha^* - \alpha \zeta^*}$$

$$= \int d^2\alpha \, e^{\alpha \xi^* - \xi \alpha^*} \left\langle \alpha \left| e^{\zeta a^\dagger} e^{-a \zeta^*} \right| \alpha \right\rangle \tag{16.99}$$

in which we recognize the normally ordered form Eq. (16.16) of the displacement operator

$$\delta(\xi - \zeta) = \int d^2\alpha \, e^{\alpha \xi^* - \xi \alpha^*} \, \langle \alpha | D_N(\zeta) | \alpha \rangle \, . \tag{16.100}$$

By using the trace formula Eq. (16.70), we may write this δ function as the trace

$$\delta(\xi - \zeta) = \int d^2\alpha \, e^{\alpha \xi^* - \xi \alpha^*} \, \mathrm{Tr}\left[D_N(\zeta) | \alpha \rangle \langle -\alpha | \right]$$

$$= \mathrm{Tr}\left[D_N(\zeta) E_A(-\xi) \right] \tag{16.101}$$

of the product of the normally ordered displacement operator $D_N(\zeta)$ with an even operator $E_A(\xi)$ defined as the Fourier transform

$$E_A(\xi) = \int d^2\alpha\, e^{\xi\alpha^* - \alpha\xi^*} |\alpha\rangle\langle -\alpha| \tag{16.102}$$

of the coherent-statest dyadic $|\alpha\rangle\langle -\alpha|$. As intimated by its subscript, the operator $E_A(\xi)$ will turn out to be useful for dealing with antinormally ordered operators.

We may now use the completeness Eqns. (16.57)–(16.60) of the displacement operators and the trace identity Eq. (16.101) to expand an arbitrary operator F in terms of the normally ordered displacement operators $D_N(\xi)$,

$$F = \int d^2\xi\, f(\xi)\, D_N(-\xi) . \tag{16.103}$$

We may solve for the function $f(\xi)$ by multiplying on the right by the operator $E_A(\zeta)$ and forming the trace:

$$\text{Tr}[F E_A(\zeta)] = \int d^2\xi\, f(\xi)\, \text{Tr}[D_N(-\xi) E_A(\zeta)]$$
$$= \int d^2\xi\, f(\xi)\, \delta(\zeta - \xi) = f(\zeta) . \tag{16.104}$$

The full expansion is thus

$$F = \int d^2\xi\, \text{Tr}[F E_A(\xi)] D_N(-\xi) . \tag{16.105}$$

By using the Grassmann calculus, one may compute the Fourier transform Eq. (16.102) of the coherent-state dyadic $|\alpha\rangle\langle -\alpha|$ and find for the operator $E_A(\xi)$ the formulas

$$E_A(\xi) = |0\rangle\langle 0| - (\xi^* + a^\dagger)|0\rangle\langle 0|(\xi + a) \tag{16.106}$$
$$= 2(\tfrac{1}{2} - a^\dagger a) + \xi\xi^* a a^\dagger + \xi a^\dagger - \xi^* a , \tag{16.107}$$

with which it is easy to exhibit the completeness of the operators $E_A(\xi)$:

$$I = \int d^2\xi\, 2(1 + \xi^*\xi) E_A(\xi) , \tag{16.108}$$

$$a = \int d^2\xi\, (-\xi) E_A(\xi) , \tag{16.109}$$

$$a^\dagger = \int d^2\xi\, (-\xi^*) E_A(\xi) , \tag{16.110}$$

$$\tfrac{1}{2} - a^\dagger a = \int d^2\xi\, \tfrac{1}{2}\xi\xi^* E_A(\xi) . \tag{16.111}$$

Since the operators $E_A(\xi)$ are complete, we may expand an arbitrary operator G in terms of them,

$$G = \int d^2\xi\, g(\xi)\, E_A(-\xi) \tag{16.112}$$

and then use the trace formula Eq. (16.101) and the evenness of the displacement operators to evaluate the weight function $g(\xi)$,

$$\mathrm{Tr}[D_N(\zeta)G] = \int d^2\xi\, g(\xi)\, \mathrm{Tr}[D_N(\zeta)E_A(-\xi)]$$

$$= \int d^2\xi\, g(\xi)\, \delta(\xi - \zeta) = g(\zeta)\,. \tag{16.113}$$

The full expansion is thus

$$G = \int d^2\xi\, \mathrm{Tr}[GD_N(\xi)]E_A(-\xi)\,. \tag{16.114}$$

16.8
Characteristic Functions

For a system described by the density operator ρ, we define the characteristic function $\chi(\xi)$ of Grassmann argument ξ (and ξ^*) as the mean value

$$\chi(\xi) = \mathrm{Tr}\left[\rho \exp\left(\sum_n (\xi_n a_n^\dagger - a_n \xi_n^*)\right)\right]\,. \tag{16.115}$$

It is thus a species of Fourier transform of the density operator ρ. Because $\xi_i^2 = \xi_i^{*2} = 0$, we may expand the exponential as

$$\chi(\xi) = \mathrm{Tr}\left[\rho \prod_n [1 + \xi_n a_n^\dagger - a_n \xi_n^* + \xi_n^* \xi_n (a_n^\dagger a_n - \tfrac{1}{2})]\right]\,. \tag{16.116}$$

We may also define the normally ordered characteristic function $\chi_N(\xi)$ as

$$\chi_N(\xi) = \mathrm{Tr}\left[\rho \exp\left(\sum_n \xi_n a_n^\dagger\right) \exp\left(-\sum_m a_m \xi_m^*\right)\right]\,, \tag{16.117}$$

with the expansion

$$\chi_N(\xi) = \mathrm{Tr}\left[\rho \prod_n (1 + \xi_n a_n^\dagger - a_n \xi_n^* + \xi_n^* \xi_n a_n^\dagger a_n)\right]\,. \tag{16.118}$$

The antinormally ordered characteristic function $\chi_A(\xi)$ is

$$\chi_A(\xi) = \mathrm{Tr}\left[\rho \exp\left(-\sum_m a_m \xi_m^*\right) \exp\left(\sum_n \xi_n a_n^\dagger\right)\right] \tag{16.119}$$

$$= \mathrm{Tr}\left[\rho \prod_n [1 + \xi_n a_n^\dagger - a_n \xi_n^* + \xi_n^* \xi_n (a_n^\dagger a_n - 1)]\right]\,. \tag{16.120}$$

Because the density operator ρ is an even operator and because the displacement operators are constructed from bilinear forms in fermionic quantities, it follows that the characteristic functions are doubly even in the sense that they commute with Grassmann variables and also are of even parity.

16.8.1
The s-Ordered Characteristic Function

We may define a more general ordering of the annihilation operator a_n and the creation operator $a_n^\dagger$, much as we did earlier for boson field operators[4]. It is an ordering specified by a real parameter s that runs from $s = -1$ for antinormal ordering to $s = 1$ for normal ordering. For the quadratic case, the s-ordered product for fermions is

$$\{a_n^\dagger a_n\}_s = a_n^\dagger a_n + \tfrac{1}{2}(s-1), \tag{16.121}$$

to which we append the trivial definitions

$$\{a_n^\dagger\}_s = a_n^\dagger \quad \text{and} \quad \{a_n\}_s = a_n. \tag{16.122}$$

We note that the definition Eq. (16.121) differs by a crucial sign from that[4] of s ordering for bosonic operators b_n and $b_n^\dagger$:

$$\{b_n^\dagger b_n\}_s = b_n^\dagger b_n + \tfrac{1}{2}(1-s). \tag{16.123}$$

In particular, the antinormally ordered product $\{a_n^\dagger a_n\}_{-1}$ is $-a_n a_n^\dagger$, and the symmetrically ordered product $\{a_n^\dagger a_n\}_0$ is half the commutator,

$$\{a_n^\dagger a_n\}_0 = \tfrac{1}{2}\left[a_n, a_n^\dagger\right]. \tag{16.124}$$

We define the s-ordered characteristic function $\chi(\xi, s)$ as

$$\chi(\xi, s) = \text{Tr}\left[\rho\left\{\exp\left(\sum_n (\xi_n a_n^\dagger - a_n \xi_n^*)\right)\right\}_s\right] \tag{16.125}$$

$$= \text{Tr}\left[\rho \prod_n \left(1 + \xi_n a_n^\dagger - a_n \xi_n^* + \xi_n^* \xi_n \{a_n^\dagger a_n\}_s\right)\right] \tag{16.126}$$

$$= \text{Tr}\left[\rho \prod_n \left\{1 + \xi_n a_n^\dagger - a_n \xi_n^* + \xi_n^* \xi_n [a_n^\dagger a_n + \tfrac{1}{2}(s-1)]\right\}\right] \tag{16.127}$$

$$= \text{Tr}\left\{\rho \exp\left[\sum_n \left(\xi_n a_n^\dagger - a_n \xi_n^* + \tfrac{s}{2}\xi_n^* \xi_n\right)\right]\right\} \tag{16.128}$$

$$= \chi(\xi) \exp\left(\tfrac{s}{2}\sum_n \xi_n^* \xi_n\right), \tag{16.129}$$

which, incidentally, shows it to be an even function and of even parity,

$$\chi(-\xi, s) = \chi(\xi, s). \tag{16.130}$$

A particularly useful example of these characteristic functions is the case of the antinormally ordered function $\chi_A(\xi) = \chi(\xi, -1)$. We see by inserting the resolution Eq. (16.63) of the identity between the exponential functions in its definition Eq. (16.119) that

$$\chi(\xi,-1) = \text{Tr}\left[\rho \exp\left(-\sum_m \beta_m \xi_m^*\right) \int d^2\beta \, |\beta\rangle \langle\beta| \exp\left(\sum_n \xi_n \beta_n^*\right)\right], \qquad (16.131)$$

in which we have replaced the annihilation and creation operators by their eigenvalues in the coherent states. By using the trace formula Eq. (16.69), we find

$$\chi(\xi,-1) = \int d^2\beta \, \exp\left(\sum_n (\xi_n \beta_n^* - \beta_n \xi_n^*)\right) \langle\beta|\rho|-\beta\rangle, \qquad (16.132)$$

which expresses the antinormally ordered characteristic function $\chi(\xi,-1)$ as the Fourier transform of the matrix element $\langle\beta|\rho|-\beta\rangle$.

If we define the s-ordered displacement operator $D(\xi,s)$ as

$$D(\xi,s) = \{D(\xi)\}_s = D(\xi) \exp\left(\frac{s}{2}\sum_n \xi_n^* \xi_n\right), \qquad (16.133)$$

then we may write the s-ordered characteristic function Eq. (16.129) as the trace

$$\chi(\xi,s) = \text{Tr}[\rho D(\xi,s)]. \qquad (16.134)$$

16.9
s-Ordered Expansions for Operators

A convenient extension of the definition of the operator $E_A(\xi)$ is

$$E(\xi,s) \equiv E_A(\xi) \exp\left(\frac{s+1}{2}\sum_n \xi_n^* \xi_n\right), \qquad (16.135)$$

from which we note that

$$E_A(\xi) = E(\xi,-1). \qquad (16.136)$$

This is one sense in which the operator $E_A(\xi)$ is related to antinormal ordering.

By using the s-ordered operators $D(\xi,s)$ and $E(\xi,s)$, we may generalize the expansions Eqns. (16.105) and (16.114) of the arbitrary operators F and G to

$$F = \int d^2\xi \, \text{Tr}[F E(\xi,-s)] \, D(-\xi,s), \qquad (16.137)$$

$$G = \int d^2\xi \, \text{Tr}[G D(\xi,-s)] \, E(-\xi,s). \qquad (16.138)$$

The obvious generalization

$$\delta(\xi - \zeta) = \text{Tr}[D(\xi,s)E(-\zeta,-s)] \tag{16.139}$$

of the trace formula Eq. (16.101) then gives the trace of the product FG as

$$\text{Tr}[FG] = \int d^2\xi \, \text{Tr}[FE(\xi,-s)] \, \text{Tr}[GD(-\xi,s)]. \tag{16.140}$$

We may now use the second Parseval relation Eq. (16.92) to cast the expansions Eq. (16.137) and (16.138) into forms that will prove to be quite useful. First let us define the complete sets of operators $\widetilde{D}(\alpha,s)$ and $\widetilde{E}(\alpha,s)$ as the Fourier transforms of the operators $D(\xi,s)$ and $E(\xi,s)$:

$$\widetilde{D}(\alpha,s) \equiv \int d^2\xi \, \exp\left(\sum_n (\alpha_n\xi_n^* - \xi_n\alpha_n^*)\right) D(\xi,s), \tag{16.141}$$

$$\widetilde{E}(\alpha,s) \equiv \int d^2\xi \, \exp\left(\sum_n (\alpha_n\xi_n^* - \xi_n\alpha_n^*)\right) E(\xi,s). \tag{16.142}$$

Next let us define the weight functions $F_E(\alpha,-s)$ and $G_D(\alpha,-s)$ as the Fourier transforms of the traces

$$F_E(\alpha,-s) \equiv \int d^2\xi \, \exp\left(\sum_n (\alpha_n\xi_n^* - \xi_n\alpha_n^*)\right) \text{Tr}[FE(\xi,-s)], \tag{16.143}$$

$$G_D(\alpha,-s) \equiv \int d^2\xi \, \exp\left(\sum_n (\alpha_n\xi_n^* - \xi_n\alpha_n^*)\right) \text{Tr}[GD(\xi,-s)]. \tag{16.144}$$

It follows then from the Parseval relation Eq. (16.92) and from the expansions Eqns. (16.137) and (16.138) that the operators $\widetilde{D}(\alpha,s)$ and $\widetilde{E}(\alpha,s)$ form complete sets of operators and afford us the expansions

$$F = \int d^2\alpha \, F_E(\alpha,-s) \, \widetilde{D}(\alpha,s), \tag{16.145}$$

$$G = \int d^2\alpha \, G_D(\alpha,-s) \, \widetilde{E}(\alpha,s) \tag{16.146}$$

of the arbitrary operators F and G. Applying the Parseval relation Eq. (16.92) to the trace formula Eq. (16.140), we have the trace relation

$$\text{Tr}[FG] = \int d^2\alpha \, F_E(\alpha,-s) \, G_D(\alpha,s). \tag{16.147}$$

The operators $\widetilde{E}(\alpha,s)$ are particularly simple when $s = \pm 1$. It follows from the definitions Eqns. (16.135) and (16.102) of the operators $\widetilde{E}(\alpha,s)$ and $E_A(\xi)$, and from the formula Eq. (16.86) for the δ function that the operator $\widetilde{E}(\alpha,-1)$ is just the coherent-state dyadic

$$\widetilde{E}(\alpha,-1) = |\alpha\rangle\langle -\alpha|. \tag{16.148}$$

Similarly, by using the definitions Eq. (16.135) and (16.102) and the Fourier-transform relation Eq. (16.50), one may write the operator $\widetilde{E}(\alpha,1)$ as the integral

$$\widetilde{E}(\alpha,-1) = \int \prod_i (-d^2\beta_i)\, e^{-(\alpha-\beta)\cdot(\alpha^*-\beta^*)}\, |\beta\rangle\langle -\beta| \,. \tag{16.149}$$

By performing the integration and referring to the explicit formula Eq. (16.31), we may show that the operator $\widetilde{E}(\alpha,1)$ is the dyadic of the eigenstates (16.27) of the creation operators $|\alpha\rangle'$:

$$\widetilde{E}(\alpha,1) = |\alpha\rangle'\,'\langle -\alpha| \,. \tag{16.150}$$

16.10
Quasiprobability Distributions

Among the most important of the foregoing expansions is the expansion Eq. (16.138) when the operator G is the density operator ρ,

$$\rho = \int d^2\xi \, \mathrm{Tr}[\rho D(\xi,s)]\, E(-\xi,-s) \,, \tag{16.151}$$

in which case the trace is the s-ordered characteristic function $\chi(\xi,s)$,

$$\rho = \int d^2\xi \, \chi(\xi,s)\, E(-\xi,-s) \,. \tag{16.152}$$

We may define the s-ordered quasiprobability distribution $W(\alpha,s)$ as the Fourier transform of the s-ordered characteristic function $\chi(\xi,s)$,

$$W(\alpha,s) = \int d^2\xi \, \exp\left(\sum_n (\alpha_n \xi_n^* - \xi_n \alpha_n^*)\right) \chi(\xi,s) \,. \tag{16.153}$$

It follows now from the expansion Eq. (16.146) that the s-ordered quasiprobability distribution $W(\alpha,s)$ is the weight function for the density operator ρ in the expansion

$$\rho = \int d^2\alpha \, W(\alpha,s)\, \widetilde{E}(\alpha,-s) \,. \tag{16.154}$$

Because density operators must be physical operators, $W(\alpha,s)$ like $\chi(\xi,s)$ is an even function of even parity,

$$W(-\alpha,s) = W(\alpha,s) \,. \tag{16.155}$$

When the density operator possesses the additional phase symmetry Eq. (16.82), then both $\chi(\xi,s)$ and $W(\alpha,s)$ are invariant under the rotation of all the variables α_i by the same angle θ,

$$W(e^{i\theta}\alpha,s) = W(\alpha,s) \quad \text{and} \quad \chi(e^{i\theta}\xi,s) = \chi(\xi,s) \,. \tag{16.156}$$

The functions $W(\alpha,s)$ for different values of the order parameter s are intimately related to one another because the characteristic functions obey the identity

$$\chi(\xi,s) = \exp\left(\frac{s}{2}\xi^* \cdot \xi\right)\chi(\xi) = \exp\left(\frac{s-t}{2}\xi^* \cdot \xi\right)\chi(\xi,t). \tag{16.157}$$

The function $W(\alpha,s)$ is therefore the Fourier transform of the product of $\exp\{[(s-t)/2]\xi^* - \xi\}$ with the characteristic function $\chi(\xi,t)$

$$W(\alpha,s) = \int d^2\xi \exp\left(\sum_n (\alpha_n \xi_n^* - \xi_n \alpha_n^*)\right) \exp\left(\frac{s-t}{2}\xi^* \cdot \xi\right)\chi(\xi,t). \tag{16.158}$$

The Fourier transform of the characteristic function $\chi(\xi,t)$ is $W(\alpha,t)$, while that of $\exp([(s-t)/2]\xi^* - \xi)$ according to Eq. (16.50) is

$$\int d^2\xi\, e^{\sum_n (\gamma_n \xi_n^* - \xi_n \gamma_n^*)} e^{[(s-t)/2]\xi^* \cdot \xi} = \prod_n \left[\frac{(t-s)}{2} e^{[2/(t-s)]\gamma_n \gamma_n^*}\right]. \tag{16.159}$$

The convolution theorem Eq. (16.93) now gives $W(\alpha,s)$ as

$$W(\alpha,s) = \int \prod_j \left[\frac{t-s}{2} d^2\beta_j\right] \exp\left[\frac{2}{t-s}\sum_i (\alpha_i - \beta_i)(\alpha_i^* - \beta_i^*)\right] W(\beta,t). \tag{16.160}$$

A useful example of $W(\alpha,s)$ is the function

$$W(\alpha,-1) = \int d^2\xi \exp\left(\sum_n (\alpha_n \xi_n^* - \xi_n \alpha_n^*)\right)\chi(\xi,-1), \tag{16.161}$$

which according to Eq. (16.132) is the Fourier transform

$$W(\alpha,-1) = \int d^2\xi\, d^2\beta \exp\left[\sum_n [(\alpha_n - \beta_n)\xi_n^* - \xi_n(\alpha_n^* - \beta_n^*)]\right] \langle \beta|\rho|-\beta\rangle. \tag{16.162}$$

By using the δ-function identity Eq. (16.86), we see that this expression reduces to

$$W(\alpha,-1) = \int d^2\beta\, \delta(\alpha - \beta) \langle \beta|\rho|-\beta\rangle = \langle \alpha|\rho|-\alpha\rangle. \tag{16.163}$$

This function is the fermionic analog of the function $Q(\beta) = \langle \beta|\rho|\beta\rangle$ which is often used to represent the density operator ρ in terms of the bosonic coherent states $|\beta\rangle$. It is the weight function that gives the mean values of antinormally ordered products of creation and annihilation operators in terms of integrals of the corresponding products of Grassmann numbers.

16.11
Mean Values of Operators

We shall here be concerned with computing the mean values of the products of s-ordered monomials,

$$\prod_i \left\{ (a_i^\dagger)^{n_i} a_i^{m_i} \right\}_s, \tag{16.164}$$

in which the exponents n_i and m_i take the value 0 or 1. The ordering of the modes labeled by the index i is arbitrary but fixed. We shall show that we may express the mean values of such products of monomials as integrals of the s-ordered weight function $W(\boldsymbol{\alpha}, s)$ multiplied by the monomials in the same order. By using the definition Eq. (16.153) of $W(\boldsymbol{\alpha}, s)$, we may write these integrals in the form

$$\int d^2\boldsymbol{\alpha} \prod_i (\alpha_i^*)^{n_i} \alpha_i^{m_i} W(\boldsymbol{\alpha}, s) = \int d^2\boldsymbol{\alpha} \prod_i (\alpha_i^*)^{n_i} \alpha_i^{m_i}$$
$$\times \int d^2\boldsymbol{\xi} \exp\left(\sum_j (\alpha_j \xi_j^* - \xi_j \alpha_j^*) \right) \chi(\boldsymbol{\xi}, s). \tag{16.165}$$

It is now easy to write the monomial as a multiple derivative,

$$\int d^2\boldsymbol{\alpha} \prod_i (\alpha_i^*)^{n_i} \alpha_i^{m_i} W(\boldsymbol{\alpha}, s) = \int d^2\boldsymbol{\alpha} \, d^2\boldsymbol{\xi}$$
$$\times \prod_i \left[\frac{\partial^{n_i}}{\partial(-\xi_i)^{n_i}} e^{-\xi_i \alpha_i^*} \frac{\partial^{m_i}}{\partial(-\xi_i^*)^{m_i}} e^{-\xi_i^* \alpha_i} \right] \chi(\boldsymbol{\xi}, s). \tag{16.166}$$

On using our formula Eq. (16.54) for integration by parts, we have

$$\int d^2\boldsymbol{\alpha} \prod_i (\alpha_i^*)^{n_i} \alpha_i^{m_i} W(\boldsymbol{\alpha}, s) = \int d^2\boldsymbol{\xi} \, d^2\boldsymbol{\alpha} \exp\left(\sum_j (\alpha_j \xi_j^* - \xi_j \alpha_j^*) \right)$$
$$\times \prod_i \left[\frac{\partial^{n_i}}{\partial(\xi_i)^{n_i}} \frac{\partial^{m_i}}{\partial(\xi_i^*)^{m_i}} \right] \chi(\boldsymbol{\xi}, s) \tag{16.167}$$

in which we recognize the δ-function formula Eq. (16.87) which gives

$$\int d^2\boldsymbol{\alpha} \prod_i (\alpha_i^*)^{n_i} \alpha_i^{m_i} W(\boldsymbol{\alpha}, s) = \int d^2\boldsymbol{\xi} \, \delta(\boldsymbol{\xi}) \prod_i \frac{\partial^{n_i}}{\partial(\xi_i)^{n_i}} \frac{\partial^{m_i}}{\partial(\xi_i^*)^{m_i}} \chi(\boldsymbol{\xi}, s) \tag{16.168}$$

$$= \prod_i \frac{\partial^{n_i}}{\partial(\xi_i)^{n_i}} \frac{\partial^{m_i}}{\partial(\xi_i^*)^{m_i}} \chi(\boldsymbol{\xi}, s) \bigg|_{\boldsymbol{\xi}=0} \tag{16.169}$$

$$= \text{Tr}\left[\rho \prod_i \frac{\partial^{n_i}}{\partial(\xi_i)^{n_i}} \frac{\partial^{m_i}}{\partial(\xi_i^*)^{m_i}} \right.$$
$$\left. \times \left(1 + \xi_i a_i^\dagger + \xi_i^* a_i + \xi_i^* \xi_i \left\{ a_i^\dagger a_i \right\}_s \right) \bigg|_{\boldsymbol{\xi}=0} \right]. \tag{16.170}$$

If we recall the definitions of s ordering in Eqns. (16.121) and (16.122), we then find

$$\int d^2\alpha \prod_i (\alpha_i^*)^{n_i} \alpha_i^{m_i} W(\boldsymbol{\alpha},s) = \text{Tr}\left[\rho \prod_i \{(a_i^\dagger)^{n_i} a_i^{m_i}\}_s\right] . \tag{16.171}$$

In particular, by taking $n_i = m_i = 0$, we see that the weight function $W(\boldsymbol{\alpha},s)$ is normalized,

$$\int d^2\alpha\, W(\boldsymbol{\alpha},s) = \text{Tr}\,\rho = 1 . \tag{16.172}$$

16.12
P Representation

Of the representations Eq. (16.154) for the density operator ρ, by far the most important is the one for $s = 1$ with the normally ordered weight function $P(\boldsymbol{\alpha}) = W(\boldsymbol{\alpha},1)$. By Eq. (16.148) it takes the simple form

$$\rho = \int d^2\alpha\, P(\boldsymbol{\alpha}) |\alpha\rangle \langle -\alpha| , \tag{16.173}$$

which recalls the P representation[2,3] for boson fields. Since the function $P(\boldsymbol{\alpha})$ is even, we may also write

$$\rho = \int d^2\alpha\, P(\boldsymbol{\alpha}) |-\alpha\rangle \langle \alpha| . \tag{16.174}$$

Because Grassmann integration is differentiation, the fermionic P representation is not affected by the mathematical limitations[2–4,7] that restricted somewhat the use of the bosonic P representation.

The P representation may be used directly to compute the mean values of normally ordered products

$$\text{Tr}\left[\rho a_k^{\dagger n} a_l^m\right] = \int d^2\alpha\, P(\boldsymbol{\alpha}) \langle \alpha | (a_k^\dagger)^n a_l^m | \alpha \rangle$$
$$= \int d^2\alpha\, P(\boldsymbol{\alpha}) (\alpha_k^*)^n \alpha_l^m . \tag{16.175}$$

This extremely useful relation is just a special case of Eq. (16.171) for $s = 1$.

Since the operator $\widetilde{E}(\boldsymbol{\alpha},1)$ is the dyadic Eq. (16.150) of the eigenstates of the creation operators, it follows from the expansion Eq. (16.154) that the weight function Eq. (16.163)

$$Q(\boldsymbol{\alpha}) \equiv W(\boldsymbol{\alpha},-1) = \langle \alpha | \rho | -\alpha \rangle \tag{16.176}$$

is the weight function in the representation

$$\rho = \int d^2\alpha\, Q(\boldsymbol{\alpha}) |\alpha\rangle' \,'\langle -\alpha| , \tag{16.177}$$

which affords the simple way of computing the mean values of antinormally ordered products that corresponds to Eq. (16.171) for $s = -1$.

Another use of the weight function $Q(\alpha) = W(\alpha, -1) = \langle \alpha | \rho | -\alpha \rangle$, however, is that it allows us to compute the weight function $P(\alpha) = W(\alpha, 1)$ of the P representation as the simple convolution

$$P(\alpha) = \int \prod_m (-d^2 \beta_m) \exp\left[-\sum_n (\alpha_n - \beta_n)(\alpha_n^* - \beta_n^*)\right] \langle \beta | \rho | -\beta \rangle , \qquad (16.178)$$

as follows from the general convolution formula Eq. (16.160) with $s = 1$ and $t = -1$. Although the analogous relation for bosons often is singular[2–4,7], this result holds for all fermionic density operators ρ.

16.13
Correlation Functions for Fermions

A principal use of the P representation for bosonic fields has been the evaluation of the normally ordered correlation functions, which play an important role in the theory of coherence and of the statistics of photon-counting experiments[2]. The analogously defined correlation functions for fields of fermionic atoms can be shown to play a similar role in the description of atom-counting experiments[8]. If we use $\psi(x)$ to denote the positive-frequency part of the Fermi field as a function of a space-time variable x, then the first two of these correlation functions may be defined as

$$G^{(1)}(x, y) = \text{Tr}\left[\rho \psi^\dagger(x) \psi(y)\right] , \qquad (16.179)$$

$$G^{(2)}(x_1, x_2, y_2, y_1) = \text{Tr}\left[\rho \psi^\dagger(x_1) \psi^\dagger(x_2) \psi(y_2) \psi(y_1)\right] . \qquad (16.180)$$

The n-th-order correlation function is

$$G^{(n)}(x_1, \ldots, x_n, y_n, \ldots, y_1) = \text{Tr}\left[\rho \psi^\dagger(x_1) \ldots \psi^\dagger(x_n) \psi(y_n) \ldots \psi(y_1)\right] . \qquad (16.181)$$

If we expand the positive-frequency part of the Fermi field in terms of its mode functions $\phi_k(x)$ as

$$\psi(x) = \sum_k a_k \phi_k(x) , \qquad (16.182)$$

then its eigenvalue in the coherent state $|\alpha\rangle$,

$$\psi(x)|\alpha\rangle = \varphi(x)|\alpha\rangle , \qquad (16.183)$$

is the Grassmann field

$$\varphi(x) = \sum_k \alpha_k \phi_k(x) \qquad (16.184)$$

in which the annihilation operators in Eq. (16.182) are replaced by the Grassmann variables $\alpha = \{\alpha_k\}$.

We may use the P representation to evaluate the n-th-order correlation function $G^{(n)}$ as the integral

$$G^{(n)}(x_1, \ldots, x_n, y_n, \ldots, y_1)$$
$$= \int d^2\alpha\, P(\alpha) \left\langle \alpha \left| \psi^\dagger(x_1) \ldots \psi^\dagger(x_n) \psi(y_n) \ldots \psi(y_1) \right| \alpha \right\rangle \quad (16.185)$$
$$= \int d^2\alpha\, P(\alpha) \varphi^\dagger(x_1) \ldots \varphi^\dagger(x_n) \varphi(y_n) \ldots \varphi(y_1) \,. \quad (16.186)$$

16.14
Chaotic States of the Fermion Field

The reduced density operator for a single mode of the fermion field can be represented by a 2×2 matrix for the states with occupation numbers 0 and 1. If the matrix is diagonal, it is specified completely by the mean number of quanta $\langle n \rangle$ in the mode. The density operator for the k-th mode, in other words, must take the form

$$\rho_k = (1 - \langle n_k \rangle) |0\rangle\langle 0| + \langle n_k \rangle |1\rangle\langle 1| \,. \quad (16.187)$$

We shall speak of this density operator as representing a chaotic state of the k-th mode. A chaotic state of the entire field will then be represented as a direct product of such density operators for all the modes of the field,

$$\rho_c = \prod_k \rho_k \,. \quad (16.188)$$

It is specified by the complete set of mean occupation numbers $\{\langle n_k \rangle\}$.

The total number of fermions, $N = \sum_k a_k^\dagger a_k$, present in chaotic states will in general be indefinite. Indeed it is easily seen that in the state specified by Eq. (16.188) we have

$$\langle N^2 \rangle - \langle N \rangle^2 = \sum_k \langle n_k \rangle (1 - \langle n_k \rangle) \quad (16.189)$$

so that N cannot be fixed unless all the $\langle n_k \rangle$ take the value 0 or 1. The indefiniteness of the number of particles present is a feature that the chaotic states of the fermion and boson fields have in common. For sufficiently large values of N, however, the fluctuations of $N/\langle N \rangle$ may be quite small so the specification of N in these relative terms may be quite precise. Fluctuations of this type in the number of particles present are a familiar property of the grand-canonical ensemble in statistical mechanics, and that ensemble, as we shall see, represents a special class of chaotic states.

The single-mode density operator Eq. (16.187) can also be written as

$$\rho_k = (1 - \langle n_k \rangle) \left(\frac{\langle n_k \rangle}{1 - \langle n_k \rangle} \right)^{a_k^\dagger a_k} (|0\rangle\langle 0| + |1\rangle\langle 1|) \quad (16.190)$$

in which we recognize the unit operator I_k for the subspace of the k-th mode. Within this subspace we have

$$\rho_k = (1 - \langle n_k \rangle) \left(\frac{\langle n_k \rangle}{1 - \langle n_k \rangle} \right)^{a_k^\dagger a_k}. \tag{16.191}$$

This expression can be used quite directly to evaluate the weight function $W(\alpha, -1) = Q(\alpha)$.

We first note that for any real number v

$$v^{a^\dagger a} |\alpha\rangle = e^{\alpha \alpha^* (1 - v^2)/2} |\alpha v\rangle, \tag{16.192}$$

so that we have

$$\left\langle \alpha \left| v^{a^\dagger a} \right| - \alpha \right\rangle = e^{\alpha \alpha^* (1 - v^2)/2} \langle \alpha | -\alpha v \rangle = e^{\alpha \alpha^* (1 + v)}. \tag{16.193}$$

Then if we let $v = \langle n_k \rangle / (1 - \langle n_k \rangle)$, we see that

$$\langle \alpha_k | \rho_k | -\alpha_k \rangle = (1 - \langle n_k \rangle) \exp\left(\frac{\alpha_k \alpha_k^*}{1 - \langle n_k \rangle} \right) \tag{16.194}$$

and

$$Q(\alpha) = W(\alpha, -1) = \prod_k \langle \alpha_k | \rho_k | -\alpha_k \rangle$$

$$= \prod_k (1 - \langle n_k \rangle) \exp\left(\frac{\alpha_k \alpha_k^*}{1 - \langle n_k \rangle} \right). \tag{16.195}$$

This product is the weight function appropriate to averaging antinormally ordered operator products in chaotic states.

We may find the weight functions corresponding to all the other ordering schemes by using the convolution Eq. (16.160) with $t = -1$ and carrying out the required integration with sufficient attention to the implicit minus signs. The result for the k-th mode is

$$W_k(\alpha_k, s) = -\frac{s + 2\langle n_k \rangle - 1}{2} \exp\left(-\frac{2\alpha_k \alpha_k^*}{s + 2\langle n_k \rangle - 1} \right), \tag{16.196}$$

and the weight function for the multimode field is simply the product

$$W(\alpha, s) = \prod_k W_k(\alpha_k, s). \tag{16.197}$$

Thus the function $W_k(\alpha_k, 0)$, which is analogous to the Wigner function for boson fields, is given by

$$W_k(\alpha_k, 0) = -(\langle n_k \rangle - \tfrac{1}{2}) \exp\left(-\frac{\alpha_k \alpha_k^*}{\langle n_k \rangle - \tfrac{1}{2}} \right), \tag{16.198}$$

and the function $W_k(\alpha_k, 1)$, which is the analog of the function $P_k(\alpha_k)$ for boson fields, is

$$W_k(\alpha_k, 1) \equiv P_k(\alpha_k) = -\langle n_k \rangle \exp\left(-\frac{\alpha_k \alpha_k^*}{\langle n_k \rangle}\right). \tag{16.199}$$

The latter result is a particularly useful one since there are many physical contexts that call for the averaging of normally ordered products of annihilation and creation operators. For chaotic fields one may calculate all such averages as Grassmann integrals by making use of the fermionic P representation with $P(\alpha)$ given by Eq. (16.199).

The minus signs in front of the expressions (16.198) and (16.199) may be somewhat surprising since these functions are the fermionic analogs of quasiprobability densities that are predominantly positive for boson fields. It is worth pointing out, therefore, that these signs result from our convention that defines $d^2\alpha$ as $d\alpha^* \, d\alpha$. Had we chosen the differential instead to be $d\alpha \, d\alpha^*$, the signs would have been positive.

For a chaotically excited boson field, the P representation expresses the density operator as a Gaussian integral of a diagonal coherent-state dyadic. For fermion fields the corresponding expression of ρ_k for a single mode is

$$\rho_k = -\langle n_k \rangle \int d^2\alpha_k \, e^{-\alpha_k \alpha_k^*/\langle n_k \rangle} |\alpha_k\rangle \langle -\alpha_k|. \tag{16.200}$$

According to Eq. (16.190), the density operator ρ_k can also be written as a sum over the m-fermion states as

$$\rho_k = (1 - \langle n_k \rangle) \sum_{m_k=0}^{1} \left(\frac{\langle n_k \rangle}{1 - \langle n_k \rangle}\right)^{m_k} |m_k\rangle \langle m_k|. \tag{16.201}$$

What we have shown, in effect, is that the two expressions are identical and that statistical averages can be evaluated by means of Gaussian integrations for fermions as well as for bosons. The multimode density operator is represented, of course, by the product of the single-mode density operators, $\rho = \prod_k \rho_k$.

Fields in thermal equilibrium with a suitable particle reservoir represent particular examples of the kind of chaotic excitation we have been describing. If it is appropriate to describe such fields by means of the grand-canonical ensemble, then their overall density operator may be written as

$$\rho = \frac{1}{\Xi(\beta, \mu)} e^{-\beta(H - \mu N)}, \tag{16.202}$$

where $\beta = 1/k_B T$, μ is the chemical potential, H is the Hamiltonian for the system, N is the particle number, and the normalizing factor $\Xi(\beta, \mu)$ is the grand partition function. For a field with dynamically independent mode functions labeled by the index k, we can write

$$H = \sum_k \varepsilon_k a_k^\dagger a_k, \qquad N = \sum_k a_k^\dagger a_k, \tag{16.203}$$

where ε_k is the energy of a particle in the k-th mode.

Under these circumstances the equilibrium number of fermions in the k-th mode is

$$\langle n_k \rangle = \frac{1}{e^{\beta(\varepsilon_k - \mu)} + 1} \,. \tag{16.204}$$

In this case the ratio $\langle n_k \rangle / (1 - \langle n_k \rangle)$ is simply the generalized Boltzmann factor

$$\frac{\langle n_k \rangle}{1 - \langle n_k \rangle} = e^{\beta(\varepsilon_k - \mu)} \,. \tag{16.205}$$

We then find that the product of the ρ_k given by Eq. (16.200) is precisely equal to the grand-canonical density operator Eq. (16.202),

$$\int \prod_k \left(-\langle n_k \rangle \, d^2\alpha_k \, e^{-\alpha_k \alpha_k^* / \langle n_k \rangle} \right) |\alpha\rangle \langle -\alpha| = \frac{1}{\Xi(\beta, \mu)} e^{-\beta(H - \mu N)} \,. \tag{16.206}$$

There are many examples of thermal equilibria for which the P representation on the left should furnish a useful computational tool.

16.15
Correlation Functions for Chaotic Field Excitations

We have introduced a succession of normally ordered correlation functions $G^{(n)}(x_1, \ldots, x_n, y_n, \ldots, y_1)$ in Sect. 16.13 and shown how they can be expressed as integrals over the Grassmann variables $\alpha = \{\alpha_k\}$. For the case of chaotic fields, the appropriate weight function is

$$P(\alpha) = \prod_k P_k(\alpha_k) \,, \tag{16.207}$$

the product of the Gaussian functions in Eq. (16.199). The first-order correlation function is thus given by

$$G^{(1)}(x, y) = \int \prod_k \left(-\langle n_k \rangle \, d^2\alpha_k \, e^{-\alpha_k \alpha_k^* / \langle n_k \rangle} \right) \langle \alpha | \psi^\dagger(x) \psi(y) | \alpha \rangle \,. \tag{16.208}$$

The fields ψ and $\psi^\dagger$ may now be replaced by their Grassmann field eigenvalues defined by Eqns. (16.183) and (16.184). Their product is a quadratic form in the variables α_k and α_k^*, which is easily integrated:

$$G^{(1)}(x, y) = \int \prod_k \left(-\langle n_k \rangle \, d^2\alpha_k \, e^{-\alpha_k \alpha_k^* / \langle n_k \rangle} \right) \sum_{l,m} \alpha_l^* \alpha_m \phi_l^\dagger(x) \phi_m(y)$$

$$= \sum_k \langle n_k \rangle \, \phi_k^\dagger(x) \phi_k(y) \,. \tag{16.209}$$

To find the higher-order correlation functions, we can make use of a species of generating functional. We first define the Grassmann fields

$$\zeta(x) = \sum_k \beta_k \phi_k(x) \,, \tag{16.210}$$

$$\eta(y) = \sum_k \gamma_k \phi_k(y) \,, \tag{16.211}$$

and use them to construct the normally ordered expectation value

$$\Gamma[\zeta,\eta] \equiv \text{Tr}\left[\rho \exp\left(\int \zeta(x)\psi^\dagger(x)\,d^4x\right) \exp\left(\int \psi(y)\eta^*(y)\,d^4y\right)\right]. \quad (16.212)$$

If we form the variational derivative of Γ with respect to $\zeta(x_1)$ from the left and with respect to $\eta^*(y_1)$ from the right, subsequently setting ζ and η to zero, then we find an alternative expression for the first-order correlation function,

$$\frac{\delta}{\delta_L \zeta(x_1)} \frac{\delta}{\delta_R \eta^*(y_1)} \Gamma \bigg|_{\zeta=\eta=0} = \text{Tr}\left[\rho \psi^\dagger(x_1)\psi(y_1)\right] = G^{(1)}(x_1, y_1), \quad (16.213)$$

where left and right differentiation have been indicated explicitly in the subscripts.

It is evident then that one may generate all of the higher-order correlation functions by performing further differentiations,

$$G^{(n)}(x_1,\ldots,x_n, y_n,\ldots,y_1) = \frac{\delta}{\delta_L \zeta(x_1)} \cdots \frac{\delta}{\delta_L \zeta(x_n)} \frac{\delta}{\delta_R \eta^*(y_n)} \cdots \frac{\delta}{\delta_R \eta^*(y_1)} \Gamma \bigg|_{\zeta=\eta=0}. \quad (16.214)$$

To evaluate the generating functional Γ for a chaotic field, we make use of the orthonormality of the mode functions ϕ_k and then carry out the Grassmann integration

$$\Gamma = \int \prod_k \left(-\langle n_k\rangle \, d^2\alpha_k\, e^{-\alpha_k \alpha_k^*/\langle n_k\rangle}\right) \exp\left(\sum_l (\beta_l \alpha_l^* + \alpha_l \gamma_l^*)\right)$$

$$= \prod_k (1 + \langle n_k\rangle \beta_k \gamma_k^*) = \exp\left(\sum_k \langle n_k\rangle \beta_k \gamma_k^*\right)$$

$$= \exp\left(\int \zeta(x)\, G^{(1)}(x,y)\, \eta^*(y)\, d^4x\, d^4y\right). \quad (16.215)$$

If we begin performing the variational differentiations to find the second-order correlation function, we may write

$$\frac{\delta}{\delta_R \eta^*(y_2)} \frac{\delta}{\delta_R \eta^*(y_1)} \Gamma \bigg|_{\eta=0} = \int \zeta(x)\, G^{(1)}(x,y_2)\, d^4x \int \zeta(x')\, G^{(1)}(x',y_1)\, d^4x'. \quad (16.216)$$

We then find

$$G^{(2)}(x_1 x_2 y_2 y_1) = \frac{\delta}{\delta_L \zeta(x_1)} \frac{\delta}{\delta_L \zeta(x_2)} \int \zeta(x)\, G^{(1)}(x,y_2)\, d^4x \int \zeta(x')\, G^{(1)}(x',y_1)\, d^4x' \quad (16.217)$$

and since $\zeta(x)$ and $\zeta(x')$ anticommute,

$$G^{(2)}(x_1, x_2, y_2, y_1) = G^{(1)}(x_1, y_1)\, G^{(1)}(x_2, y_2) - G^{(1)}(x_1, y_2)\, G^{(1)}(x_2, y_1). \quad (16.218)$$

The generalization to *n*-th order is immediate. It expresses the *n*-th-order correlation function for chaotic fields as a sum of products of first-order correlation functions with permuted arguments,

$$G^{(n)}(x_1,\ldots,x_n,y_n,\ldots,y_1) = \sum_P (-1)^P \prod_{j=1}^n G^{(1)}(x_j,y_{P_j}) . \qquad (16.219)$$

This expression is summed over the $n!$ permutations of the indices $1,\ldots,n$. The factor $(-1)^P$ is the parity of the permutation, and the index P_j is the index that replaces j in the permutation.

The expression of the *n*-th-order correlation function in terms of first-order correlation functions is characteristic of chaotic fields. Such fields are completely specified by the set of mean occupation numbers $\langle n_k \rangle$, and these are already contained in the first-order correlation function.

16.16
Fermion-Counting Experiments

The use of photon-counting techniques has for many years been the most direct means of investigating the statistical properties of light beams. Experiments of this type began with that of Hanbury Brown and Twiss[9] in 1956 and expanded greatly in scope with the development of the laser. The theory[3] underlying these experiments is based on the evaluation of quantum-mechanical expectation values of normally ordered products of electromagnetic field operators. The coherent states of the field[2] thus play a special role in the formulation of that theory. The application of the theory, furthermore, extends to boson fields of much more general sorts, including, for example, beams of heavy atoms[8].

In the case of the electromagnetic field, it has been shown[3] that the probability of detecting n photons in a given interval of time can be expressed as the n-th derivative with respect to a parameter λ of a certain generating function $Q(\lambda)$,

$$p(n) = \frac{(-1)^n}{n!} \frac{d^n}{d\lambda^n} Q(\lambda) \bigg|_{\lambda=1} . \qquad (16.220)$$

The generating function $Q(\lambda)$ for the electromagnetic field is the expectation value of a normally ordered exponential function of the form

$$Q(\lambda) = \mathrm{Tr}\left(\rho : e^{-\lambda I} :\right) , \qquad (16.221)$$

in which the symbols $: \ :$ stand for normal ordering, and the operator I is a space-time integral of the product of the positive-frequency and negative-frequency parts of the field, $E^{(+)}$ and $E^{(-)}$, respectively.

For the case of fermion fields, it can easily be shown[8] that the probability of counting n fermions in a given interval of time falls into precisely the same general form. In the simplest instance, for detectors that respond to the density rather than the flux of the particles, the integral I takes the form

$$I = \kappa \int \psi^\dagger(\mathbf{r},t)\psi(\mathbf{r},t)\,d^3\mathbf{r}\,dt, \qquad (16.222)$$

where the constant κ is a measure of the sensitivity of the counter and the integration is carried out over the counting-time interval and over the volume being observed.

To obtain the expectation value of the normally ordered exponential function in Eq. (16.221), we may use the P representation for the density operator ρ. In that case the field operators $\psi(\mathbf{r},t)$ and $\psi^\dagger(\mathbf{r},t)$ are, in effect, always applied to their eigenstates, coherent states such as $|\alpha\rangle$ and $\langle\alpha|$. They can then be replaced by their Grassmann-field eigenvalue functions defined by Eq. (16.184) and its adjoint, so that we have

$$Q(\lambda) = \int d^2\alpha\, P(\alpha)\, e^{-\lambda \mathcal{J}}, \qquad (16.223)$$

where

$$\mathcal{J} = \kappa \int \varphi^*(\mathbf{r},t)\,\varphi(\mathbf{r},t)\,d^3\mathbf{r}\,dt. \qquad (16.224)$$

The expression $\mathcal{J}$ is a quadratic form that we can write as

$$\mathcal{J} = \sum_{k,k'} \alpha_k^* B_{kk'} \alpha_{k'}, \qquad (16.225)$$

so the evaluation of the generating function $Q(\lambda)$ reduces to the calculation of the integral

$$Q(\lambda) = \int d^2\alpha\, P(\alpha)\, \exp\!\left(-\lambda \sum_{k,k'} \alpha_k^* B_{kk'} \alpha_{k'}\right), \qquad (16.226)$$

in which the normal ordering symbols are no longer necessary because of the simple anticommutation properties of the Grassmann variables α_k.

For the case of the chaotic fields defined in Sect. 16.14 this integral takes the form

$$Q(\lambda) = \int \prod_k \left(-\langle n_k\rangle\, d^2\alpha_k\, e^{-\alpha_k \alpha_k^*/\langle n_k\rangle}\right) \exp\!\left(-\lambda \sum_{k,k'} \alpha_k^* B_{kk'} \alpha_{k'}\right). \qquad (16.227)$$

If we define a new set of variables $\beta_k = \alpha_k/\sqrt{\langle n_k\rangle}$, we find according to the rule Eq. (16.48) that the integral can be written as

$$Q(\lambda) = \int \prod_k (-d^2\beta_k)\, \exp\!\left(\sum_{k,k'} \beta_k^*(\delta_{kk'} - \lambda M_{kk'})\beta_{k'}\right), \qquad (16.228)$$

where the matrix M is

$$M_{kk'} = \sqrt{\langle n_k\rangle}\, B_{kk'}\, \sqrt{\langle n_{k'}\rangle}. \qquad (16.229)$$

A unitary linear transformation on the variables β_k can then be used to diagonalize the quadratic form in brackets. If the eigenvalues of the matrix $1 - \lambda M$ are μ_l, then the integral is easily seen, according to the formula Eq. (16.50) for $\alpha = 0$, to be

$$Q(\lambda) = \prod_l \mu_l = \det(1 - \lambda M) . \tag{16.230}$$

This result may be used directly to find the various probabilities given by Eq. (16.220). It contrasts quite interestingly with the generating function for boson-counting distributions, which with closely corresponding definitions takes the form[3]

$$Q_B(\lambda) = \frac{1}{\det(1 + \lambda M)} . \tag{16.231}$$

16.17
Some Elementary Examples

16.17.1
The Vacuum State

For the density operator

$$\rho = |0 \cdots 0\rangle \langle 0 \cdots 0| , \tag{16.232}$$

which represents the multimode vacuum state, the normally ordered characteristic function $\chi_N(\xi)$ is

$$\chi_N(\xi) = \operatorname{Tr}\left[\rho \exp\left(\sum_n \xi_n a_n^\dagger\right) \exp\left(-\sum_n a_n \xi_n^*\right)\right]$$
$$= \left\langle 0 \cdots 0 \left| \exp\left(\sum_n \xi_n a_n^\dagger\right) \exp\left(-\sum_n a_n \xi_n^*\right) \right| 0 \cdots 0 \right\rangle$$
$$= 1 . \tag{16.233}$$

The weight function of the P representation is then

$$P(\alpha) = \int d^2\xi \exp\left(\sum_i (\alpha_i \xi_i^* - \xi_i \alpha_i^*)\right) = \delta(\alpha) . \tag{16.234}$$

The mean values of the normally ordered products of creation and annihilation operators all vanish,

$$\operatorname{Tr}\left[\rho \prod_i (a_i^\dagger)^{n_i} a_i^{m_i}\right] = \int d^2\alpha \prod_i (\alpha_i^*)^{n_i} \alpha_i^{m_i} \delta(\alpha) = 0 , \tag{16.235}$$

except for the trace

$$\operatorname{Tr}[\rho] = \int d^2\alpha \, \delta(\alpha) = 1 . \tag{16.236}$$

The general weight function $W(\alpha, s)$ of the vacuum is given by

$$W(\alpha, s) = \frac{1}{2}(1-s) \exp\left(\frac{2\alpha \cdot \alpha^*}{1-s}\right). \qquad (16.237)$$

16.17.2
A Physical Two-Mode Density Operator

Let us consider the most general physical two-mode fermionic density operator

$$\rho = r|00\rangle\langle00| + u|10\rangle\langle10| + v|01\rangle\langle01| + w|10\rangle\langle01| + w^*|01\rangle\langle10|$$
$$+ x|00\rangle\langle11| + x^*|11\rangle\langle00| + t|11\rangle\langle11| \qquad (16.238)$$

in which $|10\rangle = a_1^\dagger|00\rangle$, $|11\rangle = a_2^\dagger a_1^\dagger|00\rangle$, etc., and the Latin letters r, t, u, and v represent non-negative real numbers that sum to unity, while x and w may be complex. The mean values of the s-ordered products are

$$\mathrm{Tr}\,\rho = r + u + v + t = 1, \qquad (16.239)$$

$$\mathrm{Tr}\,\rho\{a_1^\dagger a_1\}_s = u + t + \tfrac{1}{2}(s-1), \qquad (16.240)$$

$$\mathrm{Tr}\,\rho\{a_2^\dagger a_2\}_s = v + t + \tfrac{1}{2}(s-1), \qquad (16.241)$$

$$\mathrm{Tr}\,\rho\{a_2^\dagger a_1\}_s = w, \qquad (16.242)$$

$$\mathrm{Tr}\,\rho\{a_1^\dagger a_2\}_s = w^*, \qquad (16.243)$$

$$\mathrm{Tr}\,\rho\{a_1 a_2\}_s = x^*, \qquad (16.244)$$

$$\mathrm{Tr}\,\rho\{a_1^\dagger a_2^\dagger\}_s = -x, \qquad (16.245)$$

$$\mathrm{Tr}\,\rho\{a_2^\dagger a_1^\dagger a_1 a_2\}_s = st + \tfrac{1}{2}(s-1)(u+v) + \tfrac{1}{4}(s-1)^2. \qquad (16.246)$$

If the fermion number N commutes with the density operator ρ, then $x = x^* = 0$. The normally ordered characteristic function $\chi_N(\xi)$ is

$$\chi_N(\xi) = \mathrm{Tr}\left[\rho(1 + \xi_1 a_1^\dagger - a_1 \xi_1^* + \xi_1^* \xi_1 a_1^\dagger a_1)(1 + \xi_2 a_2^\dagger - a_2 \xi_2^* + \xi_2^* \xi_2 a_2^\dagger a_2)\right]$$
$$= 1 + w\xi_1^*\xi_2 + w^*\xi_2^*\xi_1 + (u+t)\xi_1^*\xi_1 + (v+t)\xi_2^*\xi_2$$
$$+ x\xi_1\xi_2 + x^*\xi_2^*\xi_1^* + t\xi_1^*\xi_1\xi_2^*\xi_2. \qquad (16.247)$$

Because the density operator ρ is physical, $\chi_N(\xi)$ is even. But unless $x = 0$, the phase transformation $\xi_1 \to e^{i\theta}\xi_1$, $\xi_2 \to e^{i\theta}\xi_2$ changes $\chi_N(\xi)$.

By its definition Eq. (16.153), the weight function $W(\alpha,s)$ is the Fourier transform of the s-ordered characteristic function $\chi(\xi,s) = \exp[\frac{1}{2}(s-1)\xi^*\xi]\chi_N(\xi)$,

$$W(\alpha,s) = \int d^2\xi_1 \, d^2\xi_2 \left[1 + \frac{1}{2}(s-1)(\xi_1^*\xi_1 + \xi_2^*\xi_2) + \frac{1}{4}(s-1)^2 \xi_1^*\xi_1\xi_2^*\xi_2\right]$$

$$\times (1 + \alpha_1\xi_1^* + \alpha_1^*\xi_1 + \alpha_1^*\alpha_1\xi_1^*\xi_1)(1 + \alpha_2\xi_2^* + \alpha_2^*\xi_2 + \alpha_2^*\alpha_2\xi_2^*\xi_2)$$

$$\times \left[1 + w\xi_1^*\xi_2 + w^*\xi_2^*\xi_1 + (u+t)\xi_1^*\xi_1 + (v+t)\xi_2^*\xi_2 + x\xi_1\xi_2 + x^*\xi_2^*\xi_1^* + t\xi_1^*\xi_1\xi_2^*\xi_2\right],$$

and after following the rules from Eqns. (16.40) and (16.42), we find

$$W(\alpha,s) = st + \frac{1}{2}(s-1)(u+v) + \frac{1}{4}(s-1)^2 + w\alpha_2\alpha_1^* + w^*\alpha_1\alpha_2^*$$

$$+ [v+t+\frac{1}{2}(s-1)]\alpha_1^*\alpha_1 + [u+t+\frac{1}{2}(s-1)]\alpha_2^*\alpha_2$$

$$+ x\alpha_1\alpha_2 + x^*\alpha_2^*\alpha_1^* + \alpha_1^*\alpha_1\alpha_2^*\alpha_2 . \quad (16.248)$$

We may now use this weight function to compute the mean values

$$\int d^2\alpha_1 \, d^2\alpha_2 \, W(\alpha,s) = \text{Tr}\,\rho = 1, \quad (16.249)$$

$$\int d^2\alpha_1 \, d^2\alpha_2 \, \alpha_1^*\alpha_1 \, W(\alpha,s) = \text{Tr}\,\rho\{a_1^\dagger a_1\}_s = u + t + \frac{1}{2}(1-s), \quad (16.250)$$

$$\int d^2\alpha_1 \, d^2\alpha_2 \, \alpha_2^*\alpha_2 \, W(\alpha,s) = \text{Tr}\,\rho\{a_2^\dagger a_2\}_s = v + t + \frac{1}{2}(1-s), \quad (16.251)$$

$$\int d^2\alpha_1 \, d^2\alpha_2 \, \alpha_2^*\alpha_1 \, W(\alpha,s) = \text{Tr}\,\rho\{a_2^\dagger a_1\}_s = w, \quad (16.252)$$

$$\int d^2\alpha_1 \, d^2\alpha_2 \, \alpha_1^*\alpha_2 \, W(\alpha,s) = \text{Tr}\,\rho\{a_1^\dagger a_2\}_s = w^*, \quad (16.253)$$

$$\int d^2\alpha_1 \, d^2\alpha_2 \, \alpha_1\alpha_2 \, W(\alpha,s) = \text{Tr}\,\rho\{a_1 a_2\}_s = x^*, \quad (16.254)$$

$$\int d^2\alpha_1 \, d^2\alpha_2 \, \alpha_1^*\alpha_2^* \, W(\alpha,s) = \text{Tr}\,\rho\{a_1^\dagger a_2^\dagger\}_s = -x, \quad (16.255)$$

$$\int d^2\alpha_1 \, d^2\alpha_2 \, \alpha_2^*\alpha_1^*\alpha_1\alpha_2 \, W(\alpha,s) = \text{Tr}\,\rho\{a_2^\dagger a_1^\dagger a_1 a_2\}_s$$

$$= st + \frac{1}{2}(s-1)(u+v) + \frac{1}{4}(s-1)^2, \quad (16.256)$$

which agree with the results from Eqns. (16.239) and (16.246).

With $P(\alpha) = W(\alpha, 1)$ as given by Eq. (16.248) with $s = 1$, we may write the density operator Eq. (16.238) in the form of the fermionic P representation

$$\rho = \int d^2\alpha \, P(\alpha)|-\alpha\rangle\langle\alpha| = \int d^2\alpha \, P(\alpha)|\alpha\rangle\langle-\alpha| . \quad (16.257)$$

References

1. H. C. Stoof, M. Houbiers, C. A. Sackett, R. G. Hulet, *Phys. Rev. Lett.* **76**, 10 (1996); C. J. Myatt, E. A. Burt, R. W. Ghrist, E. A. Cornell, C. E. Wieman, *Phys. Rev. Lett.* **78**, 586 (1997).
2. R. J. Glauber, *Phys. Rev.* **130**, 2529 (1963), reprinted as Chapter 1 in this volume; *Phys. Rev.* **131**, 2766 (1963), reprinted as Sect. 2.9 in this volume.
3. R. J. Glauber, in *Quantum Optics and Electronics*, C. deWitt, A. Blandin, C. Cohen-Tannoudji (Eds.), Gordon & Breach, New York 1965, pp. 65–185; reprinted as Chapter 2 in this volume.
4. K. Cahill, R. J. Glauber, *Phys. Rev.* **177**, 1857 (1969), reprinted as Chapter 9 in this volume; *Phys. Rev.* **177**, 1882 (1969), reprinted as Chapter 10 in this volume.
5. J. Schwinger, *Phys. Rev.* **92**, 1283 (1953).
6. Grassmann algebra is conventionally used in the formulation of the path integral for fermion fields. It is discussed in a number of texts: F. A. Berezin, *The Method of Second Quantization*, Academic Press, New York 1966; S. Weinberg, *The Quantum Theory of Fields, Volume I: Foundations*, Cambridge University Press, Cambridge/England 1995; J. Zinn-Justin, *Quantum Field Theory and Critical Phenomena*, Oxford University Press, New York 1989; Lowell Brown, *Quantum Field Theory*, Cambridge University Press, Cambridge/England 1992.
7. K. Cahill, *Phys. Rev. B* **138**, 1566 (1965); *Phys. Rev.* **180**, 1239 (1969); *Phys. Rev.* **180**, 1244 (1969).
8. R. J. Glauber, *Bull. Am. Phys. Soc.* **41**, 1056 (1996).
9. R. Hanbury Brown, R. Q. Twiss, *Nature* **177**, 27 (1956).

Index

a

amplification   542, 554
– and uncertainty principle   557
amplitude operators   79, 291
– expectation values   219
annihilation operators   4, 36, 185, 198, 217, 292, 434, 472, 581
– eigenstates   199, 207, 208, 596
– properties of eigenstates   209
attenuation   542
autocorrelation function   162
average photon number   100, 103

b

Baker–Hausdorff identity   598
Bayes' theorem   561, 566, 567, 573
Bernoulli distribution   326
Bessel functions   180, 534
blackbody radiation   103, 114
Bloch equation   452, 453
Bloch theorem   104
Bose–Einstein condensation   595
Bose–Einstein statistics   526

c

c-number   471
Cauchy theorem   55
causality   56
central limit theorem   141
chaos   321, 514
chaotic states   242

characteristic function   225, 314, 394, 505
– s-ordered   507, 613
– antinormally ordered   315, 401
– boundedness condition   394
– for system of modes   258
– in Grassmann algebra   612
– normally ordered   234, 392, 438
classical electrodynamics   27
coherence   1, 10, 487, 489
– and monochromaticity   14, 63, 150, 433, 490
– and noise   61, 74
– classical theory   186
– conditions   18, 300, 489
– conditions for first-order   12, 204
– definition   59, 490
– first-order   431, 433
– full   17, 70, 190, 198, 207, 298, 299, 494, 495
– higher-order   205, 494
– higher-order analogs   67
– m-th-order   183, 186, 197
– orders of   183
– range   150
– relativistic   18
– traditional interpretation   135
coherent states   74, 307
– nonorthogonality   307
– normalized   599
coincidence detection   1

Quantum Theory of Optical Coherence. Selected Papers and Lectures. Roy J. Glauber.
Copyright © 2007 WILEY-VCH Verlag GmbH & Co. KGaA, Weinheim
ISBN: 978-3-527-40687-6

coincidence rate 144
– average 140
– measurement 296
– time dependence 152
communication theory 299
commutation relation 331, 333
– of creation and annihilation operators 434, 457, 472
– of electric field components 3
– of position and momentum 253
completeness relation 37
complex signal 28
convolution theorem 101
– fermionic analog 609
cooperation length 533
correlation function 429, 432, 481
– n-th-order 8, 389
– as scalar product 186
– classical 9, 471
– factorization 17, 70, 431
– first-order 30, 77, 145, 201, 480, 516
– for incoherent fields 115
– for stationary light beams 134
– higher-order 189, 487
– measurement 50
– measurement of n-th order 49
– measurement of second order 66
– (n,m)-th-order 483
– normalized 11
– n-th-order 38, 185, 197, 480
– properties 51
correspondence principle 473
creation operators 36, 185, 292, 472
cross correlation function 517

d

delayed coincidence 287, 321
– counting rate 146, 150
delta function 124
– integral representation 315, 507

density operator 77, 92, 391, 427, 497
– associated weight function 384
– chaotic 521
– definition 7
– diagonal matrix elements 124, 314
– expansion in n-quantum states 310
– factorization property 215
– for chaotic states 241, 242
– for the radiation field 437
– Fourier transform 313
– Gaussian 102, 113, 126
– matrix elements 92, 96, 264
– normalization condition 499
– occupation-number representation 203, 207
– physical 608
– projection on coherent states 499
– reduced 233, 445, 459
– stationary 97, 107
– superposition of projection operators 95
– thermal 104
– two-mode fermionic 629
detector
– based on stimulated emission 480
– Compton-effect 481
– quantum-mechanical 28
– single-quantum 23
diffraction 56
displacement operators 82, 108, 336, 505, 597
– antinormally ordered 353
– completeness 340
– matrix elements 378
– multiplication law 336
– normally ordered 353
– orthogonality rule 340
– symmetrically ordered 353

dissipation   542
disturbance field   10
Doppler effect   577
dyadic product   *see* outer product
Dyson–Wick expansion   389

e

electric dipole approximation   39, 46
electric dipole transition   455
electromagnetic spectrum   463
energy conservation   578
ensemble averages   29, 35, 37, 408
– quantum-mechanical   420
entire function   349, 351, 498
entropy   513
– maximum   319
EPR paradox   538–540
ergodicity   31, 35
expectation value   427
– Heisenberg picture   482

f

factorial moments   172
factorization condition   298, 490, 495
– higher order   298
field
– n-th-order coherent   12
– chaotic   408, 522
– chaotically generated   175
– coherent   408
– complex   28
– expansion in orthogonal modes   29
– first-order coherent   199, 204
– Fourier coefficients   469
– ground state   472
– Hamiltonian   27
– monochromatic   200
– operator   471
– polarized   71
– probability distribution   470

– quantum-mechanical characterization   26, 426
– stationary   31, 54
field operators   77, 185
– expectation value   498
– normal ordering   477
– wave equation   3
field vectors   27
Floquet form   585, 588
Floquet solution   586
Fock space   292, 473
Fokker–Planck equation   450
Fourier amplitudes   29, 79
Fourier coefficients   30, 74
Fourier integral theory   317
Fresnel number   533
fringe contrast   297, 431
– maximum   298
functional   136
functional differentiation   136

g

Gauss' law   434
Gaussian distribution   100, 102, 103
Gaussian units   434
gedanken experiment   537, 541
generating function   323
Grassmann algebra   596
Grassmann calculus   601, 604
Grassmann field   620
Grassmann integration   603, 619
Grassmann variables   596, 597
Green's function   450

h

Hölder inequality   193
Hamiltonian
– electromagnetic   79
– for freely oscillating mode   223
– free-field   533
– interaction   219
– time dependent interaction   39

Hanbury Brown–Twiss effect 145,
  190, 492, 517, 519, 521, 526
Hanbury Brown–Twiss experiment
  10, 12, 66, 68, 190, 192, 288, 426,
  466, 468, 492
– Rebka–Pound modification 69
– result 69, 151
Hanbury Brown–Twiss interferometer
  34
Hankel function 180
heat bath 543, 544, 550
heat diffusion
– equation 385, 417, 450
– time-reversed 418
Heaviside–Lorentz units 434
Heisenberg picture 85, 108, 443,
  482, 580
Helmholtz equation 289, 434
Hermite functions 334
Hilbert transform 56
Hilbert–Schmidt norm 340, 343,
  356
Hill's equation 585
hohlraum 16
Huygens' principle 25, 488

i

integration by parts 604
intensity distribution measurement
  183
interference
– between different photons 59
– between independent light beams
  164
– quantum-mechanical interpretation
  58
interference pattern 25
irreversibility 441, 448

j

Jacobi polynomials 249

l

Laguerre polynomials 239, 261,
  378, 393, 411
– expansion 260
– generating function 131, 246, 327,
  378
laser 468
– amplification 242
– ideally monochromatic 153, 155
– realistic 156
laser gain tube 540, 561, 565
least upper bound 343
Lebesgue class of functions 339
Legendre polynomials 416
light
– corpuscular behavior 425
– frequency-splitting 228
– from chaotic source 191
light cone 3
Liouville equation 271, 445, 450
Lorentz transformation 464
lowering operator 455

m

Mach–Zehnder interferometer 573,
  574
Markoff process 446
– continuous 159, 446
master equation 453
Maxwell equations 2
– source-free 27
measurement process 50, 214
– quantum theory 541
Michelson interferometer 32–35, 59
micromotion 586
minimum uncertainty condition 436
minimum uncertainty states 255
mode functions 28, 78, 434, 469
– non-monochromatic 494
– plane-wave 78
Morse code 539

n

near-diagonality, asymptotic   555
Nicol prism   539
noise   75, 321
– generators   23
– temperature   24
noise theory   24, 288, 470
– quantum mechanical   290
nonlocality   538
normal mode expansion   134
number operator   198

o

one-photon experiments   468
one-photon optics   468
operator–function correspondence   368
operators
– bounded   340
– expansion in coherent states   90
– finite   343
– matrix elements   90
– multiplication   90
– multiplication of exponential   83
– normally ordered form   226
– nuclear   344
– ordering   331
– trace-class   344, 345
– unbounded   90
– variance of a non-Hermitian   210
optical cooling   595
optical disturbance function   25
optical instrumentation   24
order parameter   332
orthonormality condition   78, 469
oscillating electromagnetic fields   577
oscillators
– coherent states   217
– damped   426
– damped harmonic   441, 542
– Hamiltonian of harmonic   335
– infinite set   472
– inverted harmonic   548, 557
– one-dimensional harmonic   79
– quantized harmonic   436
– single mode   80
– statistical properties   426
– variances of coordinate and momentum   217
– weight function for damped   449
outer product   350, 476

p

P representation   96, 222, 225, 311, 384, 390, 438, 499, 507
– as phase-space distribution   122
– connection with classical theory   116
– fermionic   619
– positive definiteness condition   107
– regularized   393
parametric amplifier   221, 228, 474, 484, 503
– characteristic function   273
– Hamiltonian   265
– Heisenberg equations of motion   271
– mode correlation   283
– noise   244
– quantum-mechanical model   222
– state vectors   232
– statistical behavior   263
– Wigner function   267, 270
Parseval relation   609, 615
Paul trap   577, 578
Pauli equation   453
Pauli exclusion principle   595
phase space density   122, 126
photoabsorption probability   293
photoabsorption process   5
photodetector   *see* photon counter
photoionization rate   8

photomultiplier   466
photon
– cloning   540
– polarization   539, 558, 561
– statistics   287
photon coincidence experiment   517
photon counter   26, 288, 293
– counting probability   49
– counting rate   46, 170
– ideal   28, 38, 46, 293, 429, 474
– predicting the response   453
– realistic   170
– sensitivity function   42
Planck density operator   513
Planck distribution   103, 114, 179
Planck formula   444
plane wave   516
Poincaré recurrences   545
Poincaré sphere   559
Poisson distribution
– compound   454
– for average occupation numbers   81
– for number of photons   494
polarization   45, 71, 148
– density   153, 533
– vector   472
polarizing filter   539
Poynting vector   25
probability density   500
– conditioned   147, 328
– reduced   171
projection operator   94, 307, 438
– on coherent states   401
– superposition of   95

q

q-number   471
quadrupole fields   578
quantum detection   466, 467
quantum electrodynamics   27, 292, 463, 467, 474

quantum fluctuations   532, 535
– amplified   531
quantum noise   531
quarter-wave plate   562
quasi-energy   587, 589
quasiprobability density   146, 312, 315, 384, 393, 446, 501, 504
– conditioned   147, 159, 545, 551, 552

r

radiation damping   578, 592
raising operator   455
rate equation   453
residue theorem   149
rotating wave approximation   456

s

s ordering   353
scattering matrix   389
Schrödinger equation   108, 581
Schrödinger picture   85, 218, 233, 443, 580
Schrödinger state   219, 444
– coherent   232
– time dependence   230
Schrödinger's cat   541, 570
Schwarz inequality   187, 342, 489
semi-classical approach   26
signal-to-noise ratio   326
spectral intensity function   427
spectroscopy   577
spinors   559
state vectors   472, 476
statistical thermodynamics   29
stimulated emission   480
stochastic processes   471
– Gaussian   35
– stationary   481
sum rules   590
superfluid   204
superfluorescence   531, 558

superfluorescence time   533
superluminal communication   540, 558, 565
superposition principle   389

t
Tchebycheff inequality   192
tempered distributions   375, 377, 439
time translation operator   233
trace operation   77, 359, 427, 476
– symmetry   482
transition amplitude   294, 475, 588
transition frequency   532
transition probability   37, 43, 44, 475, 478
– rate of increase   45
transversality condition   78, 289, 434, 469
trapping   578, 584
trapping condition   584
trapping field   587

u
ultradistributions   377
uncertainty principle   85, 212, 253, 306
uncertainty product   209

v
vacuum fluctuations   474
vacuum state   292, 596, 628

w
wave–particle duality   75
wavepacket
– coordinate and momentum   305
– Gaussian   217, 584
– Kennard   306, 436
– minimum-uncertainty   224
– trajectories in phase space   219
Weisskopf–Wigner approximation   447, 448, 453
Weyl correspondence   396
Wiener–Khintchine theorem   112, 113, 163
Wigner function   96, 127, 263, 267, 312, 319, 384, 395, 397, 398, 501, 504, 510, 596
– n-th excited state of oscillator   131
– in terms of decoupled variables   276
– time dependence   281
Wollaston prism   539, 542
Wronskian identity   579, 580

y
Young experiment   25, 30, 56, 58, 297, 429, 487, 569, 570
– fringe contrast   60
Young interferometer   25

z
Zeeman levels   557
zero-point fluctuations   85, 100, 237, 477

Related Titles

Cohen-Tannoudji, C., Dupont-Roc, J., Grynberg, G.

Photons and Atoms

Introduction to Quantum Electrodynamics

486 pages
1989, Softcover
ISBN-13: 978-0-471-18433-1
ISBN-10: 0-471-18433-0

Saleh, B. E. A., Teich, M. C.

Fundamentals of Photonics, 2nd Edition

ca. 1200 pages
2007, Hardcover
ISBN-13: 978-0-471-35832-9
ISBN-10: 0-471-35832-0

Schleich, Wolfgang P.

Quantum Optics in Phase Space

716 pages with approx. 220 figures
2001, Hardcover
ISBN-13: 978-3-527-29435-0
ISBN-10: 3-527-29435-X